Cyclostationary Processes and Time Series

Cyclostationary Processes and Time Series

Theory, Applications, and Generalizations

Antonio Napolitano
University of Napoli "Parthenope"
Department of Engineering
Napoli, Italy

ELSEVIER
ACADEMIC PRESS
An imprint of Elsevier

Academic Press is an imprint of Elsevier
125 London Wall, London EC2Y 5AS, United Kingdom
525 B Street, Suite 1650, San Diego, CA 92101, United States
50 Hampshire Street, 5th Floor, Cambridge, MA 02139, United States
The Boulevard, Langford Lane, Kidlington, Oxford OX5 1GB, United Kingdom

Library of Congress Cataloging-in-Publication Data
A catalog record for this book is available from the Library of Congress

British Library Cataloguing-in-Publication Data
A catalogue record for this book is available from the British Library

ISBN: 978-0-08-102708-0

For information on all Academic Press publications
visit our website at https://www.elsevier.com/books-and-journals

Publisher: Mara Conner
Acquisition Editor: Tim Pitts
Editorial Project Manager: Marianna L. Kuhl
Production Project Manager: R. Vijay Bharath
Designer: Matthew Limbert

Typeset by VTeX

To Ubalda and Asha

Contents

About the Author

Antonio Napolitano was born in Napoli, Italy, in 1964. He received the Ph.D. in Electronic Engineering and Computer Science in 1994, from the University of Napoli Federico II.

In 1997, he was with the Department of Electrical and Computer Engineering at the University of California, Davis, USA, as a Postdoctorate Research Associate. From 1995 to 2001 he has been Assistant Professor and from 2001 to 2005 Associate Professor at the University of Napoli Federico II. Since 2005 he is Full Professor of Telecommunications at the University of Napoli "Parthenope". In 2005, he was Visiting Professor at the Institute de Recherche Mathematique de Rennes, France. In 2010 and 2013 he was Visiting Professor at Laboratoire d'Analyse des Signaux & des Processus Industriels, Universite Jean-Monnet, Roanne, France. In 2014 and 2015 he was Visiting Professor at Ecole Nationale Superieure d'Electronique, d'Electrotechnique, d'Informatique et d'Hydraulique et Telecommunications, Toulouse, France. He held visiting appointments at Universite de Nice–Sophia Antipolis, France (1994); University of North Carolina, Chapel Hill, NC, USA (1997, 2001, 2002); Technical University of Wroclaw, Poland (1999, 2001); Centro de Investigacion en Matematicas, Guanajuato, Mexico (2000–2002); Econometric Department, Wyzsza Szkola Biznesu, Nowy Sacz, Poland (2000–2007); School of Electrical and Information Engineering, University of South Australia (2010).

In 1996 he received the Best Paper of the Year Award from the European Association for Signal Processing (EURASIP) for a paper on higher-order cyclostationarity. In 2007 was recipient of the EURASIP Best Paper Award for a paper on the functional approach in signal analysis. In 2008 he received from Elsevier the Most Cited Paper Award for a review article on cyclostationarity. Since 2016 he is IEEE Fellow with the citation "for contributions to the statistical theory of nonstationary signal processing."

From 2006 to 2009 and from 2011 to 2015 he has been Associate Editor of the IEEE Transactions on Signal Processing. He is in the Editorial Board of Signal Processing (Elsevier) since 2008 and Digital Signal Processing (Elsevier) since 2015. From 2008 to 2013 he has been Elected Member of the Signal Processing Theory and Method Technical Committee (SPTM-TC) of the IEEE Signal Processing Society. From 2017 he has been Elected Member of the Sensor Array and Multichannel Technical Committee (SAM-TC) of the IEEE Signal Processing Society. He is EURASIP Local Liaison Officer.

He is the author of the book Generalizations of Cyclostationary Signal Processing: Spectral Analysis and Applications, John Wiley & Sons Ltd. – IEEE Press, 2012.

Preface

Ubiquity of science data generated by periodic phenomena

Many processes in nature arise from the interaction of periodic phenomena with random phenomena. The result are processes which are not periodic, but whose statistical functions are periodic functions of time. These processes are called cyclostationary (CS) (Gardner, 1987f) and are an appropriate mathematical model for signals encountered in many fields. In contrast, the classical model of stationary processes considers statistical functions which do not depend on time. More generally, if different periodicities are present in the generation mechanism of the process, the statistical functions are almost periodic and the process is called almost cyclostationary (ACS). Cyclostationary and almost-cyclostationary processes have also been referred to as periodically correlated and almost-periodically correlated, respectively (Hurd and Miamee, 2007).

The cyclostationary and almost-cyclostationary models are ubiquitous statistical models for science data. In communications, radar, sonar, and telemetry, periodicities in the statistical functions arise from the modulation by random data of carriers or pulse trains (Gardner, 1987f), (Napolitano, 2016b). Periodicities are also due to operations such as sampling, scanning, multiplexing, and coding. In monitoring and diagnosis of mechanical machinery, periodicities in the statistics of vibro-acoustic signals are due to rotations of gears, belts, and bearings (Antoni, 2007b), (Section 10.6). In econometrics, the weekly opening and closing of markets and the season-depending supply and demand of products give rise to periodicities in the statistical functions of prices and exchange rates (Parzen and Pagano, 1979), (Section 10.7). In radio astronomy, periodicities are due to the revolution and rotation of planets and pulsation of stars (Demorest, 2011). In human biological signals, cyclostationarity is due to heart pulsation (Haritopoulos et al., 2014) or alternation of day and night (Section 10.8). Hidden periodicities are present in genome sequences (Arora et al., 2008) and diffusion processes of molecular dynamics (Ho and Hamelberg, 2016). Evidence of spectral correlation, which is equivalent to cyclostationarity, is present in signals encountered in neuroscience (Aru et al., 2015), (Olhede and Ombao, 2013).

As a simple means of assessing the ubiquity and current prevalence of the cyclostationarity paradigm in scientific data processing in various fields of science and engineering, a web search using `https://scholar.google.com` was performed in April 2018, and reported in Appendix of (Gardner, 2018b). The total number of hits was about 136,000.

A plethora of applications of cyclostationarity in communications and other fields are considered in Chapters 9 and 10.

Usefulness of the cyclostationarity model

Cyclostationary (or periodically correlated) processes have been introduced in the Russian literature (Gudzenko, 1959), (Lebedev, 1959), (Gladyshev, 1961), and then extensively treated in the pioneering works of Harry L. Hurd (Hurd, 1969) and William A. Gardner (Gardner, 1972a).

Almost all modulated signals adopted in communications, radar, and sonar can be modeled as ACS. For these signals, the ACS model is much more informative than the stationary one whose statistical functions are just the zero frequency components of the corresponding periodic or almost periodic statistical functions of the ACS model.

Fundamental noise and interference immunity and signal-separability properties of CS and ACS processes discovered in the early 1980s by Gardner (e.g., (Gardner, 1985) and (Gardner, 1987f)) and theoretically developed in his seminal work performed primarily during the subsequent decade provided the foundation for the last 30 years of exploitation of almost-cyclostationarity in the design and analysis of signal processing methods and algorithms for communications, telemetry, and radar systems. This body of work has demonstrated beyond any doubt that substantial improvements in system performance can be obtained in various signal processing applications, such as detection, estimation, and classification of signals, by recognizing and modeling cyclostationarity instead of using stationary-process models (Chapter 9). The performance gain is significant in severe noise and interference environments. It is due to a proper description of the nonstationarity of the signals, that is, the periodic or almost-periodic time variability of their statistical functions. Moreover, applications benefit of the existence of estimators of statistical functions that are consistent and asymptotically complex normal.

By exploiting statistical functions (cyclic autocorrelations or cyclic spectra) at a cycle frequency of the signal-of-interest (SOI) which is not shared with the disturbance signals, signal processing algorithms can be built that are inherently immune to the effects of noise and interference, regardless of the temporal or spectral overlap of SOI and disturbance. Thus, due to the existence of consistent estimators for these statistical functions, these algorithms are highly tolerant to noise and interference in practice, provided that sufficiently long observation intervals are adopted for the estimates.

The capability to capture nonstationarity features of the observed signals is also the reason of the usefulness of the cyclostationary model in fields of applications other than communications and radar (see Chapter 10).

Generalizations of cyclostationarity

The data-record length is in principle the only limit to the possibility to obtain satisfactory performance by cyclostationarity-based algorithms in the presence of arbitrarily low signal-to-noise and signal-to-interference ratios. The data-record length can be increased provided that the ACS model for the data continues to be valid. Thus, limits to the exploitation of the cyclostationary model arise from possible irregularities or imperfections in the pace of the periodic phenomena underlying the analyzed signal, imperfections that become evident on large observation intervals. These irregularities or imperfections can be caused, for example, by (slight) time variability of timing parameters or by the relative motion between transmitter and receiver.

In the case of relative motion between transmitter and receiver, the propagation channel is time variant, not necessarily periodically or almost-periodically time variant. In such a case, if the length of the observation interval and other parameters such as the signal bandwidth and the motion parameters are such that these channel time variations can be appreciated, even if the transmitted signal is ACS, the received signal should be modeled by more general nonstationarity models.

Two extensions of the cyclostationary model have been proposed for higher fidelity modeling of communications signals in high mobility scenarios: the *generalized almost-cyclostationary* (GACS) signals and the *spectrally correlated* (SC) signals.

The class of the GACS time series is introduced and characterized in (Izzo and Napolitano, 1998a). For GACS signals, multivariate statistical functions are almost-periodic functions of time whose Fourier series expansions have both coefficients and frequencies depending on the lag parameters. ACS signals are obtained as special cases when the frequencies do not depend on the lag parameters. An extensive treatment is provided in (Napolitano, 2012, Chaps. 2 and 3).

The class of the SC processes is introduced in (Napolitano, 2001) and characterized in (Napolitano, 2003). SC processes have the Loève bifrequency spectrum concentrated on a countable set of support curves in the bifrequency plane. ACS processes are obtained as special case when the curves are lines with unit slope. An extensive treatment is provided in (Napolitano, 2012, Chaps. 4 and 5).

The way in which random and periodic or almost-periodic phenomena mix in order to generate cyclostationary or almost-cyclostationary processes or time series is such that statistical functions are exactly periodic or almost-periodic functions of time. Thus, we say that cyclostationary and almost-cyclostationary models correspond to regular statistical cyclicity in the data. There are many phenomena, however, where statistical cyclicity in the data is irregular (Gardner, 2018b). These include, for example, biological signals. In most cases the signals can be described as time-warped ACS signals (Napolitano and Gardner, 2016), (Napolitano, 2017c), (Gardner, 2018b). *Oscillatory almost-cyclostationary* (OACS) processes introduced in (Napolitano, 2016a) are a suitable model for processes exhibiting irregular cyclicity. OACS processes have autocorrelation function constituted by the superposition of amplitude- and angle-modulated sinewaves.

Two approaches for signal analysis

Two alternative approaches have been adopted for the analysis and characterization of cyclostationary signals (Gardner, 1991d). Both approaches are adopted in this book.

The first approach is the classical stochastic approach, where a signal is modeled as a realization of an ensemble of realizations, namely, the stochastic process. In such a case, the cyclostationarity or almost-cyclostationarity property corresponds to the periodicity or almost-periodicity of probabilistic functions defined in terms of ensemble averages of functions of the process and its time translates. This approach is adopted for the second-order characterization of cyclostationary stochastic processes in Chapter 1. The second approach is the *functional* or *fraction-of-time* (FOT) probability approach, which is also referred to as the *nonstochastic* approach (Gardner, 1987f), (Leśkow and Napolitano, 2006). In such a case, a signal is modeled as a single function of time or time series and all the familiar probabilistic functions are built starting from this single time series. The cyclostationarity or almost cyclostationarity property corresponds to the presence of finite-strength additive sine-wave components in functions of the single time series and its translates. In this framework, the almost-periodic component extraction operator is recognized to be the expectation operator. This approach is adopted for the second-order characterization of cyclostationary time series in Chapter 2 and for the higher-order characterization in Chapter 4.

Signal analysis was originally developed in the functional approach. In particular, a complete theory of generalized harmonic analysis for "persistent" or finite-power signals was developed in (Wiener, 1930). However, after the foundation of the theory of probability appeared in (Kolmogorov, 1933), the functional approach was abandoned in favor of the stochastic process model for signal analysis, except in few instances (Brennan, 1961), (Hofstetter, 1964), (Lee, 1967). The FOT probability framework for CS and ACS is introduced in (Gardner, 1985) and treated in depth in (Gardner, 1986c), (Gardner, 1987f, Part II), (Gardner and Brown, 1991) and (Gardner, 1991a) with reference to continuous-time signals and in (Gardner, 1994a) for both continuous- and discrete-time signals. The case of complex signals is considered in (Gardner, 1987f, Part II) and (Brown, 1987).

From the point of view of establishing a powerful mathematical theory for signal analysis, the stochastic process model has the advantage that the probability measure and the ensemble average posses amenable mathematical properties. Specifically, the probability measure is σ-additive and the ensemble average is σ-linear. In contrast, the relative measure, which is the FOT counterpart of the probability measure, is not σ-additive and the infinite time average, which is the FOT counterpart of the ensemble average, is not σ-linear (Section 2.2), (Leśkow and Napolitano, 2006). Similarly, also the almost-periodic component extraction operator is not σ-linear. Adopting a stochastic process model with amenable mathematical properties makes easier the derivation of signal processing algorithms, but requires suitable *ergodicity conditions* once the algorithms must be realized in practice, and probabilistic functions defined in terms of ensemble averages must be estimated by time averages. The ergodicity conditions are expressed in terms of mixing conditions (for example α-mixing conditions) or summability of stochastic cumulants. These conditions need to be assumed and cannot be tested if only one realization is available at the hands of the experimenter. In contrast, conditions in the FOT approach for the convergence of the statistical functions defined in terms of infinite time averages are just a property of the unique available function, and can be tested on this function. For this reason, albeit the FOT approach based on the relative measure and infinite time average possesses less amenable properties of the stochastic approach, it should be preferable since it better fits the physical situation when the experimenter makes measurements on a single time series.

The σ-additivity property of probability measure (Kolmogorov, 1933, Generalized Addition Theorem, p. 15) is an immediate consequence of the Axiom of Continuity which is a mathematical property of the sample space or field, that cannot be verified in practice. Citing (Kolmogorov, 1933, p. 15): "For infinite fields, on the other hand, the Axiom of Continuity, VI, [is] proved to be independent of Axioms I–V. Since the new axiom is essential for infinite fields of probability only, *it is almost impossible to elucidate its empirical meaning*, as has been done, for example, in the case of Axioms I–V in §2 of the first chapter. For, in describing any observable random process we can obtain only finite fields of probability. *Infinite fields of probability occur only as idealized models of real random processes. We limit ourselves, arbitrarily, to only those models that satisfy Axiom VI.* This limitation has been found *expedient* in researches of the most different sort."

The profound reason of the mathematical differences of the stochastic and FOT approaches is that the sample space for the probability measure is implicitly assumed with finite measure (Halmos, 1944), (Halmos, 1956, p. 31). Hence, the normalization that its probability is one is obtained just by a division operation. In contrast, the Lebesgue measure of the whole real line is infinite, and the normalization that the relative measure of the real line is one is obtained by a limit operation. Such a limit operation is responsible of the lack of σ-additivity. Similarly, the presence of the limit operation in the infinite time average is responsible of the lack of σ-linearity (Dehay et al., 2013, Secs. I and IV).

The existence of a FOT model for a time series is intimately linked to the existence of consistent estimators for the statistical functions of the corresponding stochastic process model. While in the stationary case it is also intuitively acceptable that under appropriate assumptions on the process (time invariant) statistical functions can be estimated by (ideally infinite) time averages, it is not obvious that time varying probabilistic functions of a nonstationary process can be estimated by infinite-time averages. Cyclostationarity is a very special kind of nonstationarity. Due to the periodicity or almost-periodicity of the statistical functions, and assuming known the frequencies of the Fourier series expansions of these functions, the Fourier coefficients can be estimated with the desired accuracy by considering sinusoidally weighted time averages over a sufficiently large observation interval. The set of frequencies and coefficients of the Fourier series provides a complete description of the periodic or almost-periodic function and, hence, a characterization of the process in terms of the considered statistical function. Such a result is obviously consequence of the special structure of nonstationarity.

A functional or FOT characterization is not available for the SC signals (Chapter 13) and the OACS signals (Chapter 14) that are two classes of signals that extend the class of the ACS signals. Thus, in Part 2, the generalizations of cyclostationarity are presented in the classical stochastic approach, even if for the GACS signals a FOT characterization is available (Izzo and Napolitano, 1998a).

The estimation problem is presented in Chapter 5 in the classical stochastic approach since most of the existing results have been established in this approach (Dehay and Hurd, 1994), (Dandawaté and Giannakis, 1994a). Moreover, results on the estimation of the cyclic spectrum of ACS signals are derived as special case of results on the estimation of the spectral density of spectrally correlated signals (Napolitano, 2012, Chaps. 4 and 5) for which no FOT characterization is available. However, in Section 5.6, the estimation of cyclic statistical functions is addressed in the FOT approach and the link between the two estimation philosophies is enlightened.

The lack of an FOT model for SC and OACS processes reflects in difficulties in the estimation of their statistical functions.

For SC processes, if the structure of the nonstationarity is unknown, the estimation of the spectral density of the Loève bifrequency spectrum can be made with a certain degree of uncertainty, depending on the "distance" of the nonstationarity from the almost cyclostationarity. In contrast, if the structure of the nonstationarity is known, namely, the location of the support curves of the Loève bifrequency spectrum is known, then the spectral density can be consistently estimated (Napolitano, 2012, Chap. 4). For the amplitude-modulated time-warped ACS processes, a subclass of the OACS processes, the statistical function estimation problem is addressed by first estimating and compensating the time warping and the amplitude modulation functions and then estimating the statistical functions of the underlying ACS process. For the GACS processes, the estimation of the cyclic statistics in the time domain is substantially similar to the case of ACS processes.

Fundamental treatments on cyclostationarity

The first fundamental treatment on cyclostationary processes in a book can be found in (Gardner, 1985, Chap. 12). Several *books* are entirely dedicated to cyclostationarity. The extensive treatment in Part II of (Gardner, 1987f) contains the foundations of the cyclic spectral analysis in the FOT approach. A collection of articles from several authors on both theoretical aspects and applications can be found in (Gardner, 1994b). A rigorous theory of cyclostationary (periodically correlated) discrete-time processes within the framework of the harmonizable processes is developed in (Hurd and Miamee, 2007). Generalizations of cyclostationary signal processing is the subject of (Napolitano, 2012), which focus on spectral analysis and applications in mobility scenarios. Mathematical models and analysis of stochastic oscillations in vibro-acoustic signals are treated in (Javorskyj, 2013).

Several *review articles* have been published on the subject of cyclostationarity. They contain complementary points of view and material. In (Gardner, 1986c), the spectral correlation theory of cyclostationary time series is presented in the FOT probability framework. In (Gardner, 1991a), the emphasis is on the spectral redundancy property of cyclostationary time series. An introduction to cyclostationary signals is provided in (Gardner, 1994a), where motivations to adopt the cyclostationary model (instead of the stationary one) are enlightened and the philosophies behind the two approaches, the classical stochastic one and the FOT one, are explained. In (Giannakis, 1998) discrete-time almost-cyclostationary stochastic processes are considered. In (Gardner et al., 2006), both stochastic and nonstochastic approaches for ACS signal analysis are treated and an extensive bibliography is included. An update can be found in (Napolitano, 2016b). The review article (Antoni, 2007a) is mainly oriented to the practitioner in mechanical engineering. ACS, GACS and SC processes are reviewed and compared in (Napolitano, 2013). In (Napolitano,

2016a), limits to the applicability of the cyclostationary model in the presence of relative motion between transmitter and receiver are derived and the GACS and SC models reviewed. In addition, the OACS model is introduced for the first time.

Extensive *bibliographies* on cyclostationarity up to year 2005 can be found in (Serpedin et al., 2005) and (Gardner et al., 2006) and up to year 2016 in (Napolitano, 2016b).

Several *PhD theses* contain fundamental contributions to the theory of cyclostationary processes and time series. The first extensive treatments on cyclostationary processes can be found in the theses (Hurd, 1969) and (Gardner, 1972a). An extensive treatment of ACS time series in the FOT probability framework is in (Brown, 1987). The theory and applications of higher-order cyclostationarity in the FOT approach is the subject of (Spooner, 1992). In (Dandawaté, 1993), this subject is treated in the classical stochastic approach. An *Habilitation à Diriger des Recherches (HDR) thesis* on the spectral analysis of cyclostationary processes is (Dehay, 1991).

Recently, freely accessible resources on cyclostationarity can be found on the Internet. Chad M. Spooner is developing and handling the blog "Cyclostationary Signal Processing, understanding and using the statistics of communication signals" (Spooner, 2015). This blog is oriented to the practitioner interested in issues on communications systems and cyclic spectral analysis. William A. Gardner is developing and handling the web site (Gardner, 2018a). This website presents an introduction to the fundamental concepts, history, basic theory, and applications of cyclostationarity and its exploitation for purposes of statistical inference or information extraction from time-series data.

Events on cyclostationarity

Several *Workshops* on cyclostationarity have been held, aimed at promoting the interaction among scholars with different backgrounds, including mathematicians, statisticians, and electrical and mechanical engineers.

The first one is the "Workshop on Cyclostationary Signals", Yountville, California, August 16–18, 1992, organized by W.A. Gardner. The second one is the "Second Workshop on Cyclostationary Signals", Monterey, CA, August 1–2, 1994, organized by S.V. Shell and C.M. Spooner. The third one is the "Workshop on Cyclostationary Processes", Marne la Vallée, France, July 1–5, 1996, organized by H.L. Hurd and P. Loubaton. Every year in February, from 2008 to 2018, the "Workshop on Cyclostationary Systems and Their Applications", has been organized by Jacek Leśkow in Gródek nad Dunajcem, Poland. In 2019, this workshop was held in Warsaw, Poland.

Several special sessions, plenary talks, and keynote speeches in international conferences have been dedicated to cyclostationarity.

Book outline

In **Part 1**, theory and applications of cyclostationary processes and time series are treated.

In **Chapter 1**, the characterization of almost-cyclostationary processes is made in the classical stochastic process framework.

In **Chapter 2**, the characterization of almost-cyclostationary time series is made in the fraction-of-time probability framework.

In **Chapter 3**, the processing of ACS signals is considered. Linear almost-periodically time variant filtering and uniform sampling are treated in detail.

In **Chapter 4**, the extension of the second-order theory to higher-orders for both continuous- and discrete-time almost-cyclostationary time series is presented in the fraction-of-time probability framework.

In **Chapter 5**, ergodicity issues are addressed. Estimators of second- and higher-order cyclic statistical functions of almost-cyclostationary processes are presented and their properties discussed. The estimation problem in the fraction-of-time approach is also addressed.

In **Chapter 6**, quadratic time-frequency distributions of second-order ACS signals are derived.

In **Chapter 7**, fundamental examples of cyclostationary communication signals are considered and their second- and higher-order cyclic statistics are derived.

In **Chapter 8**, cyclostationarity-based signal detection techniques and cycle frequency estimation methods are presented.

In **Chapter 9**, the exploitation of cyclostationarity properties of modulated signals in the analysis and design of communications systems is treated.

In **Chapter 10**, selected topics and applications are presented and discussed.

In **Part 2**, generalizations of the class of the almost-cyclostationary signals are presented.

In **Chapter 11**, the Doppler effect on ACS signals is described and limits of the ACS model are discussed. The necessity of considering generalizations of the class of the ACS signals is enlightened. For an extensive treatment see (Napolitano, 2012, Chap. 7).

In **Chapter 12**, the class of the generalized almost-cyclostationary processes is characterized. The problems of statistical function estimation and the link between continuous- and discrete-time signals is addressed. Examples of applications of the GACS model are presented. For an extensive treatment see (Napolitano, 2012, Chaps. 2 and 3).

In **Chapter 13**, the class of the spectrally correlated processes is characterized. The problems of statistical function estimation and the link between continuous- and discrete-time signals is addressed. Examples of applications of the SC model are presented. For an extensive treatment see (Napolitano, 2012, Chaps. 4 and 5).

In **Chapter 14**, the recently introduced class of the oscillatory almost-cyclostationary processes is characterized. Examples of applications of the OACS model are presented.

In **Chapter 15**, relationships among the classes of nonstationary signals considered in the previous chapters are enlightened.

In **Part 3**, six appendices contain, for easy reference, results exploited in Parts 1 and 2.

In **Appendix A**, the second-order characterization of nonstationary stochastic processes is reviewed in both time and frequency domains and for both continuous- and discrete-time cases.

In **Appendix B**, definitions and properties of uniformly almost-periodic functions and their generalizations are reviewed.

In **Appendix C**, relationships regarding sampling and replication of signals in time and frequency domains are reported.

In **Appendix D**, results on Hilbert transform, analytic signal, and complex envelope of a real valued signal are reported.

In **Appendix E**, results on complex random variables and vectors, quadratic forms of random normal vectors, and chi squared distribution are reported.

In **Appendix F**, selected references are classified into categories and listed in chronological order.

In the **Bibliography** section, 1440 references including books, journal and conference articles, patents, and web sites are listed in alphabetical order.

For each chapter or appendix, proofs of theorems and results are provided in the last section. In some cases, proofs are presented just as formal mathematical manipulations without justifying the interchange of limit, integral, infinite sum, and infinite time average operations. In such a case, when available, references where a rigorous proof can be possibly found are cited. Proofs available in (Napolitano, 2012) for GACS and SC processes that can be specialized to the case of ACS processes are only briefly sketched here and the reader is remanded to (Napolitano, 2012).

Antonio Napolitano
Napoli
May 3, 2019

Acknowledgment

Parts of this material from Elsevier are reprinted with permission:

- W. A. Gardner, A. Napolitano, and L. Paura, "Cyclostationarity: Half a century of research," *Signal Processing*, vol. 86, no. 4, pp. 639–697, April 2006.
- A. Napolitano "Cyclostationarity: New trends and applications", *Signal Processing*, vol. 120, pp. 385–408, March 2016.
- A. Napolitano "Cyclostationarity: Limits and generalizations", *Signal Processing*, vol. 120, pp. 323–347, March 2016.
- A. Napolitano and K. Doğançay, "Bandpass sampling of almost-cyclostationary signals", *Signal Processing*, vol. 153, pp. 266–274, December 2018.

A special thank is due to my parents. I gratefully acknowledge for scientific interaction and friendship Dominique Dehay, Anna E. Dudek, William A. Gardner, Harry L. Hurd, Luciano Izzo, Ihor M. Javorskyj, Giacomo Lauridia, Jacek Leśkow, Andrzej Makagon, Luigi Paura, Chad M. Spooner. Tim Pitts is acknowledged for believing in this book.

Antonio Napolitano
May 3, 2019

List of Abbreviations

a.e. almost everywhere
AM amplitude modulation
AP almost periodic
ACS almost cyclostationary
AWGN additive white Gaussian noise
CCP cyclic cross-polyspectrum
CDMA code-division multiple access
CS cyclostationary
CSCCF cyclic spectral cross-cumulant function
CSCMF cyclic spectral cross-moment function
CTCCF cyclic temporal cross-cumulant function
CTCMF cyclic temporal cross-moment function
DFT discrete Fourier transform
DSSS direct-sequence spread-spectrum
fBm fractional Brownian motion
FFT fast Fourier transform
FOT fraction-of-time
FRESH frequency shift
GACS generalized almost-cyclostationary
gcd greatest common divisor
GLLRT generalized log-likelihood ratio test
IDFT inverse discrete Fourier transform
LAPTV linear almost-periodically time-variant
lcm least common multiple
lhs left hand side
LLRT log-likelihood ratio test
LPI low-probability-of-intercept
LPTV linear periodically time-variant
LTI linear time-invariant
LTV linear time-variant
MC modulated cyclical
ML maximum likelihood
MIMO multi-input multi-output
MMSE minimum mean-squared error
OACS oscillatory almost-cyclostationary
OSC oscillatory spectrally correlated
PAM pulse amplitude modulated
pdf probability density function
PSD power spectral density
p.v. Cauchy principal value
QAM quadrature amplitude modulation
QTI quadratic time-invariant
RM relatively measurable
RD-CSCMF reduced-dimension CSCMF
RD-CTCCF reduced-dimension CTCCF
RD-CTCMF reduced-dimension CTCMF
rhs right hand side

ROC	receiver operating characteristic
RX	receiver
SC	spectrally correlated
SIR	signal-to-interference ratio
SNR	signal-to-noise ratio
SOI	signal of interest
STFT	short-time Fourier transform
TDOA	time-difference-of-arrival
t.m.s.s.	temporal mean-squared sense
TFD	time-frequency distribution
TX	transmitter
TW	time warping
w.p.1	with probability 1
w.r.t.	with respect to
WSS	wide-sense stationary

Notation

$\delta(\cdot)$	Dirac delta
$\tilde{\delta}(\cdot)$	periodic train of Dirac deltas with period 1 or unit element with respect to convolution of complex envelopes
$\delta_{(\cdot)}$	Kronecker delta
$\mathbf{u}(\cdot)$	unit step function
$\mathbf{rect}(\cdot)$	rect function
$\mathbf{sign}(\cdot)$	signum function
$\mathbf{sinc}(\cdot)$	$\sin(\pi(\cdot))/(\pi(\cdot))$
$\otimes$	convolution
$\langle\cdot\rangle_t$	continuous-time time average
$\langle\cdot\rangle_n$	discrete-time time average
$\mathrm{E}^{\{\alpha\}}\{\cdot\}$	almost-periodic component extraction operator (continuous-time)
$\mathrm{E}^{\{\tilde{\alpha}\}}\{\cdot\}$	almost-periodic component extraction operator (discrete-time)
$\boldsymbol{u}^{\mathsf{T}}$	transpose of vector $\boldsymbol{u}$
$\boldsymbol{u}^{\mathsf{H}}$	Hermitian transpose of vector $\boldsymbol{u}$
$\boldsymbol{u}'$	$[u_1, \ldots, u_{K-1}]^{\mathsf{T}}$ if $\boldsymbol{u} = [u_1, \ldots, u_K]^{\mathsf{T}}$
$\Vert\cdot\Vert_\infty$	essential supremum
$\mathbf{Re}[\cdot]$	real part
$\mathbf{Im}[\cdot]$	imaginary part
$\mathbf{1}_{\{\cdot\}}$	indicator of a set
$L^p(\mathbb{R})$	space of functions with summable pth power of magnitude
$L^1_{\mathbf{loc}}(\mathbb{R})$	space of locally summable functions
$L^\infty(\mathbb{R})$	space of bounded functions
$\stackrel{t \leftrightarrow f}{\longleftrightarrow}$	Fourier transform from variable t to f
$\stackrel{\mathcal{F}}{\longleftrightarrow}$	Fourier transform
$\stackrel{\mathcal{F}_{\mathbf{D}}}{\longleftrightarrow}$	discrete Fourier transform
$\stackrel{\mathbf{d}}{\longrightarrow}$	convergence in distribution
$\stackrel{\mathbf{p}}{\longrightarrow}$	convergence in probability
$\stackrel{\mathbf{w.p.1}}{\longrightarrow}$	convergence with probability 1

PART 1

Cyclostationarity

1

Characterization of stochastic processes

Contents

Abstract

Cyclostationary and almost-cyclostationary stochastic processes are defined and characterized in the time and frequency domains and in the continuous- and discrete-time cases. Strict-sense and second-order wide-sense characterizations are addressed. For a generic cyclostationary process, two representations in terms of elementary cyclostationary processes are considered: the harmonic series representation and the translation series representation.

Keywords

Stochastic process, Cyclic autocorrelation, Cyclic spectrum, Spectral coherence, Harmonic series representation, Translation series representation

1.1 Introduction

The first results on characterization and properties of cyclostationary stochastic processes are in the Russian literature (Gudzenko, 1959), (Lebedev, 1959), (Gladyshev, 1961),

Cyclostationary Processes and Time Series. https://doi.org/10.1016/B978-0-08-102708-0.00012-1

(Gladyshev, 1963), (Markelov, 1966b), (Kayatskas, 1968), (Dragan, 1969b), (Voychishin and Dragan, 1973), (Dragan, 1980).

The first extensive treatments of the theory of cyclostationary processes can be found in the pioneering works of H.L. Hurd (Hurd, 1969) and W.A. Gardner (Gardner, 1972a).

In (Boyles and Gardner, 1983) and (Gardner, 1985) (and (Gardner, 1990b)), the theory of second-order cyclostationary (CS) processes is developed mainly with reference to continuous-time stochastic processes, but discrete-time is the focus in (Boyles and Gardner, 1983). Discrete-time processes are treated more generally in (Giannakis, 1998). A rigorous treatment of periodically correlated processes within the framework of harmonizable processes is given in (Hurd, 1989b), (Hurd, 1991), (Dehay, 1991), (Hurd and Miamee, 2007).

General properties of cyclostationary processes are derived in terms of the Fourier series expansion of the periodic autocorrelation function. The frequencies, called *cycle frequencies*, are multiples of the reciprocal of the period of cyclostationarity and the coefficients, referred to as *cyclic autocorrelation functions*, are continuous functions of the lag parameter. Cyclostationary processes are characterized in the frequency domain by the *cyclic spectra*, which are the Fourier transforms of the cyclic autocorrelation functions. The cyclic spectrum at a given cycle frequency represents the density of correlation between two spectral components of the process that are separated by an amount equal to the cycle frequency. Almost-cyclostationary (ACS) processes, introduced in (Gardner, 1978), have autocorrelation functions that can be expressed in a (generalized) Fourier series whose frequencies are not harmonically related, that is, are incommensurate. The periodicity or almost-periodicity in the time domain reflects in the fact that the Loève bifrequency spectrum has spectral masses concentrated on lines parallel to the main diagonal in the bifrequency plane. The intercepts of these lines are the cycle frequencies and the spectral densities on these lines are the cyclic spectra.

The problem of spectral analysis is mainly treated in (Dragan, 1971), (Ogura, 1971), (Honda, 1981), (Honda, 1982), (Dragan, 1985), (Hurd, 1989a), (Alekseev, 1990), (Makagon et al., 1991), (Dehay, 1994), (Makagon et al., 1994), (Makagon, 2001), and (Hurd and Koski, 2004a). The stationarizing effects of random shifts are examined in (Hurd, 1974b), (Gardner, 1978), (Miamee, 1990) and the important impact of random shifts on cycloergodic properties is exposed in (Gardner, 1994a). The problem of filtering is considered in (Dragan, 1969b), (Gupta, 1972), (Sathe and Vaidyanathan, 1993). Wavelet analysis of cyclostationary processes is addressed in (Cambanis and Houdré, 1995), (Touati and Pesquet, 2001).

For further related references, see the general treatments in (Hurd, 1969), (Gardner, 1972a), (Dragan, 1980), (Gardner, 1985), (Gardner, 1986c), (Gardner, 1988a), (Yaglom, 1986), (Gardner, 1987f), (Gardner, 1990b), (Gardner and Chen, 1991), (Gardner, 1994a), (Hurd and Miamee, 2007), (Napolitano, 2012), (Javorskyj, 2013), and also (Boyles and Gardner, 1983), (Gardner, 1991d), (Dehay and Hurd, 1994), (Reed and Wickert, 1989), (Cartaxo and Albuquerque, 1990), (Spooner and Gardner, 1991b), (Todd, 1990), (Dragan, 1972b), (Zhang, 1997), (Akdi, 1998), (Izzo and Napolitano, 1998a), (Napolitano, 2003).

1.2 Stochastic processes

1.2.1 Continuous-time processes

1.2.1.1 Time domain characterization

Let us consider a continuous-time real-valued stochastic process $\{x(t,\omega), t \in \mathbb{R}, \omega \in \Omega\}$, with abbreviated notation $x(t)$ when this does not create ambiguity, defined on a probability space $(\Omega, \mathcal{F}, P)$, where Ω is the sample space, equipped with the σ-field $\mathcal{F}$, and P is a probability measure defined on the elements of $\mathcal{F}$.

Definition 1.1. The process $x(t)$ is said to be *Nth-order cyclostationary in the strict sense* (Gardner, 1972a), (Gardner, 1985) if its Nth-order distribution function

$$F_{x(t+\tau_1)\cdots x(t+\tau_{N-1})x(t)}(\xi_1,\ldots,\xi_{N-1},\xi_N) \triangleq P\{x(t+\tau_1) \leqslant \xi_1,\ldots,x(t+\tau_{N-1}) \leqslant \xi_{N-1}, x(t) \leqslant \xi_N\} \tag{1.1}$$

is periodic in t with some period, say T_0:

$$\begin{aligned} &F_{x(t+\tau_1+T_0)\cdots x(t+\tau_{N-1}+T_0)x(t+T_0)}(\xi_1,\ldots,\xi_{N-1},\xi_N) \\ &\quad = F_{x(t+\tau_1)\cdots x(t+\tau_{N-1})x(t)}(\xi_1,\ldots,\xi_{N-1},\xi_N) \\ &\quad\quad \forall t \in \mathbb{R},\ \forall(\tau_1,\ldots,\tau_{N-1}) \in \mathbb{R}^{N-1},\ \forall(\xi_1,\ldots,\xi_N) \in \mathbb{R}^N . \end{aligned} \tag{1.2}$$

□

Definition 1.2. The process $x(t)$ is said to be *second-order cyclostationary in the wide sense* (Gardner, 1972a), (Gardner, 1985) if its mean $\mathrm{E}\{x(t)\}$ and autocorrelation function

$$R_x(t,\tau) \triangleq \mathrm{E}\{x(t+\tau)\,x(t)\} \tag{1.3}$$

are periodic with some period, say T_0:

$$\mathrm{E}\{x(t+T_0)\} = \mathrm{E}\{x(t)\} \tag{1.4}$$

$$R_x(t+T_0,\tau) = R_x(t,\tau) \tag{1.5}$$

for all t and τ. □

Theorem 1.3. *Let $x(t)$ be a second-order wide-sense cyclostationary process with period T_0 and let us assume that the series of the Fourier coefficients*

$$m_{k/T_0} \triangleq \frac{1}{T_0}\int_{-T_0/2}^{T_0/2} \mathrm{E}\{x(t)\}\, e^{-j2\pi(k/T_0)t}\,\mathrm{d}t \tag{1.6}$$

is absolutely convergent, that is

$$\sum_{k=-\infty}^{+\infty} \left|m_{k/T_0}\right| < \infty . \tag{1.7}$$

Thus the Fourier series expansion of $\mathrm{E}\{x(t)\}$ *is convergent to* $\mathrm{E}\{x(t)\}$ *and we have*

$$\mathrm{E}\{x(t)\} = \sum_{k=-\infty}^{+\infty} m_{k/T_0}\, e^{j2\pi(k/T_0)t} . \tag{1.8}$$

□

Theorem 1.4. *Let* $x(t)$ *be a second-order wide-sense cyclostationary process with period* T_0 *and let us assume that the series of the Fourier coefficients*

$$R_x^{k/T_0}(\tau) \triangleq \frac{1}{T_0} \int_{-T_0/2}^{T_0/2} R_x(t,\tau)\, e^{-j2\pi(k/T_0)t}\, \mathrm{d}t \tag{1.9}$$

is absolutely convergent, that is

$$\sum_{k=-\infty}^{+\infty} \left| R_x^{k/T_0}(\tau) \right| < \infty . \tag{1.10}$$

Thus the Fourier series expansion of $R_x(t,\tau)$ *is convergent to* $R_x(t,\tau)$ *and we have*

$$R_x(t,\tau) = \sum_{k=-\infty}^{+\infty} R_x^{k/T_0}(\tau)\, e^{j2\pi(k/T_0)t} . \tag{1.11}$$

The functions $R_x^{k/T_0}(\tau)$ *are referred to as* cyclic autocorrelation functions *and the frequencies* $\{k/T_0\}_{n\in\mathbb{Z}}$ *are called* cycle frequencies. □

Wide-sense cyclostationary processes have also been called *periodically correlated processes* (see e.g., (Gladyshev, 1961), (Hurd, 1969), (Hurd, 1991)). As first shown in (Gardner and Franks, 1975), $x(t)$ and its frequency-shifted version $x(t)e^{j2\pi(k/T_0)t}$ are correlated.

The wide-sense stationary processes (Section A.5) are the special case of cyclostationary processes for which $R_x^{k/T_0}(\tau) \not\equiv 0$ only for $k = 0$.

Theorem 1.5. *Let* $x(t)$ *be cyclostationary with period* T_0 *and* θ *a random variable uniformly distributed in* $[0, T_0)$ *and statistically independent of* $x(t)$*. Then the stochastic process* $x(t+\theta)$ *is wide sense stationary (Hurd, 1974b), (Gardner, 1978), (Gardner, 1985, Chap. 12), (Gardner, 1987f, Chap. 10, Sec. E), (Miamee, 1990). Stationarization effects are also studied in the early work (Voychishin and Dragan, 1972a).* □

A more general class of stochastic processes is obtained if the autocorrelation function $R_x(t,\tau)$, for each τ, is *almost periodic* in t in the sense of Bohr (Bohr, 1933, par. 84–92) or, equivalently, uniformly almost periodic in t in the sense of Besicovitch (Besicovitch, 1932, Chap. 1) (Section B.2).

Almost periodic function are treated in (Besicovitch, 1932), (Bohr, 1933), (Wiener, 1933, par. 24–25), (Amerio and Prouse, 1971), (Bass, 1971, part 5), (Corduneanu, 1989). A review of the main properties can be found in Appendix B and in (Napolitano, 2012, Sec. 1.2).

Definition 1.6. The process $x(t)$ is said to be *Nth-order almost-cyclostationary in the strict sense* (Gardner, 1972a), (Gardner, 1985) if its Nth-order distribution function (1.1) is almost-periodic in t. □

Definition 1.7. A continuous-time real-valued stochastic process $x(t)$ is said to be *almost-cyclostationary (ACS) in the wide sense* (Gardner, 1972a), (Gardner, 1978), (Gardner, 1985) if its mean $\mathrm{E}\{x(t)\}$ is an almost periodic function of t and its autocorrelation function $R_x(t, \tau)$ is an almost periodic function of t with frequencies not depending on τ. That is, it is an almost-periodic function depending on a parameter τ, according to the definition of (Corduneanu, 1989, Chap. II). □

Definition 1.8. The Fourier coefficient

$$R_x^\alpha(\tau) \triangleq \lim_{T\to\infty} \frac{1}{T} \int_{-T/2}^{T/2} R_x(t, \tau)\, e^{-j2\pi\alpha t}\, \mathrm{d}t \tag{1.12}$$

is referred to as *cyclic autocorrelation function* at *cycle frequency* α. □

For $\alpha = 0$, the cyclic autocorrelation reduces to the time averaged autocorrelation $R_x^0(\tau)$ (A.31) which is also denoted by $R_x(\tau)$.

Accounting for the properties of the almost-periodic functions (Section B.2), we have the following result.

Theorem 1.9. *The autocorrelation function of an ACS process in the wide sense is the limit of a uniformly convergent sequence of trigonometric polynomials in t. If the series of the Fourier coefficients is absolutely convergent*

$$\sum_{\alpha\in A} \left|R_x^\alpha(\tau)\right| < \infty \tag{1.13}$$

where A is the countable set, not depending on τ, of possibly incommensurate cycle frequencies α, then the Fourier series is uniformly convergent to the autocorrelation

$$R_x(t, \tau) = \sum_{\alpha\in A} R_x^\alpha(\tau)\, e^{j2\pi\alpha t}\,. \tag{1.14}$$

□

As first shown in (Gardner, 1978), if $x(t)$ is an ACS process, then the process $x(t)$ and its frequency-shifted version $x(t)\, e^{j2\pi\alpha t}$ are correlated when $\alpha \in A$. Wide-sense almost-cyclostationary processes have also been called *almost-periodically correlated* processes (see e.g., (Gladyshev, 1963), (Hurd, 1969), (Hurd, 1991)).

Remark 1.10. In the special case where the almost-periodic autocorrelation is poly-periodic (Definition B.6), that is, $A \equiv \{\ell_1/T_1\} \cup \cdots \cup \{\ell_M/T_M\}$, $\ell_p \in \mathbb{Z}$, with $T_1, \ldots, T_M$ incommensurate, then the process is said *poly-cyclostationary* (Gardner, 1987f), (Gardner, 1994a).

In the special case where the almost-periodic autocorrelation is periodic, that is, for some T_0, $A \equiv \{k/T_0\}$, $k \in \mathbb{Z}$, then the process is wide-sense cyclostationary. □

Theorem 1.11. Characterization of ACS Processes (Hurd, 1991). *Let A_τ be the set*

$$A_\tau \triangleq \left\{\alpha \in \mathbb{R} : R_x^\alpha(\tau) \neq 0\right\} . \tag{1.15}$$

The ACS processes are characterized by the following conditions:

1. *The set*

$$A \triangleq \bigcup_{\tau \in \mathbb{R}} A_\tau \tag{1.16}$$

is countable.

2. *The autocorrelation function $R_x(t,\tau)$ is uniformly continuous in t and τ.*
3. *The time-averaged autocorrelation function*

$$R_x^0(\tau) \triangleq \lim_{T\to\infty} \frac{1}{T} \int_{-T/2}^{T/2} R_x(t,\tau)\,\mathrm{d}t \tag{1.17}$$

is continuous for $\tau = 0$ (and, hence, for every τ).

4. *The process is mean-square continuous, that is*

$$\sup_{t\in\mathbb{R}} \mathrm{E}\left\{|x(t+\tau) - x(t)|^2\right\} \to 0 \quad \text{as } \tau \to 0 . \tag{1.18}$$

□

More generally, a stochastic process $x(t)$ is said to *exhibit cyclostationarity* at cycle frequency α if $R_x^\alpha(\tau) \not\equiv 0$ (Gardner, 1985). If $x(t)$ exhibits cyclostationarity for $\alpha \in A$, the autocorrelation function can be expressed as

$$R_x(t,\tau) = \sum_{\alpha\in A} R_x^\alpha(\tau)\, e^{j2\pi\alpha t} + r_x(t,\tau) \tag{1.19}$$

where, in the right-hand side, the first term is the Fourier series of an almost-periodic function possibly of generalized type (Section B.4), not necessarily continuous in t, and the term $r_x(t,\tau)$ does not contain any finite-strength additive sine-wave component

$$\lim_{T\to\infty} \frac{1}{T} \int_{-T/2}^{T/2} r_x(t,\tau)\, e^{-j2\pi\alpha t}\,\mathrm{d}t = 0 \quad \forall \alpha \in \mathbb{R} . \tag{1.20}$$

In the special case where $\lim_{|t|\to\infty} r_x(t,\tau) = 0$, $x(t)$ is said to be *asymptotically almost cyclostationary* (Gardner, 1978).

1.2.1.2 Frequency domain characterization

A covariance (or autocorrelation) function $\mathrm{E}\{x(t_1)\, x(t_2)\}$ is said to be *harmonizable* if it can be expressed as

$$\mathrm{E}\{x(t_1)\, x(t_2)\} = \int_{\mathbb{R}^2} e^{j2\pi(f_1 t_1 - f_2 t_2)}\mathrm{d}\gamma(f_1, f_2) \tag{1.21}$$

where $\gamma(f_1, f_2)$ is a covariance of bounded variation on $\mathbb{R} \times \mathbb{R}$

$$\int_{\mathbb{R}^2} \left|\mathrm{d}\gamma(f_1, f_2)\right| < \infty \tag{1.22}$$

and the integral is a Fourier-Stieltjes transform (Definition A.4) (Loève, 1963). Moreover, a second-order stochastic process $x(t)$ is said to be harmonizable if there exists a second-order stochastic process $\chi(f)$ with covariance function $\mathrm{E}\{\chi(f_1)\,\chi^*(f_2)\} = \gamma(f_1, f_2)$ of bounded variation on $\mathbb{R} \times \mathbb{R}$ such that

$$x(t) = \int_{\mathbb{R}} e^{j2\pi f t}\,\mathrm{d}\chi(f) \tag{1.23}$$

with probability one (Definition A.5).

Theorem 1.12. (Dragan, 1978), (Hurd, 1969), (Hurd, 1991). *If the stochastic process $x(t)$ is harmonizable and ACS, then we have that*

$$\mathrm{d}\gamma(f_1, f_2) = \sum_{\alpha\in A} \delta(f_2 - f_1 + \alpha)\,\mathrm{d}\mu_x^\alpha(f_1)\,\mathrm{d}f_2 \tag{1.24}$$

where $\delta(\cdot)$ is the Dirac delta and $\mu_x^\alpha(f_1)$, $\alpha \in A$, is a family of complex measures. In addition, it results that

$$R_x^\alpha(\tau) = \int_{\mathbb{R}} e^{j2\pi f\tau}\,\mathrm{d}\mu_x^\alpha(f) \tag{1.25a}$$

$$= \int_{\mathbb{R}} e^{j2\pi f\tau}\,S_x^\alpha(f)\,\mathrm{d}f \tag{1.25b}$$

where the second equality holds if $\mu_x^\alpha(f)$ does not contain a singular component and, hence $\mathrm{d}\mu_x^\alpha(f) = S_x^\alpha(f)\,\mathrm{d}f$.

Proof. See Section 1.9. □

The condition of bounded variation for the spectral covariance can be written as

$$\int_{\mathbb{R}^2} \left|\mathrm{d}\gamma(f_1, f_2)\right| = \sum_{\alpha\in A} \int_{\mathbb{R}} \left|\mathrm{d}\mu_x^\alpha(f)\right| < \infty \tag{1.26a}$$

$$\int_{\mathbb{R}^2} \left|\mathrm{d}\gamma(f_1, f_2)\right| = \sum_{\alpha\in A} \int_{\mathbb{R}} \left|S_x^\alpha(f)\right|\,\mathrm{d}f < \infty \tag{1.26b}$$

where the second expression holds if $\mu_x^\alpha(f)$ does not contain singular component $\forall\alpha \in A$.

The function $S_x^\alpha(f)$

$$S_x^\alpha(f) = \int_{\mathbb{R}} R_x^\alpha(\tau)\,e^{-j2\pi f\tau}\,\mathrm{d}\tau \tag{1.27}$$

is the *cyclic spectrum* of $x(t)$ at cycle frequency α.

Let us formally denote by

$$X(f) \triangleq \int_{\mathbb{R}} x(t)\, e^{-j2\pi f t}\, \mathrm{d}t \tag{1.28}$$

the stochastic process obtained by Fourier transformation of the ACS process $x(t)$, where the Fourier transform is assumed to exist, at least in the sense of distributions (generalized functions) (Gel'fand and Vilenkin, 1964, Chap. 3), (Henniger, 1970), (Napolitano, 2012, Secs. 1.1.2, 4.2.1), with probability 1.

If a stochastic process is harmonizable and $\chi(f)$ does not contain singular component w.p.1, in the sense of distributions (generalized functions) we have $\mathrm{d}\chi(f) = X(f)\mathrm{d}f$ and $\chi(f)$ is the indefinite integral of $X(f)$ or the integrated Fourier transform of $x(t)$ (Wiener, 1933) (Section A.3). In addition, if also $\gamma(f_1, f_2)$ does not contain singular component, we have $\mathrm{d}\gamma(f_1, f_2) = \mathrm{E}\{\mathrm{d}\chi(f_1)\, \mathrm{d}\chi^*(f_2)\} = \mathrm{E}\{X(f_1)\, X^*(f_2)\}\mathrm{d}f_1 \mathrm{d}f_2$.

By using (1.14), in the sense of distributions (Gel'fand and Vilenkin, 1964, Chap. 3), (Henniger, 1970), (Napolitano, 2012, Secs. 1.1.2, 4.2.1) we obtain the *Loève bifrequency spectrum*

$$\mathrm{E}\left\{X(f_1)\, X^*(f_2)\right\} = \sum_{\alpha \in A} S_x^{\alpha}(f_1)\, \delta(f_2 - f_1 + \alpha) \tag{1.29}$$

Therefore, for an ACS process, correlation exists between spectral components that are separated by amounts equal to the cycle frequencies. The support in the (f_1, f_2) plane of the spectral correlation function $\mathrm{E}\{X(f_1)\, X^*(f_2)\}$ consists of parallel lines with unity slope. The density of spectral correlation on this support is described by the cyclic spectra $S_x^{\alpha}(f)$, $\alpha \in A$.

Alternatively, the cyclic spectrum $S_x^{\alpha}(f)$ at cycle frequency α can be defined as (Gardner, 1985)

$$S_x^{\alpha}(f) \triangleq \lim_{\Delta f \to 0} \lim_{T \to \infty} \frac{1}{T} \int_{-T/2}^{T/2} \mathrm{E}\left\{\Delta f\, X_{1/\Delta f}(t, f)\, X^*_{1/\Delta f}(t, f - \alpha)\right\} \mathrm{d}t \tag{1.30}$$

where the order of the two limits, provided that they exist, cannot be reversed, and

$$X_Z(t, f) \triangleq \int_{t-Z/2}^{t+Z/2} x(s)\, e^{-j2\pi f s}\, \mathrm{d}s \tag{1.31}$$

is the *short-time Fourier transform* (STFT) of $x(t)$. Therefore, the cyclic spectrum $S_x^{\alpha}(f)$ is also called the *spectral correlation density function*. It represents the time-averaged statistical correlation (with zero lag) of two spectral components at frequencies f and $f - \alpha$, as the bandwidth approaches zero. For $\alpha = 0$, the cyclic spectrum reduces to the *power spectrum* or *power spectral density* (PSD) function $S_x^0(f)$, which is also denoted as $S_x(f)$, and (1.27) reduces to the *Wiener-Khinchin relation* (A.28b).

If $S_x^{\alpha}(f)$ is *defined* as in (1.30), then the Fourier transform relationship (1.27) can be *proved* (Gardner, 1985, Chap. 12) and is dubbed *Gardner relation* which is also referred to as *cyclic Wiener-Khinchin relation*.

For wide-sense stationary processes the autocorrelation function does not depend on t,

$$\mathrm{E}\{x(t+\tau)\,x(t)\} = R_x^0(\tau) \tag{1.32}$$

and, equivalently, no correlation exists between distinct spectral components,

$$\mathrm{E}\left\{X(f_1)\,X^*(f_2)\right\} = S_x^0(f_1)\,\delta(f_2 - f_1)\,. \tag{1.33}$$

Finally, note that symmetric definitions of cyclic autocorrelation function and cyclic spectrum have been widely used (see, e.g., (Gardner, 1985), (Gardner, 1987f), (Gardner, 1990b)). They are linked to the asymmetric definitions (1.12) and (1.30) by the relationships

$$\lim_{T\to\infty} \frac{1}{T} \int_{-T/2}^{T/2} \mathrm{E}\{x(t+\tau/2)\,x(t-\tau/2)\}\, e^{-j2\pi\alpha t}\,\mathrm{d}t = R_x^\alpha(\tau)\, e^{-j\pi\alpha\tau} \tag{1.34}$$

and

$$\lim_{\Delta f\to 0}\lim_{T\to\infty} \frac{1}{T} \int_{-T/2}^{T/2} \Delta f\, \mathrm{E}\left\{X_{1/\Delta f}(t, f+\alpha/2)\, X^*_{1/\Delta f}(t, f-\alpha/2)\right\} \mathrm{d}t = S_x^\alpha(f+\alpha/2) \tag{1.35}$$

respectively.

In Fig. 1.1, the magnitude of the cyclic autocorrelation function as function of α and τ and in Fig. 1.2, the magnitude of the bifrequency spectral correlation density $\bar{\mathbb{S}}_x(f_1, f_2)$ (obtained by replacing Dirac deltas with Kronecker deltas in the expression of the Loève bifrequency spectrum (1.33) and (1.29)), as function of f_1 and f_2, are reported for a WSS and an ACS signal with the same time-averaged autocorrelation function and power spectrum. The WSS signal has nonzero cyclic autocorrelation function only for $\alpha = 0$ (Fig. 1.1 (top)). The ACS signal is a cyclostationary pulse-amplitude modulated (PAM) signal with rectangular pulse. The cyclic autocorrelation function is nonzero for values of cycle frequency α belonging to the countable set $\{k/T_0\}_{k\in\mathbb{Z}}$, where T_0 is the symbol period (Fig. 1.1 (bottom)). The bifrequency spectral correlation density of the WSS signal is nonzero only on the main diagonal of the bifrequency plane (Fig. 1.2 (top)). In contrast, the ACS signal has the bifrequency spectral correlation density nonzero on lines which are parallel to the main diagonal (Fig. 1.2 (bottom)). Spectral components that are correlated are separated by quantities equal to the cycle frequencies. According to (1.29), on each support line $f_2 = f_1 - \alpha$ the spectral correlation density is the cyclic spectrum $S_x^\alpha(f)$.

1.2.2 Discrete-time processes

Let us consider a discrete-time real-valued stochastic process $\{x(n,\omega), n \in \mathbb{Z}, \omega \in \Omega\}$, with abbreviated notation $x(n)$ when this does not create ambiguity.

Definitions of almost cyclostationarity in the strict sense and in the wide sense for discrete-time processes are obtained straightforwardly from the continuous-time counterparts.

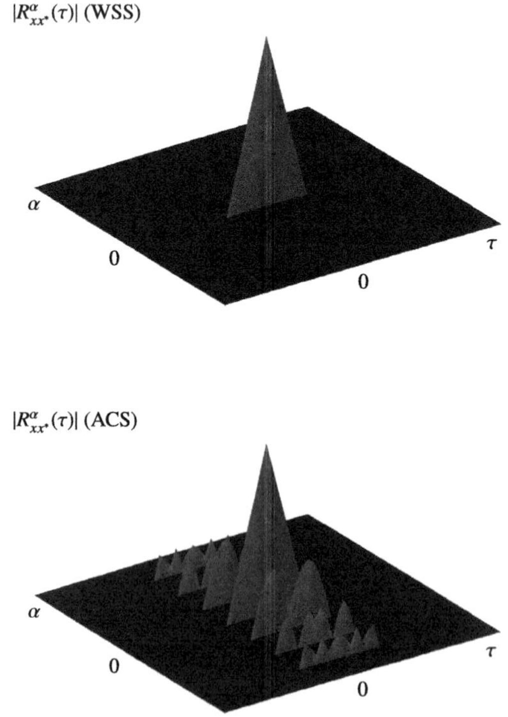

FIGURE 1.1 Magnitude of the cyclic autocorrelation function $R^{\alpha}_{xx^*}(\tau)$ of (top) a WSS signal and (bottom) an ACS signal as function of α and τ. *Source: (Napolitano, 2016b) Copyright of Elsevier.*

Definition 1.13. The stochastic process $x(n)$ is said to be *second-order almost-cyclostationary in the wide sense* (Gardner, 1972a), (Boyles and Gardner, 1983), (Gardner, 1985) if its mean $\mathrm{E}\{x(n)\}$ and its autocorrelation function

$$\widetilde{R}_x(n,m) \triangleq \mathrm{E}\{x(n+m)\,x(n)\} \tag{1.36}$$

are almost-periodic functions of the discrete-time parameter n. □

Definition 1.14. The Fourier coefficient

$$\widetilde{R}^{\widetilde{\alpha}}_x(m) \triangleq \lim_{N\to\infty} \frac{1}{2N+1} \sum_{n=-N}^{N} \widetilde{R}_x(n,m)\, e^{-j2\pi\widetilde{\alpha}n} \tag{1.37}$$

is referred to as *cyclic autocorrelation function* at *cycle frequency* $\widetilde{\alpha}$. □

Theorem 1.15. *The autocorrelation function of a second-order almost-cyclostationary process in the wide sense is the limit of a sequence of discrete-time trigonometric polynomials in n. If the series of the Fourier coefficients is absolutely convergent*

$$\sum_{\widetilde{\alpha}\in\widetilde{A}} \left|\widetilde{R}^{\widetilde{\alpha}}_x(m)\right| < \infty \tag{1.38}$$

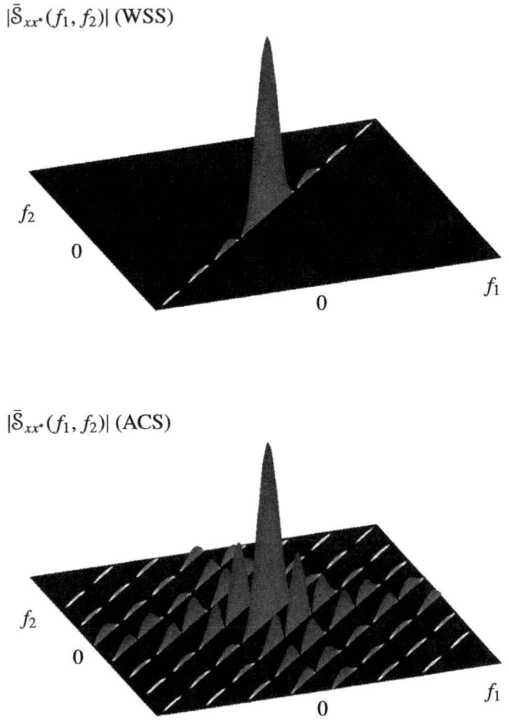

FIGURE 1.2 Magnitude of the bifrequency spectral correlation density $\bar{\mathcal{S}}_{xx^*}(f_1, f_2)$ of (top) a WSS signal and (bottom) an ACS signal as function of f_1 and f_2. *Source: (Napolitano, 2016b) Copyright of Elsevier.*

where

$$\widetilde{A} \triangleq \{\widetilde{\alpha} \in [-1/2, 1/2) \; : \; \widetilde{R}_x^{\widetilde{\alpha}}(m) \not\equiv 0\} \tag{1.39}$$

is the countable set of cycle frequencies, then the autocorrelation can be expressed as

$$\widetilde{R}_x(n, m) = \sum_{\widetilde{\alpha} \in \widetilde{A}} \widetilde{R}_x^{\widetilde{\alpha}}(m) \, e^{j2\pi \widetilde{\alpha} n} \tag{1.40}$$

□

The cyclic autocorrelation function $\widetilde{R}_x^{\widetilde{\alpha}}(m)$ is periodic in $\widetilde{\alpha}$ with period 1. Thus, the sum in (1.40) can be equivalently extended to the set $\widetilde{A}_1 \triangleq \{\widetilde{\alpha} \in [0, 1) \; : \; \widetilde{R}_x^{\widetilde{\alpha}}(m) \not\equiv 0\}$.

In general, the set $\widetilde{A}$ (or $\widetilde{A}_1$) contains possibly incommensurate cycle frequencies $\widetilde{\alpha}$. In the special case where $\widetilde{A}_1 \equiv \{0, 1/N_0, \ldots, (N_0 - 1)/N_0\}$ for some integer N_0, the autocorrelation function $\widetilde{R}_x(n, m)$ is periodic in n with period N_0 and the process $x(n)$ is said to be *cyclostationary in the wide sense.* If $N_0 = 1$ then $x(n)$ is *wide-sense stationary.*

Let us formally denote by

$$\widetilde{X}(\nu) \triangleq \sum_{n \in \mathbb{Z}} x(n) \, e^{-j2\pi \nu n} \tag{1.41}$$

the stochastic process obtained by Fourier transformation (in a generalized sense) of the ACS process $x(n)$. By using (1.40), in the sense of distributions we have that (Hurd, 1991)

$$\mathrm{E}\left\{\widetilde{X}(\nu_1)\, \widetilde{X}^*(\nu_2)\right\} = \sum_{\widetilde{\alpha}\in\widetilde{A}} \widetilde{S}_x^{\widetilde{\alpha}}(\nu_1) \sum_{\ell\in\mathbb{Z}} \delta(\nu_2 - \nu_1 + \widetilde{\alpha} - \ell) \tag{1.42}$$

where

$$\widetilde{S}_x^{\widetilde{\alpha}}(\nu) \triangleq \sum_{m\in\mathbb{Z}} \widetilde{R}_x^{\widetilde{\alpha}}(m)\, e^{-j2\pi\nu m} \tag{1.43}$$

is the *cyclic spectrum* at cycle frequency $\widetilde{\alpha}$. The cyclic spectrum $\widetilde{S}_x^{\widetilde{\alpha}}(\nu)$ is periodic in both ν and $\widetilde{\alpha}$ with period 1.

A covariance (or autocorrelation) is said to be *harmonizable* if it can be expressed as

$$\mathrm{E}\{x(n_1)\, x(n_2)\} = \int_{[-1/2,1/2]^2} e^{j2\pi[\nu_1 n_1 - \nu_2 n_2]}\, \mathrm{d}\widetilde{\gamma}(\nu_1, \nu_2) \tag{1.44}$$

where $\widetilde{\gamma}(\nu_1, \nu_2)$ is a spectral covariance of bounded variation in $[-1/2, 1/2] \times [-1/2, 1/2]$, that is,

$$\int_{[-1/2,1/2]^2} \left|\mathrm{d}\widetilde{\gamma}(\nu_1, \nu_2)\right| < \infty\,. \tag{1.45}$$

Moreover, a second-order stochastic process $x(n)$ is said to be harmonizable if there exists a second-order stochastic process $\chi(\nu)$ with covariance function $\mathrm{E}\{\chi(\nu_1)\, \chi(\nu_2)\} = \widetilde{\gamma}(\nu_1, \nu_2)$ of bounded variation in $[-1/2, 1/2] \times [-1/2, 1/2]$ such that

$$x(n) = \int_{-1/2}^{1/2} e^{j2\pi\nu n}\, \mathrm{d}\chi(\nu) \tag{1.46}$$

with probability 1.

Theorem 1.16. (Gladyshev, 1961), (Dehay and Hurd, 1994). *Discrete-time cyclostationary processes are harmonizable.* □

1.3 Complex signals

The case of complex-valued processes and time series, first treated in depth in (Brown, 1987), is extensively treated in (Gardner, 1986c), (Gardner, 1987f), (Gardner, 1987c), (Spooner and Gardner, 1994a), (Napolitano, 1995), (Izzo and Napolitano, 1997). Several results with reference to higher-order statistics are reported in Chapter 4.

Let $x(t)$ be a zero-mean complex-valued continuous-time process. Its wide-sense characterization can be made in terms of the two second-order moments

$$R_{xx^*}(t, \tau) \triangleq \mathrm{E}\{x(t+\tau)\, x^*(t)\}$$

$$= \sum_{\alpha \in A_{xx^*}} R_{xx^*}^{\alpha}(\tau)\, e^{j2\pi\alpha t} \tag{1.47}$$

$$R_{xx}(t,\tau) \triangleq \mathrm{E}\{x(t+\tau)\, x(t)\}$$
$$= \sum_{\beta \in A_{xx}} R_{xx}^{\beta}(\tau)\, e^{j2\pi\beta t} \tag{1.48}$$

which are called the *autocorrelation function* and *conjugate autocorrelation function*, respectively. The conjugate autocorrelation function $R_{xx}(t,\tau)$ is also referred to as *relation function* (Picinbono and Bondon, 1997) or *complementary correlation* (Schreier and Scharf, 2003b). The Fourier coefficients

$$R_{xx^*}^{\alpha}(\tau) \triangleq \lim_{T\to\infty} \frac{1}{T} \int_{-T/2}^{T/2} \mathrm{E}\{x(t+\tau)\, x^*(t)\}\, e^{-j2\pi\alpha t}\, \mathrm{d}t \tag{1.49}$$

$$R_{xx}^{\beta}(\tau) \triangleq \lim_{T\to\infty} \frac{1}{T} \int_{-T/2}^{T/2} \mathrm{E}\{x(t+\tau)\, x(t)\}\, e^{-j2\pi\beta t}\, \mathrm{d}t \tag{1.50}$$

are referred to as the *cyclic autocorrelation function* at *cycle frequency* α and the *conjugate cyclic autocorrelation function* at *conjugate cycle frequency* β, respectively. The Fourier transforms of the cyclic autocorrelation function and conjugate cyclic autocorrelation function

$$S_{xx^*}^{\alpha}(f) \triangleq \lim_{\Delta f\to 0} \lim_{T\to\infty} \frac{1}{T} \int_{-T/2}^{T/2} \Delta f\, \mathrm{E}\left\{X_{1/\Delta f}(t,f)\, X_{1/\Delta f}^{*}(t,f-\alpha)\right\} \mathrm{d}t$$
$$= \int_{\mathbb{R}} R_{xx^*}^{\alpha}(\tau)\, e^{-j2\pi f\tau} \mathrm{d}\tau \tag{1.51}$$

$$S_{xx}^{\beta}(f) \triangleq \lim_{\Delta f\to 0} \lim_{T\to\infty} \frac{1}{T} \int_{-T/2}^{T/2} \Delta f\, \mathrm{E}\left\{X_{1/\Delta f}(t,f)\, X_{1/\Delta f}(t,\beta-f)\right\} \mathrm{d}t$$
$$= \int_{\mathbb{R}} R_{xx}^{\beta}(f)\, e^{-j2\pi f\tau} \mathrm{d}\tau \tag{1.52}$$

are called the *cyclic spectrum* and the *conjugate cyclic spectrum*, respectively.

Let us denote by the superscript $(*)$ an *optional complex conjugation*. In the following, when it does not create ambiguity, both cyclic autocorrelation function (1.49) and conjugate cyclic autocorrelation function (1.50) are simultaneously represented by

$$R_{xx^{(*)}}^{\alpha}(\tau) \triangleq \lim_{T\to\infty} \frac{1}{T} \int_{-T/2}^{T/2} \mathrm{E}\{x(t+\tau)\, x^{(*)}(t)\}\, e^{-j2\pi\alpha t}\, \mathrm{d}t \qquad \alpha \in A_{xx^{(*)}}\,. \tag{1.53}$$

Analogously, both cyclic spectrum (1.51) and conjugate cyclic spectrum (1.52) are simultaneously represented by

$$S_{xx^{(*)}}^{\alpha}(f) \triangleq \lim_{\Delta f\to 0} \lim_{T\to\infty} \frac{1}{T} \int_{-T/2}^{T/2} \Delta f\, \mathrm{E}\left\{X_{1/\Delta f}(t,f)\, X_{1/\Delta f}^{(*)}(t,{}_{(-)}(\alpha-f))\right\} \mathrm{d}t$$

$$= \int_{\mathbb{R}} R^{\alpha}_{xx^{(*)}}(\tau)\, e^{-j2\pi f\tau}\, \mathrm{d}\tau \tag{1.54}$$

where $(-)$ is an optional minus sign linked to $(*)$.

Both autocorrelation (1.47) and conjugate autocorrelation (1.48) are simultaneously represented by

$$\begin{aligned} R_{xx^{(*)}}(t,\tau) &\triangleq \mathrm{E}\left\{x(t+\tau)\, x^{(*)}(t)\right\} \\ &= \sum_{\alpha \in A_{xx^{(*)}}} R^{\alpha}_{xx^{(*)}}(\tau)\, e^{j2\pi\alpha t}\,. \end{aligned} \tag{1.55}$$

The Loève bifrequency spectrum with and without conjugation for the second term is denoted by

$$\begin{aligned} \mathcal{S}_{xx^{(*)}}(f_1, f_2) &= \mathrm{E}\left\{X(f_1)\, X^{(*)}(f_2)\right\} \\ &= \sum_{\alpha \in A_{xx^{(*)}}} S^{\alpha}_{xx^{(*)}}(f_1)\, \delta(f_2 - (-)(\alpha - f_1)) \end{aligned} \tag{1.56}$$

It should be noted that, in (Gardner, 1986c), (Gardner, 1987f), and (Brown, 1987), for example, a different conjugation notation in the subscripts is adopted to denote the autocorrelation and the conjugate autocorrelation function. Specifically, $R_{xx}(t,\tau) = \mathrm{E}\{x(t+\tau/2)\, x^*(t-\tau/2)\}$ and $R_{xx^*}(t,\tau) = \mathrm{E}\{x(t+\tau/2)\, x(t-\tau/2)\}$. Moreover, an analogous notation is adopted in these references and others for the (conjugate) cyclic autocorrelation function and the (conjugate) cyclic spectrum.

By Fourier transforming with respect to τ both sides of (1.55), we obtain the almost-periodically *time-variant spectrum*

$$\mathcal{S}_{xx^{(*)}}(t, f) = \sum_{\alpha \in A_{xx^{(*)}}} S^{\alpha}_{xx^{(*)}}(f)\, e^{j2\pi\alpha t}\,. \tag{1.57}$$

According to (1.57), the Wigner spectrum (A.24a) of an ACS signal is (see Section 1.9)

$$\begin{aligned} \mathcal{W}_{xx^{(*)}}(t', f) &\triangleq \int_{\mathbb{R}} \mathrm{E}\left\{x\left(t' + \frac{\tau}{2}\right) x^{(*)}\left(t' - \frac{\tau}{2}\right)\right\} e^{-j2\pi f\tau}\, \mathrm{d}\tau \\ &= \sum_{\alpha \in A_{xx^{(*)}}} S^{\alpha}_{xx^{(*)}}\left(f + \frac{\alpha}{2}\right) e^{j2\pi\alpha t'}\,. \end{aligned} \tag{1.58}$$

It is an almost-periodic function of $t' = t + \tau/2$ with frequencies $\alpha \in A_{xx^{(*)}}$ and Fourier-series coefficients $S^{\alpha}_{xx^{(*)}}(f + \alpha/2)$, where the frequency shift $\alpha/2$ is due to the symmetric definition of the Wigner spectrum in terms of the time and lag variables t' and τ (see (A.24a)) (Izzo and Napolitano, 2005).

The (conjugate) *Rihaczek distribution* (A.23) of ACS processes is given by

$$\mathcal{S}_{xx^{(*)}}(t, f)\, \mathrm{d}f = \mathrm{E}\left\{\mathrm{d}\chi(f)\, x^{(*)}(t)\right\} e^{j2\pi f t} \tag{1.59a}$$

$$= \sum_{\alpha \in A_{xx^{(*)}}} \mathrm{d}\mu^{\alpha}_{xx^{(*)}}(f)\, e^{j2\pi\alpha t} \tag{1.59b}$$

$$= \sum_{\alpha \in A_{xx^{(*)}}} S^{\alpha}_{xx^{(*)}}(f)\, e^{j2\pi\alpha t}\, \mathrm{d}f \tag{1.59c}$$

where the last equality holds provided that the functions $\mu^{\alpha}_{xx^{(*)}}(f)$ do not contain singular component.

Finally, note that in some cases, when it does not create ambiguity, subscript $xx^{(*)}$ is replaced by $\boldsymbol{x}$.

Definition 1.17. Let $x(t)$ be a zero-mean ACS stochastic process. The *(conjugate) spectral coherence function* of $x(t)$ at cycle frequency α is defined as (Gardner, 1987f, Chap. 11), (Gardner, 1992)

$$\gamma^{\alpha}_{xx^{(*)}}(f) \triangleq \frac{S^{\alpha}_{xx^{(*)}}(f)}{\left[S^{0}_{xx^*}(f)\, S^{0}_{xx^*}((-)(\alpha - f))\right]^{1/2}} . \tag{1.60}$$

It results

$$0 \leqslant |\gamma^{\alpha}_{xx^{(*)}}(f)| \leqslant 1 . \tag{1.61}$$

□

The magnitude of the spectral coherence is invariant to linear time-invariant (LTI) filtering (Example 3.9).

1.4 Nonzero-mean signals

For a process $x(t)$ with finite or practically finite memory, $x(t+\tau)$ and $x^{(*)}(t)$ become uncorrelated when $|\tau|$ is large. That is

$$\mathrm{E}\{x(t+\tau)\, x^{(*)}(t)\} \to \mathrm{E}\{x(t+\tau)\}\, \mathrm{E}\{x^{(*)}(t)\} \quad \text{as } |\tau| \to \infty . \tag{1.62}$$

Thus, if the process is zero mean, we have that $|R^{\alpha}_{xx^{(*)}}(\tau)| \to 0$ as $|\tau| \to \infty$, $\forall \alpha \in A$. In contrast, if the process has nonzero almost-periodic expected value $\mathrm{E}\{x(t)\}$, then some (conjugate) cyclic autocorrelation functions $R^{\alpha}_{xx^{(*)}}(\tau)$ contain additive sinusoidal functions of τ which arise from products of finite-strength sinusoidal terms contained in $\mathrm{E}\{x(t)\}$ (Gardner and Spooner, 1994a).

Let us consider the decomposition of $x(t)$ into a (possibly zero) almost-periodic component and a residual term $x_r(t)$ not containing any finite-strength additive sine-wave component:

$$x(t) = \sum_{\eta \in E_1} x_\eta\, e^{j2\pi\eta t} + x_r(t) \tag{1.63}$$

where the coefficients x_η are random variables, E_1 is a countable set of possibly incommensurate frequencies, and the residual term $x_r(t)$ is assumed independent of x_η $\forall \eta$, with $\mathrm{E}\{x_r(t)\} = 0$, and such that with probability 1 it results

$$\lim_{T\to\infty} \frac{1}{T} \int_{-T/2}^{T/2} x_r(t)\, e^{-j2\pi\eta t}\, \mathrm{d}t = 0 \qquad \forall \eta \in \mathbb{R} . \tag{1.64}$$

If the almost-periodic component is nonzero, the process $x(t)$ is said to exhibit *first-order cyclostationarity*. The expected value of $x(t)$ is given by

$$\mathrm{E}\{x(t)\} = \sum_{\eta\in E_1} \mathrm{E}\left\{x_\eta\right\} e^{j2\pi\eta t} . \tag{1.65}$$

ACS signals in communications are zero mean unless pilot tones are present for synchronization purposes. Vibroacoustic ACS signals arising in mechanical applications generally have an almost-periodic expected value (Antoni, 2009). The most general case of $x_r(t)$ and x_η statistically dependent is considered in (Napolitano, 2012, Sec. 4.2) in the context of the spectrally correlated processes.

From decomposition (1.63) and accounting for (1.64), it follows that the (conjugate) cyclic autocorrelation function of $x(t)$ can be written as

$$R^{\alpha}_{xx^{(*)}}(\tau) = R^{\alpha}_{x_r x_r^{(*)}}(\tau) + \sum_{\eta\in E_1} \mathrm{E}\left\{x_\eta\, x^{(*)}_{(-)(\alpha-\eta)}\right\} e^{j2\pi\eta\tau} \tag{1.66}$$

where $(-)$ is an optional minus sign linked to $(*)$ and in the sum over E_1 only the values $\eta \in E_1$ such that $(-)(\alpha-\eta) \in E_1$ give nonzero contribution. The (conjugate) cyclic autocorrelation function $R^{\alpha}_{x_r x_r^{(*)}}(\tau)$ approaches zero as $|\tau| \to \infty$, provided that the residual term $x_r(t)$ has finite or practically finite memory.

If $R^{\alpha}_{x_r x_r^{(*)}}(\tau) \in L^1(\mathbb{R})$, by Fourier transforming both sides of (1.66) one obtains

$$S^{\alpha}_{xx^{(*)}}(f) = S^{\alpha}_{x_r x_r^{(*)}}(f) + \sum_{\eta\in E_1} \mathrm{E}\left\{x_\eta\, x^{(*)}_{(-)(\alpha-\eta)}\right\} \delta(f-\eta) \tag{1.67}$$

where $\delta(\cdot)$ denotes Dirac delta and the (conjugate) cyclic spectrum $S^{\alpha}_{x_r x_r^{(*)}}(f)$ is an ordinary Fourier transform and does not contain impulses.

From (1.66) and (1.67) it follows that the presence of an almost-periodic component in $x(t)$ reflects into the lack of summability of the (conjugate) cyclic autocorrelation functions and the presence of Dirac deltas in the (conjugate) cyclic spectra. Better behaved functions can be obtained by considering the *(conjugate) covariance function*

$$C_{xx^{(*)}}(t,\tau) \triangleq \mathrm{E}\left\{\left[x(t+\tau) - \mathrm{E}\{x(t+\tau)\}\right]\left[x(t) - \mathrm{E}\{x(t)\}\right]^{(*)}\right\} \tag{1.68a}$$

$$= \sum_{\beta\in B_{xx^{(*)}}} C^{\beta}_{xx^{(*)}}(\tau)\, e^{j2\pi\beta t} \tag{1.68b}$$

where $B_{xx^{(*)}}$ is a countable set of (conjugate) covariance cycle frequencies and the second equality holds provided that the series of the *(conjugate) cyclic covariance functions*

$$C^{\beta}_{xx^{(*)}}(\tau) \triangleq \lim_{T\to\infty} \frac{1}{T} \int_{-T/2}^{T/2} C_{xx^{(*)}}(t,\tau)\, e^{-j2\pi\beta t}\, dt = R^{\beta}_{x_r x_r^{(*)}}(\tau) \tag{1.69}$$

is absolutely summable, that is,

$$\sum_{\beta\in B_{xx^{(*)}}} \left| C^{\beta}_{xx^{(*)}}(\tau) \right| < \infty . \tag{1.70}$$

For a finite or practically finite memory process, accounting for (1.62) we have

$$C_{xx^{(*)}}(t,\tau) \to 0 \text{ as } |\tau| \to \infty . \tag{1.71}$$

If the rate of decay to zero is sufficiently fast so that $C^{\beta}_{xx^{(*)}}(\tau) \in L^1(\mathbb{R})$, the Fourier transform

$$P^{\beta}_{xx^{(*)}}(f) \triangleq \int_{\mathbb{R}} C^{\beta}_{xx^{(*)}}(\tau)\, e^{-j2\pi f\tau}\, d\tau = S^{\beta}_{x_r x_r^{(*)}}(f) \tag{1.72}$$

which is referred to as *second-order (conjugate) cyclic polyspectrum* (Theorem 4.18), (Gardner and Spooner, 1994a), is defined in ordinary sense and does not contain Dirac impulses.

From (1.69) and (1.72) it follows that the (conjugate) cyclic covariance and the second-order (conjugate) cyclic polyspectrum of $x(t)$ are coincident with the (conjugate) cyclic autocorrelation and the (conjugate) cyclic spectrum, respectively, of $x_r(t)$.

By specializing to the case $N = 2$ the results of Section 4.2.3.1, we have that

$$B_{xx^{(*)}} \subseteq A_{xx^{(*)}} . \tag{1.73}$$

1.5 Jointly ACS signals

Let $x_i(t)$, $i = 1, 2$, $t \in \mathbb{R}$, be two complex-valued continuous-time processes.

Definition 1.18. The processes $x_1(t)$ and $x_2(t)$ are said to be *jointly almost-cyclostationary* for a given choice of conjugation $(*)$, if their (conjugate) cross-correlation function can be expressed as

$$\begin{aligned} R_{x_1x_2^{(*)}}(t,\tau) &\triangleq \mathrm{E}\left\{x_1(t+\tau)\, x_2^{(*)}(t)\right\} \\ &= \sum_{\alpha\in A_{12}} R^{\alpha}_{x_1x_2^{(*)}}(\tau)\, e^{j2\pi\alpha t} \end{aligned} \tag{1.74}$$

where

$$R^{\alpha}_{x_1x_2^{(*)}}(\tau) \triangleq \lim_{T\to\infty} \frac{1}{T} \int_{-T/2}^{T/2} \mathrm{E}\left\{x_1(t+\tau)\, x_2^{(*)}(t)\right\} e^{-j2\pi\alpha t}\, dt \tag{1.75}$$

is the *(conjugate) cyclic cross-correlation function* of x_1 and x_2 at (conjugate) cycle frequency α and

$$A_{12} \triangleq \{\alpha \in \mathbb{R} : R^{\alpha}_{x_1 x_2^{(*)}}(\tau) \not\equiv 0\} \tag{1.76}$$

is a countable set. □

Note that in Definition 1.18 no assumption is made on the possible marginal almost-cyclostationarity of $x_1(t)$ and $x_2(t)$. In addition, in general, the set A_{12} depends on whether $(*)$ is conjugation or not and can be different from the sets A_{11} and A_{22} of the (conjugate) cycle frequencies of $x_1(t)$ and $x_2(t)$.

If the set A_{12} contains only the element $\alpha = 0$, then the processes $x_1(t)$ and $x_2(t)$ are jointly WSS. Since no assumption is made on the possible wide-sense stationarity of $x_1(t)$ and $x_2(t)$, this definition of jointly WSS processes is different from that given in (Papoulis, 1991, p. 299) where each $x_1(t)$ and $x_2(t)$ is WSS and their cross-correlation depends only on τ.

Finally, note that two processes can be jointly ACS for a choice of conjugation and not jointly ACS for the other choice.

The *(conjugate) cyclic cross-spectrum* of x_1 and x_2 at (conjugate) cycle frequency α is defined as

$$S^{\alpha}_{x_1 x_2^{(*)}}(f) \triangleq \lim_{\Delta f \to 0} \lim_{T \to \infty} \frac{1}{T} \int_{-T/2}^{T/2} \Delta f \, \mathrm{E}\left\{X_{1,1/\Delta f}(t, f)\, X^{(*)}_{2,1/\Delta f}(t, (-)(\alpha - f))\right\} \mathrm{d}t \tag{1.77}$$

and is related to the (conjugate) cyclic cross-correlation function by the *Gardner relation* (also referred to as *cyclic Wiener-Khinchin* relation)

$$S^{\alpha}_{x_1 x_2^{(*)}}(f) = \int_{\mathbb{R}} R^{\alpha}_{x_1 x_2^{(*)}}(\tau)\, e^{-j2\pi f \tau} \mathrm{d}\tau \,. \tag{1.78}$$

In (1.77), $X_{i,1/\Delta f}(t, f)$ is the STFT of $x_i(t)$ defined according to (1.31).

In the case of distinct processes (or processes denoted as distinct), the use of optional conjugation can be avoided without lack of generality. In fact, if $x_i(t) = x_0^*(t)$, then $X_i(f) = X_0^*(-f)$. However, the following results are useful for future reference. Let $\boldsymbol{x} \triangleq x_1^{(*)_1}\, x_2^{(*)_2}$, where superscript $(*)_i$ denotes ith optional complex conjugation.

Proposition 1.19. *If*

$$\mathrm{E}\left\{x_1^{(*)_1}(t+\tau)\, x_2^{(*)_2}(t)\right\} = \sum_{\alpha \in A_{\boldsymbol{x}}} R^{\alpha}_{\boldsymbol{x}}(\tau)\, e^{j2\pi\alpha t} \tag{1.79}$$

with

$$R^{\alpha}_{\boldsymbol{x}}(\tau) \triangleq \lim_{T \to \infty} \frac{1}{T} \int_{-T/2}^{T/2} \mathrm{E}\left\{x_1^{(*)_1}(t+\tau)\, x_2^{(*)_2}(t)\right\} e^{-j2\pi\alpha t} \, \mathrm{d}t \tag{1.80}$$

then

$$\mathrm{E}\left\{X_1^{(*)_1}(f_1)\, X_2^{(*)_2}(f_2)\right\} = \sum_{\alpha \in A_{\boldsymbol{x}}} S_{\boldsymbol{x}}^{\alpha}((-)_1 f_1)\, \delta(f_2 - (-)_2(\alpha - (-)_1 f_1)) \tag{1.81}$$

where $(-)_i$ *is an optional minus sign linked to* $(*)_i$ *and*

$$S_{\boldsymbol{x}}^{\alpha}(f) = \int_{\mathbb{R}} R_{\boldsymbol{x}}^{\alpha}(\tau)\, e^{-j2\pi f \tau}\, \mathrm{d}\tau\,. \tag{1.82}$$

Proof. See Section 1.9. □

By specializing (1.79) and (1.81) to the case $x_1^{(*)_1} \equiv y$ and $x_2^{(*)_2} \equiv x^{(*)}$, we have

$$\mathrm{E}\left\{y(t+\tau)\, x^{(*)}(t)\right\} = \sum_{\alpha \in A_{yx^{(*)}}} R_{yx^{(*)}}^{\alpha}(\tau)\, e^{j2\pi\alpha t} \tag{1.83}$$

$$\mathrm{E}\left\{Y(f_1)\, X^{(*)}(f_2)\right\} = \sum_{\alpha \in A_{yx^{(*)}}} S_{yx^{(*)}}^{\alpha}(f_1)\, \delta(f_2 - (-)(\alpha - f_1))\,. \tag{1.84}$$

1.5.1 Symmetry relationships

(Conjugate) cyclic cross-correlations and spectra exhibit the following symmetries (Section 1.9)

$$R_{yx^{(*)}}^{\alpha}(\tau) = R_{x^{(*)}y}^{\alpha}(-\tau)\, e^{j2\pi\alpha\tau} \tag{1.85}$$

$$S_{yx^{(*)}}^{\alpha}(f) = S_{x^{(*)}y}^{\alpha}(\alpha - f) \tag{1.86}$$

$$R_{yx^{(*)}}^{-\alpha}(\tau) = R_{y^*x^{(*)*}}^{\alpha}(\tau)^* \tag{1.87}$$

$$S_{yx^{(*)}}^{-\alpha}(f) = S_{y^*x^{(*)*}}^{\alpha}(-f)^* \tag{1.88}$$

For $y(t) \equiv x(t)$, one obtains relationships between cyclic statistics and conjugate cyclic statistics of the same signal. For $y(t) \equiv x(t)$ real-valued, the above relationships provide symmetry properties of cyclic autocorrelation and cyclic spectra of real-valued signals.

Note that different symmetry relationships are obtained if the symmetric definition of (conjugate) cyclic cross-correlation and cross-spectrum (see (1.34) and (1.35) for the case $y \equiv x$) are adopted (Gardner, 1987f, Chap. 11, Sec. C-3). Symmetry properties are addressed in (Dragan et al., 1982), (Alekseev, 1987b), (Brown, 1987).

1.6 Representations by stationary components

1.6.1 Continuous-time

In this section, two representations of a generic cyclostationary process in terms of elementary cyclostationary processes are presented. The former is the harmonic series representation, where the elementary cyclostationary processes are (complex-valued) amplitude-modulated signals (Section 7.2). The latter is the translation series representation, where the elementary cyclostationary processes are pulse-amplitude modulated signals (Section 7.3). These representations are exploited in (Gardner and Franks, 1975) to provide general solutions for filters that minimize mean-squared error for continuous-waveform estimation.

1.6.1.1 Harmonic series representation

A continuous-time wide-sense cyclostationary process $x(t)$ can be expressed in terms of singularly and jointly wide-sense stationary processes with non-overlapping spectral bands (Dragan, 1975), (Gardner and Franks, 1975), (Gardner, 1985), (Dehay and Hurd, 1994), (Makagon, 1999a). Let T_0 be the period of cyclostationarity, and define

$$\widetilde{x}_k(t) \triangleq \left[x(t)\, e^{-j2\pi(k/T_0)t}\right] \otimes h_0(t) \tag{1.89}$$

where $h_0(t) \triangleq (1/T_0)\mathrm{sinc}(t/T_0)$ is the impulse-response function of the ideal low-pass filter $H_0(f) \triangleq \mathrm{rect}(fT_0)$ with monolateral bandwidth $1/2T_0$. The process $x(t)$ can be expressed by the *harmonic series representation*

$$x(t) = \sum_{k=-\infty}^{+\infty} \widetilde{x}_k(t)\, e^{j2\pi(k/T_0)t} \tag{1.90}$$

where (see Section 1.9)[1]

$$\begin{aligned} \mathrm{E}\{\widetilde{x}_k(t+\tau)\, \widetilde{x}_h^{(*)}(t)\} &= \int_{-1/2T_0}^{1/2T_0} S_{xx^{(*)}}^{(k+(-)h)/T_0}\left(f + \frac{k}{T_0}\right) e^{j2\pi f\tau}\, \mathrm{d}f \\ &\triangleq R_{\widetilde{x}_k\widetilde{x}_h^{(*)}}(\tau) \end{aligned} \tag{1.91}$$

which is independent of t. This result reflects the fact that the Fourier transforms (defined in a generalized sense (Section A.3)) $\widetilde{X}_k(f)$ of the signals $\widetilde{x}_k(t)$ have non-overlapping supports of width $1/T_0$

$$\widetilde{X}_k\left(f - \frac{k}{T_0}\right) = \begin{cases} X(f) & |f - k/T_0| \leqslant 1/2T_0 \\ 0 & \text{otherwise;} \end{cases} \tag{1.92}$$

that immediately follows from (1.89). Therefore, since only spectral components of $x(t)$ with frequencies separated by an integer multiple of $1/T_0$ can be correlated, then the only pairs of spectral components in $\widetilde{x}_k(t)$ and $\widetilde{x}_h(t)$ that can be (conjugate) correlated are those with the same frequency f. Thus, there is no (conjugate) spectral cross-correlation at distinct frequencies in $\widetilde{x}_k(t)$ and $\widetilde{x}_h(t)$.

It follows that a continuous-time wide-sense cyclostationary scalar signal $x(t)$ is equivalent to the infinite-dimensional vector-valued wide-sense stationary signal

$$[\ldots, \widetilde{x}_{-k}(t), \ldots, \widetilde{x}_{-1}(t), \widetilde{x}_0(t), \widetilde{x}_1(t), \ldots, \widetilde{x}_k(t), \ldots].$$

[1] For $(*) = *$, corrected version of (Gardner et al., 2006, Eq. (3.117)).

The (conjugate) autocorrelation is given by (see Section 1.9)

$$\mathrm{E}\left\{x(t_1)\,x^{(*)}(t_2)\right\} = \sum_{k_1=-\infty}^{+\infty}\sum_{k_2=-\infty}^{+\infty} R_{\widetilde{x}_{k_1}\widetilde{x}_{k_2}^{(*)}}(t_1-t_2)\,e^{j2\pi(k_1/T_0)t_1}\,e^{(-)j2\pi(k_2/T_0)t_2}\,. \tag{1.93}$$

The (conjugate) autocorrelation can also be expressed as (see Section 1.9)

$$\mathrm{E}\left\{x(t+\tau)\,x^{(*)}(t)\right\} = \sum_{k=-\infty}^{+\infty} R_{xx^{(*)}}^{k/T_0}(\tau)\,e^{j2\pi(k/T_0)t} \tag{1.94}$$

with (conjugate) cyclic autocorrelation functions (Gladyshev, 1961)

$$R_{xx^{(*)}}^{k/T_0}(\tau) = \sum_{k_1=-\infty}^{+\infty} R_{\widetilde{x}_{k_1}\widetilde{x}_{(-)(k-k_1)}^{(*)}}(\tau)\,e^{j2\pi(k_1/T_0)\tau}\,. \tag{1.95}$$

The (conjugate) cyclic spectra are obtained by Fourier transforming both sides of (1.95):

$$S_{xx^{(*)}}^{k/T_0}(f) = \sum_{k_1=-\infty}^{+\infty} S_{\widetilde{x}_{k_1}\widetilde{x}_{(-)(k-k_1)}^{(*)}}\left(f-\frac{k_1}{T_0}\right) \tag{1.96}$$

where

$$S_{\widetilde{x}_k\widetilde{x}_h^{(*)}}(f) \triangleq \int_{\mathbb{R}} R_{\widetilde{x}_k\widetilde{x}_h^{(*)}}(\tau)\,e^{-j2\pi f\tau}\,\mathrm{d}\tau\,. \tag{1.97}$$

A harmonic series representation can also be obtained for an ACS process $x(t)$, provided that it belongs to the sub-class of the almost-periodically unitary processes (Hurd, 1992). In this case,

$$x(t) = \sum_{k=-\infty}^{+\infty} \widetilde{x}_k(t)\,e^{j2\pi\lambda_k t} \tag{1.98}$$

where the frequencies λ_k are possibly incommensurate and the processes $\widetilde{x}_k(t)$ are singularly and jointly wide-sense stationary but not necessarily band limited.

1.6.1.2 Translation series representation

Let $x(t)$ be a cyclostationary process with period of cyclostationarity T_0.

Let $\{\phi_i(t)\}_{i\in\mathbb{N}}$ a complete orthonormal set of basis functions in $L^2([0,T_0])$. It results $\phi_i(t)=0$ for $t\notin[0,T_0]$ and

$$\int_0^{T_0} \phi_{i_1}(t)\,\phi_{i_2}^*(t)\,\mathrm{d}t = \delta_{i_1-i_2}\,. \tag{1.99}$$

The cyclostationary signal $x(t)$ can be expanded on the intervals $t \in [nT_0, (n+1)T_0)$ for all integers n in terms of translates of the basis functions. That is,

$$x(t) = \sum_{i=1}^{+\infty} a_i(0)\, \phi_i(t) \qquad t \in [0, T_0] \tag{1.100}$$

and

$$x(t) = \sum_{i=1}^{+\infty} a_i(n)\, \phi_i(t - nT_0) \qquad t \in [nT_0, (n+1)T_0] \tag{1.101}$$

where the random variables $a_i(n)$ are defined as

$$a_i(n) \triangleq \int_{nT_0}^{(n+1)T_0} x(t)\, \phi_i^*(t - nT_0)\, \mathrm{d}t \tag{1.102a}$$

$$= \int_0^{T_0} x(t' + nT_0)\, \phi_i^*(t')\, \mathrm{d}t' \,. \tag{1.102b}$$

The translates $\phi_i(t - nT_0)$ for different values of n do not overlap and we have

$$x(t) = \sum_{n=-\infty}^{+\infty} \left[\sum_{i=1}^{+\infty} a_i(n)\, \phi_i(t - nT_0) \right]$$

$$= \sum_{i=1}^{+\infty} \left[\sum_{n=-\infty}^{+\infty} a_i(n)\, \phi_i(t - nT_0) \right] \qquad t \in \mathbb{R} \tag{1.103}$$

which is referred to as *translation series representation (TSR)*.

The sequences $a_{i_1}(n)$ and $a_{i_2}(n)$ are jointly WSS (see Section 1.9):

$$\mathrm{E}\{a_{i_1}(n_1)\, a_{i_2}^{(*)}(n_2)\}$$

$$= \sum_{k=-\infty}^{+\infty} \int_{-T_0}^{T_0} R_{xx^{(*)}}^{k/T_0}(\tau + (n_1 - n_2)T_0)\, r_{\phi_{i_1}\phi_{i_2}^{(*)}}^{k/T_0}(\tau)^*\, \mathrm{d}\tau \tag{1.104a}$$

$$= \sum_{k=-\infty}^{+\infty} \int_{\mathbb{R}} S_{xx^{(*)}}^{k/T_0}(f)\, e^{j2\pi(n_1-n_2)fT_0}\, \Phi_{i_1}^*(f)\, \Phi_{i_2}^{(*)*}\left((-)\left(\frac{k}{T_0} - f\right)\right) \mathrm{d}f \tag{1.104b}$$

$$\triangleq r_{i_1 i_2 (*)}(n_1 - n_2)$$

where

$$r_{\phi_{i_1}\phi_{i_2}^{(*)}}^{k/T_0}(\tau) \triangleq \int_0^{T_0} \phi_{i_1}(s + \tau)\, \phi_{i_2}^{(*)}(s)\, e^{-j2\pi(k/T_0)s}\, \mathrm{d}s \tag{1.105}$$

$$\overset{\mathcal{F}}{\longleftrightarrow} \quad \Phi_{i_1}(f)\, \Phi_{i_2}^{(*)}\left((-)\left(\frac{k}{T_0} - f\right)\right) \tag{1.106}$$

and

$$\Phi_i(f) \triangleq \int_{\mathbb{R}} \phi_i(t)\, e^{-j2\pi f t}\, \mathrm{d}t\,. \tag{1.107}$$

In (1.103), the cyclostationary process $x(t)$ is expressed as the superposition of pulse-amplitude-modulated (PAM) processes with jointly WSS sequences (see (1.104a)–(1.104b)) and orthogonal pulses (see (1.99)) of finite duration T_0.

By using the TSR (1.103) for the cyclostationary process $x(t)$, one obtains the following expression for the (conjugate) autocorrelation function (see Section 1.9)

$$\begin{aligned} &\mathrm{E}\left\{x(t_1)\, x^{(*)}(t_2)\right\} \\ &\quad = \sum_{n_1=-\infty}^{+\infty} \sum_{n_2=-\infty}^{+\infty} \sum_{i_1=1}^{+\infty} \sum_{i_2=1}^{+\infty} r_{i_1 i_2 (*)}(n_1 - n_2)\, \phi_{i_1}(t_1 - n_1 T_0)\, \phi_{i_2}^{(*)}(t_2 - n_2 T_0)\,. \end{aligned} \tag{1.108}$$

Equivalently, by replacing $n_1 = n_2 + m$ and $t_1 = t_2 + \tau$ into (1.108) and then renaming $t_2 = t$ and $n_2 = n$, the (conjugate) autocorrelation can be expressed as the periodic replication with period T_0

$$\mathrm{E}\left\{x(t+\tau)\, x^{(*)}(t)\right\} = \sum_{n=-\infty}^{+\infty} \rho_{xx^{(*)}}(t - nT_0, \tau) \tag{1.109}$$

where the generator signal is

$$\rho_{xx^{(*)}}(t,\tau) = \sum_{i_1=1}^{+\infty} \sum_{i_2=1}^{+\infty} \sum_{m=-\infty}^{+\infty} r_{i_1 i_2 (*)}(m)\, \phi_{i_1}(t + \tau - mT_0)\, \phi_{i_2}^{(*)}(t) \tag{1.110}$$

which is nonzero only for $t \in [0, T_0]$.

The (conjugate) cyclic autocorrelation functions are the Fourier coefficients of the periodic function of t (1.109). Thus, they are given by the samples of the Fourier transform (with respect to the variable t) of the generator (1.110) divided by T_0 (Section C.2):

$$R_{xx^{(*)}}^{k/T_0}(\tau) = \frac{1}{T_0} \int_{\mathbb{R}} \rho_{xx^{(*)}}(t,\tau)\, e^{-j2\pi\alpha t}\, \mathrm{d}t \Bigg|_{\alpha = k/T_0} \tag{1.111a}$$

$$= \frac{1}{T_0} \sum_{i_1=1}^{+\infty} \sum_{i_2=1}^{+\infty} \sum_{m=-\infty}^{+\infty} r_{i_1 i_2 (*)}(m)\, r_{\phi_{i_1} \phi_{i_2}^{(*)}}^{k/T_0}(\tau - mT_0)\,. \tag{1.111b}$$

By Fourier transforming expression (1.111b) of the (conjugate) cyclic autocorrelation function one obtains the (conjugate) cyclic spectrum (Section 1.9)

$$S_{xx^{(*)}}^{k/T_0}(f) = \frac{1}{T_0} \sum_{i_1=1}^{+\infty} \sum_{i_2=1}^{+\infty} \Phi_{i_1}(f)\, \Phi_{i_2}^{(*)}\left((-)\left(\frac{k}{T_0} - f\right)\right) S_{i_1 i_2 (*)}(f T_0) \tag{1.112}$$

where

$$S_{i_1 i_2(*)}(\nu) \triangleq \sum_{m=-\infty}^{+\infty} r_{i_1 i_2(*)}(m)\, e^{-j2\pi \nu m} \,. \tag{1.113}$$

By double Fourier transforming both sides of (1.108) one obtains the Loève bifrequency spectrum (see Section 1.9)

$$\mathrm{E}\left\{X(f_1)\, X^{(*)}(f_2)\right\} = \sum_{k=-\infty}^{+\infty} S^{k/T_0}_{xx^{(*)}}(f_1)\, \delta\!\left(f_2 - {}_{(-)}\!\left(\frac{k}{T_0} - f_1\right)\right). \tag{1.114}$$

No translation series representation with stationary components has been demonstrated to exist for ACS processes.

1.6.2 Discrete-time

1.6.2.1 Harmonic series representation

A discrete-time wide-sense cyclostationary process $x(n)$ can be expressed in terms of a finite number of singularly and jointly wide-sense stationary processes $\widetilde{x}_k(n)$ with non-overlapping bands (Hurd, 1969), (Pagano, 1978), (Gardner, 1985), (Sakai, 1991), (Dehay and Hurd, 1994), (Giannakis, 1998). Let N_0 be the period of cyclostationarity, and define

$$\widetilde{x}_k(n) \triangleq \left[x(n)\, e^{-j2\pi(k/N_0)n}\right] \otimes h_0(n) \tag{1.115}$$

where $h_0(n) \triangleq (1/N_0)\mathrm{sinc}(n/N_0)$ is the ideal low-pass filter with monolateral bandwidth $1/2N_0$. The process $x(n)$ can be expressed by the *harmonic series representation*

$$x(n) = \sum_{k=0}^{N_0-1} \widetilde{x}_k(n)\, e^{j2\pi(k/N_0)n} \,. \tag{1.116}$$

Therefore, a discrete-time wide-sense cyclostationary scalar process $x(n)$ is equivalent to the N_0-dimensional vector-valued wide-sense stationary process

$$[\widetilde{x}_0(n), \widetilde{x}_1(n), \ldots, \widetilde{x}_{N_0-1}(n)] \,.$$

1.6.2.2 Polyphase decomposition

A further decomposition of a discrete-time cyclostationary process can be obtained in terms of sub-sampled components. Let $x(n)$ be a discrete-time real-valued wide-sense cyclostationary process with cyclostationarity period N_0. The sub-sampled (or decimated) processes

$$x_i(n) \triangleq x(nN_0 + i) \qquad i = 0, \ldots, N_0 - 1 \tag{1.117}$$

constitute what is called the *polyphase decomposition* of $x(n)$. Given the set of processes $x_i(n), i = 0, \ldots, N_0 - 1$, the original process $x(n)$ can be reconstructed by using the synthesis formula

$$x(n) = \sum_{i=0}^{N_0-1} \sum_{\ell \in \mathbb{Z}} x_i(\ell) \delta_{n-i-\ell N_0} \tag{1.118}$$

where δ_γ is the Kronecker delta ($\delta_\gamma = 1$ for $\gamma = 0$ and $\delta_\gamma = 0$ for $\gamma \neq 0$). The process $x(n)$ is wide-sense cyclostationary with period N_0 if and only if the set of sub-sampled $x_i(n)$ are jointly wide-sense stationary (Gladyshev, 1961). In fact, due to the cyclostationarity of $x(n)$,

$$\begin{aligned} \mathrm{E}\{x_i(n+m)\, x_k(n)\} &= \mathrm{E}\{x((n+m)N_0+i)\, x(nN_0+k)\} \\ &= \mathrm{E}\{x(mN_0+i)\, x(k)\}\,, \end{aligned} \tag{1.119}$$

which is independent of n.

1.7 Special topics

Special topics related to cyclostationary and almost-cyclostationary processes are:

- *General properties of stochastic processes*
  (Jordan, 1961), (Stratonovich, 1967), (Franks, 1969), (Dragan, 1969a), (Yaglom, 1986), (Rytov et al., 1988), (Franks, 1997);
- *Characterization of periodically correlated (cyclostationary) processes*
  (Voychishin and Dragan, 1971), (Dragan, 1972a), (Lugannani, 1973), (Dragan and Yavorskii, 1974b), (van Der Wurf, 1974), (Piranashvili and Kavtaradze, 1975), (Berezin and Krasheninnikov, 1978), (Porotskii and Tsymbalyuk, 1979), (Dragan et al., 1980b), (Honda, 1980), (Yuran, 1980), (Kapustinskas, 1981), (Gartshtein and Smelyakov, 1982), (Ogura and Yoshida, 1982), (Kuznetsov and Chernenko, 1984), (Martin, 1990), (Hurd and Kallianpur, 1991), (Kim et al., 1996), (Lambert, 1996), (De Nicolao and Ferrari-Trecate, 1999), (Alpay et al., 2001), (Lacaze and Roviras, 2002), (Miamee and Shahkar, 2002), (Hurd, 2005), (Lambert-Lacroix, 2005), (Soltani and Parvardeh, 2005), (Wang et al., 2005), (Soltani and Shishebor, 2007), (Makagon, 2011), (Thomson, 2011), (Shevgunov, 2019);
- *Characterization of almost-periodically correlated (almost-cyclostationary) processes*
  (Kayatskas, 1970), (Lenart and Pipien, 2013);
- *Variance analysis*
  (Herbst, 1963), (Herbst, 1965);
- *Extreme values*
  (Ballerini and McCormick, 1989), (Konstantinides and Pieterbarg, 1993);
- *Trend extraction*
  (Alekseev, 1987a);
- *Markov models*
  (Kochel, 1980), (Osteyee, 1981);

- *Non-Gaussian and long memory models*
  (Willis, 1964), (Renger, 1977), (Melititskiy and Mosionzhik, 1987), (Konstantinides et al., 2004), (Philippe et al., 2006), (Tan and Hashorva, 2013b), (Dehay, 2015b);
- *Stochastic difference and differential equations*
  (Dorogovtsev and Tkhuan, 1988), (Dorogovtsev, 1990b), (Dorogovtsev, 1990a), (Dorogovtsev, 1991);
- *Discrete scale invariance*
  (Borgnat et al., 2002), (Borgnat et al., 2005);
- *Other subjects*
  (Rootenberg and Ghozati, 1978), (Sakai, 1983), (Kamolor, 1984), (Dragan, 1988), (Miamee and Hardin, 1990), (McLeod, 1993), (Dorogovtsev, 2002), (Woerdeman, 2002), (Tan and Hashorva, 2013a).

1.8 Summary

Cyclostationary processes in the wide sense have mean and autocorrelation function that are periodic functions of time (Definition 1.2). Under suitable regularity conditions, they can be expanded in Fourier series. The frequencies and coefficients of the Fourier series of the autocorrelation are called *cycle frequencies* and *cyclic autocorrelation functions*, respectively (Theorem 1.4).

More generally, the process is called *almost-cyclostationary* (Definition 1.7) if mean and autocorrelation are almost-periodic functions of time. Under suitable regularity conditions they can be expressed as Fourier series where the frequencies are incommensurate (Theorem 1.9). That is, cycle frequencies are not integer multiple of a fundamental frequency.

In the frequency domain, assuming that the process is harmonizable, the Loève bifrequency spectrum is concentrated on lines parallel to the main diagonal of the bifrequency plane that are separated by quantities equal to the cycle frequencies (Theorem 1.12). The densities of spectral correlation along these lines are the *cyclic spectra* that are the Fourier transforms of the cyclic autocorrelation functions.

Alternatively, the cyclic spectrum at cycle frequency α can be defined as the time-averaged statistical correlation (with zero lag) of two spectral components at frequencies f and $f - \alpha$, as the bandwidth approaches zero. It is proved to be the Fourier transform of the cyclic autocorrelation function at cycle frequency α and the Fourier pair is dubbed *Gardner relation*.

Almost-cyclostationary processes can be defined at both continuous and discrete time. In addition, complex-valued almost-cyclostationary processes can be considered (Section 1.3). In such a case, both autocorrelation (the second term is conjugated) and conjugate autocorrelation (the second term is not conjugated) are almost-periodic functions of time. For the conjugate autocorrelation the frequencies of the Fourier series are called conjugate cycle frequencies and the coefficients are called conjugate cyclic autocorrelation functions.

Almost-cyclostationary processes with almost-periodic mean function have the cyclic autocorrelation functions that contain finite-strength additive sinewave components. This reflects into the presence of Dirac impulses in the cyclic spectra (Section 1.4).

Jointly almost-cyclostationary processes have the (conjugate) cross-correlation function that is an almost-periodic function of time (Definition 1.18). The frequencies and coefficients of the Fourier series are called (conjugate) cycle frequencies and (conjugate) cyclic cross-correlation functions, respectively. The Fourier transforms of the (conjugate) cyclic cross-correlation functions are the (conjugate) cyclic cross-spectra.

A generic cyclostationary process can be expressed in terms of elementary cyclostationary processes. Two representations are considered. The former is the *harmonic series representation* (Section 1.6.1.1), where the elementary cyclostationary processes are amplitude modulated signals with jointly wide-sense stationary modulating signals with non-overlapping spectra. The latter is the *translation series representation* (Section 1.6.1.2), where the elementary cyclostationary processes are pulse-amplitude modulated signals with jointly wide-sense stationary modulating sequences. At discrete time, the harmonic series representation allows one to establish the equivalence between a scalar cyclostationary process and a finite-dimension vector-valued wide-sense stationary process.

1.9 Proofs

In this section, proofs are reported for results of the present chapter. Unless otherwise specified, only formal manipulations are reported, without explicitly justifying inversion of linear operators such as time averages, infinite sums, integrals, and ensemble average (expectation). In general, these manipulations are justified at least in the sense of generalized functions (distributions) (Champeney, 1990), (Zemanian, 1987).

Proof of Theorem 1.12

By replacing $t+\tau \curvearrowright t_1$ and $t \curvearrowright t_2$ into (1.21) with (1.24) substituted into, we have

$$\begin{aligned}
&\mathrm{E}\{x(t+\tau)\,x(t)\} \\
&\quad = \int_{\mathbb{R}^2} e^{j2\pi(f_1(t+\tau)-f_2 t)} \sum_{\alpha\in A} \delta(f_2 - f_1 + \alpha)\,\mathrm{d}\mu_x^{\alpha}(f_1)\,\mathrm{d}f_2 \\
&\quad = \sum_{\alpha\in A} \int_{\mathbb{R}} e^{j2\pi f_1(t+\tau)}\,\mathrm{d}\mu_x^{\alpha}(f_1) \underbrace{\int_{\mathbb{R}} \delta(f_2 - f_1 + \alpha)\,e^{-j2\pi f_2 t}\,\mathrm{d}f_2}_{e^{-j2\pi(f_1-\alpha)t}} \\
&\quad = \sum_{\alpha\in A} \int_{\mathbb{R}} e^{j2\pi f_1 \tau}\,\mathrm{d}\mu_x^{\alpha}(f_1)\,e^{j2\pi\alpha t} \qquad (1.120)
\end{aligned}$$

Proof of (1.58)

According to (1.55) we have

$$\mathrm{E}\left\{x(t_1)\,x^{(*)}(t_2)\right\} = \sum_{\alpha\in A} R^{\alpha}_{xx^{(*)}}(t_1 - t_2)\, e^{j2\pi\alpha t_2}\,. \tag{1.121}$$

Thus,

$$\begin{aligned}
\mathcal{W}_{xx^{(*)}}(t', f) &\triangleq \int_{\mathbb{R}} \mathrm{E}\left\{x\left(t' + \frac{\tau}{2}\right) x^{(*)}\left(t' - \frac{\tau}{2}\right)\right\} e^{-j2\pi f\tau}\, \mathrm{d}\tau \\
&= \int_{\mathbb{R}} \sum_{\alpha\in A} R^{\alpha}_{xx^{(*)}}(\tau)\, e^{j2\pi\alpha(t'-\tau/2)}\, e^{-j2\pi f\tau}\, \mathrm{d}\tau \\
&= \sum_{\alpha\in A} \underbrace{\int_{\mathbb{R}} R^{\alpha}_{xx^{(*)}}(\tau)\, e^{-j2\pi(\alpha/2)\tau}\, e^{-j2\pi f\tau}\, \mathrm{d}\tau}_{S^{\alpha}_{xx^{(*)}}(f+\alpha/2)}\, e^{j2\pi\alpha t'}\,.
\end{aligned} \tag{1.122}$$

Proof of Proposition 1.19

By making the variable change $t_1 = t + \tau$, $t_2 = t$ into (1.79), it results

$$\mathrm{E}\left\{x_1^{(*)_1}(t_1)\, x_2^{(*)_2}(t_2)\right\} = \sum_{\alpha\in A_{\boldsymbol{x}}} R^{\alpha}_{\boldsymbol{x}}(t_1 - t_2)\, e^{j2\pi\alpha t_2} \tag{1.123}$$

and in the sense of distributions we formally have

$$\begin{aligned}
&\mathrm{E}\left\{X_1^{(*)_1}(f_1)\, X_2^{(*)_2}(f_2)\right\} \\
&= \mathrm{E}\left\{\left[\int_{\mathbb{R}} x_1(t_1)\, e^{-j2\pi f_1 t_1} \mathrm{d}t_1\right]^{(*)_1} \left[\int_{\mathbb{R}} x_2(t_2)\, e^{-j2\pi f_2 t_2} \mathrm{d}t_2\right]^{(*)_2}\right\} \\
&= \int_{\mathbb{R}}\int_{\mathbb{R}} \mathrm{E}\left\{x_1^{(*)_1}(t_1)\, x_2^{(*)_2}(t_2)\right\} e^{-j2\pi[(-)_1 f_1 t_1 + (-)_2 f_2 t_2]} \mathrm{d}t_1\, \mathrm{d}t_2 \\
&= \int_{\mathbb{R}}\int_{\mathbb{R}} \sum_{\alpha\in A_{\boldsymbol{x}}} R^{\alpha}_{\boldsymbol{x}}(t_1 - t_2)\, e^{j2\pi\alpha t_2}\, e^{-j2\pi[(-)_1 f_1 t_1 + (-)_2 f_2 t_2]} \mathrm{d}t_1\, \mathrm{d}t_2 \\
&= \sum_{\alpha\in A_{\boldsymbol{x}}} \int_{\mathbb{R}} R^{\alpha}_{\boldsymbol{x}}(\tau)\, e^{-(-)_1 j2\pi f_1\tau} \mathrm{d}\tau \int_{\mathbb{R}} e^{j2\pi[\alpha - ((-)_1 f_1 + (-)_2 f_2)]t}\, \mathrm{d}t
\end{aligned} \tag{1.124}$$

from which (1.81) follows since

$$\begin{aligned}
\int_{\mathbb{R}} e^{j2\pi[\alpha - ((-)_1 f_1 + (-)_2 f_2)]t}\, \mathrm{d}t &= \delta(\alpha - ((-)_1 f_1 + (-)_2 f_2)) \\
&= \delta(f_2 - (-)_2(\alpha - (-)_1 f_1))\,.
\end{aligned} \tag{1.125}$$

Proof of (1.85)–(1.88)

$$
\begin{aligned}
R^{\alpha}_{yx^{(*)}}(\tau) &\triangleq \left\langle \mathrm{E}\left\{ y(t+\tau)\, x^{(*)}(t) \right\} e^{-j2\pi\alpha t} \right\rangle_t \\
&= \left\langle \mathrm{E}\left\{ x^{(*)}(t'-\tau)\, y(t') \right\} e^{-j2\pi\alpha t'} \right\rangle_{t'} e^{j2\pi\alpha\tau} \\
&\triangleq R^{\alpha}_{x^{(*)}y}(-\tau)\, e^{j2\pi\alpha\tau}
\end{aligned} \tag{1.126}
$$

$$
\stackrel{\mathcal{F}}{\longleftrightarrow} \; S^{\alpha}_{yx^{(*)}}(f) = S^{\alpha}_{x^{(*)}y}(-(f-\alpha)) \tag{1.127}
$$

where in the second equality in (1.126) the variable change $t' = t + \tau$ is made.

$$
\begin{aligned}
R^{-\alpha}_{yx^{(*)}}(\tau) &\triangleq \left\langle \mathrm{E}\left\{ y(t+\tau)\, x^{(*)}(t) \right\} e^{j2\pi\alpha t} \right\rangle_t \\
&= \left\langle \mathrm{E}\left\{ y^*(t+\tau)\, x^{(*)*}(t) \right\} e^{-j2\pi\alpha t} \right\rangle_t^* \\
&\triangleq R^{\alpha}_{y^*x^{(*)*}}(\tau)^*
\end{aligned} \tag{1.128}
$$

$$
\stackrel{\mathcal{F}}{\longleftrightarrow} \; S^{-\alpha}_{yx^{(*)}}(f) = S^{\alpha}_{y^*x^{(*)*}}(-f)^* \tag{1.129}
$$

Proof of (1.91)

Accounting for (1.92) and (1.56), we have

$$
\begin{aligned}
&\mathrm{E}\{\widetilde{x}_k(t+\tau)\,\widetilde{x}_h^{(*)}(t)\} \\
&= \mathrm{E}\left\{ \int_{\mathbb{R}} \widetilde{X}_k(f_1)\, e^{j2\pi f_1(t+\tau)}\, \mathrm{d}f_1 \int_{\mathbb{R}} \widetilde{X}_h^{(*)}(f_2)\, e^{(-)j2\pi f_2 t}\, \mathrm{d}f_2 \right\} \\
&= \mathrm{E}\left\{ \int_{\mathbb{R}} X\left(f_1 + \frac{k}{T_0}\right) \mathrm{rect}(f_1 T_0)\, e^{j2\pi f_1(t+\tau)}\, \mathrm{d}f_1 \right. \\
&\qquad \left. \int_{\mathbb{R}} X^{(*)}\left(f_2 + \frac{h}{T_0}\right) \mathrm{rect}(f_2 T_0)\, e^{(-)j2\pi f_2 t}\, \mathrm{d}f_2 \right\} \\
&= \int_{-1/(2T_0)}^{1/(2T_0)} \int_{-1/(2T_0)}^{1/(2T_0)} \mathrm{E}\left\{ X\left(f_1 + \frac{k}{T_0}\right) X^{(*)}\left(f_2 + \frac{h}{T_0}\right) \right\} e^{j2\pi[f_1 + (-)f_2]t}\, e^{j2\pi f_1 \tau}\, \mathrm{d}f_1\, \mathrm{d}f_2 \\
&= \int_{-1/(2T_0)}^{1/(2T_0)} \int_{-1/(2T_0)}^{1/(2T_0)} \sum_{\ell \in \mathbb{Z}} S_{\boldsymbol{x}}^{\ell/T_0}\left(f_1 + \frac{k}{T_0}\right) \delta\left(f_2 + \frac{h}{T_0} - (-)\left(\frac{\ell}{T_0} - f_1 - \frac{k}{T_0}\right)\right) \\
&\qquad e^{j2\pi[f_1 + (-)f_2]t}\, e^{j2\pi f_1 \tau}\, \mathrm{d}f_1\, \mathrm{d}f_2 \\
&= \sum_{\ell \in \mathbb{Z}} \int_{-1/2}^{1/2} S_{\boldsymbol{x}}^{\ell/T_0}\left(f_1 + \frac{k}{T_0}\right) e^{j2\pi f_1 \tau} \\
&\qquad \int_{-1/(2T_0)}^{1/(2T_0)} \delta\left(f_2 + \frac{h}{T_0} - (-)\left(\frac{\ell}{T_0} - f_1 - \frac{k}{T_0}\right)\right) e^{j2\pi[f_1 + (-)f_2]t}\, \mathrm{d}f_2\, \mathrm{d}f_1
\end{aligned} \tag{1.130}
$$

For $f_1 \in [-1/(2T_0), 1/(2T_0)]$ and $(*) = *$ (that is, $(-) = -$), we have

$$\begin{aligned}
&\int_{-1/(2T_0)}^{1/(2T_0)} \delta\Big(f_2 + \frac{h}{T_0} + \Big(\frac{\ell}{T_0} - f_1 - \frac{k}{T_0}\Big)\Big)\, e^{j2\pi[f_1 - f_2]t}\, \mathrm{d}f_2 \\
&\quad = \int_{-1/(2T_0)+(h+\ell-k)/T_0}^{1/(2T_0)+(h+\ell-k)/T_0} \delta(f_2' - f_1)\, e^{j2\pi[f_1 - f_2' + (h+\ell-k)/T_0]t}\, \mathrm{d}f_2' \\
&\quad = \begin{cases} 1 & \ell = k - h \\ 0 & \text{otherwise} \end{cases}
\end{aligned} \tag{1.131}$$

where in the first equality the variable change $f_2' = f_2 + (h + \ell - k)/T_0$ is made and the second equality is obtained observing that the integral is nonzero only if the integration interval for f_2' is $[-1/(2T_0), 1/(2T_0)]$.

For $f_1 \in [-1/(2T_0), 1/(2T_0)]$ and $(*)$ absent (that is, $(-) = +$), we have

$$\begin{aligned}
&\int_{-1/(2T_0)}^{1/(2T_0)} \delta\Big(f_2 + \frac{h}{T_0} - \Big(\frac{\ell}{T_0} + f_1 + \frac{k}{T_0}\Big)\Big)\, e^{j2\pi[f_1 + f_2]t}\, \mathrm{d}f_2 \\
&\quad = \int_{1/(2T_0)+(h-\ell+k)/T_0}^{-1/(2T_0)+(h-\ell+k)/T_0} \delta(-f_2' + f_1)\, e^{j2\pi[f_1 - f_2' + (h-\ell+k)/T_0]t}\, \mathrm{d}(-f_2') \\
&\quad = \begin{cases} 1 & \ell = k + h \\ 0 & \text{otherwise} \end{cases}
\end{aligned} \tag{1.132}$$

where in the first equality the variable change $f_2' = -f_2 + (h - \ell + k)/T_0$ is made and the second equality is obtained observing that the integral is nonzero only if the integration interval for f_2' is $[-1/(2T_0), 1/(2T_0)]$.

Proof of (1.93)

Using (1.90), we have

$$\begin{aligned}
&\mathrm{E}\Big\{x(t_1)\, x^{(*)}(t_2)\Big\} \\
&\quad = \mathrm{E}\left\{ \sum_{k_1=-\infty}^{+\infty} \widetilde{x}_{k_1}(t_1)\, e^{j2\pi(k_1/T_0)t_1} \sum_{k_2=-\infty}^{+\infty} \widetilde{x}_{k_2}^{(*)}(t_2)\, e^{(-)j2\pi(k_2/T_0)t_2} \right\} \\
&\quad = \sum_{k_1=-\infty}^{+\infty} \sum_{k_2=-\infty}^{+\infty} \mathrm{E}\Big\{\widetilde{x}_{k_1}(t_1)\, \widetilde{x}_{k_2}^{(*)}(t_2)\Big\}\, e^{j2\pi(k_1/T_0)t_1}\, e^{(-)j2\pi(k_2/T_0)t_2} .
\end{aligned} \tag{1.133}$$

Eq. (1.93) follows accounting for definition (1.91).

Proof of (1.94)–(1.95)

Using (1.90) we have

$$
\begin{aligned}
&\mathrm{E}\left\{x(t+\tau)\, x^{(*)}(t)\right\} \\
&= \mathrm{E}\left\{ \sum_{k_1=-\infty}^{+\infty} \widetilde{x}_{k_1}(t+\tau)\, e^{j2\pi(k_1/T_0)(t+\tau)} \sum_{k_2=-\infty}^{+\infty} \widetilde{x}_{k_2}^{(*)}(t)\, e^{(-)j2\pi(k_2/T_0)t} \right\} \\
&= \sum_{k_1=-\infty}^{+\infty} \sum_{k_2=-\infty}^{+\infty} \mathrm{E}\left\{\widetilde{x}_{k_1}(t+\tau)\, \widetilde{x}_{k_2}^{(*)}(t)\right\} e^{j2\pi(k_1/T_0)(t+\tau)}\, e^{(-)j2\pi(k_2/T_0)t} \\
&= \sum_{k=-\infty}^{+\infty} \sum_{k_1=-\infty}^{+\infty} R_{\widetilde{x}_{k_1}\widetilde{x}^{(*)}_{(-)(k-k_1)}}(\tau)\, e^{j2\pi(k_1/T_0)\tau}\, e^{j2\pi(k/T_0)t}
\end{aligned}
\tag{1.134}
$$

where in the third equality the variable change $k = k_1 + (-)k_2$ is made.

Proof of (1.104a)–(1.104b)

$$
\begin{aligned}
&\mathrm{E}\left\{a_{i_1}(n_1)\, a_{i_2}^{(*)}(n_2)\right\} \\
&= \mathrm{E}\left\{ \int_0^{T_0} x(s_1+n_1T_0)\, \phi_{i_1}^*(s_1)\, \mathrm{d}s_1 \int_0^{T_0} x^{(*)}(s_2+n_2T_0)\, \phi_{i_2}^{(*)*}(s_2)\, \mathrm{d}s_2 \right\} \\
&= \int_0^{T_0}\int_0^{T_0} \mathrm{E}\left\{x(s_1+n_1T_0)\, x^{(*)}(s_2+n_2T_0)\right\} \phi_{i_1}^*(s_1)\, \phi_{i_2}^{(*)*}(s_2)\, \mathrm{d}s_1\, \mathrm{d}s_2 \\
&= \int_0^{T_0}\int_0^{T_0} \sum_{k=-\infty}^{+\infty} R_{xx^{(*)}}^{k/T_0}(s_1-s_2+(n_1-n_2)T_0)\, e^{j2\pi(k/T_0)(s_2+n_2T_0)} \\
&\qquad \phi_{i_1}^*(s_1)\, \phi_{i_2}^{(*)*}(s_2)\, \mathrm{d}s_1\, \mathrm{d}s_2 \\
&= \sum_{k=-\infty}^{+\infty} \int_{-T_0}^{T_0} R_{xx^{(*)}}^{k/T_0}(\tau+(n_1-n_2)T_0) \left[\int_0^{T_0} \phi_{i_1}(s_2+\tau)\, \phi_{i_2}^{(*)}(s_2)\, e^{-j2\pi(k/T_0)s_2}\, \mathrm{d}s_2\right]^* \mathrm{d}\tau
\end{aligned}
\tag{1.135}
$$

where in the fourth equality the variable change $s_1 - s_2 = \tau$ is made (when $s_1 \in [0, T_0]$ and $s_2 \in [0, T_0]$ we have that $\tau \in [-T_0, T_0]$).

Thus we have (1.104a) accounting for the definition (1.105). Eq. (1.104b) follows accounting for the Parseval theorem (Champeney, 1990, Theorem 7.5) and the Fourier transform in (1.106).

Proof of (1.108)

By using the TSR (1.103) one has

$$\begin{aligned}
&\mathrm{E}\left\{x(t_1)\, x^{(*)}(t_2)\right\} \\
&\quad = \mathrm{E}\left\{\sum_{i_1=1}^{+\infty}\sum_{n_1=-\infty}^{+\infty} a_{i_1}(n_1)\,\phi_{i_1}(t_1 - n_1 T_0) \sum_{i_2=1}^{+\infty}\sum_{n_2=-\infty}^{+\infty} a_{i_2}^{(*)}(n_2)\,\phi_{i_2}^{(*)}(t_2 - n_2 T_0)\right\} \\
&\quad = \sum_{i_1=1}^{+\infty}\sum_{n_1=-\infty}^{+\infty}\sum_{i_2=1}^{+\infty}\sum_{n_2=-\infty}^{+\infty} \mathrm{E}\left\{a_{i_1}(n_1)\, a_{i_2}^{(*)}(n_2)\right\}\,\phi_{i_1}(t_1 - n_1 T_0)\,\phi_{i_2}^{(*)}(t_2 - n_2 T_0)\,. \qquad (1.136)
\end{aligned}$$

Accounting for the definition of $r_{i_1 i_2 (*)}(n_1 - n_2)$ for (1.104a)–(1.104b), one obtains (1.108).

Proof of (1.112)

By Fourier transforming both sides of (1.111b) and accounting for (1.106), we have

$$\begin{aligned}
S_{xx^{(*)}}^{k/T_0}(f) &= \int_{\mathbb{R}} R_{xx^{(*)}}^{k/T_0}(\tau)\, e^{-j2\pi f\tau}\,\mathrm{d}\tau \\
&= \frac{1}{T_0}\sum_{i_1=1}^{+\infty}\sum_{i_2=1}^{+\infty}\sum_{m=-\infty}^{+\infty} r_{i_1 i_2 (*)}(m) \int_{\mathbb{R}} r_{\phi_{i_1}\phi_{i_2}^{(*)}}^{k/T_0}(\tau - mT_0)\, e^{-j2\pi f\tau}\,\mathrm{d}\tau \\
&= \frac{1}{T_0}\sum_{i_1=1}^{+\infty}\sum_{i_2=1}^{+\infty} \Phi_{i_1}(f)\,\Phi_{i_2}^{(*)}\left((-)\left(\frac{k}{T_0} - f\right)\right) \sum_{m=-\infty}^{+\infty} r_{i_1 i_2 (*)}(m)\, e^{-j2\pi f m T_0} \qquad (1.137)
\end{aligned}$$

Eq. (1.112) immediately follows accounting for definition (1.113).

Proof of (1.114)

By using (1.108) one has

$$\begin{aligned}
&\mathrm{E}\left\{X(f_1)\, X^{(*)}(f_2)\right\} \\
&= \int_{\mathbb{R}^2} \mathrm{E}\left\{x(t_1)\, x^{(*)}(t_2)\right\}\, e^{-j2\pi[f_1 t_1 + (-) f_2 t_2]}\,\mathrm{d}t_1\,\mathrm{d}t_2 \\
&= \sum_{n_1=-\infty}^{+\infty}\sum_{n_2=-\infty}^{+\infty}\sum_{i_1=1}^{+\infty}\sum_{i_2=1}^{+\infty} r_{i_1 i_2 (*)}(n_1 - n_2)\,\Phi_{i_1}(f_1)\, e^{-j2\pi f_1 n_1 T_0}\,\Phi_{i_2}^{(*)}(f_2)\, e^{-(-)j2\pi f_2 n_2 T_0} \\
&= \sum_{i_1=1}^{+\infty}\sum_{i_2=1}^{+\infty}\left[\sum_{m=-\infty}^{+\infty} r_{i_1 i_2 (*)}(m)\, e^{-j2\pi f_1 m T_0}\right]\Phi_{i_1}(f_1)\,\Phi_{i_2}^{(*)}(f_2) \sum_{n_2=-\infty}^{+\infty} e^{-j2\pi f_1 n_2 T_0}\, e^{-(-)j2\pi f_2 n_2 T_0} \\
&= \sum_{i_1=1}^{+\infty}\sum_{i_2=1}^{+\infty} S_{i_1 i_2 (*)}(f_1 T_0)\,\Phi_{i_1}(f_1)\,\Phi_{i_2}^{(*)}(f_2)\,\frac{1}{T_0}\sum_{k=-\infty}^{+\infty}\delta\left(f_2 - (-)\left(\frac{k}{T_0} - f_1\right)\right)
\end{aligned}$$

$$= \sum_{i_1=1}^{+\infty} \sum_{i_2=1}^{+\infty} \sum_{k=-\infty}^{+\infty} \frac{1}{T_0} S_{i_1 i_2 (*)}(f_1 T_0)\, \Phi_{i_1}(f_1)\, \Phi_{i_2}^{(*)}\left({}_{(-)}\left(\frac{k}{T_0} - f_1\right)\right) \delta\left(f_2 - {}_{(-)}\left(\frac{k}{T_0} - f_1\right)\right) \tag{1.138}$$

where, in the second equality, the Fourier transform

$$\phi^{(*)}(t - T) \overset{\mathcal{F}}{\longleftrightarrow} \Phi^{(*)}((-)f)\, e^{-j2\pi f T} \tag{1.139}$$

and relation

$$\Phi^{(*)}((-)\lambda)\, e^{-j2\pi\lambda T}|_{\lambda=(-)f} = \Phi^{(*)}(f)\, e^{-(-)j2\pi f T} \tag{1.140}$$

are used; in the third equality, the variable change $n_1 - n_2 = m$ is made; in the fourth equality, the Poisson formula (C.8b)

$$\sum_{n=-\infty}^{+\infty} e^{-j2\pi n f/f_s} = f_s \sum_{k=-\infty}^{+\infty} \delta(f - k f_s), \tag{1.141}$$

with $f = f_1 + {}_{(-)}f_2$ and $f_s = 1/T_0$ is used, the property $\delta(\cdot) = \delta(-(\cdot))$ is exploited, and $S_{i_1 i_2 (*)}(\nu)$ is defined in (1.113); in the fifth equality the sampling property of Dirac delta is used.

The densities of Loève bifrequency spectrum are the (conjugate) cyclic spectra whose expression is in agreement with (1.112).

2 Characterization of time-series

Contents

Abstract

Cyclostationary and almost-cyclostationary time series are defined and characterized in the time and frequency domains and in the continuous- and discrete-time cases. The characterization is made in the functional or fraction-of-time approach, where a signal is modeled as a single function of time, without introducing the concept of stochastic process. In such a framework, probabilistic functions are built starting from this single function of time and the expectation operator is the almost-periodic component extraction operator. The link with the characterization in the classical stochastic approach is enlightened.

Keywords

Fraction-of-time probability, Relative measure, Nonstochastic approach, Time series, Spectral line regeneration, Almost-periodic component extraction operator, Synchronized averaging

2.1 Introduction

In the classical stochastic approach, signals are modeled as realizations (or sample paths) of stochastic processes (Doob, 1953). For this purpose, a sample space Ω equipped with a σ-field $\mathcal{F}$ and a probability measure P defined on $\mathcal{F}$ are introduced. In Section 1.2, such an approach is adopted, and cyclostationary and almost-cyclostationary stochastic processes are defined and characterized. Probabilistic functions are defined as ensemble averages of functions of stochastic processes and their time translates. Estimates of the Fourier coefficients of these probabilistic functions are time averages, over finite time intervals, of sinusoidally weighted functions of a single realization and its time translates. Then some mixing assumptions assure the convergence of the estimates to the corresponding probabilistic functions as the time interval becomes infinite (Chapter 5).

In several practical cases, only one realization of the time series exists. This is the case, for example, of a biological signal of a specific living being, the vibro-acoustic signal registered from a specific machinery, the signal coming from a specific astronomic source, a communication signal arising from a specific transmitter, the temperature registered in a specific place, and the exchange rate between two specific currencies. In such cases, the ensemble of realizations does not exist. The stochastic process, that is the ensemble of realizations, is an abstraction introduced just to create a mathematical model for signal analysis. Such a model can be unnecessarily abstract when there is only a single time series at hand.

An alternative approach for signal analysis consists in modeling the observed signal or time series as a single function of time. Such a signal is modeled as a segment of an idealized one existing on the whole real line (that is, a "persistent" signal). Such an approach is referred to as *functional approach* (Brillinger, 1981, Sec. 2.11) or *fraction-of-time (FOT) probability* approach (Gardner, 1985, Chap. 8), (Gardner, 1987f, Chap. 5) and was first adopted in (Wiener, 1930) for correlation and spectral analysis (see (Leśkow and Napolitano, 2006) for an extensive historical perspective). In the FOT approach, the only abstraction consists in considering the signal existing also outside the observation interval. In contrast, in the classical stochastic approach, a double process of abstraction is needed since the observed signal segment is first extended on the whole real line and then considered as one of an ensemble of realizations.

The key concept in the FOT approach is the relative measure μ_R of sets introduced in (Kac and Steinhaus, 1938). Once the concept of relative measurability is defined for a signal $x(t)$, a valid distribution function can be constructed whose expectation is the infinite time average (Leśkow and Napolitano, 2006). Using such distribution and expectation, familiar time-invariant probabilistic functions as mean, moments and cumulants can be constructed. Moreover, considering two or more time-shifted versions of the same signal, concepts can be extended to the multivariate case leading to the definition of time-invariant joint distributions, autocorrelation, cross moments, and cumulants (Gardner, 1987f, Chap. 5), (Leśkow and Napolitano, 2006). Within such a framework, independence of functions can be defined (Kac and Steinhaus, 1938), (Kac, 1959), and central limit theorems (CLTs) proved (Dehay et al., 2013), (Dehay et al., 2018b).

The FOT probability approach based on the infinite time average as expectation operator leads to a stationary probabilistic model built form a single function of time. It is extended in (Gardner, 1987f, Chap. 15), (Brown, 1987), (Gardner and Brown, 1991), (Gardner, 1994a) to accommodate periodically and almost-periodically time variant probabilistic functions. In such a case, the expectation operator is the almost-periodic component extraction operator. It extracts the additive almost-periodic component of its argument.

Higher-order almost-periodically time variant statistics are considered in (Gardner and Spooner, 1994a), (Spooner and Gardner, 1994a), (Izzo and Napolitano, 1998a) and a link with the relative measure is established in (Napolitano, 2012, Chap. 6). Further results are in (Hurd and Koski, 2004a), (Hurd and Koski, 2004b), (Dehay et al., 2018b).

2.2 Fraction-of-time probability

In this section, basic concepts of the fraction-of-time probability are presented for the stationary model for real-valued time series. Specifically, for a single time series, time invariant probabilistic parameters are built starting from the concept of relative measure (Kac and Steinhaus, 1938). For more extensive treatments, see (Gardner, 1987f, Chap. 5), (Leśkow and Napolitano, 2006), (Dehay et al., 2013), (Dehay et al., 2018b).

2.2.1 Continuous time

Definition 2.1. Let us consider the set $A \in \mathcal{B}_{\mathbb{R}}$, where $\mathcal{B}_{\mathbb{R}}$ is the σ-field of the Borel subsets and μ is the Lebesgue measure on the real line $\mathbb{R}$. The *relative measure* of A is defined as (Kac and Steinhaus, 1938)

$$\mu_R(A) \triangleq \lim_{T\to\infty} \frac{1}{T} \mu(A \cap [t_0 - T/2, t_0 + T/2]) \tag{2.1}$$

provided that the limit exists. In such a case, the limit does not depend on t_0 and the set A is said to be relatively measurable (RM). □

The relative measure μ_R plays an equivalent role to the theoretical probability P in the stochastic approach where the object $(\Omega, \mathcal{F}, P)$ is considered. However, the relative measure is not closed under union and intersection operations (Leśkow and Napolitano, 2006, Fact 2.2). In addition, it is additive (Leśkow and Napolitano, 2006, Fact 2.4) but not σ-additive (Leśkow and Napolitano, 2006, Fact 2.5). The lack of σ-additivity is consequence of the normalization $\mu_R(\mathbb{R}) = 1$ obtained by a limit operation (see (2.1)). In contrast, in the classical probability theory, the sample space Ω is implicitly assumed to have a finite measure (Halmos, 1944), (Halmos, 1956) and the normalization $P(\Omega) = 1$ is obtained just dividing a measure by the (finite) measure of the whole space. Thus, the classical probability measure turns out to be σ-additive.

Definition 2.2. Let $x(t)$ be a real-valued Lebesgue measurable function. The function $x(t)$ is said to be *relatively measurable* (RM) if and only if the set $\{t \in \mathbb{R} : x(t) \leqslant \xi\}$ is RM for every $\xi \in \mathbb{R} - \Xi_0$, where Ξ_0 is at most a countable set of points. □

Examples of RM functions are the almost-periodic functions (Appendix B), (Leśkow and Napolitano, 2006, Secs. 6.1–6.2) and some pseudo-random functions (Bass, 1974a), (Leśkow and Napolitano, 2006, Sec. 6.3), (Dehay et al., 2018b, Sec. 4). Functions that are not RM can be easily constructed, in contrast to functions that are not Lebesgue measurable (Leśkow and Napolitano, 2006), (Leśkow and Napolitano, 2007).

Let $\mathbf{1}_A$ be the indicator function of the set A, that is,

$$\mathbf{1}_A = \begin{cases} 1 & t \in A \\ 0 & t \notin A \end{cases} \tag{2.2}$$

and let us denote by

$$\langle \phi(t) \rangle_t \triangleq \lim_{T\to\infty} \frac{1}{T} \int_{t_0-T/2}^{t_0+T/2} \phi(t)\, \mathrm{d}t \tag{2.3}$$

the infinite time average of the function $\phi(t)$, where the limit, if it exists, is independent of t_0.

For every Lebesgue measurable function $x(t)$, the function

$$\begin{aligned} F_{T,t}(\xi) &\triangleq \frac{1}{T}\, \mu\big(\{u \in [t - T/2, t + T/2] \,:\, x(u) \leqslant \xi\}\big) \\ &= \frac{1}{T} \int_{t-T/2}^{t+T/2} \mathbf{1}_{\{x(u)\leqslant\xi\}}\, \mathrm{d}u \end{aligned} \tag{2.4}$$

represents the *fraction-of-time* that the function $x(u)$ is below the threshold ξ when $u \in [t - T/2, t + T/2]$ (Gardner, 1987f, Chap. 5). The function $\xi \mapsto F_{T,t}(\xi)$ has all the properties of a valid distribution function including the right continuity in the discontinuity points (Leśkow and Napolitano, 2006) and is referred to as finite-time FOT distribution. The expectation induced by $F_{T,t}(\xi)$ is the finite time average on the interval $[t - T/2, t + T/2]$.

If the function $x(t)$ is RM, then, according to Definition 2.2, the limit as $T \to \infty$ of $F_{T,t}(\xi)$ exists.

Definition 2.3. Each RM function $x(t)$ generates a function

$$F_x(\xi) \triangleq \mu_R(\{t \in \mathbb{R} \,:\, x(t) \leqslant \xi\}) = \big\langle \mathbf{1}_{\{t\in\mathbb{R}\,:\,x(t)\leqslant\xi\}} \big\rangle_t \tag{2.5}$$

in all points ξ where the limit (2.1) exists.

The function $F_x(\xi)$ is called *FOT distribution* of $x(t)$. It has all the properties of a valid distribution function except for the right continuity in the discontinuity points (Leśkow and Napolitano, 2006). □

The expectation induced by the FOT distribution (2.5) is the infinite time average (2.3) (Leśkow and Napolitano, 2006, Corollary 3.1) (Napolitano, 2012, Sec. 6.2). That is

$$\langle x(t) \rangle_t = \int_{\mathbb{R}} \xi \, \mathrm{d}F_x(\xi)\,. \tag{2.6}$$

Thus, the infinite time average (2.3) plays, in the FOT probability framework, the same role played by the ensemble average $\mathrm{E}\{\cdot\}$ in the classical stochastic framework. For example, for a stationary stochastic process $X(t)$ we have that the cumulative distribution function can be expressed as

$$F_X(\xi) = \mathrm{E}\big\{\mathbf{1}_{\{X(t)\leqslant\xi\}}\big\} \tag{2.7}$$

which is the stochastic counterpart of (2.5). However, the infinite time average is linear but not σ-linear (Leśkow and Napolitano, 2006, Fact 4.4). In contrast, the expectation operator of the classical probability theory, that is the ensemble average, is σ-linear.

Since the relative measure is not closed under union and intersection operations, it follows that the set of RM functions is not closed under addition and multiplication (Leśkow and Napolitano, 2006, Theorem 3.5).

The joint characterization of k signals is made starting from the following definition.

Definition 2.4. The Lebesgue measurable functions $x_1(t), \ldots, x_k(t)$ are said to be *jointly relatively measurable* if the limit

$$F_{x_1 \ldots x_k}(\xi_1, \ldots, \xi_k) \triangleq \mu_R(\{t \in \mathbb{R} : x_1(t) \leqslant \xi_1\} \cap \cdots \cap \{t \in \mathbb{R} : x_k(t) \leqslant \xi_k\})$$
$$= \left\langle \mathbf{1}_{\{t \in \mathbb{R} : x_1(t) \leqslant \xi_1\}} \cdots \mathbf{1}_{\{t \in \mathbb{R} : x_k(t) \leqslant \xi_k\}} \right\rangle_t \tag{2.8}$$

exists for all $(\xi_1, \ldots, \xi_k) \in \mathbb{R}^k - \Xi_0$, where Ξ_0 is at most a countable set of $(k-1)$-dimensional manifolds of $\mathbb{R}^k$. □

By introducing the concept of conditioned relative measure (Leśkow and Napolitano, 2006, Def. 5.1), the independence of sets can be defined (Leśkow and Napolitano, 2006, Def. 5.3), (Dehay et al., 2013, Def. 2.4). Then, the independence of functions is defined in (Leśkow and Napolitano, 2006, Def. 5.4), (Dehay et al., 2013, Def. 2.5), and the following result (that is assumed as definition of independence in (Kac and Steinhaus, 1938)), is proved.

Theorem 2.5. (Dehay et al., 2013, Theorem 2.6). *The jointly RM functions* $x_1(t), \ldots, x_k(t)$ *are* FOT-independent *if and only if*

$$F_{x_1 \cdots x_k}(\xi_1, \ldots, \xi_k) = F_{x_1}(\xi_1) \cdots F_{x_k}(\xi_k) \qquad \forall (\xi_1, \ldots, \xi_k) \in \mathbb{R}^k - \Xi_0 \tag{2.9}$$

where Ξ_0 *is at most a countable set of* $(k-1)$*-dimensional manifolds of* $\mathbb{R}^k$. □

Note that in the FOT approach the independence condition (2.9) is a theorem. In contrast, in the classical stochastic approach, the factorization of the kth-order joint distribution into the product of the k marginal distributions is assumed as definition of independence for k stochastic processes (Doob, 1953, Sec. 1.3).

Definition 2.6. Let $x_1(t), \ldots, x_k(t)$ be jointly RM functions. Their joint *FOT cumulant* is defined as

$$\operatorname{cum}_t \{x_1(t), \ldots, x_k(t)\} \triangleq (-j)^k \frac{\partial^k}{\partial \omega_1 \ldots \partial \omega_k} \log \left\langle e^{j\omega_1 x_1(t)} \cdots e^{j\omega_k x_k(t)} \right\rangle_t \Big|_{\boldsymbol{\omega} = \mathbf{0}} \tag{2.10}$$

where $\boldsymbol{\omega} \triangleq [\omega_1, \ldots, \omega_k]$. □

For $k = 2$ in Definition 2.6, we have the FOT covariance

$$\operatorname{cov}_t[x_1(t+s), x_2(t)] \triangleq \langle x_1(t+s)\, x_2(t) \rangle_t - \langle x_1(t) \rangle_t \, \langle x_2(t) \rangle_t \, . \tag{2.11}$$

The lack of σ-additivity of the relative measure μ_R and the lack of σ-linearity of the infinite time average $\langle\cdot\rangle_t$ make the FOT model less attractive with respect to the classical stochastic-process framework, from a mathematical point of view. However, it is worthwhile to underline that such less attractive properties are those of the signals at the hands of the experimenter. Thus, more amenable properties of a stochastic process could not have a counterpart in their sample paths and hence could not be exploitable in practice.

Note that, for stochastic processes $\{X(t,\omega),\ t \in \mathbb{R},\ \omega \in \Omega\}$, the usual assumption $X(t,\omega) \in L^2(\Omega)\ \forall t$, that is, $\mathrm{E}\{|X(t,\omega)|^2\} < \infty\ \forall t$, where $\mathrm{E}\{\cdot\}$ denotes ensemble average, allows one to build a vector space of random variables $\forall t$. In contrast, for single functions of time $\{x(t),\ t \in \mathbb{R}\}$, the analogous assumption $\langle|x(t)|^2\rangle_t < \infty$, does not allow one to build a vector space since finite-power signals do not constitute a vector space (Mari, 1996), (Mäkilä et al., 1998), (Mäkilä, 2006). Therefore, in some cases, linear combinations of finite-power signals should be carefully handled (Leśkow and Napolitano, 2007).

2.2.2 Discrete time

Definition 2.7. Let $\#[\cdot]$ denote the counting measure. That is, for every finite or countable set F, $\#[F]$ is the number of elements (cardinality) of F. The *relative measure* of the countable set $A \subseteq \mathbb{Z}$ is defined as

$$\mu_R(A) \triangleq \lim_{N\to\infty} \frac{1}{2N+1} \#(A \cap [n_0 - N, n_0 + N]) \tag{2.12}$$

provided that the limit exists. In such a case, the limit does not depend on n_0 and the set A is said to be relatively measurable. □

Definition 2.8. Let $x(n)$ be a real-valued sequence. The sequence $x(n)$ is said to be *relatively measurable* (RM) if and only if the set $\{n \in \mathbb{Z} : x(n) \leqslant \xi\}$ is RM for every $\xi \in \mathbb{R} - \Xi_0$, where Ξ_0 is at most a countable set of points. □

Examples of RM sequences are the almost-periodic sequences (Chapter B) and some pseudo-random sequences (Leśkow and Napolitano, 2006, Sec. 6.3), (Dehay et al., 2018b, Sec. 4). Examples of non-RM sequences are provided in (Leśkow and Napolitano, 2006), (Leśkow and Napolitano, 2007).

Let us denote by

$$\langle\phi(n)\rangle_n \triangleq \lim_{N\to\infty} \frac{1}{2N+1} \sum_{n=n_0-N}^{n_0+N} \phi(n) \tag{2.13}$$

the infinite discrete-time time average of the sequence $\phi(n)$, where the limit, if it exists, is independent of n_0.

Definition 2.9. Each RM sequence $x(n)$ generates a function

$$F_x(\xi) \triangleq \mu_R(\{n \in \mathbb{Z} : x(n) \leqslant \xi\}) = \left\langle \mathbf{1}_{\{n\in\mathbb{Z}\,:\,x(n)\leqslant\xi\}} \right\rangle_n \tag{2.14}$$

in all points ξ where the limit (2.12) exists.

The function $F_x(\xi)$ is called *FOT distribution* of $x(n)$. It has all the properties of a valid distribution function except for the right continuity in the discontinuity points. □

The expectation induced by the discrete-time FOT distribution is the discrete-time time average (2.13). That is,

$$\langle x(n)\rangle_n = \int_{\mathbb{R}} \xi \, \mathrm{d}F_x(\xi) . \tag{2.15}$$

2.3 Almost-cyclostationary time series

2.3.1 Continuous-time

In this section, the almost-periodic FOT probability model is presented for real-valued almost-cyclostationary time series. Specifically, for a single time series, periodically time variant, poly-periodically time variant, and almost-periodically time variant probabilistic parameters are built by adopting, for each case, the appropriate expectation operator. Extensive treatments of the FOT probability framework for almost-cyclostationary time series are in (Gardner, 1987f, Chap. 15), (Brown, 1987), (Gardner and Brown, 1991), (Gardner, 1994a), (Gardner and Spooner, 1994a), (Napolitano, 2012, Sec. 6.3). Higher-order statistics of complex-valued time series are treated (Spooner and Gardner, 1994a), (Napolitano, 1995), and Chapter 4. In this section, limits are assumed to exist and infinite sums and integrals are assumed to be well defined without explicitly giving the sufficient assumptions.

2.3.1.1 Expectation operators and time series decomposition

Definition 2.10. The *periodic component extraction operator* $\mathrm{E}^{\{T_0\}}\{\cdot\}$ is defined to be the operator that extracts the periodic component with period T_0 of its argument. That is, it extracts all the finite-strength additive sine-wave components of its argument with frequencies k/T_0, $k \in \mathbb{Z}$. According to the synchronized averaging theorem (Theorem B.19), we have

$$\mathrm{E}^{\{T_0\}}\{z(t)\} \triangleq \lim_{N\to\infty} \frac{1}{2N+1} \sum_{n=-N}^{N} z(t+nT_0) \tag{2.16}$$

provided that the limit exists. □

Proposition 2.11. *Let $\mathcal{Z}^{T_0}$ be the class of functions $z(t)$ that can be expressed as the sum of a (possibly identically zero) periodic component $\mathrm{E}^{\{T_0\}}\{z(t)\}$ for some period T_0 and a residual term $z_\varrho(t)$ not containing any finite-strength sine wave component with frequency k/T_0, $k \in \mathbb{Z}$. For every $z(t) \in \mathcal{Z}^{T_0}$ it results*

$$z(t) = \mathrm{E}^{\{T_0\}}\{z(t)\} + z_\varrho(t) \tag{2.17}$$

where

$$\left\langle z_\varrho(t)\, e^{-j2\pi\eta t} \right\rangle_t = 0 \qquad \forall \eta = \frac{k}{T_0},\ k \in \mathbb{Z} . \tag{2.18}$$

Note that $z_Q(t)$ possibly contains finite-strength sine wave components with frequencies that are not integer multiples of $1/T_0$. □

Definition 2.12. The *poly-periodic component extraction operator* $\mathrm{E}^{\{T_1,\ldots,T_P\}}\{\cdot\}$ is defined to be the operator that extracts the periodic components with periods $T_1, \ldots, T_P$ of its argument. That is, it extracts all the finite-strength additive sine-wave components of its argument with frequencies $k/T_1, \ldots, k/T_P$, $k \in \mathbb{Z}$.

$$\mathrm{E}^{\{T_1,\ldots,T_P\}}\{z(t)\} \triangleq \sum_{p=1}^{P} \Big[\mathrm{E}^{\{T_p\}}\{z(t)\} - \langle z(t)\rangle_t\Big] + \langle z(t)\rangle_t \tag{2.19}$$

where each periodic component extraction operator $\mathrm{E}^{\{T_p\}}\{\cdot\}$ is defined according to Definition 2.10. □

In (2.19), the time average must be considered only once. Thus, it is subtracted by each periodic component and then added.

Proposition 2.13. *Let $\mathcal{Z}^{\mathrm{pp}}$ be the class of functions $z(t)$ that can be expressed as the sum of a (possibly identically zero) poly-periodic component with periods $T_1, \ldots, T_P$ and a residual term $z_Q(t)$ not containing any finite-strength sine wave component with frequencies $k/T_1, \ldots, k/T_P$, $k \in \mathbb{Z}$. For every $z(t) \in \mathcal{Z}^{\mathrm{pp}}$ it results*

$$z(t) = \sum_{p=1}^{P} \Big[\mathrm{E}^{\{T_p\}}\{z(t)\} - \langle z(t)\rangle_t\Big] + \langle z(t)\rangle_t + z_Q(t) \tag{2.20}$$

where $\mathrm{E}^{\{T_p\}}\{z(t)\}$ is the (possibly identically zero) periodic component with period T_p of $z(t)$ and

$$\Big\langle z_Q(t)\, e^{-j2\pi\eta t}\Big\rangle_t = 0 \qquad \forall \eta = \frac{k}{T_1}, \ldots, \frac{k}{T_P}, \qquad k \in \mathbb{Z}\,. \tag{2.21}$$

Note that $z_Q(t)$ possibly contains finite-strength sine wave components with frequencies that are not integer multiples of $1/T_1, \ldots, 1/T_P$. □

Definition 2.14. The *almost-periodic component extraction operator* $\mathrm{E}^{\{\alpha\}}\{\cdot\}$ is defined to be the operator that extracts the additive almost-periodic component of its argument or, equivalently, all the finite-strength additive sine-wave components of its argument. It is denoted by

$$\mathrm{E}^{\{\alpha\}}\{z(t)\} \equiv z_{\mathrm{ap}}(t)\,. \tag{2.22}$$

□

In some cases, the notation $\mathrm{E}_t^{\{\alpha\}}\{\cdot\}$ is adopted to clarify that the almost-periodic component is taken with respect to the variable t.

Proposition 2.15. *Let $\mathcal{Z}^{\text{ap}}$ be the class of functions $z(t)$ that can be expressed as the sum of a (possibly identically zero) almost-periodic component $z_{\text{ap}}(t)$ and a residual term $z_r(t)$ not containing any finite-strength sine wave component. For every $z(t) \in \mathcal{Z}^{\text{ap}}$ it results*

$$z(t) = z_{\text{ap}}(t) + z_r(t) \tag{2.23}$$

where

$$\left\langle z_r(t)\, e^{-j2\pi\eta t} \right\rangle_t = 0 \qquad \forall \eta \in \mathbb{R}\,. \tag{2.24}$$

□

The function $z_{\text{ap}}(t)$ can be a uniformly almost-periodic function (Section B.2) or an almost-periodic function defined in one of the generalized senses considered in Section B.4. In this latter case, equality in (2.23) must be intended in the appropriate norm. That is, with the notation introduced in Definition B.15, equality in (2.23) means that

$$\left\| z(t) - [z_{\text{ap}}(t) + z_{\text{r}}(t)] \right\|_{G^p} = 0\,. \tag{2.25}$$

In (Napolitano, 2012, Sec. 6.3.1), decomposition (2.23) is considered in the sense of Hartman and Ryll-Nardzewski, where the almost-periodic function is B^p almost-periodic (B.4.1) and the residual term is of finite power.

Obviously it results that $\mathcal{Z}^{T_0} \subset \mathcal{Z}^{\text{pp}} \subset \mathcal{Z}^{\text{ap}}$.

In Section 2.3.1.3, it is shown that the periodic component extraction operator, the poly-periodic component extraction operator, and the almost-periodic component extraction operator are the expectation operators of valid distribution functions that are built by a single time series.

The *choice* of the expectation operator determines the nonstationarity model that is adopted to characterize the time series (see Remark 2.27).

2.3.1.2 Periodic, poly-periodic, and almost-periodic components of nonnegative functions

In this section, three results on the nonnegativeness of the periodic, poly-periodic, and almost-periodic components of nonnegative functions are presented. These results are fundamental in the construction of FOT periodically, poly-periodically, and almost-periodically time variant distributions.

Lemma 2.16. *Let $z(t) \in \mathcal{Z}^{T_0}$ be a nonnegative function. The periodic component with period T_0 of $z(t)$ (Proposition 2.11) is nonnegative.*

Proof. See Section 2.6. □

Lemma 2.17. *Let $z(t) \in \mathcal{Z}^{\text{pp}}$ be a nonnegative function. The poly-periodic component with periods $T_1, \ldots, T_P$ of $z(t)$ (Proposition 2.13) is nonnegative, provided that the possible finite-strength sinewave components of the residual term have frequencies incommensurate with $h_1/T_1 + \cdots + h_P/T_P$, for all integers $h_1, \ldots, h_P$.*

Proof. See Section 2.6. □

Lemma 2.18. *Let $z(t) \in \mathcal{Z}^{\mathrm{ap}}$ be a nonnegative function. The almost-periodic component $z_{\mathrm{ap}}(t)$ of $z(t)$ (Proposition 2.15) is nonnegative.*

Proof. See Section 2.6. □

2.3.1.3 First-order characterization

Let $x(t)$, $t \in \mathbb{R}$, be a real-valued continuous-time time series (a single function of time) and let $\mathbf{1}_{\{x(t) \leqslant \xi\}}$ be the indicator function of the set $\{t \in \mathbb{R} : x(t) \leqslant \xi\}$ defined according to (2.2). It results

$$\mathbf{1}_{\{x(t) \leqslant \xi\}} = \begin{cases} 1, & t : x(t) \leqslant \xi \\ 0, & t : x(t) > \xi \end{cases} \tag{2.26}$$

Assumption 2.19.

a. *The set Γ_ξ of frequencies (depending on ξ) of the almost-periodic component of the function $t \mapsto \mathbf{1}_{\{x(t) \leqslant \xi\}}$ is countable.*
b. *The set*

$$\Gamma^{(1)} \triangleq \bigcup_{\xi \in \mathbb{R}} \Gamma_\xi \tag{2.27}$$

is countable. □

Definition 2.20. The RM function $x(t)$ is said to be T_0*-relatively measurable* (T_0-RM) if and only if the sequence $n \mapsto \mathbf{1}_{\{x(t+nT_0) \leqslant \xi\}}$ is RM for every $t \in \mathbb{R}$ and for every $\xi \in \mathbb{R} - \Xi_0$, where Ξ_0 is at most a countable set of points. □

Note that, in general, not all RM functions are T_0-RM for some T_0.

Theorem 2.21. *Every T_0-RM function $x(t)$ generates a function*

$$F_x^{\{T_0\}}(\xi; t) \triangleq \mathrm{E}^{\{T_0\}}\left\{\mathbf{1}_{\{x(t) \leqslant \xi\}}\right\} = \lim_{N \to \infty} \frac{1}{2N+1} \sum_{n=n_0-N}^{n_0+N} \mathbf{1}_{\{x(t+nT_0) \leqslant \xi\}} . \tag{2.28}$$

For every fixed t, the function $\xi \mapsto F_x^{\{T_0\}}(\xi; t)$ has all the properties of a valid distribution function except for the right continuity in the discontinuity points. That is, it has values in $[0, 1]$, is nondecreasing, $F_x^{\{T_0\}}(-\infty; t) = 0$, and $F_x^{\{T_0\}}(+\infty; t) = 1$. For every fixed ξ where the limit exists, the function $t \mapsto F_x^{\{T_0\}}(\xi; t)$ is periodic in t with period T_0.

Proof. See Section 2.6. □

In the classical stochastic approach the right-continuity property in the discontinuity points of the distribution function is proved starting from the Axiom of Continuity which is at the basis of the proof of the σ-additivity property of the probability measure in the case of infinite probability fields (Kolmogorov, 1933, Chap. II). The lack of σ-additivity of the relative measure (2.1) does not allow one to prove the right-continuity property for the stationary FOT distribution. However this fact does not influence the values of moments (Leśkow and Napolitano, 2006, Sec. 3). Thus, it is not relevant in most, if not all, applications. A similar consideration holds for the periodic FOT distributions.

In the original work (Kolmogorov, 1933), the probability distribution is left continuous. In fact, for a random variable X, in (Kolmogorov, 1933) the probability distribution is defined as $F_X(\xi) \triangleq \mathrm{P}[X < \xi]$. In contrast, the right continuity is obtained when the probability distribution is defined as $F_X(\xi) \triangleq \mathrm{P}[X \leqslant \xi]$.

The expectation induced by the periodic FOT distribution (2.28) is the operator that extracts the periodic component of $x(t)$ with period T_0 (a formal proof can be obtained as a special case of the proof of Theorem 2.24 with $g(\xi) = \xi$). Accounting for the synchronized averaging theorem (Theorem B.19), we have

$$\mathrm{E}^{\{T_0\}}\{x(t)\} \triangleq \lim_{N\to\infty} \frac{1}{2N+1} \sum_{n=-N}^{N} x(t+nT_0) \tag{2.29a}$$

$$= \int_{\mathbb{R}} \xi \, \mathrm{d}F_x^{\{T_0\}}(\xi;t) \tag{2.29b}$$

Theorem 2.22. *Let $x(t)$ be T_1-RM, ..., T_P-RM, with (incommensurate) periods $T_1, \ldots, T_P$. It generates the function*

$$F_x^{\{T_1,\ldots,T_P\}}(\xi;t) \triangleq F_x^{\{0\}}(\xi) + \sum_{p=1}^{P} \left[F_x^{\{T_p\}}(\xi;t) - F_x^{\{0\}}(\xi) \right] \tag{2.30}$$

where $F_x^{\{0\}}(\xi) \equiv F_x(\xi)$ is defined in (2.5) and $F_x^{\{T_p\}}(\xi;t)$, $p = 1, \ldots, P$, are defined according to (2.28).

For every fixed t, the function $\xi \mapsto F_x^{\{T_1,\ldots,T_P\}}(\xi;t)$ has all the properties of a valid distribution function except for the right continuity in the discontinuity points. For every continuity point ξ, the function $t \mapsto F_x^{\{T_1,\ldots,T_P\}}(\xi;t)$ is poly-periodic in t with (incommensurate) periods $T_1, \ldots, T_P$.

Proof. The proof is the same as that of Theorem 2.21 by replacing the periodic component extraction operator $\mathrm{E}^{\{T_0\}}\{\cdot\}$ with the poly-periodic component extraction operator $\mathrm{E}^{\{T_1,\ldots,T_P\}}\{\cdot\}$ and using Lemma 2.17 on the nonnegativeness of a poly-periodic component of a nonnegative function. □

Note that in Theorem 2.22, not necessarily all the additive periodic components present in the function $t \mapsto \mathbf{1}_{\{x(t)\leqslant\xi\}}$ are considered.

The expectation induced by the poly-periodic FOT distribution (2.30) is the operator that extracts the poly-periodic component of $x(t)$ with periods $T_1, \ldots, T_P$ (a formal proof can be obtained as a special case of the proof of Theorem 2.24 with $g(\xi) = \xi$). Accounting for the synchronized averaging theorem (Theorem B.19), we have

$$\mathrm{E}^{\{T_1,\ldots,T_P\}}\{x(t)\}$$
$$\triangleq \sum_{p=1}^{P}\left[\lim_{N\to\infty}\frac{1}{2N+1}\sum_{n=-N}^{N}x(t+nT_p)-\langle x(t)\rangle_t\right]+\langle x(t)\rangle_t \tag{2.31a}$$
$$=\sum_{p=1}^{P}\left[\int_{\mathbb{R}}\xi\,\mathrm{d}F_x^{\{T_p\}}(\xi;t)-\int_{\mathbb{R}}\xi\,\mathrm{d}F_x^{\{0\}}(\xi)\right]+\int_{\mathbb{R}}\xi\,\mathrm{d}F_x^{\{0\}}(\xi)\,. \tag{2.31b}$$

Theorem 2.23. (Gardner and Brown, 1991). *Under Assumption 2.19a, for every fixed t, the function of ξ*

$$F_{x(t)}^{\{\alpha\}}(\xi)\triangleq \mathrm{E}^{\{\alpha\}}\left\{\mathbf{1}_{\{x(t)\leqslant\xi\}}\right\} \tag{2.32}$$

for any t is a valid cumulative distribution function except for the right-continuity property (in the discontinuity points). That is, it has values in $[0,1]$, is nondecreasing, $F_{x(t)}^{\{\alpha\}}(-\infty)=0$, and $F_{x(t)}^{\{\alpha\}}(+\infty)=1$. In every continuity point of the function $\xi\mapsto F_{x(t)}^{\{\alpha\}}(\xi)$, the function $t\mapsto F_{x(t)}^{\{\alpha\}}(\xi)$ is almost periodic.

Proof. The proof is the same as that of Theorem 2.21 by replacing the periodic component extraction operator $\mathrm{E}^{\{T_0\}}\{\cdot\}$ with the almost-periodic component extraction operator $\mathrm{E}^{\{\alpha\}}\{\cdot\}$ and using Lemma 2.18 on the nonnegativeness of the almost-periodic component of a nonnegative function. □

In the following, several results are reported for almost-periodic FOT statistical functions. That is, by adopting the almost-periodic component extraction operator as expectation operator. Similar results can be obtained with the obvious changes by adopting the periodic or the poly-periodic component extraction operators as expectation operators. It is worth to underline that, in general, the almost-periodic component extraction operator should be adopted for a complete statistical characterization of a time series.

Under Assumption 2.19b, we have

$$F_{x(t)}^{\{\alpha\}}(\xi)=\sum_{\gamma\in\Gamma^{(1)}}F_x^{\gamma}(\xi)\,e^{j2\pi\gamma t} \tag{2.33}$$

where the Fourier coefficients are given by

$$F_x^{\gamma}(\xi)\triangleq\left\langle F_{x(t)}^{\{\alpha\}}(\xi)\,e^{-j2\pi\gamma t}\right\rangle_t=\left\langle\mathbf{1}_{\{x(t)\leqslant\xi\}}\,e^{-j2\pi\gamma t}\right\rangle_t \tag{2.34}$$

and the convergence of the Fourier series in (2.33) is uniform provided that

$$\sum_{\gamma\in\Gamma^{(1)}}\left|F_x^{\gamma}(\xi)\right|<\infty\,. \tag{2.35}$$

From Theorem 2.23, it follows that the almost-periodic component extraction operator $\mathrm{E}^{\{\alpha\}}\{\cdot\}$ plays, in the FOT probability framework, the same role played by the ensemble average $\mathrm{E}\{\cdot\}$ in the classical stochastic framework. For example, for an almost cyclostationary stochastic process $X(t)$ we have that the almost-periodically time-variant cumulative distribution function can be expressed as

$$F_X(\xi;t) = \mathrm{E}\left\{\mathbf{1}_{\{X(t)\leqslant\xi\}}\right\} \tag{2.36}$$

which is the stochastic counterpart of (2.32). However, similarly to the infinite time average, the almost-periodic component extraction operator is linear but not σ-linear.

If the derivative with respect to ξ

$$f_{x(t)}^{\{\alpha\}}(\xi) \triangleq \frac{\mathrm{d}F_{x(t)}^{\{\alpha\}}(\xi)}{\mathrm{d}\xi} = \mathrm{E}^{\{\alpha\}}\left\{\delta(\xi - x(t))\right\} \tag{2.37}$$

exists, then it is a valid almost-periodically time variant probability density function, where the second formal equality is derived in Section 2.6.

Theorem 2.24. Gardner Theorem of Temporal Expectation (Gardner, 1987f), (Gardner and Brown, 1991). *Let $g(\cdot)$ be a continuous and bounded function such that for any $\ell \in \mathbb{R}$ the equation $g(\xi) = \ell$ admits at most a finite number of solutions for ξ belonging to any finite interval. If condition (2.35) holds and the Fourier series of the left-hand side of (2.38) is convergent, under Assumption 2.19b, it results*

$$\mathrm{E}^{\{\alpha\}}\left\{g(x(t))\right\} = \int_{\mathbb{R}} g(\xi)\, \mathrm{d}F_{x(t)}^{\{\alpha\}}(\xi) \tag{2.38}$$

Proof. See Section 2.6. □

This result is in accordance with the fact that $\mathrm{E}^{\{\alpha\}}\{\cdot\}$ is the expectation operator with respect to the distribution function $F_{x(t)}^{\{\alpha\}}(\xi)$ for the time series $x(t)$. Theorem 2.24 is the analogous of the corresponding fundamental theorem of expectation from probability theory.

Fact 2.25. *For an almost-periodic function $x(t)$, we have $x(t) \equiv x_{\mathrm{ap}}(t)$ and, hence,*

$$\mathrm{E}^{\{\alpha\}}\{x(t)\} = x(t)\,. \tag{2.39}$$

That is, the almost-periodic functions are the deterministic signals in the almost-periodic FOT probability framework. □

All the other signals are the random signals. Note that *the term "random" here is not intended to be synonymous with "stochastic"*. In fact, the adjective stochastic is adopted, as usual, when an ensemble of realizations or sample paths exists, whereas the adjective random is used in reference to a single function of time.

Fact 2.26. *As in the case of the infinite time average (Leśkow and Napolitano, 2006, Fact 4.4), the periodic component extraction operator $\mathrm{E}^{\{T_0\}}\{\cdot\}$, the poly-periodic component extraction operator $\mathrm{E}^{\{T_1,\ldots,T_P\}}\{\cdot\}$, and the almost-periodic component extraction operator $\mathrm{E}^{\{\alpha\}}\{\cdot\}$ are linear but not σ-linear. In contrast, in the stochastic approach, the ensemble average is also σ-linear.* □

Remark 2.27. For a given time series $x(t)$, *one can choose which expectation operator to adopt* among the infinite time average $\langle\cdot\rangle_t$, the periodic component extraction operator $\mathrm{E}^{\{T_0\}}\{\cdot\}$, the poly-periodic component extraction operator $\mathrm{E}^{\{T_1,\ldots,T_P\}}\{\cdot\}$, or the almost-periodic component extraction operator $\mathrm{E}^{\{\alpha\}}\{\cdot\}$. This choice determines the kind on non-stationarity that is possibly evidenced in the analyzed time series. For a complete characterization of a time series, all the periodicities should be considered. Hence, the operator $\mathrm{E}^{\{\alpha\}}\{\cdot\}$ should be adopted.

In contrast, in the stochastic approach, the expectation operator is the ensemble average and the kind of nonstationarity exhibited by the process is not consequence of any choice for this operator. However, in the stochastic approach the kind of nonstationarity of the process depends on how parameters present in the process are modeled. Specifically, if some parameters are modeled as random variables, statistical functions of the process depend on the distribution and, more generally, on the probabilistic characteristics of these parameters (Gardner, 1987f, Chap. 10, Sec. E). In particular, in some cases the process can be stationary (Theorem 1.5). □

2.3.1.4 Basic results on time averages

In this section, some basic results on time averages are reported.

Theorem 2.28. Sinusoidally Weighted Time Average. *Let be $x(t) \in L^1_{\mathrm{loc}}(\mathbb{R})$ and bounded. The sinusoidally weighted time average*

$$x^\alpha \triangleq \left\langle x(t)\, e^{-j2\pi\alpha t} \right\rangle_t \tag{2.40}$$

exists provided that $x(t)\, e^{-j2\pi\alpha t}$ is relatively measurable. Alternatively, the time average exists if $x(t)$ is bounded and jointly relatively measurable with $\cos(2\pi\alpha t)$ and $\sin(2\pi\alpha t)$.

Proof. See (Leśkow and Napolitano, 2006, Corollary 4.2), (Napolitano, 2012, Sec. 6.3.2). See also (Leśkow and Napolitano, 2006, Example 6.1). □

Theorem 2.29. Interchange of Integral and Time-Average Operators. *Under the assumptions*

1. $h(t) \in L^1(\mathbb{R})$;
2. $x(t) \in L^1_{\mathrm{loc}}(\mathbb{R}) \cap L^\infty(\mathbb{R})$, *and* $\langle x(t)\rangle_t$ *exists (see Theorem 2.28);*

it results that

$$\langle x(t) \otimes h(t)\rangle_t = \int_{\mathbb{R}} h(t)\, \mathrm{d}t \; \langle x(t)\rangle_t \,. \tag{2.41}$$

Proof. See Section 2.6. □

Remark 2.30. Note that by making the variable change $u = t - \tau$ in the left-hand side of (2.155) (which is equivalent to (2.41)), one has

$$\lim_{Z\to\infty} \frac{1}{Z} \int_{-Z/2}^{+Z/2} \left[\int_{\mathbb{R}} h(\tau)\, x(t-\tau)\, \mathrm{d}\tau \right] \mathrm{d}t = \lim_{Z\to\infty} \frac{1}{Z} \int_{-Z/2}^{+Z/2} \left[\int_{\mathbb{R}} h(t-u)\, x(u)\, \mathrm{d}u \right] \mathrm{d}t \tag{2.42}$$

However, unlike equality in (2.155), the equality

$$\lim_{Z\to\infty} \frac{1}{Z} \int_{-Z/2}^{+Z/2} \left[\int_{\mathbb{R}} h(t-u)\, x(u)\, \mathrm{d}u \right] \mathrm{d}t = \int_{\mathbb{R}} x(u) \left[\lim_{Z\to\infty} \frac{1}{Z} \int_{-Z/2}^{+Z/2} h(t-u)\, \mathrm{d}t \right] \mathrm{d}u \tag{2.43}$$

is not true since in the assumptions of Theorem 2.29, $h(t)$ and $x(t)$ cannot be interchanged one another. Specifically, even if $x(t) \in L^1_{\mathrm{loc}}(\mathbb{R})$, in most cases of interest $x(t)$ is a persistent time series and hence $x(t) \notin L^1(\mathbb{R})$. Moreover, since $h(t) \in L^1(\mathbb{R})$, its time average is zero. □

Observing that $x(t) \in L^\infty(\mathbb{R}) \Rightarrow x(t)\, e^{j2\pi\alpha t} \in L^\infty(\mathbb{R})$, from Theorem 2.29, the following result immediately follows.

Corollary 2.31. *Under the assumptions*

1. $h(t) \in L^1(\mathbb{R})$;
2. $x(t) \in L^1_{\mathrm{loc}}(\mathbb{R}) \cap L^\infty(\mathbb{R})$, *and* $x^\alpha \triangleq \langle x(t)\, e^{-j2\pi\alpha t}\rangle_t$ *exists (see Theorem 2.28);*

it results that

$$\left\langle [x(t) \otimes h(t)]\, e^{-j2\pi\alpha t} \right\rangle_t = H(\alpha)\, x^\alpha \tag{2.44}$$

where

$$H(f) \triangleq \int_{\mathbb{R}} h(t)\, e^{-j2\pi f t}\, \mathrm{d}t \tag{2.45}$$

Proof. See Section 2.6. □

Theorem 2.32. Interchange of Integral and Almost-Periodic Component Extraction Operators. *Under the assumptions*

1. $h(t) \in L^1(\mathbb{R})$;
2. $x(t) \in L^1_{\mathrm{loc}}(\mathbb{R}) \cap L^\infty(\mathbb{R})$, *and* $x^\alpha \triangleq \langle x(t)\, e^{-j2\pi\alpha t}\rangle_t$ *exists* $\forall \alpha \in A$ *(see Theorem 2.28);*
3. $\sum_{\alpha\in A} |x^\alpha| < \infty$;

it results

$$\mathrm{E}^{\{\alpha\}}\left\{ \int_{\mathbb{R}} h(\tau)\, x(t-\tau)\, \mathrm{d}\tau \right\} = \int_{\mathbb{R}} h(\tau)\, \mathrm{E}^{\{\alpha\}}\{x(t-\tau)\}\, \mathrm{d}\tau \tag{2.46a}$$

$$= \sum_{\alpha\in A} H(\alpha)\, x^\alpha\, e^{j2\pi\alpha t}\,. \tag{2.46b}$$

Proof. See Section 2.6. □

2.3.1.5 Second-order characterization

Analogously to the first-order case, a second-order characterization for the real-valued time series $x(t)$ can be obtained by using the almost-periodic component extraction operator as the expectation operator. Similar results can be obtained with the obvious changes by adopting the periodic or the poly-periodic component extraction operators as expectation operators. Note that, in general, the almost-periodic component extraction operator should be adopted for a complete statistical characterization of a time series.

Assumption 2.33.

a. *The set* $\Gamma_{\xi_1,\xi_2,\tau}$ *of frequencies (depending on* ξ_1*,* ξ_2*, and* τ*) of the almost-periodic component of the function* $t \mapsto \mathbf{1}_{\{x(t+\tau)\leqslant\xi_1\}}\,\mathbf{1}_{\{x(t)\leqslant\xi_2\}}$ *is countable.*

b. *The set*

$$\Gamma_\tau^{(2)} \triangleq \bigcup_{\xi_1\in\mathbb{R}}\bigcup_{\xi_2\in\mathbb{R}} \Gamma_{\xi_1,\xi_2,\tau} \tag{2.47}$$

is countable.

c. *The set*

$$\Gamma^{(2)} \triangleq \bigcup_{\tau\in\mathbb{R}} \Gamma_\tau^{(2)} \tag{2.48}$$

is countable. □

Theorem 2.34. (Gardner and Brown, 1991). *Under Assumption 2.33b, the function of* ξ_1 *and* ξ_2

$$F^{\{\alpha\}}_{x(t+\tau)\,x(t)}(\xi_1,\xi_2) \triangleq \mathrm{E}^{\{\alpha\}}\left\{\mathbf{1}_{\{x(t+\tau)\leqslant\xi_1\}}\,\mathbf{1}_{\{x(t)\leqslant\xi_2\}}\right\} \tag{2.49}$$

is a valid second-order joint cumulative distribution function for every fixed t *and* τ*, except for the right-continuity property (in the discontinuity points) with respect to* ξ_1 *and* ξ_2*. It is an almost-periodic function of* t*.*

Proof. The proof is the same as the proof of Theorem 2.23 with $\mathbf{1}_{\{x(t)\leqslant\xi\}}$ replaced by $\mathbf{1}_{\{x(t+\tau)\leqslant\xi_1\}}\,\mathbf{1}_{\{x(t)\leqslant\xi_2\}}$. The second-order distribution is consistent in the sense that the first-order distribution is obtained when $\xi_1 \to +\infty$ or $\xi_2 \to +\infty$. In fact, $\mathbf{1}_{\{x(t)\leqslant+\infty\}} = 1$. □

If the second-order derivative with respect to ξ_1 and ξ_2

$$f^{\{\alpha\}}_{x(t+\tau)\,x(t)}(\xi_1,\xi_2) \triangleq \frac{\partial^2 F^{\{\alpha\}}_{x(t+\tau)\,x(t)}(\xi_1,\xi_2)}{\partial\xi_1\,\partial\xi_2} = \mathrm{E}^{\{\alpha\}}\left\{\delta(\xi_1 - x(t+\tau))\,\delta(\xi_2 - x(t))\right\} \tag{2.50}$$

exists, then it is a valid second-order joint probability density function (Gardner and Brown, 1991), where the second formal equality is proved in Section 2.6.

Theorem 2.35. *Under Assumption 2.33c, the function*

$$R_x(t,\tau) \triangleq \mathrm{E}^{\{\alpha\}}\{x(t+\tau)\,x(t)\} \tag{2.51}$$

is a valid autocorrelation function and it results that

$$R_x(t,\tau) = \int_{\mathbb{R}^2} \xi_1\xi_2\,\mathrm{d}F^{\{\alpha\}}_{x(t+\tau)\,x(t)}(\xi_1,\xi_2)$$

$$= \int_{\mathbb{R}^2} \xi_1 \xi_2 \, f^{\{\alpha\}}_{x(t+\tau)\,x(t)}(\xi_1, \xi_2)\, \mathrm{d}\xi_1\, \mathrm{d}\xi_2$$

$$= \sum_{\alpha \in A} R_x^\alpha(\tau)\, e^{j2\pi\alpha t} \tag{2.52}$$

where the second equality holds provided that the second-order joint probability density function (2.50) exists at least in the sense of generalized functions. In (2.52), A is a countable set and

$$R_x^\alpha(\tau) \triangleq \left\langle x(t+\tau)\, x(t)\, e^{-j2\pi\alpha t} \right\rangle_t \tag{2.53}$$

is the (nonstochastic) cyclic autocorrelation function *at cycle frequency α, ($\alpha \in \mathbb{R}$).*

Proof. See Section 2.6. □

Theorem 2.35 can be proved also under Assumption 2.33b. In such a case, if Assumption 2.33c does not hold, then the time series $x(t)$ is second-order generalized almost cyclostationary (Chapter 12).

The classification of the kind of nonstationarity of a time series is made on the basis of the elements contained in the set A. In general, the set A can contain incommensurate cycle frequencies α and, in such a case, the time series is said to be *wide-sense almost cyclostationary*. In the special case where $A \equiv \{k/T_0\}_{k\in\mathbb{Z}}$ the time series $x(t)$ is said to be *wide-sense cyclostationary*. If the set A contains only the element $\alpha = 0$, then the time series $x(t)$ is said to be *wide-sense stationary*.

2.3.1.6 Spectral line regeneration

In this section it is shown how, for an ACS signal, finite-strength additive sine-wave components are generated in the second-order lag-product even if they are not present in the signal.

Let us consider the decomposition (2.23) for the signal $x(t)$ in the special case where the AP component is uniformly AP in the sense of Besicovitch (Section B.2)

$$x(t) = x_{\mathrm{uap}}(t) + x_r(t) \tag{2.54}$$

and the residual term $x_r(t)$ does not contain any finite-strength additive sine-wave component and has possibly finite power.

The second-order lag-product waveform is given by

$$\begin{aligned} x(t+\tau)\, x(t) &= [x_{\mathrm{uap}}(t+\tau) + x_r(t+\tau)][x_{\mathrm{uap}}(t) + x_r(t)] \\ &= x_{\mathrm{uap}}(t+\tau)\, x_{\mathrm{uap}}(t) + x_r(t+\tau)\, x_{\mathrm{uap}}(t) \\ &\quad + x_{\mathrm{uap}}(t+\tau)\, x_r(t) + x_r(t+\tau)\, x_r(t)\,. \end{aligned} \tag{2.55}$$

In (2.55) it results that

a. $x_{\mathrm{uap}}(t+\tau)\, x_{\mathrm{uap}}(t)$ is a uniformly AP function since the product of uniformly AP functions is a uniformly AP function (Theorem B.9);

b. $x_r(t+\tau)\,x_{\text{uap}}(t)$ and $x_{\text{uap}}(t+\tau)\,x_r(t)$ do not contain any additive finite-strength sine-wave component;
c. $x_r(t+\tau)\,x_r(t)$ can contain additive finite-strength sine-wave components.

The (possible) almost-periodic component contained in $x_r(t+\tau)\,x_r(t)$ is constituted by the so called second-order pure sine waves (Gardner and Spooner, 1994a). That is, it is given by the superposition of the finite-strength additive sine-wave components present in the second-order lag-product $x(t+\tau)\,x(t)$ that are not generated by beats of (first-order) finite-strength sine waves present in the time series (that is, term $x_{\text{uap}}(t+\tau)\,x_{\text{uap}}(t)$). In particular, if $x(t)$ does not contain any additive finite-strength sine-wave component, that is, $x_{\text{uap}}(t) \equiv 0$ in (2.54), then the possible AP component in the lag product waveform $x(t+\tau)\,x(t)$ is only due to the product of the residual term $x_r(t)$ and its time-shifted version $x_r(t+\tau)$. In such a case, we have that periodicities that are hidden at first order are regenerated in the second-order lag-product waveform. If the second-order lag-product waveform $x(t+\tau)\,x(t)$ is uniformly continuous and bounded, and possibly possesses a uniformly AP component, according to (Kahane, 1962), (Napolitano, 2012, Theorem 1.2.20), this uniformly AP component is contained in the product of $x(t+\tau) \equiv x_r(t+\tau)$ and $x(t) \equiv x_r(t)$.

Analogously to the signal $x(t)$ that can be decomposed as in (2.23), the second-order lag-product waveform $x(t+\tau_1)x(t+\tau_2)$, for every fixed τ_1, τ_2, is a function of t that can be decomposed into the sum of its almost-periodic component and a residual term

$$x(t+\tau_1)\,x(t+\tau_2) = \mathrm{E}^{\{\alpha\}}\,\{x(t+\tau_1)\,x(t+\tau_2)\} + \ell_x(t,\tau_1,\tau_2) \tag{2.56a}$$

$$= \sum_{\alpha\in A_\tau} R_x^\alpha(\tau_1-\tau_2)\,e^{j2\pi\alpha(t+\tau_2)} + \ell_x(t,\tau_1,\tau_2) \tag{2.56b}$$

where $\ell_x(t,\tau_1,\tau_2)$ does not contain any finite-strength additive sine-wave component and $\tau \triangleq \tau_1 - \tau_2$.

In the special case of ACS time series ($A \triangleq \cup_{\tau\in\mathbb{R}} A_\tau$ countable), any homogeneous quadratic time-invariant (QTI) transformation of the signal $x(t)$ has the form

$$\begin{aligned} y(t) &\triangleq \int_{\mathbb{R}^2} k(\tau_1,\tau_2)\,x(t+\tau_1)\,x(t+\tau_2)\,\mathrm{d}\tau_1\,\mathrm{d}\tau_2 \\ &= \sum_{\alpha\in A} \int_{\mathbb{R}^2} k(\tau_1,\tau_2)\,R_x^\alpha(\tau_1-\tau_2)\,e^{j2\pi\alpha\tau_2}\,\mathrm{d}\tau_1\,\mathrm{d}\tau_2\,e^{j2\pi\alpha t} \\ &\quad + \int_{\mathbb{R}^2} k(\tau_1,\tau_2)\,\ell_x(t,\tau_1,\tau_2)\,\mathrm{d}\tau_1\,\mathrm{d}\tau_2 \end{aligned} \tag{2.57}$$

where $k(\tau_1,\tau_2)$ is the kernel of the QTI transformation. That is, finite-strength additive sine-wave components can be regenerated by homogeneous QTI transformations (Gardner, 1987f), (Gardner and Brown, 1991), (Gardner, 1994a). In contrast, for stationary time series (A containing the only element $\alpha = 0$), no spectral line at nonzero frequency can be regenerated in the lag-product or by QTI transformations.

More generally, a time series $x(t)$ is said to exhibit *higher-order cyclostationarity* if finite-strength additive sine-wave components can be regenerated by homogeneous nonlinear time-invariant transformations of $x(t)$ of order greater than two (Gardner and Spooner, 1994a), (Spooner and Gardner, 1994a). In such a case, almost-periodically time-variant higher-order moment and cumulant functions can be defined by the almost-periodic component extraction operator (Chapter 4), (Gardner and Spooner, 1994a), (Spooner and Gardner, 1994a), (Izzo and Napolitano, 1998a). For communications ACS signals, cycle frequencies of second- and higher-order statistical functions are related to parameters such as sine-wave carrier frequency, pulse rate, symbol rate, frame rate, sampling frequency. Therefore, spectral line regeneration by second- or higher-order time-invariant transformations leads to signals suitable for synchronization purposes (Section 9.4) (Gardner, 1987f).

In (Bass, 1974b, pp. 497–502), conditions are derived such that linear combinations of powers of pseudo-random functions can be decomposed into the sum of a periodic function and a pseudo-random function.

2.3.1.7 Synchronized averaging for the autocorrelation

Theorem 2.36. Synchronized Averaging Theorem for the Autocorrelation (Gardner, 1985), (Gardner, 1987f). *The periodic component with period T_0 contained in the lag product time series $x(t+\tau)\,x(t)$ can be extracted by exploiting the synchronized averaging theorem (Theorem B.19):*

$$\sum_{k=-\infty}^{+\infty} R_x^{k/T_0}(\tau)\, e^{j2\pi(k/T_0)t} = \lim_{N\to\infty} \frac{1}{2N+1} \sum_{n=-N}^{N} x(t+\tau-nT_0)\, x(t-nT_0) \tag{2.58}$$

provided that the series in the lhs is absolutely convergent and the limit in the rhs exists. □

Proposition 2.37. *Let the lag product $x(t+\tau)\,x(t)$ be the sum of a poly-periodic function with incommensurate periods $T_1,\ldots,T_P$ and a residual term not containing any finite-strength additive sine-wave component with frequencies commensurate with $1/T_1,\ldots,1/T_P$. Let*

$$\mathrm{E}^{\{T_p\}}\{x(t+\tau)\,x(t)\} \triangleq \lim_{N\to\infty} \frac{1}{2N+1} \sum_{n=-N}^{N} x(t+\tau-nT_p)\, x(t-nT_p) \tag{2.59}$$

denote the periodic component with period T_p of the lag product.

The poly-periodic component of $x(t+\tau)\,x(t)$ with periods $T_1,\ldots,T_P$ can be expressed as

$$\begin{aligned}&\mathrm{E}^{\{T_1,\ldots,T_P\}}\{x(t+\tau)\,x(t)\}\\ &\quad = \sum_{p=1}^{P} \Big[\mathrm{E}^{\{T_p\}}\{x(t+\tau)\,x(t)\} - \langle x(t+\tau)\,x(t)\rangle_t\Big] + \langle x(t+\tau)\,x(t)\rangle_t\,.\end{aligned} \tag{2.60}$$

□

If two periods are commensurate, then the two periodic functions can be merged into a unique periodic function.

2.3.1.8 Frequency domain

ACS time series are characterized in the spectral domain by the (nonstochastic) *cyclic spectrum* or *spectral correlation density function* at cycle frequency α:

$$S_x^\alpha(f) \triangleq \lim_{\Delta f \to 0} \lim_{T \to \infty} \frac{1}{T} \int_{-T/2}^{T/2} \Delta f \, X_{1/\Delta f}(t, f) \, X_{1/\Delta f}^*(t, f - \alpha) \, \mathrm{d}t \tag{2.61}$$

where $X_{1/\Delta f}(t, f)$ is defined according to (1.31) and the order of the two limits, assuming that they exist, cannot be reversed. The *Gardner relation* introduced in (Gardner, 1987f) also dubbed *cyclic Wiener relation*

$$S_x^\alpha(f) = \int_{\mathbb{R}} R_x^\alpha(\tau) \, e^{-j2\pi f \tau} \, \mathrm{d}\tau \tag{2.62}$$

links the cyclic autocorrelation function to the cyclic spectrum. This relation generalizes that for $\alpha = 0$, which was first dubbed the *Wiener relation* in (Gardner, 1985) to distinguish it from the *Khinchin relation* (1.27), which is frequently called the *Wiener-Khinchin relation* (see also Section 2.3.1.10).

Definition 2.38. Let $x(t)$ a locally summable function. The FOT Loève bifrequency spectrum is defined as

$$\begin{aligned}
\mathcal{S}_x(f_1, f_2) &\triangleq \lim_{T \to \infty} \left\langle X_T(t, f_1) \, X_T^*(t, f_2) \right\rangle_t \\
&= \lim_{T \to \infty} \lim_{Z \to \infty} \frac{1}{Z} \int_{-Z/2}^{Z/2} X_T(t, f_1) \, X_T^*(t, f_2) \, \mathrm{d}t
\end{aligned} \tag{2.63}$$

provided that the limits exist, where $X_T(t, f)$ is the STFT of $x(t)$ defined in (1.31). □

2.3.1.9 Nonzero-mean time series

For a signal containing an additive almost-periodic component, the following decomposition holds (see (2.23))

$$x(t) = \sum_{\eta \in E_1} x_\eta \, e^{j2\pi \eta t} + x_r(t) \tag{2.64}$$

which is the analogous for time series of decomposition (1.63) for stochastic processes.

Here, results are presented in the general case of complex-valued time series. See Chapter 4 for notation and definitions.

The condition that the constants x_η and the signal $x_r(t)$ are uncorrelated is automatically verified in the FOT approach. In fact, since $x_r(t)$ does not contain any finite-strength additive sine wave component, we have

$$\mathrm{E}^{\{\alpha\}} \left\{ x_\eta^{(*)} \, x_r(t) \right\} = 0 \, . \tag{2.65}$$

Thus, unlike as in the stochastic approach, such a condition does not need to be *assumed* to derive the FOT counterparts of the results of Section 1.4. In particular, the FOT counterparts of (1.66) and (1.67) are

$$R^{\alpha}_{xx^{(*)}}(\tau) = R^{\alpha}_{x_r x_r^{(*)}}(\tau) + \sum_{\eta \in E_1} x_\eta \, x^{(*)}_{(-)(\alpha-\eta)} \, e^{j2\pi\eta\tau} \tag{2.66}$$

$$S^{\alpha}_{xx^{(*)}}(f) = S^{\alpha}_{x_r x_r^{(*)}}(f) + \sum_{\eta \in E_1} x_\eta \, x^{(*)}_{(-)(\alpha-\eta)} \, \delta(f-\eta) \tag{2.67}$$

2.3.1.10 Gardner relation

In this section, the Gardner relation (2.62) is presented for the general case of cross-statistics of complex-valued time series. Proofs are reported in Section 2.6.

Lemma 2.39. (Gardner, 1987f, Chap. 13). *Let $x(t)$ and $y(t)$ be complex-valued functions in $L^1_{\text{loc}}(\mathbb{R}) \cap L^\infty(\mathbb{R})$. The* (conjugate) cyclic cross-periodogram *(see Definition 5.21)*

$$I^{(T)}_{yx^{(*)}}(\alpha, f; t_0) \triangleq \frac{1}{T} Y_T(t_0, f) \, X^{(*)}_T(t_0, (-)(\alpha - f)) \tag{2.68}$$

where $X_T(t, f)$ and $Y_T(t, f)$ are STFTs defined according to (1.31), and the (conjugate) cyclic cross-correlogram *(see Definition 5.1)*

$$R^{(T)}_{yx^{(*)}}(\alpha, \tau; t_0) \triangleq \int_{t_0 - T/2}^{t_0 + T/2} \operatorname{rect}\left(\frac{t + \tau - t_0}{T}\right) y(t+\tau) \, x^{(*)}(t) \, e^{-j2\pi\alpha t} \, \mathrm{d}t \tag{2.69}$$

are linked by the Fourier transform relationship

$$I^{(T)}_{yx^{(*)}}(\alpha, f; t_0) = \int_{\mathbb{R}} R^{(T)}_{yx^{(*)}}(\alpha, \tau; t_0) \, e^{-j2\pi f\tau} \, \mathrm{d}\tau \, . \tag{2.70}$$

Proof. See Section 2.6. □

Theorem 2.40. Gardner Relation (Gardner, 1987f, Chap. 13). *Let $x(t)$ and $y(t)$ be complex-valued functions in $L^1_{\text{loc}}(\mathbb{R}) \cap L^\infty(\mathbb{R})$. The* (conjugate) cyclic cross-spectrum

$$\begin{aligned} S^{\alpha}_{yx^{(*)}}(f) &\triangleq \lim_{T\to\infty} \lim_{Z\to\infty} \frac{1}{Z} \int_{-Z/2}^{Z/2} I^{(T)}_{yx^{(*)}}(\alpha, f; t) \, \mathrm{d}t \\ &= \lim_{T\to\infty} \lim_{Z\to\infty} \frac{1}{Z} \int_{-Z/2}^{Z/2} \frac{1}{T} Y_T(t, f) \, X^{(*)}_T(t, (-)(\alpha - f)) \, \mathrm{d}t \end{aligned} \tag{2.71}$$

and the (conjugate) cyclic cross-correlation function

$$\begin{aligned} R^{\alpha}_{yx^{(*)}}(\tau) &\triangleq \lim_{T\to\infty} R^{(T)}_{yx^{(*)}}(\alpha, \tau; t) \\ &= \lim_{T\to\infty} \frac{1}{T} \int_{-T/2}^{T/2} y(t+\tau) \, x^{(*)}(t) \, e^{-j2\pi\alpha t} \, \mathrm{d}t \end{aligned} \tag{2.72}$$

are linked by the Fourier transform relationship

$$S^{\alpha}_{yx^{(*)}}(f) = \int_{\mathbb{R}} R^{\alpha}_{yx^{(*)}}(\tau)\, e^{-j2\pi f\tau}\, \mathrm{d}\tau \tag{2.73}$$

provided that the limits in (2.71) and (2.72) exist and

$$\int_{\mathbb{R}} |\tau| \left| R^{\alpha}_{yx^{(*)}}(\tau) \right| \mathrm{d}\tau < \infty\,. \tag{2.74}$$

The order of the two limits in (2.71) cannot be inverted.

Proof. See Section 2.6. □

2.3.2 Discrete-time

The characterization of discrete-time time series (sequences) is similar to that of continuous-time time series. We consider here only the wide-sense second-order characterization.

Let $x(n)$, $n \in \mathbb{Z}$, be a real-valued discrete-time time series. Let us assume that the set $\widetilde{\Gamma}_2$ of frequencies (for every ξ_1, ξ_2 and m) of the almost-periodic component of the function $n \mapsto \mathbf{1}_{\{x(n+m)\leqslant\xi_1\}}\, \mathbf{1}_{\{x(n)\leqslant\xi_2\}}$ is countable. Then, the function of ξ_1 and ξ_2

$$F^{\{\widetilde{\alpha}\}}_{x(n+m)\,x(n)}(\xi_1, \xi_2) \triangleq \mathrm{E}^{\{\widetilde{\alpha}\}} \left\{ \mathbf{1}_{\{x(n+m)\leqslant\xi_1\}}\, \mathbf{1}_{\{x(n)\leqslant\xi_2\}} \right\} \tag{2.75}$$

is a valid second-order joint cumulative distribution function for every fixed n and m, except for the right-continuity property (in the discontinuity points) with respect to ξ_1 and ξ_2. In (2.75), $\mathrm{E}^{\{\widetilde{\alpha}\}}\{\cdot\}$ denotes the discrete-time almost-periodic component extraction operator, which is defined analogously to its continuous-time counterpart. Thus, $F^{\{\widetilde{\alpha}\}}_{x(n+m)\,x(n)}(\xi_1, \xi_2)$ is almost-periodic in n (Section B.3). The second-order derivative with respect to ξ_1 and ξ_2 of $F^{\{\widetilde{\alpha}\}}_{x(n+m)\,x(n)}(\xi_1, \xi_2)$, denoted by $f^{\{\widetilde{\alpha}\}}_{x(n+m)\,x(n)}(\xi_1, \xi_2)$, if it exists, is a valid second-order joint probability density function (Gardner and Brown, 1991). Furthermore, it can be shown that the function

$$\widetilde{R}_x(n, m) \triangleq \mathrm{E}^{\{\widetilde{\alpha}\}} \{x(n+m)\, x(n)\} \tag{2.76}$$

is a valid autocorrelation function and can be characterized by

$$\begin{aligned} \widetilde{R}_x(n, m) &= \int_{\mathbb{R}^2} \xi_1 \xi_2\, f^{\{\widetilde{\alpha}\}}_{x(n+m)\,x(n)}(\xi_1, \xi_2)\, \mathrm{d}\xi_1\, \mathrm{d}\xi_2 \\ &= \sum_{\widetilde{\alpha} \in \widetilde{A}} \widetilde{R}^{\widetilde{\alpha}}_x(m)\, e^{j2\pi\widetilde{\alpha}n} \end{aligned} \tag{2.77}$$

where

$$\widetilde{A} \triangleq \{\widetilde{\alpha} \in [-1/2, 1/2) \;:\; \widetilde{R}^{\widetilde{\alpha}}_x(m) \not\equiv 0\} \tag{2.78}$$

is a countable set and

$$\widetilde{R_x^{\widetilde{\alpha}}}(m) \triangleq \lim_{N\to\infty} \frac{1}{2N+1} \sum_{n=-N}^{N} x(n+m)x(n)\, e^{-j2\pi\widetilde{\alpha}n} \tag{2.79}$$

is the (nonstochastic) discrete-time *cyclic autocorrelation function* at cycle frequency $\widetilde{\alpha}$.

Obviously, as in the stochastic framework, the cyclic autocorrelation function $\widetilde{R_x^{\widetilde{\alpha}}}(m)$ is periodic in $\widetilde{\alpha}$ with period 1. Thus, the sum in (2.77) can be equivalently extended to the set $\widetilde{A}_1 \triangleq \{\widetilde{\alpha} \in [0, 1) : \widetilde{R_x^{\widetilde{\alpha}}}(m) \not\equiv 0\}$. Moreover, as in the continuous-time case, the classification of the kind of nonstationarity of discrete-time time series is made on the basis of the elements contained in the set $\widetilde{A}$. That is, in general, the set $\widetilde{A}$ can contain incommensurate cycle frequencies $\widetilde{\alpha}$ and, in such a case, the time series is said to be *wide-sense almost cyclostationary*. In the special case where $\widetilde{A}_1 \equiv \{0, 1/N_0, \ldots, (N_0-1)/N_0\}$ for some integer N_0, the time series $x(n)$ is said to be *wide-sense cyclostationary*. If the set $\widetilde{A}$ contains only the element $\widetilde{\alpha} = 0$, then the time series $x(n)$ is said to be *wide-sense stationary*.

The discrete-time counterpart of the synchronized averaging theorem for the autocorrelation (Theorem 2.36) holds only in the cyclostationary case (period N_0 integer):

$$\sum_{k=0}^{N_0-1} R_x^{k/N_0}(m)\, e^{j2\pi(k/N_0)n} = \lim_{N\to\infty} \frac{1}{2N+1} \sum_{p=-N}^{N} x(n+m-pN_0)\, x(n-pN_0) \tag{2.80}$$

provided that the limit in the rhs exists.

The sum of discrete-time periodic functions can be expressed as a unique periodic function whose period is at most equal to the least common multiple (lcm) of the integer periods of the individual periodic functions. Thus, no discrete-time counterpart of (2.60) exists.

Discrete-time ACS time series are characterized in the spectral domain by the (nonstochastic) *cyclic spectrum* or *spectral correlation density function* at cycle frequency $\widetilde{\alpha}$:

$$\widetilde{S_x^{\widetilde{\alpha}}}(\nu) \triangleq \lim_{\Delta\nu\to 0} \lim_{N\to\infty} \frac{1}{2N+1} \sum_{n=-N}^{N} \Delta\nu\, X_{\lfloor 1/\Delta\nu \rfloor}(n,\nu)\, X^*_{\lfloor 1/\Delta\nu \rfloor}(n,\nu-\widetilde{\alpha}) \tag{2.81}$$

where

$$X_{2M+1}(n,\nu) \triangleq \sum_{k=n-M}^{n+M} x(k)\, e^{-j2\pi\nu k} \tag{2.82}$$

and $\lfloor\cdot\rfloor$ denotes here operation that provides the closest odd integer. The *Cyclic Wiener relation* (Gardner, 1985)

$$\widetilde{S_x^{\widetilde{\alpha}}}(\nu) = \sum_{m\in\mathbb{Z}} \widetilde{R_x^{\widetilde{\alpha}}}(m)\, e^{-j2\pi\nu m} \tag{2.83}$$

links the cyclic autocorrelation function to the cyclic spectrum.

2.3.3 Nonstationarity classification in the FOT approach

In the stochastic approach, signals are modeled as stochastic processes, that is, ensemble of sample paths or realizations. In such a framework, nonstationarity is the property that statistical functions defined by ensemble averages depend on the time parameter t. In particular, the nonstationarity at first or second order in the strict sense is consequence of the possible dependence on t of the first- and second-order distribution functions.

In the functional approach, once the almost-periodically time-variant model is adopted, the kind of nonstationarity at first or second order in the strict sense depends on the elements contained in the countable sets $\Gamma^{(1)}$ or $\Gamma^{(2)}$ defined in (2.27) and (2.48), respectively.

If $\Gamma^{(2)} = \{0\}$, the only finite-strength additive sine-wave component contained in the function

$$t \mapsto \mathbf{1}_{\{x(t+\tau)\leqslant\xi_1\}}\,\mathbf{1}_{\{x(t)\leqslant\xi_2\}}$$

is that at frequency $\alpha = 0$. Thus, the time series is said to be second-order stationary in the strict sense and

$$\mathrm{E}^{\{\alpha\}}\{g(x(t+\tau), x(t))\} \equiv \langle g(x(t+\tau), x(t))\rangle_t \ . \tag{2.84}$$

If the set $\Gamma^{(2)}$ contains frequencies that are all integer multiple of a fundamental one, then the time series is said to be second-order cyclostationary in the strict sense. If the set $\Gamma^{(2)}$ contains incommensurate frequencies, then the time series is said to be second-order almost cyclostationary in the strict sense.

A similar classification can be made for the second-order nonstationarity in the wide sense. Let us consider, for the sake of simplicity, a zero mean time series. If the countable set A defined in Theorem 2.35 contains incommensurate cycle frequencies α, then the time series $x(t)$ is said to be *wide-sense almost cyclostationary*. In the special case where $A \equiv \{k/T_0\}_{k\in\mathbb{Z}}$, the time series is said to be *wide-sense cyclostationary*. If the set A contains only the element $\alpha = 0$, then the time series is said to be *wide-sense stationary*. In this classification, WSS time series are a subclass of the cyclostationary time series which are a subclass of the ACS time series.

In the FOT approach, a different classification can be made on the basis of the *existence or not* of the stationary and almost-cyclostationary probabilistic models. Specifically, time series for which a probabilistic model based on almost-periodic distributions (Theorem 2.34) exists are a subclass of time series for which a probabilistic model based on stationary distributions (Definition 2.4 with $k = 2$, $x_1 \equiv x_2$) exists. In fact, the existence of the almost-periodic distribution (2.49) requires more stringent conditions on the time series $x(t)$ with respect to those for the existence of the stationary distribution (2.8) (with $k = 2$, $x_1 \equiv x_2$). See also (Leśkow and Napolitano, 2006, Ex. 6.1).

Therefore, from this point of view, almost-cyclostationary time series are a subclass of the stationary time series. See (Napolitano, 2012, Sec. 6.4) for further discussion on this subject.

2.4 Stochastic versus fraction-of-time approach

In this section, the link between the stochastic and FOT approaches is analyzed. Moreover, it is shown how results derived in one approach should be interpreted in the other approach. The link between the two approaches is treated in depth in (Wold, 1948), (Gardner, 1985), (Gardner, 1987f), (Brown, 1987), (Gardner, 1990b), (Gardner and Brown, 1991), (Gardner, 1994a), (Hurd and Koski, 2004b), (Leśkow and Napolitano, 2006), (Napolitano, 2012, Chap. 6), (Dehay et al., 2013), (Dehay et al., 2018b).

The statistical expectation or ensemble mean $\mathrm{E}\{\cdot\}$ acts on a stochastic process, say $\{\xi(t,\omega), t \in \mathbb{R}, \omega \in \Omega\}$, and is an average over the variable ω. In the time-variant nonstochastic framework (for ACS time series), the expectation operator is the almost-periodic component extraction operator $\mathrm{E}^{\{\alpha\}}\{\cdot\}$. It acts on a single time-series and operates sinusoidally weighted averages over the time variable $t \in \mathbb{R}$.

Therefore, results derived in the FOT probability approach for time series can be interpreted in the classical stochastic process framework by replacing time series with stochastic processes and the almost-periodic component extraction operator $\mathrm{E}^{\{\alpha\}}\{\cdot\}$ with the statistical expectation $\mathrm{E}\{\cdot\}$. Vice versa, in the case of processes exhibiting suitable ergodicity properties, results in the stochastic approach can be reinterpreted in the FOT approach by replacing $\mathrm{E}\{\cdot\}$ with $\mathrm{E}^{\{\alpha\}}\{\cdot\}$ and the stochastic process with one of its typical (i.e., satisfying the weak law of large numbers) realizations.

It is worth to underline that a stochastic process model corresponding to a given ACS time series should be carefully built. In fact, modeling unknown parameters as random variables can modify or even destroy the cyclostationarity or almost-cyclostationarity of the process (Gardner, 1978), (Gardner, 1987f, Chap. 10, Sec. E). A typical case is the phase randomization (Theorem 1.5).

Let us denote by $\zeta(t)$ a generic stochastic process (possibly obtained as lag product of another process $\xi(t)$ or as a nonlinear function of $\xi(t+\tau_1), \ldots, \xi(t+\tau_N)$) and by $z(t)$ one of its typical realizations. Assuming that the following quantities exist

$$\zeta^{\beta} \triangleq \left\langle \mathrm{E}\{\zeta(t)\}\, e^{-j2\pi\beta t} \right\rangle_t \tag{2.85}$$

$$z^{\beta} \triangleq \left\langle \mathrm{E}^{\{\alpha\}}\{z(t)\}\, e^{-j2\pi\beta t} \right\rangle_t = \left\langle z(t)\, e^{-j2\pi\beta t} \right\rangle_t \tag{2.86}$$

we have

$$\zeta^{\beta} = z^{\beta}\,. \tag{2.87}$$

In particular, by assuming that $\{\xi(t,\omega), t \in \mathbb{R}, \omega \in \Omega\}$ is a stochastic process satisfying appropriate ergodicity properties (see Chapter 5), and denoting by $x(t) = \xi(t,\bar{\omega})$, $\bar{\omega}$ fixed, one of its typical realizations, we have the following equality of the FOT and stochastic autocorrelation functions

$$\mathcal{R}_{\xi}(t,\tau) \triangleq \mathrm{E}\{\xi(t+\tau)\,\xi(t)\} = \mathrm{E}^{\{\alpha\}}\{x(t+\tau)\,x(t)\} \triangleq R_x(t,\tau)\,. \tag{2.88}$$

Consequently, for their cyclic components and frequency-domain counterparts we also have

$$\mathcal{R}_{\xi}^{\beta}(\tau) \triangleq \left\langle \mathcal{R}_{\xi}(t,\tau)\, e^{-j2\pi\beta t} \right\rangle_t = \left\langle x(t+\tau)\, x(t)\, e^{-j2\pi\beta t} \right\rangle_t \triangleq R_x^{\beta}(\tau) \tag{2.89}$$

$$\mathcal{S}_{\xi}^{\beta}(f) \triangleq \int_{\mathbb{R}} \mathcal{R}_{\xi}^{\beta}(\tau)\, e^{-j2\pi f\tau}\, \mathrm{d}\tau = \int_{\mathbb{R}} R_x^{\beta}(\tau)\, e^{-j2\pi f\tau}\, \mathrm{d}\tau \triangleq S_x^{\beta}(f)\,. \tag{2.90}$$

Analogous equivalences hold in the discrete-time case.

A rigorous link between the FOT and stochastic approaches is established at discrete-time by the *Wold's isomorphism*. It was introduced in (Wold, 1948) for ergodic stationary processes. It is a one-to-one correspondence between a stationary stochastic process and translates of a single time series. The time series can be interpreted as a typical realization of the process. Specifically, the correspondence is

$$\xi(n,\omega) \quad \omega \in \Omega \longleftrightarrow x(n-\ell_\omega) \quad \ell_\omega \in \mathbb{Z} \tag{2.91}$$

It is an isomorphism in the sense that it is a bijection (invertible) and preserves inner products, and hence is also linear. It results

$$\mathrm{E}\{\xi(n+m,\omega)\,\xi(n,\omega)\} = \lim_{L\to\infty} \frac{1}{2L+1} \sum_{\ell=-L}^{L} x(n+m+\ell)\, x(n+\ell) \tag{2.92}$$

with both sides independent of n.

The extension of the Wold's isomorphism to cyclostationary sequences was first introduced in (Gardner, 1985), treated more in depth in (Gardner and Brown, 1991), and finally –with mathematical rigor– in (Hurd and Koski, 2004b). In such case, for ergodic cyclostationary processes the correspondence is

$$\xi(n,\omega) \quad \omega \in \Omega \longleftrightarrow x(n-\ell_\omega N_0) \quad \ell_\omega \in \mathbb{Z} \tag{2.93}$$

where $N_0 \in \mathbb{N}$ is the period of cyclostationarity. It results

$$\mathrm{E}\{\xi(n+m,\omega)\,\xi(n,\omega)\} = \lim_{L\to\infty} \frac{1}{2L+1} \sum_{\ell=-L}^{L} x(n+m+\ell N_0)\, x(n+\ell N_0) \tag{2.94}$$

with both sides periodic functions of n with period N_0.

2.4.1 Almost-periodic component of a pulse-amplitude modulated time series

In this section, the almost-periodic component of pulse-amplitude modulated (PAM) time series is derived. Then, a comparison with the analogous derivation of the expected value of a PAM stochastic process is made. Differences between the two procedures are enlightened.

Let

$$x_\delta(t) = \sum_{n=-\infty}^{+\infty} \widetilde{x}_n \, \delta(t - nT_0) \tag{2.95}$$

be a modulated Dirac-delta train with period T_0, where $\{\widetilde{x}_n\}_{n\in\mathbb{Z}}$ is a complex-valued numerical sequence. Let

$$\widetilde{x}^{\widetilde{\alpha}} \triangleq \left\langle \widetilde{x}_n \, e^{-j2\pi\widetilde{\alpha}n} \right\rangle_n \tag{2.96}$$

be the Fourier coefficient at frequency $\widetilde{\alpha}$ of the possible additive almost-periodic component contained in $\widetilde{x}_n$.

Let us consider the PAM time series $x(t)$ with pulse $q(t) \in L^1(\mathbb{R})$

$$x(t) = x_\delta(t) \otimes q(t) \tag{2.97a}$$

$$= \sum_{n=-\infty}^{+\infty} \widetilde{x}_n \, q(t - nT_0) \tag{2.97b}$$

The almost-periodic component of $x(t)$ is given by (Section 2.6)

$$\mathrm{E}^{\{\alpha\}}\{x(t)\} = \mathrm{E}^{\{\alpha\}}\{x_\delta(t)\} \otimes q(t) \tag{2.98a}$$

$$= \sum_{n=-\infty}^{+\infty} \mathrm{E}^{\{\widetilde{\alpha}\}}\{\widetilde{x}_n\} \, q(t - nT_0) \tag{2.98b}$$

The Fourier coefficients at frequency α of $x_\delta(t)$ and $x(t)$ are given by (Section 2.6)

$$x_\delta^\alpha \triangleq \left\langle x_\delta(t) \, e^{-j2\pi\alpha t} \right\rangle_t = \frac{1}{T_0} \, \widetilde{x}^{\alpha T_0} \tag{2.99}$$

$$x^\alpha \triangleq \left\langle x(t) \, e^{-j2\pi\alpha t} \right\rangle_t = \frac{1}{T_0} \, \widetilde{x}^{\alpha T_0} \, Q(\alpha) \tag{2.100}$$

where

$$Q(f) \triangleq \int_{\mathbb{R}} q(t) \, e^{-j2\pi f t} \, \mathrm{d}t \,. \tag{2.101}$$

Let us consider now a PAM stochastic process

$$\xi(t) \triangleq \sum_{n=-\infty}^{+\infty} \widetilde{\xi}_n \, q(t - nT_0) \tag{2.102}$$

where $\{\widetilde{\xi}_n\}_{n\in\mathbb{Z}}$ is a sequence of complex-valued random variables. Under the assumption

$$\sum_{n=-\infty}^{+\infty} \mathrm{E}\left\{|\widetilde{\xi}_n|\right\} < \infty \tag{2.103}$$

due to the σ-linearity of $\mathrm{E}\{\cdot\}$, we have

$$\mathrm{E}\{\xi(t)\} = \sum_{n=-\infty}^{+\infty} \mathrm{E}\left\{\widetilde{\xi}_n\right\} q(t - nT_0) . \tag{2.104}$$

Eq. (2.98b) is the FOT counterpart of (2.104). In contrast, even if

$$\sum_{n=-\infty}^{+\infty} |\widetilde{x}_n| < \infty \tag{2.105}$$

for the time series $x(t)$ in general we have

$$\langle x(t)\rangle_t \neq \sum_{n=-\infty}^{+\infty} \widetilde{x}_n \, \langle q(t - nT_0)\rangle_t \equiv 0 \tag{2.106}$$

$$\mathrm{E}^{\{\alpha\}}\{x(t)\} \neq \sum_{n=-\infty}^{+\infty} \widetilde{x}_n \, \mathrm{E}^{\{\alpha\}}\{q(t - nT_0)\} \equiv 0 \tag{2.107}$$

since $q(t) \in L^1(\mathbb{R})$. This is in accordance with the fact that the infinite time average $\langle\cdot\rangle_t$ and the almost-periodic component extraction operator $\mathrm{E}^{\{\alpha\}}\{\cdot\}$ are not σ-linear.

2.5 Summary

An approach for signal analysis which is alternative to the classical stochastic approach consists in modeling the observed signal or time series as a single function of time. From this single function, resorting to the concept of *relatively measurable function* (Definition 2.2), a *fraction-of-time* distribution function (Definition 2.3) is defined, whose expectation operator is the time average. Hence, all the other familiar probabilistic parameters are built. In addition, resorting to the concept of *jointly relatively measurable functions* (Definition 2.4), two or more functions are jointly characterized in terms of a joint distribution function and the independence between functions is defined and characterized (Theorem 2.5).

The approach based on relative measurability provides a stationary probabilistic (but not stochastic) model for the time series. Models with time variant probabilistic functions are obtained by considering the *periodic component extraction operator*, the *poly-periodic component extraction operator*, and the *almost-periodic component extraction operator* (Section 2.3.1.1) since they are recognized to be valid expectation operators corresponding to valid periodically, poly-periodically, and almost-periodically time variant distribution functions built from a single time series (Section 2.3.1.3).

By using the almost-periodic component extraction operator, a joint second-order distribution of a time series and its time-shifted version is build and, then, a valid almost-periodically time-variant autocorrelation function is defined. It is coincident with the almost-periodic component of the second-order lag product (Theorem 2.35). Analogously to the stochastic case, the frequencies and coefficients of the Fourier series of the autocorrelation function are referred to as *cycle frequencies* and *cyclic autocorrelation functions*. If the set of cycle frequencies contains only the element zero, then the time series is said *wide-sense stationary*, if it contains only frequencies that are integer multiples of the same fundamental frequency, then the time series is said *wide-sense cyclostationary*, if it contains incommensurate frequencies, then the time series is said *wide-sense almost cyclostationary*.

Almost-cyclostationary time series in the wide sense are characterized by the fact that spectral lines can be regenerated by a homogeneous quadratic time-invariant transformation of the time series (Section 2.3.1.6). In addition, correlation exists between spectral components of the time series that are separated by quantities equal to one of the cycle frequencies. The density of spectral correlation is described by the *cyclic spectrum* (Section 2.3.1.8). The *Gardner Relation* states that the cyclic spectrum at a given cycle frequency is the Fourier transform of the cyclic autocorrelation function at that cycle frequency (Section 2.3.1.10).

2.6 Proofs

In this Section, proofs are reported for results of the present chapter. Unless otherwise specified, only formal manipulations are reported, without explicitly justifying inversion of linear operators such as time averages, infinite sums, integrals, and ensemble average (expectation). In general, these manipulations are justified at least in the sense of generalized functions (distributions) (Champeney, 1990), (Zemanian, 1987).

Proof of Lemma 2.16

Let $z(t) \in \mathcal{Z}^{T_0}$ be a nonnegative function.

For every finite N the sum in the right-hand side of (2.16) is nonnegative. Thus, it is nonnegative also in the limit as $N \to \infty$, provided that the limit exists.

Proof of Lemma 2.17

Let $z(t) \in \mathcal{Z}^{\text{pp}}$ be a nonnegative function decomposed as a poly-periodic component and a residual term (Proposition 2.13). Let as assume, for the sake of simplicity, that the poly-periodic component of interest is constituted by the sum of two periodic components with incommensurate periods T_1 and T_2. The case with $P > 2$ incommensurate periods can be treated similarly.

We have

$$z(t) = z_{T_1}(t) + z_{T_2}(t) - \langle z(t) \rangle_t + z_{\varrho}(t) \geqslant 0 \tag{2.108}$$

and let us define

$$z_T(t) \triangleq z_{T_1}(t) + z_{T_2}(t) - \langle z(t) \rangle_t \, . \tag{2.109}$$

Let $\mathrm{u}[\cdot]$ denote the unit step function.

Let us consider the decomposition

$$z_T(t) \triangleq z_T^+(t) - z_T^-(t) \tag{2.110}$$

where

$$z_T^+(t) \triangleq z_T(t)\,\mathrm{u}[z_T(t)] = \begin{cases} z_T(t) & z_T(t) \geqslant 0 \\ 0 & z_T(t) < 0 \end{cases} \tag{2.111}$$

$$z_T^-(t) \triangleq -z_T(t)\,\mathrm{u}[-z_T(t)] = \begin{cases} 0 & z_T(t) \geqslant 0 \\ -z_T(t) & z_T(t) < 0 \end{cases} \tag{2.112}$$

denote the positive and negative parts of $z_T(t)$, respectively.

For every continuous and infinitely differentiable approximation of the unit step function depending on a parameter ϵ, say $\mathrm{u}_\epsilon(x)$, such that $\lim_{\epsilon\to 0}\mathrm{u}_\epsilon(x) = \mathrm{u}(x)$, we have the following Taylor series expansion

$$\mathrm{u}_\epsilon(x) = \mathrm{u}_\epsilon(0) + \frac{\mathrm{d}}{\mathrm{d}x}\mathrm{u}_\epsilon(x)\Big|_{x=0} x + \frac{1}{2!}\frac{\mathrm{d}^2}{\mathrm{d}x^2}\mathrm{u}_\epsilon(x)\Big|_{x=0} x^2 + \frac{1}{3!}\frac{\mathrm{d}^3}{\mathrm{d}x^3}\mathrm{u}_\epsilon(x)\Big|_{x=0} x^3 + \cdots \tag{2.113}$$

The function $\mathrm{u}_\epsilon[\pm z_T(t)]$ is a zero-memory nonlinear transformation of a poly-periodic function with periods T_1 and T_2. Thus, by substituting this poly-periodic function in place of x into (2.113), and accounting for the binomial formula

$$\begin{aligned}&\left(\sum_{h_1} z_{1,h_1}\, e^{j2\pi(h_1/T_1)t} + \sum_{h_2} z_{2,h_2}\, e^{j2\pi(h_2/T_2)t}\right)^n \\ &\quad = \sum_{k=0}^{n}\binom{n}{k}\left(\sum_{h_1} z_{1,h_1}\, e^{j2\pi(h_1/T_1)t}\right)^k \left(\sum_{h_2} z_{2,h_2}\, e^{j2\pi(h_2/T_2)t}\right)^{n-k}\end{aligned} \tag{2.114}$$

we have that $\mathrm{u}_\epsilon[\pm z_T(t)]$ can be expressed by a trigonometric series of Bohl's class (Definition B.7). That is, it can be expressed as a (generalized) Fourier series containing the frequencies that are linear combinations with integer coefficients of $1/T_1$ and $1/T_2$, i.e., the frequencies $(h_1/T_1 + h_2/T_2)$, $h_1, h_2 \in \mathbb{Z}$. The same frequencies are present in the limit as $\epsilon \to 0$, even if the Taylor series expansion does not hold for $\mathrm{u}(x)$.

Therefore, since $z_T(t)$ is poly-periodic with periods T_1 and T_2, accounting for (2.111) and (2.112), both $z_T^+(t)$ and $z_T^-(t)$ can be expressed by a trigonometric series of Bohl's class with frequencies $(h_1/T_1 + h_2/T_2)$. That is,

$$z_T^\pm(t) = \sum_{h_1}\sum_{h_2} z^\pm_{h_1 h_2}\, e^{j2\pi(h_1/T_1 + h_2/T_2)t}\,. \tag{2.115}$$

In order to prove the lemma, we prove that $z_T^-(t)$ is identically zero.

We have

$$\mathrm{u}\left[z_T^-(t)\right] = \begin{cases} 1 & z_T^-(t) \geqslant 0 \qquad \text{(that is } z_T(t) < 0) \\ 0 & z_T^-(t) < 0 \qquad \text{(that is } z_T(t) \geqslant 0) \end{cases} \tag{2.116}$$

Multiplying both terms of inequality

$$0 \leqslant z(t) = z_T^+(t) - z_T^-(t) + z_\varrho(t) \tag{2.117}$$

times $\mathrm{u}\left[z_T^-(t)\right]$ one has

$$\begin{aligned} 0 &\leqslant \mathrm{u}\left[z_T^-(t)\right] z(t) \\ &= \mathrm{u}\left[z_T^-(t)\right] z_T^+(t) - \mathrm{u}\left[z_T^-(t)\right] z_T^-(t) + \mathrm{u}\left[z_T^-(t)\right] z_\varrho(t) \\ &= -z_T^-(t) + \mathrm{u}\left[z_T^-(t)\right] z_\varrho(t)\,. \end{aligned} \tag{2.118}$$

In the last equality we used the facts that

$$\mathrm{u}\left[z_T^-(t)\right] z_T^+(t) = 0 \tag{2.119}$$

since the functions $z_T^+(t)$ and $z_T^-(t)$ have disjoint supports, and

$$\mathrm{u}\left[z_T^-(t)\right] z_T^-(t) = z_T^-(t) \tag{2.120}$$

that immediately follows from (2.116) since $z_T^-(t) \geqslant 0$.

The function in the right-hand side of (2.118) is nonnegative. Therefore, also its time average, if it exists, is nonnegative. That is

$$-\left\langle z_T^-(t)\right\rangle_t + \left\langle \mathrm{u}\left[z_T^-(t)\right] z_\varrho(t)\right\rangle_t \geqslant 0 \tag{2.121}$$

and, hence

$$\left\langle \mathrm{u}\left[z_T^-(t)\right] z_\varrho(t)\right\rangle_t \geqslant \left\langle z_T^-(t)\right\rangle_t \geqslant 0 \tag{2.122}$$

where the last inequality is consequence of the fact that $z_T^-(t) \geqslant 0$ and hence also its time average is nonnegative.

By reasoning as above, we see that the Taylor series expansion of $\mathrm{u}_\epsilon\left[z_T^-(t)\right]$ contains the terms

$$\left(z_T^-(t)\right)^n = \left(\sum_{h_1}\sum_{h_2} z^-_{h_1 h_2}\, e^{j2\pi(h_1/T_1 + h_2/T_2)t}\right)^n \tag{2.123}$$

that is, frequencies of the kind $(h_1/T_1 + h_2/T_2)$. The same frequencies are present in $\mathrm{u}\left[z_T^-(t)\right]$.

Since, by assumption, the frequencies of the possible finite-strength sine wave components in $z_\varrho(t)$ are incommensurate with all frequencies $(h_1/T_1 + h_2/T_2)$, $h_1, h_2 \in \mathbb{Z}$, we have

$$\left\langle \mathrm{u}\left[z_T^-(t)\right] z_\varrho(t)\right\rangle_t = 0\,. \tag{2.124}$$

From (2.122) and (2.124) it follows that

$$\langle z_T^-(t)\rangle_t = 0\,. \tag{2.125}$$

Consequently $z_T^-(t)$ must be identically zero since it is nonnegative by definition (see (2.112)) and we have

$$z_T(t) = z_T^+(t) \geqslant 0\,. \tag{2.126}$$

Proof of Lemma 2.18

Let $z(t) \in \mathcal{Z}^{\text{ap}}$ be a nonnegative function. According to (2.23), we have

$$z(t) = z_{\text{ap}}(t) + z_r(t) \geqslant 0\,. \tag{2.127}$$

Let us consider the decomposition

$$z_{\text{ap}}(t) \triangleq z_{\text{ap}}^+(t) - z_{\text{ap}}^-(t) \tag{2.128}$$

where

$$z_{\text{ap}}^+(t) \triangleq \begin{cases} z_{\text{ap}}(t) & z_{\text{ap}}(t) \geqslant 0 \\ 0 & z_{\text{ap}}(t) < 0 \end{cases} \tag{2.129}$$

$$z_{\text{ap}}^-(t) \triangleq \begin{cases} 0 & z_{\text{ap}}(t) \geqslant 0 \\ -z_{\text{ap}}(t) & z_{\text{ap}}(t) < 0 \end{cases} \tag{2.130}$$

In order to prove the lemma, we prove that $z_{\text{ap}}^-(t)$ is identically zero.
Let $\mathrm{u}[\cdot]$ denote the unit step function. We have

$$\mathrm{u}\left[z_{\text{ap}}^-(t)\right] = \begin{cases} 1 & z_{\text{ap}}^-(t) \geqslant 0 & (\text{that is } z_{\text{ap}}(t) < 0) \\ 0 & z_{\text{ap}}^-(t) < 0 & (\text{that is } z_{\text{ap}}(t) \geqslant 0) \end{cases} \tag{2.131}$$

Multiplying both terms of inequality

$$0 \leqslant z(t) = z_{\text{ap}}^+(t) - z_{\text{ap}}^-(t) + z_r(t) \tag{2.132}$$

times $\mathrm{u}\left[z_{\text{ap}}^-(t)\right]$ one has

$$\begin{aligned} 0 &\leqslant \mathrm{u}\left[z_{\text{ap}}^-(t)\right]\, z(t) \\ &= \mathrm{u}\left[z_{\text{ap}}^-(t)\right]\, z_{\text{ap}}^+(t) - \mathrm{u}\left[z_{\text{ap}}^-(t)\right]\, z_{\text{ap}}^-(t) + \mathrm{u}\left[z_{\text{ap}}^-(t)\right]\, z_r(t) \\ &= -z_{\text{ap}}^-(t) + \mathrm{u}\left[z_{\text{ap}}^-(t)\right]\, z_r(t)\,. \end{aligned} \tag{2.133}$$

In the last equality we used the facts that

$$\mathrm{u}\left[z_{\text{ap}}^-(t)\right]\, z_{\text{ap}}^+(t) = 0 \tag{2.134}$$

since the functions $z^+_{\text{ap}}(t)$ and $z^-_{\text{ap}}(t)$ have disjoint supports, and

$$\mathrm{u}\left[z^-_{\text{ap}}(t)\right] z^-_{\text{ap}}(t) = z^-_{\text{ap}}(t) \tag{2.135}$$

that immediately follows from (2.131) since $z^-_{\text{ap}}(t) \geqslant 0$.

The function in the right-hand side of (2.133) is nonnegative. Therefore, also its time average is nonnegative. That is

$$-\left\langle z^-_{\text{ap}}(t)\right\rangle_t + \left\langle \mathrm{u}\left[z^-_{\text{ap}}(t)\right] z_r(t)\right\rangle_t \geqslant 0 \tag{2.136}$$

and, hence

$$\left\langle \mathrm{u}\left[z^-_{\text{ap}}(t)\right] z_r(t)\right\rangle_t \geqslant \left\langle z^-_{\text{ap}}(t)\right\rangle_t \geqslant 0 \tag{2.137}$$

where the last inequality is consequence of the fact that $z^-_{\text{ap}}(t) \geqslant 0$ and hence also its time average is nonnegative.

The function $\mathrm{u}\left[z^-_{\text{ap}}(t)\right]$ is a zero-memory nonlinear transformation of an almost-periodic function. Thus, it is in turn almost-periodic in a generalized sense that accounts for the presence of discontinuities (as result of a generalization of Theorem B.10). It can be expressed by a (generalized) Fourier series with possibly incommensurate frequencies (Theorem B.16).

Therefore, since $z_r(t)$ is orthogonal to all sine waves $e^{j2\pi\eta t}$, $\eta \in \mathbb{R}$ (see (2.24)), we have

$$\left\langle \mathrm{u}\left[z^-_{\text{ap}}(t)\right] z_r(t)\right\rangle_t = 0\,. \tag{2.138}$$

From (2.137) and (2.138) it follows that

$$\left\langle z^-_{\text{ap}}(t)\right\rangle_t = 0\,. \tag{2.139}$$

Consequently $z^-_{\text{ap}}(t)$ must be identically zero since it is nonnegative by definition (see (2.130)) and we have

$$z_{\text{ap}}(t) = z^+_{\text{ap}}(t) \geqslant 0\,. \tag{2.140}$$

Proof of Theorem 2.21

The proof consists in showing that (i) $F_x^{\{T_0\}}(\xi;t)$ is nonnegative, (ii) nondecreasing, and (iii) with values in [0, 1]. The right continuity property (iv) in general cannot be proved.

i. The function $F_x^{\{T_0\}}(\xi;t)$ is obtained by the synchronized averaging with period T_0 of the nonnegative function $t \mapsto \mathbf{1}_{\{x(t)\leqslant\xi\}}$. Thus, it is periodic with period T_0 (Theorem B.19) and nonnegative (Lemma 2.16).

ii. For every $\xi_2 > \xi_1$ it results that

$$\mathbf{1}_{\{x(t)\leqslant\xi_2\}} = \mathbf{1}_{\{x(t)\leqslant\xi_1\}} + \mathbf{1}_{\{\xi_1<x(t)\leqslant\xi_2\}}\,. \tag{2.141}$$

Since the periodic component extraction operator is linear, we have

$$\mathrm{E}^{\{T_0\}}\left\{\mathbf{1}_{\{x(t)\leqslant \xi_2\}}\right\} = \mathrm{E}^{\{T_0\}}\left\{\mathbf{1}_{\{x(t)\leqslant \xi_1\}}\right\} + \mathrm{E}^{\{T_0\}}\left\{\mathbf{1}_{\{\xi_1 < x(t)\leqslant \xi_2\}}\right\} \tag{2.142}$$

Both functions in the right-hand side are the periodic component with period T_0 of a nonnegative function. Thus, these functions are both nonnegative (Lemma 2.16). Therefore, it results

$$\xi_2 > \xi_1 \Rightarrow F_x^{\{T_0\}}(\xi_2; t) \triangleq \mathrm{E}^{\{T_0\}}\left\{\mathbf{1}_{\{x(t)\leqslant \xi_2\}}\right\} \geqslant \mathrm{E}^{\{T_0\}}\left\{\mathbf{1}_{\{x(t)\leqslant \xi_1\}}\right\} \triangleq F_x^{\{T_0\}}(\xi_1; t) \tag{2.143}$$

iii. For $\xi = -\infty$ we have $\mathbf{1}_{\{x(t)\leqslant -\infty\}} = 0 \;\forall t \in \mathbb{R}$ and hence $F_x^{\{T_0\}}(-\infty; t) = 0$. For $\xi = +\infty$ we have $\mathbf{1}_{\{x(t)\leqslant +\infty\}} = 1 \;\forall t \in \mathbb{R}$ and hence $F_x^{\{T_0\}}(+\infty; t) = 1$. $F_x^{\{T_0\}}(\xi; t)$ is nondecreasing and, hence has values in $[0, 1]$.

iv. Similarly to the case of stationary FOT distributions (Leśkow and Napolitano, 2006, Lemma 6.6), the right continuity property in the discontinuity points of periodic FOT distributions cannot be proved.

Proof of (2.37)

It results that

$$\mathbf{1}_{\{x(t)\leqslant \xi\}} \equiv \mathrm{u}(\xi - x(t)) = \begin{cases} 1 & \xi \geqslant x(t) \\ 0 & \text{otherwise} \end{cases} \tag{2.144}$$

where $\mathrm{u}(\xi)$ is the unit step function. Thus, using the fact that $\delta(\xi) = \mathrm{d}\mathrm{u}(\xi)/\mathrm{d}\xi$, we formally have

$$\begin{aligned} f_{x(t)}^{\{\alpha\}}(\xi) &\triangleq \frac{\mathrm{d}F_{x(t)}^{\{\alpha\}}(\xi)}{\mathrm{d}\xi} \\ &= \frac{\mathrm{d}}{\mathrm{d}\xi}\mathrm{E}^{\{\alpha\}}\left\{\mathbf{1}_{\{x(t)\leqslant \xi\}}\right\} \\ &= \frac{\mathrm{d}}{\mathrm{d}\xi}\mathrm{E}^{\{\alpha\}}\left\{\mathrm{u}(\xi - x(t))\right\} \\ &= \mathrm{E}^{\{\alpha\}}\left\{\frac{\mathrm{d}}{\mathrm{d}\xi}\,\mathrm{u}(\xi - x(t))\right\} \\ &= \mathrm{E}^{\{\alpha\}}\{\delta(\xi - x(t))\} \end{aligned} \tag{2.145}$$

Proof of Theorem 2.24 "Gardner Theorem of Temporal Expectation"

Eq. (2.38) can be written as

$$\sum_{\gamma\in\Gamma^{(1)}} \left\langle g(x(s))\, e^{-j2\pi\gamma s}\right\rangle_s e^{j2\pi\gamma t} = \sum_{\gamma\in\Gamma^{(1)}} \int_{\mathbb{R}} g(\xi)\, \mathrm{d}F_x^{\gamma}(\xi)\, e^{j2\pi\gamma t}\,. \tag{2.146}$$

For every $\gamma \in \Gamma^{(1)}$, the proof that

$$\mathrm{Re}\left[\left\langle g(x(s))\, e^{-j2\pi\gamma s}\right\rangle_s\right] = \mathrm{Re}\left[\int_{\mathbb{R}} g(\xi)\, \mathrm{d}F_x^{\gamma}(\xi)\right] \tag{2.147}$$

$$\mathrm{Im}\left[\left\langle g(x(s))\, e^{-j2\pi\gamma s}\right\rangle_s\right] = \mathrm{Im}\left[\int_{\mathbb{R}} g(\xi)\, \mathrm{d}F_x^{\gamma}(\xi)\right] \tag{2.148}$$

is similar to the proof of (Leśkow and Napolitano, 2006, Theorem 3.2) for the stationary case. Condition (2.35) and the boundedness of $g(\cdot)$ assure the convergence of the series in the right-hand side of (2.146).

By assuming that the FOT probability density function (2.37) exists, (2.38) can be formally derived as follows:

$$\begin{aligned}
&\int_{\mathbb{R}} g(\xi)\, \mathrm{d}F_{x(t)}^{\{\alpha\}}(\xi) \\
&\quad = \int_{\mathbb{R}} g(\xi)\, f_{x(t)}^{\{\alpha\}}(\xi)\, \mathrm{d}\xi \\
&\quad = \int_{\mathbb{R}} g(\xi)\, \mathrm{E}^{\{\alpha\}}\{\delta(\xi - x(t))\}\, \mathrm{d}\xi \\
&\quad = \mathrm{E}^{\{\alpha\}}\Big\{\underbrace{\int_{\mathbb{R}} g(\xi)\, \delta(\xi - x(t))\, \mathrm{d}\xi}_{g(x(t))}\Big\}
\end{aligned} \tag{2.149}$$

Proof of Theorem 2.29

For any finite Z it results

$$\int_{-Z/2}^{+Z/2} \int_{\mathbb{R}} h(\tau)\, x(t-\tau)\, \mathrm{d}\tau\, \mathrm{d}t = \int_{\mathbb{R}} h(\tau) \int_{-Z/2}^{+Z/2} x(t-\tau)\, \mathrm{d}t\, \mathrm{d}\tau \tag{2.150}$$

where the second integral exists since $x(t) \in L^1_{\mathrm{loc}}(\mathbb{R})$ and $h(t) \in L^1(\mathbb{R})$. The interchange of the order of integrals in (2.150) is allowed by the Fubini and Tonelli theorem (Champeney, 1990, Chap. 3). In fact, we have

$$\begin{aligned}
\int_{-Z/2}^{+Z/2} \int_{\mathbb{R}} |h(\tau)\, x(t-\tau)|\, \mathrm{d}\tau\, \mathrm{d}t &\leqslant \int_{-Z/2}^{+Z/2} \|x\|_\infty \int_{\mathbb{R}} |h(\tau)|\, \mathrm{d}\tau\, \mathrm{d}t \\
&= Z\, \|x\|_\infty \int_{\mathbb{R}} |h(\tau)|\, \mathrm{d}\tau < \infty
\end{aligned} \tag{2.151}$$

where $\|x\|_\infty \triangleq \operatorname{ess\,sup}_{t\in\mathbb{R}} |x(t)|$ is the essential supremum of $x(t)$ (Champeney, 1990, pp. 6, 15). Since (2.150) holds for any finite Z, it also holds in the limit for $Z \to \infty$:

$$\lim_{Z\to\infty} \frac{1}{Z} \int_{-Z/2}^{+Z/2} \int_{\mathbb{R}} h(\tau)\, x(t-\tau)\, \mathrm{d}\tau\, \mathrm{d}t = \lim_{Z\to\infty} \frac{1}{Z} \int_{\mathbb{R}} h(\tau) \int_{-Z/2}^{+Z/2} x(t-\tau)\, \mathrm{d}t\, \mathrm{d}\tau \tag{2.152}$$

where equality holds provided that at least one of the two limits exists.

The function family

$$F_Z(\tau) \triangleq h(\tau) \frac{1}{Z} \int_{-Z/2}^{+Z/2} x(t-\tau)\, \mathrm{d}t \tag{2.153}$$

is dominated by a summable function of τ not depending on Z:

$$\begin{aligned} |F_Z(\tau)| &= \left| h(\tau) \frac{1}{Z} \int_{-Z/2}^{+Z/2} x(t-\tau)\, \mathrm{d}t \right| \\ &\leqslant \left| h(\tau) \frac{1}{Z} Z \, \|x\|_\infty \right| = |h(\tau)| \; \|x\|_\infty \in L^1(\mathbb{R}) \end{aligned} \tag{2.154}$$

Therefore, the dominated convergence theorem (Champeney, 1990, Chap. 4) can be applied to interchange the limit and integral operations in the right-hand side of (2.152) leading to

$$\lim_{Z\to\infty} \frac{1}{Z} \int_{-Z/2}^{+Z/2} \left[\int_{\mathbb{R}} h(\tau)\, x(t-\tau)\, \mathrm{d}\tau \right] \mathrm{d}t = \int_{\mathbb{R}} h(\tau) \left[\lim_{Z\to\infty} \frac{1}{Z} \int_{-Z/2}^{+Z/2} x(t-\tau)\, \mathrm{d}t \right] \mathrm{d}\tau \tag{2.155}$$

which is equivalent to (2.41).

The existence of the limit in (2.155) is assured by the assumptions $h(\tau) \in L^1(\mathbb{R})$ and that $\langle x(t) \rangle_t$ exists. Then also limits in (2.152) exist.

Proof of Corollary 2.31

We have

$$\begin{aligned} &\lim_{Z\to\infty} \frac{1}{Z} \int_{-Z/2}^{+Z/2} \left[\int_{\mathbb{R}} h(\tau)\, x(t-\tau)\, \mathrm{d}\tau \right] e^{-j2\pi\alpha t}\, \mathrm{d}t \\ &\quad = \int_{\mathbb{R}} h(\tau) \left[\lim_{Z\to\infty} \frac{1}{Z} \int_{-Z/2}^{+Z/2} x(t-\tau)\, e^{-j2\pi\alpha(t-\tau)}\, \mathrm{d}t \right] e^{-j2\pi\alpha\tau}\, \mathrm{d}\tau \end{aligned} \tag{2.156}$$

which is equivalent to (2.44).

Proof of Theorem 2.32

Let $\{A_n\}$ an increasing sequence of finite subsets of A such that $\lim_n A_n \triangleq \cup_n A_n = A$. Since A_n is finite, accounting for Corollary 2.31, we have

$$\mathrm{E}^{\{A_n\}} \left\{ \int_{\mathbb{R}} h(\tau)\, x(t-\tau)\, \mathrm{d}\tau \right\}$$

$$\triangleq \sum_{\alpha \in A_n} \lim_{Z\to\infty} \frac{1}{Z} \int_{-Z/2}^{+Z/2} \left[\int_{\mathbb{R}} h(\tau)\, x(t'-\tau)\, \mathrm{d}\tau \right] e^{-j2\pi\alpha t'}\, \mathrm{d}t'\, e^{j2\pi\alpha t}$$

$$= \int_{\mathbb{R}} h(\tau) \sum_{\alpha \in A_n} \lim_{Z\to\infty} \frac{1}{Z} \int_{-Z/2}^{+Z/2} x(t'-\tau)\, e^{-j2\pi\alpha(t'-\tau)}\, \mathrm{d}t'\, e^{-j2\pi\alpha\tau}\, e^{j2\pi\alpha t}\, \mathrm{d}\tau$$

$$= \int_{\mathbb{R}} h(\tau) \sum_{\alpha \in A_n} x^{\alpha}\, e^{-j2\pi\alpha\tau}\, e^{j2\pi\alpha t}\, \mathrm{d}\tau\,. \tag{2.157}$$

The integrand function in the right-hand side

$$F_n(\tau) \triangleq h(\tau) \sum_{\alpha \in A_n} x^{\alpha}\, e^{-j2\pi\alpha\tau}\, e^{j2\pi\alpha t} \tag{2.158}$$

is dominated by a summable function of τ not depending on n:

$$|F_n(\tau)| \leqslant |h(\tau)| \sum_{\alpha \in A_n} \left|x^{\alpha}\right| \leqslant |h(\tau)| \sum_{\alpha \in A} \left|x^{\alpha}\right| \in L^1(\mathbb{R}) \tag{2.159}$$

Thus, for the dominated convergence theorem (Champeney, 1990, Chap. 4)

$$\begin{aligned} \mathrm{E}^{\{A\}}\left\{ \int_{\mathbb{R}} h(\tau)\, x(t-\tau)\, \mathrm{d}\tau \right\} &\triangleq \lim_n \mathrm{E}^{\{A_n\}}\left\{ \int_{\mathbb{R}} h(\tau)\, x(t-\tau)\, \mathrm{d}\tau \right\} \\ &= \lim_n \int_{\mathbb{R}} h(\tau) \sum_{\alpha \in A_n} x^{\alpha}\, e^{-j2\pi\alpha\tau}\, e^{j2\pi\alpha t}\, \mathrm{d}\tau \\ &= \int_{\mathbb{R}} h(\tau) \lim_n \sum_{\alpha \in A_n} x^{\alpha}\, e^{j2\pi\alpha(t-\tau)}\, \mathrm{d}\tau \\ &= \int_{\mathbb{R}} h(\tau)\, \mathrm{E}^{\{A\}}\{x(t-\tau)\}\, \mathrm{d}\tau \end{aligned} \tag{2.160}$$

which proves (2.46a).

Similarly, Eq. (2.46b) follows from (2.157) in the limit as $n \to \infty$.

Proof of (2.50)

$$\begin{aligned} f^{\{\alpha\}}_{x(t+\tau)\,x(t)}(\xi_1, \xi_2) &\triangleq \frac{\partial^2 F^{\{\alpha\}}_{x(t+\tau)\,x(t)}(\xi_1, \xi_2)}{\partial \xi_1\, \partial \xi_2} \\ &= \frac{\partial^2}{\partial \xi_1\, \partial \xi_2} \mathrm{E}^{\{\alpha\}}\left\{ \mathrm{u}(\xi_1 - x(t+\tau))\, \mathrm{u}(\xi_2 - x(t)) \right\} \\ &= \mathrm{E}^{\{\alpha\}}\left\{ \frac{\partial}{\partial \xi_1} \mathrm{u}(\xi_1 - x(t+\tau))\, \frac{\partial}{\partial \xi_2} \mathrm{u}(\xi_2 - x(t)) \right\} \end{aligned}$$

$$= \mathrm{E}^{\{\alpha\}}\left\{\delta(\xi_1 - x(t+\tau))\,\delta(\xi_2 - x(t))\right\} \tag{2.161a}$$

$$= \sum_{\alpha \in \Gamma^{(2)}} f^{\alpha}(\xi_1, \xi_2; \tau)\, e^{j2\pi\alpha t} \tag{2.161b}$$

where

$$f^{\alpha}(\xi_1, \xi_2; \tau) \triangleq \left\langle \delta(\xi_1 - x(t+\tau))\,\delta(\xi_2 - x(t))\, e^{-j2\pi\alpha t} \right\rangle_t \tag{2.162}$$

Proof of Theorem 2.35

Using (2.161b) and (2.162) we have

$$\begin{aligned}
&\int_{\mathbb{R}^2} \xi_1 \xi_2 \, \mathrm{d}F^{\{\alpha\}}_{x(t+\tau)\,x(t)}(\xi_1, \xi_2) \\
&= \int_{\mathbb{R}^2} \xi_1 \xi_2 \, f^{\{\alpha\}}_{x(t+\tau)\,x(t)}(\xi_1, \xi_2)\, \mathrm{d}\xi_1 \mathrm{d}\xi_2 \\
&= \int_{\mathbb{R}^2} \xi_1 \xi_2 \left[\sum_{\alpha \in \Gamma^{(2)}} f^{\alpha}(\xi_1, \xi_2; \tau)\, e^{j2\pi\alpha t} \right] \mathrm{d}\xi_1 \mathrm{d}\xi_2 \\
&= \sum_{\alpha \in \Gamma^{(2)}} \int_{\mathbb{R}^2} \xi_1 \xi_2 \left\langle \delta(\xi_1 - x(s+\tau))\,\delta(\xi_2 - x(s))\, e^{-j2\pi\alpha s} \right\rangle_s \mathrm{d}\xi_1 \mathrm{d}\xi_2 \, e^{j2\pi\alpha t} \\
&= \sum_{\alpha \in \Gamma^{(2)}} \left\langle \int_{\mathbb{R}^2} \xi_1 \xi_2 \, \delta(\xi_1 - x(s+\tau))\,\delta(\xi_2 - x(s))\, \mathrm{d}\xi_1 \mathrm{d}\xi_2 \, e^{-j2\pi\alpha s} \right\rangle_s e^{j2\pi\alpha t} \\
&= \sum_{\alpha \in \Gamma^{(2)}} \left\langle x(s+\tau)\, x(s)\, e^{-j2\pi\alpha s} \right\rangle_s e^{j2\pi\alpha t}
\end{aligned} \tag{2.163}$$

In (2.52) we have $A \subseteq \Gamma^{(2)}$.

Proof of Lemma 2.39

Using definition (2.68) for the (conjugate) cyclic cross-correlogram and accounting for (5.11), we have

$$\begin{aligned}
&\mathcal{F}\left[R^{(T)}_{yx^{(*)}}(\alpha, \tau; t_0) \right] \\
&= \int_{-T}^{T} R^{(T)}_{yx^{(*)}}(\alpha, \tau; t_0)\, e^{-j2\pi f \tau} \, \mathrm{d}\tau \\
&= \int_{-T}^{T} \frac{1}{T} \int_{t_0 - T/2}^{t_0 + T/2} \mathrm{rect}\left(\frac{t + \tau - t_0}{T} \right) y(t+\tau)\, x^{(*)}(t)\, e^{-j2\pi\alpha t} \, \mathrm{d}t \, e^{-j2\pi f \tau} \, \mathrm{d}\tau \\
&= \frac{1}{T} \int_{t_0 - T/2}^{t_0 + T/2} x^{(*)}(t)\, e^{-j2\pi\alpha t} \int_{-T}^{T} \mathrm{rect}\left(\frac{t + \tau - t_0}{T} \right) y(t+\tau) e^{-j2\pi f \tau} \, \mathrm{d}\tau \, \mathrm{d}t \\
&= \frac{1}{T} \int_{t_0 - T/2}^{t_0 + T/2} x^{(*)}(t)\, e^{-j2\pi\alpha t} \int_{t-T}^{t+T} \mathrm{rect}\left(\frac{t' - t_0}{T} \right) y(t') e^{-j2\pi f (t'-t)} \, \mathrm{d}t' \, \mathrm{d}t
\end{aligned}$$

$$= \frac{1}{T} \int_{t_0-T/2}^{t_0+T/2} x^{(*)}(t)\, e^{-j2\pi(\alpha-f)t}\, \mathrm{d}t \int_{t_0-T/2}^{t_0+T/2} y(t')\, e^{-j2\pi f t'}\, \mathrm{d}t'$$
$$= \frac{1}{T} \left[\int_{t_0-T/2}^{t_0+T/2} y(t')\, e^{-j2\pi f t'}\, \mathrm{d}t' \right] \left[\int_{t_0-T/2}^{t_0+T/2} x(t)\, e^{-(-)j2\pi(\alpha-f)t}\, \mathrm{d}t \right]^{(*)}$$
$$= \frac{1}{T} Y_T(t_0, f)\, X_T^{(*)}(t_0, (-)(\alpha - f)) \tag{2.164}$$

where in the fourth equality the variable change $t' = t + \tau$ is made in the inner integral, and the fifth equality is consequence of the fact that $t \in [t_0 - T/2, t_0 + T/2]$.

Under that assumption that $x(t)$ and $y(t)$ are in $L^1_{\mathrm{loc}}(\mathbb{R}) \cap L^\infty(\mathbb{R})$, the order of integrals is interchanged due to the Fubini and Tonelli theorem (Champeney, 1990, Chap. 3).

Proof of Theorem 2.40 "Gardner Relation"

Accounting for (2.164), and the fact that $R^{(T)}_{yx^{(*)}}(\alpha, \tau)$ is zero for $|\tau| > T$ (Remark 5.2), we have

$$\frac{1}{Z} \int_{-Z/2}^{Z/2} I^{(T)}_{yx^{(*)}}(\alpha, f; t)\, \mathrm{d}t$$
$$= \frac{1}{Z} \int_{-Z/2}^{Z/2} \int_{-T}^{T} R^{(T)}_{yx^{(*)}}(\alpha, \tau; t)\, e^{-j2\pi f\tau}\, \mathrm{d}\tau\, \mathrm{d}t$$
$$= \frac{1}{Z} \int_{-Z/2}^{Z/2} \int_{-T}^{T} \frac{1}{T} \int_{t-T/2}^{t+T/2} \mathrm{rect}\Big(\frac{s+\tau-t}{T}\Big)\, y(s+\tau)\, x^{(*)}(s)\, e^{-j2\pi\alpha s}\, \mathrm{d}s\, e^{-j2\pi f\tau}\, \mathrm{d}\tau\, \mathrm{d}t$$
$$= \frac{1}{Z} \int_{-Z/2}^{Z/2} \int_{-T}^{T} \frac{1}{T} \int_{-T/2}^{T/2} \mathrm{rect}\Big(\frac{v+\tau}{T}\Big)\, y(t+v+\tau)$$
$$\mathrm{rect}\Big(\frac{v}{T}\Big)\, x^{(*)}(t+v)\, e^{-j2\pi\alpha(t+v)}\, \mathrm{d}v\, e^{-j2\pi f\tau}\, \mathrm{d}\tau\, \mathrm{d}t$$
$$= \int_{-T}^{T} \frac{1}{T} \int_{-T/2}^{T/2} \frac{1}{Z} \int_{-Z/2}^{Z/2} y(t+v+\tau)\, x^{(*)}(t+v)\, e^{-j2\pi\alpha(t+v)}\, \mathrm{d}t$$
$$\mathrm{rect}\Big(\frac{v+\tau}{T}\Big)\, \mathrm{rect}\Big(\frac{v}{T}\Big)\, \mathrm{d}v\, e^{-j2\pi f\tau}\, \mathrm{d}\tau \tag{2.165}$$

where in the third equality the variable change $v = s - t$ is made in the integral in $\mathrm{d}s$.

In the limit as $Z \to \infty$ we have

$$\lim_{Z\to\infty} \frac{1}{Z} \int_{-Z/2}^{Z/2} I^{(T)}_{yx^{(*)}}(\alpha, f; t)\, \mathrm{d}t$$
$$= \int_{-T}^{T} \frac{1}{T} \int_{-T/2}^{T/2} \lim_{Z\to\infty} \frac{1}{Z} \int_{-Z/2}^{Z/2} y(t+v+\tau)\, x^{(*)}(t+v)\, e^{-j2\pi\alpha(t+v)}\, \mathrm{d}t$$
$$\mathrm{rect}\Big(\frac{v+\tau}{T}\Big)\, \mathrm{rect}\Big(\frac{v}{T}\Big)\, \mathrm{d}v\, e^{-j2\pi f\tau}\, \mathrm{d}\tau$$
$$= \int_{-T}^{T} \frac{1}{T} \int_{-T/2}^{T/2} R^{\alpha}_{yx^{(*)}}(\tau)\, \mathrm{rect}\Big(\frac{v+\tau}{T}\Big)\, \mathrm{rect}\Big(\frac{v}{T}\Big)\, \mathrm{d}v\, e^{-j2\pi f\tau}\, \mathrm{d}\tau$$

$$
\begin{aligned}
&= \int_{-T}^{T} R^{\alpha}_{yx^{(*)}}(\tau) \, \frac{1}{T} \int_{-T/2}^{T/2} \operatorname{rect}\Big(\frac{v+\tau}{T}\Big) \operatorname{rect}\Big(\frac{v}{T}\Big) \, \mathrm{d}v \, e^{-j2\pi f\tau} \, \mathrm{d}\tau \\
&= \int_{\mathbb{R}} R^{\alpha}_{yx^{(*)}}(\tau) \left(1 - \frac{|\tau|}{T}\right) \operatorname{rect}\Big(\frac{\tau}{2T}\Big) \, e^{-j2\pi f\tau} \, \mathrm{d}\tau
\end{aligned}
\tag{2.166}
$$

In the limit as $T \to \infty$ we have

$$
\begin{aligned}
&\lim_{T\to\infty} \lim_{Z\to\infty} \frac{1}{Z} \int_{-Z/2}^{Z/2} I^{(T)}_{yx^{(*)}}(\alpha, f; t) \, \mathrm{d}t \\
&\quad = \lim_{T\to\infty} \int_{\mathbb{R}} R^{\alpha}_{yx^{(*)}}(\tau) \left(1 - \frac{|\tau|}{T}\right) \operatorname{rect}\Big(\frac{\tau}{2T}\Big) \, e^{-j2\pi f\tau} \, \mathrm{d}\tau \\
&\quad = \int_{\mathbb{R}} R^{\alpha}_{yx^{(*)}}(\tau) \, e^{-j2\pi f\tau} \, \mathrm{d}\tau
\end{aligned}
\tag{2.167}
$$

since under condition (2.74) we have

$$
\begin{aligned}
&\lim_{T\to\infty} \left| \int_{\mathbb{R}} \frac{|\tau|}{T} R^{\alpha}_{yx^{(*)}}(\tau) \operatorname{rect}\Big(\frac{\tau}{2T}\Big) \, e^{-j2\pi f\tau} \, \mathrm{d}\tau \right| \\
&\qquad \leqslant \lim_{T\to\infty} \frac{1}{T} \int_{\mathbb{R}} |\tau| \left| R^{\alpha}_{yx^{(*)}}(\tau) \right| \, \mathrm{d}\tau = 0
\end{aligned}
\tag{2.168}
$$

In (2.165), in the fourth equality the order of integrals is interchanged due to the Fubini and Tonelli theorem (Champeney, 1990, Chap. 3). In fact, from the assumption that $x(t)$ and $y(t)$ are in $L^1_{\text{loc}}(\mathbb{R}) \cap L^{\infty}(\mathbb{R})$, we have the summability condition

$$
(t, \tau, v) \mapsto y(t+v+\tau) \, x(t+v) \in L^1([-Z/2, Z/2] \times [-T, T] \times [-T/2, T/2]) \, .
$$

In (2.166), the limit as $Z \to \infty$ and integral operations in $\mathrm{d}\tau \, \mathrm{d}v$ can be interchanged due to the dominated convergence theorem (Champeney, 1990, Chap. 4). In fact, the family of functions labeled by Z

$$
\begin{aligned}
(\tau, v) \mapsto \frac{1}{T} \frac{1}{Z} \int_{-Z/2}^{Z/2} & y(t+v+\tau) \, x^{(*)}(t+v) \, e^{-j2\pi\alpha(t+v)} \, \mathrm{d}t \\
& \operatorname{rect}\Big(\frac{v+\tau}{T}\Big) \operatorname{rect}\Big(\frac{v}{T}\Big) \, e^{-j2\pi f\tau}
\end{aligned}
\tag{2.169}
$$

is bounded by the function

$$
(\tau, v) \mapsto \frac{1}{T} \, \|y\|_{\infty} \, \|x\|_{\infty} \operatorname{rect}\Big(\frac{v+\tau}{T}\Big) \operatorname{rect}\Big(\frac{v}{T}\Big)
\tag{2.170}
$$

that is independent of Z and in $L^1([-T, T] \times [-T/2, T/2])$.

Proof of (2.98a)–(2.98b)

The Fourier coefficient at frequency α of the almost-periodic component of the modulated Dirac-delta train $x_\delta(t)$ defined in (2.95), assuming that it exists in a generalized sense (Champeney, 1990, Chap. 16), is given by

$$\begin{aligned}
x_\delta^\alpha &\triangleq \lim_{T\to\infty} \frac{1}{T} \int_{-T/2}^{T/2} x_\delta(t)\, e^{-j2\pi\alpha t}\, dt \\
&= \lim_{T\to\infty} \frac{1}{T} \int_{-T/2}^{T/2} \sum_{n=-\infty}^{+\infty} \widetilde{x}_n\, \delta(t-nT_0)\, e^{-j2\pi\alpha t}\, dt \\
&= \lim_{N\to\infty} \frac{1}{2NT_0+2\epsilon} \int_{-NT_0-\epsilon}^{NT_0+\epsilon} \sum_{n=-N}^{N} \widetilde{x}_n\, \delta(t-nT_0)\, e^{-j2\pi\alpha t}\, dt \\
&= \lim_{N\to\infty} \frac{1}{2NT_0+2\epsilon} \sum_{n=-N}^{N} \widetilde{x}_n \int_{-NT_0-\epsilon}^{NT_0+\epsilon} \delta(t-nT_0)\, e^{-j2\pi\alpha t}\, dt \\
&= \lim_{N\to\infty} \frac{(2N+1)T_0}{2NT_0+2\epsilon}\, \frac{1}{2N+1}\, \frac{1}{T_0} \sum_{n=-N}^{N} \widetilde{x}_n\, e^{-j2\pi\alpha n T_0} \\
&= \frac{1}{T_0} \lim_{N\to\infty} \frac{1}{2N+1} \sum_{n=-N}^{N} \widetilde{x}_n\, e^{-j2\pi\alpha n T_0} \\
&= \frac{1}{T_0}\, \widetilde{x}^{\widetilde{\alpha}}\big|_{\widetilde{\alpha}=\alpha T_0}
\end{aligned} \tag{2.171}$$

In the third equality we set $[-T/2, T/2] \equiv [-NT_0-\epsilon, NT_0+\epsilon]$ with $0 \leqslant \epsilon < T_0$ and N is largest integer such that $2N \leqslant T/T_0$. In addition, third equality holds since the limit in the left-hand side is assumed to exist. In fact, if the limit as $T \to \infty$ of $f(T)$ exists, then it is equal to the limit as $N \to \infty$ of $f(T_N)$, where $\{T_N\}$ is every sequence such that $T_N \to \infty$ as $N \to \infty$ (Johnsonbaugh and Pfaffenberger, 2002, Theorem 31.2). The fifth equality holds since the point $t = nT_0$ is always inside the interval $[-NT_0-\epsilon, NT_0+\epsilon]$ when $n \in \{-N, \ldots, N\}$. In the right-hand side of (2.171),

$$\widetilde{x}^{\widetilde{\alpha}} \triangleq \lim_{N\to\infty} \frac{1}{2N+1} \sum_{n=-N}^{N} \widetilde{x}_n\, e^{-j2\pi\widetilde{\alpha} n} \tag{2.172}$$

denotes the Fourier coefficient at frequency $\widetilde{\alpha}$ of the sequence $\widetilde{x}_n$. Let $\widetilde{A}$ be the set of frequencies of the almost-periodic component contained in the sequence $\widetilde{x}_n$, that is

$$\widetilde{A} \triangleq \left\{\widetilde{\alpha} \in [-1/2, 1/2) : \widetilde{x}^{\widetilde{\alpha}} \neq 0\right\}. \tag{2.173}$$

Thus, the Fourier coefficient of $x_\delta(t)$ at frequency α in (2.171) is zero if $\alpha T_0 \notin \widetilde{A}$.

We have

$$
\begin{aligned}
&\sum_{n=-\infty}^{+\infty} \mathrm{E}^{\{\widetilde{\alpha}\}}\{\widetilde{x}_n\}\,\delta(t-nT_0) \\
&\quad = \sum_{n=-\infty}^{+\infty} \left[\sum_{\widetilde{\alpha}\in\widetilde{A}} \widetilde{x}^{\widetilde{\alpha}}\, e^{j2\pi\widetilde{\alpha}n} \right] \delta(t-nT_0) \\
&\quad = \sum_{\widetilde{\alpha}\in\widetilde{A}} \widetilde{x}^{\widetilde{\alpha}} \left[\sum_{n=-\infty}^{+\infty} e^{j2\pi\widetilde{\alpha}n}\, \frac{1}{T_0}\,\delta\!\left(\frac{t}{T_0}-n\right) \right] \\
&\quad = \sum_{\widetilde{\alpha}\in\widetilde{A}} \widetilde{x}^{\widetilde{\alpha}}\, e^{j2\pi\widetilde{\alpha}t/T_0} \sum_{n=-\infty}^{+\infty} \delta(t-nT_0) \\
&\quad = \sum_{\widetilde{\alpha}\in\widetilde{A}} \widetilde{x}^{\widetilde{\alpha}}\, e^{j2\pi\widetilde{\alpha}t/T_0}\, \frac{1}{T_0} \sum_{p=-\infty}^{+\infty} e^{j2\pi(t/T_0)p} \\
&\quad = \sum_{\widetilde{\alpha}\in\widetilde{A}} \widetilde{x}^{\widetilde{\alpha}}\, \frac{1}{T_0} \sum_{p=-\infty}^{+\infty} e^{j2\pi(\widetilde{\alpha}+p)(t/T_0)} \\
&\quad = \frac{1}{T_0} \sum_{\widetilde{\alpha}\in\widetilde{A}} \sum_{p=-\infty}^{+\infty} \widetilde{x}^{\widetilde{\alpha}+p}\, e^{j2\pi(\widetilde{\alpha}+p)(t/T_0)} \\
&\quad = \frac{1}{T_0} \sum_{\alpha\in A} \widetilde{x}^{\alpha T_0}\, e^{j2\pi\alpha t}
\end{aligned}
\tag{2.174}
$$

In the fourth equality, the Poisson formula (C.8a) is used. In the sixth equality we used the fact that $\widetilde{x}^{\widetilde{\alpha}} = \widetilde{x}^{\widetilde{\alpha}+p}$ for any integer p. In the seventh equality the set

$$
A \triangleq \left\{\alpha\in\mathbb{R} : \alpha = (\widetilde{\alpha}+p)/T_0,\ \widetilde{\alpha}\in\widetilde{A},\ p\in\mathbb{Z}\right\}
\tag{2.175}
$$

is used.

From (2.171) and (2.174), we have

$$
\begin{aligned}
\mathrm{E}^{\{\alpha\}}\{x_\delta(t)\} &= \mathrm{E}^{\{\alpha\}}\left\{ \sum_{n=-\infty}^{+\infty} \widetilde{x}_n\,\delta(t-nT_0) \right\} \\
&= \sum_{n=-\infty}^{+\infty} \mathrm{E}^{\{\widetilde{\alpha}\}}\{\widetilde{x}_n\}\,\delta(t-nT_0)\,.
\end{aligned}
\tag{2.176}
$$

Eqs. (2.98a)–(2.98b) follow accounting for Theorem 2.32.

Proof of (2.99) and (2.100)

From (2.171) or from the right-hand side of (2.174), Eq. (2.99) immediately follows.

Accounting for (2.98a)–(2.98b) and (2.174), we have the following Fourier transform

$$\begin{aligned}
\mathrm{E}^{\{\alpha\}}\{x(t)\} &= \mathrm{E}^{\{\alpha\}}\{x_\delta(t)\} \otimes q(t) \\
&\overset{\mathcal{F}}{\longleftrightarrow} \sum_{\alpha \in A} \frac{1}{T_0} \widetilde{x}^{\alpha T_0}\, \delta(f-\alpha)\, Q(f) = \sum_{\alpha \in A} \frac{1}{T_0} \widetilde{x}^{\alpha T_0}\, Q(\alpha)\, \delta(f-\alpha) \\
&\overset{\mathcal{F}}{\longleftrightarrow} \sum_{\alpha \in A} \frac{1}{T_0} \widetilde{x}^{\alpha T_0}\, Q(\alpha)\, e^{j2\pi\alpha t}
\end{aligned} \tag{2.177}$$

from which (2.100) follows.

3

Almost-cyclostationary signal processing

Contents

Abstract

The processing of almost-cyclostationary signals is addressed. The linear almost-periodically time-variant filtering is considered and input/output relations are derived in terms of cyclic cross-correlations and cyclic cross-spectra. The cyclic statistics of the signals involved in the Rice's representation are obtained as special cases. The discrete-time signal obtained by uniformly sampling a continuous-time almost-cyclostationary signal is considered. The cyclic statistical functions of the discrete-time signal are linked to those of the continuous-time signal and aliasing issues are addressed. Multirate processing of discrete-time almost-cyclostationary signals are also considered.

Keywords

Linear almost-periodically time variant (LAPTV) filtering, Frequency shift (FRESH) filtering, Uniform sampling, Bandpass sampling, Rice's representation, Multirate systems

Cyclostationary Processes and Time Series. https://doi.org/10.1016/B978-0-08-102708-0.00014-5

3.1 Introduction

Cyclostationarity properties of almost-cyclostationary (ACS) signals are transformed by signal processing. In particular, the linear almost-periodically time-variant (LAPTV) transformations are of great interest. In fact, many man-made signals are obtained by LAPTV transformations of random (possibly wide-sense stationary) data. In addition, the class of the ACS processes is closed under LAPTV transformations. The linear time-invariant (LTI) filtering is as a special case of LAPTV filtering.

The LAPTV filtering of ACS signals is introduced in (Dragan, 1969b), (Gardner, 1972a), (Pourahmadi and Salehi, 1983), (Gardner, 1985), (Gardner, 1987f), and (Brown, 1987). It is also considered in follow-on work in (McLernon, 1998), (Franks, 1994), (Spooner, 1992), (Spooner and Gardner, 1994a), and (Napolitano, 1995). Properties of linear periodically time-varying systems are analyzed in (Voychishin and Dragan, 1972b), (Meyer and Burrus, 1975), (Ding and Fahmy, 1990), (Chen and Qiu, 1997), (Mehr and Chen, 2001), and (Mehr and Chen, 2002). Applications are considered in (Ericson, 1981).

More general kinds of linear time-variant (LTV) filtering are addressed in (Izzo and Napolitano, 2002a), (Izzo and Napolitano, 2002b), where linear systems that are deterministic in the fraction-of-time probability sense are considered. In general, LTV transformations of ACS signals produce oscillatory ACS signals (Section 14.2.1), (Napolitano, 2016a, Sec. 6.2.1).

The ACS signal obtained by the product of ACS signals are characterized in (Gardner, 1987f), (Gardner et al., 2006). The Rice's representation models real signals as linear periodically time-variant filtering of the complex envelope or, equivalently, of the in-phase and in-quadrature components. See (Gardner, 1987c) for the second-order statistics and (Izzo and Napolitano, 1997) for the higher-order statistics.

The problem of sampling a continuous-time ACS signal is treated in (Napolitano, 1995), (Izzo and Napolitano, 1996), where sufficient conditions on the sampling frequency are derived to reconstruct cyclic statistics of the continuous-time signal starting from the discrete-time sequence obtained by uniformly sampling the continuous-time signal.

Multirate processing of discrete-time jointly ACS signals is addressed in (Sathe and Vaidyanathan, 1993), (Izzo and Napolitano, 1995), (Izzo and Napolitano, 1998b), (Akkarakaran and Vaidyanathan, 2000), (Aach, 2007), (Napolitano, 2012, Sec. 4.10). In particular, expansion (upsampling) and decimation (downsampling) operations are considered. Note that such operations are liner time variant but not periodically or almost-periodically time variant.

3.2 Linear filtering

3.2.1 Linear almost-periodically time-variant systems

Definition 3.1. A *linear time-variant* (LTV) system with input $x(t)$, output $y(t)$, impulse-response function $h(t, u)$, and input-output relation

$$y(t) = \int_{\mathbb{R}} h(t, u)\, x(u)\, \mathrm{d}u \tag{3.1}$$

is said to be *linear almost-periodically time-variant* (LAPTV) if the impulse-response function admits the Fourier series expansion

$$h(t, u) = \sum_{\sigma \in J} h_\sigma(t - u)\, e^{j2\pi\sigma u} \tag{3.2}$$

where J is a countable set of possibly incommensurate frequency shifts. □

By substituting (3.2) into (3.1) we see that the output $y(t)$ can be expressed in the two equivalent forms (Gardner, 1987f), (Franks, 1994)

$$y(t) = \sum_{\sigma \in J} h_\sigma(t) \otimes \left[x(t)\, e^{j2\pi\sigma t}\right] \tag{3.3a}$$

$$= \sum_{\sigma \in J} \left[g_\sigma(t) \otimes x(t)\right] e^{j2\pi\sigma t} \tag{3.3b}$$

where

$$g_\sigma(t) \triangleq h_\sigma(t)\, e^{-j2\pi\sigma t} \,. \tag{3.4}$$

Eqs. (3.3a) and (3.3b) can be re-expressed in the frequency domain:

$$Y(f) = \sum_{\sigma \in J} H_\sigma(f)\, X(f - \sigma) \tag{3.5a}$$

$$= \sum_{\sigma \in J} G_\sigma(f - \sigma)\, X(f - \sigma) \tag{3.5b}$$

where $H_\sigma(f)$ and $G_\sigma(f)$ are the Fourier transforms of $h_\sigma(t)$ and $g_\sigma(t)$, respectively.

From (3.3a) it follows that a LAPTV system combines LTI filtered versions of frequency-shifted versions of the input signal. For this reason LAPTV filtering is also referred to as *frequency-shift* (FRESH) filtering (Gardner, 1993). Equivalently, from (3.3b) it follows that a LAPTV system combines frequency shifted versions of LTI filtered versions of the input (Fig. 3.1).

In the special case where $J \equiv \{k/T_0\}_{k \in \mathbb{Z}}$ for some period T_0, the system is said to be linear periodically time-variant (LPTV). If J contains only the element $\sigma = 0$, then the system is linear time-invariant.

3.2.2 Input/output relations in terms of cyclic statistics

Let us consider two LAPTV systems (Fig. 3.2) with input/output relationships

$$y_i(t) = \int_{\mathbb{R}} h_i(t, u)\, x_i(u)\, \mathrm{d}u \qquad i = 1, 2 \tag{3.6}$$

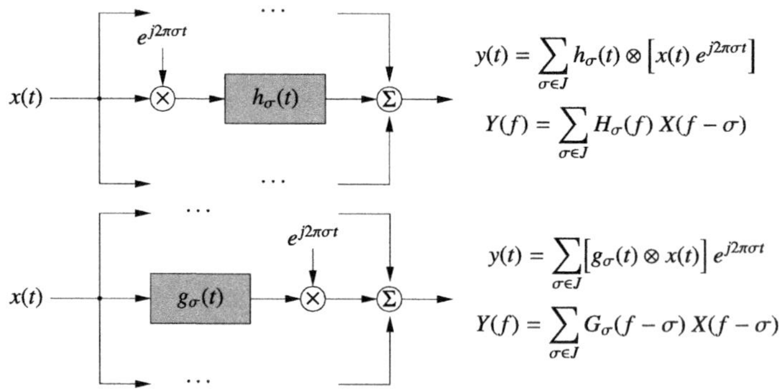

FIGURE 3.1 LAPTV systems: Two equivalent realizations and corresponding input/output relations in time and frequency domains (see (3.3a)–(3.5b)).

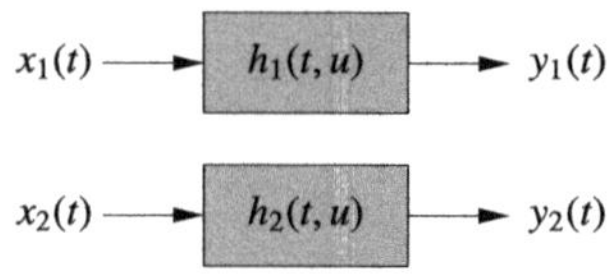

FIGURE 3.2 LAPTV systems.

whose impulse-response functions admit the Fourier series expansions

$$h_i(t,u) = \sum_{\sigma_i \in J_i} h_{\sigma_i}(t-u)\, e^{j2\pi\sigma_i u} \qquad i = 1, 2\,. \tag{3.7}$$

Assumption 3.2. *The LAPTV systems (3.6) are such that*

$$\text{a.} \qquad \sum_{\sigma_i \in J_i} \left|h_{\sigma_i}(\tau)\right| \in L^1(\mathbb{R}) \qquad i = 1, 2 \tag{3.8a}$$

$$\text{b.} \qquad \sum_{\sigma_i \in J_i} \left\|h_{\sigma_i}\right\|_\infty < \infty \qquad i = 1, 2\,. \tag{3.8b}$$

□

Assumption 3.3. *The input signals $x_1(t)$ and $x_2(t)$ have almost-periodic expected values*

$$\mathrm{E}\{x_i(t)\} = \sum_{\eta_i \in E_{1i}} \overline{x}_{\eta_i}\, e^{j2\pi\eta_i t} \qquad i = 1, 2 \tag{3.9}$$

with absolutely convergent Fourier series. □

Assumption 3.4. *The input signals $x_1(t)$ and $x_2(t)$ are jointly ACS with (conjugate) cyclic cross-correlation functions such that*

$$\text{a.} \qquad \sum_{\alpha \in A_{12}} \left| R^{\alpha}_{x_1 x_2^{(*)}}(\tau) \right| \in L^1(\mathbb{R}) \tag{3.10a}$$

$$\text{b.} \qquad \sum_{\alpha \in A_{12}} \left\| R^{\alpha}_{x_1 x_2^{(*)}} \right\|_{\infty} < \infty \tag{3.10b}$$

□

Assumptions 3.2, 3.3, and 3.4 are used to allow the interchange of integral and sum operations in the derivations of the following input/output relationships.

Theorem 3.5. *Under Assumptions 3.2a and 3.3, the expected value of* $y_i(t)$ *is given by*

$$\mathrm{E}\{y_i(t)\} = \sum_{\sigma_i \in J_i} \sum_{\eta_i \in E_{1i}} H_{\sigma_i}(\eta_i + \sigma_i)\, \overline{x}_{\eta_i}\, e^{j2\pi(\eta_i+\sigma_i)t} \qquad i = 1, 2 \tag{3.11}$$

where

$$H_{\sigma_i}(f) \triangleq \int_{\mathbb{R}} h_{\sigma_i}(\tau)\, e^{-j2\pi f \tau}\, \mathrm{d}\tau\,. \tag{3.12}$$

Proof. See Section 3.10. □

Theorem 3.6. *Under Assumptions 3.2a and 3.4b or under Assumptions 3.2b and 3.4a, the (conjugate) cross-correlation of the outputs* $y_i(t)$ *is given by*

$$\begin{aligned} R_{y_1 y_2^{(*)}}(t,\tau) &\triangleq \mathrm{E}\left\{ y_1(t+\tau)\, y_2^{(*)}(t) \right\} \\ &= \sum_{\alpha \in A_{12}} \sum_{\sigma_1 \in J_1} \sum_{\sigma_2 \in J_2} \left[R^{\alpha}_{x_1 x_2^{(*)}}(\tau)\, e^{j2\pi\sigma_1\tau} \right] \\ &\qquad \underset{\tau}{\otimes}\, r^{\alpha+\sigma_1+(-)\sigma_2}_{\sigma_1\sigma_2(*)}(\tau)\, e^{j2\pi(\alpha+\sigma_1+(-)\sigma_2)t} \end{aligned} \tag{3.13}$$

where $\underset{\tau}{\otimes}$ *denotes convolution with respect to* τ, $(-)$ *is an optional minus sign that is linked to* $(*)$, *and*

$$r^{\gamma}_{\sigma_1\sigma_2(*)}(\tau) \triangleq \int_{\mathbb{R}} h_{\sigma_1}(s+\tau)\, h^{(*)}_{\sigma_2}(s)\, e^{-j2\pi\gamma s}\, \mathrm{d}s\,. \tag{3.14}$$

Proof. See Section 3.10. □

From Theorem 3.5 it follows that $\mathrm{E}\{y_i(t)\}$ is an almost-periodic function of t.

From Theorem 3.6 it follows that $R_{y_1 y_2^{(*)}}(t,\tau)$ is an almost-periodic function of t whose (conjugate) cycle frequencies are obtained by sum and differences of those of $R_{x_1 x_2^{(*)}}(t,\tau)$ and the frequencies of the Fourier series of the two impulse-response functions $h_1(t,u)$ and $h_2(t,u)$.

Therefore, from Theorems 3.5 and 3.6 considered for $x_1(t) \equiv x_2(t)$, we have that LAPTV systems transform wide-sense ACS processes into wide-sense ACS processes (with different cyclic features). That is, *wide-sense ACS processes are closed under LAPTV transformations.*

Corollary 3.7. *Under Assumptions 3.2 and 3.4a, the (conjugate) cyclic cross-correlation and the (conjugate) cyclic cross-spectrum of the outputs $y_i(t)$ are given by*

$$R^{\alpha}_{y_1 y_2^{(*)}}(\tau) \triangleq \left\langle \mathrm{E}\left\{ y_1(t+\tau)\, y_2^{(*)}(t) \right\} e^{-j2\pi\alpha t} \right\rangle_t$$

$$= \sum_{\sigma_1 \in J_1} \sum_{\sigma_2 \in J_2} \left[R^{\alpha-\sigma_1-(-)\sigma_2}_{x_1 x_2^{(*)}}(\tau)\, e^{j2\pi\sigma_1\tau} \right] \underset{\tau}{\otimes} r^{\alpha}_{\sigma_1\sigma_2(*)}(\tau) \tag{3.15}$$

$$S^{\alpha}_{y_1 y_2^{(*)}}(f) \triangleq \int_{\mathbb{R}} R^{\alpha}_{y_1 y_2^{(*)}}(\tau)\, e^{-j2\pi f\tau}\, \mathrm{d}\tau$$

$$= \sum_{\sigma_1 \in J_1} \sum_{\sigma_2 \in J_2} S^{\alpha-\sigma_1-(-)\sigma_2}_{x_1 x_2^{(*)}}(f-\sigma_1)\, H_{\sigma_1}(f)\, H^{(*)}_{\sigma_2}((-)(\alpha-f)) \tag{3.16}$$

where in the sums in (3.15) and (3.16), only those $\sigma_1 \in G_1$ and $\sigma_2 \in G_2$ such that $\alpha - \sigma_1 - (-)\sigma_2 \in A_{12}$ give nonzero contribution.

Proof. See Section 3.10. □

The results of Corollary 3.7 can be specialized to several cases of interest.

a. By specializing (3.15) and (3.16) to the case of two LTI systems, we have

$$R^{\alpha}_{y_1 y_2^{(*)}}(\tau) = R^{\alpha}_{x_1 x_2^{(*)}}(\tau) \otimes h_1(\tau) \otimes \left[h_2^{(*)}(-\tau)\, e^{j2\pi\alpha\tau} \right] \tag{3.17}$$

$$\overset{\mathcal{F}}{\longleftrightarrow} \quad S^{\alpha}_{y_1 y_2^{(*)}}(f) = S^{\alpha}_{x_1 x_2^{(*)}}(f)\, H_1(f)\, H_2^{(*)}((-)(\alpha-f))\,. \tag{3.18}$$

b. If in (3.15) and (3.16) $x_1 = x_2 = x$, $h_1 = h_2 = h$, $y_1 = y_2 = y$, and $(*)$ is conjugation, then we obtain the input-output relations for LAPTV systems in terms of cyclic autocorrelation functions and cyclic spectra:

$$R^{\alpha}_{yy^*}(\tau) = \sum_{\sigma_1 \in J} \sum_{\sigma_2 \in J} \left[R^{\alpha-\sigma_1+\sigma_2}_{xx^*}(\tau)\, e^{j2\pi\sigma_1\tau} \right] \underset{\tau}{\otimes} r^{\alpha}_{\sigma_1\sigma_2 *}(\tau) \tag{3.19}$$

$$S^{\alpha}_{yy^*}(f) = \sum_{\sigma_1 \in J} \sum_{\sigma_2 \in J} S^{\alpha-\sigma_1+\sigma_2}_{xx^*}(f-\sigma_1)\, H_{\sigma_1}(f)\, H^*_{\sigma_2}(f-\alpha)\,. \tag{3.20}$$

c. By specializing (3.16) to the case $x_1 = x_2 = x$, $h_1 = h_2 = h$ LTI system, and $y_1 = y_2 = y$, we obtain

$$y(t) = x(t) \otimes h(t) \tag{3.21}$$

and

$$S^{\alpha}_{yy^{(*)}}(f) = S^{\alpha}_{xx^{(*)}}(f)\, H(f)\, H^{(*)}((-)(\alpha-f))\,. \tag{3.22}$$

Analogously, by specializing (3.15), we obtain the inverse Fourier transform of both sides of (3.22)

$$R^{\alpha}_{yy^{(*)}}(\tau) = R^{\alpha}_{xx^{(*)}}(\tau) \otimes r^{\alpha}_{hh^{(*)}}(\tau) \tag{3.23}$$

where the *asymmetric (conjugate) ambiguity function* $r^{\alpha}_{hh^{(*)}}(\tau)$ can be written in one of the following equivalent forms (see Section 3.10)

$$r^{\alpha}_{hh^{(*)}}(\tau) \triangleq \int_{\mathbb{R}} h(s+\tau)\, h^{(*)}(s)\, e^{-j2\pi\alpha s}\, \mathrm{d}s$$
$$= h(\tau) \otimes \left[h^{(*)}(-\tau)\, e^{j2\pi\alpha\tau} \right] \tag{3.24a}$$
$$= \int_{\mathbb{R}} H(\lambda)\, H^{(*)}((-)(\alpha-\lambda))\, e^{j2\pi\lambda\tau}\, \mathrm{d}\lambda \tag{3.24b}$$

For $\alpha = 0$ and $(*) = *$, (3.22) and (3.23) reduce to the input/output relations for LTI systems in terms of power spectral densities and averaged autocorrelations:

$$S^{0}_{yy^*}(f) = S^{0}_{xx^*}(f)\, |H(f)|^2 \tag{3.25}$$
$$R^{0}_{yy^*}(\tau) = R^{0}_{xx^*}(\tau) \otimes r^{0}_{hh^*}(\tau)\,. \tag{3.26}$$

d. By specializing (3.15) and (3.16) to the case $x_1 = x_2 = x$, $h_2(t+\tau, t) = \delta(\tau)$, $y_1 = y$, $y_2 = x$, so that $J_2 = \{0\}$ and

$$r^{\alpha}_{\sigma_1\sigma_2(*)}(\tau) = \int_{\mathbb{R}} h_{\sigma_1}(s+\tau)\, \delta(s)\, e^{-j2\pi\alpha s}\, \mathrm{d}s = h_{\sigma_1}(\tau) \tag{3.27}$$

we obtain the cyclic cross-statistics of the output and input signals

$$R^{\alpha}_{yx^{(*)}}(\tau) = \sum_{\sigma_1 \in J_1} \left[R^{\alpha-\sigma_1}_{xx^{(*)}}(\tau)\, e^{j2\pi\sigma_1\tau} \right] \underset{\tau}{\otimes} h_{\sigma_1}(\tau) \tag{3.28}$$
$$S^{\alpha}_{yx^{(*)}}(f) = \sum_{\sigma_1 \in J_1} S^{\alpha-\sigma_1}_{xx^{(*)}}(f-\sigma_1)\, H_{\sigma_1}(f)\,. \tag{3.29}$$

e. By specializing (3.16) to the case h_1 and h_2 LTI and $(*)$ absent, for $\alpha = 0$ we obtain

$$S^{0}_{y_1y_2}(f) = S^{0}_{x_1x_2}(f)\, H_1(f)\, H_2(-f) \tag{3.30}$$
$$\overset{\mathcal{F}}{\longleftrightarrow}\; R^{0}_{y_1y_2}(\tau) = R^{0}_{x_1x_2}(\tau) \otimes h_1(\tau) \otimes h_2(-\tau) \tag{3.31}$$

Example 3.8. *Time Delay.* Let us consider the LTI system that introduces a time delay d:

$$y(t) = x(t-d) = x(t) \otimes \delta(t-d) \tag{3.32}$$

The (conjugate) cyclic statistics of the output signal can be expressed in terms of those of the input signal:

$$R^{\alpha}_{yy^{(*)}}(\tau) = R^{\alpha}_{xx^{(*)}}(\tau)\, e^{-j2\pi\alpha d} \tag{3.33}$$
$$S^{\alpha}_{yy^{(*)}}(f) = S^{\alpha}_{xx^{(*)}}(f)\, e^{-j2\pi\alpha d} \tag{3.34}$$

$$R^{\alpha}_{yx^{(*)}}(\tau) = R^{\alpha}_{xx^{(*)}}(\tau - d) \tag{3.35}$$

$$S^{\alpha}_{yx^{(*)}}(f) = S^{\alpha}_{xx^{(*)}}(f)\, e^{-j2\pi f d} \tag{3.36}$$

From (3.33) it follows that the (conjugate) cyclic autocorrelation of $y(t)$ for $\alpha \neq 0$ contains a phase term depending on d. In contrast, for $\alpha = 0$ the time averaged (conjugate) autocorrelation $R^{0}_{yy^{(*)}}(\tau)$ is equal to $R^{0}_{xx^{(*)}}(\tau)$ and is independent of the time shift d.

Relationships (3.33)–(3.36) are exploited in (Gardner and Chen, 1988a), (Gardner and Chen, 1992), (Gardner and Chen, 1992) to design signal-selective time-difference-of arrival (TDOA) estimation algorithms.

Example 3.9. *Spectral Coherence and LTI Filtering.* The magnitude of the spectral coherence (1.60) is invariant to LTI filtering, provided that the spectral components are not annihilated by the filter harmonic response. Let $y(t)$ be given by (3.21). Accounting for (3.22), we have

$$\begin{aligned}
\gamma^{\alpha}_{yy^{(*)}}(f) &\triangleq \frac{S^{\alpha}_{yy^{(*)}}(f)}{\left[S^{0}_{yy^*}(f)\, S^{0}_{yy^*}((-)(\alpha - f))\right]^{1/2}} \\
&= \frac{S^{\alpha}_{xx^{(*)}}(f)\, H(f)\, H^{(*)}((-)(\alpha - f))}{\left[S^{0}_{xx^*}(f)\, |H(f)|^2\, S^{0}_{xx^*}((-)(\alpha - f))\, |H((-)(\alpha - f))|^2\right]^{1/2}} \\
&= \gamma^{\alpha}_{xx^{(*)}}(f)\, \frac{H(f)}{|H(f)|}\, \frac{H^{(*)}((-)(\alpha - f))}{|H((-)(\alpha - f))|}
\end{aligned} \tag{3.37}$$

Thus,

$$\left|\gamma^{\alpha}_{yy^{(*)}}(f)\right| = \left|\gamma^{\alpha}_{xx^{(*)}}(f)\right| \tag{3.38}$$

for all (α, f) such that

$$H(f)\, H^{(*)}((-)(\alpha - f)) \neq 0\,. \tag{3.39}$$

3.3 Products of ACS signals

Let $x_1(t)$, $x_2(t)$, $c_1(t)$, and $c_2(t)$ be jointly ACS complex-valued signals with (conjugate) cross-correlation functions

$$R_{x_1 x_2^{(*)}}(t, \tau) = \sum_{\alpha_x \in A_{x_1 x_2^{(*)}}} R^{\alpha_x}_{x_1 x_2^{(*)}}(\tau)\, e^{j2\pi \alpha_x t} \tag{3.40}$$

$$R_{c_1 c_2^{(*)}}(t, \tau) = \sum_{\alpha_c \in A_{c_1 c_2^{(*)}}} R^{\alpha_c}_{c_1 c_2^{(*)}}(\tau)\, e^{j2\pi \alpha_c t}\,. \tag{3.41}$$

If $x_1(t)$ and $x_2(t)$ are statistically independent of $c_1(t)$ and $c_2(t)$, then the eighth-order joint probability density function of their real and imaginary parts factors into the product of the two fourth-order joint probability densities of real and imaginary parts of x_1, x_2 and of c_1, c_2, (Gardner, 1987f), (Brown, 1987), (Gardner, 1994a). Moreover, the (conjugate) cross-correlation function of the product waveforms

$$y_i(t) = c_i(t)\, x_i(t) \qquad i = 1, 2 \tag{3.42}$$

also factors:

$$R_{y_1 y_2^{(*)}}(t, \tau) = R_{c_1 c_2^{(*)}}(t, \tau)\, R_{x_1 x_2^{(*)}}(t, \tau)\,. \tag{3.43}$$

Theorem 3.10. *Under Assumption 3.4 (and a similar assumption on the (conjugate) cyclic cross-correlations of c_1 and c_2) the (conjugate) cyclic cross-correlation function and the (conjugate) cyclic cross-spectrum of $y_1(t)$ and $y_2(t)$ are given by*

$$R^{\alpha}_{y_1 y_2^{(*)}}(\tau) = \sum_{\alpha_x \in A_{x_1 x_2^{(*)}}} R^{\alpha_x}_{x_1 x_2^{(*)}}(\tau)\, R^{\alpha-\alpha_x}_{c_1 c_2^{(*)}}(\tau) \tag{3.44}$$

$$S^{\alpha}_{y_1 y_2^{(*)}}(f) = \sum_{\alpha_x \in A_{x_1 x_2^{(*)}}} \int_{\mathbb{R}} S^{\alpha_x}_{x_1 x_2^{(*)}}(\lambda)\, S^{\alpha-\alpha_x}_{c_1 c_2^{(*)}}(f-\lambda)\mathrm{d}\lambda \tag{3.45}$$

where, in the sums, only those (conjugate) cycle frequencies α_x such that $\alpha - \alpha_x \in A_{c_1 c_2^{(*)}}$ give nonzero contribution.

Proof. See Section 1.9. □

From (3.44) and (3.45), it follows that the set of (conjugate) cycle frequencies of the (conjugate) cross-correlation of $y_1(t)$ and $y_2(t)$ is

$$\{\alpha\} = \left\{\alpha = \alpha_c + \alpha_x,\ \alpha_c \in A_{c_1 c_2^{(*)}}, \alpha_x \in A_{x_1 x_2^{(*)}}\right\}. \tag{3.46}$$

In the special case where $c_1(t)$ and $c_2(t)$ are almost-periodic functions with (generalized) Fourier series

$$c_i(t) = \sum_{\gamma_i \in G_i} c_{i,\gamma_i}\, e^{j2\pi\gamma_i t} \qquad i = 1, 2 \tag{3.47}$$

then

$$\begin{aligned} R_{c_1 c_2^{(*)}}(t, \tau) &= c_1(t+\tau)\, c_2^{(*)}(t) \\ &= \sum_{\gamma_1 \in G_1} \sum_{\gamma_2 \in G_2} c_{1,\gamma_1}\, c^{(*)}_{2,\gamma_2}\, e^{j2\pi\gamma_1\tau}\, e^{j2\pi(\gamma_1 + (-)\gamma_2)t} \end{aligned} \tag{3.48}$$

and

$$R^{\alpha}_{c_1 c_2^{(*)}}(\tau) = \sum_{\gamma_1 \in G_1} c_{1,\gamma_1}\, c^{(*)}_{2,(-)(\alpha-\gamma_1)}\, e^{j2\pi\gamma_1\tau} \tag{3.49}$$

$$S^{\alpha}_{c_1 c_2^{(*)}}(f) = \sum_{\gamma_1 \in G_1} c_{1,\gamma_1}\, c^{(*)}_{2,(-)(\alpha-\gamma_1)}\, \delta(f-\gamma_1) \tag{3.50}$$

where in the sum give nonzero contribution only those frequencies γ_1 such that $(-)(\alpha - \gamma_1) \in G_2$. From (3.49) and (3.50) it follows that the set of the (conjugate) cycle-frequencies of the (conjugate) cross-correlation of $c_1(t)$ and $c_2(t)$ is

$$\{\alpha\} = \{\alpha = \gamma_1 + (-)\gamma_2,\ \gamma_1 \in G_1, \gamma_2 \in G_2\}\ . \tag{3.51}$$

In the special case $x_1 \equiv x_2$ and $c_1 \equiv c_2$ (3.48)–(3.50) reduce to (corrected versions of) (3.93)–(3.95) in (Gardner et al., 2006). By substituting (3.49) and (3.50) into (3.44) and (3.45), respectively, and making the variable change $\gamma_2 = (-)(\alpha - \alpha_x - \gamma_1)$, we get (see (Napolitano, 1995, Eqs. (46)–(48)))

$$R^{\alpha}_{y_1 y_2^{(*)}}(\tau) = \sum_{\gamma_1 \in G_1} \sum_{\gamma_2 \in G_2} c_{1,\gamma_1}\, c^{(*)}_{2,\gamma_2}\, R^{\alpha-(\gamma_1+(-)\gamma_2)}_{x_1 x_2^{(*)}}(\tau)\, e^{j2\pi\gamma_1\tau} \tag{3.52}$$

$$S^{\alpha}_{y_1 y_2^{(*)}}(f) = \sum_{\gamma_1 \in G_1} \sum_{\gamma_2 \in G_2} c_{1,\gamma_1}\, c^{(*)}_{2,\gamma_2}\, S^{\alpha-(\gamma_1+(-)\gamma_2)}_{x_1 x_2^{(*)}}(f-\gamma_1)\,, \tag{3.53}$$

where, in the sums, give nonzero contribution only those values of γ_1 and γ_2 such that $\alpha - (\gamma_1 + (-)\gamma_2) \in A_{x_1 x_2^{(*)}}$. From (3.52) and (3.53), it follows that the set of the (conjugate) cycle-frequencies of the (conjugate) cross-correlation of $y_1(t)$ and $y_2(t)$ is

$$\{\alpha\} = \left\{\alpha = \gamma_1 + (-)\gamma_2 + \alpha_x,\ \alpha_x \in A_{x_1 x_2^{(*)}}, \gamma_1 \in G_1, \gamma_2 \in G_2\right\}\ . \tag{3.54}$$

Example 3.11. *Modulation.* Let us consider the system that performs the modulation of a complex carrier:

$$y(t) = x(t)\, e^{j2\pi f_0 t}\ . \tag{3.55}$$

The (conjugate) cyclic statistics of the output signal can be expressed in terms of those of the input signal:

$$R^{\alpha}_{yy^{(*)}}(\tau) = R^{\alpha-\kappa f_0}_{xx^{(*)}}(\tau)\, e^{j2\pi f_0\tau} \tag{3.56}$$

$$S^{\alpha}_{yy^{(*)}}(f) = S^{\alpha-\kappa f_0}_{xx^{(*)}}(f-f_0) \tag{3.57}$$

$$R^{\alpha}_{yx^{(*)}}(\tau) = R^{\alpha-f_0}_{xx^{(*)}}(\tau)\, e^{j2\pi f_0\tau} \tag{3.58}$$

$$S^{\alpha}_{yx^{(*)}}(f) = S^{\alpha-f_0}_{xx^{(*)}}(f-f_0) \tag{3.59}$$

where $\kappa \triangleq 1 + (-)1$, that is, $\kappa = 0$ if $(*) = *$ and $\kappa = 2$ if $(*)$ is absent.

3.4 Supports of cyclic spectra of band limited signals

Definition 3.12. A signal $x(t)$ is said to be a *strictly band-limited* low-pass signal with monolateral bandwidth B if and only if $x(t)$ passes unaltered through the ideal low-pass filter with monolateral bandwidth B. That is,

$$x(t) \equiv x(t) \otimes h_{\text{LPF}}(t) \tag{3.60}$$

where $h_{\text{LPF}}(t)$ is the impulse-response function of the ideal low-pass filter with harmonic-response function

$$H_{\text{LPF}}(f) = \text{rect}\left(\frac{f}{2B}\right) \triangleq \begin{cases} 1 & |f| \leqslant B \\ 0 & |f| > B \, . \end{cases} \tag{3.61}$$

□

Fact 3.13. *Accounting for (3.25), for a strictly band-limited low-pass signal we have*

$$S^0_{xx^*}(f) = S^0_{xx^*}(f)\,|H(f)|^2 \tag{3.62}$$

that is, $S^0_{xx^*}(f) \equiv 0$ for $f \notin (-B, B)$, which is commonly used as definition of strictly band-limited low-pass process in the stationary case. □

From (3.22) it follows that the support in the (α, f) plane of the (conjugate) cyclic spectrum of $y(t)$ is such that

$$\begin{aligned} \text{supp}\left[S^{\alpha}_{yy^{(*)}}(f)\right] &\triangleq \left\{(\alpha, f) \in \mathbb{R} \times \mathbb{R} \, : \, S^{\alpha}_{yy^{(*)}}(f) \neq 0\right\} \\ &\subseteq \left\{(\alpha, f) \in \mathbb{R} \times \mathbb{R} \, : \, H(f)\, H^{(*)}((-)(\alpha - f)) \neq 0\right\} \, . \end{aligned} \tag{3.63}$$

For the LTI filtering (3.60), accounting for the fact that rect(f) is real and even, the input/output relation (3.22) specializes to

$$S^{\alpha}_{xx^{(*)}}(f) = H_{\text{LPF}}(f)\, H_{\text{LPF}}(f - \alpha)\, S^{\alpha}_{xx^{(*)}}(f) \, . \tag{3.64}$$

Hence, (3.63) specializes into

$$\text{supp}\left[S^{\alpha}_{xx^{(*)}}(f)\right] \subseteq \{(\alpha, f) \in \mathbb{R} \times \mathbb{R} \, : \, H_{\text{LPF}}(f)\, H_{\text{LPF}}(f - \alpha) \neq 0\}$$

$$\equiv \{(\alpha, f) \in \mathbb{R} \times \mathbb{R} \, : \, |f| \leqslant B, \ |\alpha - f| \leqslant B\} \tag{3.65a}$$

$$\subseteq \{(\alpha, f) \in \mathbb{R} \times \mathbb{R} \, : \, |f| \leqslant B, \ |\alpha| \leqslant 2B\} \, . \tag{3.65b}$$

In Fig. 3.3, set (3.65a) is represented. By common convention, f and α variables are on the abscissa and ordinate axis, respectively of the (α, f)-plane. From the conditions characterizing the bigger set (3.65b) it results that

$$S^{\alpha}_{xx^{(*)}}(f) = 0 \quad |f| > B, \ |\alpha| > 2B \, . \tag{3.66}$$

Let $x(t)$ be a strictly band-limited high-pass signal, that is,

$$x(t) \equiv x(t) \otimes h_{\text{HPF}}(t) \tag{3.67}$$

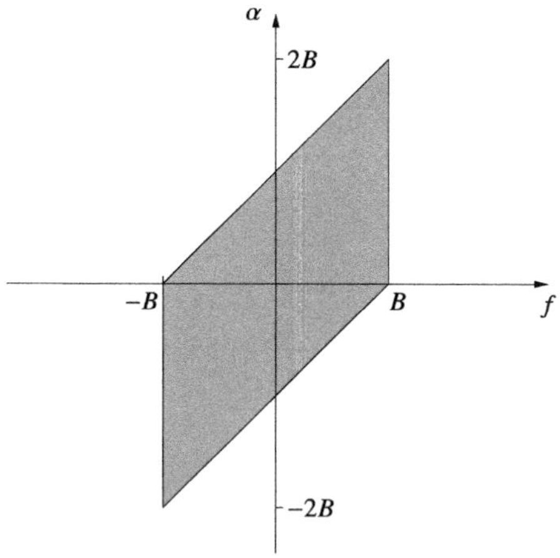

FIGURE 3.3 Cyclic-spectrum support of a low-pass signal. *Source: (Gardner et al., 2006) Copyright of Elsevier.*

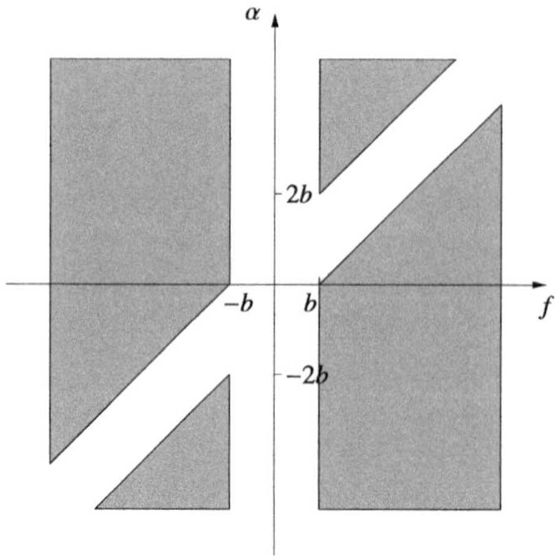

FIGURE 3.4 Cyclic-spectrum support of a high-pass signal. *Source: (Gardner et al., 2006) Copyright of Elsevier.*

where $h_{\text{HPF}}(t)$ is the impulse-response function of the ideal high-pass filter with harmonic-response function

$$H_{\text{HPF}}(f) = 1 - \text{rect}\left(\frac{f}{2b}\right) . \tag{3.68}$$

In such a case, $S^0_{xx^*}(f) \equiv 0$ for $f \in (-b, b)$.

From (3.63), we have (see Fig. 3.4)

$$\text{supp}\left[S^{\alpha}_{xx^{(*)}}(f)\right] \subseteq \{(\alpha, f) \in \mathbb{R} \times \mathbb{R} \, : \, H_{\text{HPF}}(f)\, H_{\text{HPF}}(f - \alpha) \neq 0\} . \tag{3.69}$$

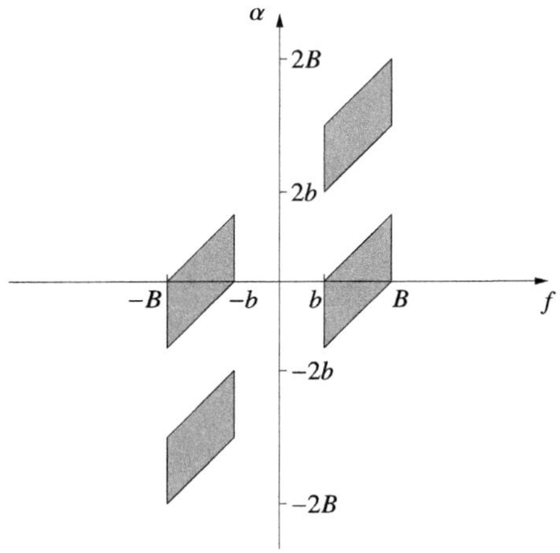

FIGURE 3.5 Cyclic-spectrum support of a band-pass signal. *Source: (Gardner et al., 2006) Copyright of Elsevier.*

Finally, let $x(t)$ be a strictly band-limited band-pass signal, that is,

$$x(t) \equiv x(t) \otimes h_{\text{BPF}}(t) \tag{3.70}$$

where $h_{\text{BPF}}(t)$ is the impulse-response function of the ideal band-pass filter with harmonic-response function

$$H_{\text{BPF}}(f) = H_{\text{LPF}}(f)\, H_{\text{HPF}}(f) \tag{3.71}$$

where $H_{\text{LPF}}(f)$ and $H_{\text{HPF}}(f)$ are given by (3.61) and (3.68), respectively. In such a case, $S^0_{xx^*}(f) \equiv 0$ for $f \notin (-B, -b) \cup (b, B)$, where $0 < b < B$.

Accounting for (3.63), we have (see Fig. 3.5)

$$\begin{aligned} \text{supp}\left[S^{\alpha}_{xx^{(*)}}(f)\right] &\subseteq \{(\alpha, f) \in \mathbb{R} \times \mathbb{R} : H_{\text{BPF}}(f)\, H_{\text{BPF}}(f-\alpha) \neq 0\} \\ &= \{(\alpha, f) \in \mathbb{R} \times \mathbb{R} : H_{\text{LPF}}(f)\, H_{\text{LPF}}(f-\alpha) \neq 0\} \\ &\quad \cap \{(\alpha, f) \in \mathbb{R} \times \mathbb{R} : H_{\text{HPF}}(f)\, H_{\text{HPF}}(f-\alpha) \neq 0\}\ . \end{aligned} \tag{3.72}$$

It is noted that supports for symmetric definitions of cyclic spectra (1.35) are "diamond shaped" and are given in (Gardner, 1986c), (Gardner, 1987f, Chap. 11), (Gardner, 1991a).

3.5 Rice's representation

In this section, cyclic statistical functions of real signal, analytic signal, complex envelope, and in-phase and in-quadrature components are linked among them (Gardner, 1987c), (Izzo and Napolitano, 1997). For related results see also (Rybakov, 1981), (Rybakov, 1986), (Gurevich and Korshunov, 1987). Expressions specialized for $\alpha = 0$ reveal that cyclostationarity cannot be neglected even if only the time-averaged autocorrelation or the power spectrum is computed.

Let $x(t)$ be a real-valued signal.

Assumption 3.14. $x_{\mathrm{dc}} \triangleq \langle \mathrm{E}\{x(t)\}\rangle_t = 0$, *that is, the power spectrum* $S_{xx}(f)$ *does not contain the additive term* $x_{\mathrm{dc}}\,\delta(f)$. □

Assumption 3.15. *The signal* $x(t)$ *does not contain any finite-strength additive sine-wave component. That is, cyclic spectra* $S^{\alpha}_{xx}(f)$ *do not contain Dirac deltas (Section 1.4).* □

Assumption 3.15 avoids in the following expressions the presence of undetermined products of sign(·), u(·), and rect(·) functions with Dirac deltas centered in the discontinuity points of these functions. In the special case $\alpha = 0$, only Assumption 3.14 is needed.

Hilbert transform

The Hilbert transform of a signal $x(t)$ is defined by the LTI transformation (Section D.1)

$$\widehat{x}(t) \triangleq x(t) \otimes h_{\mathrm{H}}(t) \tag{3.73}$$

where

$$h_{\mathrm{H}}(t) \triangleq \frac{1}{\pi t} \stackrel{\mathcal{F}}{\longleftrightarrow} H_{\mathrm{H}}(f) = -j\,\mathrm{sign}(f)\,. \tag{3.74}$$

Under Assumption 3.15, and using (3.22) and (3.23), we have

$$\begin{aligned} S^{\alpha}_{\widehat{x}\widehat{x}}(f) &= S^{\alpha}_{xx}(f)\,\big[-j\,\mathrm{sign}(f)\big]\,\big[-j\,\mathrm{sign}(\alpha-f)\big] \\ &= S^{\alpha}_{xx}(f)\,\mathrm{sign}(f)\,\mathrm{sign}(f-\alpha) \\ &= S^{\alpha}_{xx}(f)\left[1-2\,\mathrm{rect}\left(\frac{f-\alpha/2}{|\alpha|}\right)\right] \end{aligned} \tag{3.75}$$

$$\begin{aligned} \stackrel{\mathcal{F}}{\longleftrightarrow}\; R^{\alpha}_{\widehat{x}\widehat{x}}(\tau) &= R^{\alpha}_{xx}(\tau) \otimes \frac{1}{\pi\tau} \otimes \left[-\frac{1}{\pi\tau}\,e^{j2\pi\alpha\tau}\right] \\ &= R^{\alpha}_{xx}(\tau) \otimes \left[\delta(\tau) - 2|\alpha|\,\mathrm{sinc}(|\alpha|\tau)\,e^{j\pi\alpha\tau}\right] \end{aligned} \tag{3.76}$$

$$S^{\alpha}_{\widehat{x}x}(f) = S^{\alpha}_{xx}(f)\,\big[-j\,\mathrm{sign}(f)\big] \tag{3.77}$$

$$\stackrel{\mathcal{F}}{\longleftrightarrow}\; R^{\alpha}_{\widehat{x}x}(\tau) = R^{\alpha}_{xx}(\tau) \otimes \frac{1}{\pi\tau} \triangleq \widehat{R}^{\alpha}_{xx}(\tau) \tag{3.78}$$

$$S^{\alpha}_{x\widehat{x}}(f) = S^{\alpha}_{xx}(f)\,\big[-j\,\mathrm{sign}(\alpha-f)\big] \tag{3.79}$$

$$\stackrel{\mathcal{F}}{\longleftrightarrow}\; R^{\alpha}_{x\widehat{x}}(\tau) = R^{\alpha}_{xx}(\tau) \otimes \left[-\frac{1}{\pi\tau}\,e^{j2\pi\alpha\tau}\right] \tag{3.80}$$

In the special case $\alpha = 0$, under Assumption 3.14 it results

$$S_{\widehat{x}\widehat{x}}(f) = S_{xx}(f)\,\big[-j\,\mathrm{sign}(f)\big]\,\big[-j\,\mathrm{sign}(-f)\big] = S_{xx}(f) \tag{3.81}$$

$$\stackrel{\mathcal{F}}{\longleftrightarrow} \quad R_{\widehat{x}\widehat{x}}(\tau) = R_{xx}(\tau) \tag{3.82}$$

$$S_{\widehat{x}x}(f) = S_{xx}(f)\left[-j\,\operatorname{sign}(f)\right] \tag{3.83}$$

$$\stackrel{\mathcal{F}}{\longleftrightarrow} \quad R_{\widehat{x}x}(\tau) = R_{xx}(\tau) \otimes \frac{1}{\pi\tau} \triangleq \widehat{R}_{xx}(\tau) \tag{3.84}$$

$$S_{x\widehat{x}}(f) = S_{xx}(f)\left[-j\,\operatorname{sign}(-f)\right] = j\,\operatorname{sign}(f)\, S_{xx}(f) \tag{3.85}$$

$$\stackrel{\mathcal{F}}{\longleftrightarrow} \quad R_{x\widehat{x}}(\tau) = -\, R_{xx}(\tau) \otimes \frac{1}{\pi\tau} \triangleq -\widehat{R}_{xx}(\tau) \tag{3.86}$$

Analytic signal

Let us consider the LTI system with impulse response function

$$h_{\mathrm{A}}(t) \triangleq \mathring{\delta}(t) = \frac{1}{k}\left[\delta(t) + j\frac{1}{\pi t}\right] \stackrel{\mathcal{F}}{\longleftrightarrow} H_{\mathrm{A}}(f) = \frac{2}{k}\,\mathrm{u}(f) \tag{3.87}$$

where k is a normalizing factor which is chosen differently from different authors ($k = \sqrt{2}$ in (Van Trees, 1971b), $k = 1$ in (Papoulis, 1983), (Gardner, 1987c), (Gardner, 1987f, Chap. 11, Sec. D-2), (Papoulis, 1991), and $k = 2$ in (Izzo and Napolitano, 1997)). The analytic signal of a signal $x(t)$ is defined by the LTI transformation (Section D.1)

$$\mathring{x}(t) \triangleq x(t) \otimes h_{\mathrm{A}}(t) = \frac{1}{k}\left[x(t) + j\widehat{x}(t)\right]. \tag{3.88}$$

Under Assumption 3.15, and using (3.22) and (3.23) and $\mathring{\delta}^*(-t) = \mathring{\delta}(t)$, we have

$$R^{\alpha}_{\mathring{x}\mathring{x}^*}(\tau) \triangleq \left\langle \mathrm{E}\left\{\mathring{x}(t+\tau)\,\mathring{x}^*(t)\right\}\, e^{-j2\pi\alpha t}\right\rangle_t$$

$$= R^{\alpha}_{xx}(\tau) \otimes \mathring{\delta}(\tau) \otimes \left[\mathring{\delta}(\tau)\, e^{j2\pi\alpha\tau}\right] \tag{3.89}$$

$$\stackrel{\mathcal{F}}{\longleftrightarrow} \quad S^{\alpha}_{\mathring{x}\mathring{x}^*}(f) = S^{\alpha}_{xx}(f)\,\frac{4}{k^2}\,\mathrm{u}(f)\,\mathrm{u}(f-\alpha) \tag{3.90}$$

$$= S^{\alpha}_{xx}(f)\,\frac{4}{k^2}\begin{cases}\mathrm{u}(f-\alpha) & \alpha \geqslant 0\\ \mathrm{u}(f) & \alpha < 0\end{cases} \tag{3.91}$$

$$\stackrel{\mathcal{F}}{\longleftrightarrow} \quad R^{\alpha}_{\mathring{x}\mathring{x}^*}(\tau) = \frac{2}{k}\, R^{\alpha}_{xx}(\tau) \otimes \begin{cases}\mathring{\delta}(\tau)\, e^{j2\pi\alpha\tau} & \alpha \geqslant 0\\ \mathring{\delta}(\tau) & \alpha < 0\end{cases} \tag{3.92}$$

$$R^{\alpha}_{\mathring{x}\mathring{x}}(\tau) \triangleq \left\langle \mathrm{E}\left\{\mathring{x}(t+\tau)\,\mathring{x}(t)\right\}\, e^{-j2\pi\alpha t}\right\rangle_t$$

$$= R^{\alpha}_{xx}(\tau) \otimes \mathring{\delta}(\tau) \otimes \left[\mathring{\delta}(-\tau)\, e^{j2\pi\alpha\tau}\right] \tag{3.93}$$

$$\stackrel{\mathcal{F}}{\longleftrightarrow} \quad S^{\alpha}_{\mathring{x}\mathring{x}}(f) = S^{\alpha}_{xx}(f)\,\frac{4}{k^2}\,\mathrm{u}(f)\,\mathrm{u}(\alpha-f) \tag{3.94}$$

$$= S^{\alpha}_{xx}(f)\,\frac{4}{k^2}\begin{cases}\operatorname{rect}\left(\frac{f-\alpha/2}{\alpha}\right) & \alpha \geqslant 0\\ 0 & \alpha < 0\end{cases} \tag{3.95}$$

$$\stackrel{\mathcal{F}}{\longleftrightarrow}\; R^{\alpha}_{\overset{\circ}{x}\overset{\circ}{x}}(\tau) = \frac{4}{k^2}\,R^{\alpha}_{xx}(\tau)\otimes\begin{cases}\alpha\,\operatorname{sinc}(\alpha\tau)\,e^{j\pi\alpha\tau} & \alpha \geqslant 0\\ 0 & \alpha < 0\end{cases} \tag{3.96}$$

Using the symmetry relationships (Section 1.5.1) and the fact that $x(t)$ is real, we have

$$\begin{aligned} S^{\alpha}_{\overset{\circ}{x}^*\overset{\circ}{x}}(f) &= S^{\alpha}_{\overset{\circ}{x}\overset{\circ}{x}^*}(\alpha-f)\\ &= S^{\alpha}_{xx}(\alpha-f)\,\frac{4}{k^2}\,\mathrm{u}(\alpha-f)\,\mathrm{u}(-f)\\ &= S^{\alpha}_{xx}(f)\,\frac{4}{k^2}\,\mathrm{u}(\alpha-f)\,\mathrm{u}(-f) \end{aligned} \tag{3.97}$$

$$\begin{aligned} S^{\alpha}_{\overset{\circ}{x}^*\overset{\circ}{x}^*}(f) &= S^{-\alpha}_{\overset{\circ}{x}\overset{\circ}{x}}(-f)^*\\ &= S^{-\alpha}_{xx}(-f)^*\,\frac{4}{k^2}\,\mathrm{u}(-f)\,\mathrm{u}(-\alpha+f)\\ &= S^{\alpha}_{xx}(f)\,\frac{4}{k^2}\,\mathrm{u}(-f)\,\mathrm{u}(-\alpha+f) \end{aligned} \tag{3.98}$$

From (3.88) we have

$$x(t) = k\,\mathrm{Re}\left[\overset{\circ}{x}(t)\right] = \frac{k}{2}\,\left[\overset{\circ}{x}(t) + \overset{\circ}{x}^*(t)\right]. \tag{3.99}$$

Thus,

$$R^{\alpha}_{xx}(\tau) = \frac{k^2}{4}\left[R^{\alpha}_{\overset{\circ}{x}\overset{\circ}{x}^*}(\tau) + R^{\alpha}_{\overset{\circ}{x}^*\overset{\circ}{x}}(\tau) + R^{\alpha}_{\overset{\circ}{x}\overset{\circ}{x}}(\tau) + R^{\alpha}_{\overset{\circ}{x}^*\overset{\circ}{x}^*}(\tau)\right] \tag{3.100}$$

$$\stackrel{\mathcal{F}}{\longleftrightarrow}\; S^{\alpha}_{xx}(f) = \frac{k^2}{4}\left[S^{\alpha}_{\overset{\circ}{x}\overset{\circ}{x}^*}(f) + S^{\alpha}_{\overset{\circ}{x}^*\overset{\circ}{x}}(f) + S^{\alpha}_{\overset{\circ}{x}\overset{\circ}{x}}(f) + S^{\alpha}_{\overset{\circ}{x}^*\overset{\circ}{x}^*}(f)\right] \tag{3.101}$$

In (3.101), the cyclic spectra in the sum have non-overlapping supports in the (α, f) plane (Gardner, 1986c), (Gardner, 1987f, Chap. 11, Sec. D-2), (Izzo and Napolitano, 1997). In fact, $S^{\alpha}_{\overset{\circ}{x}\overset{\circ}{x}^*}(f)$ is nonzero for $f \geqslant 0$ and $f \geqslant \alpha$ (see (3.90)), $S^{\alpha}_{\overset{\circ}{x}\overset{\circ}{x}}(f)$ is nonzero for $f \geqslant 0$ and $f \leqslant \alpha$ (see (3.94)), $S^{\alpha}_{\overset{\circ}{x}^*\overset{\circ}{x}}(f)$ is nonzero for $f \leqslant 0$ and $f \leqslant \alpha$ (see (3.97)), and $S^{\alpha}_{\overset{\circ}{x}^*\overset{\circ}{x}^*}(f)$ is nonzero for $f \leqslant 0$ and $f \geqslant \alpha$ (see (3.98)).

In the special case $\alpha = 0$, under Assumption 3.14 it results

$$R_{\overset{\circ}{x}\overset{\circ}{x}^*}(\tau) \triangleq \left\langle \mathrm{E}\left\{\overset{\circ}{x}(t+\tau)\,\overset{\circ}{x}^*(t)\right\}\right\rangle_t \tag{3.102}$$

$$\stackrel{\mathcal{F}}{\longleftrightarrow}\; S_{\overset{\circ}{x}\overset{\circ}{x}^*}(f) = S_{xx}(f)\,\frac{4}{k^2}\,\mathrm{u}(f)\,\mathrm{u}(-(-f)) = S_{xx}(f)\,\frac{4}{k^2}\,\mathrm{u}(f) \tag{3.103}$$

$$\stackrel{\mathcal{F}}{\longleftrightarrow}\; R_{\overset{\circ}{x}\overset{\circ}{x}^*}(\tau) = \frac{2}{k}\,R_{xx}(\tau)\otimes\frac{1}{k}\left[\delta(t) + j\,\frac{1}{\pi t}\right] = \frac{2}{k}\,\overset{\circ}{R}_{xx}(\tau) \tag{3.104}$$

$$R_{\overset{\circ}{x}\overset{\circ}{x}}(\tau) \triangleq \left\langle \mathrm{E}\left\{\overset{\circ}{x}(t+\tau)\,\overset{\circ}{x}(t)\right\}\right\rangle_t \tag{3.105}$$

$$\overset{\mathcal{F}}{\longleftrightarrow} \quad S_{\overset{\circ}{x}\overset{\circ}{x}}(f) = S_{xx}(f)\,\frac{4}{k^2}\,\mathrm{u}(f)\,\mathrm{u}(-f) = 0 \tag{3.106}$$

$$\overset{\mathcal{F}}{\longleftrightarrow} \quad R_{\overset{\circ}{x}\overset{\circ}{x}}(\tau) = 0 \tag{3.107}$$

$$S_{\overset{\circ}{x}^*\overset{\circ}{x}}(f) = S_{xx}(f)\,\frac{4}{k^2}\,\mathrm{u}(-f) \tag{3.108}$$

$$S_{\overset{\circ}{x}^*\overset{\circ}{x}^*}(f) = 0 \tag{3.109}$$

Complex envelope

The complex envelope of a signal $x(t)$ is defined as (Section D.1)

$$\widetilde{x}(t) \triangleq \overset{\circ}{x}(t)\,e^{-j2\pi f_0 t}\,. \tag{3.110}$$

Under Assumption 3.15 we have

$$\begin{aligned}
R^{\alpha}_{\widetilde{x}\widetilde{x}^*}(\tau) &\triangleq \left\langle \mathrm{E}\left\{\widetilde{x}(t+\tau)\,\widetilde{x}^*(t)\right\}\,e^{-j2\pi\alpha t}\right\rangle_t \\
&= \left\langle \mathrm{E}\left\{\overset{\circ}{x}(t+\tau)\,\overset{\circ}{x}^*(t)\right\}\,e^{-j2\pi\alpha t}\right\rangle_t\,e^{-j2\pi f_0\tau} \\
&= R^{\alpha}_{\overset{\circ}{x}\overset{\circ}{x}^*}(\tau)\,e^{-j2\pi f_0\tau}
\end{aligned} \tag{3.111}$$

$$\begin{aligned}
\overset{\mathcal{F}}{\longleftrightarrow} \quad S^{\alpha}_{\widetilde{x}\widetilde{x}^*}(f) &= S^{\alpha}_{\overset{\circ}{x}\overset{\circ}{x}^*}(f+f_0) \\
&= S^{\alpha}_{xx}(f+f_0)\,\frac{4}{k^2}\,\mathrm{u}(f+f_0)\,\mathrm{u}(f+f_0-\alpha) \qquad (3.112)\\
&= S^{\alpha}_{xx}(f+f_0)\,\frac{4}{k^2}\begin{cases}\mathrm{u}(f+f_0-\alpha) & \alpha \geqslant 0\\ \mathrm{u}(f+f_0) & \alpha < 0\end{cases} \qquad (3.113)
\end{aligned}$$

$$\begin{aligned}
R^{\alpha}_{\widetilde{x}\widetilde{x}}(\tau) &\triangleq \left\langle \mathrm{E}\{\widetilde{x}(t+\tau)\,\widetilde{x}(t)\}\,e^{-j2\pi\alpha t}\right\rangle_t \\
&= \left\langle \mathrm{E}\left\{\overset{\circ}{x}(t+\tau)\,\overset{\circ}{x}(t)\right\}\,e^{-j2\pi[\alpha+2f_0]t}\right\rangle_t\,e^{-j2\pi f_0\tau} \\
&= R^{\alpha+2f_0}_{\overset{\circ}{x}\overset{\circ}{x}}(\tau)\,e^{-j2\pi f_0\tau}
\end{aligned} \tag{3.114}$$

$$\begin{aligned}
\overset{\mathcal{F}}{\longleftrightarrow} \quad S^{\alpha}_{\widetilde{x}\widetilde{x}}(f) &= S^{\alpha+2f_0}_{\overset{\circ}{x}\overset{\circ}{x}}(f+f_0) \\
&= S^{\alpha+2f_0}_{xx}(f+f_0)\,\frac{4}{k^2}\,\mathrm{u}(f+f_0)\,\mathrm{u}(\alpha+2f_0-(f+f_0)) \qquad (3.115)\\
&= S^{\alpha+2f_0}_{xx}(f+f_0)\,\frac{4}{k^2}\begin{cases}\mathrm{rect}\left(\frac{f-\alpha/2}{\alpha+2f_0}\right) & \alpha \geqslant -2f_0\\ 0 & \alpha < -2f_0\end{cases} \qquad (3.116)
\end{aligned}$$

Using the symmetry relationships (Section 1.5.1) and the fact that $x(t)$ is real, we have

$$S^{\alpha}_{\widetilde{x}^*\widetilde{x}}(f) = S^{\alpha}_{\widetilde{x}\widetilde{x}^*}(\alpha - f)$$

$$
\begin{aligned}
&= S^{\alpha}_{xx}(\alpha - f + f_0)\,\frac{4}{k^2}\,\mathrm{u}(\alpha - f + f_0)\,\mathrm{u}(\alpha - f + f_0 - \alpha)\\
&= S^{\alpha}_{xx}(f - f_0)\,\frac{4}{k^2}\,\mathrm{u}(\alpha - f + f_0)\,\mathrm{u}(-f + f_0)
\end{aligned} \tag{3.117}
$$

$$
\begin{aligned}
S^{\alpha}_{\widetilde{x}^*\widetilde{x}^*}(f) &= S^{-\alpha}_{\widetilde{x}\widetilde{x}}(-f)^*\\
&= S^{-\alpha+2f_0}_{xx}(-f + f_0)^*\,\frac{4}{k^2}\,\mathrm{u}(-f + f_0)\,\mathrm{u}(-\alpha + f_0 + f)\\
&= S^{\alpha-2f_0}_{xx}(f - f_0)\,\frac{4}{k^2}\,\mathrm{u}(-f + f_0)\,\mathrm{u}(-\alpha + f_0 + f)
\end{aligned} \tag{3.118}
$$

From (3.88) and (3.110) we have

$$
x(t) = k\,\mathrm{Re}\left[\widetilde{x}(t)\,e^{j2\pi f_0 t}\right] = \frac{k}{2}\left[\widetilde{x}(t)\,e^{j2\pi f_0 t} + \widetilde{x}^*(t)\,e^{-j2\pi f_0 t}\right]. \tag{3.119}
$$

Thus (see Example 3.11),

$$
\begin{aligned}
R^{\alpha}_{xx}(\tau) = \frac{k^2}{4}\Big[&R^{\alpha}_{\widetilde{x}\,\widetilde{x}^*}(\tau)\,e^{j2\pi f_0 \tau} + R^{\alpha}_{\widetilde{x}^*\,\widetilde{x}}(\tau)\,e^{-j2\pi f_0 \tau}\\
&+ R^{\alpha-2f_0}_{\widetilde{x}\,\widetilde{x}}(\tau)\,e^{j2\pi f_0 \tau} + R^{\alpha+2f_0}_{\widetilde{x}^*\,\widetilde{x}^*}(\tau)\,e^{-j2\pi f_0 \tau}\Big]
\end{aligned} \tag{3.120}
$$

$$
\begin{aligned}
\stackrel{\mathcal{F}}{\longleftrightarrow}\quad S^{\alpha}_{xx}(f) = \frac{k^2}{4}\Big[&S^{\alpha}_{\widetilde{x}\,\widetilde{x}^*}(f - f_0) + S^{\alpha}_{\widetilde{x}^*\,\widetilde{x}}(f + f_0)\\
&+ S^{\alpha-2f_0}_{\widetilde{x}\,\widetilde{x}}(f - f_0) + S^{\alpha+2f_0}_{\widetilde{x}^*\,\widetilde{x}^*}(f + f_0)\Big]
\end{aligned} \tag{3.121}
$$

In the special case $\alpha = 0$, under Assumption 3.15 it results

$$
\begin{aligned}
R_{\widetilde{x}\widetilde{x}^*}(\tau) &\triangleq \left\langle \mathrm{E}\left\{\widetilde{x}(t+\tau)\widetilde{x}^*(t)\right\}\right\rangle_t\\
&= \left\langle \mathrm{E}\left\{\mathring{x}(t+\tau)\,\mathring{x}^*(t)\right\}\right\rangle_t\,e^{-j2\pi f_0 \tau}\\
&= R_{\mathring{x}\mathring{x}^*}(\tau)\,e^{-j2\pi f_0 \tau}
\end{aligned} \tag{3.122}
$$

$$
\begin{aligned}
\stackrel{\mathcal{F}}{\longleftrightarrow}\quad S_{\widetilde{x}\widetilde{x}^*}(f) &= S_{\mathring{x}\,\mathring{x}^*}(f + f_0)\\
&= S_{xx}(f + f_0)\,\frac{4}{k^2}\,\mathrm{u}(f + f_0)
\end{aligned} \tag{3.123}
$$

$$
\begin{aligned}
R_{\widetilde{x}\widetilde{x}}(\tau) &\triangleq \langle \mathrm{E}\{\widetilde{x}(t+\tau)\widetilde{x}(t)\}\rangle_t\\
&= \left\langle \mathrm{E}\left\{\mathring{x}(t+\tau)\,\mathring{x}(t)\right\}\,e^{-j2\pi 2 f_0 t}\right\rangle_t\,e^{-j2\pi f_0 \tau}\\
&= R^{2f_0}_{\mathring{x}\mathring{x}}(\tau)\,e^{-j2\pi f_0 \tau}
\end{aligned} \tag{3.124}
$$

$$
\begin{aligned}
\stackrel{\mathcal{F}}{\longleftrightarrow}\quad S_{\widetilde{x}\widetilde{x}}(f) &= S^{2f_0}_{\mathring{x}\,\mathring{x}}(f + f_0)\\
&= S^{2f_0}_{xx}(f + f_0)\,\frac{4}{k^2}\,\mathrm{u}(f + f_0)\,\mathrm{u}(f_0 - f)
\end{aligned}
$$

$$= S_{xx}^{2f_0}(f+f_0)\,\frac{4}{k^2}\,\text{rect}\left(\frac{f}{2f_0}\right) \tag{3.125}$$

$$S_{\widetilde{x}^*\widetilde{x}}(f) = S_{xx}(f-f_0)\,\frac{4}{k^2}\,\text{u}(-f+f_0) \tag{3.126}$$

$$S_{\widetilde{x}^*\widetilde{x}^*}(f) = S_{xx}^{-2f_0}(f-f_0)\,\frac{4}{k^2}\,\text{rect}\left(\frac{f}{2f_0}\right) \tag{3.127}$$

Supports of cyclic spectra of band pass signals

Definition 3.16. The signal $x(t)$ is bandpass with central frequency f_0 and strictly band-limited with monolateral bandwidth W, with $f_0 > W$, if and only if it passes unaltered through the ideal bandpass filter with harmonic response

$$H_{\text{BPF}}(f) \triangleq \text{rect}\Big(\frac{f-f_0}{2W}\Big) + \text{rect}\Big(\frac{f+f_0}{2W}\Big) = \begin{cases} 1 & ||f|-f_0| \leqslant W \\ 0 & \text{otherwise.} \end{cases} \tag{3.128}$$

That is,

$$x(t) = x(t) \otimes h_{\text{BPF}}(t) \tag{3.129}$$

where the impulse-response function $h_{\text{BPF}}(t)$ is the inverse Fourier transform of $H_{\text{BPF}}(f)$. □

If $x(t)$ is ACS, then accounting for (3.128) and (3.129), and the input/output relation in terms of cyclic spectra for linear time-invariant (LTI) filters (3.22), the support in the (α, f)-plane of its cyclic spectra is given by

$$\begin{aligned} \text{supp}\left\{S_{xx}^{\alpha}(f)\right\} &\subseteq \{(\alpha,f)\in\mathbb{R}\times\mathbb{R} : H_{\text{BPF}}(f)\,H_{\text{BPF}}(\alpha-f)\neq 0\} \\ &= \{(\alpha,f)\in\mathbb{R}\times\mathbb{R} : ||f|-f_0|\leqslant W,\ ||\alpha-f|-f_0|\leqslant W\}\,. \end{aligned} \tag{3.130}$$

Remember that f and α variables are on the abscissa and ordinate axis, respectively of the (α, f)-plane.

From (3.112) and (3.130) it follows that

$$\begin{aligned} \text{supp}\left\{S_{\widetilde{x}\widetilde{x}^*}^{\alpha}(f)\right\} &= \text{supp}\left\{S_{xx}^{\alpha}(f+f_0)\right\} \cap \{(\alpha,f)\in\mathbb{R}\times\mathbb{R} : f\geqslant -f_0,\ \alpha\leqslant f+f_0\} \\ &\subseteq \mathcal{D}_1(\alpha,f) \end{aligned} \tag{3.131}$$

where $\mathcal{D}_1(\alpha, f)$ is the domain defined as (Fig. 3.3, Eq. (3.65a) with $B = W$)

$$\mathcal{D}_1(\alpha,f) \triangleq \{(\alpha,f)\in\mathbb{R}\times\mathbb{R} : |f|\leqslant W,\ |\alpha-f|\leqslant W\} \tag{3.132}$$

which is obtained when the asymmetric definition for the cyclic spectrum (1.30) is used.

Note that a diamond-shaped domain is obtained for $\mathcal{D}_1(\alpha, f)$ (Gardner, 1991a, Fig. 2b) if the symmetric definition for the cyclic spectrum (1.35) (Gardner, 1991a, Eqs. (44)–(48)) is adopted.

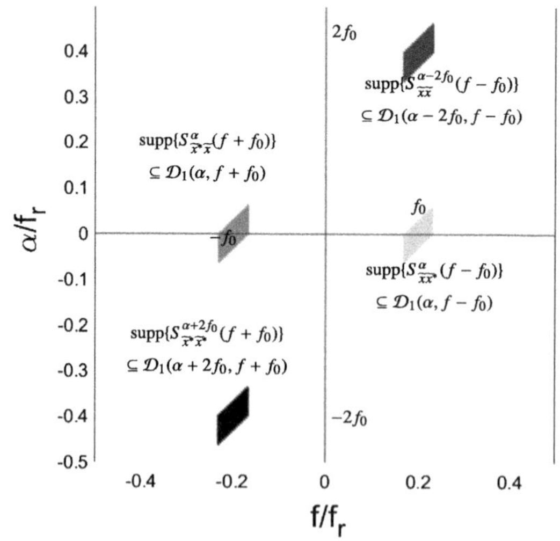

FIGURE 3.6 Set $\mathcal{D}_2(\alpha, f)$ defined in (3.137) containing the support of the cyclic spectrum of a bandpass ACS signal. *Source: (Napolitano and Doğançay, 2018) Copyright of Elsevier.*

Analogously, it can be shown that

$$\operatorname{supp}\left\{S^{\alpha}_{\widetilde{x}^*\widetilde{x}}(f)\right\} \subseteq \mathcal{D}_1(\alpha, f) \tag{3.133}$$

$$\operatorname{supp}\left\{S^{\alpha}_{\widetilde{x}\widetilde{x}}(f)\right\} \subseteq \mathcal{D}_1(\alpha, f) \tag{3.134}$$

$$\operatorname{supp}\left\{S^{\alpha}_{\widetilde{x}^*\widetilde{x}^*}(f)\right\} \subseteq \mathcal{D}_1(\alpha, f)\,. \tag{3.135}$$

Therefore, from (3.121) and (3.131)–(3.135) we have (Fig. 3.6)

$$\operatorname{supp}\left\{S^{\alpha}_{xx}(f)\right\} \subseteq \mathcal{D}_2(\alpha, f) \tag{3.136}$$

where

$$\begin{aligned}\mathcal{D}_2(\alpha, f) \triangleq\ & \mathcal{D}_1(\alpha, f - f_0) \cup \mathcal{D}_1(\alpha, f + f_0)\\ & \cup \mathcal{D}_1(\alpha - 2f_0, f - f_0) \cup \mathcal{D}_1(\alpha + 2f_0, f + f_0)\,.\end{aligned} \tag{3.137}$$

In Fig. 3.6, the set $\mathcal{D}_2(\alpha, f)$ defined in (3.137) is represented with axes α and f normalized to a reference frequency f_r. It is coincident with the set represented in Fig. 3.5 (Eq. (3.72) with $b = f_0 - W$ and $B = f_0 + W$).

In (3.121), the cyclic spectrum of $x(t)$ is expressed as the superposition of the contributions of cyclic spectra of the complex envelope $\widetilde{x}(t)$ with different conjugation choices. In Fig. 3.6 the location of such contributions in the (α, f) plane is evidenced.

In-phase and in-quadrature components

The in-phase and in-quadrature components are defined as (Section D.1)

$$x_{\mathrm{c}}(t) \triangleq k\,\mathrm{Re}[\widetilde{x}(t)] = \frac{k}{2}\left[\widetilde{x}(t) + \widetilde{x}^*(t)\right] \tag{3.138}$$

$$x_{\mathrm{s}}(t) \triangleq k\,\mathrm{Im}[\widetilde{x}(t)] = -j\frac{k}{2}\left[\widetilde{x}(t) - \widetilde{x}^*(t)\right] \tag{3.139}$$

respectively.

Under Assumption 3.15 we have

$$\begin{aligned}
R^{\alpha}_{x_{\mathrm{c}}x_{\mathrm{c}}}(\tau) &\triangleq \left\langle \mathrm{E}\{x_{\mathrm{c}}(t+\tau)\,x_{\mathrm{c}}(t)\}\; e^{-j2\pi\alpha t}\right\rangle_t \\
&= \frac{k^2}{4}\left\langle \mathrm{E}\left\{\left[\widetilde{x}(t+\tau) + \widetilde{x}^*(t+\tau)\right]\left[\widetilde{x}(t) + \widetilde{x}^*(t)\right]\right\}\; e^{-j2\pi\alpha t}\right\rangle_t \\
&= \frac{k^2}{4}\left[R^{\alpha}_{\widetilde{x}\widetilde{x}^*}(\tau) + R^{\alpha}_{\widetilde{x}^*\widetilde{x}}(\tau) + R^{\alpha}_{\widetilde{x}\widetilde{x}}(\tau) + R^{\alpha}_{\widetilde{x}^*\widetilde{x}^*}(\tau)\right]
\end{aligned} \tag{3.140}$$

$$\overset{\mathcal{F}}{\longleftrightarrow}\quad S^{\alpha}_{x_{\mathrm{c}}x_{\mathrm{c}}}(f) = \frac{k^2}{4}\left[S^{\alpha}_{\widetilde{x}\widetilde{x}^*}(f) + S^{\alpha}_{\widetilde{x}^*\widetilde{x}}(f) + S^{\alpha}_{\widetilde{x}\widetilde{x}}(f) + S^{\alpha}_{\widetilde{x}^*\widetilde{x}^*}(f)\right] \tag{3.141}$$

$$\begin{aligned}
R^{\alpha}_{x_{\mathrm{s}}x_{\mathrm{s}}}(\tau) &\triangleq \left\langle \mathrm{E}\{x_{\mathrm{s}}(t+\tau)\,x_{\mathrm{s}}(t)\}\; e^{-j2\pi\alpha t}\right\rangle_t \\
&= -\frac{k^2}{4}\left\langle \mathrm{E}\left\{\left[\widetilde{x}(t+\tau) - \widetilde{x}^*(t+\tau)\right]\left[\widetilde{x}(t) - \widetilde{x}^*(t)\right]\right\}\; e^{-j2\pi\alpha t}\right\rangle_t \\
&= \frac{k^2}{4}\left[R^{\alpha}_{\widetilde{x}\widetilde{x}^*}(\tau) + R^{\alpha}_{\widetilde{x}^*\widetilde{x}}(\tau) - R^{\alpha}_{\widetilde{x}\widetilde{x}}(\tau) - R^{\alpha}_{\widetilde{x}^*\widetilde{x}^*}(\tau)\right]
\end{aligned} \tag{3.142}$$

$$\overset{\mathcal{F}}{\longleftrightarrow}\quad S^{\alpha}_{x_{\mathrm{s}}x_{\mathrm{s}}}(f) = \frac{k^2}{4}\left[S^{\alpha}_{\widetilde{x}\widetilde{x}^*}(f) + S^{\alpha}_{\widetilde{x}^*\widetilde{x}}(f) - S^{\alpha}_{\widetilde{x}\widetilde{x}}(f) - S^{\alpha}_{\widetilde{x}^*\widetilde{x}^*}(f)\right] \tag{3.143}$$

$$\begin{aligned}
R^{\alpha}_{x_{\mathrm{c}}x_{\mathrm{s}}}(\tau) &\triangleq \left\langle \mathrm{E}\{x_{\mathrm{c}}(t+\tau)\,x_{\mathrm{s}}(t)\}\; e^{-j2\pi\alpha t}\right\rangle_t \\
&= -j\,\frac{k^2}{4}\left\langle \mathrm{E}\left\{\left[\widetilde{x}(t+\tau) + \widetilde{x}^*(t+\tau)\right]\left[\widetilde{x}(t) - \widetilde{x}^*(t)\right]\right\}\; e^{-j2\pi\alpha t}\right\rangle_t \\
&= j\,\frac{k^2}{4}\left[R^{\alpha}_{\widetilde{x}\widetilde{x}^*}(\tau) - R^{\alpha}_{\widetilde{x}^*\widetilde{x}}(\tau) - R^{\alpha}_{\widetilde{x}\widetilde{x}}(\tau) + R^{\alpha}_{\widetilde{x}^*\widetilde{x}^*}(\tau)\right]
\end{aligned} \tag{3.144}$$

$$\overset{\mathcal{F}}{\longleftrightarrow}\quad S^{\alpha}_{x_{\mathrm{c}}x_{\mathrm{s}}}(f) = j\,\frac{k^2}{4}\left[S^{\alpha}_{\widetilde{x}\widetilde{x}^*}(f) - S^{\alpha}_{\widetilde{x}^*\widetilde{x}}(f) - S^{\alpha}_{\widetilde{x}\widetilde{x}}(f) + S^{\alpha}_{\widetilde{x}^*\widetilde{x}^*}(f)\right] \tag{3.145}$$

$$\begin{aligned}
R^{\alpha}_{x_{\mathrm{s}}x_{\mathrm{c}}}(\tau) &\triangleq \left\langle \mathrm{E}\{x_{\mathrm{s}}(t+\tau)\,x_{\mathrm{c}}(t)\}\; e^{-j2\pi\alpha t}\right\rangle_t \\
&= -j\,\frac{k^2}{4}\left\langle \mathrm{E}\left\{\left[\widetilde{x}(t+\tau) - \widetilde{x}^*(t+\tau)\right]\left[\widetilde{x}(t) + \widetilde{x}^*(t)\right]\right\}\; e^{-j2\pi\alpha t}\right\rangle_t \\
&= -j\,\frac{k^2}{4}\left[R^{\alpha}_{\widetilde{x}\widetilde{x}^*}(\tau) - R^{\alpha}_{\widetilde{x}^*\widetilde{x}}(\tau) + R^{\alpha}_{\widetilde{x}\widetilde{x}}(\tau) - R^{\alpha}_{\widetilde{x}^*\widetilde{x}^*}(\tau)\right]
\end{aligned} \tag{3.146}$$

$$\overset{\mathcal{F}}{\longleftrightarrow}\quad S^{\alpha}_{x_{\mathrm{s}}x_{\mathrm{c}}}(f) = -j\,\frac{k^2}{4}\left[S^{\alpha}_{\widetilde{x}\widetilde{x}^*}(f) - S^{\alpha}_{\widetilde{x}^*\widetilde{x}}(f) + S^{\alpha}_{\widetilde{x}\widetilde{x}}(f) - S^{\alpha}_{\widetilde{x}^*\widetilde{x}^*}(f)\right] \tag{3.147}$$

$$R^{\alpha}_{\tilde{x}\,\tilde{x}^{(*)}}(\tau) = \frac{1}{k^2}\Big\langle \mathrm{E}\{(x_c(t+\tau) + j x_s(t+\tau))\,(x_c(t) + {\scriptstyle(-)} j x_s(t))\}\, e^{-j2\pi\alpha t}\Big\rangle_t$$
$$= \frac{1}{k^2}\Big[R^{\alpha}_{x_c x_c}(\tau) + j R^{\alpha}_{x_s x_c}(\tau) + {\scriptstyle(-)} j R^{\alpha}_{x_c x_s}(\tau) - {\scriptstyle(-)} R^{\alpha}_{x_s x_s}(\tau)\Big] \tag{3.148}$$
$$\overset{\mathcal{F}}{\longleftrightarrow}\; S^{\alpha}_{\tilde{x}\,\tilde{x}^{(*)}}(f) = \frac{1}{k^2}\Big[S^{\alpha}_{x_c x_c}(f) + j S^{\alpha}_{x_s x_c}(f) + {\scriptstyle(-)} j S^{\alpha}_{x_c x_s}(f) - {\scriptstyle(-)} S^{\alpha}_{x_s x_s}(f)\Big] \tag{3.149}$$

In the special case $\alpha = 0$, under Assumption 3.15 and accounting for (D.17) and (D.18), it results

$$S_{x_c x_c}(f) = S_{xx}(f+f_0)\,\mathrm{u}(f+f_0) + S_{xx}(f-f_0)\,\mathrm{u}(-f+f_0)$$
$$+ S^{2f_0}_{xx}(f+f_0)\,\mathrm{rect}\Big(\frac{f}{2f_0}\Big) + S^{-2f_0}_{xx}(f-f_0)\,\mathrm{rect}\Big(\frac{f}{2f_0}\Big) \tag{3.150}$$
$$\overset{\mathcal{F}}{\longleftrightarrow}\; R_{x_c x_c}(\tau) = \big[R_{xx}(\tau)\big]_{\mathrm{c}} + \Big[R^{2f_0}_{xx}(\tau)\, e^{-j2\pi f_0\tau} + R^{-2f_0}_{xx}(\tau)\, e^{j2\pi f_0\tau}\Big]$$
$$\otimes\, 2f_0\,\mathrm{sinc}(2f_0\tau) \tag{3.151}$$

$$S_{x_s x_s}(f) = S_{xx}(f+f_0)\,\mathrm{u}(f+f_0) + S_{xx}(f-f_0)\,\mathrm{u}(-f+f_0)$$
$$- S^{2f_0}_{xx}(f+f_0)\,\mathrm{rect}\Big(\frac{f}{2f_0}\Big) - S^{-2f_0}_{xx}(f-f_0)\,\mathrm{rect}\Big(\frac{f}{2f_0}\Big) \tag{3.152}$$
$$\overset{\mathcal{F}}{\longleftrightarrow}\; R_{x_s x_s}(\tau) = \big[R_{xx}(\tau)\big]_{\mathrm{c}} - \Big[R^{2f_0}_{xx}(\tau)\, e^{-j2\pi f_0\tau} + R^{-2f_0}_{xx}(\tau)\, e^{j2\pi f_0\tau}\Big]$$
$$\otimes\, 2f_0\,\mathrm{sinc}(2f_0\tau) \tag{3.153}$$

$$S_{x_c x_s}(f) = j\,\Big[S_{xx}(f+f_0)\,\mathrm{u}(f+f_0) - S_{xx}(f-f_0)\,\mathrm{u}(-f+f_0)$$
$$- S^{2f_0}_{xx}(f+f_0)\,\mathrm{rect}\Big(\frac{f}{2f_0}\Big) + S^{-2f_0}_{xx}(f-f_0)\,\mathrm{rect}\Big(\frac{f}{2f_0}\Big)\Big] \tag{3.154}$$
$$\overset{\mathcal{F}}{\longleftrightarrow}\; R_{x_c x_s}(\tau) = -\big[R_{xx}(\tau)\big]_{\mathrm{s}} - j\,\Big[R^{2f_0}_{xx}(\tau)\, e^{-j2\pi f_0\tau} - R^{-2f_0}_{xx}(\tau)\, e^{j2\pi f_0\tau}\Big]$$
$$\otimes\, 2f_0\,\mathrm{sinc}(2f_0\tau) \tag{3.155}$$

$$S_{x_s x_c}(f) = -\,j\,\Big[S_{xx}(f+f_0)\,\mathrm{u}(f+f_0) - S_{xx}(f-f_0)\,\mathrm{u}(-f+f_0)$$
$$+ S^{2f_0}_{xx}(f+f_0)\,\mathrm{rect}\Big(\frac{f}{2f_0}\Big) - S^{-2f_0}_{xx}(f-f_0)\,\mathrm{rect}\Big(\frac{f}{2f_0}\Big)\Big] \tag{3.156}$$
$$\overset{\mathcal{F}}{\longleftrightarrow}\; R_{x_s x_c}(\tau) = \big[R_{xx}(\tau)\big]_{\mathrm{s}} - j\,\Big[R^{2f_0}_{xx}(\tau)\, e^{-j2\pi f_0\tau} - R^{-2f_0}_{xx}(\tau)\, e^{j2\pi f_0\tau}\Big]$$
$$\otimes\, 2f_0\,\mathrm{sinc}(2f_0\tau) \tag{3.157}$$

where $[\cdot]_{\mathrm{c}}$ and $[\cdot]_{\mathrm{s}}$ denote the operators that provide the in-phase and in-quadrature components of their argument according to (D.15) and (D.16), respectively.

From (3.150)–(3.156) it follows that the power spectra and cross-power spectra ($\alpha = 0$) of in-phase and in-quadrature components of a cyclostationary signal depend not only on the power spectrum of the real signal, but also from cyclic spectra at cycle frequencies $\alpha = \pm 2 f_0$ of the real signal. Thus, neglecting these cyclic spectra leads to wrong results.

Wide-sense stationary signals

Let $x(t)$ be a zero mean wide-sense stationary signal. Under the zero-mean assumption we have

$$\mathrm{E}\{x(t)\} = 0 \quad \Rightarrow \quad \mathrm{E}\{\mathring{x}(t)\} = \mathrm{E}\{\widetilde{x}(t)\} = \mathrm{E}\{x_\mathrm{c}(t)\} = \mathrm{E}\{x_\mathrm{c}(t)\} = 0 \,. \tag{3.158}$$

From the wide-sense stationarity assumption it follows that

$$S_{xx}^{\pm 2 f_0}(f) \equiv 0 \overset{\mathcal{F}}{\longleftrightarrow} R_{xx}^{\pm 2 f_0}(\tau) \equiv 0 \,. \tag{3.159}$$

In addition, under Assumption 3.14, it results that

$$S_{\widetilde{x}\widetilde{x}}(f) = S_{\widetilde{x}^*\widetilde{x}^*}(f) \equiv 0 \overset{\mathcal{F}}{\longleftrightarrow} R_{\widetilde{x}\widetilde{x}}(\tau) = R_{\widetilde{x}^*\widetilde{x}^*}(\tau) \equiv 0 \tag{3.160}$$

$$\begin{aligned} R_{x_\mathrm{c}x_\mathrm{c}}(\tau) = R_{x_\mathrm{s}x_\mathrm{s}}(\tau) &= \big[R_{xx}(\tau)\big]_\mathrm{c} \\ &= R_{xx}(\tau)\, \cos(2\pi f_0 \tau) + \widehat{R}_{xx}(\tau)\, \sin(2\pi f_0 \tau) \end{aligned} \tag{3.161}$$

$$\begin{aligned} \overset{\mathcal{F}}{\longleftrightarrow} S_{x_\mathrm{c}x_\mathrm{c}}(f) = S_{x_\mathrm{s}x_\mathrm{s}}(f) = {} & S_{xx}(f + f_0)\, \mathrm{u}(f + f_0) \\ & + S_{xx}(f - f_0)\, \mathrm{u}(-f + f_0) \end{aligned} \tag{3.162}$$

$$\begin{aligned} R_{x_\mathrm{s}x_\mathrm{c}}(\tau) = -R_{x_\mathrm{c}x_\mathrm{s}}(\tau) &= \big[R_{xx}(\tau)\big]_\mathrm{s} \\ &= -\, R_{xx}(\tau)\, \sin(2\pi f_0 \tau) + \widehat{R}_{xx}(\tau)\, \cos(2\pi f_0 \tau) \end{aligned} \tag{3.163}$$

$$\begin{aligned} \overset{\mathcal{F}}{\longleftrightarrow} S_{x_\mathrm{s}x_\mathrm{c}}(f) = -S_{x_\mathrm{c}x_\mathrm{s}}(f) = {} & -\, j\, \big[S_{xx}(f + f_0)\, \mathrm{u}(f + f_0) \\ & - S_{xx}(f - f_0)\, \mathrm{u}(-f + f_0)\big] \end{aligned} \tag{3.164}$$

$$\widehat{R}_{xx}(-\tau) = \widehat{R}_{xx}(\tau) \tag{3.165}$$

$$R_{x_\mathrm{s}x_\mathrm{c}}(\tau) = R_{x_\mathrm{c}x_\mathrm{s}}(-\tau) \tag{3.166}$$

$$R_{xx}(\tau) = \big[R_{xx}(\tau)\big]_\mathrm{c} \cos(2\pi f_0 \tau) - \big[R_{xx}(\tau)\big]_\mathrm{s} \sin(2\pi f_0 \tau) \tag{3.167}$$

Stationary white noise

In this section, the complex envelope signal

$$\widetilde{w}(t) = \frac{1}{k}\, \big[w(t) + j\widehat{w}(t)\big]\, e^{-j2\pi f_0 t} = \frac{1}{k}\, \big[w_c(t) + j\, w_s(t)\big] \tag{3.168}$$

is analyzed, where $w(t)$ is wide-sense stationary white noise (Kailath, 1966). For related results see also (Viswanathan, 2006).

It results that

$$R_{ww}(\tau) = \mathrm{E}\{w(t+\tau)\, w(t)\} = \frac{N_0}{2}\, \delta(\tau) \tag{3.169}$$

$$\stackrel{\mathcal{F}}{\longleftrightarrow} \quad S_{ww}(f) = \frac{N_0}{2} \tag{3.170}$$

$$R_{\widehat{w}\widehat{w}}(\tau) = R_{ww}(\tau) = \frac{N_0}{2}\,\delta(\tau) \tag{3.171}$$

$$\stackrel{\mathcal{F}}{\longleftrightarrow} \quad S_{\widehat{w}\widehat{w}}(f) = \frac{N_0}{2} \tag{3.172}$$

$$R_{\widehat{w}w}(\tau) = -\,R_{w\widehat{w}}(\tau) = \widehat{R}_{ww}(\tau) = \frac{N_0}{2}\,\delta(\tau) \otimes \frac{1}{\pi\tau} = \frac{N_0}{2}\,\frac{1}{\pi\tau} \tag{3.173}$$

$$\stackrel{\mathcal{F}}{\longleftrightarrow} \quad S_{\widehat{w}w}(f) = -\,S_{w\widehat{w}}(f) = -j\,\mathrm{sign}(f)\,\frac{N_0}{2} \tag{3.174}$$

$$R_{w_c w_c}(\tau) = \left[R_{ww}(\tau)\right]_c = R_{ww}(\tau)\,\cos(2\pi f_0\tau) + \widehat{R}_{ww}(\tau)\,\sin(2\pi f_0\tau) \tag{3.175a}$$

$$= \frac{N_0}{2}\delta(\tau)\,\cos(2\pi f_0\tau) + \frac{N_0}{2}\,\frac{1}{\pi\tau}\,\sin(2\pi f_0\tau)$$

$$= \frac{N_0}{2}\delta(\tau) + \frac{N_0}{2}\,\frac{1}{\pi\tau}\,\sin(2\pi f_0\tau) \tag{3.175b}$$

$$R_{w_s w_c}(\tau) = -\,R_{w_c w_s}(\tau) = \left[R_{ww}(\tau)\right]_s$$

$$= -\,R_{ww}(\tau)\,\sin(2\pi f_0\tau) + \widehat{R}_{ww}(\tau)\,\cos(2\pi f_0\tau) \tag{3.176a}$$

$$= -\,\frac{N_0}{2}\delta(\tau)\,\sin(2\pi f_0\tau) + \frac{N_0}{2}\,\frac{1}{\pi\tau}\,\cos(2\pi f_0\tau)$$

$$= \frac{N_0}{2}\,\frac{1}{\pi\tau}\,\cos(2\pi f_0\tau) \tag{3.176b}$$

$$R_{\widetilde{w}\widetilde{w}^*}(\tau) = \mathrm{E}\{\widetilde{w}(t+\tau)\,\widetilde{w}^*(t)\} = \frac{2}{k^2}\left\{\left[R_{ww}(\tau)\right]_c + j\left[R_{ww}(\tau)\right]_s\right\}$$

$$= \frac{2}{k^2}\left\{\frac{N_0}{2}\delta(\tau) + \frac{N_0}{2}\,\frac{1}{\pi\tau}\,\sin(2\pi f_0\tau) + j\,\frac{N_0}{2}\,\frac{1}{\pi\tau}\,\cos(2\pi f_0\tau)\right\}$$

$$= \frac{2}{k^2}\,\frac{N_0}{2}\left\{\delta(\tau) + j\frac{1}{\pi\tau}\right\}\,e^{-j2\pi f_0\tau} \triangleq \frac{2}{k}\left[R_{ww}(\tau)\right]_{\mathrm{lp}} \tag{3.177}$$

$$\stackrel{\mathcal{F}}{\longleftrightarrow} \quad S_{\widetilde{w}\widetilde{w}^*}(f) = \frac{2}{k^2}\,\frac{N_0}{2}\,\{1 + j(-j\,\mathrm{sign}(f))\} \otimes \delta(f+f_0)$$

$$= \frac{4}{k^2}\,\frac{N_0}{2}\,\mathrm{u}(f+f_0) \tag{3.178}$$

where $[\cdot]_{\mathrm{lp}}$ denotes the operator that provides the complex envelope of its argument according to (D.11).

$$S_{w_s w_c}(f) = -\,j\left[S_{ww}(f+f_0)\,\mathrm{u}(f+f_0) - S_{ww}(f-f_0)\,\mathrm{u}(-f+f_0)\right]$$

$$= -\,j\,\frac{N_0}{2}\left[\mathrm{u}(f+f_0) - \mathrm{u}(-f+f_0)\right] \tag{3.179}$$

$$\lim_{\tau\to 0} R_{w_s w_c}(\tau) = \mathrm{p.v.}\int_{-\infty}^{+\infty} S_{w_s w_c}(f)\,\mathrm{d}f = 0\,. \tag{3.180}$$

From the wide-sense stationarity of $w(t)$ it follows that

$$\mathrm{E}\{\widetilde{w}(t+\tau)\,\widetilde{w}(t)\} = 0\,. \tag{3.181}$$

3.6 Sampling and aliasing

Let

$$x(n) \triangleq x_a(t)|_{t=nT_s} \qquad n \in \mathbb{Z} \tag{3.182}$$

be the discrete-time process obtained by uniformly sampling with period $T_s = 1/f_s$ the continuous-time ACS signal $x_a(t)$ with (conjugate) autocorrelation

$$\mathrm{E}\left\{x_a(t+\tau)\,x_a^{(*)}(t)\right\} = \sum_{\alpha \in A_{\boldsymbol{x}_a}} R_{\boldsymbol{x}_a}^{\alpha}(\tau)\,e^{j2\pi\alpha t} \tag{3.183}$$

and Loève bifrequency spectrum (see (1.56))

$$\mathrm{E}\left\{X_a(f_1)\,X_a^{(*)}(f_2)\right\} = \sum_{\alpha \in A_{\boldsymbol{x}_a}} S_{\boldsymbol{x}_a}^{\alpha}(f_1)\,\delta(f_2 - (-)(\alpha - f_1))\,, \tag{3.184}$$

where subscript $\boldsymbol{x}_a \equiv x_a\,x_a^{(*)}$. The relationship between the Fourier transform $X_a(f)$ of the continuous-time signal $x_a(t)$ and the Fourier transform $X(\nu)$ of the discrete-time sampled signal $x(n)$ is (Section C.1) (Lathi, 2002, Sec. 9.5)

$$X(\nu) = \frac{1}{T_s} \sum_{k=-\infty}^{+\infty} X_a((\nu - k)\,f_s)\,. \tag{3.185}$$

If the signal $x_a(t)$ is strictly band limited, that is $X_a(f) = 0$ for $|f| > B$, the following two conditions are equivalent, provided that $f_s \geqslant 2B$:

1. Replicas in (3.185) do not overlap;
2. In the principal frequency domain $\nu \in [-1/2, 1/2)$, the Fourier transform $X(\nu)$ of the discrete-time sampled signal is an amplitude and frequency scaled version of only one replica (the one with $k = 0$) of the Fourier transform $X_a(f)$ of the continuous time signal. That is

$$X(\nu) = \frac{1}{T_s}\,X_a(f)|_{f=\nu f_s} \qquad \nu \in [-1/2, 1/2)\,. \tag{3.186}$$

In the following, it is shown that in aliasing formulas for cyclic spectra, conditions for not overlap of replicas and conditions for the presence of a unique replica in the principal cycle- and spectral-frequency domain $(\widetilde{\alpha}, \nu) \in [-1/2, 1/2) \times [-1/2, 1/2)$ are not equivalent (see the proofs of Theorem 3.22 and 3.25 in Section 3.10).

3.6.1 Cyclic statistics of the sampled signal

Lemma 3.17. Aliasing Formula for the Loève Bifrequency Spectrum of a Sampled ACS Signal. *Let $x_a(t)$ be a continuous-time ACS signal with Loève bifrequency spectrum (3.184). The*

Loève bifrequency spectrum of the sampled signal $x(n)$ *(3.182) is given by*

$$\mathrm{E}\left\{X(\nu_1)\, X^{(*)}(\nu_2)\right\} = \frac{1}{T_s} \sum_{\alpha \in A_{\boldsymbol{x}_a}} \sum_{k_1 \in \mathbb{Z}} S^{\alpha}_{\boldsymbol{x}_a}\left((\nu_1 - k_1) f_s\right)$$

$$\sum_{k_2 \in \mathbb{Z}} \delta\left((\nu_2 - k_2) - (-)(\alpha/f_s - (\nu_1 - k_1))\right) \tag{3.187}$$

Proof. See Section 3.10. □

Corollary 3.18. *From (3.187), it follows that the discrete-time signal* $x(n) \triangleq x_a(t)|_{t=nT_s}$ *obtained by uniformly sampling a continuous-time ACS signal* $x_a(t)$ *is a discrete-time ACS signal. In fact, its Loève bifrequency spectrum has support concentrated on lines with slope* ± 1 *in the bifrequency plane* (ν_1, ν_2) *and can be written in the form (1.42).*

The countable set $\widetilde{A}$ *in (1.42) is linked to* $A_{\boldsymbol{x}_a}$ *by the relationships*

$$\widetilde{A} = \bigcup_{\alpha \in A_{\boldsymbol{x}_a}} \left\{\widetilde{\alpha} \in (-1/2, 1/2] \,:\, \widetilde{\alpha} = (\alpha/f_s) \bmod 1\right\} \tag{3.188}$$

$$A_{\boldsymbol{x}_a} \subseteq \bigcup_{\widetilde{\alpha} \in \widetilde{A}} \bigcup_{p \in \mathbb{Z}} \left\{\alpha \in \mathbb{R} \,:\, \alpha = \widetilde{\alpha} f_s - p f_s\right\} \tag{3.189}$$

with mod 1 *denoting the modulo 1 operation with values in* $[-1/2, 1/2)$. □

The (conjugate) cyclic spectrum and the (conjugate) cyclic autocorrelation function of $x(n)$ are given in the following result.

Lemma 3.19. Aliasing Formulas for Cyclic Statistics of ACS Signals (Napolitano, 1995), (Izzo and Napolitano, 1996). *Let* $x_a(t)$ *be a continuous-time ACS signal with (conjugate) autocorrelation (3.183) and Loève bifrequency spectrum (3.184). The (conjugate) autocorrelation, cyclic autocorrelation, and cyclic spectrum of the sampled signal* $x(n)$ *(3.182) are given by*

$$\mathrm{E}\{x(n+m)\, x^{(*)}(n)\} = \mathrm{E}\{x_a(t+\tau)\, x_a(t)\}|_{t=nT_s, \tau=mT_s}\,, \tag{3.190}$$

$$\widetilde{R}^{\widetilde{\alpha}}_{\boldsymbol{x}}(m) = \sum_{p \in \mathbb{Z}} R^{\alpha - p f_s}_{\boldsymbol{x}_a}(\tau)\Big|_{\tau = mT_s, \alpha = \widetilde{\alpha} f_s} \tag{3.191}$$

$$\widetilde{S}^{\widetilde{\alpha}}_{\boldsymbol{x}}(\nu) = \frac{1}{T_s} \sum_{p \in \mathbb{Z}} \sum_{q \in \mathbb{Z}} S^{\alpha - p f_s}_{\boldsymbol{x}_a}(f - q f_s)\Big|_{f = \nu f_s, \alpha = \widetilde{\alpha} f_s}\,, \tag{3.192}$$

where subscript $\boldsymbol{x} \equiv x\, x^{(*)}$.

Proof. See Section 3.10. □

Assumption 3.20. *The continuous-time process* $x_a(t)$ *is strictly bandlimited, i.e.,* $S^0_{x_a x_a^*}(f) = 0$ *for* $|f| > B$ *(Fact 3.13).* □

According to (3.65a)–(3.66), we have the following result.

Lemma 3.21. Support of the (Conjugate) Cyclic Spectrum. *Under Assumption 3.20 ($x_a(t)$ strictly band-limited), it results*

$$\operatorname{supp}\left\{S^{\alpha}_{\boldsymbol{x}_a}(f)\right\} \subseteq \{(\alpha, f) \in \mathbb{R} \times \mathbb{R} : |f| \leqslant B, \ |\alpha - f| \leqslant B\} \tag{3.193a}$$

$$\subseteq \{(\alpha, f) \in \mathbb{R} \times \mathbb{R} : |f| \leqslant B, \ |\alpha| \leqslant 2B\} . \tag{3.193b}$$

Therefore,

$$S^{\alpha}_{\boldsymbol{x}_a}(f) = 0 \quad |f| > B, \ |\alpha| > 2B . \tag{3.194}$$

Moreover, in (3.184) it results

$$\operatorname{supp}\left\{S^{\alpha}_{\boldsymbol{x}_a}(f)\delta(f_2 - {}_{(-)}(\alpha - f_1))\right\}$$

$$\subseteq \{(f_1, f_2) \in \mathbb{R} \times \mathbb{R} : |f_1| \leqslant B, \ |\alpha - f_1| \leqslant B, \ f_2 = {}_{(-)}(\alpha - f_1)\} \tag{3.195a}$$

$$= \{(f_1, f_2) \in \mathbb{R} \times \mathbb{R} : |f_1| \leqslant B, \ |f_2| \leqslant B, \ f_2 = {}_{(-)}(\alpha - f_1)\} . \tag{3.195b}$$

□

Theorem 3.22. *Let $x_a(t)$ be a continuous-time ACS signal with Loève bifrequency spectrum (3.184). Under Assumption 3.20 and for $f_s \geqslant 4B$, it results*

$$\mathrm{E}\left\{X(\nu_1)\, X^{(*)}(\nu_2)\right\} = \frac{1}{T_s} \sum_{\alpha \in A_{\boldsymbol{x}_a}} S^{\alpha}_{\boldsymbol{x}_a}(\nu_1 f_s)\, \delta\left(\nu_2 - {}_{(-)}(\alpha/f_s - \nu_1)\right)$$

$$|\nu_1| \leqslant \frac{1}{2}, \ |\nu_2| \leqslant \frac{1}{2} . \tag{3.196}$$

Moreover,

$$\mathrm{E}\left\{X(\nu_1)\, X^{(*)}(\nu_2)\right\} = 0 \ \textit{for} \ 1/4 \leqslant |\nu_1| \leqslant 1/2 \ \textit{and} \ 1/4 \leqslant |\nu_2| \leqslant 1/2 . \tag{3.197}$$

Proof. See Section 3.10. □

Theorem 3.22 provides a lower bound to the sampling frequency f_s such that the Loève bifrequency spectrum of the discrete-time signal $x(n)$ in the principal frequency domain $(\nu_1, \nu_2) \in [-1/2, 1/2] \times [-1/2, 1/2]$ is identical, but for frequency scaling factors, to the Loève bifrequency spectrum of the continuous-time signal $x_a(t)$.

Note that the condition $f_s \geqslant 4B$ is more restrictive with respect to $f_s \geqslant 2B$ that is sufficient for having non-overlapping replicas in (3.187) or for the mean-square reconstruction of the process by the Shannon interpolation formula (Gardner, 1972a) (see also the proof of Theorem 3.22 in Section 3.10).

Theorem 3.23. Sampling Theorem for (Conjugate) Cyclic Autocorrelation of ACS Signals (Napolitano, 1995), (Izzo and Napolitano, 1996). *Let $x_a(t)$ be a continuous-time ACS signal* *with (conjugate) autocorrelation (3.183) and Loève bifrequency spectrum (3.184). Under Assumption 3.20 and for $f_s \geqslant 4B$, it results:*

$$\widetilde{R}^{\widetilde{\alpha}}_{\boldsymbol{x}}(m) = R^{\alpha}_{\boldsymbol{x}_a}(\tau)\big|_{\tau=mT_s,\alpha=\widetilde{\alpha} f_s} \qquad |\widetilde{\alpha}| \leqslant \frac{1}{2} \tag{3.198a}$$

$$R^{\alpha}_{\boldsymbol{x}_a}(\tau)\big|_{\tau=mT_s} = \begin{cases} \widetilde{R}^{\widetilde{\alpha}}_{\boldsymbol{x}}(m)\big|_{\widetilde{\alpha}=\alpha/f_s} & |\alpha| \leqslant \dfrac{f_s}{2} \\ 0 & \text{otherwise} \end{cases} \tag{3.198b}$$

$$R^{\alpha}_{\boldsymbol{x}_a}(\tau) = \sum_{m\in\mathbb{Z}} \widetilde{R}^{\widetilde{\alpha}}_{\boldsymbol{x}}(m) \operatorname{sinc}\left(\frac{\tau}{T_s} - m\right) \qquad \forall \tau \in \mathbb{R},\ \alpha = \widetilde{\alpha} f_s \tag{3.199}$$

where subscripts $\boldsymbol{x}_a \equiv x_a\, x_a^{(*)}$ *and* $\boldsymbol{x} \equiv x\, x^{(*)}$.

Proof. The results are consequence of (3.191) and (3.193b). □

Theorem 3.24. Sampling Theorem for (Conjugate) Cyclic Spectrum of ACS Signals (Napolitano, 1995), (Izzo and Napolitano, 1996). *Let $x_a(t)$ be a continuous-time ACS signal with (conjugate) autocorrelation (3.183) and Loève bifrequency spectrum (3.184). Under Assumption 3.20 and for $f_s \geqslant 4B$, it results:*

$$\widetilde{S}^{\widetilde{\alpha}}_{\boldsymbol{x}}(\nu) = \frac{1}{T_s}\, S^{\alpha}_{\boldsymbol{x}_a}(f)\big|_{f=\nu f_s,\alpha=\widetilde{\alpha} f_s} \qquad |\widetilde{\alpha}| \leqslant \frac{1}{2},\ |\nu| \leqslant \frac{1}{2} \tag{3.200a}$$

$$S^{\alpha}_{\boldsymbol{x}_a}(f) = \begin{cases} T_s\, \widetilde{S}^{\widetilde{\alpha}}_{\boldsymbol{x}}(\nu)\big|_{\nu=f/f_s,\widetilde{\alpha}=\alpha/f_s} & |\alpha| \leqslant \dfrac{f_s}{2},\ |f| \leqslant \dfrac{f_s}{2} \\ 0 & \text{otherwise} \end{cases} \tag{3.200b}$$

By Fourier transforming both sides in (3.199) we get the following expression for the cyclic spectra of a strictly band-limited signal

$$S^{\alpha}_{\boldsymbol{x}_a}(f) = \sum_{m\in\mathbb{Z}} \widetilde{R}^{\widetilde{\alpha}}_{\boldsymbol{x}}(m)\, e^{-j2\pi f m T_s}\, T_s \operatorname{rect}(f T_s)\,, \qquad \alpha = \widetilde{\alpha} f_s \tag{3.201}$$

which is equivalent to (3.200b).

Proof. The results are consequence of (3.192) and (3.193b). □

3.6.2 (Almost-)cyclostationarity of a sampled cyclostationary signal

Let $x_a(t)$ be cyclostationary with period T_0. It results that (Theorem 1.4)

$$S^{\alpha}_{\boldsymbol{x}_a}(f) \not\equiv 0 \ \text{ for } \alpha = m/T_0,\ m \in \mathbb{Z}\,. \tag{3.202}$$

In such a case, in the double sum in (3.192), only terms such that $(\widetilde{\alpha} - p) f_s$ is an integer multiple of $1/T_0$ give nonzero contribution. Thus, (3.192) specializes into

$$\widetilde{S}^{\widetilde{\alpha}}_{\boldsymbol{x}}(\nu) = \frac{1}{T_s} \sum_{p\in\mathbb{Z}} \sum_{q\in\mathbb{Z}} S^{(\widetilde{\alpha}-p) f_s}_{\boldsymbol{x}_a}((\nu - q) f_s)\, \delta_{(\widetilde{\alpha}-p) \bmod (T_s/T_0)}$$

$$= \frac{1}{T_s} \sum_{m \in \mathbb{Z}} \sum_{q \in \mathbb{Z}} S_{x_a}^{m/T_0}((\nu - q) f_s)\, \delta_{(\widetilde{\alpha} - m T_s/T_0) \bmod 1} \tag{3.203}$$

where δ_α is the Kronecker delta, that is, $\delta_\alpha = 1$ for $\alpha = 0$ and $\delta_\alpha = 0$ for $\alpha \neq 0$.

From (3.203) it follows that, for a given conjugation configuration (xx^* or xx), the cyclostationarity properties of the discrete-time signal $x(n)$ depend on the ratio T_s/T_0.

Let T_s/T_0 be a rational number, say $q_0 \triangleq \mathrm{n}(q_0)/\mathrm{d}(q_0)$ with numerator $\mathrm{n}(q_0)$ and denominator $\mathrm{d}(q_0)$ relatively prime integers. We have that

- if $\mathrm{d}(q_0) = 1$ (i.e., T_s is an integer multiple of T_0), then $x(n)$ is WSS, i.e., $\widetilde{S}_{\boldsymbol{x}}^{\widetilde{\alpha}}(\nu) \not\equiv 0$ only for $\widetilde{\alpha} \in \mathbb{Z}$;
- if $\mathrm{d}(q_0) > 1$, then $x(n)$ is cyclostationary with period $\mathrm{d}(q_0)$ i.e., all cycle frequencies are multiple of the rational number $1/\mathrm{d}(q_0)$. That is, $\widetilde{\alpha} = k/\mathrm{d}(q_0)$, $k = 0, 1, \ldots, \mathrm{d}(q_0) - 1$.

If T_s and T_0 are incommensurate, the discrete-time signal $x(n)$ exhibits cyclostationarity, for the considered conjugation configuration, with cycle frequencies $\widetilde{\alpha} = (mT_s/T_0) \bmod 1$, $m \in \mathbb{Z}$. Therefore the cycle frequencies are not all integer multiple of a fundamental rational cycle frequency, that is, $x(n)$ is not cyclostationary but, rather, almost-cyclostationary.

If the cyclostationary continuous-time signal $x_a(t)$ is strictly band limited to $[-B, B]$, then it exhibits cyclostationarity only in correspondence of the finite number of cycle frequencies $\alpha = m/T_0$, $m \in \mathbb{Z}$, such that $|\alpha| \leqslant 2B$ (see Fig. 3.3).

Therefore, if $x(n)$ is obtained by sampling $x_a(t)$ at rate $f_s \geqslant 4B$, then from (3.200a) we have that $x(n)$ exhibits cyclostationarity with cycle frequencies (considered in the interval $[0, 1[$)

$$\widetilde{\alpha} = \frac{\alpha}{f_s} = m\,\frac{T_s}{T_0}, \qquad 0 < m \leqslant \mathrm{int}(4BT_0) \leqslant \frac{T_0}{T_s} \tag{3.204}$$

where $\mathrm{int}(\cdot)$ denotes integer part. As already observed, $x(n)$ is cyclostationary if T_s/T_0 is rational and is almost-cyclostationary otherwise.

In general, no information on the cyclostationarity properties of a continuous-time signal $x_a(t)$ can be achieved by the knowledge of the cyclostationarity properties of the sampled signal $x(n)$. However, if $x(n)$ is obtained by sampling a strictly band limited signal at a rate $f_s \geqslant 4B$, then (3.200b) holds. Consequently, if $x(n)$ exhibits cyclostationarity with cycle frequencies given by (3.204), then the continuous-time signal $x_a(t)$ is wide-sense cyclostationary with period T_0, whether T_0/T_s is a rational number or not.

3.6.3 Bandpass sampling

From Theorems 3.23 and 3.24 it follows that the continuous-time signal should be sampled at twice the Nyquist frequency in order to avoid aliasing in both cycle and spectral frequency domains. Thus, for bandpass signals in communications and optics applications, very high sampling rates are required.

The problem of sub-Nyquist sampling for cyclic spectrum reconstruction can be addressed resorting to compressive sampling (Section 10.2). An alternative way to reduce the sampling rate without demodulating the bandpass signal is to resort to bandpass sampling (Napolitano and Doğançay, 2018). It consists in sampling a bandpass signal with a sampling frequency significantly lower than the Nyquist frequency of the bandpass signal and comparable with the Nyquist frequency of the complex envelope. Sufficient conditions involving sampling frequency and carrier frequency and the bandwidth of the bandpass signal must be satisfied in order to avoid the overlap of replicas in the Fourier transform of the sampled signal and to allow one to reconstruct the continuous-time complex envelope. Conditions for the reconstruction of the cyclic spectrum can be derived under assumptions less restrictive than those made for compressive sampling based techniques (Napolitano and Doğançay, 2018).

By substituting the expression (3.121) (with x replaced by x_a) of the cyclic spectrum of $x_a(t)$ expressed in terms of the (conjugate) cyclic spectra of $\widetilde{x}_a(t)$ into the expression (3.192) of the cyclic spectrum of $x(n)$ expressed in terms of replicas of cyclic spectra of $x_a(t)$ leads to

$$\widetilde{S}_{xx}^{\widetilde{\alpha}}(\nu) = \frac{k^2}{4} \sum_{p\in\mathbb{Z}} \sum_{q\in\mathbb{Z}} \Big[S_{\widetilde{x}_a\widetilde{x}_a^*}^{\alpha-pf_s}(f - f_0 - qf_s) + S_{\widetilde{x}_a^*\widetilde{x}_a}^{\alpha-pf_s}(f + f_0 - qf_s)$$
$$+ S_{\widetilde{x}_a\widetilde{x}_a}^{\alpha-2f_0-pf_s}(f - f_0 - qf_s) + S_{\widetilde{x}_a^*\widetilde{x}_a^*}^{\alpha+2f_0-pf_s}(f + f_0 - qf_s) \Big]_{f=\nu f_s, \alpha=\widetilde{\alpha} f_s} . \tag{3.205}$$

In the following, sufficient conditions on the sampling frequency f_s, the frequency f_0, and the signal bandwidth W are obtained such that the double sum in (3.205) involves terms with non-overlapping supports in the $(\widetilde{\alpha}, \nu)$ plane. In addition, more restrictive conditions are derived such that, for each $\widetilde{\alpha} \in [-1/2, 1/2)$, *only one replica* in (3.205) gives nonzero contribution for $\nu \in [-1/2, 1/2)$.

Theorem 3.25. Band Pass Sampling (Napolitano and Doğançay, 2018). *Let $x_a(t)$ be a band pass ACS signal with central frequency f_0 and strictly band limited with monolateral bandwidth W, with $f_0 > W$ (Definition 3.16).*

1. *For every integer m and sampling frequency f_s such that*

$$\begin{cases} m \leqslant f_0/(2W) - 1/2 \\ f_s = 2f_0/(m + 1/2) \end{cases} \tag{3.206}$$

is verified, replicas in (3.205) do not overlap.

2. *For every integer m and sampling frequency f_s such that*

$$\begin{cases} m \leqslant f_0/(4W) - 1/2 \\ f_s = 2f_0/(m + 1/2) \end{cases} \tag{3.207}$$

is verified, only one replica *in (3.205), for each $\widetilde{\alpha} \in [-1/2, 1/2)$, gives nonzero contribution for $\nu \in [-1/2, 1/2)$.*

Proof. See Section 3.10. □

The smallest sampling frequency satisfying (3.206) is obtained by taking

$$m = m_{\max} \triangleq \text{floor}(f_0/(2W) - 1/2) \tag{3.208}$$

and the admissible values for f_s that allow bandpass sampling are

$$f_s = 2f_0/(m + 1/2) \qquad m = 1, 2, \ldots, m_{\max}\,. \tag{3.209}$$

The smallest sampling frequency satisfying (3.207) is obtained by

$$m = m'_{\max} \triangleq \text{floor}(f_0/(4W) - 1/2)\,. \tag{3.210}$$

The set of all admissible sampling frequencies is obtained by replacing $m_{\max}$ with $m'_{\max}$ in (3.209).

If condition (3.207) is satisfied, then for each fixed $\widetilde{\alpha} \in [-1/2, 1/2)$, the cyclic spectrum $\widetilde{S}^{\widetilde{\alpha}}_{xx}(\nu)$, with $\nu \in [-1/2, 1/2)$, is proportional to only one of the four (frequency scaled) cyclic spectra of the complex envelope $\widetilde{x}_a(t)$ in the right-hand side of (3.205).

In Fig. 3.7, an example is illustrated corresponding to two different values of sampling frequency. Let f_r be a reference frequency. In the first case (Fig. 3.7 (top)), $f_s = f_r/4$ and condition (3.207) is satisfied. That is, in the principal domain $(\widetilde{\alpha}, \nu) = (\alpha/f_s, f/f_s) \in [-1/2, 1/2)^2$, for each $\widetilde{\alpha}$ only one cyclic spectrum in (3.205) gives nonzero contribution. In the second case (Fig. 3.7 (bottom)), $f_s = f_r/6$ and condition (3.206) (but not (3.207)) is satisfied. In such a case, replicas in (3.205) do not overlap, but for each $\widetilde{\alpha} \in [-1/2, 1/2)$ either one or two cyclic spectra in (3.205) can give nonzero contribution for $\nu \in [-1/2, 1/2)$.

Note that the reciprocal positions of the replicas of the four parallelograms depend on the value of f_s and are not necessarily coincident with the reciprocal positions as in Fig. 3.6.

In practice, since $f_0 \gg W$, the above results on bandpass sampling allow one to choose a sampling frequency f_s for the continuous-time signal $x_a(t)$ which is significantly smaller than the sampling frequency $f_{\text{N}} = 4(f_0 + W)$ (i.e., twice the Nyquist frequency) which is necessary to have no aliasing in both f and α domains (Theorem 3.24) when the bandpass nature of $x_a(t)$ is not exploited. In particular, for $m = m_{\max}$ we have

$$f_s = \frac{2f_0}{\text{floor}(f_0/(2W) - 1/2) + 1/2} \simeq 4\,W \tag{3.211}$$

which is the sampling frequency to avoid aliasing in both α and f for a low-pass ACS signal with monolateral bandwidth W (Theorem 3.24).

3.7 Multirate processing of discrete-time ACS signals

In multirate digital signal processing, the main building blocks are constituted by expansors (upsamplers) and decimators (downsamplers). These devices are linear time-variant systems that can be classified as *deterministic* in the FOT probability framework (Izzo

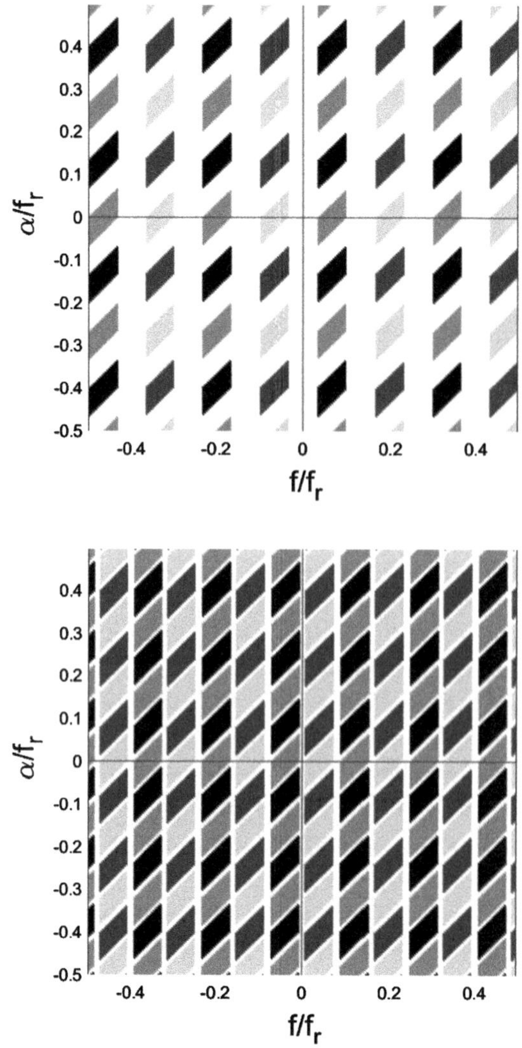

FIGURE 3.7 Set $\mathcal{D}_3(\alpha, f)$ defined in (3.274) containing the support of the cyclic spectrum of the bandpass sampled bandpass signal. (Top) sampling frequency $f_s = f_r/4$ (condition (3.207) satisfied). (Bottom) $f_s = f_r/6$ (condition (3.206) satisfied and (3.207) not satisfied). *Source: (Napolitano and Doğançay, 2018) Copyright of Elsevier.*

and Napolitano, 2002a), (Napolitano, 2012, Sec. 6.3.8). In particular, these systems are not LAPTV.

The effects of multirate systems on second-order cyclostationary processes are analyzed in (Sathe and Vaidyanathan, 1993) and (Akkarakaran and Vaidyanathan, 2000). A cyclostationary analysis of multirate filter banks is made in (Aach, 2007). Higher-order statistics of expanded or decimated ACS time series are considered in (Izzo and Napolitano, 1998b).

In (Sathe and Vaidyanathan, 1993), (Izzo and Napolitano, 1998b), and (Akkarakaran and Vaidyanathan, 2000), multirate operations on a single signal are addressed and no cross-statistics of a signal and its expanded or decimated version are considered. Moreover, no cross-statistics are considered between signals obtained by multirate elaborations with different rates of the same signal or of jointly ACS signals. In these cases, since the elaborations are not LAPTV, the output signals are not jointly ACS but, rather, jointly SC (Section 13.4).

In (Napolitano, 2012, Sec. 4.10), multirate processing of jointly SC processes is considered. Results for jointly ACS processes reported here are obtained as special case of the results in (Napolitano, 2012, Sec. 4.10).

Let $x_1(n)$ and $x_2(n)$ be discrete-time jointly ACS processes with (conjugate) cross-correlation function

$$\mathrm{E}\left\{x_1(n+m)\, x_2^{(*)}(n)\right\} = \sum_{\widetilde{\alpha}\in\widetilde{A}} \widetilde{R}_{\boldsymbol{x}}^{\widetilde{\alpha}}(m)\, e^{j2\pi\widetilde{\alpha}n} \tag{3.212}$$

where subscript $\boldsymbol{x} \equiv x_1 x_2^{(*)}$, the Fourier coefficients

$$\widetilde{R}_{\boldsymbol{x}}^{\widetilde{\alpha}}(m) \triangleq \lim_{N\to\infty} \frac{1}{2N+1} \sum_{n=-N}^{N} \mathrm{E}\left\{x_1(n+m)\, x_2^{(*)}(n)\right\} e^{-j2\pi\widetilde{\alpha}n} \tag{3.213}$$

are the (conjugate) cyclic cross-correlation functions, and

$$\widetilde{A} \triangleq \left\{\widetilde{\alpha} \in [-1/2, 1/2) \, : \, \widetilde{R}_{\boldsymbol{x}}^{\widetilde{\alpha}}(m) \not\equiv 0\right\} \tag{3.214}$$

is the countable set of (conjugate) cycle frequencies $\widetilde{\alpha}$ in the principal domain $[-1/2, 1/2)$. By double Fourier transforming both sides of (3.212) with $n_1 = n+m$ and $n_2 = n$, one obtains the Loève bifrequency cross-spectrum

$$\mathrm{E}\left\{X_1(\nu_1)\, X_2^{(*)}(\nu_2)\right\} = \sum_{\widetilde{\alpha}\in\widetilde{A}} \widetilde{S}_{\boldsymbol{x}}^{\widetilde{\alpha}}(\nu_1)\, \widetilde{\delta}(\nu_2 - {\scriptstyle(-)}(\widetilde{\alpha} - \nu_1)) \tag{3.215}$$

where $\widetilde{\delta}(\cdot)$ denotes the periodic delta train with period 1 (C.18) and

$$\widetilde{S}_{\boldsymbol{x}}^{\widetilde{\alpha}}(\nu) = \sum_{m\in\mathbb{Z}} \widetilde{R}_{\boldsymbol{x}}^{\widetilde{\alpha}}(m)\, e^{-j2\pi\nu m} \tag{3.216}$$

are the (conjugate) cyclic cross-spectra.

3.7.1 Expansion (upsampling)

Let $x_{Ii}(n)$ be the L_i-fold expanded (upsampled) version of $x_i(n)$, $(i = 1, 2)$ (see Fig. 3.8). That is

$$x_{Ii}(n) = x_i\left[\frac{n}{L_i}\right] \triangleq \begin{cases} x_i\left(\dfrac{n}{L_i}\right) & n = kL_i, k \in \mathbb{Z} \\ 0 & \text{otherwise} \end{cases} \tag{3.217}$$

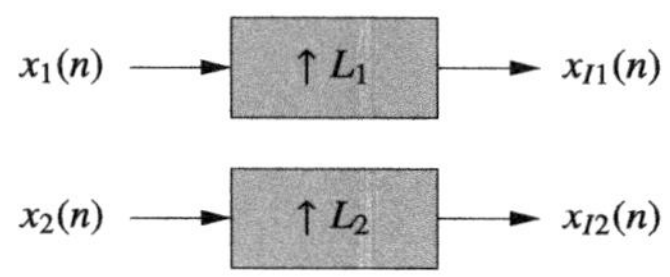

FIGURE 3.8 Expansion of two processes with different expansion factors.

whose Fourier transform is

$$X_{Ii}(\nu) = X_i(\nu L_i)\,. \tag{3.218}$$

Theorem 3.26. (Napolitano, 2012, Sec. 4.10.1.2). *Let $x_1(n)$ and $x_2(n)$ be jointly ACS with Loève bifrequency cross-spectrum (3.215). The Loève bifrequency cross-spectrum of the expanded processes $x_{I1}(n)$ and $x_{I2}(n)$ is given by*

$$\begin{aligned}
&\mathrm{E}\left\{X_{I1}(\nu_1)\, X_{I2}^{(*)}(\nu_2)\right\} \\
&\quad = \mathrm{E}\left\{X_1(\nu_1 L_1)\, X_2^{(*)}(\nu_2 L_2)\right\} \\
&\quad = \sum_{\widetilde{\alpha}\in\widetilde{A}} \widetilde{S_{\boldsymbol{x}}^{\widetilde{\alpha}}}(\nu_1 L_1)\, \widetilde{\delta}(\nu_2 L_2 - {}_{(-)}(\widetilde{\alpha} - \nu_1 L_1)) && (3.219a)\\
&\quad = \sum_{\widetilde{\alpha}\in\widetilde{A}} \widetilde{S_{\boldsymbol{x}}^{\widetilde{\alpha}}}(\nu_1 L_1)\, \frac{1}{L_2} \sum_{h\in\mathbb{Z}} \delta\left(\nu_2 - \frac{h}{L_2} - {}_{(-)}\left(\frac{\widetilde{\alpha}}{L_2} - \frac{L_1}{L_2}\nu_1\right)\right) && (3.219b)\\
&\quad = \sum_{\widetilde{\alpha}\in\widetilde{A}} \widetilde{S_{\boldsymbol{x}}^{\widetilde{\alpha}}}(\nu_1 L_1)\, \frac{1}{L_2} \sum_{p=0}^{L_2-1} \widetilde{\delta}_{\nu_2}\left(\nu_2 - \frac{p}{L_2} - {}_{(-)}\left(\frac{\widetilde{\alpha}}{L_2} - \frac{L_1}{L_2}\nu_1\right)\right). && (3.219c)
\end{aligned}$$

Proof. It is a consequence of the scaling property of the Dirac delta (Zemanian, 1987, Sec. 1.7). See also (Napolitano, 2012, Sec. 5.11). □

In (3.219a)–(3.219c), the cyclic spectra $\widetilde{S_{\boldsymbol{x}}^{\widetilde{\alpha}}}(\nu_1 L_1)$ are periodic in ν_1 with period $1/L_1$, and the delta train is periodic in ν_1 with period $1/L_1$ and is periodic in ν_2 with period $1/L_2$. Notation $\widetilde{\delta}_{\nu_2}(\cdot)$ clarifies that the periodicity with period 1 must be considered with respect to the variable ν_2.

Therefore, $\mathrm{E}\{X_{I1}(\nu_1)\, X_{I2}^{(*)}(\nu_2)\}$ is periodic in ν_1 with period $1/L_1$ and in ν_2 with period $1/L_2$. In addition, from (3.219c) it follows that the Loève bifrequency cross-spectrum of the expanded processes $x_{I1}(n)$ and $x_{I2}(n)$ in the principal domain $(\nu_1, \nu_2) \in [-1/2, 1/2)^2$ has support (for the case ${}_{(-)}$ present)

$$\nu_2 = (L_1/L_2)\nu_1 - \widetilde{\alpha}/L_2 + p/L_2 \bmod 1 \qquad \widetilde{\alpha}\in\widetilde{A},\ p\in\{0,1,\ldots,L_2-1\} \tag{3.220}$$

that is, lines with slope L_1/L_2. Consequently, $x_{I1}(n)$ and $x_{I2}(n)$ are jointly SC (Section 13.4). In particular, for a single process $x_1(n) \equiv x_2(n) \equiv x(n)$, by taking $L_1 = 1$, that is $x_{I1}(n) \equiv x(n)$, it follows that the ACS process $x(n)$ and its expanded version $x_{I2}(n)$ are jointly SC with support lines with slope $1/L_2$. In the special case of $x(n)$ WSS, $x(n)$ and $x_{I2}(n)$ are jointly SC with support constituted by a unique line.

Theorem 3.27. (Napolitano, 2012, Sec. 4.10.1.2). *Let $x_1(n)$ and $x_2(n)$ be jointly ACS with (conjugate) cross-correlation (3.212). The (conjugate) cross-correlation function of the expanded processes $x_{I1}(n)$ and $x_{I2}(n)$ is given by*

$$\begin{aligned}
\mathrm{E}&\left\{x_{I1}(n+m)\, x_{I2}^{(*)}(n)\right\} \\
&= \int_{[-1/2,1/2]^2} \mathrm{E}\left\{X_{I1}(\nu_1)\, X_{I2}^{(*)}(\nu_2)\right\} e^{j2\pi[\nu_1(n+m)+(-)\nu_2 n]} \mathrm{d}\nu_1 \mathrm{d}\nu_2 \\
&= \sum_{\widetilde{\alpha}\in\widetilde{A}} e^{j2\pi(\widetilde{\alpha}/L_2)n}\, \widetilde{\widetilde{R}}{}_{\boldsymbol{x}}^{\widetilde{\alpha}}\left[\frac{(1-L_1/L_2)n+m}{L_1}\right] \delta_{n \bmod L_2}
\end{aligned} \tag{3.221}$$

where the Kronecker delta $\delta_{n \bmod L_2}$ is defined in (C.17) and $\widetilde{\widetilde{R}}{}_{\boldsymbol{x}}^{\widetilde{\alpha}}[\ell/L_2]$ is defined according to (3.217), that is,

$$\widetilde{\widetilde{R}}{}_{\boldsymbol{x}}^{\widetilde{\alpha}}\left[\frac{\ell}{L}\right] \triangleq \begin{cases} \widetilde{R}_{\boldsymbol{x}}^{\widetilde{\alpha}}(\ell/L) & \ell = kL, k\in\mathbb{Z} \\ 0 & \textit{otherwise} \end{cases} \tag{3.222}$$

Proof. See (Napolitano, 2012, Sec. 5.11.2). □

When $L_1 \neq L_2$, the argument of the cyclic cross-correlation function $\widetilde{\widetilde{R}}{}_{\boldsymbol{x}}^{\widetilde{\alpha}}(\cdot)$ contains the time variable n. Consequently, under the mild assumption that $\widetilde{\widetilde{R}}{}_{\boldsymbol{x}}^{\widetilde{\alpha}}(\cdot) \in \ell_1(\mathbb{Z})$, it follows that, for every fixed m, the cross-correlation function in (3.221) as a function of n does not contain any finite-strength additive sinewave component. That is, $x_{I1}(n)$ and $x_{I2}^{(*)}(n)$ do not exhibit joint cyclostationarity at any cycle frequency:

$$\widetilde{R}^{\widetilde{\alpha}}_{x_{I1}x_{I2}^{(*)}}(m) \triangleq \lim_{N\to\infty} \frac{1}{2N+1} \sum_{n=-N}^{N} \mathrm{E}\left\{x_{I1}(n+m)\, x_{I2}^{(*)}(n)\right\} e^{-j2\pi\widetilde{\alpha}n} = 0 \quad \forall \widetilde{\alpha} \in [-1/2, 1/2)\,. \tag{3.223}$$

In particular, their time-averaged (conjugate) cross-correlation $\widetilde{R}^{0}_{x_{I1}x_{I2}^{(*)}}(m)$ is identically zero. Moreover, from (3.221) it follows that $x_{I1}(n)$ and $x_{I2}(n)$ are jointly ACS if and only if $L_1 = L_2 = L$. In such a case, their cyclic cross-correlation function is (Napolitano, 2012, Sec. 5.11.3)

$$\widetilde{R}^{\widetilde{\alpha}}_{x_{I1}x_{I2}^{(*)}}(m) = \frac{1}{L}\, \widetilde{\widetilde{R}}{}_{\boldsymbol{x}}^{\widetilde{\alpha}L}\left[\frac{m}{L}\right] \qquad \widetilde{\alpha}L \bmod 1 \in \widetilde{A}\,. \tag{3.224}$$

Furthermore, since from (3.224) it follows

$$\widetilde{R}_{\boldsymbol{x}}^{\widetilde{\alpha}}(m) = L\, \widetilde{R}^{\widetilde{\alpha}/L}_{x_{I1}x_{I2}^{(*)}}(mL) \tag{3.225}$$

in accordance with the Fourier-transform pair (3.217) and (3.218), the cyclic cross-spectra of $x_{I1}(n)$ and $x_{I2}(n)$ are

$$\widetilde{S}^{\widetilde{\alpha}}_{x_{I1}x^{(*)}_{I2}}(\nu) = \frac{1}{L}\widetilde{S}^{\widetilde{\alpha}L}_{\boldsymbol{x}}(\nu L) \qquad \widetilde{\alpha}L \bmod 1 \in \widetilde{A}\,. \tag{3.226}$$

In the special case of $x_1 \equiv x_2$, (3.224) and (3.226) reduce to (Izzo and Napolitano, 1998b, eqs. (32) and (34)) specialized to second order.

The problem of image suppression in (3.226) is addressed in (Napolitano, 2012, Sec. 4.10.6).

3.7.2 Sampling

Let $x_{\delta i}(n)$ be the sampled version of $x_i(n)$, with sampling period M_i $(i = 1, 2)$. That is

$$x_{\delta i}(n) \triangleq x_i(n) \sum_{\ell=-\infty}^{+\infty} \delta_{n-\ell M_i} = \sum_{\ell=-\infty}^{+\infty} x_i(\ell M_i)\, \delta_{n-\ell M_i}\,. \tag{3.227}$$

Accounting for (C.17), the Fourier transforms of the sampled processes are

$$X_{\delta i}(\nu) = X_i(\nu) \otimes \frac{1}{M_i}\sum_{p=0}^{M_i-1} \widetilde{\delta}\left(\nu - \frac{p}{M_i}\right) = \frac{1}{M_i}\sum_{p=0}^{M_i-1} X_i\left(\nu - \frac{p}{M_i}\right) \tag{3.228}$$

where $\otimes$ denotes periodic convolution with period 1.

Theorem 3.28. (Napolitano, 2012, Sec. 4.10.2). *Let $x_1(n)$ and $x_2(n)$ be jointly ACS, with Loève bifrequency cross-spectrum (3.215). The Loève bifrequency cross-spectrum of the sampled processes $x_{\delta 1}(n)$ and $x_{\delta 2}(n)$ is given by*

$$\mathrm{E}\left\{X_{\delta 1}(\nu_1)\, X^{(*)}_{\delta 2}(\nu_2)\right\} = \frac{1}{M_1 M_2} \sum_{p_1=0}^{M_1-1} \sum_{p_2=0}^{M_2-1} \sum_{\widetilde{\alpha}\in\widetilde{A}} \widetilde{S}^{\widetilde{\alpha}}_{\boldsymbol{x}}\left(\nu_1 - \frac{p_1}{M_1}\right) \widetilde{\delta}\left(\nu_2 - \frac{p_2}{M_2} - (-)\left(\widetilde{\alpha} - \nu_1 + \frac{p_1}{M_1}\right)\right). \tag{3.229}$$

Proof. It is an immediate consequence of (3.215) and (3.228). □

From (3.229) it follows that the sampled processes are in turn jointly ACS for every value of M_1 and M_2. Such a result is not surprising since sampling is a linear periodically time-variant transformation (Napolitano, 1995).

Theorem 3.29. (Napolitano, 2012, Sec. 4.10.2). *For $M_1 = M_2 = M$, the (conjugate) cyclic cross-correlation function of $x_{\delta 1}(n)$ and $x_{\delta 2}(n)$ is given by*

$$\widetilde{R}^{\widetilde{\alpha}}_{x_{\delta 1}x^{(*)}_{\delta 2}}(m) = \frac{1}{M}\sum_{q=0}^{M-1} \widetilde{R}^{\widetilde{\alpha}-q/M}_{\boldsymbol{x}}(m)\, \delta_{m \bmod M}\,. \tag{3.230}$$

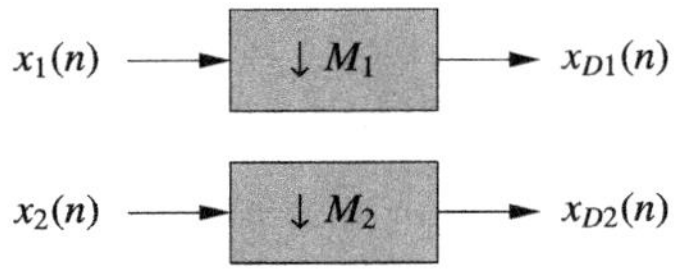

FIGURE 3.9 Decimation of two processes with different decimation factors.

Proof. See (Napolitano, 2012, Sec. 5.11.4). □

Theorem 3.30. (Napolitano, 2012, Sec. 4.10.2). *For $M_1 = M_2 = M$, the (conjugate) cyclic cross-spectrum of the sampled processes $x_{\delta 1}(n)$ and $x_{\delta 2}(n)$ is given by*

$$\widetilde{S}^{\widetilde{\alpha}}_{x_{\delta 1} x^{(*)}_{\delta 2}}(\nu) = \frac{1}{M} \sum_{q=0}^{M-1} \widetilde{S}_{\boldsymbol{x}}^{\widetilde{\alpha} - q/M}(\nu) \otimes \frac{1}{M} \sum_{p=0}^{M-1} \widetilde{\delta}\left(\nu - \frac{p}{M}\right)$$

$$= \frac{1}{M^2} \sum_{p=0}^{M-1} \sum_{q=0}^{M-1} \widetilde{S}_{\boldsymbol{x}}^{\widetilde{\alpha} - q/M}\left(\nu - \frac{p}{M}\right). \tag{3.231}$$

Proof. In (3.231), $\otimes$ denotes periodic convolution with period 1. By Fourier transforming both sides of (3.230), using (C.17) and the fact that $\widetilde{\delta}(\nu)$ is the unit element for the periodic convolution with period 1, we obtain (3.231). □

3.7.3 Decimation (downsampling)

Let $x_{Di}(n)$ be the decimated (downsampled) version of $x_i(n)$, with decimation factor M_i $(i = 1, 2)$ (Fig. 3.9). That is,

$$x_{Di}(n) = x_i(nM_i) = x_{\delta i}(nM_i). \tag{3.232}$$

The (conjugate) cross-correlation functions of decimated, sampled, and original signals are linked by the relationship

$$\mathrm{E}\left\{x_{D1}(n+m)\, x^{(*)}_{D2}(n)\right\} = \mathrm{E}\left\{x_1((n+m)M_1)\, x_2^{(*)}(nM_2)\right\}$$

$$= \mathrm{E}\left\{x_{\delta 1}((n+m)M_1)\, x^{(*)}_{\delta 2}(nM_2)\right\}. \tag{3.233}$$

Since the sampled process is an expanded version of the decimated process, that is $x_{\delta i}(n) = x_{Di}[n/M_i]$, accounting for (3.218), the Fourier transform of the sampled and decimated processes are linked by $X_{\delta i}(\nu) = X_{Di}(\nu M_i)$. Thus, using (3.228), the Fourier transforms of the decimated processes are

$$X_{Di}(\nu) = X_{\delta i}\left(\frac{\nu}{M_i}\right) = \frac{1}{M_i} \sum_{p=0}^{M_i - 1} X_i\left(\frac{\nu - p}{M_i}\right). \tag{3.234}$$

Theorem 3.31. (Napolitano, 2012, Sec. 4.10.3.2). *Let $x_1(n)$ and $x_2(n)$ be jointly ACS, with Loève bifrequency cross-spectrum (3.215). The Loève bifrequency cross-spectrum of the decimated processes $x_{D1}(n)$ and $x_{D2}(n)$ is given by*

$$
\begin{aligned}
&\mathrm{E}\left\{X_{D1}(\nu_1)\, X_{D2}^{(*)}(\nu_2)\right\} \\
&\quad = \frac{1}{M_1} \sum_{p_1=0}^{M_1-1} \sum_{\widetilde{\alpha}\in\widetilde{A}} \widetilde{S_{\boldsymbol{x}}^{\alpha}}\left(\frac{\nu_1 - p_1}{M_1}\right) \widetilde{\delta}_{\nu_2}\left(\nu_2 - {}_{(-)}\left(\widetilde{\alpha} M_2 - (\nu_1 - p_1)\frac{M_2}{M_1}\right)\right). \qquad (3.235)
\end{aligned}
$$

Proof. It is consequence of (3.234), the scaling property of Dirac delta (Zemanian, 1987, Sec. 1.7), and identity (Napolitano, 2012, Eq. (5.192)). □

The cyclic spectra $\widetilde{S_{\boldsymbol{x}}^{\alpha}}((\nu_1 - p_1)/M_1)$ are periodic in ν_1 with period 1 and the delta train is periodic in both ν_1 and ν_2 with period 1. In addition, from (3.235) it follows that the Loève bifrequency cross-spectrum of the decimated processes $x_{D1}(n)$ and $x_{D2}(n)$ in the principal domain $(\nu_1, \nu_2) \in [-1/2, 1/2)^2$ has support (in the case ${}_{(-)}$ present)

$$
\begin{aligned}
\nu_2 = (M_2/M_1)\nu_1 - \widetilde{\alpha} M_2 - (M_2/M_1) p_1 \bmod 1 \\
\widetilde{\alpha} \in \widetilde{A},\ p_1 \in \{0, 1, \ldots, M_1 - 1\} \qquad (3.236)
\end{aligned}
$$

that is, lines with slope M_2/M_1. Consequently, unlike the case of sampled processes, the decimated processes $x_{D1}(n)$ and $x_{D2}(n)$ are jointly SC. In particular, for a single process $x_1(n) \equiv x_2(n) = x(n)$, by taking $M_1 = 1$, that is $x_{D1}(n) \equiv x(n)$, it follows that the ACS process $x(n)$ and its decimated version $x_{D2}(n)$ are jointly SC with support lines with slope M_2. In the special case of $x(n)$ WSS, the support line is unique.

Theorem 3.32. (Napolitano, 2012, Sec. 4.10.3.2). *Let $x_1(n)$ and $x_2(n)$ be jointly ACS with (conjugate) cross-correlation (3.212). The (conjugate) cross-correlation function of the decimated processes $x_{D1}(n)$ and $x_{D2}(n)$ is given by*

$$
\begin{aligned}
&\mathrm{E}\left\{x_{D1}(n+m)\, x_{D2}^{(*)}(n)\right\} \\
&\quad = \int_{[-1/2,1/2]^2} \mathrm{E}\left\{X_{D1}(\nu_1)\, X_{D2}^{(*)}(\nu_2)\right\} e^{j2\pi[\nu_1(n+m) + {}_{(-)}\nu_2 n]} \mathrm{d}\nu_1 \mathrm{d}\nu_2 \\
&\quad = \sum_{\widetilde{\alpha}\in\widetilde{A}} e^{j2\pi\widetilde{\alpha} M_2 n} \widetilde{R_{\boldsymbol{x}}^{\alpha}}\left((M_1 - M_2)n + M_1 m\right). \qquad (3.237)
\end{aligned}
$$

Proof. See (Napolitano, 2012, Sec. 5.11.6). □

By reasoning as for (3.221), we have that if $\widetilde{R_{\boldsymbol{x}}^{\alpha}}(\cdot) \in \ell^1(\mathbb{Z})$ then for $M_1 \neq M_2$ the right-hand side of (3.237), as a function of n, does not contain any finite-strength additive sinewave component. That is, for $M_1 \neq M_2$ the processes $x_{D1}(n)$ and $x_{D2}^{(*)}(n)$ do not exhibit joint cyclostationarity at any cycle frequency. From (3.235) and (3.237) it follows that $x_{D1}(n)$ and $x_{D2}(n)$ are jointly ACS if and only if $M_1 = M_2 = M$.

Theorem 3.33. (Napolitano, 2012, Sec. 4.10.3.2). *For $M_1 = M_2 = M$, the (conjugate) cyclic cross-correlation function of the decimated processes $x_{D1}(n)$ and $x_{D2}(n)$ is given by*

$$\widetilde{R}^{\widetilde{\alpha}}_{x_{D1}x^{(*)}_{D2}}(m) = \sum_{q=0}^{M-1} \widetilde{R}_{\boldsymbol{x}}^{(\widetilde{\alpha}-q)/M}(mM) \,. \tag{3.238}$$

Proof. See (Napolitano, 2012, Sec. 5.11.7). □

Theorem 3.34. (Napolitano, 2012, Sec. 4.10.3.2). *For $M_1 = M_2 = M$, the (conjugate) cyclic cross-spectrum of the decimated processes $x_{D1}(n)$ and $x_{D2}(n)$ is given by*

$$\widetilde{S}^{\widetilde{\alpha}}_{x_{D1}x^{(*)}_{D2}}(\nu) = \frac{1}{M} \sum_{p=0}^{M-1} \sum_{q=0}^{M-1} \widetilde{S}_{\boldsymbol{x}}^{(\widetilde{\alpha}-q)/M}\left(\frac{\nu-p}{M}\right) . \tag{3.239}$$

Proof. Eq. (3.239) is obtained by Fourier transforming both sides of (3.238) and accounting for the Fourier-transform pair (3.232) and (3.234). □

In the special case $x_1 \equiv x_2$, (3.238) and (3.239) reduce to (Izzo and Napolitano, 1998b, eqs. (22) and (24)) specialized to second order. In addition, by comparing (3.230) with (3.238) the following relationship between (conjugate) cyclic cross-correlations of decimated and sampled signals can be established

$$\widetilde{R}^{\widetilde{\alpha}}_{x_{D1}x^{(*)}_{D2}}(m) = M \, \widetilde{R}^{\widetilde{\alpha}/M}_{x_{\delta1}x^{(*)}_{\delta2}}(mM) \tag{3.240}$$

or, equivalently

$$\widetilde{R}^{\widetilde{\alpha}}_{x_{\delta1}x^{(*)}_{\delta2}}(m) = \frac{1}{M} \widetilde{R}^{\widetilde{\alpha}M}_{x_{D1}x^{(*)}_{D2}}\left[\frac{m}{M}\right] . \tag{3.241}$$

Moreover, by comparing (3.231) with (3.239) it results

$$\widetilde{S}^{\widetilde{\alpha}}_{x_{D1}x^{(*)}_{D2}}(\nu) = M \, \widetilde{S}^{\widetilde{\alpha}/M}_{x_{\delta1}x^{(*)}_{\delta2}}\left(\frac{\nu}{M}\right) . \tag{3.242}$$

3.8 Special topics

Special topics related to filtering and sampling of cyclostationary and almost-cyclostationary signals are:

- *Sampling with jitter*
  (Dehay and Monsan, 2007);
- *Models for time variant systems*
  (Duverdier et al., 1999), (Duverdier and Lacaze, 2000), (Huang et al., 2002b), (Guidorzi and Diversi, 2003), (Sadeghi and Yu, 2016);

- *Digital-to-analog conversion*
  (Mandyam, 2002);
- *Cyclic spectral analysis with missing observations*
  (Dunsmuir, 1984), (Dandawaté and Giannakis, 1993), (Giannakis and Zhou, 1994), (Drake et al., 2014), (Drake et al., 2015).

3.9 Summary

LAPTV systems, also referred to as FRESH filters, transform ACS processes into ACS processes with different cyclic features. That is, the class of the ACS processes is closed under LAPTV transformations. Input/output relations in terms of (conjugate) cross-correlation (Theorem 3.6), (conjugate) cyclic cross-correlations and (conjugate) cyclic cross-spectra (Corollary 3.7) are derived for two LAPTV systems with two jointly ACS inputs. Input/output relations for a single system with a single input are obtained a special case.

Cyclic statistics of the product of ACS processes are also derived (Theorem 3.10).

Cyclic statistics of signals involved in the Rice's representation of a real-valued signal are derived (Section 3.5). Specifically, cyclic statistics of real-valued signal, analytic signal, complex envelope, and in-phase and in-quadrature components are linked among them.

Cyclic statistics of the discrete-time signal obtained by uniformly sampling with sampling frequency f_s a continuous-time ACS signal are derived. It is shown that the discrete-time signal is ACS (Corollary 3.18) and its cyclic statistics are obtained by the superposition of replicas of those of the continuous-time ACS signal (Lemmas 3.17 and 3.19).

For a continuous-time signal with bandwidth B, the sufficient condition $f_s \geqslant 4B$ is derived such that the Loève bifrequency spectrum and the (conjugate) cyclic spectrum of the discrete-time sampled signal are amplitude- and frequency-scaled versions of the corresponding functions of the continuous-time signal in the whole main cycle and spectral frequency domains (Theorems 3.22 and 3.24). A similar sampling theorem is proved for the (conjugate) cyclic autocorrelation function (Theorem 3.23). Condition $f_s \geqslant 4B$ is more restrictive than condition $f_s \geqslant 2B$ which is sufficient to have non-overlapping replicas in the aliasing formulas or to reconstruct the process by the Shannon interpolation formula.

For bandpass signals adopted in communications, the anti-aliasing condition requires very high sampling rates. This problem can be circumvented by bandpass sampling. It consists in sampling the continuous-time bandpass signal by a sampling frequency significantly lower than the Nyquist frequency $2B$ and comparable with the Nyquist frequency of the complex envelope. Admissible values of the sampling frequency for bandpass sampling are derived such that (conjugate) cyclic spectra of the complex envelope can be reconstructed by samples of the continuous-time signal (Theorem 3.25).

The effects of expansors (upsamplers) (Theorems 3.26 and 3.27) and decimators (downsamplers) (Theorems 3.31, 3.32, 3.33, 3.34) on ACS processes are analyzed since these devices are the main building blocks of multirate systems. These devices are linear time-variant systems that are not LAPTV. The expanded (upsampled) version of an ACS process is ACS. Similarly, the decimated (downsampled) version of an ACS process is ACS. However, expanded or decimated versions of the same ACS signal with different expansion or decimation factors are not jointly ACS but, rather, jointly spectrally correlated.

3.10 Proofs

Proof of Theorem 3.5

Here the proof of the FOT counterpart of Theorem 3.5 is reported, since the proof in the FOT approach is more challenging than the analogous proof in the stochastic process framework.

$$
\begin{aligned}
\mathrm{E}^{\{\alpha\}}\left\{y_i(t)\right\} &= \mathrm{E}^{\{\alpha\}}\left\{\int_{\mathbb{R}} h_i(t,u)\, x_i(u)\,\mathrm{d}u\right\} \\
&= \mathrm{E}^{\{\alpha\}}\left\{\int_{\mathbb{R}} \sum_{\sigma_i\in J_i} h_{\sigma_i}(t-u)\, e^{j2\pi\sigma_i u}\, x_i(u)\,\mathrm{d}u\right\} \\
&= \mathrm{E}^{\{\alpha\}}\left\{\int_{\mathbb{R}} \sum_{\sigma_i\in J_i} h_{\sigma_i}(s)\, e^{j2\pi\sigma_i (t-s)}\, x_i(t-s)\,\mathrm{d}s\right\} \\
&= \int_{\mathbb{R}} \sum_{\sigma_i\in J_i} h_{\sigma_i}(s)\, e^{j2\pi\sigma_i (t-s)}\, \mathrm{E}^{\{\alpha\}}\{x_i(t-s)\}\,\mathrm{d}s \\
&= \mathrm{E}^{\{\alpha\}}\left\{\int_{\mathbb{R}} \sum_{\sigma_i\in J_i} h_{\sigma_i}(s)\, e^{j2\pi\sigma_i (t-s)} \sum_{\eta_i\in E_{1i}} \overline{x}_{\eta_i}\, e^{j2\pi\eta_i (t-s)}\,\mathrm{d}s\right\} \\
&= \sum_{\sigma_i\in J_i}\sum_{\eta_i\in E_{1i}} \overline{x}_{\eta_i} \int_{\mathbb{R}} h_{\sigma_i}(s)\, e^{-j2\pi(\eta_i+\sigma_i)s}\,\mathrm{d}s\; e^{j2\pi(\eta_i+\sigma_i)t}\,.
\end{aligned}
\tag{3.243}
$$

In the third equality the variable change $t-u=s$ is made. This allows to interchange the order of integral and almost-periodic component extraction operator (Theorem 2.32 and Remark 2.30).

Eq. (3.5) follows accounting for (3.12).

Assumption 3.2a assures that

$$\sum_{\sigma_i\in J_i} h_{\sigma_i}(s)\, e^{-j2\pi\sigma_i s} \in L^1(\mathbb{R}) \tag{3.244}$$

and Assumption 3.3 assures that

$$\sum_{\eta_i\in E_{1i}} \left|\overline{x}_{\eta_i}\right| < \infty\,. \tag{3.245}$$

Thus the assumptions for Theorem 2.32 are satisfied.

Proof of Theorem 3.6

Here the proof of the FOT counterpart of Theorem 3.6 is reported, since the proof in the FOT approach is more challenging than the analogous proof in the stochastic process framework.

By substituting (3.7) into (3.6) we have

$$
\begin{aligned}
&\mathrm{E}^{\{\alpha\}}\{y_1(t+\tau)\, y_2^{(*)}(t)\} \\
&\quad = \mathrm{E}^{\{\alpha\}}\left\{\int_{\mathbb{R}} h_1(t+\tau, u_1)\, x_1(u_1)\, \mathrm{d}u_1 \int_{\mathbb{R}} h_2^{(*)}(t, u_2)\, x_2^{(*)}(u_2)\, \mathrm{d}u_2\right\} \\
&\quad = \mathrm{E}^{\{\alpha\}}\left\{\int_{\mathbb{R}} \sum_{\sigma_1 \in J_1} h_{\sigma_1}(t+\tau-u_1)\, e^{j2\pi\sigma_1 u_1}\, x_1(u_1)\, \mathrm{d}u_1\right. \\
&\qquad\qquad \left.\int_{\mathbb{R}} \sum_{\sigma_2 \in J_2} h_{\sigma_2}^{(*)}(t-u_2)\, e^{(-)j2\pi\sigma_2 u_2}\, x_2^{(*)}(u_2)\, \mathrm{d}u_2\right\} \\
&\quad = \mathrm{E}^{\{\alpha\}}\left\{\int_{\mathbb{R}} \sum_{\sigma_1 \in J_1} h_{\sigma_1}(s_1+\tau)\, e^{j2\pi\sigma_1 (t-s_1)}\, x_1(t-s_1)\, \mathrm{d}s_1\right. \\
&\qquad\qquad \left.\int_{\mathbb{R}} \sum_{\sigma_2 \in J_2} h_{\sigma_2}^{(*)}(s_2)\, e^{(-)j2\pi\sigma_2 (t-s_2)}\, x_2^{(*)}(t-s_2)\, \mathrm{d}s_2\right\}
\end{aligned} \tag{3.246}
$$

where, in the last equality, the variable changes $s_1 = t - u_1$ and $s_2 = t - u_2$ are made.

Under Assumptions 3.2a and 3.4b, or alternatively under Assumptions 3.2b and 3.4a, the order of integration and summation operations can be interchanged. In fact, the Fubini and Tonelli theorem (Champeney, 1990, Chapter 3) and the dominated convergence theorem (Champeney, 1990, Chapter 4) can be applied. In addition, under the above assumptions also the almost-periodic component extraction operator can be interchanged with the integration and summation operations (Theorem 2.32). Finally, by observing that almost-periodic functions can be led outside the almost-periodic component extraction operator (see also Remark 2.30), we get

$$
\begin{aligned}
&\mathrm{E}^{\{\alpha\}}\{y_1(t+\tau)\, y_2^{(*)}(t)\} \\
&\quad = \sum_{\sigma_1 \in J_1} \sum_{\sigma_2 \in J_2} \int_{\mathbb{R}} \int_{\mathbb{R}} h_{\sigma_1}(s_1+\tau)\, h_{\sigma_2}^{(*)}(s_2)\, e^{-j2\pi\sigma_1 s_1}\, e^{-(-)j2\pi\sigma_2 s_2} \\
&\qquad \mathrm{E}^{\{\alpha\}}\left\{x_1(t-s_1)\, x_2^{(*)}(t-s_2)\right\}\, \mathrm{d}s_1\, \mathrm{d}s_2\, e^{j2\pi(\sigma_1 + (-)\sigma_2)t} \\
&\quad = \sum_{\alpha \in A_{12}} \sum_{\sigma_1 \in J_1} \sum_{\sigma_2 \in J_2} \int_{\mathbb{R}} \int_{\mathbb{R}} R^{\alpha}_{x_1 x_2^{(*)}}(s_2 - s_1)\, h_{\sigma_1}(s_1+\tau)\, h_{\sigma_2}^{(*)}(s_2) \\
&\qquad e^{-j2\pi\sigma_1 s_1}\, e^{-(-)j2\pi\sigma_2 s_2}\, e^{-j2\pi\alpha s_2}\, \mathrm{d}s_1\, \mathrm{d}s_2\, e^{j2\pi(\alpha + \sigma_1 + (-)\sigma_2)t}
\end{aligned}
$$

$$= \sum_{\alpha \in A_{12}} \sum_{\sigma_1 \in J_1} \sum_{\sigma_2 \in J_2} \int_{\mathbb{R}} R^{\alpha}_{x_1 x_2^{(*)}}(u)\, e^{j2\pi\sigma_1 u}$$

$$\int_{\mathbb{R}} h_{\sigma_1}(s_2 - u + \tau)\, h_{\sigma_2}^{(*)}(s_2)\, e^{-j2\pi(\alpha+\sigma_1+(-)\sigma_2)s_2} \mathrm{d}s_2 \,\mathrm{d}u\, e^{j2\pi(\alpha+\sigma_1+(-)\sigma_2)t} \tag{3.247}$$

where in the second equality, according to (1.74) (with the replacements $t - s_1 \curvearrowright t + \tau$ and $t - s_2 \curvearrowright t$), we put

$$\mathrm{E}^{\{\alpha\}}\left\{x_1(t-s_1)\, x_2^{(*)}(t-s_2)\right\} = \sum_{\alpha \in A_{12}} R^{\alpha}_{x_1 x_2^{(*)}}(s_2 - s_1)\, e^{j2\pi\alpha(t-s_2)} \tag{3.248}$$

and in the last equality the variable change $s_2 - s_1 = u$ is made.

By substituting (3.14) into (3.247), we get the FOT counterpart of (3.13) which is formally the same as (3.13).

By replacing (3.14) into (3.247), for the generic term of the multiple sum in (3.247) it results that

$$\begin{aligned}
&\left|\left[R^{\alpha}_{x_1 x_2^{(*)}}(\tau)\, e^{j2\pi\sigma_1\tau}\right] \underset{\tau}{\otimes} r^{\alpha+\sigma_1+(-)\sigma_2}_{\sigma_1\sigma_2(*)}(\tau)\right| \\
&\quad = \left|\int_{\mathbb{R}} R^{\alpha}_{x_1 x_2^{(*)}}(s)\, e^{j2\pi\sigma_1 s}\, r^{\alpha+\sigma_1+(-)\sigma_2}_{\sigma_1\sigma_2(*)}(\tau - s)\,\mathrm{d}s\right| \\
&\quad \leqslant \int_{\mathbb{R}} \left|R^{\alpha}_{x_1 x_2^{(*)}}(s)\, r^{\alpha+\sigma_1+(-)\sigma_2}_{\sigma_1\sigma_2(*)}(\tau - s)\right| \mathrm{d}s \\
&\quad \leqslant \int_{\mathbb{R}} \left|R^{\alpha}_{x_1 x_2^{(*)}}(s)\right| \mathrm{d}s\, \left\|r^{\alpha+\sigma_1+(-)\sigma_2}_{\sigma_1\sigma_2(*)}\right\|_{\infty}
\end{aligned} \tag{3.249}$$

Using Assumption 3.2 we have

$$\begin{aligned}
\left\|r^{\alpha+\sigma_1+(-)\sigma_2}_{\sigma_1\sigma_2(*)}\right\|_{\infty} &\leqslant \int_{\mathbb{R}} \left|h_{\sigma_1}(\tau + s)\, h_{\sigma_2}^{(*)}(s)\right| \mathrm{d}s \\
&\leqslant \int_{\mathbb{R}} \left|h_{\sigma_2}(s)\right| \mathrm{d}s\, \left\|h_{\sigma_1}\right\|_{\infty} < \infty\,.
\end{aligned} \tag{3.250}$$

Therefore, accounting for Assumptions 3.2 and 3.4, the series in the right-hand side of (3.13) is convergent.

In addition, the assumptions for (a bi-dimensional version of) Theorem 2.32 hold. In fact, accounting for Assumptions 3.2 and 3.4 we have

1. $\sum_{\sigma_1} h_{\sigma_1}(s_1) \sum_{\sigma_2} h_{\sigma_2}(s_2) \in L^1(\mathbb{R}^2)$.
2. The function $t \mapsto x_1(t-s_1)\, x_2^{(*)}(t-s_2) \in L^1_{\mathrm{loc}}(\mathbb{R}) \cap L^{\infty}(\mathbb{R})$ and $R^{\alpha}_{x_1 x_2^{(*)}}(\tau)$ exists $\forall \alpha \in A$.
3. $\sum_{\alpha} \|R^{\alpha}_{x_1 x_2^{(*)}}\|_{\infty} < \infty$.

Proof of Corollary 3.7

By taking the Fourier coefficient at frequency α of the almost periodic-function in the FOT counterpart of (3.13), we have

$$\begin{aligned} R^{\alpha}_{y_1 y_2^{(*)}}(\tau) &\triangleq \left\langle y_1(t+\tau)\, y_2^{(*)}(t)\, e^{-j2\pi\alpha t} \right\rangle_t \\ &= \sum_{\beta\in A_{12}} \sum_{\sigma_1\in J_1} \sum_{\sigma_2\in J_2} \left[R^{\beta}_{x_1 x_2^{(*)}}(\tau)\, e^{j2\pi\sigma_1\tau} \right] \underset{\tau}{\otimes} r^{\beta+\sigma_1+(-)\sigma_2}_{\sigma_1\sigma_2(*)}(\tau) \\ &\quad \left\langle e^{j2\pi(\beta+\sigma_1+(-)\sigma_2)t}\, e^{-j2\pi\alpha t} \right\rangle_t . \end{aligned} \tag{3.251}$$

Since

$$\left\langle e^{j2\pi(\beta+\sigma_1+(-)\sigma_2)t}\, e^{-j2\pi\alpha t} \right\rangle_t = \delta_{\alpha-(\beta+\sigma_1+(-)\sigma_2)} \tag{3.252}$$

the FOT counterpart of (3.15), which is formally identical to (3.15), easily follows.

Under Assumptions 3.2 (a and b) and 3.4a, accounting for inequalities (3.249), (3.250), we have that $R^{\alpha}_{y_1 y_2^{(*)}}(\tau) \in L^1(\mathbb{R})$. Thus, its Fourier transform $S^{\alpha}_{y_1 y_2^{(*)}}(f)$ exists in ordinary sense.

Proof of (3.24a) and (3.24b)

Accounting for the Fourier pairs

$$\begin{aligned} &h(-\tau) \overset{\mathcal{F}}{\longleftrightarrow} H(-f) \\ &h^{(*)}(-\tau) \overset{\mathcal{F}}{\longleftrightarrow} H^{(*)}((-)(-f)) \\ &h^{(*)}(-\tau)\, e^{j2\pi\alpha\tau} \overset{\mathcal{F}}{\longleftrightarrow} H^{(*)}((-)(-(f-\alpha))) = H^{(*)}((-)(\alpha-f)) \end{aligned}$$

we have

$$\begin{aligned} r^{\alpha}_{hh^{(*)}}(\tau) &\triangleq \int_{\mathbb{R}} h(s+\tau)\, h^{(*)}(s)\, e^{-j2\pi\alpha s}\, \mathrm{d}s \\ &= \int_{\mathbb{R}} h(\tau-s')\, h^{(*)}(-s')\, e^{j2\pi\alpha s'}\, \mathrm{d}s' \\ &= h(\tau) \otimes \left[h^{(*)}(-\tau)\, e^{j2\pi\alpha\tau} \right] \end{aligned} \tag{3.253a}$$

$$\begin{aligned} r^{\alpha}_{hh^{(*)}}(\tau) &\triangleq \int_{\mathbb{R}} h(s+\tau)\, h^{(*)}(s)\, e^{-j2\pi\alpha s}\, \mathrm{d}s \\ &= \underset{s\to\alpha}{\mathcal{F}} \left[h(\tau+s)\, h^{(*)}(s) \right] \\ &= \left[H(\alpha)\, e^{j2\pi\alpha\tau} \right] \underset{\alpha}{\otimes} H^{(*)}((-)\alpha) \end{aligned}$$

$$= \int_{\mathbb{R}} H(\lambda)\, H^{(*)}((-)(\alpha - \lambda))\, e^{j2\pi\lambda\tau}\, d\lambda \tag{3.253b}$$

Therefore, we have the Fourier transform pair

$$r^{\alpha}_{hh^{(*)}}(\tau) \triangleq \int_{\mathbb{R}} h(t+\tau)\, h^{(*)}(t)\, e^{-j2\pi\alpha t}\, dt \overset{\mathcal{F}}{\longleftrightarrow} H(f)\, H^{(*)}((-)(\alpha - f)) \tag{3.254}$$

Proof of Theorem 3.10

By using the FOT counterparts of (3.40) and (3.41), which are formally identical to (3.40) and (3.41), we get

$$\begin{aligned} R^{\alpha}_{y_1 y_2^{(*)}}(\tau) &\triangleq \left\langle y_1(t+\tau)\, y_2^{(*)}(t)\, e^{-j2\pi\alpha t} \right\rangle_t \\ &= \left\langle \mathrm{E}^{\{\alpha\}}\left\{ y_1(t+\tau)\, y_2^{(*)}(t) \right\} e^{-j2\pi\alpha t} \right\rangle_t \\ &= \left\langle \mathrm{E}^{\{\alpha\}}\left\{ c_1(t+\tau)\, c_2^{(*)}(t) \right\} \mathrm{E}^{\{\alpha\}}\left\{ x_1(t+\tau)\, x_2^{(*)}(t) \right\} e^{-j2\pi\alpha t} \right\rangle_t \\ &= \left\langle \sum_{\alpha_c \in A_{c_1 c_2^{(*)}}} R^{\alpha_c}_{c_1 c_2^{(*)}}(\tau)\, e^{j2\pi\alpha_c t} \sum_{\alpha_x \in A_{x_1 x_2^{(*)}}} R^{\alpha_x}_{x_1 x_2^{(*)}}(\tau)\, e^{j2\pi\alpha_x t}\, e^{-j2\pi\alpha t} \right\rangle_t \\ &= \sum_{\alpha_x \in A_{x_1 x_2^{(*)}}} \sum_{\alpha_c \in A_{c_1 c_2^{(*)}}} R^{\alpha_x}_{x_1 x_2^{(*)}}(\tau)\, R^{\alpha_c}_{c_1 c_2^{(*)}}(\tau) \\ &\quad \left\langle e^{j2\pi\alpha_x t}\, e^{j2\pi\alpha_c t}\, e^{-j2\pi\alpha t} \right\rangle_t \end{aligned} \tag{3.255}$$

from which the FOT counterpart of (3.44), formally identical to (3.44), follows observing that

$$\left\langle e^{j2\pi\alpha_x t}\, e^{j2\pi\alpha_c t}\, e^{-j2\pi\alpha t} \right\rangle_t = \delta_{\alpha - \alpha_x - \alpha_c}\,. \tag{3.256}$$

Under Assumption 3.4a for $R^{\alpha}_{x_1 x_2^{(*)}}(\tau)$ and Assumption 3.4b for $R^{\alpha}_{c_1 c_2^{(*)}}(\tau)$, we have

$$\left| R^{\alpha}_{y_1 y_2^{(*)}}(\tau) \right| \leqslant \sum_{\alpha_x \in A_{x_1 x_2^{(*)}}} \left| R^{\alpha_x}_{x_1 x_2^{(*)}}(\tau) \right| \sup_{\alpha_c \in A_{c_1 c_2^{(*)}}} \left\| R^{\alpha_c}_{c_1 c_2^{(*)}} \right\|_{\infty} \in L^1(\mathbb{R})\,. \tag{3.257}$$

Thus, the Fourier transform of $R^{\alpha}_{y_1 y_2^{(*)}}(\tau)$ exists in the ordinary sense and is given by (3.45).

The FOT counterpart of (3.45) easily follows by Fourier transforming both sides of the FOT counterpart of (3.44).

Proof of Lemma 3.17

Accounting for (3.185) and (3.184) we have

$$
\begin{aligned}
\mathrm{E}\left\{X(\nu_1)\, X^{(*)}(\nu_2)\right\} &= \mathrm{E}\left\{\frac{1}{T_s}\sum_{k_1\in\mathbb{Z}} X_a((\nu_1-k_1)f_s)\,\frac{1}{T_s}\sum_{k_2\in\mathbb{Z}} X_a^{(*)}((\nu_2-k_2)f_s)\right\}\\
&= \frac{1}{T_s^2}\sum_{k_1\in\mathbb{Z}}\sum_{k_2\in\mathbb{Z}} \mathrm{E}\left\{X_a((\nu_1-k_1)f_s)\, X_a^{(*)}((\nu_2-k_2)f_s)\right\}\\
&= \frac{1}{T_s^2}\sum_{k_1\in\mathbb{Z}}\sum_{k_2\in\mathbb{Z}}\sum_{\alpha\in A_{\boldsymbol{x}_a}} S^{\alpha}_{\boldsymbol{x}_a}((\nu_1-k_1)f_s)\\
&\qquad \delta\left((\nu_2-k_2)f_s-(-)(\alpha-(\nu_1-k_1)f_s)\right)\\
&= \frac{1}{T_s}\sum_{\alpha\in A_{\boldsymbol{x}_a}}\sum_{k_1\in\mathbb{Z}} S^{\alpha}_{\boldsymbol{x}_a}((\nu_1-k_1)f_s)\\
&\qquad \sum_{k_2\in\mathbb{Z}}\delta\left((\nu_2-k_2)-(-)(\alpha/f_s-(\nu_1-k_1))\right)
\end{aligned}
\tag{3.258}
$$

where, in the last equality, the scaling property of the Dirac delta $f_s\delta(\nu f_s)=\delta(\nu)$ is used (Zemanian, 1987, Sec. 1.7).

Proof of Lemma 3.19

The proof of (3.190) in the stochastic approach is straightforward. In contrast, the proof of the FOT Counterpart of (3.190) is more elaborated. In the following, a formal derivation of the FOT Counterpart of (3.190) is reported. Moreover, the general case of continuous-time GACS signals (Chapter 12) is considered which include the ACS signals as special case. For continuous-time GACS signals, the set of cycle frequencies depends on the lag parameter.

The FOT counterpart of (3.190) is

$$
\mathrm{E}^{\{\widetilde{\alpha}\}}\{x(n+m)\,x^{(*)}(n)\} = \mathrm{E}^{\{\alpha\}}\{x_a(t+\tau)\,x_a(t)\}\Big|_{t=nT_s,\tau=mT_s}
\tag{3.259}
$$

where $\mathrm{E}^{\{\alpha\}}\{\cdot\}$ is the *continuous-time* almost-periodic component extraction operator (see (2.22)) and $\mathrm{E}^{\{\widetilde{\alpha}\}}\{\cdot\}$ is the *discrete-time* almost-periodic component extraction operator which is defined similarly to its continuous-time counterpart.

The second-order lag product $x_a(t+\tau)\,x_a^{(*)}(t)$ can be decomposed into the sum of its almost-periodic component and a residual term $\ell_{\boldsymbol{x}_a}(t,\tau)$ not containing any finite-strength additive sine-wave component (Section 2.3.1.6).

For a GACS signal it results

$$
\begin{aligned}
x_a(t+\tau)\,x_a^{(*)}(t) &= \mathrm{E}^{\{\alpha\}}\left\{x_a(t+\tau)\,x_a^{(*)}(t)\right\} + \ell_{\boldsymbol{x}_a}(t,\tau)\\
&= \sum_{\alpha\in A_\tau} R^{\alpha}_{\boldsymbol{x}_a}(\tau)\,e^{j2\pi\alpha t} + \ell_{\boldsymbol{x}_a}(t,\tau)
\end{aligned}
\tag{3.260}
$$

where subscript $\boldsymbol{x}_a \equiv x_a\, x_a^{(*)}$, $A_\tau \triangleq A_{\boldsymbol{x}_a,\tau}$, and

$$\left\langle \ell_{\boldsymbol{x}_a}(t,\tau)\, e^{-j2\pi\alpha t} \right\rangle_t \equiv 0 \quad \forall \alpha \in \mathbb{R}\,. \tag{3.261}$$

For any $0 < \epsilon < T_s$ we formally have

$$\begin{aligned}
&\frac{1}{2N+1} \sum_{n=-N}^{N} \ell_{\boldsymbol{x}_a}(t,\tau)\big|_{t=nT_s,\tau=mT_s}\; e^{-j2\pi\widetilde{\alpha}n} \\
&= \frac{1}{2N+1} \int_{-NT_s-\epsilon}^{NT_s+\epsilon} \ell_{\boldsymbol{x}_a}(t,mT_s) \sum_{n=-\infty}^{+\infty} \delta(t-nT_s)\, e^{-j2\pi\widetilde{\alpha}t/T_s}\, \mathrm{d}t \\
&= \sum_{p=-\infty}^{+\infty} \frac{1}{(2N+1)T_s} \int_{-NT_s-\epsilon}^{NT_s+\epsilon} \ell_{\boldsymbol{x}_a}(t,mT_s)\, e^{-j2\pi(\widetilde{\alpha}-p)t/T_s}\, \mathrm{d}t
\end{aligned} \tag{3.262}$$

where, in the second equality, the Poisson formula (C.8a)

$$\sum_{n=-\infty}^{+\infty} \delta(t-nT_s) = \frac{1}{T_s} \sum_{p=-\infty}^{+\infty} e^{j2\pi pt/T_s} \tag{3.263}$$

is accounted for. Taking the limit for $N \to \infty$ in (3.262) and accounting for (3.261) one has

$$\lim_{N\to\infty} \frac{1}{2N+1} \sum_{n=-N}^{N} \ell_{\boldsymbol{x}_a}(t,\tau)\big|_{t=nT_s,\tau=mT_s}\; e^{-j2\pi\widetilde{\alpha}n} = 0\,. \tag{3.264}$$

That is, if the continuous-time residual term $\ell_{\boldsymbol{x}_a}(t,\tau)$ does not contain any (continuous-time) finite-strength additive sine-wave component (see (3.261)), then also the sampled residual term $\ell_{\boldsymbol{x}_a}(nT_s,mT_s)$ does not contain any discrete-time finite-strength additive sine-wave component.

Thus,

$$\begin{aligned}
&\mathrm{E}^{\{\widetilde{\alpha}\}}\{x(n+m)\,x^{(*)}(n)\} \\
&= \mathrm{E}^{\{\widetilde{\alpha}\}}\left\{ x_a(t+\tau)\,x_a^{(*)}(t)\Big|_{t=nT_s,\tau=mT_s} \right\} \\
&= \mathrm{E}^{\{\widetilde{\alpha}\}}\left\{ \mathrm{E}^{\{\alpha\}}\{x_a(t+\tau)\,x_a^{(*)}(t)\}\Big|_{t=nT_s,\tau=mT_s} \right\} \\
&\quad + \mathrm{E}^{\{\widetilde{\alpha}\}}\left\{ \ell_{\boldsymbol{x}_a}(t,\tau)\big|_{t=nT_s,\tau=mT_s} \right\} \\
&= \mathrm{E}^{\{\widetilde{\alpha}\}}\left\{ \sum_{\alpha\in A_\tau} R_{\boldsymbol{x}_a}^{\alpha}(\tau)\, e^{j2\pi\alpha t}\Big|_{t=nT_s,\tau=mT_s} \right\} \\
&= \sum_{\alpha\in A_\tau} R_{\boldsymbol{x}_a}^{\alpha}(\tau)\, e^{j2\pi\alpha t}\Big|_{t=nT_s,\tau=mT_s}
\end{aligned} \tag{3.265}$$

where in the third equality (3.264) is accounted for and in the last equality the fact that the right-hand side of (3.265) is a discrete-time almost periodic function is exploited.

Eq. (3.259) is then proved observing that the right-hand side of (3.265) is coincident with the right-hand side of (3.259).

Note that uniformly sampling a continuous-time GACS signal leads to a discrete-time ACS signal (Section 12.4).

Proof of the FOT counterpart of (3.191)
Accounting for (3.265), we have

$$
\begin{aligned}
\widetilde{R}_{\boldsymbol{x}}^{\widetilde{\alpha}}(m) &= \left\langle x(n+m)\, x(n)\, e^{-j2\pi\widetilde{\alpha}n} \right\rangle_n \\
&= \left\langle \sum_{\alpha\in A_{mT_s}} R_{\boldsymbol{x}_a}^{\alpha}(mT_s)\, e^{j2\pi\alpha nT_s}\, e^{-j2\pi\widetilde{\alpha}n} \right\rangle_n \\
&= \sum_{\alpha\in A_{mT_s}} R_{\boldsymbol{x}_a}^{\alpha}(mT_s) \underbrace{\left\langle e^{j2\pi(\alpha T_s-\widetilde{\alpha})n} \right\rangle_n}_{\sum_{p\in\mathbb{Z}} \delta_{\alpha T_s-\widetilde{\alpha}+p}} \\
&= \sum_{p\in\mathbb{Z}} R_{\boldsymbol{x}_a}^{(\widetilde{\alpha}-p)f_s}(mT_s)
\end{aligned}
\tag{3.266}
$$

where in the last equality we use the fact that there is nonzero contribution only for those cycle frequencies α such that $\alpha = (\widetilde{\alpha}-p)/T_s$, $p\in\mathbb{Z}$.

Proof of the FOT counterpart of (3.192)
Using (3.266) we have

$$
\begin{aligned}
\widetilde{S}_{\boldsymbol{x}}^{\widetilde{\alpha}}(\nu) &= \sum_{m\in\mathbb{Z}} \widetilde{R}_{\boldsymbol{x}}^{\widetilde{\alpha}}(m)\, e^{-j2\pi\nu m} \\
&= \sum_{m\in\mathbb{Z}} \left[\sum_{p\in\mathbb{Z}} R_{\boldsymbol{x}_a}^{(\widetilde{\alpha}-p)f_s}(mT_s) \right] e^{-j2\pi\nu m} \\
&= \sum_{p\in\mathbb{Z}} \left[\sum_{m\in\mathbb{Z}} R_{\boldsymbol{x}_a}^{(\widetilde{\alpha}-p)f_s}(mT_s)\, e^{-j2\pi\nu f_s mT_s} \right] \\
&= \frac{1}{T_s} \sum_{p\in\mathbb{Z}} \sum_{q\in\mathbb{Z}} S_{\boldsymbol{x}_a}^{(\widetilde{\alpha}-p)f_s}((\nu-q)f_s)
\end{aligned}
\tag{3.267}
$$

where, in the last equality, the second Poisson's summation formula (C.12) is used.

Proof of Theorem 3.22

From (3.195a) and (3.195b), under Assumption 3.20 ($x_a(t)$ strictly band-limited), and accounting for (3.193a), the support of the replica with $k_1 = k_2 = 0$ in the aliasing formula (3.187) is given by

$$\begin{aligned}
&\operatorname{supp}\left\{S^{\alpha}_{\boldsymbol{x}_a}(\nu_1 f_s)\,\delta(\nu_2 - (-)(\alpha/f_s - \nu_1))\right\} \\
&\quad\subseteq \{(\nu_1,\nu_2) \in \mathbb{R}\times\mathbb{R} : |\nu_1 f_s| \leqslant B,\ |\alpha - \nu_1 f_s| \leqslant B, \\
&\qquad \nu_2 - (-)(\alpha/f_s - \nu_1) = 0\} &(3.268a)\\
&\quad= \{(\nu_1,\nu_2) \in \mathbb{R}\times\mathbb{R} : |\nu_1| \leqslant B/f_s,\ |\nu_2| \leqslant B/f_s,\ \alpha = (\nu_1 + (-)\nu_2) f_s\} &(3.268b)
\end{aligned}$$

Replicas in (3.187) are separated by 1 in both ν_1 and ν_2 variables. Thus, from (3.268b) it follows that $B/f_s \leqslant 1/2$, that is,

$$f_s \geqslant 2B \tag{3.269}$$

is a *sufficient condition such that replicas do not overlap.* Note that, however, (3.269) does not assure that the mappings $f_1 = \nu_1 f_s$ and $\alpha = \widetilde{\alpha} f_s$ in

$$\widetilde{S}^{\widetilde{\alpha}}_{\boldsymbol{x}}(\nu_1) = \frac{1}{T_s}\, S^{\alpha}_{\boldsymbol{x}_a}(f_1)\Big|_{f_1=\nu_1 f_s,\alpha=\widetilde{\alpha} f_s} \tag{3.270}$$

holds $\forall\widetilde{\alpha} \in [-1/2, 1/2)$ and $\forall\nu_1 \in [-1/2, 1/2)$. For example, for $(*)$ present and $\alpha = -B$, equality (3.270) holds only for $\nu_1 \in [-1/2, 0]$ and not for $\nu_1 \in [0, 1/2]$. In fact, for $\nu_1 \in [0, 1/2]$ the density of the replica with $n_1 = 0$, $n_2 = 1$ is present in the right-hand side of (3.270).

In contrast, condition $B/f_s \leqslant 1/4$, that is,

$$f_s \geqslant 4B \tag{3.271}$$

assures in (3.268b)

$$|\nu_1| \leqslant 1/4, \quad |\nu_2| \leqslant 1/4, \quad |\alpha/f_s| = |\nu_1 + (-)\nu_2| \leqslant |\nu_1| + |\nu_2| \leqslant 1/2 \tag{3.272}$$

and, consequently, the mappings $f_1 = \nu_1 f_s$ and $\alpha = \widetilde{\alpha} f_s$ in (3.270) hold $\forall\widetilde{\alpha} \in [-1/2, 1/2)$ and $\forall\nu_1 \in [-1/2, 1/2)$. Moreover, (3.196) holds. Finally, note that the effect of sampling at two times the Nyquist rate (see (3.271)) leads also to (3.197).

Proof of Theorem 3.25

From (3.205), it follows that the support of the rescaled cyclic spectrum of the sampled signal is such that

$$\operatorname{supp}\left\{\widetilde{S}^{\widetilde{\alpha}}_{xx}(\nu)|_{\nu=f/f_s,\widetilde{\alpha}=\alpha/f_s}\right\} \subseteq \mathcal{D}_3(\alpha, f) \tag{3.273}$$

where

$$\mathcal{D}_3(\alpha, f) \triangleq \bigcup_{p\in\mathbb{Z}}\bigcup_{q\in\mathbb{Z}} \mathcal{D}_2(\alpha - pf_s, f - qf_s)$$

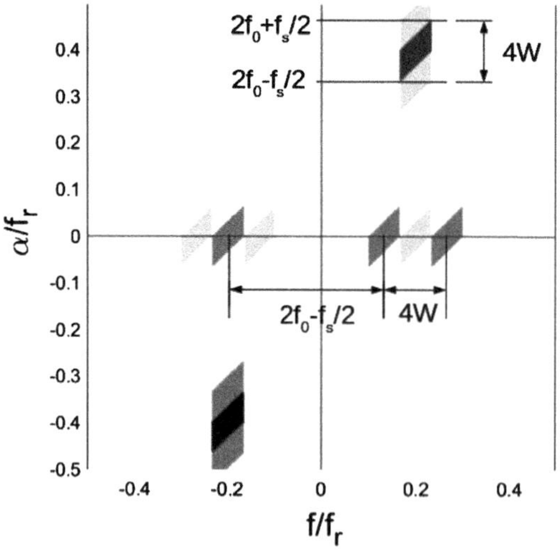

FIGURE 3.10 Reciprocal positions of replicas of $\mathcal{D}_1(\alpha, f)$ in $\mathcal{D}_3(\alpha, f)$ defined in (3.274). Only the replicas closest to $\mathcal{D}_2(\alpha, f)$ defined in (3.137) are represented. In this example, $2f_0 - f_s/2 = mf_s$, $f_s = 4W$. *Source: (Napolitano and Doğançay, 2018) Copyright of Elsevier.*

$$= \bigcup_{p\in\mathbb{Z}} \bigcup_{q\in\mathbb{Z}} \Big\{ \mathcal{D}_1(\alpha - pf_s, f - f_0 - qf_s) \cup \mathcal{D}_1(\alpha - pf_s, f + f_0 - qf_s)$$

$$\cup\, \mathcal{D}_1(\alpha - 2f_0 - pf_s, f - f_0 - qf_s) \cup \mathcal{D}_1(\alpha + 2f_0 - pf_s, f + f_0 - qf_s) \Big\} \quad (3.274)$$

with $\mathcal{D}_1(\alpha, f)$ and $\mathcal{D}_2(\alpha, f)$ defined in (3.132) and (3.137), respectively.

In the following, sufficient conditions are derived such that the set $\mathcal{D}_3(\alpha, f)$ in (3.274) is constituted by the union of non-overlapping translations of the set $\mathcal{D}_1(\alpha, f)$.

In (3.274), in order to avoid overlap along the α direction, it is sufficient that there exists an integer $m \geqslant 1$ such that $2f_0 - f_s/2 = mf_s$ and, moreover (Fig. 3.10),

$$\mathcal{D}_1(\alpha - mf_s, f - f_0) \underset{\alpha}{\preceq} \mathcal{D}_1(\alpha - 2f_0, f - f_0) \underset{\alpha}{\preceq} \mathcal{D}_1(\alpha - (m+1)f_s, f - f_0) \quad (3.275)$$

or, equivalently,

$$\mathcal{D}_1(\alpha - mf_s, f + f_0) \underset{\alpha}{\preceq} \mathcal{D}_1(\alpha + 2f_0, f + f_0) \underset{\alpha}{\preceq} \mathcal{D}_1(\alpha - (m+1)f_s, f + f_0)\,. \quad (3.276)$$

In (3.275) and (3.276), the notation $\mathcal{D}' \underset{\alpha}{\preceq} \mathcal{D}''$ means that the domain $\mathcal{D}'$ is contained in a horizontal strip of the (α, f) plane with cycle frequencies less than those of the domain $\mathcal{D}''$. Consequently, accounting for (3.132), we must have $f_s \geqslant 4W$. Analogously, in order to avoid overlap along the f direction, it is again sufficient that there exists an integer $m \geqslant 1$ such that $2f_0 - f_s/2 = mf_s$ and, moreover (Fig. 3.10),

$$\mathcal{D}_1(\alpha, f - mf_s - f_0) \underset{f}{\preceq} \mathcal{D}_1(\alpha, f - f_0) \underset{f}{\preceq} \mathcal{D}_1(\alpha, f - (m+1)f_s - f_0) \quad (3.277)$$

or, equivalently,

$$\mathcal{D}_1(\alpha, f - mf_s + f_0) \underset{f}{\preceq} \mathcal{D}_1(\alpha, f + f_0) \underset{f}{\preceq} \mathcal{D}_1(\alpha, f - (m+1)f_s + f_0)\,. \tag{3.278}$$

In (3.277) and (3.278), the notation $\mathcal{D}' \underset{f}{\preceq} \mathcal{D}''$ means that the domain $\mathcal{D}'$ is contained in a vertical strip of the (α, f) plane with spectral frequencies less than those of the domain $\mathcal{D}''$. Thus we have no aliasing in α and f provided that (Fig. 3.10)

$$\begin{cases} 2f_0 - f_s/2 = mf_s \\ f_s \geqslant 4W \end{cases} \tag{3.279}$$

which is equivalent to (3.206).

If, for the purpose of cyclic spectral analysis, it is required that for each $\widetilde{\alpha} \in [-1/2, 1/2)$ only one replica (i.e., only one cyclic spectrum) in (3.205) gives nonzero contribution if $\nu \in [-1/2, 1/2)$, then, in all the above relations involving supports, the domain $\mathcal{D}_1(\alpha, f)$ must be replaced by (see (3.65a) and (3.65b) with $B = W$)

$$\mathcal{D}_1'(\alpha, f) \triangleq \{(\alpha, f) \in \mathbb{R} \times \mathbb{R} : |f| \leqslant W,\ |\alpha| \leqslant 2W\}\,. \tag{3.280}$$

In such a case, the condition to avoid the overlap of domains along the f direction remains unaltered. In contrast, in order to avoid the overlap of domains along the α direction, the inequality in (3.279) must be replaced by the more stringent condition $f_s \geqslant 8W$. Consequently, (3.206) must be replaced by (3.207).

4 Higher-order cyclostationarity

Contents

Abstract

Almost-cyclostationary signals are characterized in terms of their higher-order statistics: characteristic functions, moments, and cumulants in both time and frequency domains and in the continuous- and discrete-time cases. Input/output relations in terms of higher-order cyclic statistics are derived for multi-input/multi-output linear almost-periodically time variant systems. The discrete-time signal obtained by uniformly sampling a continuous-time almost-cyclostationary signal is considered. Cyclic higher-order statistics of the discrete-time signal are linked to those of the continuous-time signal and aliasing issues are addressed. Results are presented in the functional or fraction-of-time approach and easily interpreted in the classical stochastic approach.

Keywords

Higher-order statistics, Cyclic cumulant, Cyclic moment, Cyclic polyspectrum, Characteristic function, Submanifold

4.1 Introduction

As first defined in (Gardner, 1990c), in the fraction-of-time probability framework a signal $x(t)$ is said to exhibit *higher-order cyclostationarity* (HOCS) if there exists a homogeneous

Cyclostationary Processes and Time Series. https://doi.org/10.1016/B978-0-08-102708-0.00015-7

nonlinear time-invariant transformation of $x(t)$ of order greater than two such that the output of this transformation contains finite-strength additive sinewave components.

Alternatively, in the classical stochastic approach, a stochastic process is said to be higher-order almost-cyclostationary if its moments and cumulants of order greater than two are almost-periodic functions of time.

Motivations to study HOCS properties of signals include the following:

1. Signals not exhibiting second-order cyclostationarity can exhibit HOCS (Gardner and Spooner, 1994a).
2. Narrow-band filtering can destroy second-order (wide-sense) cyclostationarity. In fact, let us consider the input/output relationship for LTI systems in terms of cyclic spectra (3.22). If the bandwidth of $H(f)$ is smaller than the smallest nonzero second-order cycle frequency of the input signal $x(t)$, then the output signal $y(t)$ does not exhibit second-order wide-sense cyclostationarity. The signal $y(t)$, however, can exhibit HOCS (see also (Gupta, 1972)).
3. The exploitation of HOCS can be useful for signal classification (Spooner, 1995). Specifically, different communication signals, even if they exhibit the same second-order cyclostationarity properties, can exhibit different cyclic features of higher order. Moreover, different behaviors can be obtained for different conjugation configurations (Spooner, 1992), (Spooner and Gardner, 1994a), (Spooner, 1994).
4. Exploitation of HOCS can be useful for many estimation problems.

The theory of higher-order cyclostationarity in the FOT probability framework is introduced for continuous time in (Gardner, 1990c), (Gardner and Spooner, 1990) and developed in (Spooner, 1992), (Gardner and Spooner, 1994a), (Spooner and Gardner, 1994a), (Spooner, 1994), and, for discrete and continuous time, in (Napolitano, 1995). Treatments within the stochastic framework are presented in (Dandawaté, 1993), (Dandawaté and Giannakis, 1994a), (Gardner and Spooner, 1994b), (Dandawaté and Giannakis, 1995), (Lacoume et al., 1997).

In this chapter, cyclic higher-order (joint) statistics for both continuous-time and discrete-time time series are presented in the FOT probability framework by following (Gardner and Spooner, 1994a), (Spooner and Gardner, 1994a), and (Napolitano, 1995).

All the presented results can be immediately interpreted in the stochastic process framework by interpreting signals as stochastic processes and replacing the almost-periodic component extraction operator $\mathrm{E}^{\{\alpha\}}\{\cdot\}$ with the ensemble average $\mathrm{E}\{\cdot\}$ (Section 2.4).

4.2 Continuous-time signals

4.2.1 Moments

Let $x_1(t), \ldots, x_N(t)$ be N not necessarily distinct complex-valued continuous-time time series and let $(*)_k$ represent optional complex conjugation of the kth signal $x_k(t)$.

All the results in this chapter are derived under the following fundamental assumptions.

Assumption 4.1. *The complex-valued time series* $x_1(t), \ldots, x_N(t)$ *are such that for every* $(\tau_1, \ldots, \tau_N) \in \mathbb{R}^N$ *the* $2N$ *real-valued time series*

$$\mathrm{Re}\left[x_1(t+\tau_1)\right], \mathrm{Im}\left[x_1(t+\tau_1)\right], \ldots, \mathrm{Re}\left[x_N(t+\tau_N)\right], \mathrm{Im}\left[x_N(t+\tau_N)\right]$$

are jointly relatively measurable (Definition 2.4). □

Assumption 4.1 assures the existence of the $2N$th-order joint FOT (stationary) distribution function. The existence of the $2N$th-order joint FOT almost-periodic distribution function is assured by the following assumption.

Assumption 4.2. *The complex-valued time series* $x_1(t), \ldots, x_N(t)$ *are jointly* (T_0) *relatively measurable for every* $T_0 \in \mathbb{R}$ *(Definition 2.20). That is, the sequence*

$$\begin{aligned} n \mapsto \mathbf{1}_{\{\mathrm{Re}[x(t+\tau_1+nT_0)] \leqslant \xi_{1\mathrm{r}}\}} & \mathbf{1}_{\{\mathrm{Im}[x(t+\tau_1+nT_0)] \leqslant \xi_{1\mathrm{i}}\}} \\ \cdots \mathbf{1}_{\{\mathrm{Re}[x(t+\tau_N+nT_0)] \leqslant \xi_{N\mathrm{r}}\}} & \mathbf{1}_{\{\mathrm{Im}[x(t+\tau_N+nT_0)] \leqslant \xi_{N\mathrm{i}}\}} \end{aligned} \tag{4.1}$$

is relatively measurable for every $T_0 \in \mathbb{R}$*, for every* $(\tau_1, \ldots, \tau_N) \in \mathbb{R}^N$ *(and, hence, for every* $t \in \mathbb{R}$*), and for every* $(\xi_{1\mathrm{r}}, \xi_{1\mathrm{i}}, \ldots, \xi_{N\mathrm{r}}, \xi_{N\mathrm{i}}) \in \mathbb{R}^{2N} - \Xi_0$*, where* Ξ_0 *is at most a countable set of* $(2N-1)$*-dimensional manifolds of* $\mathbb{R}^{2N}$. □

Furthermore, even if not explicitly specified, the (generalized) Fourier series of the almost-periodic functions are assumed to be absolutely convergent.

Definition 4.3. The N time series $x_1(t), \ldots, x_N(t)$ are said to exhibit joint Nth-order wide-sense cyclostationarity with cycle frequency $\alpha \neq 0$ for a given conjugation configuration $(*)_1, \ldots, (*)_N$, if the Nth-order *cyclic temporal cross-moment function* (CTCMF)

$$\mathcal{R}_{\boldsymbol{x}}^{\alpha}(\boldsymbol{\tau}) \triangleq \left\langle \prod_{k=1}^{N} x_k^{(*)_k}(t+\tau_k) e^{-j2\pi\alpha t} \right\rangle_t \tag{4.2}$$

exists and is not identically zero. In (4.2), $\boldsymbol{\tau} \triangleq [\tau_1, \ldots, \tau_N]^{\mathrm{T}}$ and subscript $\boldsymbol{x} \equiv x_1^{(*)_1} \cdots x_N^{(*)_N}$. □

Thus, N time series exhibit wide-sense joint Nth-order cyclostationarity with cycle frequency $\alpha \neq 0$ for a given conjugation configuration if, for some $\boldsymbol{\tau}$, the *lag product waveform*

$$L_{\boldsymbol{x}}(t, \boldsymbol{\tau}) \triangleq \prod_{k=1}^{N} x_k^{(*)_k}(t+\tau_k) \tag{4.3}$$

contains a finite-strength additive sinewave component with frequency α, whose amplitude and phase are the magnitude and phase of $\mathcal{R}_{\boldsymbol{x}}^{\alpha}(\boldsymbol{\tau})$, respectively.

Definition 4.4. The Nth-oder *temporal cross-moment function* (TCMF) of the time series $x_1^{(*)_1}(t), \ldots, x_N^{(*)_N}(t)$ is defined as

$$\mathcal{R}_{\boldsymbol{x}}(t, \boldsymbol{\tau}) \triangleq \mathrm{E}^{\{\alpha\}}\left\{L_{\boldsymbol{x}}(t, \boldsymbol{\tau})\right\} \tag{4.4a}$$

$$= \sum_{\alpha \in A_x} \mathcal{R}_x^{\alpha}(\boldsymbol{\tau})\, e^{j2\pi\alpha t} \tag{4.4b}$$

where $\mathrm{E}^{\{\alpha\}}\{\cdot\}$ is the almost-periodic component extraction operator (Definition 2.14) and A_x is the countable set (not depending on $\boldsymbol{\tau}$) of the Nth-order cycle frequencies of the time series $x_1^{(*)_1}(t+\tau_1), \ldots, x_N^{(*)_N}(t+\tau_N)$ (for the given conjugation configuration) and (4.4b) holds provided that the series of the Fourier coefficients $\mathcal{R}_x^{\alpha}(\boldsymbol{\tau})$ is absolutely convergent. □

Note that both A_x and $\mathcal{R}_x^{\alpha}(\boldsymbol{\tau})$ depend on the conjugation configuration $(*)_1, \ldots, (*)_N$. Even for a single time series $x_1 \equiv \cdots \equiv x_N \equiv x$, different conjugation configurations can give rise to different sets of cycle frequencies and different behaviors of CTCMFs. Such a property is exploited for modulation format classification (Section 8.12), (Spooner, 1995), (Spooner and Nicholls, 2009).

In (4.4a), the almost-periodic component extraction operator $\mathrm{E}^{\{\alpha\}}\{\cdot\}$ extracts all the frequencies of the Nth-order lag product (4.3). These frequencies are a subset of those of the $2N$-variate fraction-of-time joint cumulative distribution of the real and imaginary parts of the time-series $x_k(t)$, $k = 1, \ldots, N$:

$$\begin{aligned} &F^{\{\alpha\}}_{x_{1\mathrm{r}}(t+\tau_1)\, x_{1\mathrm{i}}(t+\tau_1) \cdots x_{N\mathrm{r}}(t+\tau_N)\, x_{N\mathrm{i}}(t+\tau_N)}(\xi_{1\mathrm{r}}, \xi_{1\mathrm{i}}, \ldots, \xi_{N\mathrm{r}}, \xi_{N\mathrm{i}}) \\ &\quad \triangleq \mathrm{E}^{\{\alpha\}}\left\{\mathbf{1}_{\{x_{1\mathrm{r}}(t+\tau_1) \leqslant \xi_{1\mathrm{r}}\}}\, \mathbf{1}_{\{x_{1\mathrm{i}}(t+\tau_1) \leqslant \xi_{1\mathrm{i}}\}} \cdots \mathbf{1}_{\{x_{N\mathrm{r}}(t+\tau_N) \leqslant \xi_{N\mathrm{r}}\}}\, \mathbf{1}_{\{x_{N\mathrm{i}}(t+\tau_N) \leqslant \xi_{N\mathrm{i}}\}}\right\} \end{aligned} \tag{4.5}$$

where $x_{k\mathrm{r}}(t) \triangleq \mathrm{Re}[x_k(t)]$, $x_{k\mathrm{i}}(t) \triangleq \mathrm{Im}[x_k(t)]$, $\xi_{k\mathrm{r}} \triangleq \mathrm{Re}[\xi_k]$, $\xi_{k\mathrm{i}} \triangleq \mathrm{Im}[\xi_k]$.

Let us assume that the set $\Gamma^{(N)}$, defined as the generalization to Nth-order and for complex-valued time series of the set $\Gamma^{(2)}$ in (2.48), is countable. $\Gamma^{(N)}$ is the set of cycle frequencies for the $2N$th-order temporal joint distribution function of the real and imaginary parts of the complex-valued time series $x_1(t), \ldots, x_N(t)$. If the almost-periodic function (4.5) has a convergent (generalized) Fourier series, we have

$$\begin{aligned} &F^{\{\alpha\}}_{x_{1\mathrm{r}}(t+\tau_1)\, x_{1\mathrm{i}}(t+\tau_1) \cdots x_{N\mathrm{r}}(t+\tau_N)\, x_{N\mathrm{i}}(t+\tau_N)}(\xi_{1\mathrm{r}}, \xi_{1\mathrm{i}}, \ldots, \xi_{N\mathrm{r}}, \xi_{N\mathrm{i}}) \\ &\quad = \sum_{\gamma \in \Gamma^{(N)}} F^{\gamma}_{x_{1\mathrm{r}}\, x_{1\mathrm{i}} \cdots x_{N\mathrm{r}}\, x_{N\mathrm{i}}}(\xi_{1\mathrm{r}}, \xi_{1\mathrm{i}}, \ldots, \xi_{N\mathrm{r}}, \xi_{N\mathrm{i}}; \boldsymbol{\tau})\, e^{j2\pi\gamma t} \end{aligned} \tag{4.6}$$

with

$$\begin{aligned} &F^{\gamma}_{x_{1\mathrm{r}}\, x_{1\mathrm{i}} \cdots x_{N\mathrm{r}}\, x_{N\mathrm{i}}}(\xi_{1\mathrm{r}}, \xi_{1\mathrm{i}}, \ldots, \xi_{N\mathrm{r}}, \xi_{N\mathrm{i}}; \boldsymbol{\tau}) \\ &\quad \triangleq \left\langle \mathbf{1}_{\{x_{1\mathrm{r}}(t+\tau_1) \leqslant \xi_{1\mathrm{r}}\}}\, \mathbf{1}_{\{x_{1\mathrm{i}}(t+\tau_1) \leqslant \xi_{1\mathrm{i}}\}} \cdots \mathbf{1}_{\{x_{N\mathrm{r}}(t+\tau_N) \leqslant \xi_{N\mathrm{r}}\}}\, \mathbf{1}_{\{x_{N\mathrm{i}}(t+\tau_N) \leqslant \xi_{N\mathrm{i}}\}}\, e^{-j2\pi\gamma t} \right\rangle_t \end{aligned} \tag{4.7}$$

If the $2N$th-order partial derivative of the function in (4.5) exists, then the $2N$-variate fraction-of-time joint probability density function of the real and imaginary parts of the time-series $x_k(t)$, $k = 1, \ldots, N$, is given by

$$f^{\{\alpha\}}_{x_{1\mathrm{r}}(t+\tau_1)\, x_{1\mathrm{i}}(t+\tau_1) \cdots x_{N\mathrm{r}}(t+\tau_N)\, x_{N\mathrm{i}}(t+\tau_N)}(\xi_{1\mathrm{r}}, \xi_{1\mathrm{i}}, \ldots, \xi_{N\mathrm{r}}, \xi_{N\mathrm{i}})$$

$$\triangleq \frac{\partial^{2N}}{\partial \xi_{1\mathrm{r}}\, \partial \xi_{1\mathrm{i}} \cdots \partial \xi_{N\mathrm{r}}\, \partial \xi_{N\mathrm{i}}} F^{\{\alpha\}}_{x_{1\mathrm{r}}(t+\tau_1)\, x_{1\mathrm{i}}(t+\tau_1) \cdots x_{N\mathrm{r}}(t+\tau_N)\, x_{N\mathrm{i}}(t+\tau_N)}(\xi_{1\mathrm{r}}, \xi_{1\mathrm{i}}, \ldots, \xi_{N\mathrm{r}}, \xi_{N\mathrm{i}}) . \tag{4.8}$$

In addition, if the almost-periodic function (4.8) has a convergent Fourier series, we have

$$f^{\{\alpha\}}_{x_{1\mathrm{r}}(t+\tau_1)\, x_{1\mathrm{i}}(t+\tau_1) \cdots x_{N\mathrm{r}}(t+\tau_N)\, x_{N\mathrm{i}}(t+\tau_N)}(\xi_{1\mathrm{r}}, \xi_{1\mathrm{i}}, \ldots, \xi_{N\mathrm{r}}, \xi_{N\mathrm{i}}) = \sum_{\gamma \in \Gamma^{(N)}} f^{\gamma}_{x_{1\mathrm{r}}\, x_{1\mathrm{i}} \cdots x_{N\mathrm{r}}\, x_{N\mathrm{i}}}(\xi_{1\mathrm{r}}, \xi_{1\mathrm{i}}, \ldots, \xi_{N\mathrm{r}}, \xi_{N\mathrm{i}}; \boldsymbol{\tau})\, e^{j2\pi\gamma t} \tag{4.9}$$

with

$$\begin{aligned} f^{\gamma}_{x_{1\mathrm{r}}\, x_{1\mathrm{i}} \cdots x_{N\mathrm{r}}\, x_{N\mathrm{i}}}&(\xi_{1\mathrm{r}}, \xi_{1\mathrm{i}}, \ldots, \xi_{N\mathrm{r}}, \xi_{N\mathrm{i}}; \boldsymbol{\tau}) \\ &= \frac{\partial^{2N}}{\partial \xi_{1\mathrm{r}}\, \partial \xi_{1\mathrm{i}} \cdots \partial \xi_{N\mathrm{r}}\, \partial \xi_{N\mathrm{i}}} F^{\gamma}_{x_{1\mathrm{r}}\, x_{1\mathrm{i}} \cdots x_{N\mathrm{r}}\, x_{N\mathrm{i}}}(\xi_{1\mathrm{r}}, \xi_{1\mathrm{i}}, \ldots, \xi_{N\mathrm{r}}, \xi_{N\mathrm{i}}; \boldsymbol{\tau}) \\ &= \Big\langle \delta(\xi_{1\mathrm{r}} - x_{1\mathrm{r}}(t+\tau_1))\, \delta(\xi_{1\mathrm{i}} - x_{1\mathrm{i}}(t+\tau_1)) \\ &\qquad \cdots \delta(\xi_{N\mathrm{r}} - x_{N\mathrm{r}}(t+\tau_N))\, \delta(\xi_{N\mathrm{i}} - x_{N\mathrm{i}}(t+\tau_N))\, e^{-j2\pi\gamma t} \Big\rangle_t \end{aligned} \tag{4.10}$$

where the second equality is obtained by following the guidelines of the formal proof of (2.37) in Section 2.6.

The time-series $x_k(t+\tau_k)$, $k = 1, \ldots, N$ are *independent* (in the FOT sense for ACS time-series) if the $2N$th-order cumulative distribution (4.5) factorizes into the product of the N 2nd-order joint distributions of $\mathrm{Re}[x_k(t+\tau_k)]$ and $\mathrm{Im}[x_k(t+\tau_k)]$, (Gardner and Brown, 1991), (Gardner, 1994a).

We have the following generalization to higher-orders and complex signals of the Gardner theorem of temporal expectation (Theorem 2.24):

Theorem 4.5. Gardner Theorem of Temporal Expectation (Higher Order). *Let* $g(\cdot) : \mathbb{C}^N \mapsto \mathbb{C}$ *be a continuous and bounded function such that for any* $\ell \in \mathbb{C}$ *the equation* $g(\xi_1, \ldots, \xi_N) = \ell$ *admits at most a finite number of solutions* $(\xi_1, \ldots, \xi_N)$ *belonging to any finite parallelepiped of* $\mathbb{C}^N$. *If the Fourier series (4.6) is convergent, it results*

$$\begin{aligned} \mathrm{E}^{\{\alpha\}}&\left\{ g\left(x_1^{(*)_1}(t+\tau_1), \ldots, x_N^{(*)_N}(t+\tau_N) \right) \right\} \\ &= \int_{\mathbb{R}^{2N}} g\left(\xi_1^{(*)_1}, \ldots, \xi_N^{(*)_N} \right) \mathrm{d}F^{\{\alpha\}}_{x_{1\mathrm{r}}(t+\tau_1)\, x_{1\mathrm{i}}(t+\tau_1) \cdots x_{N\mathrm{r}}(t+\tau_N)\, x_{N\mathrm{i}}(t+\tau_N)}(\xi_{1\mathrm{r}}, \xi_{1\mathrm{i}}, \ldots, \xi_{N\mathrm{r}}, \xi_{N\mathrm{i}}) \qquad (4.11\mathrm{a}) \\ &= \int_{\mathbb{R}^{2N}} g\left(\xi_1^{(*)_1}, \ldots, \xi_N^{(*)_N} \right) f^{\{\alpha\}}_{x_{1\mathrm{r}}(t+\tau_1)\, x_{1\mathrm{i}}(t+\tau_1) \cdots x_{N\mathrm{r}}(t+\tau_N)\, x_{N\mathrm{i}}(t+\tau_N)}(\xi_{1\mathrm{r}}, \xi_{1\mathrm{i}}, \ldots, \xi_{N\mathrm{r}}, \xi_{N\mathrm{i}}) \end{aligned}$$

$$\mathrm{d}\xi_{1\mathrm{r}}\,\mathrm{d}\xi_{1\mathrm{i}}\cdots\mathrm{d}\xi_{N\mathrm{r}}\,\mathrm{d}\xi_{N\mathrm{i}} \tag{4.11b}$$

where the second equality holds provided that the fraction-of-time joint probability density function (4.8) exists. □

Eq. (4.4a) is obtained as special case of Theorem 4.5 when

$$g\left(\xi_1^{(*)_1},\ldots,\xi_N^{(*)_N}\right) \triangleq \prod_{k=1}^{N} \xi_k^{(*)_k}\,. \tag{4.12}$$

By considering Fourier transforms in a generalized sense (Zemanian, 1987), (Champeney, 1990, Chaps. 13 and 14), the frequency domain statistical characterization of the N time series $x_k^{(*)_k}(t)$, $k=1,\ldots,N$, for a given conjugation configuration, is provided by the following results (Gardner and Spooner, 1994a).

Definition 4.6. The N-dimensional Fourier transform of the CTCMF

$$\mathcal{S}_{\boldsymbol{x}}^{\alpha}(\boldsymbol{f}) \triangleq \int_{\mathbb{R}^N} \mathcal{R}_{\boldsymbol{x}}^{\alpha}(\boldsymbol{\tau})\, e^{-j2\pi \boldsymbol{f}^{\mathrm{T}}\boldsymbol{\tau}}\,\mathrm{d}\boldsymbol{\tau} \tag{4.13}$$

where $\boldsymbol{f} \triangleq [f_1,\ldots,f_N]^{\mathrm{T}}$, is called the Nth-order *cyclic spectral cross-moment function* (CSCMF). □

Theorem 4.7. *The CSCMF (4.13) can be written as*

$$\mathcal{S}_{\boldsymbol{x}}^{\alpha}(\boldsymbol{f}) = S_{\boldsymbol{x}}^{\alpha}(\boldsymbol{f}')\,\delta(\boldsymbol{f}^{\mathrm{T}}\mathbf{1}-\alpha) \tag{4.14}$$

where $\mathbf{1}$ *is the vector* $[1,\ldots,1]^{\mathrm{T}}$, *and prime denotes the operator that transforms a vector* $\boldsymbol{u} \triangleq [u_1,\ldots,u_K]^{\mathrm{T}}$ *into the reduced-dimension version* $\boldsymbol{u}' \triangleq [u_1,\ldots,u_{K-1}]^{\mathrm{T}}$. *The function* $S_{\boldsymbol{x}}^{\alpha}(\boldsymbol{f}')$, *referred to as the* reduced-dimension CSCMF *(RD-CSCMF), can be expressed as*

$$S_{\boldsymbol{x}}^{\alpha}(\boldsymbol{f}') = \int_{\mathbb{R}^{N-1}} R_{\boldsymbol{x}}^{\alpha}(\boldsymbol{\tau}')\, e^{-j2\pi \boldsymbol{f}'^{\mathrm{T}}\boldsymbol{\tau}'}\,\mathrm{d}\boldsymbol{\tau}' \tag{4.15}$$

where

$$R_{\boldsymbol{x}}^{\alpha}(\boldsymbol{\tau}') \triangleq \mathcal{R}_{\boldsymbol{x}}^{\alpha}(\boldsymbol{\tau})|_{\tau_N=0} \tag{4.16}$$

is the reduced-dimension CTCMF *(RD-CTCMF).*

Proof. See Section 4.11. □

Proposition 4.8. (Gardner and Spooner, 1994a). *The RD-CSCMF (4.15) can be expressed as*

$$S_{\boldsymbol{x}}^{\alpha}(\boldsymbol{f}') = \lim_{T\to\infty}\lim_{Z\to\infty}\frac{1}{Z}\int_{t-Z/2}^{t+Z/2}\frac{1}{T}X_{N,T}^{(*)_N}(u,(-)_N(\alpha-\boldsymbol{f}'^{\mathrm{T}}\mathbf{1}))\prod_{k=1}^{N-1}X_{k,T}^{(*)_k}(u,(-)_k f_k)\,\mathrm{d}u \tag{4.17}$$

where

$$X_{k,T}(t, f_k) \triangleq \int_{t-T/2}^{t+T/2} x_k(s) e^{-j2\pi f_k s} \mathrm{d}s \tag{4.18}$$

is the short-time Fourier transform (STFT) of $x_k(t)$ *and* $_{(-)_k}$ *denotes an optional minus sign that is linked to the optional conjugation* $(*)_k$.

Proof. It is a generalization of the proof of Theorem 2.40. □

Eq. (4.17) reveals that N time-series exhibit joint Nth-order wide-sense cyclostationarity with cycle frequency α for a given conjugation configuration (i.e., $\mathcal{R}_{\boldsymbol{x}}^{\alpha}(\boldsymbol{\tau}) \not\equiv 0$ or, equivalently, $S_{\boldsymbol{x}}^{\alpha}(\boldsymbol{f}') \not\equiv 0$) if and only if the Nth-order temporal cross-moment of their spectral components at frequencies f_k, whose sum is equal to α, is nonzero. In fact, by using

$$h_B(t) \triangleq B \operatorname{rect}(Bt) \tag{4.19}$$

with $B \triangleq 1/T$ and $\operatorname{rect}(t) = 1$ if $|t| \leqslant 1/2$ and $\operatorname{rect}(t) = 0$ if $|t| > 1/2$, Eq. (4.17) can be rewritten as

$$S_{\boldsymbol{x}}^{\alpha}(\boldsymbol{f}') = \lim_{B \to 0} \lim_{Z \to \infty} \frac{1}{Z} \int_{t-Z/2}^{t+Z/2} \frac{1}{B^{N-1}} \left[\left(x_N^{(*)_N}(u)\, e^{-j2\pi(\alpha - \boldsymbol{f}'^{\mathrm{T}}\mathbf{1})u} \right) \otimes h_B(u) \right]$$
$$\prod_{k=1}^{N-1} \left[\left(x_k^{(*)_k}(u)\, e^{-j2\pi f_k u} \right) \otimes h_B(u) \right] \mathrm{d}u \tag{4.20}$$

which is the Nth-order temporal cross-moment of low-pass-filtered versions of the frequency-shifted signals $x_k^{(*)_k}(t)\, e^{-j2\pi f_k t}$ when the sum of the frequency shifts is equal to α and the bandwidth B approaches zero. Such a property is the generalization to the order $N > 2$ of the spectral correlation property of signals that exhibit second-order cyclostationarity. Moreover, the relation (4.13) between the Nth-order cyclic spectral cross-moment function (4.14), (4.17) and Nth-order cyclic temporal cross-moment (4.2) is, as first pointed out in (Gardner, 1990c), the Nth-order generalization of the *Gardner Relation* or *Cyclic Wiener Relation*.

The time series $t \mapsto X_{k,T}(t, f_k)$, when $T \to \infty$, are jointly stationary in the FOT sense (Izzo and Napolitano, 2005, App. B). Thus, the expectation operator $\mathrm{E}^{\{\alpha\}}\{\cdot\}$ reduces to the infinite time average $\langle \cdot \rangle_t$ and the following definition can be given.

Definition 4.9. The Nth-order *spectral cross-moment function* (SCMF) of the time series $x_1^{(*)_1}(t), \ldots, x_N^{(*)_N}(t)$ is defined as

$$\mathcal{S}_{\boldsymbol{x}}(\boldsymbol{f}) \triangleq \lim_{T \to \infty} \left\langle \prod_{k=1}^{N} X_{k,T}^{(*)_k}(t, {}_{(-)_k} f_k) \right\rangle_t \tag{4.21}$$

where $X_{k,T}(t, f_k)$ are the STFTs defined in (4.18). □

Theorem 4.10. *The* N*th-order SCMF of the time series* $x_1^{(*)_1}(t), \ldots, x_N^{(*)_N}(t)$ *can be expressed as*

$$\mathcal{S}_{\boldsymbol{x}}(\boldsymbol{f}) = \int_{\mathbb{R}^N} \mathcal{R}_{\boldsymbol{x}}(0, \boldsymbol{\tau})\, e^{-j2\pi \boldsymbol{f}^{\mathsf{T}} \boldsymbol{\tau}}\, \mathrm{d}\boldsymbol{\tau} = \sum_{\alpha \in A_{\boldsymbol{x}}} S_{\boldsymbol{x}}^{\alpha}(\boldsymbol{f}')\, \delta(\boldsymbol{f}^{\mathsf{T}} \mathbf{1} - \alpha)\,. \tag{4.22}$$

Proof. See Section 4.11. □

Remark 4.11. *Conjugation Convention.* Note that a different convention is adopted in defining the Nth-order spectral cross-moment function (4.21) or its stochastic counterpart

$$\mathrm{E}\left\{\prod_{k=1}^{N} X_k^{(*)_k}((-)_k f_k)\right\}$$

and, at second order, the Loève bifrequency spectrum (1.56) or (1.81) or its FOT counterpart (2.63). In fact, in (4.21) considered with $N = 2$,

$$\mathcal{S}_{\boldsymbol{x}}(f_1, f_2) = \lim_{T \to \infty} \left\langle X_{1,T}^{(*)_1}(t, (-)_1 f_1)\, X_{2,T}^{(*)_2}(t, (-)_2 f_2) \right\rangle_t \tag{4.23}$$

unlike in (1.81), an optional minus sign $(-)_k$ multiplies each spectral frequency variable f_k.

If the stochastic spectral cross-moment function is defined according to the convention adopted for the Loève bifrequency spectrum (1.81), one formally has

$$\mathcal{S}_{\boldsymbol{x}}(\boldsymbol{f}) = \mathrm{E}\left\{\prod_{k=1}^{N} X_k^{(*)_k}(f_k)\right\} \tag{4.24}$$

and the Fourier transform relationship with the stochastic temporal cross-moment function is

$$\mathrm{E}\left\{\prod_{k=1}^{N} x_k^{(*)_k}(t + \tau_k)\right\} = \int_{\mathbb{R}^N} \mathrm{E}\left\{\prod_{k=1}^{N} X_k^{(*)_k}(f_k)\right\} e^{j2\pi \boldsymbol{f}^{(-)\mathsf{T}}(\mathbf{1}t + \boldsymbol{\tau})}\, \mathrm{d}\boldsymbol{\tau} \tag{4.25}$$

where $\boldsymbol{f}^{(-)} \triangleq [(-)_1 f_1, \ldots, (-)_N f_N]^{\mathsf{T}}$. In such a case the RD-CTCMF is given by

$$S_{\boldsymbol{x}}^{\alpha}(\boldsymbol{f}') = \int_{\mathbb{R}^{N-1}} R_{\boldsymbol{x}}^{\alpha}(\boldsymbol{\tau}')\, e^{-j2\pi \boldsymbol{f}'^{(-)\mathsf{T}} \boldsymbol{\tau}'}\, \mathrm{d}\boldsymbol{\tau}' \tag{4.26}$$

and (4.22) becomes

$$\mathcal{S}_{\boldsymbol{x}}(\boldsymbol{f}) = \sum_{\alpha \in A_{\boldsymbol{x}}} S_{\boldsymbol{x}}^{\alpha}(\boldsymbol{f}')\, \delta(\boldsymbol{f}^{(-)\mathsf{T}} \mathbf{1} - \alpha)\,. \tag{4.27}$$

Note that convention (4.21) is adopted for complex-valued ACS signals in (Spooner, 1992), (Gardner and Spooner, 1994a), (Spooner and Gardner, 1994a), (Spooner, 1994), (Napolitano, 1995), (Izzo and Napolitano, 1996), (Izzo and Napolitano, 1997), (Izzo and Napolitano, 1998b), (Flagiello et al., 2000), (Napolitano and Spooner, 2000) and for

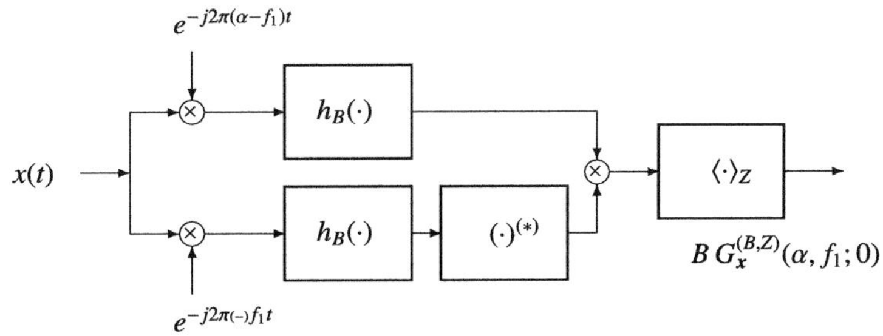

FIGURE 4.1 Second-oder spectral correlation analyzer ($(*)_2$ absent, notation of Section 5.2.4).

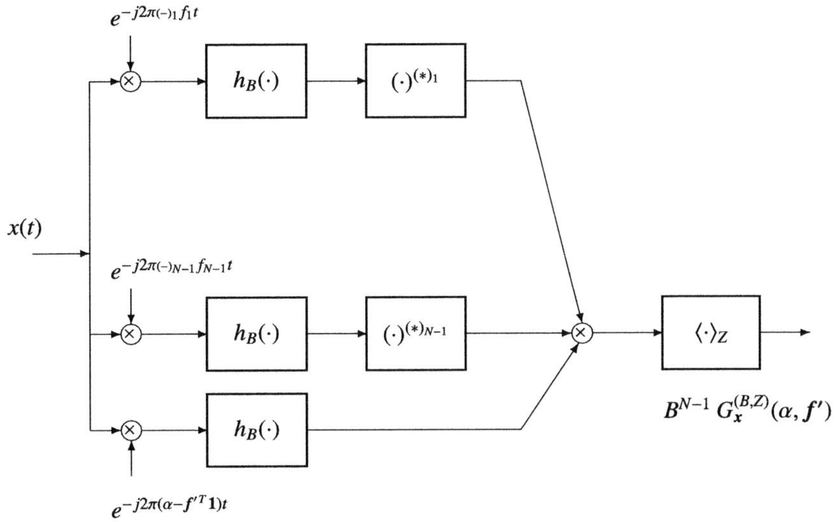

FIGURE 4.2 Nth-oder spectral correlation analyzer ($(*)_N$ absent, notation of Section 5.7).

complex-valued GACS signals in (Izzo and Napolitano, 1998a), (Izzo and Napolitano, 2002a), (Izzo and Napolitano, 2002b), (Izzo and Napolitano, 2003). In contrast, convention (4.24) is adopted in (Lacoume et al., 1997) and for complex-valued SC processes in (Napolitano, 2003), (Napolitano, 2011), (Napolitano, 2012). Thus, the existence of different conventions must be accounted for when results established in (Napolitano, 2003), (Napolitano, 2011), (Napolitano, 2012) for complex-valued SC processes are specialized to complex-valued ACS processes and used jointly with results derived using convention (4.21).

Similar considerations hold for the spectral cumulant and cyclic polyspectra defined in Section 4.2.2. □

In Figs. 4.1 and 4.2, the schemes of a second- and Nth-order spectral correlation analyzers are reported (see Chapter 5 for notation).

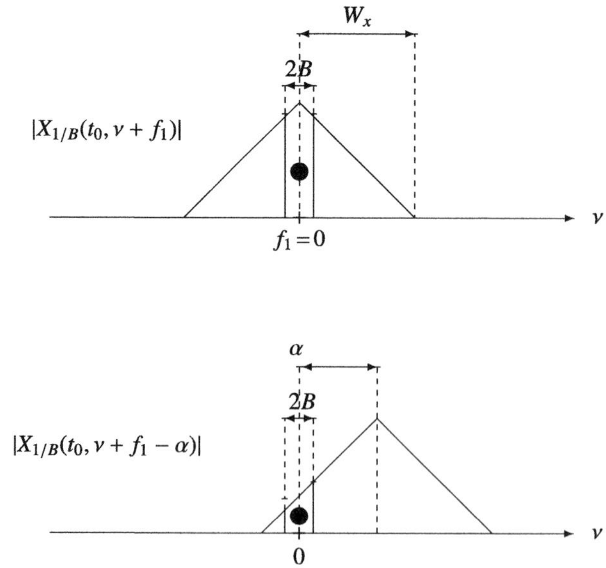

FIGURE 4.3 Second-oder spectral correlation for a wide-band cyclostationary signal ($2W_x > \alpha$). Bullets "•" denote correlated spectral bands.

Example 4.12. Let α be a candidate cycle frequency (e.g., the symbol rate of a PAM signal (Section 7.3)). A cyclostationary signal $x(t)$ with monolateral bandwidth $2W_x > \alpha$ can exhibit 2nd-order spectral correlation at cycle frequency α. In fact, spectral bands of the STFTs $X_{1/B}(t_0, \cdot)$ (defined according to (4.18)) whose central frequencies are separated by α can be correlated (Fig. 4.3). In contrast, if $2W_x < \alpha$, then no 2nd-order correlation exists between spectral bands separated by α (Fig. 4.4) However, in the case $2W_x < \alpha$, if 4th-order correlation is considered among the STFTs, more degrees of freedom are available for the frequency shifts of the STFTs so that the signal can exhibit 4th-oder cyclostationarity (Fig. 4.5, where $f_1 = f_2 = f_3$ and $\alpha = f_1 + f_2 + f_3$).

Low-pass filtering can destroy or significantly reduce the degree of second-order cyclostationarity if the bandwidth of the filter is smaller than the smallest cycle frequency (see (3.22)). Note that very narrow-band filtering makes the output almost Gaussian (Papoulis, 1972). Thus cumulants, and hence cyclic polyspectra, are negligible (Section E.1.4). □

4.2.2 Cumulants

In signal processing algorithms based on higher-order statistics, some advantages can be gained using cumulants rather than moments due to the following properties of cumulants:

1. The cumulant of the sum of two independent signals is the sum of the cumulants of the two signals, while this is not the case for the moments;

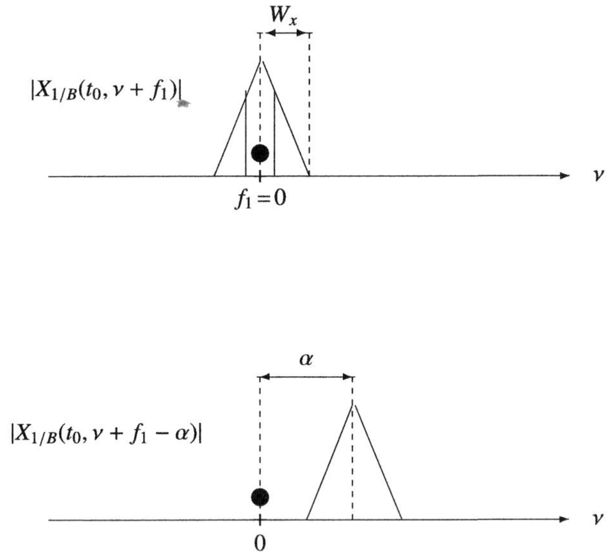

FIGURE 4.4 Absence of second-order spectral correlation for a narrow-band cyclostationary signal ($2W_x < \alpha$). Bullets "•" denote correlated spectral bands.

2. In the case of asymptotic independence of signals, cumulants, unlike moments, are summable functions of the lag parameters;
3. The cumulants of order greater than two of Gaussian signals is zero.

Definition 4.13. The Nth-order *temporal joint characteristic function* of the time series $x_1^{(*)_1}(t), \ldots, x_N^{(*)_N}(t)$ is defined as

$$\Phi_{\boldsymbol{x}}(t, \boldsymbol{\omega}, \boldsymbol{\tau}) \triangleq \mathrm{E}^{\{\alpha\}}\left\{\exp\left[j \sum_{k=1}^{N} \omega_k \, x_k^{(*)_k}(t + \tau_k)\right]\right\} \tag{4.28}$$

where $\boldsymbol{\omega} \triangleq [\omega_1, \ldots, \omega_N]^{\mathrm{T}}$, $\boldsymbol{\tau} \triangleq [\tau_1, \ldots, \tau_N]^{\mathrm{T}}$, and subscript $\boldsymbol{x} \equiv x_1^{(*)_1} \cdots x_N^{(*)_N}$. □

In (4.28), the almost-periodic component extraction operator $\mathrm{E}^{\{\alpha\}}\{\cdot\}$ extracts the frequencies of the $2N$-variate fraction-of-time joint probability density function of the real and imaginary parts of the time-series $x_k(t)$, $k = 1, \ldots, N$ according to (4.11a), (4.11b) with

$$g\left(\xi_1^{(*)_1}, \ldots, \xi_N^{(*)_N}\right) \triangleq \exp\left[j \sum_{k=1}^{N} \omega_k \, \xi_k^{(*)_k}\right]. \tag{4.29}$$

Using Theorem 4.5, we have

$$\Phi_{\boldsymbol{x}}(t, \boldsymbol{\omega}, \boldsymbol{\tau}) = \int_{\mathbb{R}^{2N}} \exp\left[j \sum_{k=1}^{N} \omega_k \left(\xi_{k\mathrm{r}} + (-)_k \, j \, \xi_{k\mathrm{i}}\right)\right] \mathrm{d}F^{\{\alpha\}}_{x_{1\mathrm{r}}(t+\tau_1)\, x_{1\mathrm{i}}(t+\tau_1) \cdots x_{N\mathrm{r}}(t+\tau_N)\, x_{N\mathrm{i}}(t+\tau_N)}(\xi_{1\mathrm{r}}, \xi_{1\mathrm{i}}, \ldots, \xi_{N\mathrm{r}}, \xi_{N\mathrm{i}}) \tag{4.30}$$

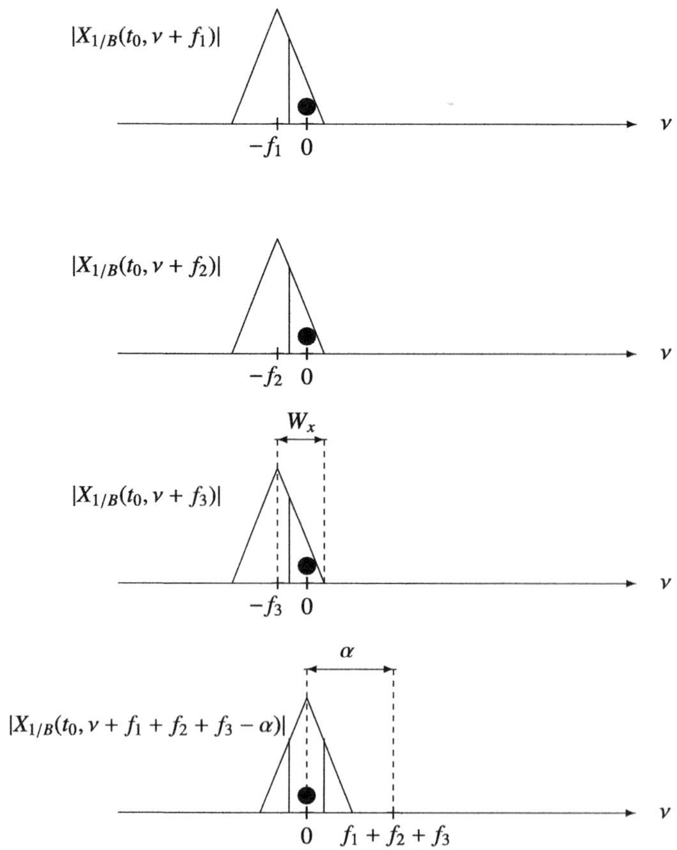

FIGURE 4.5 Fourth-oder spectral correlation for a narrow-band cyclostationary signal ($2W_x < \alpha$). Bullets "•" denote correlated spectral bands.

Definition 4.14. The coefficient of the Fourier series expansion of the almost-periodic function (4.28)

$$\Phi_{\boldsymbol{x}}^{\gamma}(\boldsymbol{\omega}, \boldsymbol{\tau}) \triangleq \left\langle \Phi_{\boldsymbol{x}}(t, \boldsymbol{\omega}, \boldsymbol{\tau})\, e^{-j2\pi\gamma t} \right\rangle_t \tag{4.31}$$

is referred to as the Nth-order *cyclic temporal joint characteristic function.* □

Accounting for (4.6) and (4.7) and the $2N$th-order Fourier transform relation (4.30), from (4.31) it follows that

$$\Phi_{\boldsymbol{x}}^{\gamma}(\boldsymbol{\omega}, \boldsymbol{\tau})$$
$$= \left\langle \exp\left[j \sum_{k=1}^{N} \omega_k\, x_k^{(*)_k}(t + \tau_k) \right] e^{-j2\pi\gamma t} \right\rangle_t$$

$$= \int_{\mathbb{R}^{2N}} \exp\Big[j \sum_{k=1}^{N} \omega_k \, (\xi_{k\mathrm{r}} + (-)_k \, j \, \xi_{k\mathrm{i}}) \Big] \, \mathrm{d}F^{\gamma}_{x_{1\mathrm{r}} \, x_{1\mathrm{i}} \cdots x_{N\mathrm{r}} \, x_{N\mathrm{i}}}(\xi_{1\mathrm{r}}, \xi_{1\mathrm{i}}, \ldots, \xi_{N\mathrm{r}}, \xi_{N\mathrm{i}}; \boldsymbol{\tau}) \, . \tag{4.32}$$

If the series of the Fourier coefficients (4.31) is absolutely convergent, we have

$$\Phi_{\boldsymbol{x}}(t, \boldsymbol{\omega}, \boldsymbol{\tau}) = \sum_{\gamma \in \Gamma^{(N)}} \Phi^{\gamma}_{\boldsymbol{x}}(\boldsymbol{\omega}, \boldsymbol{\tau}) \, e^{j2\pi\gamma t} \, . \tag{4.33}$$

Definition 4.15. The Nth-order *temporal cross-cumulant function* (TCCF) of the time series $x_1^{(*)_1}(t), \ldots, x_N^{(*)_N}(t)$ is defined as

$$\mathcal{C}_{\boldsymbol{x}}(t, \boldsymbol{\tau}) \triangleq \operatorname{cum}\left\{ x_k^{(*)_k}(t + \tau_k), \; k = 1, \ldots, N \right\} \tag{4.34a}$$

$$= (-j)^N \, \frac{\partial^N}{\partial \omega_1 \cdots \partial \omega_N} \log_e \Phi_{\boldsymbol{x}}(t, \boldsymbol{\omega}, \boldsymbol{\tau}) \bigg|_{\boldsymbol{\omega} = \mathbf{0}} \tag{4.34b}$$

□

See Section E.1.2 for a discussion on this definition of cumulant in the complex case.

Definition 4.16. The coefficient of the Fourier series expansion of the almost-periodic function (4.34b)

$$\mathcal{C}^{\beta}_{\boldsymbol{x}}(\boldsymbol{\tau}) \triangleq \left\langle \mathcal{C}_{\boldsymbol{x}}(t, \boldsymbol{\tau}) \, e^{-j2\pi\beta t} \right\rangle_t \tag{4.35}$$

is referred to as the Nth-order *cyclic temporal cross-cumulant function* (CTCCF). □

If the series of the Fourier coefficients (4.35) is absolutely convergent, we have

$$\mathcal{C}_{\boldsymbol{x}}(t, \boldsymbol{\tau}) = \sum_{\beta \in B_{\boldsymbol{x}}} \mathcal{C}^{\beta}_{\boldsymbol{x}}(\boldsymbol{\tau}) \, e^{j2\pi\beta t} \, . \tag{4.36}$$

Definition 4.17. The N-dimensional Fourier transform of the CTCCF

$$\mathcal{P}^{\beta}_{\boldsymbol{x}}(\boldsymbol{f}) \triangleq \int_{\mathbb{R}^N} \mathcal{C}^{\beta}_{\boldsymbol{x}}(\boldsymbol{\tau}) \, e^{-j2\pi \boldsymbol{f}^{\mathsf{T}} \boldsymbol{\tau}} \, \mathrm{d}\boldsymbol{\tau} \tag{4.37}$$

is called the Nth-order *cyclic spectral cross-cumulant function* (CSCCF). □

Theorem 4.18. *The CSCCF can be written as*

$$\mathcal{P}^{\beta}_{\boldsymbol{x}}(\boldsymbol{f}) = P^{\beta}_{\boldsymbol{x}}(\boldsymbol{f}') \, \delta(\boldsymbol{f}^{\mathsf{T}} \mathbf{1} - \beta) \tag{4.38}$$

where

$$P^{\beta}_{\boldsymbol{x}}(\boldsymbol{f}') \triangleq \int_{\mathbb{R}^{N-1}} C^{\beta}_{\boldsymbol{x}}(\boldsymbol{\tau}') \, e^{-j2\pi \boldsymbol{f}'^{\mathsf{T}} \boldsymbol{\tau}'} \mathrm{d}\boldsymbol{\tau}' \tag{4.39}$$

is the Nth-order cyclic cross-polyspectrum *(CCP), and*

$$C^{\beta}_{\boldsymbol{x}}(\boldsymbol{\tau}') \triangleq \mathcal{C}^{\beta}_{\boldsymbol{x}}(\boldsymbol{\tau})|_{\tau_N = 0} \tag{4.40}$$

is called the reduced-dimension CTCCF *(RD-CTCCF).*

Proof. It is analogous to the proof of Theorem 4.7 reported in Section 4.11. □

Definition 4.19. The Nth-order *spectral cross-cumulant function* (SCCF) of the time series $x_1^{(*)_1}(t), \ldots, x_N^{(*)_N}(t)$ is defined as

$$\mathcal{P}_{\boldsymbol{x}}(\boldsymbol{f}) \triangleq \lim_{T\to\infty} \operatorname{cum}\left\{X_{k,T}^{(*)_k}(t, (-)_k f_k),\ \ k = 1, \ldots, N\right\} \tag{4.41}$$

where the STFT $X_{k,T}^{(*)_k}(u, (-)_k f_k)$ are defined in (4.18) and for the time series $t \mapsto X_{k,T}(t, f_k)$ the stationary model can be adopted as $T \to \infty$. Thus, the almost-periodic component extraction operators $\mathrm{E}^{\{\alpha\}}\{\cdot\}$ simplifies into the infinite time average $\langle\cdot\rangle_t$ in the cumulant definition (4.34b) with $x_k(t)$ replaced by $X_{k,T}(t, f_k)$. □

Theorem 4.20. *The Nth-order SCCF can be expressed as*

$$\mathcal{P}_{\boldsymbol{x}}(\boldsymbol{f}) = \int_{\mathbb{R}^N} \mathcal{C}_{\boldsymbol{x}}(0, \boldsymbol{\tau})\, e^{-j2\pi \boldsymbol{f}^{\mathsf{T}}\boldsymbol{\tau}}\, \mathrm{d}\boldsymbol{\tau} = \sum_{\beta\in B_{\boldsymbol{x}}} P_{\boldsymbol{x}}^{\beta}(\boldsymbol{f}')\, \delta(\boldsymbol{f}^{\mathsf{T}}\mathbf{1} - \beta)\,. \tag{4.42}$$

Proof. See Section 4.11. □

Proposition 4.21. *The CCP can be expressed as*

$$\begin{aligned} P_{\boldsymbol{x}}^{\beta}(\boldsymbol{f}') = \lim_{T\to\infty} \frac{1}{T} \operatorname{cum}\Big\{& X_{N,T}^{(*)_N}(t, (-)_N(\beta - \boldsymbol{f}'^{\mathsf{T}}\mathbf{1})), \\ & X_{k,T}^{(*)_k}(t, (-)_k f_k), k = 1, \ldots, N-1\Big\} \end{aligned} \tag{4.43}$$

where the STFT $X_{k,T}^{(*)_k}(u, (-)_k f_k)$ are defined in (4.18) and, in the computation of the cumulant in (4.43), the stationary FOT expectation operation (that is, the infinite time average) is adopted as $T \to \infty$. □

Eq. (4.43) reveals that the CCP of N time series is the Nth-order cross-cumulant of their spectral components at frequencies f_k whose sum is equal to β. In fact, Eq. (4.43) can be re-written as

$$\begin{aligned} P_{\boldsymbol{x}}^{\beta}(\boldsymbol{f}') = \lim_{B\to 0} \frac{1}{B^{N-1}} \operatorname{cum}\Big\{&\Big[\Big(x_N^{(*)_N}(t)\, e^{-j2\pi(\beta - \boldsymbol{f}'^{\mathsf{T}}\mathbf{1})t}\Big) \otimes h_B(t)\Big], \\ &\Big[\Big(x_k^{(*)_k}(t)\, e^{-j2\pi f_k t}\Big) \otimes h_B(t)\Big], k = 1, \ldots, N-1\Big\} \end{aligned} \tag{4.44}$$

where $h_B(t)$ is defined in (4.19) with $B = 1/T$. The rhs of (4.44) is the Nth-order temporal cross-cumulant of low-pass filtered versions of the frequency-shifted signals $x_k^{(*)_k}(t)\, e^{-j2\pi f_k t}$ when the sum of the frequency shifts is equal to β and the bandwidth B approaches zero.

4.2.3 Moments versus cumulants

In this section, relationships between temporal moment and cumulant functions are presented. Moreover, relationships among their Fourier coefficients are derived.

Theorem 4.22. *Moments and cumulants are linked by the* Leonov and Shiryaev formulas *(Leonov and Shiryaev, 1959)*

$$\mathcal{C}_{\boldsymbol{x}}(t,\boldsymbol{\tau}) = \sum_{\mathsf{P}} \left[(-1)^{p-1}(p-1)! \prod_{i=1}^{p} \mathcal{R}_{\boldsymbol{x}_{\mu_i}}(t,\boldsymbol{\tau}_{\mu_i}) \right] \tag{4.45}$$

$$\mathcal{R}_{\boldsymbol{x}}(t,\boldsymbol{\tau}) = \sum_{\mathsf{P}} \left[\prod_{i=1}^{p} \mathcal{C}_{\boldsymbol{x}_{\mu_i}}(t,\boldsymbol{\tau}_{\mu_i}) \right] \tag{4.46}$$

where P *is the set of distinct partitions of* $\{1,\dots,N\}$*, each constituted by the subsets* $\{\mu_i,\ i = 1,\dots,p\}$*,* $|\mu_i|$ *is the number of elements in* μ_i*,* $\boldsymbol{x}_{\mu_i}$ *is the* $|\mu_i|$*-dimensional vector whose components are those of* $\boldsymbol{x}$ *having indices in* μ_i*.*

In (4.45) and (4.46) we have

$$\mathcal{R}_{\boldsymbol{x}_{\mu_i}}(t,\boldsymbol{\tau}_{\mu_i}) \triangleq \mathrm{E}^{\{\alpha\}}\left\{ \prod_{\ell\in\mu_i} x_\ell^{(*)_\ell}(t+\tau_\ell) \right\} \tag{4.47a}$$

$$= \sum_{\alpha_{\mu_i}} \mathcal{R}_{\boldsymbol{x}_{\mu_i}}^{\alpha_{\mu_i}}(\boldsymbol{\tau}_{\mu_i})\, e^{j2\pi\alpha_{\mu_i}t} \tag{4.47b}$$

$$\mathcal{R}_{\boldsymbol{x}_{\mu_i}}^{\alpha_{\mu_i}}(\boldsymbol{\tau}_{\mu_i}) \triangleq \lim_{T\to\infty}\frac{1}{T}\int_{-T/2}^{T/2} \prod_{\ell\in\mu_i} x_\ell^{(*)_\ell}(t+\tau_\ell)\, e^{-j2\pi\alpha_{\mu_i}t}\,\mathrm{d}t \tag{4.48}$$

$$\mathcal{C}_{\boldsymbol{x}_{\mu_i}}(t,\boldsymbol{\tau}_{\mu_i}) \triangleq \operatorname{cum}\left\{ x_\ell^{(*)_\ell}(t+\tau_\ell),\ \ell\in\mu_i \right\} \tag{4.49a}$$

$$= \sum_{\beta_{\mu_i}} \mathcal{C}_{\boldsymbol{x}_{\mu_i}}^{\beta_{\mu_i}}(\boldsymbol{\tau}_{\mu_i})\, e^{j2\pi\beta_{\mu_i}t} \tag{4.49b}$$

$$\mathcal{C}_{\boldsymbol{x}_{\mu_i}}^{\beta_{\mu_i}}(\boldsymbol{\tau}_{\mu_i}) \triangleq \lim_{T\to\infty}\frac{1}{T}\int_{-T/2}^{T/2} \operatorname{cum}\left\{ x_\ell^{(*)_\ell}(t+\tau_\ell),\ \ell\in\mu_i \right\} e^{-j2\pi\beta_{\mu_i}t}\,\mathrm{d}t \tag{4.50}$$

□

In (4.47a) $\mathrm{E}^{\{\alpha\}}\{\cdot\}$ extracts the frequencies of the $|\mu_i|$th-order lag-product in its argument. See Section E.1.2 for a discussion on the Leonov and Shiryaev formulas in the complex case.

By replacing (4.45) into the right-hand side of (4.35) we have (Gardner and Spooner, 1994a)

$$\mathcal{C}_{\boldsymbol{x}}^{\beta}(\boldsymbol{\tau}) = \sum_{\mathsf{P}} \left[(-1)^{p-1}(p-1)! \sum_{\boldsymbol{\alpha}^{\mathsf{T}}\mathbf{1}=\beta} \prod_{i=1}^{p} \mathcal{R}_{\boldsymbol{x}_{\mu_i}}^{\alpha_{\mu_i}}(\boldsymbol{\tau}_{\mu_i}) \right] \tag{4.51}$$

where the inner sum is extended to all vectors $\boldsymbol{\alpha} \triangleq [\alpha_{\mu_1},\dots,\alpha_{\mu_p}]^{\mathsf{T}}$ such that the sum of their components is equal to β. Similarly, by replacing (4.46) into the right-hand side of (4.2) we get (Gardner and Spooner, 1994a)

$$\mathcal{R}_{\boldsymbol{x}}^{\alpha}(\boldsymbol{\tau}) = \sum_{\mathsf{P}} \left[\sum_{\boldsymbol{\beta}^{\mathsf{T}}\mathbf{1}=\alpha} \prod_{i=1}^{p} \mathcal{C}_{\boldsymbol{x}_{\mu_i}}^{\beta_{\mu_i}}(\boldsymbol{\tau}_{\mu_i}) \right] \tag{4.52}$$

where the inner sum is extended to all vectors $\boldsymbol{\beta} \triangleq [\beta_{\mu_1}, \ldots, \beta_{\mu_p}]^{\mathsf{T}}$ such that the sum of their components is equal to α.

From Theorem 4.22, an important interpretation of moments and cumulants in the FOT probability framework for time series that exhibit higher-order cyclostationarity is obtained, which does not have a counterpart in the classical stochastic process framework. As first shown in (Gardner and Spooner, 1990) and then, in more detail, in (Spooner, 1992), (Gardner and Spooner, 1994a), (Gardner and Spooner, 1994b), (Spooner, 1994), the Nth-order cyclic temporal cumulant function at cycle frequency α of a time series provides a mathematical characterization of the notion of a *pure Nth-order sine wave*. It is that part of the sine wave at frequency α present in the Nth-order lag product waveform that remains after removal of all parts that result from products of sine waves in lower order lag products obtained by factoring the Nth-order product. See Section 2.3.1.6 for the analysis in the second-order case. The cyclic temporal moment function gives the strength and phase of the entire sine wave with frequency α that is contained in the Nth-order lag product. Such a sine wave is dubbed *impure Nth-order sine wave*.

Cycle frequencies of the set $A_{\boldsymbol{x}}$ in (4.4b) are dubbed moment cycle frequencies and cycle frequencies of the set $B_{\boldsymbol{x}}$ in (4.36) are dubbed cumulant cycle frequencies. In Section 4.2.3.1 it is shown that $B_{\boldsymbol{x}} \subseteq A_{\boldsymbol{x}}$.

By considering the N-dimensional Fourier transform for both sides of (4.51) and (4.52), one obtains the frequency domain counterparts of these equations:

$$\mathcal{P}_{\boldsymbol{x}}^{\beta}(\boldsymbol{f}) = \sum_{\mathsf{P}} \left[(-1)^{p-1}(p-1)! \sum_{\boldsymbol{\alpha}^{\mathsf{T}}\mathbf{1}=\beta} \prod_{i=1}^{p} \mathcal{S}_{\boldsymbol{x}_{\mu_i}}^{\alpha_{\mu_i}}(\boldsymbol{f}_{\mu_i}) \right] \tag{4.53}$$

$$\mathcal{S}_{\boldsymbol{x}}^{\alpha}(\boldsymbol{f}) = \sum_{\mathsf{P}} \left[\sum_{\boldsymbol{\beta}^{\mathsf{T}}\mathbf{1}=\alpha} \prod_{i=1}^{p} \mathcal{P}_{\boldsymbol{x}_{\mu_i}}^{\beta_{\mu_i}}(\boldsymbol{f}_{\mu_i}) \right] . \tag{4.54}$$

The frequency-domain counterparts of Leonov and Shiryaev formulas (4.45) and (4.46) are

$$\begin{aligned} \mathcal{P}_{\boldsymbol{x}}(\boldsymbol{f}) &\triangleq \lim_{T\to\infty} \operatorname{cum}\left\{ X_{k,T}^{(*)_k}(t, (-)_k f_k), \ k = 1, \ldots, N \right\} \\ &= \sum_{\mathsf{P}} (-1)^{p-1}(p-1)! \prod_{i=1}^{p} \lim_{T\to\infty} \left\langle \prod_{k\in\mu_i} X_{k,T}^{(*)_k}(t, (-)_k f_k) \right\rangle_t \end{aligned} \tag{4.55}$$

$$\mathcal{S}_{\boldsymbol{x}}(\boldsymbol{f}) \triangleq \lim_{T\to\infty} \left\langle \prod_{k=1}^{N} X_{k,T}^{(*)_k}(t, (-)_k f_k) \right\rangle_t$$

$$= \sum_{\mathcal{P}} \prod_{i=1}^{p} \lim_{T\to\infty} \operatorname{cum}\left\{ X_{k,T}^{(*)_k}(t, (-)_k f_k),\ k \in \mu_i \right\}. \tag{4.56}$$

4.2.3.1 Submanifold

By reasoning as in the classical stochastic approach, if the time series $x_N(t)$ and $x_k(t+\tau_k)$, $k = 1, ..., N-1$, are asymptotically ($|\tau_k| \to \infty$) independent (in the FOT probability sense), then $C_{\boldsymbol{x}}^{\beta}(\boldsymbol{\tau}') \to 0$ as $\|\boldsymbol{\tau}'\| \to \infty$ (Napolitano, 2012, Sec. 1.4.1). Moreover, if $C_{\boldsymbol{x}}^{\beta}(\boldsymbol{\tau}') \in L^1_{\text{loc}}(\mathbb{R}^{N-1})$ and there exists an $\epsilon > 0$ such that $|C_{\boldsymbol{x}}^{\beta}(\boldsymbol{\tau}')| = o(\|\boldsymbol{\tau}'\|^{-N+1-\epsilon})$ as $\|\boldsymbol{\tau}'\| \to \infty$, then the RD-CTCCF $C_{\boldsymbol{x}}^{\beta}(\boldsymbol{\tau}') \in L^1(\mathbb{R}^{N-1})$ and its $(N-1)$-dimensional Fourier transform, the cyclic cross-polyspectrum $P_{\boldsymbol{x}}^{\beta}(\boldsymbol{f}')$, is an ordinary function (that is, it does not contain impulses or other generalized functions).

Considering the reduced-dimension versions ($\tau_N = 0$) of the CTCMF and CTCCFs in (4.52) we have

$$R_{\boldsymbol{x}}^{\alpha}(\boldsymbol{\tau}') = \underbrace{C_{\boldsymbol{x}}^{\alpha}(\boldsymbol{\tau}')}_{\text{summable}} + \sum_{\substack{\mathcal{P}\\ p\neq 1}} \underbrace{\sum_{\boldsymbol{\beta}^{\mathsf{T}}\mathbf{1}=\alpha} \left[C_{\boldsymbol{x}_{\mu_p}}^{\beta_{\mu_p}}(\boldsymbol{\tau}'_{\mu_p}) \prod_{i=1}^{p-1} C_{\boldsymbol{x}_{\mu_i}}^{\beta_{\mu_i}}(\boldsymbol{\tau}'_{\mu_i} - \mathbf{1}\tau_{r_i})\, e^{j2\pi\beta_{\mu_i}\tau_{r_i}} \right]}_{\text{not convergent as } \tau_{r_i}\to\infty \text{ with } \boldsymbol{\tau}'_{\mu_i} = \mathbf{1}\tau_{r_i}} \tag{4.57}$$

where, for each subset μ_i we have $\boldsymbol{\tau}_{\mu_i} = [\boldsymbol{\tau}'_{\mu_i}, \tau_{r_i}]$ for $i = 1, \dots, p-1$, $\boldsymbol{\tau}_{\mu_p} = [\boldsymbol{\tau}'_{\mu_p}, \tau_N]$, for the partition with $p = 1$ we have $\alpha \equiv \beta_{\mu_1}$, and the counterpart of (4.133) for CTCCFs and RD-CTCCFs is used.

By Fourier transforming both sides of (4.57) we have

$$S_{\boldsymbol{x}}^{\alpha}(\boldsymbol{f}') = \underbrace{P_{\boldsymbol{x}}^{\alpha}(\boldsymbol{f}')}_{\substack{\text{ordinary}\\ \text{Fourier transform}}} + \sum_{\substack{\mathcal{P}\\ p\neq 1}} \underbrace{\sum_{\boldsymbol{\beta}^{\mathsf{T}}\mathbf{1}=\alpha} \left[P_{\boldsymbol{x}_{\mu_p}}^{\beta_{\mu_p}}(\boldsymbol{f}'_{\mu_p}) \prod_{i=1}^{p-1} P_{\boldsymbol{x}_{\mu_i}}^{\beta_{\mu_i}}(\boldsymbol{f}'_{\mu_i})\, \delta(\beta_{\mu_i} - \boldsymbol{f}'^{\mathsf{T}}_{\mu_i}\mathbf{1}) \right]}_{\text{Dirac impulses on the } \alpha\text{-submanifold}} \tag{4.58}$$

From (4.57) we see that in general the function $R_{\boldsymbol{x}}^{\alpha}(\boldsymbol{\tau}')$ is not absolutely integrable because it does not in general decay as $\|\boldsymbol{\tau}'\| \to \infty$, but rather it oscillates. Thus, $S_{\boldsymbol{x}}^{\alpha}(\boldsymbol{f}')$ can contain impulses and, consequently, $\mathcal{S}_{\boldsymbol{x}}^{\alpha}(\boldsymbol{f})$ can contain products of impulses. In fact, from (4.58) it follows that the RD-CSCMF $S_{\boldsymbol{x}}^{\alpha}(\boldsymbol{f}')$ can contain impulsive terms if the vector $\boldsymbol{f}$ with $f_N = \alpha - \sum_{k=1}^{N-1} f_k$ lies on the *α-submanifold*; i.e., if there exists at least one partition $\{\mu_1, \cdots, \mu_p\}$ of $\{1, \cdots, N\}$ with $p > 1$ such that each sum $\beta_{\mu_i} = \sum_{k\in\mu_i} f_k$ is a $|\mu_i|$th-order cycle frequency of $x(t)$, where $|\mu_i|$ is the number of elements in μ_i.

Furthermore, from (4.58) it follows that except on an α-submanifold, the CCP $P_{\boldsymbol{x}}^{\alpha}(\boldsymbol{f}')$ is coincident with the RD-CSCMF $S_{\boldsymbol{x}}^{\alpha}(\boldsymbol{f}')$.

Finally, from (4.57) it follows that, for each conjugation configuration, the set of the cumulant cycle frequencies is a subset of that of the moment cycle frequencies

$$B_{\boldsymbol{x}} \subseteq A_{\boldsymbol{x}}. \tag{4.59}$$

4.3 Discrete-time signals

In accordance with the definition for the continuous-time case, N not necessarily distinct discrete-time complex-valued time series $x_k(n)$, $n \in \mathbb{Z}$, exhibit joint Nth-order wide-sense cyclostationarity with cycle frequency $\widetilde{\alpha} \neq 0$, ($\widetilde{\alpha} \in [-1/2, 1/2)$), for a given conjugation configuration $(*)_1, \ldots, (*)_N$ if the Nth-order CTCMF

$$\widetilde{\mathcal{R}}_{\boldsymbol{x}}^{\widetilde{\alpha}}(\boldsymbol{m}) \triangleq \lim_{N\to\infty} \frac{1}{2N+1} \sum_{n=-N}^{N} \prod_{k=1}^{N} x_k^{(*)_k}(n+m_k) e^{-j2\pi\widetilde{\alpha}n} \tag{4.60}$$

is not identically zero. In (4.60), $\boldsymbol{m} \triangleq [m_1, \ldots, m_N]^{\mathsf{T}}$ and subscript $\boldsymbol{x} \triangleq x_1^{(*)_1} \cdots x_N^{(*)_N}$.

The magnitude and phase of the CTCMF (4.60) are the amplitude and phase of the sinewave component with frequency $\widetilde{\alpha}$ contained in the discrete-time Nth-order lag product whose temporal expected value is the discrete-time Nth-order TCMF. That is,

$$\widetilde{\mathcal{R}}_{\boldsymbol{x}}(n, \boldsymbol{m}) \triangleq \mathrm{E}^{\{\widetilde{\alpha}\}} \left\{ \prod_{k=1}^{N} x_k^{(*)_k}(n+m_k) \right\} \tag{4.61a}$$

$$= \sum_{\widetilde{\alpha} \in \widetilde{A}_{\boldsymbol{x}}} \widetilde{\mathcal{R}}_{\boldsymbol{x}}^{\widetilde{\alpha}}(\boldsymbol{m})\, e^{j2\pi\widetilde{\alpha}n} \tag{4.61b}$$

where the equality in (4.61b) holds provided that the series is absolutely convergent.

In (4.61a), $\mathrm{E}^{\{\widetilde{\alpha}\}}\{\cdot\}$ is the *discrete-time almost-periodic component extraction operator*. It extracts all the finite-strength additive sine-vawe components with frequencies $\widetilde{\alpha} \in [-1/2, 1/2)$ contained in the discrete-time function in its argument.

The set

$$\widetilde{A}_{\boldsymbol{x}} \triangleq \left\{ \widetilde{\alpha} \in [-1/2, 1/2) : \widetilde{\mathcal{R}}_{\boldsymbol{x}}^{\widetilde{\alpha}}(\boldsymbol{m}) \not\equiv 0 \right\} \tag{4.62}$$

is the countable set, depending on the conjugation configuration but not depending on $\boldsymbol{m}$, of the Nth-order moment cycle frequencies.

The N-fold discrete Fourier transform of the CTCMF

$$\widetilde{\mathcal{S}}_{\boldsymbol{x}}^{\widetilde{\alpha}}(\boldsymbol{\nu}) \triangleq \sum_{\boldsymbol{m}\in\mathbb{Z}^N} \widetilde{\mathcal{R}}_{\boldsymbol{x}}^{\widetilde{\alpha}}(\boldsymbol{m})\, e^{-j2\pi\boldsymbol{m}^{\mathsf{T}}\boldsymbol{\nu}} \tag{4.63}$$

where $\boldsymbol{\nu} \triangleq [\nu_1, \ldots, \nu_N]^{\mathsf{T}}$, is called the Nth-order CSCMF and can be written as

$$\widetilde{\mathcal{S}}_{\boldsymbol{x}}^{\widetilde{\alpha}}(\boldsymbol{\nu}) = \widetilde{S}_{\boldsymbol{x}}^{\widetilde{\alpha}}(\boldsymbol{\nu}') \sum_{r=-\infty}^{+\infty} \delta(\widetilde{\alpha} - \boldsymbol{\nu}^{\mathsf{T}}\mathbf{1} - r) \tag{4.64}$$

where the function $\widetilde{S}_{\boldsymbol{x}}^{\widetilde{\alpha}}(\boldsymbol{\nu}')$ is the Nth-order RD-CSCMF, which can be expressed as the $(N-1)$-fold discrete Fourier transform

$$\widetilde{S}_{\boldsymbol{x}}^{\widetilde{\alpha}}(\boldsymbol{\nu}') = \sum_{\boldsymbol{m}'\in\mathbb{Z}^{N-1}} \widetilde{R}_{\boldsymbol{x}}^{\widetilde{\alpha}}(\boldsymbol{m}')\, e^{-j2\pi\boldsymbol{m}'^{\mathsf{T}}\boldsymbol{\nu}'} \tag{4.65}$$

of the Nth-order RD-CTCMF

$$\widetilde{R}_{\boldsymbol{x}}^{\widetilde{\alpha}}(\boldsymbol{m}') \triangleq \widetilde{\mathcal{R}}_{\boldsymbol{x}}^{\widetilde{\alpha}}(\boldsymbol{m})|_{m_N=0} \,. \tag{4.66}$$

Analogously to the continuous-time case, the discrete-time Nth-order TCCF is defined as

$$\widetilde{\mathcal{C}}_{\boldsymbol{x}}(n, \boldsymbol{m}) \triangleq \operatorname{cum}\left\{x_k^{(*)_k}(n+m_k),\ k=1,\ldots,N\right\} \tag{4.67a}$$

$$= (-j)^N \frac{\partial^N}{\partial\omega_1 \cdots \partial\omega_N} \log_e \mathrm{E}^{\{\widetilde{\alpha}\}}\left\{\exp\left[j \sum_{k=1}^{N} \omega_k\, x_k^{(*)_k}(n+m_k)\right]\right\}\Bigg|_{\boldsymbol{\omega}=\mathbf{0}} \tag{4.67b}$$

$$= \sum_{\mathsf{P}} \left[(-1)^{p-1}(p-1)! \prod_{i=1}^{p} \widetilde{\mathcal{R}}_{\boldsymbol{x}_{\mu_i}}(n, \boldsymbol{m}_{\mu_i})\right] \tag{4.67c}$$

$$= \sum_{\widetilde{\beta} \in \widetilde{B}_{\boldsymbol{x}}} \widetilde{\mathcal{C}}_{\boldsymbol{x}}^{\widetilde{\beta}}(\boldsymbol{m})\, e^{j2\pi\widetilde{\beta}n} \tag{4.67d}$$

where

$$\widetilde{\mathcal{C}}_{\boldsymbol{x}}^{\widetilde{\beta}}(\boldsymbol{m}) \triangleq \lim_{N\to\infty} \frac{1}{2N+1} \sum_{n=-N}^{N} \widetilde{\mathcal{C}}_{\boldsymbol{x}}(n, \boldsymbol{m})\, e^{-j2\pi\widetilde{\beta}n} \tag{4.68}$$

are the discrete-time Nth-order CTCCFs and equality (4.67d) holds provided that the series is absolutely convergent.

The discrete Fourier transform of the Nth-order CTCCF is referred to as the Nth-order CSCCF. It is given by

$$\widetilde{\mathcal{P}}_{\boldsymbol{x}}^{\widetilde{\beta}}(\boldsymbol{\nu}) \triangleq \sum_{\boldsymbol{m}\in\mathbb{Z}^N} \widetilde{\mathcal{C}}_{\boldsymbol{x}}^{\widetilde{\beta}}(\boldsymbol{m})\, e^{j2\pi\boldsymbol{\nu}^{\mathsf{T}}\boldsymbol{m}} \tag{4.69a}$$

$$= \widetilde{P}_{\boldsymbol{x}}^{\widetilde{\beta}}(\boldsymbol{\nu}') \sum_{r=-\infty}^{+\infty} \delta(\widetilde{\beta} - \boldsymbol{\nu}^{\mathsf{T}}\mathbf{1} - r) \tag{4.69b}$$

where the function $\widetilde{P}_{\boldsymbol{x}}^{\widetilde{\beta}}(\boldsymbol{\nu}')$ is referred to as the Nth-order CCP, which can be expressed as the $(N-1)$-fold discrete Fourier transform

$$\widetilde{P}_{\boldsymbol{x}}^{\widetilde{\beta}}(\boldsymbol{\nu}') = \sum_{\boldsymbol{m}'\in\mathbb{Z}^{N-1}} \widetilde{C}_{\boldsymbol{x}}^{\widetilde{\beta}}(\boldsymbol{m}')\, e^{-j2\pi\boldsymbol{m}'^{\mathsf{T}}\boldsymbol{\nu}'} \tag{4.70}$$

of the Nth-order RD-CTCCF

$$\widetilde{C}_{\boldsymbol{x}}^{\widetilde{\beta}}(\boldsymbol{m}') \triangleq \widetilde{\mathcal{C}}_{\boldsymbol{x}}^{\widetilde{\beta}}(\boldsymbol{m})|_{m_N=0} \,. \tag{4.71}$$

Finally, it can be easily shown that the following periodicity properties hold:

$$\widetilde{\mathcal{R}}_{\boldsymbol{x}}^{\widetilde{\alpha}}(\boldsymbol{m}) = \widetilde{\mathcal{R}}_{\boldsymbol{x}}^{\widetilde{\alpha}+p}(\boldsymbol{m}) \qquad p \in \mathbb{Z} \tag{4.72}$$

$$\widetilde{\mathcal{S}}_{\boldsymbol{x}}^{\widetilde{\alpha}}(\boldsymbol{\nu}) = \widetilde{\mathcal{S}}_{\boldsymbol{x}}^{\widetilde{\alpha}+p}(\boldsymbol{\nu}+\boldsymbol{q}) \qquad p \in \mathbb{Z},\ \boldsymbol{q} \in \mathbb{Z}^N \tag{4.73}$$

$$\widetilde{\mathcal{C}}_{\boldsymbol{x}}^{\widetilde{\beta}}(\boldsymbol{m}) = \widetilde{\mathcal{C}}_{\boldsymbol{x}}^{\widetilde{\beta}+p}(\boldsymbol{m}) \qquad p \in \mathbb{Z} \tag{4.74}$$

$$\widetilde{\mathcal{P}}_{\boldsymbol{x}}^{\widetilde{\beta}}(\boldsymbol{\nu}) = \widetilde{\mathcal{P}}_{\boldsymbol{x}}^{\widetilde{\beta}+p}(\boldsymbol{\nu}+\boldsymbol{q}) \qquad p \in \mathbb{Z},\ \boldsymbol{q} \in \mathbb{Z}^N . \tag{4.75}$$

Analogous periodicity properties can be stated for the reduced-dimension statistics.

The Nth-order CTCCF can be expressed in terms of Nth- and lower-order CTCMFs by the discrete-time counterpart of (4.51)

$$\widetilde{\mathcal{C}}_{\boldsymbol{x}}^{\widetilde{\beta}}(\boldsymbol{m}) = \sum_{\mathsf{P}} \left[(-1)^{p-1}(p-1)! \sum_{\widetilde{\boldsymbol{\alpha}}} \left(\prod_{i=1}^{p} \widetilde{\mathcal{R}}_{\boldsymbol{x}_{\mu_i}}^{\widetilde{\alpha}_{\mu_i}}(\boldsymbol{m}_{\mu_i}) \right) \sum_{r=-\infty}^{+\infty} \delta_{\widetilde{\beta}-\widetilde{\boldsymbol{\alpha}}^{\mathsf{T}}\mathbf{1}+r} \right] \tag{4.76}$$

where the second summation ranges over all vectors $\widetilde{\boldsymbol{\alpha}} \triangleq [\widetilde{\alpha}_{\mu_1}, \ldots, \widetilde{\alpha}_{\mu_p}]^{\mathsf{T}}$ such that $\widetilde{\alpha}_{\mu_i} \in [-1/2, 1/2)$ and $\widetilde{\mathcal{R}}_{\boldsymbol{x}_{\mu_i}}^{\widetilde{\alpha}_{\mu_i}}(\cdot) \not\equiv 0$.

Similarly, the Nth-order CTCMF can be expressed in terms of Nth- and lower-order CTCCFs by the discrete-time counterpart of (4.52)

$$\widetilde{\mathcal{R}}_{\boldsymbol{x}}^{\widetilde{\alpha}}(\boldsymbol{m}) = \sum_{\mathsf{P}} \left[\sum_{\widetilde{\boldsymbol{\beta}}} \left(\prod_{i=1}^{p} \widetilde{\mathcal{C}}_{\boldsymbol{x}_{\mu_i}}^{\widetilde{\beta}_{\mu_i}}(\boldsymbol{m}_{\mu_i}) \right) \sum_{r=-\infty}^{+\infty} \delta_{\widetilde{\alpha}-\widetilde{\boldsymbol{\beta}}^{\mathsf{T}}\mathbf{1}+r} \right] \tag{4.77}$$

where the second summation ranges over all vectors $\widetilde{\boldsymbol{\beta}} \triangleq [\widetilde{\beta}_{\mu_1}, \ldots, \widetilde{\beta}_{\mu_p}]^{\mathsf{T}}$ such that $\widetilde{\beta}_{\mu_i} \in [-1/2, 1/2)$ and $\widetilde{\mathcal{C}}_{\boldsymbol{x}_{\mu_i}}^{\widetilde{\beta}_{\mu_i}}(\cdot) \not\equiv 0$.

4.4 Input/output relations for MIMO LAPTV systems

4.4.1 Continuous-time

Let us consider a multi-input multi-output (MIMO) linear time-variant (LTV) continuous-time system with L-dimensional input vector $\boldsymbol{x}(t) = [x_1(t), \ldots, x_L(t)]^{\mathsf{T}}$ and M-dimensional output vector $\boldsymbol{y}(t) = [y_1(t), \ldots, y_M(t)]^{\mathsf{T}}$.

The output signal $y_m(t)$ due to the input signals $x_1(t), \ldots, x_L(t)$ is given by (see (3.1))

$$y_m(t) = \sum_{\ell=1}^{L} \int_{\mathbb{R}} h_{m\ell}(t,u)\, x_\ell(u)\, \mathrm{d}u . \tag{4.78}$$

In (4.78), $h_{m\ell}(t,u)$ is the impulse-response function of the path between the ℓth input and the mth output and is the (m,ℓ) element of the $M \times L$ impulse-response matrix $\boldsymbol{H}(t,u)$.

From (4.78) it follows that the input/output relation for the MIMO LTV system in matrix form is

$$\boldsymbol{y}(t) = \int_{\mathbb{R}} \boldsymbol{H}(t,u)\, \boldsymbol{x}(u)\, \mathrm{d}u . \tag{4.79}$$

Let us consider the case of *linear almost-periodically time-variant* (LAPTV) system whose elements of the impulse response matrix can be represented by the Fourier series expansion (see (3.7) with the replacements $u \curvearrowright t$ and $t \curvearrowright t + \tau_m$)

$$h_{m\ell}(t + \tau_m, t) = \sum_{\sigma_{m\ell} \in J_{m\ell}} h_{\sigma_{m\ell}}(\tau_m)\, e^{j2\pi\sigma_{m\ell} t}, \qquad m = 1, \ldots, M, \ \ \ell = 1, \ldots, L \tag{4.80}$$

where the sum is over the countable set $J_{m\ell}$ of all frequencies of the almost-periodic function $t \mapsto h_{m\ell}(t + \tau_m, t)$.

Since input signals and impulse-response functions can be chosen arbitrarily, the optional complex conjugations can be omitted without lack of generality.

Assumption 4.23. *For every $m = 1, \ldots, M$ and $\ell = 1, \ldots, L$ it results*

$$\text{a.} \qquad \sum_{\sigma_{m\ell} \in J_{m\ell}} \left| h_{\sigma_{m\ell}}(\tau) \right| \in L^1(\mathbb{R}) \tag{4.81a}$$

$$\text{b.} \qquad \sum_{\sigma_{m\ell} \in J_{m\ell}} \left\| h_{\sigma_{m\ell}} \right\|_\infty < \infty\,. \tag{4.81b}$$

□

Assumption 4.24. *For every $(\ell_1, \ldots, \ell_M) \in \{1, \ldots, L\}^M$ it results*

$$\text{a.} \qquad \sum_{\alpha \in A_{\ell_1,\ldots,\ell_M}} \left| R^{\alpha}_{x_{\ell_1},\ldots,x_{\ell_M}}(\boldsymbol{\tau}') \right| \in L^1(\mathbb{R}^{M-1}) \tag{4.82a}$$

$$\text{b.} \qquad \sum_{\alpha \in A_{\ell_1,\ldots,\ell_M}} \left\| R^{\alpha}_{x_{\ell_1},\ldots,x_{\ell_M}} \right\|_\infty < \infty \tag{4.82b}$$

where $A_{\ell_1,\ldots,\ell_M}$ is the set of cycle frequencies of the Mth-order temporal cross-moment function of $x_{\ell_1}(t), \ldots, x_{\ell_M}(t)$. □

Theorem 4.25. (Napolitano, 1995). *Under Assumptions 4.23a and 4.24b or under Assumptions 4.23b and 4.24a, the Mth-order CTCMF at cycle frequency α of the M outputs $y_m(t)$, $m = 1, \ldots, M$, is given by*

$$\mathcal{R}^{\alpha}_{\mathbf{y}}(\boldsymbol{\tau}) \triangleq \lim_{T\to\infty} \frac{1}{T} \int_{-T/2}^{T/2} \prod_{m=1}^{M} y_m(t + \tau_m)\, e^{-j2\pi\alpha t}\, \mathrm{d}t \tag{4.83a}$$

$$= \sum_{(\ell_1,\ldots,\ell_M)\in I_L^M} \sum_{\sigma_{1\,\ell_1}} \cdots \sum_{\sigma_{M\,\ell_M}} \left[\mathcal{R}^{\alpha-\sigma_{1\,\ell_1}-\cdots-\sigma_{M\,\ell_M}}_{x_{\ell_1},\ldots,x_{\ell_M}}(\boldsymbol{\tau})\, e^{j2\pi(\sigma_{1\,\ell_1}\tau_1+\cdots+\sigma_{M\,\ell_M}\tau_M)} \right]$$

$$\underset{\tau_1}{\otimes} h_{\sigma_{1\,\ell_1}}(\tau_1) \cdots \underset{\tau_M}{\otimes} h_{\sigma_{M\,\ell_M}}(\tau_M) \tag{4.83b}$$

where $I_L \triangleq \{1, \ldots, L\}$ and $\underset{\tau_i}{\otimes}$ denotes convolution with respect to the variable τ_i.

The RD-CTCMF $R^{\alpha}_{\mathbf{y}}(\boldsymbol{\tau}')$ is obtained by setting $\tau_M = 0$ in $\mathcal{R}^{\alpha}_{\mathbf{y}}(\boldsymbol{\tau})$.

Proof. See Section 4.11. □

Theorem 4.26. (Napolitano, 1995). *Under Assumptions 4.23a and 4.24b or under Assumptions 4.23b and 4.24a, the Mth-order CSCMF at cycle frequency α of the M outputs* $y_m(t)$, $m = 1, \ldots, M$, *is given by*

$$\mathcal{S}_{\boldsymbol{y}}^{\alpha}(\boldsymbol{f}) = \sum_{(\ell_1,\ldots,\ell_M)\in I_L^M} \sum_{\sigma_1 \ell_1} \cdots \sum_{\sigma_M \ell_M} \mathcal{S}_{x_{\ell_1},\ldots,x_{\ell_M}}^{\alpha-\sigma_1 \ell_1-\cdots-\sigma_M \ell_M}(\boldsymbol{f} - [\sigma_1 \ell_1, \ldots, \sigma_M \ell_M]^{\mathsf{T}}) \prod_{m=1}^{M} H_{\sigma_m \ell_m}(f_m) \tag{4.84}$$

where

$$H_{\sigma_m \ell}(f_m) \triangleq \int_{\mathbb{R}} h_{\sigma_m \ell}(\tau_m)\, e^{-j2\pi f_m \tau_m}\, \mathrm{d}\tau_m \,. \tag{4.85}$$

The input/output relation in terms of RD-CSCMFs is given by

$$S_{\boldsymbol{y}}^{\alpha}(\boldsymbol{f}') = \sum_{(\ell_1,\ldots,\ell_M)\in I_L^M} \sum_{\sigma_1 \ell_1} \cdots \sum_{\sigma_M \ell_M} S_{x_{\ell_1},\ldots,x_{\ell_M}}^{\alpha-\sigma_1 \ell_1-\cdots-\sigma_M \ell_M}(\boldsymbol{f}' - [\sigma_1 \ell_1, \ldots, \sigma_{M-1} \ell_{M-1}]^{\mathsf{T}}) H_{\sigma_M \ell_M}(\alpha - \boldsymbol{f}'^{\mathsf{T}}\mathbf{1}) \prod_{m=1}^{M-1} H_{\sigma_m \ell_m}(f_m) \,. \tag{4.86}$$

Proof. See Section 4.11. □

Theorem 4.27. (Napolitano, 1995). *Under Assumptions 4.23a and 4.24b or under Assumptions 4.23b and 4.24a, the Mth-order CTCCF at cycle frequency β of the M outputs* $y_m(t)$, $m = 1, \ldots, M$, *is given by*

$$\mathcal{C}_{\boldsymbol{y}}^{\beta}(\boldsymbol{\tau}) \triangleq \lim_{T\to\infty} \frac{1}{T} \int_{-T/2}^{T/2} \operatorname{cum}\{y_m(t+\tau_m),\ m = 1, \ldots, M\}\, e^{-j2\pi\beta t}\, \mathrm{d}t \tag{4.87a}$$

$$= \sum_{(\ell_1,\ldots,\ell_M)\in I_L^M} \sum_{\sigma_1 \ell_1} \cdots \sum_{\sigma_M \ell_M} \left[\mathcal{C}_{x_{\ell_1},\ldots,x_{\ell_M}}^{\beta-\sigma_1 \ell_1-\cdots-\sigma_M \ell_M}(\boldsymbol{\tau})\, e^{j2\pi(\sigma_1 \ell_1 \tau_1+\cdots+\sigma_M \ell_M \tau_M)}\right] \underset{\tau_1}{\otimes} h_{\sigma_1 \ell_1}(\tau_1) \cdots \underset{\tau_M}{\otimes} h_{\sigma_M \ell_M}(\tau_M) \tag{4.87b}$$

The RD-CTCCF $C_{\boldsymbol{y}}^{\beta}(\boldsymbol{\tau}')$ *is obtained by setting* $\tau_M = 0$ *in* $\mathcal{C}_{\boldsymbol{y}}^{\beta}(\boldsymbol{\tau})$.

Proof. See Section 4.11. □

Remark 4.28. Let us note that the input/output relations in terms of CTCMFs and CTCCFs are the same (compare (4.83b) and (4.87b)). This result is consequence of the adopted definition for cumulants of complex random variables and signals (see Section E.1.2) and of the fact that LAPTV transformations are nonrandom in the fraction-of-time probability framework for time-series that exhibit cyclostationarity (Izzo and Napolitano, 2002a), (Napolitano, 2012, Secs. 4.3.1 and 6.3.8). □

The result of Theorem 4.27 is in agreement with the *multilinearity property* of FOT cumulants that can be proved similarly to the case of stochastic cumulants (Mendel, 1991).

From Theorem 4.27, the following result is immediately derived.

Theorem 4.29. (Napolitano, 1995). *Under Assumptions 4.23a and 4.24b or under Assumptions 4.23b and 4.24a, the* M*th-order CSCCF at cycle frequency* β *of the* M *outputs* $y_m(t)$, $m = 1, \dots, M$, *is given by*

$$\mathcal{P}_{\boldsymbol{y}}^{\beta}(\boldsymbol{f}) = \sum_{(\ell_1,\dots,\ell_M)\in I_L^M} \sum_{\sigma_1 \ell_1} \cdots \sum_{\sigma_M \ell_M} \mathcal{P}_{x_{\ell_1},\dots,x_{\ell_M}}^{\beta-\sigma_1 \ell_1 - \dots - \sigma_M \ell_M}(\boldsymbol{f} - [\sigma_{1\,\ell_1}, \dots, \sigma_{M\,\ell_M}]^{\mathsf{T}}) \prod_{m=1}^{M} H_{\sigma_m \ell_m}(f_m) \tag{4.88}$$

The input/output relation in terms of CCPs is given by

$$P_{\boldsymbol{y}}^{\beta}(\boldsymbol{f}') = \sum_{(\ell_1,\dots,\ell_M)\in I_L^M} \sum_{\sigma_1 \ell_1} \cdots \sum_{\sigma_M \ell_M} P_{x_{\ell_1},\dots,x_{\ell_M}}^{\beta-\sigma_1 \ell_1 - \dots - \sigma_M \ell_M}(\boldsymbol{f}' - [\sigma_{1\,\ell_1}, \dots, \sigma_{M-1\,\ell_{M-1}}]^{\mathsf{T}}) H_{\sigma_M \ell_M}(\beta - \boldsymbol{f}'^{\mathsf{T}}\mathbf{1}) \prod_{m=1}^{M-1} H_{\sigma_m \ell_m}(f_m)\,. \tag{4.89}$$

□

Special cases of LAPTV systems are the LTI systems and systems that make the product of the input signal with an almost-periodic function. By replacing these impulse response functions or their Fourier transforms with respect to τ_m into (4.83b)–(4.89), one obtains the input/output relations for these special cases (Napolitano, 1995).

Example 4.30. *LTI Systems.* For MIMO LTI systems we have

$$h_{m\ell}(t + \tau_m, t) = h_{m\ell}(\tau_m) \tag{4.90}$$

It results, $J_{m\ell} = \{0\}$, that is, the only nonzero term in the sum in (4.80) is that with $\sigma_{m\ell} = 0$.

In the special case of $L = M$ single-input single-output (SISO) LTI systems,

$$h_{m\ell}(t + \tau_m, t) = h_m(\tau_m)\delta_{\ell-m} \tag{4.91}$$

and (4.86) specializes into

$$S_{\boldsymbol{y}}^{\alpha}(\boldsymbol{f}') = S_{\boldsymbol{x}}^{\alpha}(\boldsymbol{f}')\, H_M(\alpha - \boldsymbol{f}'^{\mathsf{T}}\mathbf{1}) \prod_{m=1}^{M-1} H_m(f_m) \tag{4.92}$$

where subscripts $\boldsymbol{y} \triangleq y_1 \dots y_M$, $\boldsymbol{x} \triangleq x_1 \dots x_M$ and the harmonic-response functions $H_m(f_m)$ are the Fourier transforms of the impulse-response functions $h_m(\tau_m)$. Similarly (4.89) specializes into

$$P_{\boldsymbol{y}}^{\beta}(\boldsymbol{f}') = P_{\boldsymbol{x}}^{\beta}(\boldsymbol{f}')\, H_M(\beta - \boldsymbol{f}'^{\mathsf{T}}\mathbf{1}) \prod_{m=1}^{M-1} H_m(f_m) \tag{4.93}$$

Starting from (4.92) and (4.93), LTI system identification formulas are derived in Section 9.5.2.2. □

Example 4.31. *Product Modulation.* Let us consider a continuous-time MIMO memoryless LAPTV system characterized by the input/output relation

$$\boldsymbol{y}(t) = \boldsymbol{C}(t)\boldsymbol{x}(t) \tag{4.94}$$

where $\boldsymbol{C}(t)$ is an $M \times L$ matrix whose elements $c_{m\ell}(t)$ are almost-periodic functions with absolutely convergent (generalized) Fourier series

$$c_{m\ell}(t) = \sum_{\sigma_{m\ell} \in J_{m\ell}} c_{\sigma_{m\ell}}\, e^{j2\pi\sigma_{m\ell}t}\,. \tag{4.95}$$

In such a case

$$h_{m\ell}(t + \tau_m, t) = c_{m\ell}(t)\, \delta(\tau_m)\,. \tag{4.96}$$

By substituting (4.95) and (4.96) into (4.83b), one has

$$\begin{aligned} \mathcal{R}_{\boldsymbol{y}}^{\alpha}(\boldsymbol{\tau}) = &\sum_{i_1,\dots,i_M=1}^{L} \sum_{\sigma_{1i_1}} \cdots \sum_{\sigma_{Mi_M}} c_{\sigma_{1i_1}} \cdots c_{\sigma_{Mi_M}} \\ &\mathcal{R}_{x_{i_1},\dots,x_{i_M}}^{\alpha-\sigma_{1i_1}-\dots-\sigma_{Mi_M}}(\boldsymbol{\tau})\, e^{j2\pi(\sigma_{1i_1}\tau_1+\dots+\sigma_{Mi_M}\tau_M)} \end{aligned} \tag{4.97}$$

which leads to

$$\begin{aligned} S_{\boldsymbol{y}}^{\alpha}(\boldsymbol{f}') = &\sum_{i_1,\dots,i_M=1}^{L} \sum_{\sigma_{1i_1}} \cdots \sum_{\sigma_{Mi_M}} c_{\sigma_{1i_1}} \cdots c_{\sigma_{Mi_M}} \\ &S_{x_{i_1},\dots,x_{i_M}}^{\alpha-\sigma_{1i_1}-\dots-\sigma_{Mi_M}}(\boldsymbol{f}' - [\sigma_{1i_1}, \dots, \sigma_{M-1\, i_{M-1}}]^{\mathsf{T}})\,. \end{aligned} \tag{4.98}$$

If $\boldsymbol{C}(t)$ is a diagonal matrix, that is, an M-SISO system is considered, from (4.98) it follows that

$$S_{\boldsymbol{y}}^{\alpha}(\boldsymbol{f}') = \sum_{\sigma_1} \cdots \sum_{\sigma_M} c_{\sigma_1} \cdots c_{\sigma_M}\, S_{x_1,\dots,x_M}^{\alpha-\sigma_1-\dots-\sigma_M}(\boldsymbol{f}' - \boldsymbol{\sigma}') \tag{4.99}$$

where $\sigma_i \triangleq \sigma_{ii}$, $c_{\sigma_i} \triangleq c_{\sigma_{ii}}$, and $\boldsymbol{\sigma}' \triangleq [\sigma_1, \dots, \sigma_{M-1}]^{\mathsf{T}}$. □

Example 4.32. *Modulation.* Let us consider the M-SISO system

$$y_m(t) = x_m(t)\, e^{j2\pi f_{0m} t} \qquad m = 1, \ldots, M, \quad L = M\,. \tag{4.100}$$

It is a special case of the MIMO system (4.96).

Considering (for subsequent reference) also the presence of optional complex conjugations, we have

$$\begin{aligned} \mathcal{R}_{\boldsymbol{y}}^{\alpha}(\boldsymbol{\tau}) &= \left\langle \prod_{m=1}^{M} x_m^{(*)_m}(t+\tau_m)\, e^{(-)_m j2\pi f_{0m}(t+\tau_m)}\, e^{-j2\pi\alpha t} \right\rangle_t \\ &= \mathcal{R}_{\boldsymbol{x}}^{\alpha - \boldsymbol{f}_0^{(-)\mathsf{T}}\mathbf{1}}(\boldsymbol{\tau})\, e^{j2\pi \boldsymbol{f}_0^{(-)\mathsf{T}}\boldsymbol{\tau}} \end{aligned} \tag{4.101}$$

where $\boldsymbol{f}_0^{(-)} \triangleq [(-)_1 f_{01}, \ldots, (-)_M f_{0M}]^{\mathsf{T}}$. Therefore

$$\mathcal{S}_{\boldsymbol{y}}^{\alpha}(\boldsymbol{f}) = \mathcal{S}_{\boldsymbol{x}}^{\alpha - \boldsymbol{f}_0^{(-)\mathsf{T}}\mathbf{1}}(\boldsymbol{f} - \boldsymbol{f}_0^{(-)}) \tag{4.102}$$

$$\mathcal{C}_{\boldsymbol{y}}^{\beta}(\boldsymbol{\tau}) = \mathcal{C}_{\boldsymbol{x}}^{\beta - \boldsymbol{f}_0^{(-)\mathsf{T}}\mathbf{1}}(\boldsymbol{\tau})\, e^{j2\pi \boldsymbol{f}_0^{(-)\mathsf{T}}\boldsymbol{\tau}} \tag{4.103}$$

$$\mathcal{P}_{\boldsymbol{y}}^{\beta}(\boldsymbol{f}) = \mathcal{P}_{\boldsymbol{x}}^{\beta - \boldsymbol{f}_0^{(-)\mathsf{T}}\mathbf{1}}(\boldsymbol{f} - \boldsymbol{f}_0^{(-)})\,. \tag{4.104}$$

□

Example 4.33. *Man-Made Signals.* Periodicities in the second- and higher-order statistical functions of man-made signals used in communications, radar, sonar, and telemetry arise from operations such as modulation, scanning, sampling, multiplexing, and coding. These operations can be modeled as LAPTV transformations of random information data. In Section 7.5, higher-order cyclic statistics of several man-made modulated signals are derived by exploiting the input/output relationships derived in this section. □

4.4.2 Discrete-time

Let us consider now a discrete-time L-input M-output linear system whose elements of the impulse response matrix can be represented as

$$\widetilde{h}_{k\ell}(n+m_k, n) = \sum_{\widetilde{\sigma}_{k\ell} \in \widetilde{J}_{k\ell}} \widetilde{h}_{\widetilde{\sigma}_{k\ell}}(m_k) e^{j2\pi\widetilde{\sigma}_{k\ell} n}, \qquad k = 1, \ldots, M, \;\; \ell = 1, \ldots, L \tag{4.105}$$

where $\widetilde{J}_{k\ell}$ is the set $\{\widetilde{\sigma}_{k\ell} \in [-1/2, 1/2) : \widetilde{h}_{\widetilde{\sigma}_{k\ell}}(\cdot) \not\equiv 0\}$ whose elements are possibly incommensurate.

The Mth-order CTCMF at cycle frequency $\widetilde{\alpha}$ of the M outputs

$$y_k(n) = \sum_{\ell=1}^{L} \sum_{p=-\infty}^{+\infty} \widetilde{h}_{k\ell}(n, p)\, x_\ell(p) \qquad k = 1, \ldots, M, \tag{4.106}$$

is given by the relation (which is the discrete-time counterpart of (4.83b))

$$\begin{aligned}
\widetilde{\mathcal{R}}_{\boldsymbol{y}}^{\widetilde{\alpha}}(\boldsymbol{m}) &\triangleq \lim_{N\to\infty} \frac{1}{2N+1} \sum_{n=-N}^{N} \prod_{k=1}^{M} y_k(n+m_k)\, e^{-j2\pi\widetilde{\alpha}n} \\
&= \sum_{(\ell_1,\ldots,\ell_M)\in I_L^M} \sum_{\widetilde{\sigma}_{1\ell_1}} \cdots \sum_{\widetilde{\sigma}_{M\ell_M}} \left[\widetilde{\mathcal{R}}_{x_{\ell_1},\ldots,x_{\ell_M}}^{\widetilde{\alpha}-\widetilde{\sigma}_{1\ell_1}-\cdots-\widetilde{\sigma}_{M\ell_M}}(\boldsymbol{m})\, e^{j2\pi(\widetilde{\sigma}_{1\ell_1}m_1+\cdots+\widetilde{\sigma}_{M\ell_M}m_M)} \right] \\
&\underset{m_1}{\otimes} \widetilde{h}_{\widetilde{\sigma}_{1\ell_1}}(m_1) \cdots \underset{m_M}{\otimes} \widetilde{h}_{\widetilde{\sigma}_{M\ell_M}}(m_M)
\end{aligned} \tag{4.107}$$

where $\underset{m_i}{\otimes}$ denotes discrete convolution with respect to the variable m_i.

The RD-CTCMF $R_{\boldsymbol{y}}^{\alpha}(\boldsymbol{m}')$ is obtained by setting $m_M = 0$ in $\mathcal{R}_{\boldsymbol{y}}^{\alpha}(\boldsymbol{m})$.

The input/output relation in terms of CTCMFs (4.107) for discrete-time signals is analogous to the input/output relation (4.83b) for continuous-time signals. Therefore, the input/output relations in terms of CSCMFs, RD-CSCMFs, CTCCFs, CSCCFs, and CCPs for discrete-time signals are obtained with the obvious modifications for discrete time from their continuous-time counterparts (4.84), (4.86), (4.87b), (4.88), and (4.89), respectively.

4.5 Sampling

Let $x_i(n)$, $i = 1, \ldots, N$, be the discrete-time time series obtained by uniformly sampling the continuous-time time series $x_{a,i}(t)$ with rate $f_s \triangleq 1/T_s$:

$$x_i(n) \triangleq x_{a,i}(t)\,|_{t=nT_s}, \qquad n \in \mathbb{Z},\ i = 1, \ldots, N\,. \tag{4.108}$$

Lemma 4.34. (Napolitano, 1995). *The relationship between the Nth-order CTCMF of the discrete-time signals $x_i(n)$ and the Nth-order CTCMFs of the continuous-time signals $x_{a,i}(t)$ is given by*

$$\begin{aligned}
\widetilde{\mathcal{R}}_{\boldsymbol{x}}^{\widetilde{\alpha}}(\boldsymbol{m}) &\triangleq \lim_{K\to\infty} \frac{1}{2K+1} \sum_{n=-K}^{K} \prod_{i=1}^{N} x_{a,i}(t+\tau_i)\,|_{t=nT_s,\boldsymbol{\tau}=\boldsymbol{m}T_s}\, e^{-j2\pi\widetilde{\alpha}n} \\
&= \sum_{p\in\mathbb{Z}} \mathcal{R}_{\boldsymbol{x}_a}^{(\widetilde{\alpha}+p)f_s}(\boldsymbol{m}T_s)\,.
\end{aligned} \tag{4.109}$$

□

Eq. (4.109) reveals that the set of the Nth-order cycle frequencies of the cross-moments of the N sampled signals $x_i(n)$ is given by

$$\{\widetilde{\alpha}\} \subseteq \left\{ \widetilde{\alpha} \in [-1/2, 1/2) : \widetilde{\alpha} = (\alpha/f_s \bmod 1),\ \mathcal{R}_{\boldsymbol{x}_a}^{\alpha}(\boldsymbol{\tau}) \not\equiv 0 \right\} \tag{4.110}$$

where mod 1 operation has values in $[-1/2, 1/2)$ and the strict inclusion can occur only if the ratio α/f_s is an integer number for some α.

By taking the N-dimensional (discrete-time) Fourier transform of the first and last sides of (4.109) and using the N-dimensional version of the second Poisson's summation formula (C.12) one obtains the following result.

Lemma 4.35. (Napolitano, 1995). *The relationship between the Nth-order CSCMF of the discrete-time signals $x_i(n)$ and the Nth-order CSCMFs of the continuous-time signals $x_{a,i}(t)$ is given by*

$$\widetilde{\mathbb{S}}_{\boldsymbol{x}}^{\widetilde{\alpha}}(\boldsymbol{\nu}) = f_s^N \sum_{p\in\mathbb{Z}} \sum_{\boldsymbol{q}\in\mathbb{Z}^N} \mathbb{S}_{\boldsymbol{x}_a}^{(\widetilde{\alpha}+p)f_s}(\boldsymbol{\nu} f_s - \boldsymbol{q} f_s) \tag{4.111}$$

whose reduced-dimension version is

$$\widetilde{S}_{\boldsymbol{x}}^{\widetilde{\alpha}}(\boldsymbol{\nu}') = f_s^{N-1} \sum_{p\in\mathbb{Z}} \sum_{\boldsymbol{q}'\in\mathbb{Z}^{N-1}} S_{\boldsymbol{x}_a}^{(\widetilde{\alpha}+p)f_s}(\boldsymbol{\nu}' f_s - \boldsymbol{q}' f_s)\,. \tag{4.112}$$

□

Analogously, accounting for (4.109) and (4.76), one has the following result.

Lemma 4.36. (Napolitano, 1995). *The relationship between the Nth-order CTCCF of the discrete-time signals $x_i(n)$ and the Nth-order CTCCFs of the continuous-time signals $x_{a,i}(t)$ is given by*

$$\widetilde{\mathcal{C}}_{\boldsymbol{x}}^{\widetilde{\beta}}(\boldsymbol{m}) = \sum_{p=-\infty}^{+\infty} \mathcal{C}_{\boldsymbol{x}_a}^{(\widetilde{\beta}+p)f_s}(\boldsymbol{m}T_s) \tag{4.113}$$

from which it follows the relationship between the Nth-order CCP of the discrete-time signals and the CCPs of the continuous-time signals:

$$\widetilde{P}_{\boldsymbol{x}}^{\widetilde{\beta}}(\boldsymbol{\nu}') = f_s^{N-1} \sum_{p=-\infty}^{+\infty} \sum_{\boldsymbol{q}'\in\mathbb{Z}^{N-1}} P_{\boldsymbol{x}_a}^{(\widetilde{\beta}+p)f_s}(\boldsymbol{\nu}' f_s - \boldsymbol{q}' f_s)\,. \tag{4.114}$$

□

Theorem 4.37. (Napolitano, 1995). *Let $x_{a,i}(t)$ $i = 1, \dots, N$ be N strictly band-limited continuous-time signals with monolateral bandwidths B_i, $i = 1, \dots, N$. The condition*

$$f_s \geqslant 2 \sum_{i=1}^{N} B_i \tag{4.115}$$

is sufficient to assure that

$$\widetilde{S}_{\boldsymbol{x}}^{\widetilde{\alpha}}(\boldsymbol{\nu}') = f_s^{N-1}\, S_{\boldsymbol{x}_a}^{\alpha}(\boldsymbol{f}')\big|_{\alpha=\widetilde{\alpha}f_s,\ \boldsymbol{f}'=\boldsymbol{\nu}' f_s} \qquad |\widetilde{\alpha}| \leqslant \frac{1}{2},\ \ |\nu_i| \leqslant \frac{1}{2}\,, i = 1, \dots, N-1 \tag{4.116}$$

$$\widetilde{P}_{\boldsymbol{x}}^{\widetilde{\beta}}(\boldsymbol{\nu}') = f_s^{N-1}\, P_{\boldsymbol{x}_a}^{\beta}(\boldsymbol{f}')\big|_{\beta=\widetilde{\beta}f_s,\ \boldsymbol{f}'=\boldsymbol{\nu}' f_s} \qquad |\widetilde{\beta}| \leqslant \frac{1}{2},\ \ |\nu_i| \leqslant \frac{1}{2}\,, i = 1, \dots, N-1\,. \tag{4.117}$$

Proof. See Section 4.11. □

In the special case of a single time series $x_{a,i} \equiv x_a$ with monolateral bandwidth B, condition (4.115) becomes

$$f_s \geqslant 2NB\,. \tag{4.118}$$

Moreover, accounting for (4.149), from (4.146) one has

$$\operatorname{supp}\left\{S^{\alpha}_{\boldsymbol{x}_a}(\boldsymbol{f}')\right\} \subseteq \left\{(\alpha, \boldsymbol{f}') \in \mathbb{R} \times \mathbb{R}^{N-1} \,:\, |\alpha| \leqslant \frac{f_s}{2}, \right.$$
$$\left. |f_i| \leqslant \frac{f_s}{2} - \sum_{\substack{k=1\\k\neq i}}^{N} B_k,\; i = 1, \ldots, N-1 \right\} \tag{4.119}$$

so that

$$\{\alpha\} \equiv \left\{\alpha \in \mathbb{R} \,:\, \mathcal{R}^{\alpha}_{\boldsymbol{x}_a}(\boldsymbol{\tau}) \not\equiv 0\right\} \equiv \left\{\alpha \in [-f_s/2, f_s/2) \,:\, \mathcal{R}^{\alpha}_{\boldsymbol{x}_a}(\boldsymbol{\tau}) \not\equiv 0\right\}\,. \tag{4.120}$$

Thus, in (4.109) only the term with $p = 0$ is not identically zero, and hence

$$\widetilde{\mathcal{R}}^{\widetilde{\alpha}}_{\boldsymbol{x}}(\boldsymbol{m}) = \mathcal{R}^{\widetilde{\alpha} f_s}_{\boldsymbol{x}_a}(\boldsymbol{m} T_s), \quad \widetilde{\alpha} \in [-1/2, 1/2) \,:\, \widetilde{\mathcal{R}}^{\widetilde{\alpha}}_{\boldsymbol{x}}(\boldsymbol{m}) \not\equiv 0\,. \tag{4.121}$$

Consequently, there is a one-to-one correspondence between cycle frequencies α of the continuous-time signal and cycle frequencies $\widetilde{\alpha}$ of the discrete-time sampled signal:

$$\alpha = \widetilde{\alpha} f_s\,. \tag{4.122}$$

Finally, let us note that relations analogous to (4.119)–(4.122), (4.145)–(4.150) can be obtained in terms of CCPs, by substituting α, $S^{\alpha}_{\boldsymbol{x}_a}(\boldsymbol{f}')$, $\mathcal{R}^{\alpha}_{\boldsymbol{x}_a}(\boldsymbol{\tau})$, and $\widetilde{\mathcal{R}}^{\widetilde{\alpha}}_{\boldsymbol{x}}(\boldsymbol{m})$ with β, $P^{\beta}_{\boldsymbol{x}_a}(\boldsymbol{f}')$, $\mathcal{C}^{\beta}_{\boldsymbol{x}_a}(\boldsymbol{\tau})$, and $\widetilde{\mathcal{C}}^{\widetilde{\beta}}_{\boldsymbol{x}}(\boldsymbol{m})$, respectively.

4.6 Rice's representation

Let us define the subscripts $\mathring{\boldsymbol{x}}^{(*)} \triangleq \mathring{x}^{(*)_1} \cdots \mathring{x}^{(*)_N}$ and $\widetilde{\boldsymbol{x}}^{(*)} \triangleq \widetilde{x}^{(*)_1} \cdots \widetilde{x}^{(*)_N}$. Using the results of Examples 4.30 (LTI systems) and 4.32 (Modulation), relationships among the RD-CSCMFs of real signal, analytic signal, and complex envelope can be derived (Izzo and Napolitano, 1997) (see Section 3.5 for notation).

$$S^{\alpha}_{\mathring{\boldsymbol{x}}^{(*)}}(\boldsymbol{f}') = \left(\frac{2}{k}\right)^{N} S^{\alpha}_{\boldsymbol{x}}(\boldsymbol{f}')\, \mathrm{u}((-)_N(\alpha - \boldsymbol{f}'^{\mathrm{T}}\mathbf{1})) \prod_{n=1}^{N-1} \mathrm{u}((-)_n f_n) \tag{4.123}$$

$$S^{\alpha}_{\widetilde{\boldsymbol{x}}^{(*)}}(\boldsymbol{f}') = S^{\alpha + \boldsymbol{f}_0^{(-)\mathrm{T}}\mathbf{1}}_{\mathring{\boldsymbol{x}}^{(*)}}(\boldsymbol{f}' + \boldsymbol{f}_0^{(-)\prime}) \tag{4.124}$$

where $\boldsymbol{f}_0^{(-)} \triangleq [(-)_1 f_0, \ldots, (-)_N f_0]^{\mathsf{T}}$ and $\mathbf{1} \triangleq [1, \ldots, 1]^{\mathsf{T}}$.

$$S_{\boldsymbol{x}}^{\alpha}(\boldsymbol{f}') = \left(\frac{k}{2}\right)^N \sum_{\{(*)\}} S_{\overset{\circ}{\boldsymbol{x}}^{(*)}}^{\alpha}(\boldsymbol{f}') \tag{4.125a}$$

$$= \left(\frac{k}{2}\right)^N \sum_{\{(*)\}} S_{\widetilde{\boldsymbol{x}}^{(*)}}^{\alpha - \boldsymbol{f}_0^{(-)\mathsf{T}} \mathbf{1}}(\boldsymbol{f}' - \boldsymbol{f}_0^{(-)\prime}) \tag{4.125b}$$

where the sums are extended to all possible conjugation configurations.

Analogous relationships are obtained in terms of CCP by replacing RD-CSCMFs with the corresponding CCPs (Theorem 4.29).

4.7 Higher-order hybrid temporal-spectral cyclic statistics

Let

$$R_{\boldsymbol{x}}^{\alpha}(\boldsymbol{\tau}') \triangleq \lim_{T\to\infty} \frac{1}{T} \int_{-T/2}^{T/2} x_1^{(*)_1}(t+\tau_1) \cdots x_{n-1}^{(*)_{n-1}}(t+\tau_{n-1})\, x_n^{(*)_n}(t)\, e^{-j2\pi\alpha t}\, \mathrm{d}t \tag{4.126}$$

be the *nth-order reduced-dimension cyclic temporal cross-moment function* (RD-CTCMF) at cycle frequency α, with $\boldsymbol{\tau}' = [\tau_1, \ldots, \tau_{n-1}]$ (see (4.16)). The $(n-1)$-dimensional Fourier transform of the RD-CTCMF is the *nth-order reduced-dimension cyclic spectral moment function* (RD-CSCMF) at cycle frequency α (see (4.15))

$$S_{\boldsymbol{x}}^{\alpha}(\boldsymbol{f}') \triangleq \int_{\mathbb{R}^{n-1}} R_{\boldsymbol{x}}^{\alpha}(\boldsymbol{\tau}')\, e^{-j2\pi \boldsymbol{f}'^{\mathsf{T}}\boldsymbol{\tau}'}\, \mathrm{d}\boldsymbol{\tau}' . \tag{4.127}$$

Let

$$C_{\boldsymbol{x}}^{\beta}(\boldsymbol{\tau}') \triangleq \lim_{T\to\infty} \frac{1}{T} \int_{-T/2}^{T/2} \operatorname{cum}\left\{x_1^{(*)_1}(t+\tau_1), \ldots, x_{n-1}^{(*)_{n-1}}(t+\tau_{n-1}), x_n^{(*)_n}(t)\right\} e^{-j2\pi\beta t}\, \mathrm{d}t \tag{4.128}$$

be the *nth-order reduced-dimension cyclic temporal cross-cumulant function* (RD-CTCCF) at cycle frequency β (see (4.40)). Its $(n-1)$-dimensional Fourier transform is the *nth-order cyclic cross-polyspectrum* (CCP) at cycle frequency β (see (4.39))

$$P_{\boldsymbol{x}}^{\beta}(\boldsymbol{f}') \triangleq \int_{\mathbb{R}^{n-1}} C_{\boldsymbol{x}}^{\beta}(\boldsymbol{\tau}')\, e^{-j2\pi \boldsymbol{f}'^{\mathsf{T}}\boldsymbol{\tau}'}\, \mathrm{d}\boldsymbol{\tau}' . \tag{4.129}$$

Higher-order cyclic features of the n signals $x_1(t), \ldots, x_n(t)$ can also be described by considering the Fourier transform of the RD-CTCMF or of the RD-CTCCF with respect to a subset of the lags $\tau_1, \ldots, \tau_{n-1}$. By assuming for the sake of simplicity that the signals $x_1(t), \ldots, x_n(t)$ are ordered so that the subset of lags of interest is $\tau_1, \ldots, \tau_p$, $p \leqslant n-2$, according to the definition in (Flagiello et al., 2000, eqs. (15), (16)), we have the following definitions.

Definition 4.38. The *nth-order hybrid temporal-spectral cyclic cross-moment function* (HTSCCMF) is defined as

$$J_{\boldsymbol{x}}^{\alpha}(\boldsymbol{\tau}^{(n-1-p)}, \boldsymbol{f}^{(p)}) \triangleq \int_{\mathbb{R}^p} R_{\boldsymbol{x}}^{\alpha}(\boldsymbol{\tau}')\, e^{-j2\pi[f_1\tau_1+\cdots+f_p\tau_p]}\, \mathrm{d}\tau_1 \cdots \mathrm{d}\tau_p \tag{4.130a}$$

$$= \int_{\mathbb{R}^{n-1-p}} S_{\boldsymbol{x}}^{\alpha}(\boldsymbol{f}')\, e^{j2\pi[f_{p+1}\tau_{p+1}+\cdots+f_{n-1}\tau_{n-1}]}\, \mathrm{d}f_{p+1} \cdots \mathrm{d}f_{n-1} \tag{4.130b}$$

where $\boldsymbol{f}^{(p)} \triangleq [f_1, \ldots, f_p]^{\mathsf{T}}$ and $\boldsymbol{\tau}^{(n-1-p)} \triangleq [\tau_{p+1}, \ldots, \tau_{n-1}]^{\mathsf{T}}$. □

The HTSCCMF $J_{\boldsymbol{x}}^{\alpha}(\boldsymbol{\tau}^{(n-1-p)}, \boldsymbol{f}^{(p)})$ is the $(p+1)$th-order RD-CSCMF at cycle frequency α of the $p+1$ signals

$$x_1^{(*)_1}(t), \ldots, x_p^{(*)_p}(t), \text{ and } x_n^{(*)_n}(t) \prod_{i=p+1}^{n-1} x_i^{(*)_i}(t+\tau_i)\,.$$

Definition 4.39. The *nth-order hybrid temporal-spectral cyclic cross-cumulant function* (HTSCCCF) is defined as

$$H_{\boldsymbol{x}}^{\beta}(\boldsymbol{\tau}^{(n-1-p)}, \boldsymbol{f}^{(p)}) \triangleq \int_{\mathbb{R}^p} C_{\boldsymbol{x}}^{\beta}(\boldsymbol{\tau}')\, e^{-j2\pi[f_1\tau_1+\cdots+f_p\tau_p]}\, \mathrm{d}\tau_1 \cdots \mathrm{d}\tau_p \tag{4.131a}$$

$$= \int_{\mathbb{R}^{n-1-p}} P_{\boldsymbol{x}}^{\beta}(\boldsymbol{f}')\, e^{j2\pi[f_{p+1}\tau_{p+1}+\cdots+f_{n-1}\tau_{n-1}]}\, \mathrm{d}f_{p+1} \cdots \mathrm{d}f_{n-1} \tag{4.131b}$$

□

The HTSCCCF $H_{\boldsymbol{x}}^{\beta}(\boldsymbol{\tau}^{(n-1-p)}, \boldsymbol{f}^{(p)})$ cannot be interpreted as CCP of $p+1$ signals.

The nth-order HTSCCMFs and HTSCCCFs can be exploited for the purpose of HOCS-based modulation classification or HOCS-based LTI system identification (Section 9.5.2.3).

4.8 Stochastic processes

Let us consider the not necessarily distinct complex-valued continuous-time stochastic processes $x_1(t), \ldots, x_N(t)$.

All the results established in the previous sections in the FOT probability framework can be reinterpreted in the classical stochastic process framework by following the guidelines given in Section 2.4. Specifically, signals $x_k(t)$ that are modeled as single time series should be modeled as stochastic processes. In addition, the almost-periodic component extraction operator $\mathrm{E}^{\{\alpha\}}\{\cdot\}$ (that acts on the time variable $t \in \mathbb{R}$) should be replaced by the ensemble average operator $\mathrm{E}\{\cdot\}$ (that acts on the variable $\omega \in \Omega$).

Under appropriate ergodicity conditions (Chapter 5), (Boyles and Gardner, 1983), (Dandawaté and Giannakis, 1994a), (Dehay and Hurd, 1994), equations involving cyclic statistics defined in the classical stochastic approach (in terms of time and ensemble averages) are formally analogous to their FOT counterparts where cyclic statistics are defined in terms of time averages.

The following clarification is necessary. In equation

$$\mathrm{E}\left\{\prod_{k=1}^{N} X_k^{(*)_k}((-)_k f_k)\right\} \triangleq \lim_{T\to\infty}\left\langle \mathrm{E}\left\{\prod_{k=1}^{N} X_{k,T}^{(*)_k}(t,(-)_k f_k)\right\}\right\rangle_t \tag{4.132}$$

the formal expression in the lhs is defined as the rhs where the limit should be in a generalized sense that accommodates the presence of Dirac deltas.

4.9 Developments and applications

Developments and applications on the subject of higher-order cyclostationarity have been considered in the following fields:

- *Estimation issues*
  (Giannakis and Zhou, 1994), (Giannakis, 1995), (Li and Cheng, 1997);
- *Signal parameter and waveform estimation*
  (Dandawaté and Giannakis, 1993), (Giannakis and Zhou, 1994), (Shamsunder and Giannakis, 1994), (Shamsunder et al., 1995), (Zhou and Giannakis, 1995b), (Zhou and Giannakis, 1995a), (Zhou, 1995), (Martone, 1999), (Xin et al., 2000), (Morelande and Zoubir, 2002), and (Gelli et al., 2005);
- *Nonlinear system identification and equalization*
  (Gardner and Archer, 1993), (Izzo et al., 1994b), (Gardner and Paura, 1995), (Liang et al., 1997), (Liang and Leyman, 1998), (Mattera and Paura, 1998), and (Mattera, 1999);
- *Signal detection*
  (Dandawaté and Giannakis, 1994b);
- *Signal classification*
  (Boiteau and Le Martret, 1998), (Marchand et al., 1995), (Marchand, 1998), (Marchand et al., 1998), (Spooner, 1995), (Spooner and Nicholls, 2009);
- *Spectral analysis of PAM signals*
  (Bilardi and Pupolin, 1984);
- *Cyclic higher-order properties of CPM signals*
  (Napolitano and Spooner, 2001);
- *Blind adaptive detection*
  (Martone, 2000);
- *Source separation*
  (Ferréol and Chevalier, 2000);
- *Wavelet decomposition*
  (Touati and Pesquet, 2001);
- *Acoustics and mechanics*
  (Raad et al., 2002), (Raad, 2003), and (Antoni et al., 2004a);

- *Higher-order characterization of generalized almost-cyclostationary signals* (Izzo and Napolitano, 1998a), (Izzo and Napolitano, 2002a), (Izzo and Napolitano, 2002b), (Izzo and Napolitano, 2003), (Napolitano and Tesauro, 2011), (Napolitano, 2012, Sec. 2.2.4);
- *Higher-order characterization of spectrally correlated processes* (Napolitano, 2012, Sec. 4.2.3).

4.10 Summary

N time series are said to exhibit joint Nth-order cyclostationarity with cycle frequency $\alpha \neq 0$ if their Nth-order lag product contains a finite-strength additive sine wave component with frequency α. The amplitude and phase of this sine wave are the magnitude and phase of the cyclic temporal cross-moment function at cycle frequency α (Definition 4.3). For complex-valued time series, these magnitude and phase depend on the possible conjugations of the terms of the lag product.

By using the almost-periodic component extraction operator as expectation operator, the temporal cross-moment function and the almost-periodically time-variant FOT joint distribution function are defined starting from these single time series without introducing stochastic process models.

The Nth-order characterization in the frequency domain is made by the Nth-order cyclic spectral cross-moment function which is the N-dimensional Fourier transform of the Nth-order cyclic temporal cross-moment function (Definition 4.6).

For both cyclic temporal and spectral cross-moment functions a reduced dimension version can be considered since one of the N lag or spectral-frequency variables is redundant. A nonzero reduced-dimension cyclic spectral cross-moment function at cycle frequency α reveals that N time series exhibit joint Nth-order cyclostationarity at cycle frequency α if and only if it is nonzero the Nth-order temporal cross-moment of their spectral components with frequencies that sum to α (Proposition 4.8). This is the generalization to higher orders of the (second-order) spectral correlation property.

The Nth-order temporal cross-cumulant function is defined through the Nth-order partial derivative of the logarithm of the characteristic function. Its Fourier coefficients are the Nth-order cyclic temporal cross-cumulant functions whose N-dimensional Fourier transforms are the Nth-order cyclic spectral cross-cumulant functions. The reduced-dimension version of these functions are the reduced-dimension cyclic temporal cross-cumulant functions and the Nth-order cyclic cross-polyspectra.

Starting from the Leonov and Shiryaev formulas that link moments and cumulants (Theorem 4.22), relationships among cyclic temporal moments and cumulants and their frequency counterparts are established ((4.51)–(4.54)).

FOT cumulant-based cyclic statistics present all the properties of cumulants for stochastic processes: the cumulant of the sum of two independent signals is the sum of the cumulants of the two signals; in the case of asymptotic independence of signals, cumulants, unlike moments, are summable functions of the lag parameters; the cumulants of order greater than two of Gaussian signals is zero. Moreover, the Nth-order cyclic temporal cumulant function of a time series at cycle frequency α provides the amplitude and phase of the pure Nth-order sine wave at frequency α contained in the Nth-order lag product. It is that part of the sine wave at frequency α present in the Nth-order lag product waveform that remains after removal of all parts that result from products of sine waves in lower order lag products obtained by factoring the Nth-order product.

The cyclic cross-polyspectrum and the reduced-dimension cyclic spectral cross-moment function at cycle frequency α are equal except on the so called α-submanifold where the latter is impulsive (Section 4.2.3.1). This difference turns out to be fundamental in the design of cyclic statistic estimators.

Higher-order cyclic statistics for discrete-time time series in both time and frequency domains are defined analogously to their counterparts for continuous-time time series (Section 4.3).

Input/output relationships in terms of higher-order cyclic statistics for multi-input multi-output linear almost-periodically time variant systems are derived in both time and frequency domains and for both continuous- and discrete-time time series (Section 4.4). Relationships in terms of moments and cumulants have the same form (Theorem 4.27).

For a continuous-time signal with bandwidth B that is uniformly sampled with sampling frequency f_s, the sufficient condition $f_s \geqslant 2NB$ is derived such that Nth-order cyclic spectral moments and cumulants of the discrete-time sampled signal are amplitude- and frequency-scaled versions of the corresponding Nth-order cyclic spectral moments and cumulants of the continuous-time signal in the whole main cycle and spectral frequency domains (Theorem 4.37).

Nth-order hybrid temporal-spectral cyclic statistics are defined by Fourier transforming the Nth-order reduced-dimension cyclic temporal cross-moments and cross-cumulant functions with respect to a subset of the $N-1$ lag parameters (Section 4.7). The adoption of these higher-order statistics can be advantageous in estimation problems in terms of computational complexity and bias and variance of the estimators (Sections 5.7.4 and 9.5.2.3).

All the results presented in the FOT approach can be immediately re-interpreted in the classical stochastic approach by replacing time series with stochastic processes and the almost-periodic component extraction operator with the statistical expectation (Section 4.8).

4.11 Proofs

In this section, proofs are reported for results of the present chapter. Unless otherwise specified, only formal manipulations are reported, without explicitly justifying inversion of linear operators such as time averages, infinite sums, integrals, and ensemble average (expectation). In general, these manipulations are justified at least in the sense of generalized functions (distributions) (Champeney, 1990), (Zemanian, 1987).

Proof of Theorem 4.7

$$\begin{aligned}
\mathcal{R}_{\boldsymbol{x}}^{\alpha}(\boldsymbol{\tau}) &\triangleq \lim_{T\to\infty} \frac{1}{T} \int_{-T/2}^{T/2} \prod_{k=1}^{N} x_k^{(*)_k}(t+\tau_k)\, e^{-j2\pi\alpha t}\, \mathrm{d}t \\
&= \lim_{T\to\infty} \frac{1}{T} \int_{\tau_N-T/2}^{\tau_N+T/2} \prod_{k=1}^{N-1} x_k^{(*)_k}(t'+\tau_k-\tau_N)\, x_N^{(*)_N}(t')\, e^{-j2\pi\alpha t'}\, \mathrm{d}t'\, e^{j2\pi\alpha\tau_N} \\
&= R_{\boldsymbol{x}}(\boldsymbol{\tau}'-\mathbf{1}\tau_N)\, e^{j2\pi\alpha\tau_N}
\end{aligned} \tag{4.133}$$

where in the second equality the variable change $t' = t + \tau_N$ is made.

$$\begin{aligned}
\mathcal{S}_{\boldsymbol{x}}^{\alpha}(\boldsymbol{f}) &\triangleq \int_{\mathbb{R}^N} \mathcal{R}_{\boldsymbol{x}}^{\alpha}(\boldsymbol{\tau})\, e^{-j2\pi \boldsymbol{f}^{\mathsf{T}}\boldsymbol{\tau}}\, \mathrm{d}\boldsymbol{\tau} \\
&= \int_{\mathbb{R}} e^{j2\pi\alpha\tau_N}\, e^{-j2\pi f_N\tau_N} \underbrace{\int_{\mathbb{R}^{N-1}} R_{\boldsymbol{x}}(\boldsymbol{\tau}'-\mathbf{1}\tau_N)\, e^{-j2\pi \boldsymbol{f}'^{\mathsf{T}}\boldsymbol{\tau}'}\, \mathrm{d}\boldsymbol{\tau}'}_{S_{\boldsymbol{x}}^{\alpha}(\boldsymbol{f}')\, e^{-j2\pi \boldsymbol{f}'^{\mathsf{T}}\mathbf{1}\tau_N}}\, \mathrm{d}\tau_N \\
&= S_{\boldsymbol{x}}^{\alpha}(\boldsymbol{f}') \int_{\mathbb{R}} e^{-j2\pi(\boldsymbol{f}^{\mathsf{T}}\mathbf{1}-\alpha)\tau_N}\, \mathrm{d}\tau_N \\
&= S_{\boldsymbol{x}}^{\alpha}(\boldsymbol{f}')\, \delta(\boldsymbol{f}^{\mathsf{T}}\mathbf{1}-\alpha)
\end{aligned} \tag{4.134}$$

Proof of Theorem 4.10

Accounting for (4.18), we have

$$\begin{aligned}
&\prod_{k=1}^{N} X_{k,T}^{(*)_k}(t, (-)_k f_k) \\
&\quad= \prod_{k=1}^{N} \left[\int_{\mathbb{R}} x_k(s_k)\, \mathrm{rect}\Big(\frac{s_k-t}{T}\Big)\, e^{-j2\pi(-)_k f_k s_k}\, \mathrm{d}s_k \right]^{(*)_k} \\
&\quad= \int_{\mathbb{R}^N} \prod_{k=1}^{N} x_k^{(*)_k}(s_k) \prod_{k=1}^{N} \mathrm{rect}\Big(\frac{s_k-t}{T}\Big) \prod_{k=1}^{N} e^{-j2\pi f_k s_k}\, \mathrm{d}s_1 \cdots \mathrm{d}s_N \\
&\quad= \int_{\mathbb{R}^N} \prod_{k=1}^{N} x_k^{(*)_k}(t+v_k) \prod_{k=1}^{N} \mathrm{rect}\Big(\frac{v_k}{T}\Big)\, e^{-j2\pi \boldsymbol{f}^{\mathsf{T}}\boldsymbol{v}}\, e^{-j2\pi \boldsymbol{f}^{\mathsf{T}}\mathbf{1}t}\, \mathrm{d}\boldsymbol{v}
\end{aligned} \tag{4.135}$$

where in the third equality the variable changes $v_k = s_k - t$ are made.

$$\begin{aligned}
&\left\langle \prod_{k=1}^{N} X_{k,T}^{(*)_k}(t, (-)_k f_k) \right\rangle_t \\
&\quad= \int_{\mathbb{R}^N} \left\langle \prod_{k=1}^{N} x_k^{(*)_k}(t+v_k)\, e^{-j2\pi \boldsymbol{f}^{\mathsf{T}}\mathbf{1}t} \right\rangle_t \prod_{k=1}^{N} \mathrm{rect}\Big(\frac{v_k}{T}\Big)\, e^{-j2\pi \boldsymbol{f}^{\mathsf{T}}\boldsymbol{v}}\, \mathrm{d}\boldsymbol{v}
\end{aligned}$$

$$= \int_{\mathbb{R}^N} \mathcal{R}_{\boldsymbol{x}}^{\boldsymbol{f}^\top \mathbf{1}}(\boldsymbol{v}) \prod_{k=1}^{N} \operatorname{rect}\left(\frac{v_k}{T}\right) e^{-j2\pi \boldsymbol{f}^\top \boldsymbol{v}} \,\mathrm{d}\boldsymbol{v} \tag{4.136}$$

Using the identity

$$\mathcal{R}_{\boldsymbol{x}}^{\boldsymbol{f}^\top \mathbf{1}}(\boldsymbol{v}) = \sum_{\alpha \in A_{\boldsymbol{x}}} \mathcal{R}_{\boldsymbol{x}}^{\alpha}(\boldsymbol{v}) \, \delta_{\alpha - \boldsymbol{f}^\top \mathbf{1}} \tag{4.137}$$

and (4.14), we have

$$\begin{aligned}
&\lim_{T\to\infty} \left\langle \prod_{k=1}^{N} X_{k,T}^{(*)_k}(t, (-)_k f_k) \right\rangle_t \\
&= \lim_{T\to\infty} \int_{\mathbb{R}^N} \mathcal{R}_{\boldsymbol{x}}^{\boldsymbol{f}^\top \mathbf{1}}(\boldsymbol{v}) \prod_{k=1}^{N} \operatorname{rect}\left(\frac{v_k}{T}\right) e^{-j2\pi \boldsymbol{f}^\top \boldsymbol{v}} \,\mathrm{d}\boldsymbol{v} \\
&= \int_{\mathbb{R}^N} \mathcal{R}_{\boldsymbol{x}}^{\boldsymbol{f}^\top \mathbf{1}}(\boldsymbol{v}) \, e^{-j2\pi \boldsymbol{f}^\top \boldsymbol{v}} \,\mathrm{d}\boldsymbol{v} \\
&= \sum_{\alpha \in A_{\boldsymbol{x}}} \int_{\mathbb{R}^N} \mathcal{R}_{\boldsymbol{x}}^{\alpha}(\boldsymbol{v}) \, e^{-j2\pi \boldsymbol{f}^\top \boldsymbol{v}} \,\mathrm{d}\boldsymbol{v} \, \delta_{\alpha - \boldsymbol{f}^\top \mathbf{1}} \\
&= \sum_{\alpha \in A_{\boldsymbol{x}}} \mathcal{S}_{\boldsymbol{x}}^{\alpha}(\boldsymbol{f}) \, \delta_{\alpha - \boldsymbol{f}^\top \mathbf{1}} \\
&= \sum_{\alpha \in A_{\boldsymbol{x}}} S_{\boldsymbol{x}}^{\alpha}(\boldsymbol{f}') \, \delta(\alpha - \boldsymbol{f}^\top \mathbf{1}) \, \delta_{\alpha - \boldsymbol{f}^\top \mathbf{1}} \, .
\end{aligned} \tag{4.138}$$

Using the formal identity $\delta(\alpha - \boldsymbol{f}^\top \mathbf{1}) \, \delta_{\alpha - \boldsymbol{f}^\top \mathbf{1}} = \delta(\alpha - \boldsymbol{f}^\top \mathbf{1})$, one obtains the right-hand side of (4.22).

Proof of Theorem 4.20

Under mild regularity conditions, the multilinearity property for FOT cumulants can be proved similarly to the case of stochastic cumulants (Mendel, 1991).

Due to the multilinearity property of FOT cumulants, we have

$$\begin{aligned}
&\operatorname{cum}\left\{ X_{k,T}^{(*)_k}(t, (-)_k f_k), \; k = 1, \ldots, N \right\} \\
&= \operatorname{cum}\left\{ \left[\int_{\mathbb{R}} x_k(s_k) \operatorname{rect}\left(\frac{s_k - t}{T}\right) e^{-j2\pi (-)_k f_k s_k} \,\mathrm{d}s_k \right]^{(*)_k}, \; k = 1, \ldots, N \right\} \\
&= \operatorname{cum}\left\{ \left[\int_{\mathbb{R}} x_k^{(*)_k}(t + v_k) \operatorname{rect}\left(\frac{v_k}{T}\right) e^{-j2\pi f_k (t+v_k)} \,\mathrm{d}v_k \right], \; k = 1, \ldots, N \right\}
\end{aligned}$$

$$= \int_{\mathbb{R}^N} \text{cum}\left\{ x_k^{(*)_k}(t + v_k)\, e^{-j2\pi f_k t},\ k = 1, \ldots, N \right\} \prod_{k=1}^{N} \text{rect}\left(\frac{v_k}{T}\right) e^{-j2\pi \boldsymbol{f}^\top \boldsymbol{v}}\, \mathrm{d}\boldsymbol{v} \tag{4.139}$$

where in the second equality the variable changes $s_k = t + v_k$ are made.

Since the periodic functions $e^{-j2\pi f_k t}$ are deterministic in the FOT probability framework (Fact 2.25), they can be put outside the cum{·} operator. Thus, replacing (4.36) into (4.139), one has

$$\begin{aligned} &\text{cum}\left\{ X_{k,T}^{(*)_k}(t, (-)_k f_k),\ k = 1, \ldots, N \right\} \\ &\quad = \int_{\mathbb{R}^N} \sum_{\beta \in B_{\boldsymbol{x}}} \mathcal{C}_{\boldsymbol{x}}^{\beta}(\boldsymbol{v})\, e^{j2\pi\beta t}\, e^{-j2\pi \boldsymbol{f}^\top \mathbf{1} t} \prod_{k=1}^{N} \text{rect}\left(\frac{v_k}{T}\right) e^{-j2\pi \boldsymbol{f}^\top \boldsymbol{v}}\, \mathrm{d}\boldsymbol{v} \end{aligned} \tag{4.140}$$

Thus

$$\begin{aligned} &\lim_{T\to\infty} \text{cum}\left\{ X_{k,T}^{(*)_k}(t, (-)_k f_k),\ k = 1, \ldots, N \right\} \\ &\quad = \sum_{\beta \in B_{\boldsymbol{x}}} \int_{\mathbb{R}^N} \mathcal{C}_{\boldsymbol{x}}^{\beta}(\boldsymbol{v})\, e^{-j2\pi \boldsymbol{f}^\top \boldsymbol{v}}\, \mathrm{d}\boldsymbol{v}\, e^{j2\pi(\beta - \boldsymbol{f}^\top \mathbf{1})t} \\ &\quad = \sum_{\beta \in B_{\boldsymbol{x}}} \mathcal{P}_{\boldsymbol{x}}^{\beta}(\boldsymbol{f})\, e^{j2\pi(\beta - \boldsymbol{f}^\top \mathbf{1})t} \\ &\quad = \sum_{\beta \in B_{\boldsymbol{x}}} P_{\boldsymbol{x}}^{\beta}(\boldsymbol{f}')\, \delta(\beta - \boldsymbol{f}^\top \mathbf{1})\, e^{j2\pi(\beta - \boldsymbol{f}^\top \mathbf{1})t} \end{aligned} \tag{4.141}$$

from which (4.42) immediately follows.

Proof of Theorem 4.25

The proof is the generalization to higher orders of the proofs of Theorem 3.6 and Corollary 3.7 for second-order cyclic statistics.

By replacing (4.80) into (4.78), interchanging the order of integration and summation operations, and then taking the expected value (almost-periodic component extraction) of the Mth-order lag product one obtains the temporal cross-moment function of the output signals $y_m(t)$. Thus, taking the Fourier coefficient (4.83a) at frequency α one has (4.83b).

The interchange of the order of sum and integrals can be justified under Assumptions 4.23a and 4.24b or, alternatively, under Assumptions 4.23b and 4.24a.

Proof of Theorem 4.26

Taking the M-dimensional Fourier transform of the left and right sides of (4.83b) one obtains (4.84).

Eq. (4.86) is derived from (4.84) accounting for (4.134).

Proof of Theorem 4.27

Accounting for (4.51) and (4.83b), one has

$$
\begin{aligned}
\mathcal{C}_{\boldsymbol{y}}^{\beta}(\boldsymbol{\tau}) &= \sum_{\mathsf{P}} \left[(-1)^{p-1}(p-1)! \sum_{\boldsymbol{\alpha}^{\mathsf{T}}\mathbf{1}=\beta} \prod_{i=1}^{p} \mathcal{R}_{\boldsymbol{y}_{\mu_i}}^{\alpha_{\mu_i}}(\boldsymbol{\tau}_{\mu_i}) \right] \\
&= \sum_{\mathsf{P}} \left[(-1)^{p-1}(p-1)! \sum_{\boldsymbol{\alpha}^{\mathsf{T}}\mathbf{1}=\beta} \prod_{i=1}^{p} \left(\sum_{\substack{\ell_k=1 \\ k\in\mu_i}}^{L} \sum_{\substack{\sigma_{k\ell_k} \\ k\in\mu_i}} \right. \right. \\
&\qquad \left. \left. \int_{\mathbb{R}^{|\mu_i|}} \mathcal{R}_{\boldsymbol{x}_{\ell,\mu_i}}^{\alpha_{\mu_i}-\boldsymbol{\sigma}_{\ell,\mu_i}^{\mathsf{T}}\mathbf{1}}(\boldsymbol{v}_{\mu_i})\, e^{j2\pi\boldsymbol{\sigma}_{\ell,\mu_i}^{\mathsf{T}}\boldsymbol{v}_{\mu_i}} \prod_{k\in\mu_i} h_{\sigma_{k\ell_k}}(\tau_k - v_k)\, \mathrm{d}\boldsymbol{v}_{\mu_i} \right) \right] \\
&= \sum_{(\ell_1,\ldots,\ell_M)\in I_L^M} \sum_{\substack{\sigma_{k\ell_k} \\ k\in\{1,\ldots,M\}}} \int_{\mathbb{R}^M} \prod_{k=1}^{M} h_{\sigma_{k\ell_k}}(\tau_k - v_k) \\
&\qquad \cdot \sum_{\mathsf{P}} \left[(-1)^{p-1}(p-1)! \sum_{\boldsymbol{\alpha}^{\mathsf{T}}\mathbf{1}=\beta} \prod_{i=1}^{p} \mathcal{R}_{\boldsymbol{x}_{\ell,\mu_i}}^{\alpha_{\mu_i}-\boldsymbol{\sigma}_{\ell,\mu_i}^{\mathsf{T}}\mathbf{1}}(\boldsymbol{v}_{\mu_i}) \right] e^{j2\pi\boldsymbol{v}^{\mathsf{T}}\boldsymbol{\sigma}_{\ell_1,\ldots,\ell_M}}\, \mathrm{d}\boldsymbol{v}
\end{aligned}
\tag{4.142}
$$

where $\boldsymbol{\sigma}_{\ell,\mu_i}$ and $\boldsymbol{x}_{\ell,\mu_i}$ are the $|\mu_i|$-dimensional vectors whose elements are $\sigma_{k\ell_k}$ and $x_{k\ell_k}$, respectively, with $k \in \mu_i$, $\boldsymbol{\sigma}_{\ell_1,\ldots,\ell_M} \triangleq [\sigma_{1\ell_1}, \ldots, \sigma_{M\ell_M}]^{\mathsf{T}}$, and the last equality has been obtained by interchanging the order of integration and summation operators and taking into account that the subsets μ_i are disjoint.

Then, (4.87b) follows by substituting (4.51) into (4.142).

Proof of Theorem 4.37

The continuous-time signals $x_{a,i}(t)$ are strictly band-limited with monolateral bandwidths B_i. Thus we have

$$
y_{a,i}(t) \triangleq x_{a,i}(t) \otimes h_{B_i}(t) = x_{a,i}(t) \qquad i = 1, \ldots, N \tag{4.143}
$$

with

$$
h_{B_i}(t) \triangleq \mathcal{F}^{-1}\left[\operatorname{rect}\left(\frac{f}{2B_i}\right)\right], \tag{4.144}
$$

impulse-response function of the ideal low-pass filter with monolateral bandwidth B_i. In (4.144), $\mathcal{F}^{-1}[\cdot]$ denotes the inverse Fourier transform operation.

From (4.92), (4.143), and (4.144), it results that

$$
S_{\boldsymbol{y}_a}^{\alpha}(\boldsymbol{f}') = S_{\boldsymbol{x}_a}^{\alpha}(\boldsymbol{f}') \operatorname{rect}\left(\frac{\alpha - \boldsymbol{f}'^{\mathsf{T}}\mathbf{1}}{2B_N}\right) \prod_{i=1}^{N-1} \operatorname{rect}\left(\frac{f_i}{2B_i}\right) = S_{\boldsymbol{x}_a}^{\alpha}(\boldsymbol{f}'). \tag{4.145}
$$

Thus, for the support of $S^{\alpha}_{\boldsymbol{x}_a}(\boldsymbol{f}')$ in the $(\alpha, \boldsymbol{f}')$ space, the following inclusion relationship holds:

$$\begin{aligned}
\operatorname{supp}\{S^{\alpha}_{\boldsymbol{x}_a}(\boldsymbol{f}')\} &= \operatorname{supp}\left\{\operatorname{rect}\left(\frac{\alpha - \boldsymbol{f}'^{\mathrm{T}}\mathbf{1}}{2B_N}\right)\prod_{i=1}^{N-1}\operatorname{rect}\left(\frac{f_i}{2B_i}\right)\right\} \\
&= \left\{(\alpha, \boldsymbol{f}') \in \mathbb{R}\times\mathbb{R}^{N-1} : |\alpha - \boldsymbol{f}'^{\mathrm{T}}\mathbf{1}| \leqslant B_N,\ |f_i| \leqslant B_i,\ i = 1, \ldots, N-1\right\} \\
&\subseteq \left\{(\alpha, \boldsymbol{f}') \in \mathbb{R}\times\mathbb{R}^{N-1} : |\alpha| \leqslant \sum_{i=1}^{N} B_i,\ |f_i| \leqslant B_i,\ i = 1, \ldots, N-1\right\}
\end{aligned} \tag{4.146}$$

where the inclusion holds because, for $|f_i| \leqslant B_i$, it results that

$$|\alpha - \boldsymbol{f}'^{\mathrm{T}}\mathbf{1}| \geqslant |\alpha| - |\boldsymbol{f}'^{\mathrm{T}}\mathbf{1}| \geqslant |\alpha| - \sum_{i=1}^{N-1} B_i. \tag{4.147}$$

The discrete-time RD-CSCMF $\widetilde{S}^{\widetilde{\alpha}}_{\boldsymbol{x}}(\boldsymbol{\nu}')$ is the superposition (see (4.112)) of the replicas $S^{(\widetilde{\alpha}+p)f_s}_{\boldsymbol{x}_a}(\boldsymbol{f}' - \boldsymbol{q}' f_s)$, $p \in \mathbb{Z}$ and $\boldsymbol{q}' \in \mathbb{Z}^{N-1}$, whose supports, accounting for (4.146), are included in the sets

$$\left\{(\widetilde{\alpha}, \boldsymbol{f}') \in \mathbb{R}\times\mathbb{R}^{N-1} : |(\widetilde{\alpha} + p) f_s| \leqslant \sum_{i=1}^{N} B_i, \quad |f_i - q_i f_s| \leqslant B_i,\ i = 1, \ldots, N-1\right\}. \tag{4.148}$$

Therefore, a sufficient condition that assures that the sets (4.148) are disjoint and, then, replicas in (4.112) do not overlap, is the following:

$$f_s \geqslant 2\sum_{i=1}^{N} B_i. \tag{4.149}$$

When a single time-series is considered ($x_{a,i}(t) \equiv x_a(t)$, $B_i \equiv B$, $i = 1, \ldots, N$), (4.149) reduces to $f_s \geqslant 2NB$, which reveals that the sampling rate assuring no aliasing grows linearly with the order N.

Let us note that, when condition (4.149) assuring no aliasing is satisfied, it results that

$$\widetilde{S}^{\widetilde{\alpha}}_{\boldsymbol{x}}(\boldsymbol{\nu}') = f_s^{N-1} S^{\alpha}_{\boldsymbol{x}_a}(\boldsymbol{f}')\,|_{\boldsymbol{f}'=\boldsymbol{\nu}' f_s,\ \alpha=\widetilde{\alpha} f_s} \qquad |\widetilde{\alpha}| \leqslant \frac{1}{2}, \quad |\nu_i| \leqslant \frac{1}{2},\ i = 1, \ldots, N-1. \tag{4.150}$$

5

Ergodic properties and measurement of characteristics

Contents

Cyclostationary Processes and Time Series. https://doi.org/10.1016/B978-0-08-102708-0.00016-9

Abstract

Measurements of statistical functions of almost-cyclostationary processes and ergodicity issues are addressed. Second- and higher-order statistical functions are considered. Sufficient conditions for mean-square consistency and asymptotic normality of estimators of cyclic statistical functions in both time and frequency domains are expressed in terms of summability of cumulants. Aliasing issues and the cycle leakage effect are addressed. Computationally efficient estimators and their Matlab/Octave implementations are presented. Numerical experiments corroborate the theoretical results. The estimation problem is presented in the classical stochastic approach and the interpretation in the functional or fraction-of-time approach is provided.

Keywords

Ergodicity, Cyclic correlogram, Cyclic periodogram, Frequency-smoothed cyclic periodogram, Time-smoothed cyclic periodogram, Mean-square consistency, Asymptotic normality, Cycle leakage

5.1 Introduction

One of the main reasons of the practical usefulness of cyclostationarity-based signal processing algorithms is the existence of consistent estimators for cyclic statistics of ACS signals. For these estimators, mean-square consistency and asymptotic complex normality of a properly normalized estimation error are proved under the assumption that the process has finite or practically finite memory. Such an assumption is generally expressed in terms of summability of cumulants.

Ergodicity issues are treated in (Boyles and Gardner, 1983), (Honda, 1990), (Jacob, 1990), (Gardner, 1991d), (Leśkow and Weron, 1992), (Cambanis et al., 1994), (Gardner, 1994a), (Makagon and Miamee, 1996).

Results on second-order cyclic spectral analysis have been presented in the classical stochastic approach in (Dragan et al., 1979), (Yavorskii, 1983), (Dragan et al., 1984a), (Dragan and Yavorskii, 1985), (Yavorskii, 1985), (Yavorskii, 1986), (Yavorskii, 1987b), (Yavorskii, 1987a), (Alekseev, 1988), (Mezentsev and Yavorskii, 1988), (Yavorskii, 1989), (Alekseev, 1991b), (Hurd and Leśkow, 1992b), (Dehay, 1994), (Dehay and Hurd, 1994), (Dandawaté and Giannakis, 1994a), (Dandawaté and Giannakis, 1994b), (Genossar et al., 1994), (Leskow, 1994), (Dandawaté and Giannakis, 1995), (Dehay, 1995), (Dehay and Leśkow, 1996a), (Dehay and Monsan, 1996), (Makagon and Miamee, 1996), (Huang et al., 2003), (Lii and Rosenblatt, 2006), (Hurd and Miamee, 2007), (Yavorskyj et al., 2014). The estimation problem is addressed in the fraction-of-time (or functional) approach in (Gardner, 1987f), (Brown, 1987), (Gardner and Brown, 1991), (Gardner et al., 2006), (Napolitano, 2012, Sec. 6.3.5), (Dehay et al., 2018b), (Section 5.6).

Bootstrap methods and subsampling-based estimators to obtain confidence intervals are presented in (Synowiecki, 2007), (Lenart et al., 2008), (Leskow and Synowiecki, 2010), (Dudek and Leskow, 2011), (Lenart, 2013), (Dudek et al., 2014b), (Dehay et al., 2014), (Dudek et al., 2014a), (Dudek and Potorski, 2014), (Gajecka-Mirek, 2014), (Dudek, 2015), (Dehay and Dudek, 2015), (Dudek and Uzar, 2015), (Dehay et al., 2018a). These methods are discussed in the context of signal detection in Section 8.7.

Computationally efficient estimators for second-order cyclic spectral analysis are presented in (Brown and Loomis, 1988), (Roberts, 1989), (Roberts et al., 1991), (Brown and Loomis, 1993), (Gardner and Roberts, 1993), (Roberts et al., 1994), (Roberts and Loomis, 1995), (Spooner et al., 2013), (Antoni et al., 2017), (Spooner and Mody, 2018), and (Borghesani and Antoni, 2018), (Section 5.4.4). A patent is described in (Anthony et al., 2017).

Higher-order cyclic statistic estimators are given in (Spooner and Gardner, 1991a), (Dandawaté and Giannakis, 1994a), (Dandawaté and Giannakis, 1994b), (Spooner and Gardner, 1994a), (Dandawaté and Giannakis, 1995), (Zhou and Giannakis, 1996), (Wu and Lev-Ari, 1997), (Sadler and Dandawaté, 1998), (Napolitano and Spooner, 2000), (Napolitano and Tesauro, 2011), (Section 5.7).

In the derivation of the above mentioned results it is assumed that the cycle frequency of the signal of interest is exactly known. Second-order cyclic statistic estimators with unknown cycle frequency estimated by the available data are given in (Napolitano, 2015a), (Napolitano, 2016c), (Napolitano, 2017a), (Section 5.5).

Further developments on the estimation of the statistical functions of ACS processes can be found in (Honda, 1983), (Tian, 1988), (Dehay and Monsan, 1996), (Dominique and Reed, 1997), (Nematollahi and Rao, 2005), (Lii and Rosenblatt, 2006), (Soltani and Azimmohseni, 2007), (Javorskyj et al., 2007), (Javorskyj et al., 2010), (Sarnaglia et al., 2010), (Javorskyj et al., 2011), (Javorskyj et al., 2012), (Napolitano, 2012), (Dehay and Waled, 2013), (Javorskyj et al., 2014a), (Javorskyj et al., 2014b), (Yavorskyj et al., 2014).

In this chapter, for the sake of generality, the estimation of cyclic cross-statistics is considered. For the estimation of cyclic statistics in the time domain, it is observed that ACS processes are a special case of GACS processes when the lag-dependent cycle frequencies are constant with respect to the lag parameter. Thus, results for the (conjugate) cyclic cross-correlogram are obtained by specializing to ACS processes results derived in (Napolitano, 2012, Chapter 2) for GACS processes. The proofs are outlined by recalling those in (Napolitano, 2012, Chapter 3). Analogously, ACS processes are a special case of SC processes when the support curves of the Loève bifrequency spectrum are lines with unit slope. Therefore, results for the time- and frequency-smoothed (conjugate) cyclic cross-periodogram are obtained by specializing to ACS processes results derived in (Napolitano, 2012, Chapter 4) for SC processes. The proofs are outlined by recalling those in (Napolitano, 2012, Chapter 5).

5.2 Second-order cyclic statistic estimators

Let $x(t)$ and $y(t)$ be continuous-time jointly ACS processes with (conjugate) cross-correlation function

$$\mathrm{E}\left\{y(t+\tau)\,x^{(*)}(t)\right\} = \sum_{\alpha \in A_{yx^{(*)}}} R^{\alpha}_{yx^{(*)}}(\tau)\,e^{j2\pi\alpha t} \tag{5.1}$$

and (conjugate) cyclic cross-correlation functions

$$R^{\alpha}_{yx^{(*)}}(\tau) \triangleq \lim_{T\to\infty} \frac{1}{T} \int_{-T/2}^{T/2} \mathrm{E}\left\{y(t+\tau)\,x^{(*)}(t)\right\}\,e^{-j2\pi\alpha t}\,\mathrm{d}t\,. \tag{5.2}$$

Under the assumption that the processes $x(t)$ and $y(t)$ are jointly harmonizable, their Loève bifrequency cross-spectrum is given by

$$\mathrm{E}\left\{Y(f_1)\,X^{(*)}(f_2)\right\} = \sum_{\alpha \in A_{yx^{(*)}}} S^{\alpha}_{yx^{(*)}}(f_1)\,\delta(f_2 - (-)(\alpha - f_1)) \tag{5.3}$$

where the functions

$$S^{\alpha}_{yx^{(*)}}(f) = \int_{\mathbb{R}} R^{\alpha}_{yx^{(*)}}(\tau)\,e^{-j2\pi f\tau}\,\mathrm{d}\tau \tag{5.4}$$

are the (conjugate) cyclic cross-spectra.

For discrete-time processes jointly ACS processes $x(n)$ and $y(n)$, the (conjugate) cross-correlation function is expressed as

$$\mathrm{E}\left\{y(n+m)\,x^{(*)}(n)\right\} = \sum_{\widetilde{\alpha} \in \widetilde{A}_{yx^{(*)}}} \widetilde{R}^{\widetilde{\alpha}}_{yx^{(*)}}(m)\,e^{j2\pi\widetilde{\alpha}n} \tag{5.5}$$

where

$$\widetilde{R}^{\widetilde{\alpha}}_{yx^{(*)}}(m) \triangleq \lim_{N\to\infty} \frac{1}{N} \sum_{n=-\lfloor N/2\rfloor}^{\lfloor N/2\rfloor} \mathrm{E}\left\{y(n+m)\,x^{(*)}(n)\right\}\,e^{-j2\pi\widetilde{\alpha}n} \tag{5.6}$$

are the (conjugate) cyclic cross-correlation functions

Under the assumption that the processes $x(n)$ and $y(n)$ are jointly harmonizable, their Loève bifrequency cross-spectrum is given by

$$\mathrm{E}\left\{Y(\nu_1)\,X^{(*)}(\nu_2)\right\} = \sum_{\widetilde{\alpha} \in \widetilde{A}_{yx^{(*)}}} \widetilde{S}^{\widetilde{\alpha}}_{yx^{(*)}}(\nu_1) \sum_{\ell \in \mathbb{Z}} \delta(\nu_2 - (-)(\widetilde{\alpha} - \nu_1) - \ell) \tag{5.7}$$

where the functions

$$\widetilde{S}^{\widetilde{\alpha}}_{yx^{(*)}}(\nu) = \sum_{m \in \mathbb{Z}} \widetilde{R}^{\widetilde{\alpha}}_{yx^{(*)}}(m)\,e^{-j2\pi\nu m} \tag{5.8}$$

are the (conjugate) cyclic cross-spectra.

5.2.1 Cyclic cross-correlogram

Let $w_T(t) \triangleq a(t/T)/T$ be a data-tapering window with $a(t)$ with unit-area and nonzero in $(-1/2, 1/2)$. Strictly speaking, $w_T(t)$ is the lag-product tapering window (see (Napolitano, 2012, Fact 3.5.3) for a discussion). The general case of complex-valued data-tapering windows is considered. In fact, in (Politis, 2005) it is shown that, by the adoption of appropriate complex-valued tapers, the bias of spectral estimators can be reduced of orders of magnitude.

Definition 5.1. The *(conjugate) cyclic cross-correlogram* of $y(t)$ and $x(t)$ at (conjugate) cycle frequency α is defined as

$$R^{(T)}_{yx^{(*)}}(\alpha, \tau; t_0) \triangleq \int_{\mathbb{R}} w_T(t - t_0)\, y(t + \tau)\, x^{(*)}(t)\, e^{-j2\pi\alpha t}\, \mathrm{d}t \tag{5.9}$$

where, in the left-hand side, when it does not create ambiguity, the dependence on t_0 is omitted in the notation. Due to the presence of the data-tapering window, the integration is over the interval $[t_0 - T/2, t_0 + T/2]$. If $y(t) \equiv x(t)$, we have the definition of *(conjugate) cyclic correlogram.* □

The (conjugate) cyclic cross-correlogram $R^{(T)}_{yx^{(*)}}(\alpha, \tau; t_0)$ is the natural estimator of the (conjugate) cyclic cross-correlation function $R^{\alpha}_{yx^{(*)}}(\tau)$ defined in (5.2) on the basis of the observation of $y(t)$ and $x(t)$ for $t \in [t_0 - T/2, t_0 + T/2]$.

Remark 5.2. If both $x(t)$ and $y(t)$ are observed for $t \in [t_0 - T/2, t_0 + T/2]$, then the correct definition for the (conjugate) cyclic cross-correlogram is

$$R^{(T)}_{yx^{(*)}}(\alpha, \tau; t_0) \triangleq \int_{\mathbb{R}} w_T(t - t_0)\, y(t + \tau)\, \mathrm{rect}\Big(\frac{t + \tau - t_0}{T}\Big)\, x^{(*)}(t)\, e^{-j2\pi\alpha t}\, \mathrm{d}t \tag{5.10}$$

where the rectangular window $\mathrm{rect}((t + \tau - t_0)/T)$ accounts for the fact that values of $y(t + \tau)$ are not available if $t + \tau \notin [t_0 - T/2, t_0 + T/2]$. However, unless otherwise specified, definition (5.9) is adopted since, for every fixed τ such that $R^{\alpha}_{yx^{(*)}}(\tau)$ is significantly nonzero, a data-record length $T \gg |\tau|$ is needed so that $R^{(T)}_{yx^{(*)}}(\alpha, \tau; t_0)$ is a reliable estimate of $R^{\alpha}_{yx^{(*)}}(\tau)$. Thus, ignoring the rectangular windows only gives rise to edge effects that become negligible asymptotically ($T \to \infty$).

From (5.10) we have that

$$R^{(T)}_{yx^{(*)}}(\alpha, \tau; t_0) = 0 \qquad \text{for } |\tau| > T\,. \tag{5.11}$$

□

Let us consider zero-mean processes $x(t)$ and $y(t)$ satisfying the following assumptions, that are obtained as special cases for ACS processes of assumptions made in (Napolitano, 2007a), (Napolitano, 2012, Chap. 2) for the generalized ACS processes.

Assumption 5.3. Uniformly Almost-Periodic Statistics.

a. *The stochastic processes $y(t)$ and $x(t)$ are singularly and jointly (second-order) ACS in the wide sense, that is, for any choice of z_1 and z_2 in $\{x, x^*, y, y^*\}$*

$$\mathrm{E}\{z_1(t+\tau_1)\, z_2(t+\tau_2)\} = \sum_{\alpha \in A_{z_1 z_2}} R^{\alpha}_{z_1 z_2}(\tau_1 - \tau_2)\, e^{j2\pi\alpha(t+\tau_2)} \tag{5.12}$$

(uniformly almost-periodic in t in the sense of Besicovitch (Besicovitch, 1932)).

b. *For any choice of z_1 in $\{y, y^*\}$ and z_2 in $\{x, x^*\}$, the fourth-order cumulant* $\mathrm{cum}\{y(t+\tau_1), x^{(*)}(t+\tau_2), z_1(t+\tau_3), z_2(t)\}$ *can be expressed as*

$$\mathrm{cum}\{y(t+\tau_1), x^{(*)}(t+\tau_2), z_1(t+\tau_3), z_2(t)\} = \sum_{\beta \in A_4} C^{\beta}_{yx^{(*)}z_1 z_2}(\tau_1, \tau_2, \tau_3)\, e^{j2\pi\beta t} \tag{5.13}$$

(uniformly almost-periodic in t in the sense of Besicovitch), where A_4 depends on z_1, z_2, and $(*)$, and cumulants of complex processes are defined according to (Spooner and Gardner, 1994a, App. A), (Napolitano, 2012, Sec. 1.4.2) (see also Section E.1.4). □

Assumption 5.4. Fourier Series Regularity.

a. *For any choice of z_1 and z_2 in $\{x, x^*, y, y^*\}$ it results*

$$\sum_{\alpha \in A_{z_1, z_2}} \left\| R^{\alpha}_{z_1 z_2} \right\|_{\infty} < \infty \tag{5.14}$$

where $\|R\|_\infty \triangleq \mathrm{ess\,sup}_{\tau \in \mathbb{R}} |R(\tau)|$ is the essential supremum of $R(\tau)$ (Champeney, 1990).

b. *For any choice of z_1 in $\{y, y^*\}$ and z_2 in $\{x, x^*\}$, it results that*

$$\sum_{\beta \in A_4} \left\| C^{\beta}_{yx^{(*)}z_1 z_2} \right\|_{\infty} < \infty\,. \tag{5.15}$$

□

From (5.14) it follows that the function series

$$\rho_{z_1 z_2}(\tau) = \sum_{\alpha \in A_{z_1 z_2}} \left| R^{\alpha}_{z_1 z_2}(\tau) \right| \tag{5.16}$$

converges uniformly with respect to τ. A similar result for the fourth-order cumulant follows from (5.15).

Assumption 5.5. Fourth-Order Moment Boundedness. *The stochastic processes $x(t)$ and $y(t)$ have uniformly bounded fourth-order absolute moments. That is, for any $z \in \{x, y\}$ there exists a positive number M_4 such that*

$$\mathrm{E}\left\{|z(t)|^4\right\} \leqslant M_4 < \infty \quad \forall t \in \mathbb{R}\,. \tag{5.17}$$

□

Assumption 5.6. Data-Tapering Window Regularity. *The data-tapering window $w_T(t)$ is a T-duration window that can be expressed as*

$$w_T(t) = \frac{1}{T} a\left(\frac{t}{T}\right) \tag{5.18}$$

with $a(t)$ Lebesgue measurable, with finite support $[-1/2, 1/2]$, bounded, continuous almost everywhere (a.e.), and such that

$$\int_{\mathbb{R}} a(t) \mathrm{d}t = 1 \tag{5.19}$$

$$\lim_{T \to \infty} a\left(\frac{t}{T}\right) = 1 \quad \forall t \in \mathbb{R}. \tag{5.20}$$

□

Assumption 5.6 is easily verified by taking $a(t) = \text{rect}(t)$.

If $a(t)$ is continuous at $t = 0$, then from (5.20) it follows $a(0) = 1$. From Assumption 5.6 we have that $a(t) \in L^1(\mathbb{R})$. Hence, its Fourier transform $A(f)$ is bounded, continuous, and infinitesimal as $|f| \to \infty$ (Champeney, 1990, Theorem 8.1). In addition, since $a(t)$ has finite support, we have that $A(f) = \mathcal{O}(|f|^{-r})$ as $|f| \to \infty$ with $r \geqslant 1$ (Papoulis, 1977, Appendix 3-B, p. 94).

In some cases it is convenient to assume $a(t)$ be an even function. For further properties of the data-tapering window and its Fourier transform see (Napolitano, 2012, Lemma 3.5.1).

Theorem 5.7. Expected Value of the (Conjugate) Cyclic Cross-Correlogram. *Let $y(t)$ and $x(t)$ be wide-sense jointly ACS stochastic processes with cross-correlation function (5.1). Under Assumptions*

5.3a (uniformly almost-periodic statistics),
5.4a (Fourier series regularity),
5.6 (data-tapering window regularity),

the expected value of the (conjugate) cyclic cross-correlogram $R^{(T)}_{yx^{(*)}}(\alpha, \tau; t_0)$ is given by

$$\mathrm{E}\left\{R^{(T)}_{yx^{(*)}}(\alpha, \tau; t_0)\right\} = \sum_{\beta \in A_{yx^{(*)}}} R^{\beta}_{yx^{(*)}}(\tau)\, W_{\frac{1}{T}}(\alpha - \beta)\, e^{-j2\pi[\alpha-\beta]t_0} \tag{5.21}$$

where $W_{\frac{1}{T}}(f) = A(f\,T)$ is the Fourier transform of $w_T(t) = a(t/T)/T$.

Proof. By taking the expected value of the (conjugate) cyclic cross-correlogram (5.9) and using (5.1), one obtains (5.21). The made assumptions allow one to interchange the order of expectation, integral, and sum operations.

The proof is obtained by specializing to jointly ACS processes the proof of (Napolitano, 2012, Theorem 2.4.6) reported in (Napolitano, 2012, Sec. 3.4) for the case of jointly generalized ACS processes. See also (Napolitano, 2007a, Theorem 3.1). □

From Theorem 5.7 it follows that the expected value of the estimate $R^{(T)}_{yx^{(*)}}(\alpha, \tau; t_0)$ depends not only on the (conjugate) cyclic cross-correlation at cycle frequency α, but also on all the other (conjugate) cyclic cross-correlations at cycle frequencies $\beta \neq \alpha$. The influence is stronger from cycle frequencies β closer to α and with larger $|R^{\beta}_{yx^{(*)}}(\tau)|$. Such a phenomenon was dubbed *cycle leakage* in (Gardner, 1987f, Chap. 13).

The function $w_T(t)$ has duration T. Thus, its Fourier transform $W_{\frac{1}{T}}(f)$ has a bandwidth of the order of $1/T$. Consequently, due to the presence of term $W_{\frac{1}{T}}(\alpha - \beta)$, we have that cycle frequencies that are closer than $1/T$ cannot be distinguished. That is, the *cycle frequency resolution* of the (conjugate) cyclic cross-correlogram (5.9) is of the order of $1/T$. Therefore, in order to capture all the cyclic features, the condition $1/T < h_\alpha$, with h_α defined in (5.26), must be satisfied.

Theorem 5.8. (Conjugate) Covariance of Cyclic Cross-Correlograms. *Let $x_1(t)$, $x_2(t)$, $y_1(t)$, and $y_2(t)$ be zero-mean wide-sense pairwise jointly ACS stochastic processes with cross-correlation functions defined according to (5.1) and almost-periodic fourth-order cross cumulants. Under Assumptions*

5.3 (uniformly almost-periodic statistics),
5.4 (Fourier series regularity),
5.5 (fourth-order moment boundedness),
5.6 (data-tapering window regularity)

valid with x replaced by x_1 or x_2 and y replaced by y_1 or y_2, the covariance

$$\operatorname{cov}\left\{R^{(T)}_{y_1x_1}(\alpha_1, \tau_1; t_1), R^{(T)}_{y_2x_2}(\alpha_2, \tau_2; t_2)\right\}$$

and the conjugate covariance

$$\operatorname{cov}\left\{R^{(T)}_{y_1x_1}(\alpha_1, \tau_1; t_1), R^{(T)\,*}_{y_2x_2}(\alpha_2, \tau_2; t_2)\right\}$$

of the cyclic cross-correlograms $R^{(T)}_{y_1x_1}(\alpha_1, \tau_1; t_1)$ and $R^{(T)}_{y_2x_2}(\alpha_2, \tau_2; t_2)$ can be computed similarly to the case of the generalized ACS processes. See (Napolitano, 2007a, Theorem 3.2), (Napolitano, 2012, Theorem 2.4.7) for the expression of the covariance and (Napolitano, 2007a, Theorem D.1), (Napolitano, 2012, Theorem 3.7.1) for the expression of the conjugate covariance. □

Note that since processes $x_1(t)$, $x_2(t)$, $y_1(t)$, and $y_2(t)$ are arbitrary, optional complex conjugations of these processes can be omitted without lack of generality.

In the following, assumptions are made to prove the mean-square consistency of the (conjugate) cyclic cross-correlogram and to derive the asymptotic covariance of cyclic cross-correlograms.

Assumption 5.9. Mixing Conditions (I).

a. *For any choice of z_1 and z_2 in $\{x, x^*, y, y^*\}$ it results*

$$\sum_{\alpha \in A_{z_1,z_2}} \int_{\mathbb{R}} \left| R^{\alpha}_{z_1 z_2}(s) \right| \mathrm{d}s < \infty . \tag{5.22}$$

b. *For any choice of z_1 in $\{y, y^*\}$ and z_2 in $\{x, x^*\}$ and $\forall \tau_1, \tau_2 \in \mathbb{R}$ it results*

$$\sum_{\beta \in A_4} \int_{\mathbb{R}} \left| C^{\beta}_{yx^{(*)}z_1 z_2}(s + \tau_1, s, \tau_2) \right| \mathrm{d}s < \infty . \tag{5.23}$$

□

Assumption 5.10. Mixing Conditions (II). *For any choice of z_1 in $\{y, y^*\}$ and z_2 in $\{x, x^*\}$ there exist nonnegative functions $\phi^{\beta}(s)$ such that*

$$\sup_{\tau_1, \tau_2 \in \mathbb{R}} \left| C^{\beta}_{yx^{(*)}z_1 z_2}(s + \tau_1, s, \tau_2) \right| \leqslant \phi^{\beta}(s) \in L^1(\mathbb{R}) \tag{5.24}$$

with

$$\sum_{\beta \in A_4} \int_{\mathbb{R}} \phi^{\beta}(s) \mathrm{d}s < \infty . \tag{5.25}$$

□

Assumption 5.11. Lack of cycle frequency clusters. *There is no cluster (accumulation point) of (conjugate) cycle frequencies except possibly at $\pm\infty$ (Napolitano, 2007a, As. 4.4), (Napolitano, 2012, As. 2.4.10). Thus, for every $\alpha \in A \equiv A_{yx^{(*)}}$ we have*

$$h_{\alpha} \triangleq \inf_{\substack{\beta \in A \\ \beta \neq \alpha}} |\beta - \alpha| > 0 . \tag{5.26}$$

□

Theorem 5.12. Asymptotic Expected Value of the (Conjugate) Cyclic Cross-Correlogram. *Let $y(t)$ and $x(t)$ be wide-sense jointly ACS stochastic processes with cross-correlation function (5.1). Under Assumptions*

5.3a (uniformly almost-periodic statistics),
5.4a (Fourier series regularity),
5.6 (data-tapering window regularity),

the asymptotic expected value of the cyclic cross-correlogram $R^{(T)}_{yx^{(*)}}(\alpha, \tau; t_0)$ is given by

$$\lim_{T \to \infty} \mathrm{E}\left\{ R^{(T)}_{yx^{(*)}}(\alpha, \tau; t_0) \right\} = R^{\alpha}_{yx^{(*)}}(\tau) . \tag{5.27}$$

Proof. The proof can be obtained specializing to the case of ACS processes the proof of (Napolitano, 2007a, Theorem 4.1), (Napolitano, 2012, Theorem 2.4.11) for GACS processes. □

Let $A(f)$ be the Fourier transform of $a(t)$ with rate of decay to zero $|f|^{-\gamma}$, $\gamma \geqslant 1$, for $|f| \to \infty$ as specified in Assumption 5.6.

Theorem 5.13. *Let $y(t)$ and $x(t)$ be wide-sense jointly ACS stochastic processes with cross-correlation function (5.1). Under Assumptions*

5.3a (uniformly almost-periodic statistics),
5.4a (Fourier series regularity),
5.6 (data-tapering window regularity),
5.11 (lack of cycle frequency clusters),

for the bias of the (conjugate) cyclic cross-correlogram it results

$$\lim_{T\to\infty} T^{\gamma} \left|\text{bias}\left\{R^{(T)}_{yx^{(*)}}(\alpha, \tau; t_0)\right\}\right| = \lim_{T\to\infty} T^{\gamma} \left|\mathrm{E}\left\{R^{(T)}_{yx^{(*)}}(\alpha, \tau; t_0)\right\} - R^{\alpha}_{yx^{(*)}}(\tau)\right| = \mathcal{O}(1)\,. \tag{5.28}$$

Proof. The proof can be obtained specializing to the case of ACS processes the proof of (Napolitano, 2007a, Theorem 4.2), (Napolitano, 2012, Theorem 2.4.12) for GACS processes. □

Theorem 5.14. Asymptotic (Conjugate) Covariance of Cyclic Cross-Correlograms. *Let $x_1(t)$, $x_2(t)$, $y_1(t)$, and $y_2(t)$ be zero-mean wide-sense pairwise jointly ACS stochastic processes with cross-correlation functions defined according to (5.1) and almost-periodic fourth-order cross cumulants. Under Assumptions*

5.3 (uniformly almost-periodic statistics),
5.4 (Fourier series regularity),
5.5 (fourth-order moment boundedness),
5.6 (data-tapering window regularity),
5.9 (mixing conditions (I)),

the asymptotic covariance of the cyclic cross-correlograms $R^{(T)}_{y_1x_1}(\alpha_1, \tau_1; t_1)$ and $R^{(T)}_{y_2x_2}(\alpha_2, \tau_2; t_2)$ is given by

$$\lim_{T\to\infty} T \operatorname{cov}\left\{R^{(T)}_{y_1x_1}(\alpha_1, \tau_1; t_1), R^{(T)}_{y_2x_2}(\alpha_2, \tau_2; t_2)\right\} = \mathcal{T}_1' + \mathcal{T}_2' + \mathcal{T}_3' \tag{5.29}$$

where

$$\mathcal{T}_1' \triangleq \mathcal{E}_a \sum_{\alpha'} \int_{\mathbb{R}} R^{\alpha'}_{y_1y_2^*}(\tau_1 - \tau_2 + s)\, R^{\alpha_1-\alpha_2-\alpha'}_{x_1x_2^*}(s)\, e^{j2\pi\alpha'\tau_2}\, e^{-j2\pi\alpha_1 s}\, \mathrm{d}s \tag{5.30}$$

$$\mathcal{T}_2' \triangleq \mathcal{E}_a \sum_{\alpha'''} \int_{\mathbb{R}} R^{\alpha'''}_{y_1x_2^*}(\tau_1 + s)\, R^{\alpha_1-\alpha_2-\alpha'''}_{x_1y_2^*}(s - \tau_2)\, e^{j2\pi(-\alpha'''+\alpha_1-\alpha_2)\tau_2}\, e^{-j2\pi\alpha_1 s}\, \mathrm{d}s \tag{5.31}$$

$$\mathcal{T}_3' \triangleq \mathcal{E}_a \int_{\mathbb{R}} C^{\alpha_1-\alpha_2}_{y_1 x_1 y_2^* x_2^*}(\tau_1 + s, s, \tau_2)\, e^{-j2\pi\alpha_1 s} \mathrm{d}s \tag{5.32}$$

with $\mathcal{E}_a \triangleq \int_{\mathbb{R}} |a(u)|^2 \mathrm{d}u$. *In (5.30), the sum on* α' *is extended to the set of cycle frequencies* $\alpha' \in A_{y_1 y_2^*}$ *such that* $\alpha_1 - \alpha_2 - \alpha' \in A_{x_1 x_2^*}$. *In (5.31), the sum on* α''' *is extended to the set of cycle frequencies* $\alpha''' \in A_{y_1 x_2^*}$ *such that* $\alpha_1 - \alpha_2 - \alpha''' \in A_{x_1 y_2^*}$.

The expression of the asymptotic conjugate covariance is obtained from that of the asymptotic covariance by removing the conjugations in the subscripts in (5.30)–(5.32).

Proof. The proof can be obtained specializing to the case of ACS processes the proofs for GACS processes (Napolitano, 2007a, Theorem 4.3), (Napolitano, 2012, Theorem 2.4.13) (covariance) and (Napolitano, 2007a, Theorem D.2), (Napolitano, 2012, Theorem 3.7.2) (conjugate covariance). Eq. (5.31) is a corrected version of (Napolitano, 2012, Eq. (2.153)). □

Theorem 5.15. Mean-Square Consistency of the (Conjugate) Cyclic Cross-Correlogram. *Under the assumptions of Theorems 5.12 and 5.14 (with* $x_1 \equiv x_2 \equiv x$ *and* $y_1 \equiv y_2 \equiv y$*), the (conjugate) cyclic cross-correlogram is a mean-square consistent estimator of the (conjugate) cyclic autocorrelation function*

$$\lim_{T\to\infty} \mathrm{E}\left\{ \left| R^{(T)}_{yx^{(*)}}(\alpha, \tau; t_0) - R^{\alpha}_{yx^{(*)}}(\tau) \right|^2 \right\} = 0 \tag{5.33}$$

Proof. See Section 5.9. □

Assumption 5.16. Mixing Conditions (III). *For every* k*, and every choice of* z_i *in* $\{x, x^*, y, y^*\}$, $i = 1, \dots, k$ *the* k*th-order cumulants are almost-periodic functions of* t

$$\operatorname{cum}\{z_1(t+\tau_1), \dots, z_{k-1}(t+\tau_{k-1}), z_k(t)\} = \sum_{\beta\in\mathbb{A}_k} C^{\beta}_{\mathbf{z}}(\tau_1, \dots, \tau_{k-1})\, e^{j2\pi\beta t} \tag{5.34}$$

with

$$\sum_{\beta\in\mathbb{A}_k} \left| C^{\beta}_{\mathbf{z}}(\tau_1, \dots, \tau_{k-1}) \right| \in L^1(\mathbb{R}^{k-1}), \tag{5.35}$$

where subscript $\mathbf{z} \triangleq z_1 \cdots z_k$. □

This assumption is equivalent to (Napolitano, 2007a, As. 5.1), (Napolitano, 2012, As. 2.4.15) specialized to ACS processes.

Assumption 5.17. Cross-Moment Boundedness. *For every* k*, and every choice of* z_i *in* $\{x, x^*, y, y^*\}$, $i = 1, \dots, k$ *there exists a positive number* M *(depending on the* z_i*) such that*

$$\mathrm{E}\{|z_1(t_1)\cdots z_k(t_k)|\} \leqslant M < \infty \qquad \forall t_1, \dots, t_k \in \mathbb{R}\,. \tag{5.36}$$

□

Theorem 5.18. Asymptotic Joint Complex Normality of (Conjugate) Cyclic Cross-Correlograms. *Let* $x(t)$ *and* $y(t)$ *be zero mean processes. Under Assumptions*

5.3 (uniformly almost-periodic statistics),
5.4 (Fourier series regularity),
5.5 (fourth-order moment boundedness),
5.6 (data-tapering window regularity),
5.9 (mixing conditions (I)),
5.11 (lack of cycle frequency clusters),
5.16 (mixing conditions (III)),
5.17 (cross-moment boundedness),

and if $\gamma > \frac{1}{2}$ *in Theorem 5.13, it results that, for every* α_i, τ_i, t_i, $i = 1, \ldots, M$, *the random variables*

$$V_i^{(T)}(\alpha_i, \tau_i) \triangleq \sqrt{T}\left[R_{yx^{(*)}}^{(T)}(\alpha_i, \tau_i; t_i) - R_{yx^{(*)}}^{\alpha_i}(\tau_i)\right] \qquad (i = 1, 2, \ldots, M) \tag{5.37}$$

are asymptotically ($T \to \infty$) *zero-mean jointly complex Normal with asymptotic covariance matrix* $\mathbf{\Sigma}$ *with entries*

$$\Sigma_{ij} = \lim_{T\to\infty} \operatorname{cov}\left\{V_i^{(T)}, V_j^{(T)}\right\} \tag{5.38}$$

given by (5.29)–(5.32) and asymptotic conjugate covariance matrix $\mathbf{\Sigma}^{(c)}$ *with entries*

$$\Sigma_{ij}^{(c)} = \lim_{T\to\infty} \operatorname{cov}\left\{V_i^{(T)}, V_j^{(T)*}\right\} \tag{5.39}$$

given by (5.29)–(5.32) without conjugations in the subscripts of (5.30)–(5.32).

Proof. See Section 5.9. □

Similar results are found in (Gardner, 1987f), (Hurd and Leśkow, 1992b), (Dehay and Hurd, 1994), (Dandawaté and Giannakis, 1995), (Hurd and Miamee, 2007), (Lii and Rosenblatt, 2006).

These results generalize to jointly ACS processes the classical results for the cross-correlogram ($\alpha = 0$ in (5.9)) and the cross-correlation function of jointly WSS processes (Brillinger, 1981).

5.2.2 Cyclic cross-periodogram

Definition 5.19. The *short-time Fourier transform* (STFT) of the process $z(t)$ is defined as

$$Z_T(t, f) \triangleq \int_{\mathbb{R}} z(u)\, b_T(u - t)\, e^{-j2\pi f u}\, \mathrm{d}u \tag{5.40a}$$

$$= \left[B_{\frac{1}{T}}(f)\, e^{-j2\pi f t}\right] \otimes Z(f)$$

$$= \int_{\mathbb{R}} Z(\nu)\, B_{1/T}(f - \nu)\, e^{-j2\pi(f-\nu)t}\, \mathrm{d}\nu \tag{5.40b}$$

with $Z(f)$ denoting the (generalized) Fourier transform of $z(t)$ (Section A.3) and $b_T(t)$ a T-duration data-tapering window whose Fourier transform $B_{\frac{1}{T}}(f)$ has bandwidth $1/T$. □

Assuming $b_T(t)$ be an even function, the STFT (5.40a) is a convolution that represents a low-pass filtering of a frequency-shifted version of $z(t)$ with approximate bandwidth $1/T$ equal to the reciprocal of the width of the data-tapering window. Thus, the function $t \mapsto Z_T(t, f)$ is the superposition of the spectral components of $z(t)$ within the spectral interval $[f - 1/2T, f + 1/2T]$ (see (5.40b)).

Lemma 5.20. (Hurd, 1973). *If $z(t)$ is harmonizable, its spectral covariance is of bounded variation (see (A.8)), and $b_T(t)$ is a Fourier-Stiltjes transform, then the STFT $Z_T(t, f)$ is harmonizable.* □

Definition 5.21. The *(conjugate) cyclic cross-periodogram* of the random processes $x(t)$ and $y(t)$ is defined as

$$I^{(T)}_{yx^{(*)}}(\alpha, f; t_0) \triangleq \frac{1}{T} Y_T(t_0, f)\, X^{(*)}_T(t_0, (-)(\alpha - f)) \tag{5.41}$$

where $X_T(t_0, f)$ and $Y_T(t_0, f)$ are the STFTs of $x(t)$ and $y(t)$, respectively, defined according to (5.40a). □

Note that, with a small abuse of notation, the same symbol is used for the (conjugate) cyclic cross-periodogram (5.41) which is a function of $(\alpha, f; t)$ and the (conjugate) cross-periodogram (13.18) which is a function of $(f_1, f_2; t)$.

In Lemma 2.39, it is proved that the (conjugate) cyclic cross-periodogram is the Fourier transform of the (conjugate) cyclic cross-correlogram $R^{(T)}_{yx^{(*)}}(\alpha, \tau; t_0)$ with $w_T(t) = \text{rect}(t/T)/T$.

Assumption 5.22. ACS Statistics (Time Domain).

a. *The second-order harmonizable stochastic processes $\{y(t),\ t \in \mathbb{R}\}$ and $\{x(t),\ t \in \mathbb{R}\}$ are singularly and jointly (second-order) almost-cyclostationary, that is, for any choice of z_1 and z_2 in $\{x, x^*, y, y^*\}$ it results*

$$\mathrm{E}\{z_1(t+\tau)\, z_2(t)\} = \sum_{\alpha \in A_{z_1 z_2}} R^{\alpha}_{z_1 z_2}(\tau)\, e^{j2\pi\alpha t} \tag{5.42}$$

with

$$\sum_{\alpha \in A_{z_1 z_2}} \left| R^{\alpha}_{z_1 z_2}(\tau) \right| \in L^1(\mathbb{R}) \tag{5.43}$$

where $A_{z_1 z_2}$ is the countable set of cycle frequencies of the second-order moment of z_1 and z_2.

b. *The fourth-order temporal cumulant* $\text{cum}\{y(t+\tau_1),\ x^{(*)}(t+\tau_2),\ y^*(t+\tau_3),\ x^{(*)*}(t)\}$ *can be expressed as*

$$\text{cum}\left\{y(t+\tau_1),\ x^{(*)}(t+\tau_2),\ y^*(t+\tau_3),\ x^{(*)*}(t)\right\} = \sum_{\beta \in A_4} C^{\beta}_{yx^{(*)}y^*x^{(*)*}}(\tau_1, \tau_2, \tau_3)\, e^{j2\pi\beta t} \tag{5.44}$$

with

$$\sum_{\beta \in A_4} \left| C^{\beta}_{yx^{(*)}y^*x^{(*)*}}(\tau_1, \tau_2, \tau_3) \right| \in L^1(\mathbb{R}) \tag{5.45}$$

where A_4 is a countable set of cycle frequencies and the cumulant of complex random variables is defined according to (E.21a). □

Lemma 5.23. ACS Statistics (Frequency Domain). *Under Assumption 5.22, we have*

a.

$$\mathrm{E}\{Z_1(f_1)\, Z_2(f_2)\} = \sum_{\alpha \in A_{z_1 z_2}} S^{\alpha}_{z_1 z_2}(f_1)\, \delta\left(f_2 - \Psi^{\alpha}_{z_1 z_2}(f_1)\right) \tag{5.46}$$

with

$$S^{\alpha}_{z_1 z_2}(f_1) = \int_{\mathbb{R}} R^{\alpha}_{z_1 z_2}(\tau)\, e^{-j2\pi f \tau}\, \mathrm{d}\tau \tag{5.47}$$

and

$$\sum_{\alpha \in A_{z_1 z_2}} \left\| S^{\alpha}_{z_1 z_2} \right\|_{\infty} < \infty \tag{5.48}$$

where $\Psi^{\alpha}_{z_1 z_2}(f_1) = (-)_2(\alpha - (-)_1 f_1)$, with $(-)_i = -$ if z_i is a conjugated term (x^* or y^*) and $(-)_i = +$ if z_i is unconjugated (x or y), and $\|S\|_{\infty}$ denotes the essential supremum of S (Champeney, 1990).

b.

$$\begin{aligned} &\mathrm{cum}\left\{Y(f_1),\, X^{(*)}(f_2),\, Y^*(f_3),\, X^{(*)*}(f_4)\right\} \\ &= \sum_{\beta \in A_4} P^{\beta}_{yx^{(*)}y^*x^{(*)*}}(f_1, f_2, f_3)\, \delta\left(f_4 - \Psi^{\beta}_{yx^{(*)}y^*x^{(*)*}}(f_1, f_2, f_3)\right) \end{aligned} \tag{5.49}$$

with

$$P^{\beta}_{yx^{(*)}y^*x^{(*)*}}(f_1, f_2, f_3) = \int_{\mathbb{R}^3} C^{\beta}_{yx^{(*)}y^*x^{(*)*}}(\tau_1, \tau_2, \tau_3)\, e^{-j2\pi(f_1\tau_1 + f_2\tau_2 + f_3\tau_3)}\, \mathrm{d}\tau_1 \mathrm{d}\tau_2 \mathrm{d}\tau_3 \tag{5.50}$$

and

$$\sum_{\beta \in A_4} \left\| P^{\beta}_{yx^{(*)}y^*x^{(*)*}} \right\|_{\infty} < \infty \tag{5.51}$$

where the function $\Psi^{\beta}_{yx^{(*)}y^*x^{(*)*}}(f_1, f_2, f_3)$ is given by sum and differences of f_1, f_2, f_3, and β, depending on the conjugated terms (see Remark 4.11).

Proof. The results follow directly from the assumptions since the Fourier transform of a summable function is continuous, bounded and infinitesimal at infinity (Champeney, 1990). □

Note that the results of Lemma 5.23 can be obtained as special case of (Napolitano, 2012, Assumptions 4.4.2 and 4.4.3) made for SC processes and specialized to the case of ACS processes.

Assumption 5.24. Data-Tapering Window Regularity. *$b_T(t)$ is a T-duration data-tapering window with Fourier transform $B_{1/T}(f)$ such that*

$$b_T(t) = w_b\left(\frac{t}{T}\right) \tag{5.52}$$

$$\lim_{T\to\infty} w_b\left(\frac{t}{T}\right) = 1 \qquad \forall t \in \mathbb{R} \tag{5.53}$$

$$B_{\frac{1}{T}}(f) = T\, W_B(fT) \tag{5.54}$$

where $W_B(f)$, the Fourier transform of $w_b(t)$, is a.e. continuous and regular as $|f| \to \infty$, $W_B(f) \in L^1(\mathbb{R}) \cap L^\infty(\mathbb{R})$, and $\int_{\mathbb{R}} W_B(f)\,\mathrm{d}f = w_b(0) = 1$. □

The function $w_b(t)$ in (5.52) is different from zero for $|t| \leqslant 1/2$. Thus, its Fourier transform $W_B(f)$ has a bandwidth of the order of 1 and, hence, $B_{\frac{1}{T}}(f)$ has a bandwidth of the order of $1/T$.

From (5.53), it follows that

$$\lim_{T\to\infty} T\, W_B(fT) = \delta(f) \tag{5.55}$$

$$\lim_{T\to\infty} W_B(fT) = W_B(0)\,\delta_f \tag{5.56}$$

where $\delta(f)$ is Dirac delta and δ_f is Kronecker delta, (that is, $\delta_f = 1$ if $f = 0$ and $\delta_f = 0$ if $f \neq 0$), and the first limit must be intended in the sense of distributions (generalized functions) (Zemanian, 1987).

In general, a rectangular data-tapering window cannot be used in the previous results since it does not satisfy Assumption 5.24. In fact,

$$w_b(t) = \mathrm{rect}(t) \overset{\mathcal{F}}{\longleftrightarrow} W_B(f) = \mathrm{sinc}(f) \notin L^1(\mathbb{R})$$

However, in Section 5.3.2 it is shown that a rectangular window can be adopted if more stringent assumptions are made on the stochastic processes x and y. For example, if the number of cycle frequencies is finite (as it happens for strictly bandlimited cyclostationary processes).

The triangular data-tapering window satisfies Assumption 5.24. In fact,

$$w_b(t) = (1 - 2|t|)\,\mathrm{rect}(t) \overset{\mathcal{F}}{\longleftrightarrow} W_B(f) = \frac{1}{2}\,\mathrm{sinc}^2(f/2) \in L^1(\mathbb{R})$$

In general, any data-tapering window continuous and with possibly discontinuous first order derivative of bounded variation has Fourier transform with rate of decay to zero which is at least $\mathcal{O}(|f|^{-2})$ and hence is summable (Papoulis, 1977, p. 95).

The general case of complex-valued data-tapering window $b_T(t)$ is considered. In fact, in (Politis, 2005) it is shown that, by the adoption of appropriate complex-valued tapers, the bias of spectral estimators can be reduced of orders of magnitude.

Lemma 5.25. Expected Value of the (Conjugate) Cyclic Cross-Periodogram. *Let $\{y(t), t \in \mathbb{R}\}$ and $\{x(t), t \in \mathbb{R}\}$ be second-order harmonizable jointly ACS stochastic processes with Loève bifrequency cross-spectrum (5.3).*

Under Assumptions

5.22a (ACS statistics and Fourier series regularity),
5.24 (data-tapering window regularity),

the expected value of the (conjugate) cyclic cross-periodogram (5.41) is given by

$$\mathrm{E}\left\{I_{yx^{(*)}}^{(T)}(\alpha, f; t_0)\right\} = \frac{1}{T} \sum_{\beta \in A_{yx^{(*)}}} \int_{\mathbb{R}} S_{yx^{(*)}}^{\beta}(\nu)\, B_{\frac{1}{T}}(f-\nu)\, B_{\frac{1}{T}}^{(*)}((-)(\alpha-\beta+\nu-f))\, \mathrm{d}\nu\, e^{-j2\pi(\alpha-\beta)t_0} . \tag{5.57}$$

Proof. From (5.43) it follows (5.48). Thus, the assumptions are those of (Napolitano, 2003, Lemma 3.1), (Napolitano, 2012, Lemma 4.4.6) specialized to ACS processes. Therefore, (5.57) follows from (Napolitano, 2003, Eq. (39)), (Napolitano, 2012, Eq. (4.106)) specialized to ACS processes. □

By expressing the covariance of the (conjugate) cyclic cross-periodogram in terms of second-order moments and a fourth-order cumulant of the STFTs of $x(t)$ and $y(t)$, the following result is proved.

Lemma 5.26. Covariance and Conjugate Covariance of the (Conjugate) Cyclic Cross-Periodogram. *Let $\{y(t), t \in \mathbb{R}\}$ and $\{x(t), t \in \mathbb{R}\}$ be second-order harmonizable zero-mean singularly and jointly SC stochastic processes with bifrequency spectra and cross-spectra (5.46).*

Under Assumptions

5.22 (ACS statistics and Fourier series regularity),
5.24 (data-tapering window regularity),

the covariance

$$\operatorname{cov}\left\{I_{yx^{(*)}}^{(T)}(\alpha_1, f_1; t_1),\; I_{yx^{(*)}}^{(T)}(\alpha_2, f_2; t_2)\right\}$$

and the conjugate covariance

$$\operatorname{cov}\left\{I_{yx^{(*)}}^{(T)}(\alpha_1, f_1; t_1),\; I_{yx^{(*)}}^{(T)*}(\alpha_2, f_2; t_2)\right\}$$

of the (conjugate) cyclic cross-periodogram (5.41) can be computed similarly to the case of spectrally correlated processes (Napolitano, 2003, Lemma 3.2), (Napolitano, 2012, Lemma 4.4.7) (covariance) and (Napolitano, 2012, Lemma 5.8.1) (conjugate covariance). □

Theorem 5.27. Asymptotic Expected Value of the (Conjugate) Cyclic Cross-Periodogram. *Let* $\{y(t),\ t \in \mathbb{R}\}$ *and* $\{x(t),\ t \in \mathbb{R}\}$ *be second-order harmonizable jointly ACS stochastic processes with Loève bifrequency cross-spectrum (5.3).*

Under Assumptions

5.22a (ACS statistics and Fourier series regularity),
5.24 (data-tapering window regularity),

the asymptotic ($T \to \infty$) *expected value (5.57) of the (conjugate) cyclic cross-periodogram (5.41) is given by*

$$\lim_{T\to\infty} \mathrm{E}\left\{ I^{(T)}_{yx^{(*)}}(\alpha, f; t_0) \right\} = \mathcal{E}_{(*)}\, S^{\alpha}_{yx^{(*)}}(f) \tag{5.58}$$

where

$$\mathcal{E}_{(*)} \triangleq \int_{\mathbb{R}} W_B(\lambda)\, W_B^{(*)}(-_{(-)}\lambda)\, \mathrm{d}\lambda\,. \tag{5.59}$$

Proof. The proof can be obtained by specializing to ACS processes the proof of (Napolitano, 2003, Theorem 3.1), (Napolitano, 2012, Theorem 4.4.8) for SC processes. □

In general, $W_B(\cdot)$ is complex valued. However, if $W_B(\cdot)$ is real and even, (5.59) specializes into

$$\mathcal{E}_{(*)} = \int_{\mathbb{R}} W_B^2(\lambda)\, \mathrm{d}\lambda \tag{5.60}$$

independent on $(*)$. Note that by taking $w_b(\cdot)$ real and even, then $W_B(\cdot)$ is Hermitian and even and, hence, real and even. In addition, if $w_b(\cdot)$ is also unit energy, so is also $W_B(\cdot)$ and we have

$$\mathcal{E}_{(*)} = \int_{\mathbb{R}} W_B^2(\lambda)\, \mathrm{d}\lambda = \int_{\mathbb{R}} w_b^2(t)\, \mathrm{d}t = 1\,. \tag{5.61}$$

Such conditions are verified, for example, if $w_b(\cdot) = \mathrm{rect}(\cdot)$.

Theorem 5.28. Asymptotic Covariance of the (Conjugate) Cyclic Cross-Periodogram. *Let* $\{y(t),\ t \in \mathbb{R}\}$ *and* $\{x(t),\ t \in \mathbb{R}\}$ *be second-order harmonizable zero-mean singularly and jointly ACS stochastic processes with Loève bifrequency cross-spectrum (5.3).*

Under Assumptions

5.22 (ACS statistics and Fourier series regularity),
5.24 (data-tapering window regularity),

the asymptotic ($T \to \infty$) *covariance of the (conjugate) cyclic cross-periodogram (5.41) is given by*

$$\begin{aligned}
&\lim_{T\to\infty} \mathrm{cov}\left\{ I^{(T)}_{yx^{(*)}}(\alpha_1, f_1; t_1),\ I^{(T)}_{yx^{(*)}}(\alpha_2, f_2; t_2) \right\} \\
&\quad = \mathcal{E}^2\, S^{f_1-f_2}_{yy^*}(f_1)\, S^{\alpha_1-f_1-\alpha_2+f_2}_{x^{(*)}x^{(*)*}}((-)(\alpha_1 - f_1))
\end{aligned}$$

$$+ \mathcal{E}\, \mathcal{E}_{(*)}\, S^{f_1-\alpha_2+f_2}_{yx^{(*)}*}(f_1)\, S^{-f_2+\alpha_1-f_1}_{x^{(*)}y^*}((-)(\alpha_1 - f_1)) \tag{5.62}$$

where $\mathcal{E}$ is given in (5.59) with $(*)$ and $(-)$ absent.

Proof. The proof can be obtained by specializing to ACS processes the proof of (Napolitano, 2003, Theorem 3.2), (Napolitano, 2012, Theorem 4.4.9) for SC processes. Eq. (5.62) is a corrected version of (Napolitano, 2003, Eq. (57)) and is coincident with (Napolitano, 2012, Eq. (4.122)). □

The (conjugate) cyclic cross-correlogram is an asymptotically unbiased (Theorem 5.27) but not consistent (Theorem 5.28) estimator of the (conjugate) cyclic cross-spectrum $S^{\alpha}_{yx^{(*)}}(f)$. As for the cross-periodogram of jointly stationary processes ($\alpha = 0$ in (5.41)), consistency can be obtained by time or frequency smoothing (Brillinger, 1981, Chap. 5), (Rosenblatt, 1985, Chap. V).

In (Mahmoudi et al., 2018), it is shown that the limiting distributions of the finite Fourier transform and of the cyclic periodogram are multivariate complex normal and complex Wishart, respectively. This result is used in (Mahmoudi et al., 2019) to test the difference between spectral densities of two independent cyclostationary time series model.

5.2.3 Frequency-smoothed cyclic cross-periodogram

Definition 5.29. Given two jointly ACS stochastic processes $\{y(t),\ t \in \mathbb{R}\}$ and $\{x(t),\ t \in \mathbb{R}\}$, their *frequency-smoothed (conjugate) cyclic cross-periodogram* at cycle frequency α is defined as

$$S^{(T,\Delta f)}_{yx^{(*)}}(\alpha, f; t_0) \triangleq I^{(T)}_{yx^{(*)}}(\alpha, f; t_0) \underset{f}{\otimes} A_{\Delta f}(f) \tag{5.63}$$

where $A_{\Delta f}(f)$ is a Δf-bandwidth frequency-smoothing window, $\underset{f}{\otimes}$ denotes convolution with respect to f, and, in the left-hand side, when it does not create ambiguity, the dependence on t_0 is omitted in the notation. □

Assumption 5.30. Frequency-Smoothing Window Regularity. *$A_{\Delta f}(f)$ is a Δf-bandwidth frequency-smoothing window such that*

$$A_{\Delta f}(f) = \frac{1}{\Delta f}\, W_A\left(\frac{f}{\Delta f}\right) \tag{5.64}$$

with $W_A(f)$ a.e. continuous, $W_A(f) \in L^1(\mathbb{R}) \cap L^\infty(\mathbb{R})$, $\int_{\mathbb{R}} W_A(f)\mathrm{d}f = 1$, and

$$\lim_{\Delta f \to 0} \frac{1}{\Delta f}\, W_A\left(\frac{f}{\Delta f}\right) = \delta(f)\,. \tag{5.65}$$

□

For example, if $A_{\Delta f}(f)$ is the rectangular window with width Δf, that is, $A_{\Delta f}(f) = \text{rect}(f/\Delta f)/\Delta f$, the frequency-smoothed (conjugate) cyclic cross-periodogram (5.63) specializes into

$$S_{yx^{(*)}}^{(T,\Delta f)}(\alpha, f; t_0) = \frac{1}{\Delta f} \int_{f-\Delta f/2}^{f+\Delta f/2} I_{yx^{(*)}}^{(T)}(\alpha, \lambda; t_0)\, \mathrm{d}\lambda\,. \tag{5.66}$$

That is, the value at spectral frequency f of the estimate $S_{yx^{(*)}}^{(T,\Delta f)}(\alpha, f; t_0)$ of the (conjugate) cyclic cross-spectrum $S_{yx^{(*)}}^{\alpha}(f)$ is obtained by averaging the values of the (conjugate) cyclic cross-periodogram $I_{yx^{(*)}}^{(T)}(\alpha, \lambda; t_0)$ over values of spectral frequencies $\lambda \in [f - \Delta f/2, f + \Delta f/2]$. As shown below, such an averaging has a beneficial effect on the variance of the estimator.

Theorem 5.31. Expected Value of the Frequency-Smoothed (Conjugate) Cyclic Cross-Periodogram. *Let $\{y(t),\ t \in \mathbb{R}\}$ and $\{x(t),\ t \in \mathbb{R}\}$ be second-order harmonizable jointly ACS stochastic processes with Loève bifrequency cross-spectrum (5.3).*

Under Assumptions

5.22a (ACS statistics and Fourier series regularity),
5.24 (data-tapering window regularity),
5.30 (frequency-smoothing window regularity),

the expected value of the frequency-smoothed (conjugate) cyclic cross-periodogram (5.63) is given by

$$\mathrm{E}\left\{S_{yx^{(*)}}^{(T,\Delta f)}(\alpha, f; t_0)\right\} = \sum_{\beta \in A_{yx^{(*)}}} \int_{\mathbb{R}} S_{yx^{(*)}}^{\beta}(\nu)\, K^{(T,\Delta f)}(\alpha, \beta, f, \nu)\, \mathrm{d}\nu\, e^{-j2\pi(\alpha-\beta)t_0} \tag{5.67}$$

where

$$K^{(T,\Delta f)}(\alpha, \beta, f, \nu) \triangleq \frac{1}{T} \int_{\mathbb{R}} B_{\frac{1}{T}}(\lambda - \nu)\, B_{\frac{1}{T}}^{(*)}((-)(\alpha - \beta + \nu - \lambda))\, A_{\Delta f}(f - \lambda)\, \mathrm{d}\lambda\,. \tag{5.68}$$

For $T\Delta f \gg 1$ the expected value (5.67) can be approximated as

$$\begin{aligned}
\mathrm{E}\left\{S_{yx^{(*)}}^{(T,\Delta f)}(\alpha, f; t_0)\right\} &\simeq \sum_{\beta \in A_{yx^{(*)}}} \left[S_{yx^{(*)}}^{\beta}(f) \underset{f}{\otimes} A_{\Delta f}(f)\right] \chi_B((\alpha-\beta)T)\, e^{-j2\pi(\alpha-\beta)t_0} \\
&= \left[S_{yx^{(*)}}^{\alpha}(f) \underset{f}{\otimes} A_{\Delta f}(f)\right] \chi_B(0) \\
&\quad + \sum_{\beta \neq \alpha} \left[S_{yx^{(*)}}^{\beta}(f) \underset{f}{\otimes} A_{\Delta f}(f)\right] \chi_B((\alpha-\beta)T)\, e^{-j2\pi(\alpha-\beta)t_0}
\end{aligned} \tag{5.69}$$

where

$$\chi_B(f) \triangleq W_B(f) \underset{f}{\otimes} W_B^{(*)}((-)f) \tag{5.70}$$

Proof. See Section 5.9. □

From Theorem 5.31 it follows that the expected value of the estimate $S^{(T,\Delta f)}_{yx^{(*)}}(\alpha, f; t_0)$ depends not only on the (conjugate) cyclic cross-spectrum at cycle frequency α, but also on all the other (conjugate) cyclic cross-spectra at cycle frequencies $\beta \neq \alpha$. The influence is stronger from cycle frequencies β closer to α and with larger $|S^{\beta}_{yx^{(*)}}(f)|$. Such a phenomenon is referred to as *cycle leakage* (Gardner, 1987f, Chap. 13) analogously to the case of the (conjugate) cyclic cross-correlogram (Theorem 5.7).

In practice, condition $T\Delta f \gg 1$ needs to be verified to have small bias (Theorems 5.36 and 5.37) and variance (Theorem 5.38) so that approximation (5.69) generally holds.

The function $W_B(f)$ has a bandwidth of the order of unity (Assumption 5.24). Consequently, the function $\chi_B(f)$ has a width of the order of unity. Thus, due to the presence of term $\chi_B((\alpha - \beta)T)$, we have that cycle frequencies that are closer than $1/T$ cannot be distinguished. That is, the *cycle frequency resolution* of the frequency-smoothed (conjugate) cyclic cross-periodogram (5.63) is of the order of $1/T$. Therefore, in order to capture all the cyclic features, the condition $1/T < h_\alpha$, with h_α defined in (5.26), must be satisfied. In addition, from (5.69) it follows that the shape of the (conjugate) cyclic cross-spectrum is smeared by the convolution in f with the frequency-smoothed window of width Δf. Thus, variations of the cyclic spectra in f smaller than Δf cannot be appreciated. That is, the *spectral frequency resolution* is of the order of Δf.

Lemma 5.32. Covariance and Conjugate Covariance of the Frequency-Smoothed (Conjugate) Cyclic Cross-Periodogram. *Let $\{y(t),\ t \in \mathbb{R}\}$ and $\{x(t),\ t \in \mathbb{R}\}$ be second-order harmonizable zero-mean singularly and jointly ACS stochastic processes with Loève bifrequency spectra and cross-spectra (5.46).*

Under Assumptions

5.22 (ACS statistics and Fourier series regularity),
5.24 (data-tapering window regularity),
5.30 (frequency-smoothing window regularity),

the covariance

$$\text{cov}\left\{S^{(T,\Delta f)}_{yx^{(*)}}(\alpha_1, f_1; t_1),\ S^{(T,\Delta f)}_{yx^{(*)}}(\alpha_2, f_2; t_2)\right\}$$

and the conjugate covariance

$$\text{cov}\left\{S^{(T,\Delta f)}_{yx^{(*)}}(\alpha_1, f_1; t_1),\ S^{(T,\Delta f)*}_{yx^{(*)}}(\alpha_2, f_2; t_2)\right\}$$

of the frequency-smoothed (conjugate) cyclic cross-periodogram (5.63) can be computed similarly to the case of spectrally correlated processes (Napolitano, 2012, Theorem 4.6.4) (covariance) and (Napolitano, 2012, Theorem 5.8.2) (conjugate covariance). □

Assumption 5.33. Data-Tapering Window Regularity. *The data-tapering window satisfies the regularity conditions of Assumption 5.24. In addition, the first-order derivative of W_B*

exists a.e. and, where it exists, is uniformly bounded, $f^p\, W_B(f) \in L^1(\mathbb{R})$, $p = 1, 2$, *and there exists* $\gamma \geqslant 1$ *such that* $W_B(f) = \mathcal{O}(|f|^{-\gamma})$ *as* $|f| \to \infty$. □

Assumption 5.34. Frequency-Smoothing Window Regularity. *The frequency-smoothing window satisfies the regularity conditions of Assumption 5.30. In addition,* $f^p\, W_A(f) \in L^1(\mathbb{R})$, $p = 1, 2$. □

Assumption 5.35. Spectral Cross-Correlation Density Regularity. *The cyclic cross-spectra* $S^{\alpha}_{yx^{(*)}}(f)$, $\alpha \in A_{yx^{(*)}}$, *are a.e. derivable with uniformly bounded derivative.* □

Theorem 5.36. Asymptotic Expected Value of the Frequency-Smoothed (Conjugate) Cyclic Cross-Periodogram. *Let* $\{y(t),\ t \in \mathbb{R}\}$ *and* $\{x(t),\ t \in \mathbb{R}\}$ *be second-order harmonizable jointly ACS stochastic processes with Loève bifrequency cross-spectrum (5.3).*

Under Assumptions

5.22a (ACS statistics and Fourier series regularity),
5.24 (data-tapering window regularity),
5.30 (frequency-smoothing window regularity),

the asymptotic expected value of the frequency-smoothed (conjugate) cyclic cross-periodogram (5.63) is given by

$$\lim_{\Delta f \to 0} \lim_{T \to \infty} \mathrm{E}\left\{ S^{(T,\Delta f)}_{yx^{(*)}}(\alpha, f; t_0) \right\} = \mathcal{E}_{(*)}\, S^{\alpha}_{yx^{(*)}}(f) \tag{5.71}$$

where the order of the two limits cannot be interchanged ($T \to \infty$, $\Delta f \to 0$, *with* $T\Delta f \to \infty$*) and* $\mathcal{E}_{(*)}$ *is defined in (5.59).*

Proof. The proof can be obtained by specializing to ACS processes the proof of (Napolitano, 2012, Theorem 4.7.5) for SC processes. □

Theorem 5.37. Rate of Convergence of the Bias of the Frequency-Smoothed (Conjugate) Cyclic Cross-Periodogram. *Let* $\{y(t),\ t \in \mathbb{R}\}$ *and* $\{x(t),\ t \in \mathbb{R}\}$ *be second-order harmonizable jointly ACS stochastic processes with Loève bifrequency cross-spectrum (5.3).*

Under Assumptions

5.22a (ACS statistics and Fourier series regularity),
5.24 (data-tapering window regularity),
5.30 (frequency-smoothing window regularity),
5.11 (lack of cycle frequency clusters),
5.33 (data-tapering window summability),
5.34 (frequency-smoothing window summability),
5.35 (spectral cross-correlation density regularity),

asymptotically ($T \to \infty$, $\Delta f \to 0$, *with* $T\Delta f \to \infty$*) it results*

$$\text{bias} \triangleq \mathrm{E}\left\{ S^{(T,\Delta f)}_{yx^{(*)}}(\alpha, f; t_0) \right\} - \mathcal{E}_{(*)}\, S^{\alpha}_{yx^{(*)}}(f)$$

$$= \mathcal{O}\left(\frac{1}{T}\right) + \mathcal{O}(\Delta f) + \mathcal{O}\left(T(\Delta f)^2\right) \tag{5.72}$$

provided that $T(\Delta f)^2 \to 0$.

Proof. The proof is obtained by specializing to ACS processes the proof of (Napolitano, 2012, Theorem 4.7.6) for SC processes. □

Theorem 5.38. Asymptotic Covariance of the Frequency-Smoothed (Conjugate) Cyclic Cross-Periodogram. *Let* $\{y(t),\ t \in \mathbb{R}\}$ *and* $\{x(t),\ t \in \mathbb{R}\}$ *be second-order harmonizable zero-mean singularly and jointly ACS stochastic processes with Loève bifrequency spectra and cross-spectra (5.46).*

Under Assumptions

5.22 (ACS statistics and Fourier series regularity),
5.24 (data-tapering window regularity),
5.30 (frequency-smoothing window regularity),

the asymptotic ($T \to \infty$, $\Delta f \to 0$, *with* $T\Delta f \to \infty$) *covariance of the frequency-smoothed (conjugate) cyclic cross-periodogram (5.63) is such that*

$$\begin{aligned}
&\lim_{\Delta f \to 0} \lim_{T \to \infty} (T\Delta f) \operatorname{cov}\left\{S_{yx^{(*)}}^{(T,\Delta f)}(\alpha_1, f_1; t_1),\ S_{yx^{(*)}}^{(T,\Delta f)}(\alpha_2, f_2; t_2)\right\} \\
&= \mathcal{I}_{11}\, \mathcal{I}_{12} S_{yy^*}^{f_1-f_2}(f_1)\, S_{x^{(*)}x^{(*)*}}^{\alpha_1-f_1-\alpha_2+f_2}((-)(\alpha_1 - f_1)) \\
&\quad + \mathcal{I}_{21}\, \mathcal{I}_{22}\, S_{yx^{(*)*}}^{f_1-\alpha_2+f_2}(f_1)\, S_{x^{(*)}y^*}^{-f_2+\alpha_1-f_1}((-)(\alpha_1 - f_1))
\end{aligned} \tag{5.73}$$

where

$$\mathcal{I}_{11} \triangleq \int_{\mathbb{R}^3} W_B^{(*)*}((-)(\nu_{y1} - \lambda_2) + \nu_{x1})\, W_B(\nu_{y1})\, W_B^{(*)}(\nu_{x1})\, W_B^*(\lambda_2)\, \mathrm{d}\nu_{y1} \mathrm{d}\nu_{x1} \mathrm{d}\lambda_2 \tag{5.74}$$

$$\mathcal{I}_{12} \triangleq \int_{\mathbb{R}} |W_A(\lambda_1)|^2\, \mathrm{d}\lambda_1 \tag{5.75}$$

$$\mathcal{I}_{21} \triangleq \int_{\mathbb{R}} W_A(\lambda_1)\, W_A^*(-\lambda_1)\, \mathrm{d}\lambda_1 \tag{5.76}$$

$$\mathcal{I}_{22} = \int_{\mathbb{R}^3} W_B^{(*)*}((-)(\nu_{y1} - \lambda_2) - \nu_{x1})\, W_B(\nu_{y1})\, W_B^{(*)}(\nu_{x1})\, W_B^*(\lambda_2)\, \mathrm{d}\nu_{y1} \mathrm{d}\nu_{x1} \mathrm{d}\lambda_2 \tag{5.77}$$

Proof. The proof is obtained by specializing to ACS processes the proof of (Napolitano, 2012, Theorem 4.7.7) for SC processes. □

In Theorems 5.36 and 5.38, the order of the two limits cannot be interchanged. Furthermore, as corollary of Theorem 5.38 we have that asymptotically ($T \to \infty$, $\Delta f \to 0$, with $T\Delta f \to \infty$)

$$\operatorname{var}\left\{S_{yx^{(*)}}^{(T,\Delta f)}(\alpha, f; t_0)\right\} = \mathcal{O}\left(\frac{1}{T\Delta f}\right). \tag{5.78}$$

In practice (T and Δf finite), T must be sufficiently large to resolve the cycle frequencies, Δf sufficiently small to capture all the spectral-frequency variations of the cyclic spectra, and, moreover the *smoothing product* $T\Delta f$ must be large in order to have reliable estimates. If Δf is taken large in order to reduce the variance (see (5.78)), then the bias increases (see (5.72)). This is a typical *bias-variance trade-off* in spectral analysis.

Theorem 5.39. Mean-Square Consistency of the Frequency-Smoothed (Conjugate) Cyclic Cross-Periodogram. *Under the assumptions of Theorems 5.36 and 5.38, the frequency-smoothed (conjugate) cyclic cross-periodogram is a mean-square consistent estimator of the (conjugate) cyclic spectrum*

$$\lim_{\Delta f \to 0} \lim_{T \to \infty} \mathrm{E}\left\{ \left| S_{yx^{(*)}}^{(T,\Delta f)}(\alpha, f; t_0) - \mathcal{E}_{(*)}\, S_{yx^{(*)}}^{\alpha}(f) \right|^2 \right\} = 0 . \tag{5.79}$$

Proof. See Section 5.9. □

Assumption 5.40. Spectral Cumulants. *For any choice of V_i, $i = 1, ..., k$ in $\{X, X^*, Y, Y^*\}$, $V_i(f)$ are the (generalized) Fourier transforms of k processes kth-order jointly almost cyclostationary. That is (Theorem 4.20),*

$$\operatorname{cum}\{V_1(f_1), \ldots, V_k(f_k)\} = \sum_{\beta} P_{V_1 \cdots V_k}^{\beta}(f_1, \ldots, f_{k-1})\, \delta\left(f_k - \Psi_{V_1 \cdots V_k}^{\beta}(f_1, \ldots, f_{k-1})\right) \tag{5.80}$$

with

$$\sum_{\beta} \| P_{V_1 \cdots V_k}^{\beta} \|_{\infty} < \infty \tag{5.81}$$

where $\Psi_{V_1 \cdots V_k}^{\beta}(f_1, \ldots, f_{k-1})$ is given by sum and differences of $f_1, \ldots, f_{k-1}$, and β, depending on the conjugated terms (see Remark 4.11). □

Theorem 5.41. Asymptotic Joint Complex Normality of the Frequency-Smoothed (Conjugate) Cyclic Cross-Periodograms. *Under the assumptions for Theorem 5.37 (rate of convergence of the bias of the frequency-smoothed (conjugate) cyclic cross-periodogram), that is, under Assumptions*

5.22a (ACS statistics and Fourier series regularity),
5.24 (data-tapering window regularity),
5.30 (frequency-smoothing window regularity),
5.11 (lack of cycle frequency clusters),
5.33 (data-tapering window summability),
5.34 (frequency-smoothing window summability),
5.35 (spectral cross-correlation density regularity),

and the further assumptions

5.40 (spectral cumulants),
$\Delta f \equiv \Delta f_T = T^{-a}$, *with* $3/5 < a < 1$,

it results that for every fixed α_i, f_i, t_i *the random variables*

$$U_i^{(T,\Delta f)}(\alpha_i, f_i) \triangleq \sqrt{T\Delta f}\left[S_{yx^{(*)}}^{(T,\Delta f)}(\alpha_i, f_i; t_i) - \mathcal{E}_{(*)}\, S_{yx^{(*)}}^{\alpha_i}(f_i)\right] \qquad (i = 1, 2, \ldots) \tag{5.82}$$

are asymptotically ($T \to \infty$ *and* $\Delta f \to 0$ *with* $T\Delta f \to \infty$) *zero-mean jointly complex normal with asymptotic covariance matrix* $\boldsymbol{\Sigma}$ *with entries*

$$\Sigma_{ij} = \lim_{\Delta f \to 0} \lim_{T \to \infty} \operatorname{cov}\left\{U_i^{(T,\Delta f)}, U_j^{(T,\Delta f)}\right\} \tag{5.83}$$

given by (5.73) and asymptotic conjugate covariance matrix $\boldsymbol{\Sigma}^{(c)}$ *with entries*

$$\Sigma_{ij}^{(c)} = \lim_{\Delta f \to 0} \lim_{T \to \infty} \operatorname{cov}\left\{U_i^{(T,\Delta f)}, U_j^{(T,\Delta f)*}\right\} \tag{5.84}$$

given by (5.73) without conjugations in the subscripts and in the definitions of $\mathcal{I}_{hk}$.

Proof. See Section 5.9. □

Results analogous to those of Theorem 5.41 can be found in (Gardner, 1987f), (Hurd and Leśkow, 1992a), (Dehay and Hurd, 1994), (Dandawaté and Giannakis, 1994a), (Dehay and Leśkow, 1996a), (Hurd and Miamee, 2007).

In the asymptotic results of Theorems 5.36 and 5.38, we have that $T \to \infty$ and $\Delta f \to 0$, in this order, so that $T\Delta f \to \infty$. Therefore, we can take Δf finite and fixed, make $T \to \infty$, and then make $\Delta f \to 0$. Analogously, we could consider $\Delta f = \Delta f_T$ such that $\Delta f_T \to 0$ and $T\Delta f_T \to \infty$ as $T \to \infty$. This approach is adopted in Theorem 5.41 and in (Dandawaté and Giannakis, 1994a) and (Sadler and Dandawaté, 1998).

5.2.4 Time-smoothed cyclic cross-periodogram

Definition 5.42. Given two stochastic processes $\{y(t),\ t \in \mathbb{R}\}$ and $\{x(t),\ t \in \mathbb{R}\}$, their *time-smoothed (conjugate) cyclic cross-periodogram* is defined as

$$G_{yx^{(*)}}^{(\Delta f,T)}(\alpha, f; t) \triangleq I_{yx^{(*)}}^{(1/\Delta f)}(\alpha, f; t) \underset{t}{\otimes} a_T(t) \tag{5.85}$$

where $I_{yx^{(*)}}^{(1/\Delta f)}(\alpha, f; t)$ is the (conjugate) cyclic cross-periodogram defined according to (5.41) (with T replaced by $1/\Delta f$) and $a_T(t)$ is a unit-area T-duration time-smoothing window. □

In the special case of rectangular time-smoothing window $a_T(t) = \operatorname{rect}(t/T)/T$, we have

$$G_{yx^{(*)}}^{(\Delta f,T)}(\alpha, f; t) = \frac{1}{T}\int_{t-T/2}^{t+T/2} I_{yx^{(*)}}^{(1/\Delta f)}(\alpha, f; s)\, \mathrm{d}s\,. \tag{5.86}$$

The STFTs $Y_{\frac{1}{\Delta f}}(t, f)$ and $X_{\frac{1}{\Delta f}}(t, f - \alpha)$ represent the spectral components of $y(t)$ and $x(t)$, respectively, contained in frequency bands of width Δf and centered around two

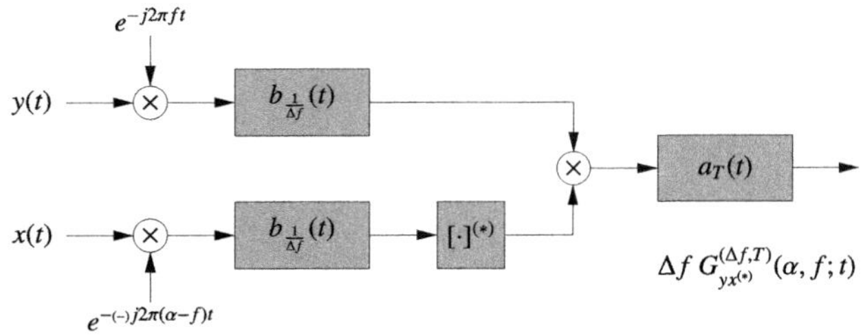

FIGURE 5.1 Spectral cross-correlation analyzer.

spectral frequencies separated by α. The time-smoothed cyclic cross-periodogram represents the temporal cross-correlation between these two spectral components (Fig. 5.1). In definition (5.41) with $T = 1/\Delta f$, according to (5.40a), we have

$$X_{1/\Delta f}^{(*)}(t, (-)(\alpha - f)) = \left[\int_{\mathbb{R}} x(u)\, b_{\frac{1}{\Delta f}}(u - t)\, e^{-(-)j2\pi f(\alpha - f)u}\, \mathrm{d}u\right]^{(*)}. \tag{5.87}$$

Assumption 5.43. Time-Smoothing Window Regularity. *$a_T(t)$ is a T-duration time-smoothing window that can be expressed as*

$$a_T(t) = \frac{1}{T} w_a\left(\frac{t}{T}\right) \tag{5.88}$$

with $w_a(t) \in L^1(\mathbb{R})$ and $\lim_{T\to\infty} w_a(t/T) = 1$. □

Theorem 5.44. Expected Value of the Time-Smoothed (Conjugate) Cyclic Cross-Periodogram. *Let $\{y(t),\ t \in \mathbb{R}\}$ and $\{x(t),\ t \in \mathbb{R}\}$ be second-order harmonizable jointly ACS stochastic processes with Loève bifrequency cross-spectrum (5.3).*

Under Assumptions

5.22a (ACS statistics and Fourier series regularity),
5.24 (data-tapering window regularity),
5.43 (time-smoothing window regularity),

the expected value of the time-smoothed (conjugate) cyclic cross-periodogram (5.85) is given by

$$\begin{aligned} &\mathrm{E}\left\{G_{yx^{(*)}}^{(\Delta f,T)}(\alpha, f; t)\right\} \\ &\quad = \Delta f \sum_{\beta \in A_{yx^{(*)}}} \int_{\mathbb{R}} S_{yx^{(*)}}^{\beta}(\nu)\, B_{\Delta f}(f - \nu)\, B_{\Delta f}^{(*)}((-)(\alpha - \beta - f + \nu))\, \mathrm{d}\nu \\ &\quad\quad A_{\frac{1}{T}}(\beta - \alpha)\, e^{-j2\pi(\alpha - \beta)t} \end{aligned} \tag{5.89}$$

where $A_{\frac{1}{T}}(f)$ is the Fourier transform of $a_T(t)$.

For $T\Delta f \gg 1$ the expected value (5.89) can be approximated as

$$\mathrm{E}\left\{G_{yx^{(*)}}^{(\Delta f,T)}(\alpha, f; t)\right\}$$
$$\simeq \Delta f \sum_{\beta \in A_{yx^{(*)}}} \left[S_{yx^{(*)}}^{\beta}(f) \underset{f}{\otimes} \left(B_{\Delta f}(f)\, B_{\Delta f}^{(*)}(-_{(-)}f)\right)\right] A_{\frac{1}{T}}(\beta - \alpha)\, e^{-j2\pi(\alpha-\beta)t}\,. \tag{5.90}$$

Proof. See Section 5.9. □

Lemma 5.45. Covariance of the Time-Smoothed (Conjugate) Cyclic Cross-Periodogram. *Let $\{y(t),\ t \in \mathbb{R}\}$ and $\{x(t),\ t \in \mathbb{R}\}$ be second-order harmonizable zero-mean singularly and jointly ACS stochastic processes with Loève bifrequency cross-spectrum (5.3).*

Under Assumptions

5.22 (ACS statistics and Fourier series regularity),
5.24 (data-tapering window regularity),
5.43 (time-smoothing window regularity),

the covariance

$$\operatorname{cov}\left\{G_{yx^{(*)}}^{(\Delta f,T)}(\alpha_1, f_1; t_1),\ G_{yx^{(*)}}^{(\Delta f,T)}(\alpha_2, f_2; t_2)\right\}$$

and the conjugate covariance

$$\operatorname{cov}\left\{G_{yx^{(*)}}^{(\Delta f,T)}(\alpha_1, f_1; t_1),\ G_{yx^{(*)}}^{(\Delta f,T)*}(\alpha_2, f_2; t_2)\right\}$$

of the time-smoothed (conjugate) cyclic cross-periodogram (5.85) can be computed similarly to the case of spectrally correlated processes (Napolitano, 2003, Lemma 4.2), (Napolitano, 2012, Lemma 4.5.6) (covariance). □

Theorem 5.46. Asymptotic Expected Value of the Time-Smoothed (Conjugate) Cyclic Cross-Periodogram. *Let $\{y(t),\ t \in \mathbb{R}\}$ and $\{x(t),\ t \in \mathbb{R}\}$ be second-order harmonizable jointly ACS stochastic processes with Loève bifrequency cross-spectrum (5.3).*

Under Assumptions

5.22a (ACS statistics and Fourier series regularity),
5.24 (data-tapering window regularity),
5.43 (time-smoothing window regularity),

the asymptotic ($T \to \infty$, $\Delta f \to 0$, with $T\Delta f \to \infty$) expected value (5.89) of the time-smoothed (conjugate) cyclic cross-periodogram (5.85) is given by

$$\lim_{\Delta f \to 0} \lim_{T \to \infty} \mathrm{E}\left\{G_{yx^{(*)}}^{(\Delta f,T)}(\alpha, f; t)\right\} = k_G\, S_{yx^{(*)}}^{\alpha}(f) \tag{5.91}$$

where

$$k_G \triangleq W_A(0) \int_{\mathbb{R}} W_B(\lambda)\, W_B^{(*)}(-_{(-)}\lambda)\, \mathrm{d}\lambda\,. \tag{5.92}$$

Proof. The proof is obtained by specializing to ACS processes the proof of (Napolitano, 2003, Theorem 4.1), (Napolitano, 2012, Theorem 4.5.7) for SC processes. □

Theorem 5.47. Asymptotic Covariance of the Time-Smoothed (Conjugate) Cyclic Cross-Periodogram. *Let $\{y(t),\ t \in \mathbb{R}\}$ and $\{x(t),\ t \in \mathbb{R}\}$ be second-order harmonizable jointly ACS stochastic processes with Loève bifrequency cross-spectrum (5.3).*

Under Assumptions

5.22 (ACS statistics and Fourier series regularity),
5.24 (data-tapering window regularity),
5.43 (time-smoothing window regularity),

the asymptotic ($T \to \infty$, $\Delta f \to 0$, with $T\Delta f \to \infty$) covariance of the time-smoothed (conjugate) cyclic cross-periodogram (5.85) is given by

$$\begin{aligned}
\lim_{\Delta f \to 0} \lim_{T \to \infty} T\Delta f \operatorname{cov}\Big\{ & G^{(\Delta f,T)}_{yx^{(*)}}(\alpha_1, f_1; t_1),\ G^{(\Delta f,T)}_{yx^{(*)}}(\alpha_2, f_2; t_2) \Big\} \\
&= W_A(0)\, \mathcal{E}_1^{(4)}\, S^{f_1-f_2}_{yy^*}(f_1)\, S^{\alpha_1-f_1-\alpha_2+f_2}_{x^{(*)}x^{(*)*}}((-)(\alpha_1 - f_1)) \\
&\quad + W_A(0)\, \mathcal{E}_2^{(4)}\, S^{f_1-\alpha_2+f_2}_{yx^{(*)*}}(f_1)\, S^{-f_2+\alpha_1-f_1}_{x^{(*)}y^*}((-)(\alpha_1 - f_1))
\end{aligned} \tag{5.93}$$

where

$$\mathcal{E}_1^{(4)} \triangleq \int_{\mathbb{R}} |W_B(-(-)\lambda)|^2\, |W_B(\lambda)|^2\, \mathrm{d}\lambda \tag{5.94}$$

$$\mathcal{E}_2^{(4)} \triangleq \int_{\mathbb{R}} W_B(-(-)\lambda)\, W_B^{(*)*}(-\lambda)\, W_B^{(*)}(\lambda)\, W_B^*((-)\lambda)\, \mathrm{d}\lambda \tag{5.95}$$

Proof. The proof is obtained by specializing to ACS processes the proof of (Napolitano, 2003, Theorem 4.2), (Napolitano, 2012, Theorem 4.5.9) for SC processes. □

The expression of the asymptotic conjugate covariance given by (5.93) without conjugations in the subscripts and in the definitions of $\mathcal{E}_h^{(4)}$.

From Theorem 5.44 it follows that for the expected value of the estimate $G^{(\Delta f,T)}_{yx^{(*)}}(\alpha, f; t)$ we have the phenomenon of *cycle leakage* (Gardner, 1987f, Chap. 13) as for the frequency-smoothed (conjugate) cyclic cross-periodogram.

In practice, condition $T\Delta f \gg 1$ needs to be verified to have small variance (Theorem 5.47) so that approximation (5.90) generally holds (Gardner, 1987f, Sec. 11-B).

The time-smoothing window $a_T(t)$ is a T-duration window. Thus, its Fourier transform $A_{\frac{1}{T}}(f)$ has a bandwidth of the order of $1/T$. Consequently, in (5.89), due to the presence of the term $A_{\frac{1}{T}}(\beta - \alpha)$, we have that cycle frequencies that are closer than $1/T$ cannot be distinguished. That is, the *cycle frequency resolution* of the time-smoothed (conjugate) cyclic cross-periodogram (5.85) is of the order of $1/T$. Therefore, in order to capture all the cyclic features, we must have $1/T < h_\alpha$, with h_α defined in (5.26). In addition, from (5.90) it follows that the shape of the cyclic spectra is smeared by the convolution in f with the function $B_{\Delta f}(f)\, B^{(*)}_{\Delta f}(-_{(-)}f)$ whose width is of the order of Δf. Thus, variations of the cyclic spectra in f smaller than Δf cannot be appreciated. That is, the *spectral frequency resolution* is of the order of Δf.

Theorem 5.48. Mean-Square Consistency of the Time-Smoothed (Conjugate) Cyclic Cross-Periodogram. *Under the assumptions of Theorems 5.46 and 5.47, the time-smoothed (conjugate) cyclic cross-periodogram is a mean-square consistent estimator of the (conjugate) cyclic spectrum*

$$\lim_{\Delta f \to 0} \lim_{T\to\infty} \mathrm{E}\left\{ \left| G^{(\Delta f,T)}_{yx^{(*)}}(\alpha, f; t_0) - k_G\, S^{\alpha}_{yx^{(*)}}(f) \right|^2 \right\} = 0\,. \tag{5.96}$$

Proof. It is analogous to the proof in Section 5.9 for Theorem 5.39. □

By following the guidelines of the proof of Theorem 5.41, the asymptotic joint complex normality of the time-smoothed (conjugate) cyclic cross-periodograms can be proved.

Theorem 5.49. Asymptotic Joint Complex Normality of the Time-Smoothed (Conjugate) Cyclic Cross-Periodograms. *Under mild suitable assumptions, for every fixed α_i, f_i, t_i the random variables*

$$W_i^{(\Delta f,T)}(\alpha_i, f_i) \triangleq \sqrt{T\Delta f}\left[G^{(\Delta f,T)}_{yx^{(*)}}(\alpha_i, f_i; t_i) - k_G\, S^{\alpha_i}_{yx^{(*)}}(f_i) \right] \qquad (i = 1, 2, \ldots) \tag{5.97}$$

with k_G defined in (5.92), are asymptotically ($T \to \infty$ and $\Delta f \to 0$ with $T\Delta f \to \infty$) zero-mean jointly complex normal with asymptotic covariance matrix $\boldsymbol{\Sigma}$ with entries

$$\Sigma_{ij} = \lim_{\Delta f \to 0} \lim_{T\to\infty} \operatorname{cov}\left\{ W_i^{(\Delta f,T)}, W_j^{(\Delta f,T)} \right\} \tag{5.98}$$

and asymptotic conjugate covariance matrix $\boldsymbol{\Sigma}^{(c)}$ with entries

$$\Sigma^{(c)}_{ij} = \lim_{\Delta f \to 0} \lim_{T\to\infty} \operatorname{cov}\left\{ W_i^{(\Delta f,T)}, W_j^{(\Delta f,T)*} \right\} \tag{5.99}$$

□

5.2.5 Median-filtering based smoothing

The existence of consistent estimators for the cyclic statistics is one of the main motivations for the success in the exploitation of cyclostationarity in several fields of application (Gardner et al., 2006). A necessary assumption to prove consistency is the zero mean and the summability of second- and fourth-order cumulants of the process. Therefore, according to the consideration made for (1.63)–(1.66), the possible additive almost-periodic component of the process must be zero or removed. In the case of periodic component with known period, this component can be estimated by resorting to the synchronized averaging identity (Theorem B.19) and then subtracted, leading to consistent estimators of the cyclic statistics of the residual part (Javorskyj et al., 2007), (Javorskyj et al., 2010).

If the period is unknown or the additive component is almost-periodic, then the (conjugate) cyclic spectrum of the zero-mean ACS residual term $x_r(t)$ can be estimated by median filtering the smoothed (conjugate) cyclic periodogram of $x(t)$ (Napolitano and Spooner, 2000). Specifically, let $Z(f) = \mathrm{Re}(Z(f)) + j\mathrm{Im}(Z(f))$ be a complex-valued function. The median value of $Z(f)$ in the interval $J(f,\delta f) \triangleq [f - \delta f/2, f + \delta f/2]$ is defined as

$$\underset{\nu \in J(f,\delta f)}{\mathrm{med}} \left[Z(\nu)\right] \triangleq \underset{\nu \in J(f,\delta f)}{\mathrm{med}} \left[\mathrm{Re}(Z(\nu))\right] + j \underset{\nu \in J(f,\delta f)}{\mathrm{med}} \left[\mathrm{Im}(Z(\nu))\right]. \tag{5.100}$$

In (Napolitano and Spooner, 2000), the following estimator of the (conjugate) cyclic spectrum of $x_r(t)$ is proposed:

$$P_{\boldsymbol{x}}^{(T,\Delta f,\delta f)}(\alpha, f) \triangleq \underset{\nu \in J(f,\delta f)}{\mathrm{med}} \left[Z_{\boldsymbol{x}}(\alpha, \nu)\right]. \tag{5.101}$$

If $Z_{\boldsymbol{x}}(\alpha, f) = S_{\boldsymbol{x}}^{(T,\Delta f)}(\alpha, f)$, the estimator $P_{\boldsymbol{x}}^{(T,\Delta f,\delta f)}(\alpha, f)$ is referred to as the median-filtered frequency-smoothed (conjugate) cyclic periodogram; if $Z_{\boldsymbol{x}}(\alpha, f) = G_{\boldsymbol{x}}^{(\Delta f,T)}(\alpha, f)$, the estimator $P_{\boldsymbol{x}}^{(T,\Delta f,\delta f)}(\alpha, f)$ is referred to as the median-filtered time-smoothed (conjugate) cyclic periodogram (Napolitano and Spooner, 2000). With probability 1 we have

$$\lim_{\delta f \to 0} \lim_{\Delta f \to 0} \lim_{T \to \infty} P_{\boldsymbol{x}}^{(T,\Delta f,\delta f)}(\alpha, f) = S_{\boldsymbol{x}_r}^{\alpha}(f) \tag{5.102}$$

where $S_{\boldsymbol{x}_r}^{\alpha}(f)$ is the (conjugate) cyclic spectrum of $x_r(t)$ and the order of the three limits cannot be interchanged (Napolitano and Spooner, 2000). In practice, conditions $T\Delta f \gg 1$ and δf sufficiently larger than Δf must be verified.

5.2.6 Loève bifrequency cross-spectrum density estimation

The density of the Loève bifrequency cross-spectrum (5.3) of jointly harmonizable ACS processes $x(t)$ and $y(t)$ observed in the time interval $t \in [-T/2, T/2]$ can be estimated as

$$\mathcal{S}_{yx^{(*)}}^{(T,\Delta t)}(f_1, f_2) = \int_{-T/2}^{T/2} \int_{-T/2}^{T/2} y(t_1)\, x^{(*)}(t_2)\, h_{\Delta t}(t_1 - t_2)\, e^{-j2\pi[f_1 t_1 + (-) f_2 t_2]}\, \mathrm{d}t_1 \mathrm{d}t_2 \tag{5.103}$$

where $h_{\Delta t}(t)$ is a Δt-width smoothing window whose effect is to taper signal samples whose distance is bigger than Δt.

In (Lenart, 2011), the discrete-time counterpart of (5.103) for the case $x(t) \equiv y(t)$ is considered. By assuming that the process is zero mean, satisfies mixing conditions, and has spectral densities along the support lines of the Loève bifrequency spectrum, it is shown that the estimator is mean-square consistent as $T \to \infty$ and $\Delta t \to \infty$ with $\Delta t/T \to 0$. In addition, it is proved that a properly normalized version of the estimation error is asymptotically complex normal. It is observed that in most cases, as in the stationary case, for the majority of smoothing windows $\Delta t = \mathcal{O}(T^{1/5})$ is the best choice.

5.3 Supplementary analysis

5.3.1 Time versus frequency smoothing

From the results of Sections 5.2.3 and 5.2.4 we have that both the frequency-smoothed (conjugate) cyclic cross-periodogram (5.63) and the time-smoothed (conjugate) cyclic cross-periodogram (5.85) have spectral frequency resolution of the order of Δf and cycle frequency resolution of the order of $1/T$. Moreover, both estimates (5.63) and (5.85) are reliable provided that $T\Delta f \gg 1$.

For both the time-smoothed and the frequency-smoothed (conjugate) cyclic cross-periodograms, T is the whole data-record length. For the time-smoothed (conjugate) cyclic cross-periodogram T is coincident with the width of the time-smoothing window adopted to smooth the (conjugate) cyclic cross-periodogram

$$I_{yx^{(*)}}^{(1/\Delta f)}(\alpha, f; t) \triangleq \Delta f \left[Y_{\frac{1}{\Delta f}}(t, f)\, X_{\frac{1}{\Delta f}}^{(*)}(t, (-)(\alpha - f)) \right] \tag{5.104}$$

which is obtained by STFTs calculated by a $(1/\Delta f)$-width temporal window. In contrast, for the frequency-smoothed (conjugate) cyclic cross-periodogram, T is the width of the temporal window adopted to calculate the STFTs

$$I_{yx^{(*)}}^{(T)}(\alpha, f; t) \triangleq \frac{1}{T} \left[Y_T(t, f)\, X_T^{(*)}(t, (-)(\alpha - f)) \right] . \tag{5.105}$$

The asymptotic equivalence of the two estimators is stated by the following result.

Theorem 5.50. (Gardner, 1986a), (Gardner, 1987f, pp. 389–390), (Napolitano, 2012, p. 230). *The time-smoothed (conjugate) cyclic cross-periodogram is asymptotically equivalent to the frequency-smoothed (conjugate) cyclic cross-periodogram in the sense that with probability 1*

$$(k_S)^{-1} \lim_{\Delta f \to 0} \lim_{T \to \infty} S_{yx^{(*)}}^{(T,\Delta f)}(\alpha, f; t) = (k_G)^{-1} \lim_{\Delta f \to 0} \lim_{T \to \infty} G_{yx^{(*)}}^{(\Delta f, T)}(\alpha, f; t) = S_{yx^{(*)}}^{\alpha}(f) \tag{5.106}$$

where the order of the two limits cannot be interchanged. The constants $k_S = \mathcal{E}_{(*)}$ and k_G given in (5.92) depend on the data-tapering, frequency-smoothing, and time-smoothing windows (see (5.71) and (5.91)). □

The estimator (5.63) for $y \equiv x$, $\alpha = 0$, and $(*)$ present reduces to the frequency smoothed periodogram that is a well known estimator of the power spectrum of WSS processes (Brillinger, 1981, Chap. 5). The estimator (5.85) for $y \equiv x$, $\alpha = 0$, $(*)$ present, and with the time average replaced by a finite sum of periodograms computed on non-overlapping blocks of data, reduces to the Bartlet estimator for the power spectrum of WSS processes (Rosenblatt, 1985, Chap. V).

5.3.2 Alternative assumptions

Alternative assumptions with respect to those considered in Sections 5.2.1, 5.2.3, and 5.2.4, can be made to derive the main results. In general, if the assumptions on the stochastic processes are relaxed, those on the data-tapering window, frequency-smoothing window or time-smoothing window must be more stringent and vice versa.

The results of Sections 5.2.3 and 5.2.4, can be obtained with minor changes by replacing (5.48) and (5.51) in Lemma 5.23 (that are an immediate consequence of (5.43) and (5.45), respectively, in Assumption 5.22) with the following assumption.

Assumption 5.51. Spectral-Density Summability.

a. *For any choice of* z_1 *and* z_2 *in* $\{x, x^*, y, y^*\}$, *the functions* $S^{\alpha}_{z_1 z_2}(f)$ *in (5.46) are almost everywhere (a.e.) continuous and such that*

$$\sum_{\alpha \in A_{z_1 z_2}} \int_{\mathbb{R}} \left| S^{\alpha}_{z_1 z_2}(f) \right| \mathrm{d}f < \infty \,. \tag{5.107}$$

b. *The functions* $P^{\beta}_{yx^{(*)}y^*x^{(*)*}}(f_1, f_2, f_3)$ *in (5.49) are a.e. continuous and such that*

$$\sum_{\beta \in A_4} \int_{\mathbb{R}^3} \left| P^{\beta}_{yx^{(*)}y^*x^{(*)*}}(f_1, f_2, f_3) \right| \mathrm{d}f_1 \mathrm{d}f_2 \mathrm{d}f_3 < \infty \,. \tag{5.108}$$

□

In (Napolitano, 2012, Sec. 4.7.3), for the more general case of SC processes, it is shown that if Assumption 5.51 is made (possibly in place of (5.43) and (5.45) that imply (5.48) and (5.51)), then $W_B(f)$ does not need to be summable and we can take $w_b(t) = \mathrm{rect}(t)$ (i.e., $W_B(f) = \mathrm{sinc}(f) \notin L^1(\mathbb{R})$).

Let us consider now the case where there is no cycle frequency cluster point (Assumption 5.11) and $W_B(f)$ and $W_A(f)$ are both rigorously bandlimited. By specializing to the ACS case the considerations made in (Napolitano, 2012, Sec. 4.7.3) for SC processes, we have that all the sums in the expressions of the expected value and covariance of the frequency-smoothed (conjugate) cyclic cross-periodogram (Theorem 5.31 and Lemma 5.32) are finite. It follows that conditions (5.48) and (5.51) are not needed. This allows, for example, to consider PAM signals with rectangular pulses (for which (5.48) is not satisfied).

5.3.3 Cyclic spectral analysis versus cyclic cross-spectral analysis

In general, there is higher variability with respect of f in cyclic cross-spectra of two signals with respect to the case of cyclic spectra of a single signal. Consider, for example, the cyclic cross-spectrum of $x(t)$ and $y(t) = x(t-d)$ (Example 3.8). In (3.36), due to the phase term $e^{-j2\pi f d}$, the bigger is $|d|$, the higher is the variability with f of the (conjugate) cyclic cross-spectrum $S^{\alpha}_{yx^{(*)}}(f)$ compared to that of $S^{\alpha}_{xx^{(*)}}(f)$.

Such a higher variability reflects into the necessity of a finer spectral resolution (smaller Δf) when $S^{\alpha}_{yx^{(*)}}(f)$ is estimated, for example, by the frequency-smoothed (conjugate) cyclic cross-periodogram. In fact, we have

$$\begin{aligned} S^{(T,\Delta f)}_{yx^{(*)}}(\alpha, f) &= I^{(T)}_{yx^{(*)}}(\alpha, f) \underset{f}{\otimes} A_{\Delta f}(f) \\ \overset{\mathcal{F}}{\longleftrightarrow} R^{(T)}_{yx^{(*)}}(\alpha, \tau)\, a_{1/\Delta f}(\tau) &\simeq R^{(T)}_{xx^{(*)}}(\alpha, \tau - d)\, a_{1/\Delta f}(\tau) \end{aligned} \tag{5.109}$$

where $I^{(T)}_{yx^{(*)}}(\alpha, f)$ is the (conjugate) cyclic cross-periodogram, the Fourier pair is consequence of Lemma 2.39 and the convolution theorem (see also (5.122a)), $A_{\Delta f}(f)$ is the frequency-smoothing window with width Δf, and $a_{1/\Delta f}(\tau)$ is its inverse Fourier transform with approximate width $1/\Delta f$.

Since the function $a_{1/\Delta f}(\tau)$ is centered in $\tau = 0$ and has width $1/\Delta f$ and the function $R^{(T)}_{xx^{(*)}}(\alpha, \tau - d)$ is centered in d, a large value of Δf can give rise to a tapering effect that reflects into a significant bias in $S^{(T,\Delta f)}_{yx^{(*)}}(\alpha, f)$ when d is large. Consequently, if Δf is selected sufficiently small (such that $1/\Delta f > |d|$), the data-record length T must be increased in order to keep a given level of reliability (large $T\Delta f$). Such an issue must be accounted for in cyclostationarity-based TDOA estimation methods (Izzo et al., 1992a, Sec. 4), (Izzo et al., 1994a, Sec. 4).

5.3.4 Cycle leakage

Cycle leakage is the influence, in cyclic statistic measurements at the considered cycle frequency α_0, of the cyclic statistics of the disturbance signals and the SOI itself at cycle frequencies different from α_0 (Gardner, 1986a), (Gardner, 1987f, Chap. 13). It is a classical effect occurring in the estimation of the parameters of a sine wave on the basis of a finite data-record length.

By considering the special cases of (5.21) with $W_{\frac{1}{T}}(t) = \mathrm{sinc}(fT)$ and of (5.69) with $\chi_B(f) = \mathrm{sinc}(f)$, both with $t_0 = 0$, for $T \gg 1/\Delta f$ and $T \gg$ width of $R^{\beta}_{\boldsymbol{x}}(\tau)$, we have

$$\mathrm{E}\left\{R^{(T)}_{\boldsymbol{x}}(\alpha, \tau)\right\} = \sum_{\beta} R^{\beta}_{\boldsymbol{x}}(\tau)\, \mathrm{sinc}((\alpha - \beta)T) \tag{5.110}$$

$$\mathrm{E}\left\{S^{(T,\Delta f)}_{\boldsymbol{x}}(\alpha, f)\right\} \simeq \sum_{\beta} \left[S^{\beta}_{\boldsymbol{x}}(f) \otimes A_{\Delta f}(f)\right] \mathrm{sinc}((\alpha - \beta)T) \tag{5.111}$$

where subscript $\boldsymbol{x} \equiv xx^*$.

In Fig. 5.2, the magnitude of the frequency-smoothed cyclic periodogram $S^{(T,\Delta f)}_{\boldsymbol{x}}(\alpha, f)$ as a function of α and f is reported, for a PAM signal $x(t)$ with baud rate $1/T_p = 0.125/T_s$ in WGN with SNR $= -10$ dB. The behavior of the sinc function along the cycle frequency dimension is evident. At $\alpha = 0$ the power spectrum of the PAM signal overlies a floor due to the WGN.

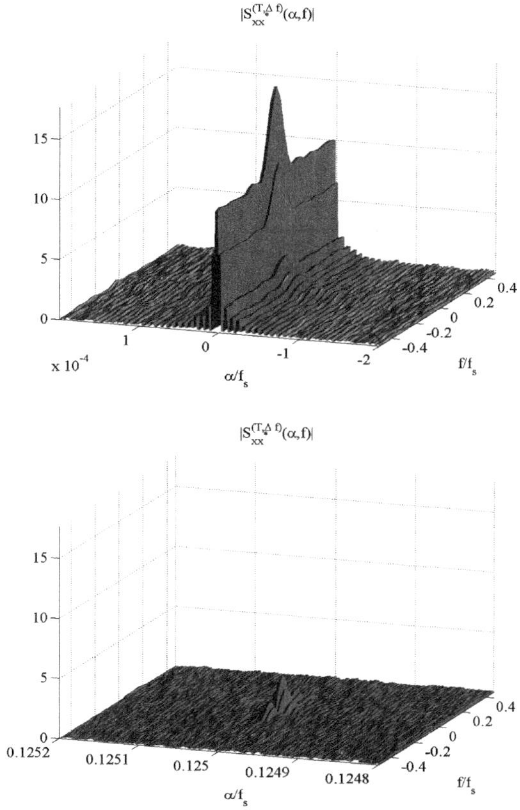

FIGURE 5.2 Magnitude of the frequency-smoothed cyclic periodogram $S_{\boldsymbol{x}}^{(T,\Delta f)}(\alpha, f)$ as a function of α and f, for a PAM signal $x(t)$ with baud rate $1/T_p = 0.125/T_s$ in WGN with SNR $= -10$ dB. Data-record length $T = 2^{17} T_s$. Zoom around (top) $\alpha = 0$ and (bottom) $\alpha = 1/T_p$.

5.3.5 Combined effects of aliasing and cycle leakage

In this section, the combined effects of aliasing and cycle leakage in the discrete-time estimator of the (conjugate) cyclic autocorrelation function of a continuous-time ACS signal are considered. The most general case of GACS signals is considered.

Let $x(n)$ be the discrete-time signal obtained by uniformly sampling with period $T_s = 1/f_s$ the complex-valued continuous-time GACS (or ACS) signal $x_a(t)$. That is,

$$x(n) = x_a(t)|_{t=nT_s} \qquad n \in \mathbb{Z}\,. \tag{5.112}$$

According to the results of Sections 2.3.1.6 and 8.2, the second-order lag product of the continuous-time signal $x_a(t)$ can be decomposed into its almost-periodic component and a residual term not containing any finite-strength additive sine-wave component

$$x_a(t+\tau)\, x_a^{(*)}(t) = \sum_{\beta \in A_\tau} R_{\boldsymbol{x}_a}^{\beta}(\tau)\, e^{j2\pi\beta t} + \ell_{\boldsymbol{x}_a}(t,\tau) \tag{5.113}$$

where subscript $\boldsymbol{x}_a \equiv x_a\, x_a^{(*)}$, $A_\tau = A_{\boldsymbol{x}_a,\tau}$ is the set of second-order (conjugate) cycle frequencies of $x_a(t)$ at lag τ, and

$$\left\langle \ell_{\boldsymbol{x}_a}(t,\tau)\, e^{-j2\pi\beta t} \right\rangle_t = 0 \qquad \forall \beta \in \mathbb{R}\,. \tag{5.114}$$

It results

$$\begin{aligned} x(n+m)\, x^{(*)}(n) &\triangleq x_a(t+\tau)\, x_a^{(*)}(t)|_{t=nT_s,\tau=mT_s} \\ &= \sum_{\beta\in A_{mT_s}} R_{\boldsymbol{x}_a}^{\beta}(mT_s)\, e^{j2\pi\beta n T_s} + \ell_{\boldsymbol{x}_a}(nT_s, mT_s) \end{aligned} \tag{5.115}$$

where the discrete-time signal $\ell_{\boldsymbol{x}_a}(nT_s, mT_s)$ does not contain any discrete-time finite-strength additive sine-wave component (see (3.264) and (Izzo and Napolitano, 2003), (Izzo and Napolitano, 2005)).

The discrete-time estimate of the (conjugate) cyclic autocorrelation function is

$$\begin{aligned} \widetilde{R}_{\boldsymbol{x}}^{(N)}(\widetilde{\alpha}, m) &\triangleq \frac{1}{N}\sum_{n=0}^{N-1} x(n+m)\, x^{(*)}(n)\, e^{-j2\pi\widetilde{\alpha}n} \\ &= \frac{1}{N}\sum_{n=0}^{N-1}\left[\sum_{\beta\in A_{mT_s}} R_{\boldsymbol{x}_a}^{\beta}(mT_s)\, e^{j2\pi\beta n T_s} + \ell_{\boldsymbol{x}_a}(nT_s, mT_s)\right] e^{-j2\pi\widetilde{\alpha}n} \\ &= \sum_{\beta\in A_{mT_s}} R_{\boldsymbol{x}_a}^{\beta}(mT_s)\left[\frac{1}{N}\sum_{n=0}^{N-1} e^{j2\pi(\beta T_s-\widetilde{\alpha})n}\right] + \frac{1}{N}\sum_{n=0}^{N-1} \ell_{\boldsymbol{x}_a}(nT_s, mT_s)\, e^{-j2\pi\widetilde{\alpha}n} \\ &= \sum_{\beta\in A_{mT_s}} R_{\boldsymbol{x}_a}^{\beta}(mT_s)\, D_{\frac{1}{N}}(\beta T_s - \widetilde{\alpha}) + \frac{1}{N}\sum_{n=0}^{N-1} \ell_{\boldsymbol{x}_a}(nT_s, mT_s)\, e^{-j2\pi\widetilde{\alpha}n} \end{aligned} \tag{5.116}$$

where

$$D_{\frac{1}{N}}(\widetilde{\gamma}) \triangleq \frac{1}{N}\sum_{n=0}^{N-1} e^{j2\pi\widetilde{\gamma}n} = \frac{1}{N}\,\frac{e^{j2\pi\widetilde{\gamma}N}-1}{e^{j2\pi\widetilde{\gamma}}-1} = \frac{1}{N}\, e^{j\pi\widetilde{\gamma}(N-1)}\,\frac{\sin[\pi\widetilde{\gamma}N]}{\sin[\pi\widetilde{\gamma}]} \tag{5.117}$$

is the Dirichlet kernel.

In the first sum in the rhs of (5.116), terms with $\beta T_s \bmod 1 \neq \widetilde{\alpha}$ describe the cycle leakage of cycle frequencies different from $\alpha = \widetilde{\alpha} f_s \bmod f_s$ on the estimate. Their effect becomes negligible as the number of samples N approaches infinite. Terms with $\beta T_s \bmod 1 = \widetilde{\alpha}$ and $\beta \neq \widetilde{\alpha} f_s$ describe the effect of aliasing due to uniform sampling. Their effect becomes negligible when the sampling period T_s approaches zero. The second sum in the rhs of (5.116) describes the randomness (in the FOT sense) of the measure. It becomes negligible as N approaches infinite.

By using the limit

$$\lim_{N\to\infty} D_{\frac{1}{N}}(\widetilde{\gamma}) = \delta_{\widetilde{\gamma} \bmod 1} = \begin{cases} 1 & \widetilde{\gamma} \in \mathbb{Z} \\ 0 & \text{otherwise} \end{cases} \tag{5.118}$$

where δ_γ is the Kronecker delta (that is, $\delta_\gamma = 1$ if $\gamma = 0$ and $\delta_\gamma = 0$ if $\gamma \neq 0$), and accounting for the fact that (see (3.264) and (Izzo and Napolitano, 2003), (Izzo and Napolitano, 2005))

$$\lim_{N\to\infty} \frac{1}{N} \sum_{n=0}^{N-1} \ell_{\boldsymbol{x}_a}(nT_s, mT_s)\, e^{-j2\pi\widetilde{\alpha}n} = 0 \qquad \forall \widetilde{\alpha} \in \mathbb{R} \tag{5.119}$$

we get the FOT counterpart of (3.191) that is

$$\begin{aligned} \widetilde{R_{\boldsymbol{x}}^{\alpha}}(m) &= \lim_{N\to\infty} \widetilde{R}_{\boldsymbol{x}}^{(N)}(\widetilde{\alpha}, m) \\ &= \sum_{\beta\in A_{mT_s}} R_{\boldsymbol{x}_a}^{\beta}(mT_s)\, \delta_{(\beta T_s - \widetilde{\alpha}) \bmod 1} \\ &= \sum_{\beta\in A_{mT_s}} R_{\boldsymbol{x}_a}^{\beta}(mT_s) \sum_{p\in\mathbb{Z}} \delta_{\beta T_s - \widetilde{\alpha} + p} \\ &= \sum_{p\in\mathbb{Z}} R_{\boldsymbol{x}_a}^{(\widetilde{\alpha}-p) f_s}(mT_s) \end{aligned} \tag{5.120}$$

which is formally identical to (3.191). In the sum in (5.120) give nonzero contribution all values of p such that $(\widetilde{\alpha} - p) f_s \in A_{mT_s}$.

5.4 Implementation of second-order cyclic statistic estimators

The continuous-time signals $x(t)$ and $y(t)$ observed in the data-record $[-T/2, T/2]$ are uniformly sampled with sampling period $T_s = 1/f_s$. The samples nT_s, $n = -N/2, \ldots, N/2 - 1$, are considered, where N is the largest even integer such that $T \geqslant NT_s$. Thus, the cycle frequency resolution of the continuous-time estimators is $1/T$ and that of the discrete-time estimators is $1/N$.

5.4.1 Cyclic cross-correlogram

The (conjugate) cyclic cross-correlogram is given by

$$R_{yx^{(*)}}^{(T)}(\alpha, \tau) = \frac{1}{T} \int_{-T/2}^{T/2} y(t+\tau)\, x^{(*)}(t)\, e^{-j2\pi\alpha t}\, \mathrm{d}t \tag{5.121a}$$

$$\simeq \frac{1}{N} \sum_{n=-N/2}^{N/2-1} y(nT_s + mT_s)\, x^{(*)}(nT_s)\, e^{-j2\pi\alpha n T_s} \tag{5.121b}$$

In the approximate equality (5.121b), samples for $\tau = mT_s$ are considered. The approximation is close if the aliasing effect is negligible (Theorem 3.23).

For every fixed m, samples of the (conjugate) cyclic cross-correlogram for $\widetilde{\alpha} = \alpha/f_s = q/N$, $q = -N/2, \dots, N/2-1$, are obtained via discrete Fourier transform (DFT) of the lag-product $y(nT_s + mT_s)\, x^{(*)}(nT_s)$. The DFT is implemented via fast Fourier transform (FFT).

5.4.2 Frequency-smoothed cyclic cross-periodogram

Let $W_a(f)$ a unit-bandwidth frequency-smoothing window with inverse Fourier transform $w_a(\tau)$ and let us denote by $\mathcal{F}[\cdot]$ the continuous-time Fourier transform operator. Accounting for the convolution theorem for Fourier transforms, we have that the frequency-smoothed (conjugate) cyclic cross-periodogram is given by (see (5.63))

$$S_{yx^{(*)}}^{(T,\Delta f)}(\alpha, f) = I_{yx^{(*)}}^{(T)}(\alpha, f) \underset{f}{\otimes} \frac{1}{\Delta f}\, W_a\Big(\frac{f}{\Delta f}\Big) \tag{5.122a}$$

$$= \mathcal{F}\Big[R_{yx^{(*)}}^{(T)}(\alpha, \tau)\, w_a(\tau \Delta f)\Big] \tag{5.122b}$$

$$\simeq \mathrm{DFT}\Big[R_{yx^{(*)}}^{(T)}(\alpha, mT_s)\, w_a(mT_s \Delta f)\Big] \tag{5.122c}$$

In the approximate equality in (5.122c), samples for $f = k/(NT_s)$, $k = -N/2, \dots, N/2-1$, of the Fourier transform in (5.122b) are obtained via DFT. The approximation is close if the aliasing effect is negligible (Theorem 3.24). In (5.122c), the DFT is implemented via FFT.

The spectral frequency resolution is Δf.

5.4.3 Time-smoothed cyclic cross-periodogram

In the case of rectangular time-smoothing window, the time-smoothed (conjugate) cyclic cross-periodogram is given by (see (5.86))

$$G_{yx^{(*)}}^{(\Delta f,T)}(\alpha, f; t) = \frac{1}{T}\int_{t-T/2}^{t+T/2} I_{yx^{(*)}}^{(1/\Delta f)}(\alpha, f; s)\, \mathrm{d}s \tag{5.123a}$$

$$\simeq \frac{1}{N}\sum_{n=0}^{N-1} I_{yx^{(*)}}^{(1/\Delta f)}(\alpha, f; t_0 + nT_s) \tag{5.123b}$$

$$\simeq \frac{1}{K}\sum_{k=1}^{K} I_{yx^{(*)}}^{(1/\Delta f)}(\alpha, f; t_0 + k\Delta t)\,, \tag{5.123c}$$

where $t_0 \triangleq t - T/2$. The approximate equality in (5.123b) is due to the discrete-time implementation of the time average in (5.123a). The approximation is close if the aliasing effect is negligible (Theorem 3.24). The approximate equality in (5.123c) is aimed at reducing the computational complexity of (5.123b). In (5.123c), the observation interval $[t-T/2, t+T/2]$ is partitioned into K possibly overlapping time intervals (blocks) of length $1/\Delta f$. Two adjacent blocks are time shifted one another by Δt. Thus, they overlap in a time interval of width $1/\Delta f - \Delta t$, provided that $\Delta t < 1/\Delta f$. K is the largest integer such that $T \geqslant K\Delta t - 2/\Delta f$. The STFTs to build samples for $f = k/(NT_s)$, $k = -N/2, \ldots, N/2-1$, of the cyclic cross-periodogram are approximated via DFT.

The spectral frequency resolution is Δf.

5.4.4 Computationally efficient estimators for cyclic spectral analysis

The (conjugate) cyclic cross-spectrum (5.8) of the discrete-time sequences $x(n)$ and $y(n)$ can be consistently estimated by the frequency-smoothed (conjugate) cyclic cross-periodogram which is defined analogously to its continuous-time counterpart (5.122a) as

$$S_{yx^{(*)}}^{(N,\Delta\nu)}(\widetilde{\alpha},\nu) \triangleq \left[\frac{1}{N}\, Y_N(n,\nu)\, X_N^{(*)}(n,{}_{(-)}(\widetilde{\alpha}-\nu))\right] \underset{\nu}{\otimes} H_{\Delta\nu}(\nu) \tag{5.124}$$

where $Y_N(n,\nu)$ and $X_N(n,\nu)$ are the STFTs of the sequences $y(m)$ and $x(m)$, respectively, with $m \in \{n-N/2, \ldots, n+N/2-1\}$, N even, and $H_{\Delta\nu}(\nu)$ is a frequency-smoothing window with bandwidth $\Delta\nu$. The condition $\Delta\nu \gg 1/N$ must be satisfied for estimate reliability (Section 5.3.1).

The estimator $S_{yx^{(*)}}^{(N,\Delta\nu)}(\widetilde{\alpha},\nu)$ has spectral frequency resolution $\Delta\nu$ and cycle frequency resolution $1/N$. Thus, assuming for simplicity $\Delta\nu = 1/M$, M even, it needs to be computed at spectral frequencies $\nu = k/M$, $k = -M/2, \ldots, M/2-1$ and cycle frequencies $\widetilde{\alpha} = q/N$, $q = -N/2, \ldots, N/2-1$.

Accounting for the above considerations, the *strip spectral correlation analyzer* (SSCA) originally proposed in (Roberts et al., 1991), (Brown and Loomis, 1993) for the estimation of the symmetric version of the (conjugate) cyclic cross-spectrum, can be used to efficiently estimate NM samples of the (conjugate) cyclic cross-spectrum in the whole principal domain $(\widetilde{\alpha},\nu) \in [-1/2,1/2] \times [-1/2,1/2]$.

One of the input signals, say $y(n)$, is channelized into narrow-band components with bandwidth $\Delta\nu = 1/M$ computed by STFT. Specifically, let $h(n)$ be an even discrete-time data-tapering window with width M whose Fourier transform is $H_{\Delta\nu}(\nu)$. The STFT of the discrete-time signal $y(m)$ with $m \in \{n-M/2, \ldots, n+M/2-1\}$ can be expressed as

$$Y_M(n,\nu) = \sum_{m'=n-M/2}^{n+M/2-1} h(n-m')\, y(m')\, e^{-j2\pi\nu m'} \tag{5.125a}$$

$$= \sum_{m=-M/2}^{M/2-1} h(m)\, y(n-m)\, e^{-j2\pi\nu(n-m)} \tag{5.125b}$$

$$= \widetilde{Y}_M(n,\nu)\, e^{-j2\pi\nu n} \tag{5.125c}$$

where

$$\widetilde{Y}_M(n,\nu) \triangleq \sum_{m=-M/2}^{M/2-1} \left[h(m)\, e^{j2\pi\nu m}\right] y(n-m) \tag{5.126a}$$

$$= y(n) \otimes \left[h(n)\, e^{j2\pi\nu n}\right] \tag{5.126b}$$

In (5.126b), $h(n)\, e^{j2\pi\nu n}$ is the impulse response function of a band-pass filter with central frequency ν and bandwidth $\Delta\nu = 1/M$.

Theorem 5.52. Strip Spectral Correlation Analyzer (Roberts et al., 1991), (Brown and Loomis, 1993). *Let $g(n)$ be a data-tapering window nonzero for $n \in \{-N/2, \ldots, N/2-1\}$. The frequency-smoothed (conjugate) cyclic cross-periodogram of $x(n)$ and $y(n)$ can be expressed as*

$$S_{yx^{(*)}}^{(N,\Delta\nu)}(\widetilde{\alpha},\nu) = \frac{1}{N} \sum_{n=-N/2}^{N/2-1} \widetilde{Y}_M(n,\nu)\, x^{(*)}(n)\, g(n)\, e^{-j2\pi\widetilde{\alpha}n} \tag{5.127}$$

where $\widetilde{Y}_M(n,\nu)$ is defined in (5.126b).

Proof. See Section 5.9. □

Samples for $\nu = k/M$, $k = -M/2, \ldots, M/2-1$ and $\widetilde{\alpha} = q/N$, $q = -N/2, \ldots, N/2-1$ are sufficient to describe $S_{yx^{(*)}}^{(N,\Delta\nu)}(\widetilde{\alpha},\nu)$. Thus, an N-point DFT at frequencies $\widetilde{\alpha} = q/N$, $q = -N/2, \ldots, N/2-1$ can be used to compute the samples of the right-hand side of (5.127).

Another computationally efficient algorithm for cyclic spectral analysis is the FFT accumulation method (FAM) (Roberts et al., 1991), (Brown and Loomis, 1993). Fast algorithms are also presented in (Spooner et al., 2013), (Antoni et al., 2017), (Spooner and Mody, 2018), and (Borghesani and Antoni, 2018).

A photonically assisted cyclostationary analyzer that eliminates the need for high-bandwidth digitization and real-time Fourier processors is presented in (Esman et al., 2017).

A cyclic spectrum analyzer with efficient memory consumption is presented in (Shevgunov et al., 2018).

5.4.5 Matlab®/Octave code for cyclic spectral analysis

In this section, Matlab (MathWorks, 1994)/Octave (GNU Octave, 1998) code is presented for cyclic cross-spectral analysis.

Signals are stored in the following Matlab/Octave vectors:

`y(1:N)` = vector of samples of $y(nT_s)$, $n = 0, 1, \ldots, N-1$

`x(1:N)` = vector of samples of $x(nT_s)$, $n = 0, 1, \ldots, N-1$

5.4.5.1 Estimation at a single cycle frequency

In this section, the Matlab/Octave functions `xcorr` and `cpsd` designed for cross-correlation and cross-power-spectrum estimation, respectively, are exploited for cyclic cross-spectral

Table 5.1 Exploitation of Matlab/Octave functions for cyclic spectral analysis.

xcorr(y, xa)	= cross-correlation of y(1 : N) and xa(1 : N)
	= cyclic cross-correlogram of y(1 : N) and x(1 : N) at cycle frequency alpha
	= estimate of the **cyclic cross-correlation function** of $y(t)$ and $x(t)$ at cycle frequency $\alpha =$ alpha T_s
cpsd(y, xa)	= cross power spectral density of y(1 : N) and xa(1 : N)
	= time-smoothed cyclic cross-periodogram of y(1 : N) and x(1 : N) at cycle frequency alpha
	= estimate of the **cyclic cross-spectrum** of $y(t)$ and $x(t)$ at cycle frequency $\alpha =$ alpha T_s

analysis. The following Matlab/Octave vector is used:

$$\begin{aligned} \texttt{xa(1:N)} &= \texttt{x(1:N).*exp(1j*2*pi*alpha*(0:1:N-1))} \\ &= \text{frequency-shifted version of } \texttt{x(1:N)} \end{aligned}$$

The second argument of the functions xcorr and cpsd is conjugated within these functions. Thus, the cyclic cross-correlation function and the cyclic cross-spectrum of $y(t)$ and $x(t)$ at cycle frequency α can be estimated as explained in Table 5.1.

The conjugate cyclic cross-correlation and conjugate cyclic cross-spectrum at conjugate cycle frequency α are obtained by considering in Table 5.1

$$\texttt{xa(1:N)} = \texttt{conj(x(1:N)).*exp(1j*2*pi*alpha*(0:1:N-1))}$$

More input arguments can be considered for the functions in Table 5.1 for a more suitable use.

If the (conjugate) cyclic cross-correlogram Ryxa has to be computed in the lag interval (−taumax : taumax), and a correct normalization is needed, we have

```
Ryxa = xcorr(y, xa, taumax)/N
```

The function cpsd computes a time-smoothed cross-periodogram (Section 5.4.3). If blocks of length Ms that overlap in Ms/4 points are exploited and a centered spectrum is needed, we have that the time-smoothed (conjugate) cyclic cross-periodogram Syxa with frequency axis ff is given by

```
[Syxa, ff] = cpsd(y, xa, Ms, round(Ms/4), [], 1, 'centered')
```

5.4.5.2 Estimation for all cycle frequencies

In this section, efficient Matlab/Octave implementations of estimators for $\widetilde{\alpha} = q/N$, $q = -N/2, \dots, N/2-1$, of the (conjugate) cyclic cross-correlation function and the (conjugate) cyclic cross-spectrum are considered.

A Matlab/Octave implementation of the cyclic cross-correlogram R is presented in Table 5.2. For the conjugate cyclic cross-correlogram, in Table 5.2 vector conj(x) must be replaced with x.

Table 5.2 Matlab/Octave implementation of the cyclic cross-correlogram.

```
R=zeros(N,2*taumax+1);
for tau=-taumax:taumax
      R(1:N, tau + taumax + 1) = ...
         fftshift(fft(wshift(1, y, tau). * conj(x)))/N;
end
```

Table 5.3 Matlab/Octave implementation of the SSCA algorithm (Section 5.4.4).

```
H = @(t) 1.0 * (abs(t) <= 1/2);
f = (-N/2 : N/2 - 1)/N;
f1 = (-M/2 : M/2 - 1)/M;
Yf = fftshift(fft(y));
S = zeros(N, M);
for k = 1 : M
      YMt = ifft(ifftshift(Yf. * H((f - f1(k)) * M) * M));
      S(1 : N, k) = fftshift(fft(YMt. * conj(x)))/N;
end
```

A Matlab/Octave implementation of the SSCA algorithm is presented in Table 5.3. Specifically, expression (5.127) for $S_{yx^*}^{(N,\Delta\nu)}(\widetilde{\alpha}, \nu)$, say S, is computed for spectral frequencies $\nu = k/M$, $k = -M/2, \ldots, M/2-1$, and cycle frequencies $\widetilde{\alpha} = q/N$, $q = -N/2 :, \ldots, N/2-1$. For the conjugate cyclic cross-spectrum estimate, in Table 5.3 vector conj(x) must be replaced with x.

5.4.6 Numerical results

In this section, numerical results are presented to illustrate the properties of the cyclic correlogram and the frequency-smoothed cyclic periodogram.

The analyzed signal $x(t)$ is a PAM signal with binary stationary white modulating sequence, full duty-cycle rectangular pulse and pulse period $T_p = 10T_s$, where $T_s = 1/f_s$ is the sampling period.

The analytic expressions of the cyclic autocorrelation and the cyclic spectrum are given by (7.19) and (7.20), respectively, with $r_{qq}^{\alpha}(\tau)$ in (7.21) and $T_0 = T_w = T_p$.

5.4.6.1 Cyclic correlogram

The cyclic correlogram $R_{xx^*}^{(T)}(\alpha, \tau)$ is computed using a rectangular data-tapering window for two values of the data-record length $T = NT_s$. In Fig. 5.3, the magnitude of $R_{xx^*}^{(T)}(\alpha, \tau)$ as function of α/f_s and τ/T_s is reported for (top) $N = 2^{10}$ and (bottom) $N = 2^{16}$.

Sample statistics are evaluated by 400 Monte Carlo trials. In Figs. 5.4 ($N = 2^{10}$) and 5.5 ($N = 2^{16}$), (top) magnitude and (bottom) phase of the cyclic correlogram $R_{xx^*}^{(T)}(\alpha, \tau)$ at cycle frequency $\alpha = 1/T_p$ are reported as function of τ/T_s. Thick line: sample mean; shaded

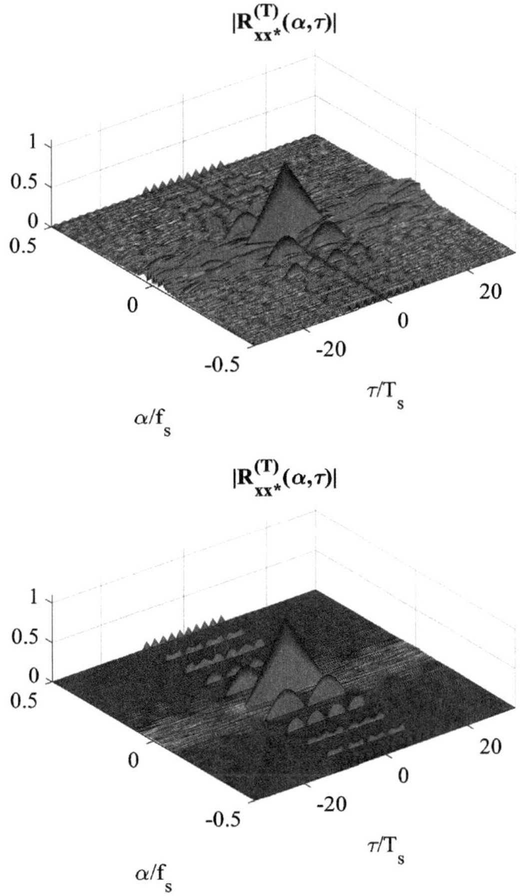

FIGURE 5.3 Magnitude of the cyclic correlogram $R^{(T)}_{xx^*}(\alpha, \tau)$ as function of α/f_s and τ/T_s. $T = NT_s$ with (top) $N = 2^{10}$ and (bottom) $N = 2^{16}$.

area: region within one sample standard deviation; thin line: numerically evaluated analytic expression of the cyclic autocorrelation. The reduction of bias and variance when N is increased confirms the mean-square consistence of the cyclic correlogram (Theorem 5.15).

The normalized cyclic correlogram $V^{(T)}(\alpha, \tau)$ of $x(t)$ defined according to (5.37) for $\alpha = 1/T_p$ as function of τ/T_s is reported in Figs. 5.6 ($N = 2^{10}$) and 5.7 ($N = 2^{16}$). The similar behavior and the same order of magnitude of the two plots for $V^{(T)}(\alpha, \tau)$ confirm the asymptotic rate of convergence proved in Theorem 5.18.

5.4.6.2 Frequency-smoothed cyclic periodogram

The frequency-smoothed cyclic periodogram $S^{(T,\Delta f)}_{xx^*}(\alpha, f)$ is computed using a rectangular data-tapering window and a triangular frequency-smoothing window for three pairs of

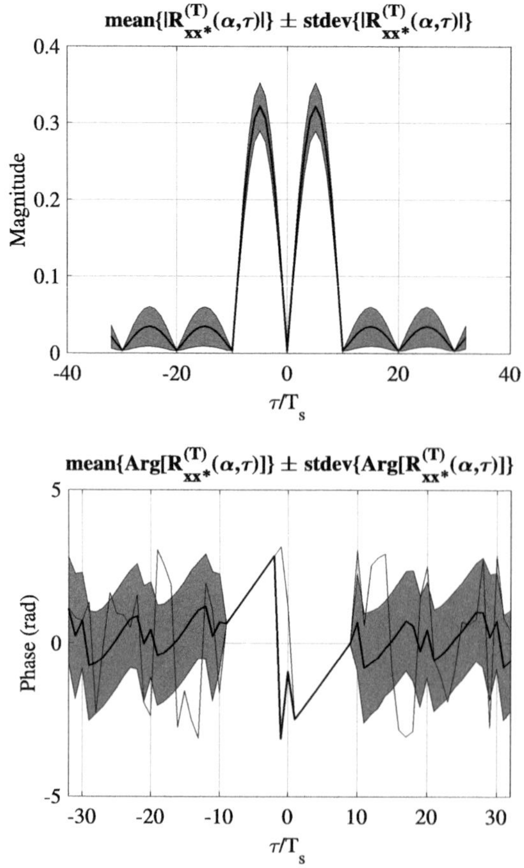

FIGURE 5.4 (Top) magnitude and (bottom) phase of the cyclic correlogram $R^{(T)}_{xx^*}(\alpha,\tau)$ as function of τ/T_s at cycle frequency $\alpha = 1/T_p$. Thick line: sample mean; shaded area: region within one sample standard deviation; thin line: numerically evaluated analytic expression of the cyclic autocorrelation. $T = NT_s$ with $N = 2^{10}$.

values $(T, \Delta f)$ of data-record length $T = NT_s$ and width Δf of the frequency-smoothing window.

The magnitude of $S^{(T,\Delta f)}_{xx^*}(\alpha, f)$ as function of α/f_s and f/f_s is reported for $N = 2^{10}$ and $\Delta f = f_s/2^4$ (Fig. 5.8), $N = 2^{10}$ and $\Delta f = f_s/2^6$ (Fig. 5.9), and $N = 2^{16}$ and $\Delta f = f_s/2^8$ (Fig. 5.10).

Sample statistics are evaluated by 400 Monte Carlo trials. In Figs. 5.11 ($N = 2^{10}$ and $\Delta f = f_s/2^4$), 5.12 ($N = 2^{10}$ and $\Delta f = f_s/2^6$), and 5.13 ($N = 2^{16}$ and $\Delta f = f_s/2^8$), (top) magnitude in dB and (bottom) phase of the frequency-smoothed cyclic periodogram $S^{(T,\Delta f)}_{xx^*}(\alpha, f)$ at cycle frequency $\alpha = 1/T_p$ are reported as functions of f/f_s. Thick line: sample mean; shaded area: region within one sample standard deviation; thin line: numerically evaluated analytic expression of the cyclic spectrum.

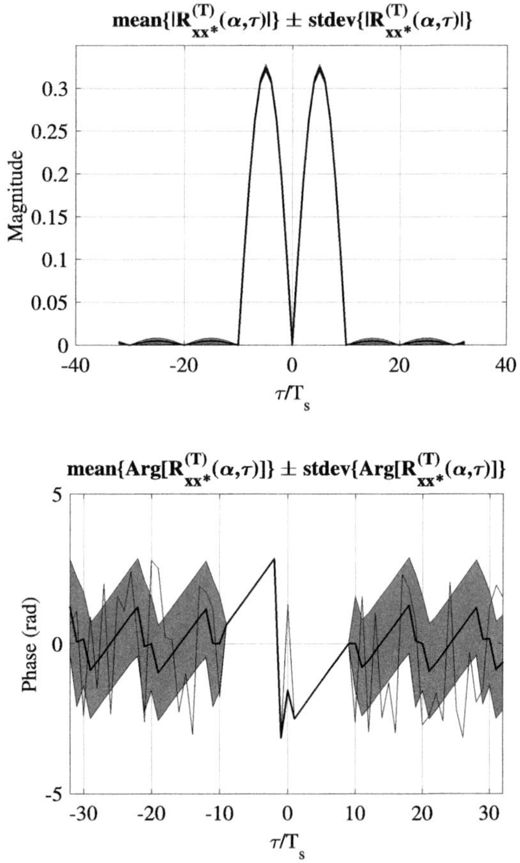

FIGURE 5.5 (Top) magnitude and (bottom) phase of the cyclic correlogram $R_{xx^*}^{(T)}(\alpha, \tau)$ as function of τ/T_s at cycle frequency $\alpha = 1/T_p$. Thick line: sample mean; shaded area: region within one sample standard deviation; thin line: numerically evaluated analytic expression of the cyclic autocorrelation. $T = NT_s$ with $N = 2^{16}$.

In Fig. 5.11, we have that the width Δf of the frequency smoothing window is large. Thus, according to (5.78), the variance of the estimator is small (the extension of the shaded area is small). However, according to (5.69), the sample mean (thick line) is smeared and is not able to capture all the variations of the true cyclic spectrum (thin line). If the width of Δf is reduced (Fig. 5.12), then the sample mean (thick line) fits better the true cyclic spectrum (thin line). However, the variance (extension of the shaded area) increases. This behavior is due to the bias-variance trade-off in spectral analysis. A better estimate is obtained by increasing the data-record length T and reducing Δf (Fig. 5.13) in accordance with the mean-square consistence proved in Theorem 5.39.

The normalized frequency-smoothed cyclic periodogram $U^{(T,\Delta f)}(\alpha, f)$ of $x(t)$ defined according to (5.82) for $\alpha = 1/T_p$ as function of f/f_s is reported in Figs. 5.14 ($N = 2^{10}$ and $\Delta f = f_s/2^4$), 5.15 ($N = 2^{10}$ and $\Delta f = f_s/2^6$), and 5.16 ($N = 2^{16}$ and $\Delta f = f_s/2^8$). The similar

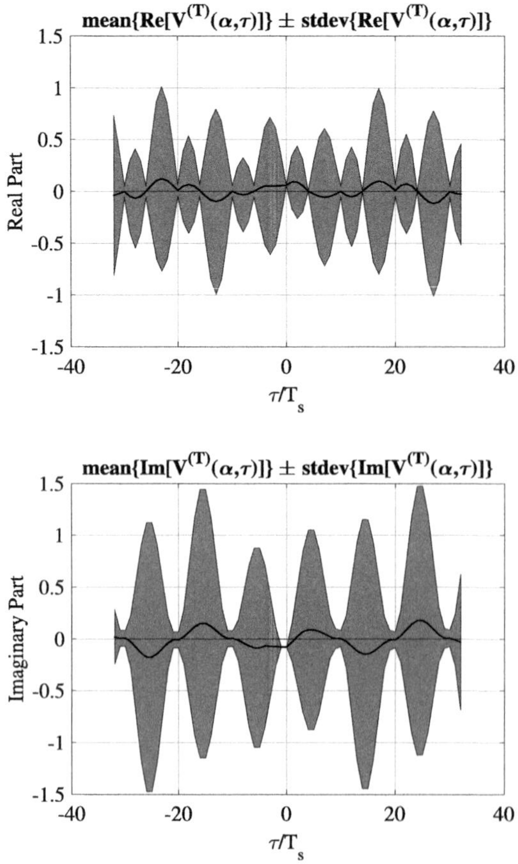

FIGURE 5.6 $V^{(T)}(\alpha, \tau)$ for $\alpha = 1/T_p$ as function of τ/T_s. (Top) real part and (bottom) imaginary part (thick line: sample mean, shaded area: region within one sample standard deviation). $T = NT_s$ with $N = 2^{10}$.

behavior and the same order of magnitude of the three plots for $U^{(T,\Delta f)}(\alpha, f)$ confirm the asymptotic rate of convergence proved in Theorem 5.41.

5.5 Estimators with estimated cycle frequencies

The consistency and asymptotic normality results of Sections 5.2.1, 5.2.3, and 5.2.4 are based on the implicit assumption that the cycle frequency is exactly known. Results on consistency and asymptotic normality of the cyclic correlogram and the frequency-smoothed cyclic periodogram with estimated cycle frequency are provided in (Napolitano, 2015a), (Napolitano, 2016c), (Napolitano, 2017a).

In the following, for the sake of simplicity, the case $y(t) \equiv x(t)$ is considered. In additions, we assume that the conditions such that $\mathcal{E}_{(*)} = 1$ are satisfied (see (5.61)).

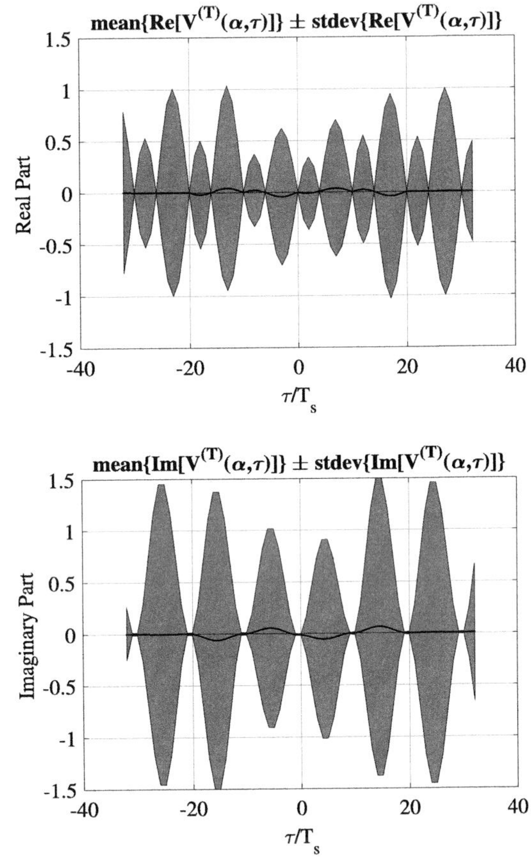

FIGURE 5.7 $V^{(T)}(\alpha, \tau)$ for $\alpha = 1/T_p$ as function of τ/T_s. (Top) real part and (bottom) imaginary part (thick line: sample mean, shaded area: region within one sample standard deviation). $T = NT_s$ with $N = 2^{16}$.

5.5.1 Cyclic correlogram with estimated cycle frequency

The function of two variables $(\alpha, \tau) \mapsto R_{\boldsymbol{x}}^{\alpha}(\tau)$ has support contained in lines parallel to the τ axis with equations $\alpha = \alpha_k$, where α_k are the (conjugate) cycle frequencies of $x(t)$. In contrast, the function $(\alpha, \tau) \mapsto R_{\boldsymbol{x}}^{(T)}(\alpha, \tau)$ is nonzero a.e. due to the cycle leakage (Section 5.3.4) which is consequence of the finite data-record length T. The cycle frequency resolution is of the order of $1/T$. That is, in the (α, τ) plane, the (conjugate) cyclic correlogram $R_{\boldsymbol{x}}^{(T)}(\alpha, \tau)$ is significantly nonzero in strips of width $1/T$ around the lines $\alpha = \alpha_k$ and its magnitude, for every τ, does not peak exactly at $\alpha = \alpha_k$ (Napolitano, 2012, Sec. 2.4.1). Instead, for T sufficiently large, it peaks in correspondence of a value of α close to α_k and that depends on τ. Techniques to obtain cycle frequency estimates not depending on τ (or not depending on the spectral frequency f) are described in Section 8.10.

In this section, the (conjugate) cyclic correlogram with estimated (conjugate) cycle frequency $\alpha^{(T)}$ is considered as estimator of the (conjugate) cyclic autocorrelation func-

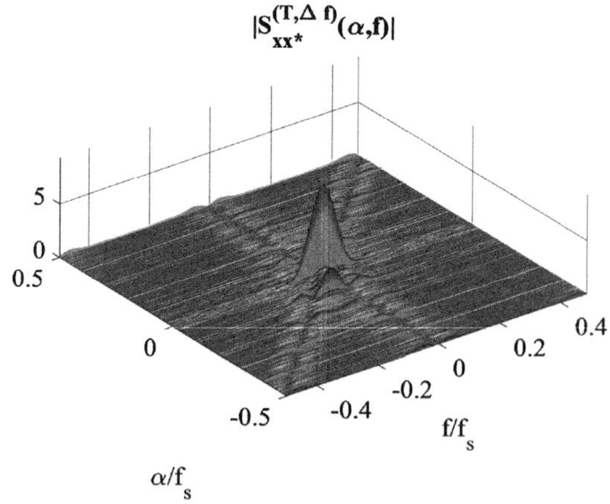

FIGURE 5.8 Magnitude of the frequency-smoothed cyclic periodogram $S_{xx^*}^{(T,\Delta f)}(\alpha, f)$ as function of α/f_s and f/f_s. $T = NT_s$ with $N = 2^{10}$ and $\Delta f = f_s/2^4$.

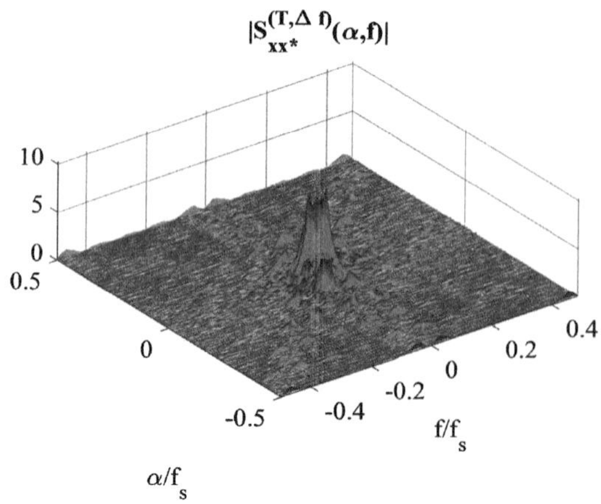

FIGURE 5.9 Magnitude of the frequency-smoothed cyclic periodogram $S_{xx^*}^{(T,\Delta f)}(\alpha, f)$ as function of α/f_s and f/f_s. $T = NT_s$ with $N = 2^{10}$ and $\Delta f = f_s/2^6$.

tion when the true (conjugate) cycle frequency α_0 is unknown. It is obtained by replacing in (5.9) with $y \equiv x$ the (conjugate) cycle frequency α with the estimate $\alpha^{(T)}$. That is,

$$R_x^{(T)}(\alpha^{(T)}, \tau) \triangleq \int_{\mathbb{R}} w_T(t - t_0)\, x(t + \tau)\, x^{(*)}(t)\, e^{-j2\pi\alpha^{(T)}t} \, dt \, . \tag{5.128}$$

Assumption 5.53. *The stochastic process $x(t)$ is such that*

$$\limsup_{T\to\infty} \frac{1}{T} \int_{-T/2}^{T/2} |x(t + \tau)\, x(t)| \, dt = M_2 < \infty \quad a.s. \tag{5.129}$$

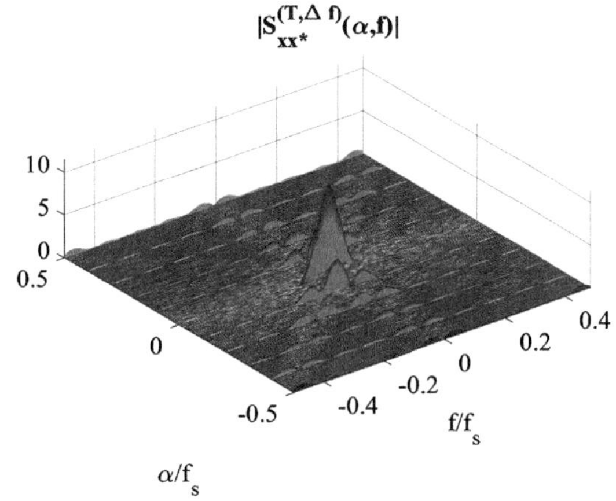

FIGURE 5.10 Magnitude of the frequency-smoothed cyclic periodogram $S_{xx^*}^{(T,\Delta f)}(\alpha, f)$ as function of α/f_s and f/f_s. $T = NT_s$ with $N = 2^{16}$ and $\Delta f = f_s/2^7$.

Assumption 5.53 is verified if $x(t;\omega)$ is a.s. bounded (uniformly w.r.t. ω) that is, if $|x(t)| \leqslant M$ a.s.

Let $\alpha^{(T)}$ be an estimate of the (conjugate) cycle frequency α_0.

Assumption 5.54. *The estimate $\alpha^{(T)}$ of the (conjugate) cycle frequency α_0 converges* almost surely (a.s.) *to α_0 with rate T^{-b}, with $b \geqslant 0$. That is,*

$$T^b\, |\alpha^{(T)} - \alpha_0| \to 0 \quad \textit{almost surely as} \ \ T \to \infty\,. \tag{5.130}$$

□

Obviously, if Assumption 5.54 holds for $b = b'$, then it also holds for every $0 \leqslant b < b'$. For example, the estimate of (Ciblat et al., 2002a), (Ciblat et al., 2002b), (see Section 8.10), is such that Assumption 5.54 is satisfied with $0 \leqslant b \leqslant 1$.

Theorem 5.55. (Napolitano, 2017a, Theorem 10). *Under Assumptions*

5.3 (uniformly almost-periodic statistics),
5.4 (Fourier series regularity),
5.5 (fourth-order moment boundedness),
5.6 (data-tapering window regularity),
5.9 (mixing conditions (I)),
5.53 (almost surely boundedness),
5.54 with $b = 1$ (almost sure convergence of cycle frequency estimate),

the (conjugate) cyclic correlogram with estimated (conjugate) cycle frequency $R_{\boldsymbol{x}}^{(T)}(\alpha^{(T)}, \tau)$ is a mean-square consistent estimator of the (conjugate) cyclic autocorrelation function $R_{\boldsymbol{x}}^{\alpha_0}(\tau)$.

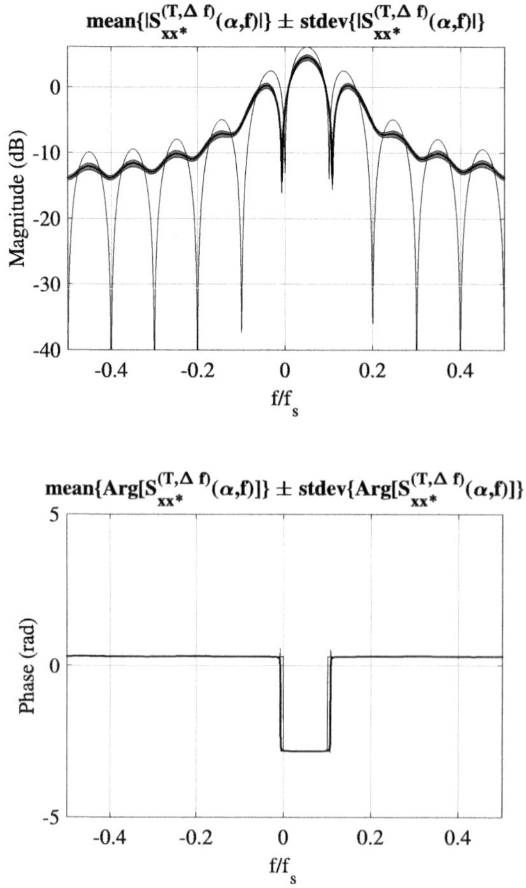

FIGURE 5.11 (Top) magnitude in dB and (bottom) phase of the frequency-smoothed cyclic periodogram $S_{xx^*}^{(T,\Delta f)}(\alpha, f)$ as function of f/f_s at cycle frequency $\alpha = 1/T_p$. Thick line: sample mean; shaded area: region within one sample standard deviation; thin line: numerically evaluated analytic expression of the cyclic spectrum. $T = NT_s$ with $N = 2^{10}$ and $\Delta f = f_s/2^4$.

Proof. See (Napolitano, 2017a). □

Assumption 5.56. *The estimation error*

$$\Delta\alpha^{(T)} \triangleq \alpha^{(T)} - \alpha_0 \tag{5.131}$$

is such that for every finite T

$$\mathrm{E}\{\Delta\alpha^{(T)}\} = 0 \tag{5.132}$$

and there exist a sequence $\{c_n\}$ *with* $c_n > 0$ *and a* $\mu > 0$ *such that as* $T \to \infty$

$$\left|\mathrm{E}\{(\Delta\alpha^{(T)})^n\}\right| \leqslant \frac{c_n}{T^{n(1+\mu)}} \tag{5.133}$$

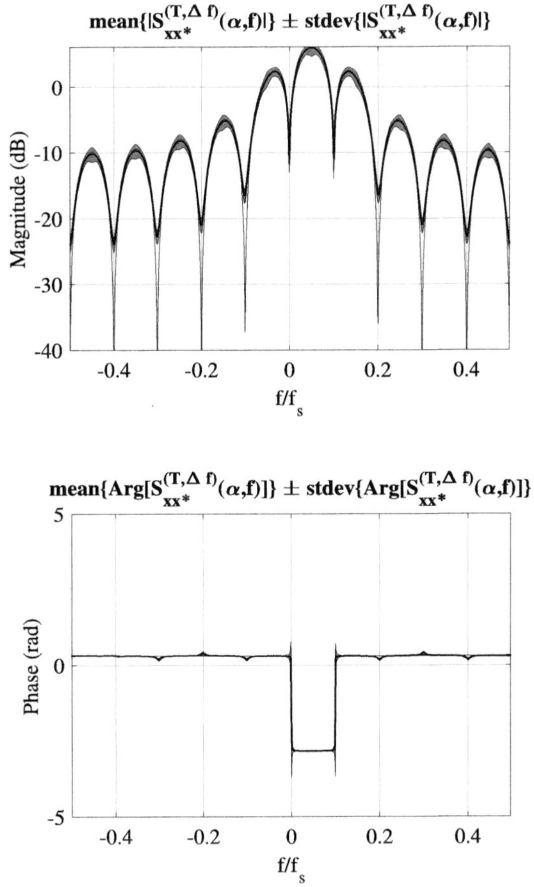

FIGURE 5.12 (Top) magnitude in dB and (bottom) phase of the frequency-smoothed cyclic periodogram $S_{xx^*}^{(T,\Delta f)}(\alpha, f)$ as function of f/f_s at cycle frequency $\alpha = 1/T_p$. Thick line: sample mean; shaded area: region within one sample standard deviation; thin line: numerically evaluated analytic expression of the cyclic spectrum. $T = NT_s$ with $N = 2^{10}$ and $\Delta f = f_s/2^6$.

and

$$\sum_{n=1}^{+\infty} \frac{a_0^n}{n!} c_n < \infty \tag{5.134}$$

for some $a_0 > 0$. □

A sufficient condition such that (5.132) is verified is that $\Delta\alpha^{(T)}$ has symmetric probability density function for finite T. In such a case, all the odd moments $\mathrm{E}\{(\Delta\alpha^{(T)})^{2k+1}\}$ are zero. In (Napolitano, 2017a), it is shown that the assumption (5.132) is not needed for the proofs of Theorems 5.58 and 5.63 if $a(t)$ is an even function.

Assumption 5.57. *Let* $\alpha_i^{(T)}$ *be estimates of M different (conjugate) cycle frequencies* $\alpha_{0,i}$, $i = 1, \ldots, M$ *such that*

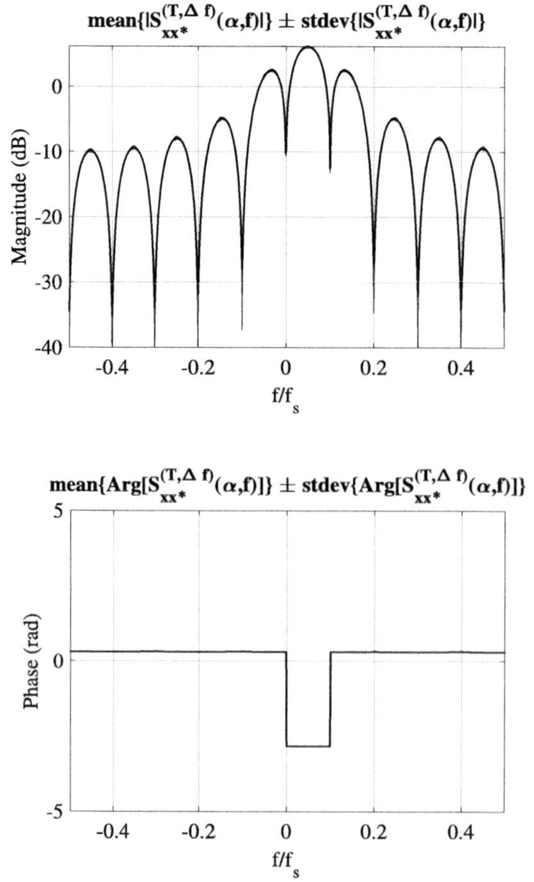

FIGURE 5.13 (Top) magnitude in dB and (bottom) phase of the frequency-smoothed cyclic periodogram $S_{xx^*}^{(T,\Delta f)}(\alpha, f)$ as function of f/f_s at cycle frequency $\alpha = 1/T_p$. Thick line: sample mean; shaded area: region within one sample standard deviation; thin line: numerically evaluated analytic expression of the cyclic spectrum. $T = NT_s$ with $N = 2^{16}$ and $\Delta f = f_s/2^8$.

$$\lim_{T\to\infty} \alpha_i^{(T)} = \alpha_{0,i} \qquad \textit{almost surely} \tag{5.135}$$

where estimates $\alpha_i^{(T)}$ *are made starting from (conjugate) cyclic correlograms* $R_{\boldsymbol{x}}^{(T)}(\alpha, \tau)$, $\alpha \in D_i$, *in non-overlapping cycle frequency intervals* D_i, *that is,* $D_i \bigcap D_j = \varnothing$ *for* $i \neq j$. □

Assumption 5.57 is verified, for example by the estimators (Section 8.10)

$$\alpha_i^{(T)} = \arg\max_{\alpha \in D_i} \int_{-\tau_M}^{\tau_M} \left| R_{\boldsymbol{x}}^{(T)}(\alpha, \tau) \right|^2 \mathrm{d}\tau \tag{5.136}$$

with non-overlapping search intervals $D_i = D(\alpha_{0,i}, \delta\alpha_i)$.

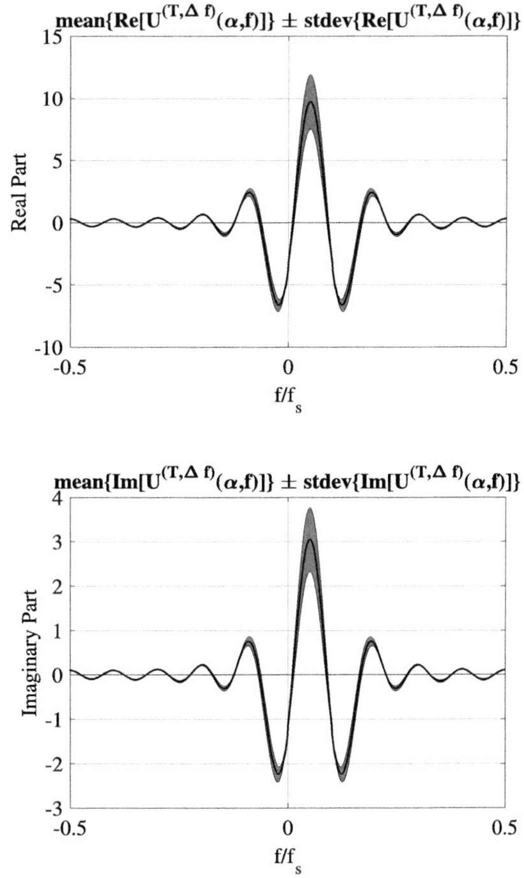

FIGURE 5.14 $U^{(T,\Delta f)}(\alpha, f)$ for $\alpha = 1/T_p$ as function of f/f_s. (Top) real part and (bottom) imaginary part (thick line: sample mean, shaded area: region within one sample standard deviation). $T = NT_s$ with $N = 2^{10}$ and $\Delta f = f_s/2^4$.

Theorem 5.58. (Napolitano, 2017a, Theorem 15). Asymptotic Joint Complex Normality of (Conjugate) Cyclic Correlograms with Estimated (Conjugate) Cycle Frequencies. *Under Assumptions*

5.3 (uniformly almost-periodic statistics),
5.4 (Fourier series regularity),
5.6 (data-tapering window regularity),
5.11 (lack of cycle frequency clusters),
5.16 (mixing conditions (III)),
5.17 (cross-moment boundedness),
5.56 (with $\mu > 1/3$) for every $\Delta\alpha_i^{(T)} = \alpha_i^{(T)} - \alpha_{0,i}$,
5.57 (almost sure convergence of cycle frequency estimates),

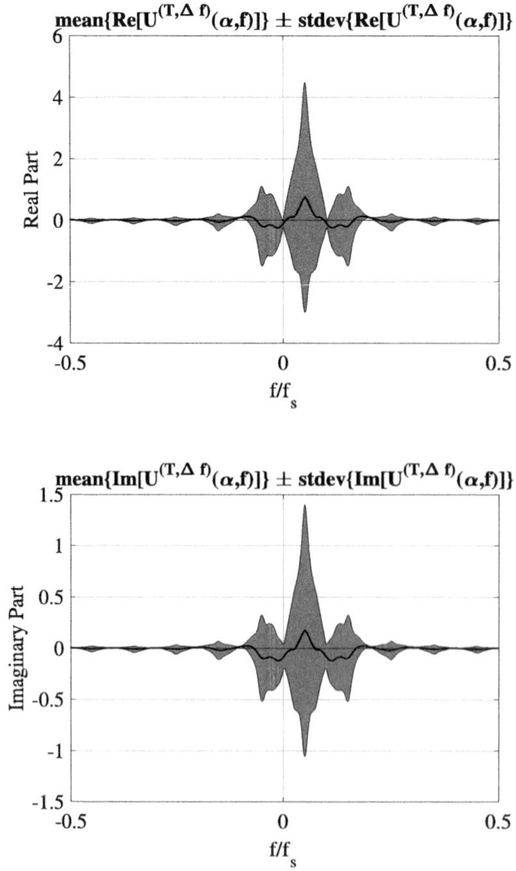

FIGURE 5.15 $U^{(T,\Delta f)}(\alpha, f)$ for $\alpha = 1/T_p$ as function of f/f_s. (Top) real part and (bottom) imaginary part (thick line: sample mean, shaded area: region within one sample standard deviation). $T = NT_s$ with $N = 2^{10}$ and $\Delta f = f_s/2^6$.

the (possibly conjugated) random variables

$$V_i^{(T)} \triangleq \sqrt{T}\Big[R_{\boldsymbol{x}}^{(T)}\big(\alpha_i^{(T)}, \tau_i\big) - R_{\boldsymbol{x}}^{\alpha_{0,i}}(\tau_i)\Big] \quad i = 1, \ldots, M \tag{5.137}$$

are asymptotically ($T \to \infty$) zero-mean jointly complex normal with asymptotic covariance matrix with entries Σ_{ij} and asymptotic conjugate covariance matrix with entries $\Sigma_{ij}^{(c)}$ having the same expressions as the case of known (conjugate) cycle frequencies (Theorem 5.18). See (5.29)–(5.32) (without conjugations in the subscripts of (5.30)–(5.32) for $\Sigma_{ij}^{(c)}$) and (Napolitano, 2012, eqs. (2.146), (2.152)–(2.154), and (3.137)).

Proof. See (Napolitano, 2017a). □

From Theorem 5.58 it follows that asymptotically the cyclic autocorrelation estimate with estimated cycle frequency has the same mean-squared error and asymptotic nor-

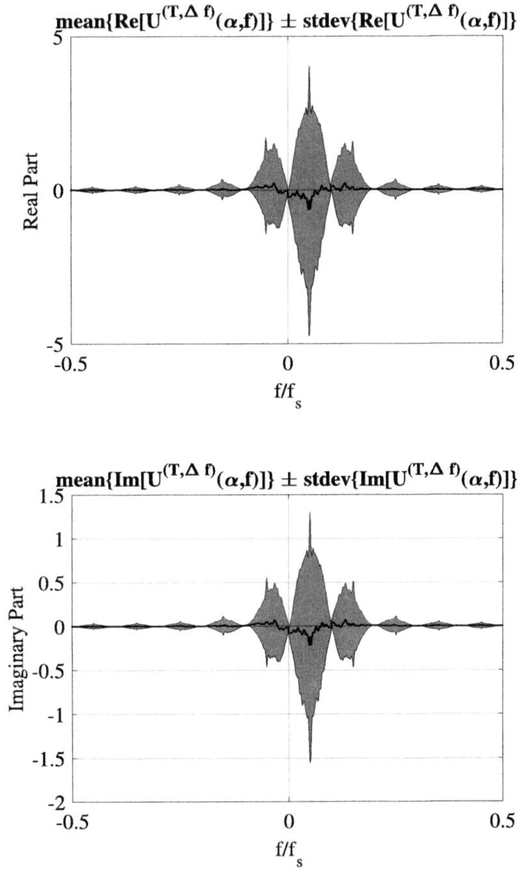

FIGURE 5.16 $U^{(T,\Delta f)}(\alpha, f)$ for $\alpha = 1/T_p$ as function of f/f_s. (Top) real part and (bottom) imaginary part (thick line: sample mean, shaded area: region within one sample standard deviation). $T = NT_s$ with $N = 2^{16}$ and $\Delta f = f_s/2^8$.

mal distribution as the cyclic autocorrelation estimate with known cycle frequency, provided that the cycle frequency estimate is sufficiently accurate, in the sense that Assumptions 5.54 and 5.56 are satisfied. An analogous result holds for the conjugate cyclic correlogram with estimated conjugate cycle frequency.

5.5.2 Frequency-smoothed cyclic periodogram with estimated cycle frequency

In this section, the frequency-smoothed (conjugate) cyclic periodogram with estimated (conjugate) cycle frequency

$$S_{\boldsymbol{x}}^{(T,\Delta f)}(\alpha^{(T)}, f) = I_{\boldsymbol{x}}^{(T)}(\alpha^{(T)}, f) \underset{f}{\otimes} \frac{1}{\Delta f} Q\left(\frac{f}{\Delta f}\right) \tag{5.138a}$$

$$= \int_{\mathbb{R}} R_{\boldsymbol{x}}^{(T)}\big(\alpha^{(T)}, \tau\big)\, q(\Delta f\ \tau)\, e^{-j2\pi f \tau}\, \mathrm{d}\tau \tag{5.138b}$$

is considered as estimator of the (conjugate) cyclic spectrum when the (conjugate) cycle frequency is unknown.

Assumption 5.59. Frequency-Smoothing Window. *The window $q(\tau)$ satisfies $q(\tau) \in L^1(\mathbb{R})$, has bounded first order derivative, its Fourier transform $Q(f)$ satisfies $Q(f) \in L^1(\mathbb{R})$, is continuous almost everywhere (a.e.), and is such that $q(0) = \int_{\mathbb{R}} Q(f)\,\mathrm{d}f = 1$.* □

Assumption 5.60. *It results $\tau\ R_{\boldsymbol{x}}^{\alpha}(\tau) \in L^1(\mathbb{R})$ and $\int_{|\tau|>T} \big|R_{\boldsymbol{x}}^{\alpha}(\tau)\big|\,\mathrm{d}\tau = \mathcal{O}(T^{-r})$ for some $r > 1/2$.* □

Fact 5.61. *The (conjugate) cyclic spectrum $S_{\boldsymbol{x}}^{\alpha}(f)$ is the Fourier transform of $R_{\boldsymbol{x}}^{\alpha}(\tau) \in L^1(\mathbb{R})$ (at least). Thus, $S_{\boldsymbol{x}}^{\alpha}(f)$ is bounded, continuous, and infinitesimal at $\pm\infty$ (Champeney, 1990, Chap. 8).*

Assumption 5.62. *The (conjugate) cyclic spectra $S_{\boldsymbol{x}}^{\alpha}(f)$ are summable.* □

Theorem 5.63. (Napolitano, 2017a, Theorem 17). Asymptotic Joint Complex Normality of Frequency-Smoothed (Conjugate) Cyclic Periodograms with Estimated (Conjugate) Cycle Frequencies. *Under Assumptions*

5.3 (uniformly almost-periodic statistics),
5.4 (Fourier series regularity),
5.6 (data-tapering window regularity),
5.11 (lack of cycle frequency clusters),
5.16 (mixing conditions (III)),
5.17 (cross-moment boundedness),
5.59 (frequency-smoothing window regularity),
5.62 ((conjugate) cyclic spectra summability),
5.56 (with $\mu > 1/3$) for every $\Delta\alpha_i^{(T)} = \alpha_i^{(T)} - \alpha_{0,i}$,
5.57 (almost sure convergence of cycle frequency estimates),
5.60 (summability conditions on $R_{\boldsymbol{x}}^{\alpha}(\tau)$),

and provided that $\Delta f = \Delta f_T = T^{-s}$ with $1/3 < s < 1$ and $s/2 + r > 1/2$, the (possibly conjugated) random variables

$$U_i^{(T,\Delta f)} \triangleq \sqrt{T\Delta f}\Big[S_{\boldsymbol{x}}^{(T,\Delta f)}\big(\alpha_i^{(T)}, f_i\big) - S_{\boldsymbol{x}}^{\alpha_{0,i}}(f_i)\Big] \quad i = 1, \ldots, M \tag{5.139}$$

are asymptotically ($T \to \infty$ and $\Delta f \to 0$ with $T\Delta f \to \infty$) zero-mean jointly complex normal with asymptotic covariance matrix with entries Σ_{ij} and asymptotic conjugate covariance matrix with entries $\Sigma_{ij}^{(c)}$, having the same expressions as the case of known (conjugate) cycle frequencies. See (5.73) and (Napolitano, 2012, eq. (4.143)) for the expression of the covariance.

Proof. See (Napolitano, 2017a). □

As a corollary, we also have the mean-square consistency of the frequency-smoothed (conjugate) cyclic periodogram with estimated (conjugate) cycle frequency. In particular, the frequency-smoothed (conjugate) cyclic periodogram is asymptotically unbiased with rate of decay to zero of the bias $o((T\Delta f)^{-1/2})$ and variance $\mathcal{O}((T\Delta f)^{-1})$.

5.6 Statistical function estimators in the functional approach

FOT probabilistic parameters of a time series $\{x(u), u \in \mathbb{R}\}$ are expressed as infinite-time averages of functions of possibly time and frequency shifted versions of this time series. Estimators of these FOT probabilistic parameters based on the observation $\{x(u), u \in [t - T/2, t + T/2]\}$ are obtained by considering finite-time averages over $[t - T/2, t + T/2]$ of the same quantities involved in the infinite-time averages (Gardner, 1987f, Chap. 15), (Brown, 1987). Thus, in both stochastic and FOT approaches, estimators of probabilistic parameters are expressed as time averages of over the finite time interval $[t - T/2, t + T/2]$ centered in t and with width T. In the stochastic approach, t is fixed (and generally assumed to be 0 or $T/2$) and the variability of the estimate is due to the sample path chosen for the estimation, that is, the estimate is a random variable (it depends on $\omega \in \Omega$). Under appropriate mixing assumptions, as $T \to \infty$ the estimate converges in some probabilistic sense to the probabilistic parameter of the process. In contrast, in the FOT approach, the variability of the estimate is due to t, the central point of the observation interval, when t ranges in a wider temporal interval, say $[-Z/2, Z/2]$, with $Z \gg T$ (Napolitano, 2012, Sec. 6.3.5). In the FOT approach, asymptotic results are expressed in terms of a double limit as $Z \to \infty$ (to average the variability with respect to t, the central point of the observation interval) and as $T \to \infty$ (to consider the whole time series), provided that $Z/T \to \infty$.

The performance of the estimator is expressed in terms of FOT bias and variance which are defined by using as expected value the almost-periodic component extraction operator acting on the time variable t viz., the central point of the finite-length time-series segment adopted for the estimation (Gardner and Brown, 1991), (Gardner et al., 2006). Thus, unlike the stochastic process framework where the variance accounts for fluctuations of the estimates over the ensemble of sample paths, in the FOT probability framework the variance accounts for the fluctuations of the estimates in the time parameter t, that is, the location on the real axis of the time segment adopted for the estimation. Mean-square consistency of the estimator means that both FOT bias and variance of the estimator approach zero as the length T of the observation interval approaches infinity. In such a case, estimates obtained by using different time segments asymptotically do not depend on the central point of the segment.

Given a time series $z(t)$ (such as a lag product of another time series with a fixed value of the lag parameter), let us define

$$z_\beta^{(T)}(t) \triangleq \frac{1}{T} \int_{t-T/2}^{t+T/2} z(u)\, e^{-j2\pi\beta u}\, \mathrm{d}u \tag{5.140}$$

to be an estimator of the parameter

$$z_\beta \triangleq \lim_{T\to\infty} \frac{1}{T} \int_{-T/2}^{T/2} z(u)\, e^{-j2\pi\beta u}\, du \,. \tag{5.141}$$

Let us assume that kth-order FOT cumulants of $z(t)$ and its time-shifted versions are summable. If $z(t)$ is the Nth-order lag product of a time series $x(t)$, then the kth-order cumulant of $z(t)$ and its time-shifted versions can be expressed in terms of cumulants up to order kN of $x(t)$ and its time shifted versions (Leonov and Shiryaev, 1959), (Brillinger, 1981, Sec. 2.3), (Napolitano, 2012, pp. 83, 84). Under such assumptions we have the convergence of the estimator in the temporal mean-square sense (t.m.s.s.) (Dehay et al., 2018b)

$$\lim_{T\to\infty} z_\beta^{(T)}(t) = z_\beta \quad \text{(t.m.s.s.)}\,, \quad \forall \beta \in \mathbb{R}\,; \tag{5.142}$$

that is,

$$\lim_{T\to\infty} \left\langle \left| z_\beta^{(T)}(t) - z_\beta \right|^2 \right\rangle_t = 0\,, \quad \forall \beta \in \mathbb{R}\,. \tag{5.143}$$

Moreover, the time series

$$t \mapsto \sqrt{T}\left[z_\beta^{(T)}(t+\tau_1) - z_\beta\right] \quad \dots \quad t \mapsto \sqrt{T}\left[z_\beta^{(T)}(t+\tau_M) - z_\beta\right] \tag{5.144}$$

are asymptotically jointly complex normal (Dehay et al., 2018b).

The t.m.s.s. convergence is a more stringent condition w.r.t. the convergence in B^2 (Section B.4.1). It can be shown that, if the time series $z(t)$ has finite-average-power (i.e., $\langle |z(t)|^2\rangle_t < \infty$), then the set $B \triangleq \{\beta \in \mathbb{R} : z_\beta \neq 0\}$ is countable, the series $\sum_{\beta\in B} |z_\beta|^2$ is summable (Brown, 1987) and, accounting for (5.142), it follows that

$$\lim_{T\to\infty} \sum_{\beta\in B} z_\beta^{(T)}(t)\, e^{j2\pi\beta t} = \sum_{\beta\in B} z_\beta\, e^{j2\pi\beta t} \quad \text{(t.m.s.s.)}\,. \tag{5.145}$$

The magnitude and phase of z_β are the amplitude and phase of the finite-strength additive complex sine wave with frequency β contained in the time series $z(t)$. Moreover, the right-hand side in (5.145) is just the almost-periodic component contained in the time series $z(t)$.

The function $z_\beta^{(T)}(t)$ is an estimator of z_β based on the observation $\{z(u),\ u \in [t-T/2, t+T/2]\}$. It is worthwhile to emphasize that, in the FOT probability framework, probabilistic functions are defined in terms of the almost-periodic component extraction operation, which plays the same role as that played by the statistical expectation operation in the stochastic process framework (Section 2.3). Therefore,

$$\begin{aligned} \text{bias}\left\{z_\beta^{(T)}(t)\right\} &\triangleq \mathrm{E}^{\{\alpha\}}\left\{z_\beta^{(T)}(t)\right\} - z_\beta \\ &\simeq \left\langle z_\beta^{(T)}(t)\right\rangle_t - z_\beta \end{aligned} \tag{5.146}$$

$$\operatorname{var}\left\{z_\beta^{(T)}(t)\right\} \triangleq \mathrm{E}^{\{\alpha\}}\left\{\left|z_\beta^{(T)}(t) - \mathrm{E}^{\{\alpha\}}\left\{z_\beta^{(T)}(t)\right\}\right|^2\right\}$$
$$\simeq \left\langle\left|z_\beta^{(T)}(t) - \left\langle z_\beta^{(T)}(t)\right\rangle_t\right|^2\right\rangle_t \tag{5.147}$$

where $\mathrm{E}^{\{\alpha\}}\{\cdot\}$ extracts the almost-periodic component of the function $t \mapsto z_\beta^{(T)}(t)$ (Definition 2.14). Approximations in (5.146) and (5.147) become exact equalities in the limit as $T \to \infty$. Therefore, the fact that the estimator asymptotically approaches the true value (the infinite-time average) in the t.m.s.s. (see (5.143)) means that the estimator is mean-square consistent in the FOT probability sense. In fact, from (5.146) and (5.147) it follows that

$$\left\langle\left|z_\beta^{(T)}(t) - z_\beta\right|^2\right\rangle_t \simeq \operatorname{var}\left\{z_\beta^{(T)}(t)\right\} + \left|\operatorname{bias}\left\{z_\beta^{(T)}(t)\right\}\right|^2 \tag{5.148}$$

and this approximation becomes exact as $T \to \infty$. In such a case, estimates obtained by using different time segments asymptotically do not depend on the central point of the segment.

In (Dehay et al., 2018b, Sec. 7), the asymptotic joint complex normality of the functions

$$t \mapsto V_i^{(T)}(\alpha_i, \tau_i; t) \triangleq \sqrt{T}\left[R_{xx^{(*)}}^{(T)}(\alpha_i, \tau_i; t) - R_{xx^{(*)}}^{\alpha_i}(\tau_i)\right] \qquad (i = 1, 2, \ldots, M) \tag{5.149}$$

is proved for a signal $x(t)$ possibly containing an additive almost-periodic component. Such a result is more general than its stochastic counterpart (Theorem 5.18) where the signal is assumed to be zero mean. Note that the dependence on t in the (conjugate) cyclic cross-correlogram is evidenced in (5.149) as in Definition 5.1.

5.7 Higher-order cyclic statistic estimators

In Chapter 4, higher order cyclic statistics are treated mostly in the FOT probability framework and results are interpreted in the classical stochastic approach in Section 4.8.

In this section, several estimators of higher-order cyclic statistics in both time and frequency domain are reviewed and presented in the classical stochastic framework, according to the approach adopted in the original sources. These results can be immediately interpreted in the FOT approach as explained in Section 5.6 where it is shown that estimators of FOT statistical functions are the same as those of their stochastic counterparts.

Estimators of temporal and spectral cyclic moments are obtained by extending to higher-order dimensions the estimators for second-order cyclic statistics. Estimators of temporal and spectral cyclic cumulants are derived from those of moments accounting for the relationships between moments and cumulants (Section 4.2.3) and, in particular, accounting for their differences on the α-submanifold (Section 4.2.3.1). Median filtering techniques are adopted for temporal and spectral cyclic cumulant estimates when the structure of the submanifold is unknown.

5.7.1 Estimators of time-domain statistics

In this section, estimators of higher-order cyclic statistical functions in the time domain are presented. In the notation, the dependence on t_0, the central point of the observation interval, is omitted.

Estimator of the Nth-order CTCMF (4.2): Nth-Order Cyclic Cross-Correlogram

$$\mathcal{R}_{\boldsymbol{x}}^{(T)}(\alpha, \boldsymbol{\tau}) = \frac{1}{T} \int_{t_0-T/2}^{t_0+T/2} \prod_{k=1}^{N} x_k^{(*)_k}(t+\tau_k)\, e^{-j2\pi\alpha t}\, \mathrm{d}t \tag{5.150}$$

Estimator of the Nth-order cyclic temporal joint characteristic function (4.31)

$$\Phi_{\boldsymbol{x}}^{(T)}(\gamma, \boldsymbol{\omega}, \boldsymbol{\tau}) = \frac{1}{T} \int_{t_0-T/2}^{t_0+T/2} \prod_{k=1}^{N} e^{j\omega_k x_k^{(*)_k}(t+\tau_k)}\, e^{-j2\pi\gamma t}\, \mathrm{d}t \tag{5.151}$$

Estimator of the Nth-order CTCCF (4.35) based on (4.51)

$$\mathcal{C}_{\boldsymbol{x}}^{(T)}(\beta, \boldsymbol{\tau}) = \mathcal{R}_{\boldsymbol{x}}^{(T)}(\beta, \boldsymbol{\tau}) + \sum_{\substack{\mathsf{P}\\ p\neq 1}} \left[(-1)^{p-1}(p-1)! \sum_{\boldsymbol{\alpha}^{\mathrm{T}}\mathbf{1}=\beta} \prod_{i=1}^{p} \mathcal{R}_{\boldsymbol{x}_{\mu_i}}^{(T)}(\alpha_{\mu_i}, \boldsymbol{\tau}_{\mu_i}) \right] \tag{5.152}$$

where, for every partition, the sum over $\boldsymbol{\alpha}$ is extended to a finite subset of the (possibly infinite) set of cycle frequencies $\{\alpha_{\mu_1}, \dots, \alpha_{\mu_p}\}$.

Estimator of the Nth-order RD-CTCMF (4.16): Nth-order Cyclic Cross-Correlogram

$$R_{\boldsymbol{x}}^{(T)}(\alpha, \boldsymbol{\tau}') = \mathcal{R}_{\boldsymbol{x}}^{(T)}(\alpha, \boldsymbol{\tau})\big|_{\tau_N=0} \tag{5.153}$$

Estimator of the Nth-order RD-CTCCF (4.40)

$$C_{\boldsymbol{x}}^{(T)}(\beta, \boldsymbol{\tau}') = \mathcal{C}_{\boldsymbol{x}}^{(T)}(\beta, \boldsymbol{\tau})\big|_{\tau_N=0} \tag{5.154}$$

The following theorem extends to higher-order the results of Section 5.2.1 for the 2nd-order cyclic cross-correlogram. An analogous result is proved in (Dandawaté and Giannakis, 1995).

Theorem 5.64. (Napolitano and Tesauro, 2011, Theorems 4.1–4.4). *For zero-mean processes, under mild assumptions on data-tapering window regularity, Fourier series regularity of the almost-periodic statistics, moment boundedness, mixing conditions, and lack of cycle frequency clusters, it results that the normalized N_ith-order cyclic cross-correlograms (with possibly different orders N_i and pairs $(\alpha_i, \boldsymbol{\tau}_i)$)*

$$V_{\mathrm{mom}}^{(T)}(\alpha_i, \boldsymbol{\tau}_i, N_i) \triangleq \sqrt{T}\left[\mathcal{R}_{\boldsymbol{x}}^{(T)}(\alpha_i, \boldsymbol{\tau}_i) - \mathcal{R}_{\boldsymbol{x}}^{\alpha_i}(\boldsymbol{\tau}_i)\right] \tag{5.155}$$

are asymptotically ($T \to \infty$) zero-mean jointly complex normal with asymptotic covariance matrix and conjugate covariance matrices given in (Napolitano and Tesauro, 2011). □

In particular, from Theorem 5.64 it follows the mean-square consistence of the Nth-order cyclic cross-correlogram and the convergence in distribution

$$\sqrt{T}\left[\mathcal{R}_{\boldsymbol{x}}^{(T)}(\alpha, \boldsymbol{\tau}) - \mathcal{R}_{\boldsymbol{x}}^{\alpha}(\boldsymbol{\tau})\right] \to \mathcal{N}(0, \Sigma_{\text{mom}}, \Sigma_{\text{mom}}^{(c)}) \qquad \text{as } T \to \infty . \tag{5.156}$$

Since all the sums in (5.152) are finite, from Theorem 5.64 we have that $\mathcal{C}_{\boldsymbol{x}}^{(T)}(\beta, \boldsymbol{\tau})$ is the sum of a finite number of products of asymptotically jointly complex normal random variables. Despite of the presence of this nonlinear operation, the following result can be proved. An analogous result is found in (Dandawaté and Giannakis, 1995).

Theorem 5.65. (Napolitano and Tesauro, 2011, Theorems 5.1–5.4). *Under the assumptions of Theorem 5.64, we have that the normalized estimators (5.152) (with possibly different orders N_i and pairs $(\beta_i, \boldsymbol{\tau}_i)$)*

$$V_{\text{cum}}^{(T)}(\beta_i, \boldsymbol{\tau}_i, N_i) \triangleq \sqrt{T}\left[\mathcal{C}_{\boldsymbol{x}}^{(T)}(\beta_i, \boldsymbol{\tau}_i) - \mathcal{C}_{\boldsymbol{x}}^{\beta_i}(\boldsymbol{\tau}_i)\right] \tag{5.157}$$

are asymptotically ($T \to \infty$) zero-mean jointly complex normal with asymptotic covariance matrix and conjugate covariance matrices given in (Napolitano and Tesauro, 2011). □

In particular, from Theorem 5.65 it follows the mean-square consistence of the estimator (5.152) and the convergence in distribution

$$\sqrt{T}\left[\mathcal{C}_{\boldsymbol{x}}^{(T)}(\beta, \boldsymbol{\tau}) - \mathcal{C}_{\boldsymbol{x}}^{\beta}(\boldsymbol{\tau})\right] \to \mathcal{N}(0, \Sigma_{\text{cum}}, \Sigma_{\text{cum}}^{(c)}) \qquad \text{as } T \to \infty . \tag{5.158}$$

In (Napolitano and Tesauro, 2011), it is shown that the results of Theorems 5.64 and 5.65 are valid also in the case of processes exhibiting joint higher-order generalized cyclostationarity (Chapter 12), that is, with TCMFs and TCCFs given by the superposition of an almost periodic term whose frequencies depend on the lag parameters and a residual term not containing any finite-strength sine wave component, provided that the residual term satisfies some mild regularity conditions.

5.7.2 Estimators of frequency-domain statistics

In this section, estimators of higher-order cyclic statistical functions in the frequency domain are presented. In the notation, in most cases the dependence on the central point of the observation interval is omitted.

Nth-order Cyclic Cross-Periodogram

$$I_{\boldsymbol{x}}^{(T)}(\alpha, \boldsymbol{f}'; t) = \frac{1}{T} X_{N,T}^{(*)_N}(t, (-)_N(\alpha - \boldsymbol{f}'^{\top}\mathbf{1})) \prod_{k=1}^{N-1} X_{k,T}^{(*)_k}(t, (-)_k f_k) \tag{5.159}$$

where $X_{k,T}(t, f)$ is the STFT of $x_k(t)$ defined according to Definition 5.19.

Estimator of the Nth-order RD-CSCMF (4.15): Frequency-Smoothed Nth-Order Cyclic Cross-Periodogram

$$S_{\boldsymbol{x}}^{(T,\Delta f)}(\alpha, \boldsymbol{f}') = \int_{\mathbb{R}^{N-1}} R_{\boldsymbol{x}}^{(T)}(\alpha, \boldsymbol{\tau}')\, w_{\frac{1}{\Delta f}}(\boldsymbol{\tau}')\, e^{-j2\pi \boldsymbol{f}'^{\mathrm{T}}\boldsymbol{\tau}'}\, \mathrm{d}\boldsymbol{\tau}' \tag{5.160a}$$

$$= I_{\boldsymbol{x}}^{(T)}(\alpha, \boldsymbol{f}'; 0) \underset{\boldsymbol{f}'}{\otimes} W_{\Delta f}(\boldsymbol{f}') \tag{5.160b}$$

where $w_{\frac{1}{\Delta f}}(\boldsymbol{\tau}')$ is an $(N-1)$-dimensional window with width $1/\Delta f$ along each dimension,

$$W_{\Delta f}(\boldsymbol{f}') = \int_{\mathbb{R}^{N-1}} w_{\frac{1}{\Delta f}}(\boldsymbol{\tau}')\, e^{-j2\pi \boldsymbol{f}'^{\mathrm{T}}\boldsymbol{\tau}'}\, \mathrm{d}\boldsymbol{\tau}' \tag{5.161}$$

is a frequency-smoothing window with bandwidth Δf along each dimension and such that

$$\int_{\mathbb{R}^{N-1}} W_{\Delta f}(\boldsymbol{f}')\, \mathrm{d}\boldsymbol{f}' = 1 \qquad\qquad \lim_{\Delta f \to 0} W_{\Delta f}(\boldsymbol{f}') = \delta(\boldsymbol{f}') \tag{5.162}$$

with $\delta(\boldsymbol{f}')$ the $(N-1)$-dimensional Dirac delta.

Theorem 5.66. (Dandawaté and Giannakis, 1994a). *Under mild assumptions on data-tapering window regularity, Fourier series regularity of the almost-periodic statistics, frequency-smoothing window regularity, mixing conditions, and lack of cycle frequency clusters, we have that the normalized frequency-smoothed cyclic cross-periodograms (with possibly different orders N_i and pairs $(\alpha_i, \boldsymbol{f}'_i)$)*

$$U^{(T,\Delta f)}(\alpha_i, \boldsymbol{f}'_i, N_i) \triangleq \sqrt{T(\Delta f)^{N_i-1}}\left[S_{\boldsymbol{x}}^{(T,\Delta f)}(\alpha_i, \boldsymbol{f}'_i) - P_{\boldsymbol{x}}^{\alpha_i}(\boldsymbol{f}'_i)\right] \tag{5.163}$$

are asymptotically ($T \to \infty$, $\Delta f \to 0$, with $T(\Delta f)^{N_i-1} \to \infty$) zero-mean complex normal outside the α_i-submanifolds. □

In particular, from Theorem 5.66 it follows the mean-square consistency of the frequency-smoothed Nth-order cyclic cross-periodogram (5.160a)–(5.160b) with asymptotic bias $o(1/\sqrt{T(\Delta f)^{N-1}})$ and asymptotic variance $\mathcal{O}(1/(T(\Delta f)^{N-1}))$, and the convergence in distribution

$$\sqrt{T(\Delta f)^{N-1}}\left[S_{\boldsymbol{x}}^{(T,\Delta f)}(\alpha, \boldsymbol{f}') - P_{\boldsymbol{x}}^{\alpha}(\boldsymbol{f}')\right] \to \mathcal{N}(0, \Sigma_f, \Sigma_f^{(c)})$$
$$\text{as } T \to \infty, \Delta f \to 0, \text{ with } T(\Delta f)^{N-1} \to \infty\,. \tag{5.164}$$

Note that outside the α-submanifold it results $P_{\boldsymbol{x}}^{\alpha}(\boldsymbol{f}') = S_{\boldsymbol{x}}^{\alpha}(\boldsymbol{f}')$ (Section 4.2.3.1). However, for the estimators based on finite data-record measurements, leakage from the α-submanifold outside the α-submanifold is present (Napolitano and Spooner, 2000).

Estimator of the Nth-order RD-CSCMF (4.15): Time-Smoothed Nth-order Cyclic Cross-Periodogram

$$G_{\boldsymbol{x}}^{(\Delta f,T)}(\alpha, \boldsymbol{f}'; t) = I_{\boldsymbol{x}}^{(1/\Delta f)}(\alpha, \boldsymbol{f}'; t) \underset{t}{\otimes} a_T(t) \tag{5.165a}$$

$$= \frac{1}{T} \int_{t-T/2}^{t+T/2} I_{\boldsymbol{x}}^{(1/\Delta f)}(\alpha, \boldsymbol{f}'; s)\, w_a\left(\frac{t-s}{T}\right) \mathrm{d}s \tag{5.165b}$$

where $a_T(t) = \frac{1}{T} w_a\left(\frac{t}{T}\right)$ is an even time-smoothing window with unit area and support $[-T/2, T/2]$.

Estimator of the CCP (4.39)

$$P_{\boldsymbol{x}}^{(T,\Delta f)}(\beta, \boldsymbol{f}') = \int_{\mathbb{R}^{N-1}} C_{\boldsymbol{x}}^{(T)}(\beta, \boldsymbol{\tau}')\, w_{\frac{1}{\Delta f}}(\boldsymbol{\tau}')\, e^{-j2\pi \boldsymbol{f}'^{\mathrm{T}}\boldsymbol{\tau}'}\, \mathrm{d}\boldsymbol{\tau}' \tag{5.166}$$

where $C_{\boldsymbol{x}}^{(T)}(\beta, \boldsymbol{\tau}')$ is given in (5.154).

The results of Theorem 5.66 hold with $S_{\boldsymbol{x}}^{(T,\Delta f)}(\alpha_i, \boldsymbol{f}_i')$ replaced by $G_{\boldsymbol{x}}^{(\Delta f,T)}(\alpha_i, \boldsymbol{f}_i')$ or $P_{\boldsymbol{x}}^{(T,\Delta f)}(\alpha_i, \boldsymbol{f}_i')$.

5.7.3 Estimators based on median filtering

The lack of knowledge of the α-submanifold can be overcome by using median-filtering based estimators (Napolitano and Spooner, 2000) (Section 5.2.5).

Estimator of the CCP (4.39): Median-Filtered Frequency-Smoothed Cyclic Cross-Periodogram (Napolitano and Spooner, 2000)

$$P_{\boldsymbol{x}}^{(T,\Delta f,\delta f)}(\beta, \boldsymbol{f}') = \underset{\boldsymbol{\nu}' \in [\boldsymbol{f}' - \mathbf{1}\delta f/2,\, \boldsymbol{f}' + \mathbf{1}\delta f/2]}{\mathrm{med}} \left[I_{\boldsymbol{x}}^{(T)}(\beta, \boldsymbol{\nu}'; t) \underset{\boldsymbol{\nu}'}{\otimes} W_{\Delta f}(\boldsymbol{\nu}') \right] \tag{5.167}$$

$T \to \infty$, $\Delta f \to 0$, $\delta f \to 0$ in this order.

Estimator of the CCP (4.39): Median-Filtered Time-Smoothed Cyclic Cross-Periodogram (Napolitano and Spooner, 2000)

$$Q_{\boldsymbol{x}}^{(\Delta f,T,\delta f)}(\beta, \boldsymbol{f}') = \underset{\boldsymbol{\nu}' \in [\boldsymbol{f}' - \mathbf{1}\delta f/2,\, \boldsymbol{f}' + \mathbf{1}\delta f/2]}{\mathrm{med}} \left[\frac{1}{T} \int_{t-T/2}^{t+T/2} I_{\boldsymbol{x}}^{(1/\Delta f)}(\beta, \boldsymbol{\nu}'; s)\, \mathrm{d}s \right] \tag{5.168}$$

$T \to \infty$, $\Delta f \to 0$, $\delta f \to 0$ in this order.

Estimators of the RD-CTCCF (4.40):

$$C_{\boldsymbol{x}}^{(T,\Delta f,\delta f)}(\beta, \boldsymbol{\tau}') = \int_{\mathbb{R}^{N-1}} P_{\boldsymbol{x}}^{(T,\Delta f,\delta f)}(\beta, \boldsymbol{f}')\, e^{j2\pi \boldsymbol{f}'^{\mathrm{T}}\boldsymbol{\tau}'}\, \mathrm{d}\boldsymbol{f}' \tag{5.169}$$

where $P_{\boldsymbol{x}}^{(T,\Delta f,\delta f)}(\beta, \boldsymbol{f}')$ can be possibly replaced by $Q_{\boldsymbol{x}}^{(\Delta f,T,\delta f)}(\beta, \boldsymbol{f}')$ (in which case the notation $C_{\boldsymbol{x}}^{(\Delta f,T,\delta f)}(\beta, \boldsymbol{\tau}')$ is preferable).

5.7.4 Estimators of hybrid temporal-spectral statistics

An estimator of the *nth-order hybrid temporal-spectral cyclic cross-moment function* (4.130a)–(4.130b) is

$$
\begin{aligned}
& J_{\boldsymbol{x}}^{(T,\Delta f)}(\alpha, \boldsymbol{\tau}^{(n-1-p)}, \boldsymbol{f}^{(p)}) \\
& \quad \triangleq \int_{\mathbb{R}^p} R_{\boldsymbol{x}}^{(T)}(\alpha, \boldsymbol{\tau}')\, w_a(\boldsymbol{\tau}^{(p)}\, \Delta f)\, e^{-j2\pi[f_1\tau_1+\cdots+f_p\tau_p]}\, \mathrm{d}\tau_1 \cdots \mathrm{d}\tau_p && (5.170a) \\
& \quad = \int_{\mathbb{R}^{n-1-p}} S_{\boldsymbol{x}}^{(T,\Delta f)}(\alpha, \boldsymbol{f}')\, e^{j2\pi[f_{p+1}\tau_{p+1}+\cdots+f_{n-1}\tau_{n-1}]}\, \mathrm{d}f_{p+1} \cdots \mathrm{d}f_{n-1} && (5.170b)
\end{aligned}
$$

where $p \leqslant n-2$, $w_a(\boldsymbol{\tau}^{(p)}\, \Delta f)$ is a p-dimensional window with width $1/\Delta f$ along each dimension, and $S_{\boldsymbol{x}}^{(T,\Delta f)}(\alpha, \boldsymbol{f}')$ can be possibly replaced by $G_{\boldsymbol{x}}^{(\Delta f,T)}(\alpha, \boldsymbol{f}')$ (in which case the notation $J_{\boldsymbol{x}}^{(\Delta f,T)}(\alpha, \boldsymbol{\tau}^{(n-1-p)}, \boldsymbol{f}^{(p)})$ is preferable).

In Section 4.7, it is observed that the HTSCCMF $J_{\boldsymbol{x}}^{\alpha}(\boldsymbol{\tau}^{(n-1-p)}, \boldsymbol{f}^{(p)})$ is the $(p+1)$th-order RD-CSCMF of the $p+1$ signals

$$
x_1^{(*)_1}(t), \ldots, x_p^{(*)_p}(t), \text{ and } \chi(t; \tau_{p+1}, \ldots, \tau_{n-1}) \triangleq x_n^{(*)_n}(t) \prod_{i=p+1}^{n-1} x_i^{(*)_i}(t+\tau_i)\,.
$$

Thus, accounting for Theorem 5.66, for estimator $J_{\boldsymbol{x}}^{\alpha}(\boldsymbol{\tau}^{(n-1-p)}, \boldsymbol{f}^{(p)})$ the asymptotic bias is $o(1/\sqrt{T(\Delta f)^p})$, the asymptotic variance is $\mathcal{O}(1/(T(\Delta f)^p)))$, and mean-square consistency is obtained as $T \to \infty$ and $\Delta f \to 0$ with $T(\Delta f)^p \to \infty$.

Thus, for a given order of magnitude of bias and variance, a finer spectral resolution Δf can be obtained by estimating a hybrid temporal-spectral cyclic statistic with respect to the frequency-smoothed nth-order cyclic cross-periodogram for which the bias and variance are $o(1/\sqrt{T(\Delta f)^{n-1}})$ and $\mathcal{O}(1/(T(\Delta f)^{n-1})))$, respectively, and $T(\Delta f)^{n-1} \to \infty$ is needed. Alternatively, for a given spectral resolution, smaller bias and variance can be obtained.

Such a more convenient asymptotic behavior is exploited in Section 9.5.2.3 for LTI system identification.

An estimator of the *nth-order hybrid temporal-spectral cyclic cross-cumulant function* (4.131a)–(4.131b) is

$$
\begin{aligned}
& H_{\boldsymbol{x}}^{(T,\Delta f,\delta f)}(\beta, \boldsymbol{\tau}^{(n-1-p)}, \boldsymbol{f}^{(p)}) \\
& \quad \triangleq \int_{\mathbb{R}^p} C_{\boldsymbol{x}}^{(T,\Delta f,\delta f)}(\beta, \boldsymbol{\tau}')\, w_a(\boldsymbol{\tau}^{(p)}\, \Delta f)\, e^{-j2\pi[f_1\tau_1+\cdots+f_p\tau_p]}\, \mathrm{d}\tau_1 \cdots \mathrm{d}\tau_p && (5.171a) \\
& \quad = \int_{\mathbb{R}^{n-1-p}} P_{\boldsymbol{x}}^{(T,\Delta f,\delta f)}(\beta, \boldsymbol{f}')\, e^{j2\pi[f_{p+1}\tau_{p+1}+\cdots+f_{n-1}\tau_{n-1}]}\, \mathrm{d}f_{p+1} \cdots \mathrm{d}f_{n-1} && (5.171b)
\end{aligned}
$$

where $P_{\boldsymbol{x}}^{(T,\Delta f,\delta f)}(\beta, \boldsymbol{f}')$ can be possibly replaced by $Q_{\boldsymbol{x}}^{(\Delta f,T,\delta f)}(\beta, \boldsymbol{f}')$ (in which case the notation $H_{\boldsymbol{x}}^{(\Delta f,T,\delta f)}(\beta, \boldsymbol{\tau}^{(n-1-p)}, \boldsymbol{f}^{(p)})$ is preferable).

The cumulants of the signals $x_1^{(*)_1}(t), \ldots, x_p^{(*)_p}(t)$ and the lag product $\chi(t; \tau_{p+1}, \ldots, \tau_{n-1})$ can be expressed in terms of cumulants of the signals $x_1^{(*)_1}(t), \ldots, x_n^{(*)_n}(t)$ (Leonov and

Shiryaev, 1959), (Brillinger, 1965), (Brillinger and Rosenblatt, 1967), (Napolitano, 2012, pp. 83-84). Thus, the assumptions of summability of cumulants to prove the mean-square consistency and asymptotic complex normality of the estimators (5.170a), (5.170b), (5.171a), and (5.171b) can be expressed in terms of summability of cumulants of the signals $x_1^{(*)_1}(t), \dots, x_n^{(*)_n}(t)$.

5.7.5 Numerical results

In this section, numerical results are presented to illustrate and discuss the cyclic higher-order statistic estimators.

Two signals are analyzed. The first one is a binary PAM signal and the second one is a 4PSK signal. Both signals have stationary white modulating sequence, full duty-cycle rectangular pulse, and pulse period $T_p = 8T_s$, where $T_s = 1/f_s$ is the sampling period. These binary PAM and 4PSK signals have the same second-order cyclic statistics. However, they can be distinguished on the basis of 4th-order cyclic features.

The 4th-order reduced-dimension cyclic spectral moment function (RD-CSMF) $S^{\alpha}_{xx^*xx^*}(f_1, f_2, f_3)$ is estimated by the time-smoothed 4th-order cyclic periodogram $G^{(\Delta f,T)}_{xx^*xx^*}(\alpha, f_1, f_2, f_3; 0)$ (5.165a) and the cyclic polyspectrum (CP) $P^{\alpha}_{xx^*xx^*}(f_1, f_2, f_3)$ is estimated by the median-filtered time-smoothed 4th-order cyclic periodogram $Q^{(\Delta f,T,\delta f)}_{xx^*xx^*}(\alpha, f_1, f_2, f_3)$ (5.168). The data-record length is $T = 2^{16}T_s$, the reciprocal of the STFT length is $\Delta f = f_s/2^7$ which is coincident with the spectral resolution, and the width of the median filtering window is $\delta f = 10\Delta f$.

Several slices of $G^{(\Delta f,T)}_{xx^*xx^*}(\alpha, f_1, f_2, f_3; 0)$ and $Q^{(\Delta f,T,\delta f)}_{xx^*xx^*}(\alpha, f_1, f_2, f_3)$ are considered.

For the PAM signal, the magnitudes of the RD-CSMF estimate $G^{(\Delta f,T)}_{xx^*xx^*}(\alpha, f_1, f_2, f_3; 0)$ and of the CP estimate $Q^{(\Delta f,T,\delta f)}_{xx^*xx^*}(\alpha, f_1, f_2, f_3)$, as functions of f_1/f_s and f_2/f_s, for $\alpha = 1/T_p$ and $f_3 = 0$, are reported in Figs. 5.17 and 5.18, respectively. In addition, the same estimators are reported as functions of α/f_s and f_1/f_s for $f_2 = 1/(2T_p)$ and $f_3 = 0$ in Figs. 5.19 and 5.20.

The same slices of the RD-CSMF and CP estimators are considered for the 4PSK signal in Figs. 5.21 and 5.22 ($\alpha = 1/T_p$ and $f_3 = 0$) and in Figs. 5.23 and 5.24 ($f_2 = 1/(2T_p)$ and $f_3 = 0$).

For the analyzed slices, the 4th-order CPs of PAM and 4PSK signals are coincident (but for an amplitude factor) while the RD-CSMFs differ on the α-submanifold. In the limit as $T \to \infty$ and $\Delta f \to 0$ with $T(\Delta f)^3 \to \infty$ the RD-CSMFs are impulsive on the α-submanifold. Thus, numerical instability can arise if a ratio of RD-CSMF estimates is considered, e.g., for the purpose of LTI system identification (Section 9.5.2.2).

As observed in Section 5.7.4, a finer spectral resolution or smaller bias and variance can be obtained by resorting to hybrid temporal-spectral cyclic higher-order statistics in order to exploit higher-order cyclostationarity features.

For the PAM signal, the magnitude of the estimator $J^{(\Delta f,T)}_{xx^*xx^*}(\alpha, \tau_1, \tau_2, f_3)$ (5.170a)–(5.170b) of the 4th-order hybrid temporal-spectral cyclic cross-moment function and of the estimator $H^{(\Delta f,T,\delta f)}_{xx^*xx^*}(\alpha, \tau_1, \tau_2, f_3)$ (5.171a)–(5.171b) of the 4th-order hybrid temporal-spectral cyclic cross-cumulant function as function of τ_1/T_s and τ_2/T_s, for $\alpha = 1/T_p$ and $f_3 = 0$, are

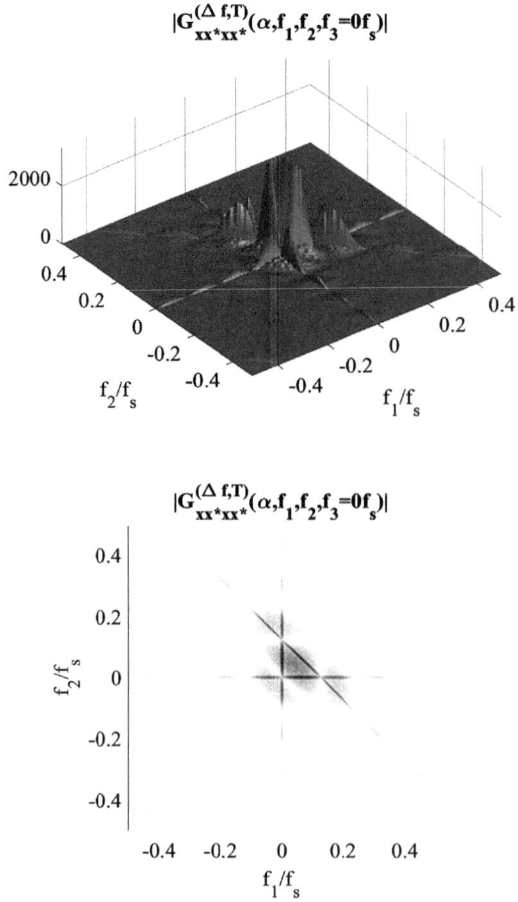

FIGURE 5.17 PAM signal. Magnitude of the time-smoothed 4th-order cyclic periodogram $G_{xx^*xx^*}^{(\Delta f,T)}(\alpha, f_1, f_2, f_3; 0)$ (5.165a) as function of f_1/f_s and f_2/f_s, for $\alpha = 1/T_p$ and $f_3 = 0$. (Top) surface and (bottom) 2-dimensional grayscale elevation map.

reported in Figs. 5.25 and 5.26, respectively. In addition, the same estimators are reported as function of α/f_s and τ_1/T_s, for $f_2 = 1/(2T_p)$ and $f_3 = 0$ in Figs. 5.27 and 5.28.

The same slices of the 4th-order hybrid temporal-spectral cyclic statistic estimators are considered for the 4PSK signal in Figs. 5.29 and 5.30 ($\alpha = 1/T_p$ and $f_3 = 0$) and in Figs. 5.31 and 5.32 ($f_2 = 1/(2T_p)$ and $f_3 = 0$).

5.8 Summary

The existence of consistent estimators for the cyclic statistics is one of the main motivations for the success in the exploitation of cyclostationarity in several fields of application.

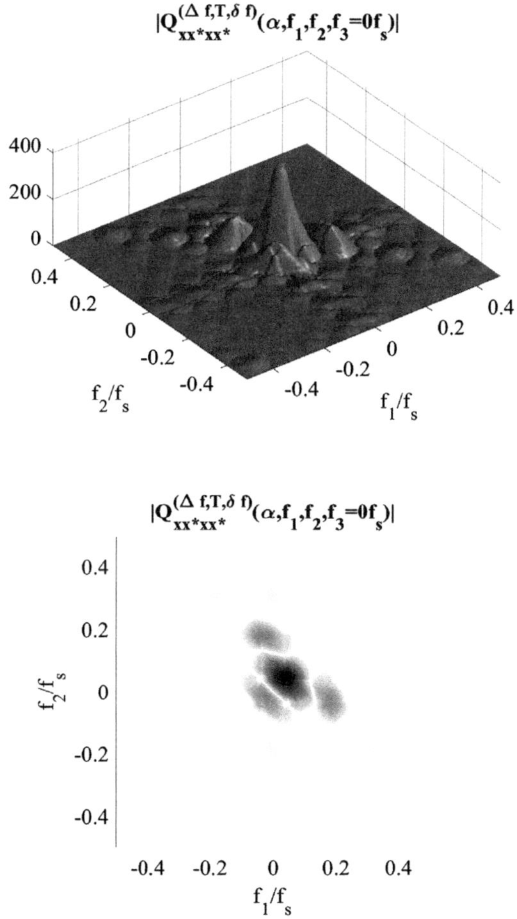

FIGURE 5.18 PAM signal. Magnitude of the median-filtered time-smoothed 4th-order cyclic periodogram $Q^{(\Delta f, T, \delta f)}_{xx^*xx^*}(\alpha, f_1, f_2, f_3)$ (5.168) as function of f_1/f_s and f_2/f_s, for $\alpha = 1/T_p$ and $f_3 = 0$. (Top) surface and (bottom) 2-dimensional grayscale elevation map.

Under the mild assumptions of zero mean and finite or practically finite memory of the processes expressed in terms of summability of cumulants, the *(conjugate) cyclic cross-correlogram* (Definition 5.1) is a mean-square consistent (as $T \to \infty$, where T is data-record length) estimator (Theorem 5.15) of the (conjugate) cyclic cross-correlation function with bias $o(T^{-1})$ and variance $\mathcal{O}(T^{-1})$. Moreover, the estimation errors –at different lags and cycle frequencies– times $\sqrt{T}$, are asymptotically jointly complex normal random variables (Theorem 5.18).

The *(conjugate) cyclic cross-periodogram* (Definition 5.21) is an asymptotically ($T \to \infty$) unbiased (Theorem 5.27) but not consistent (Theorem 5.28) estimator of the (conjugate) cyclic cross-spectrum. The mean-square consistency can be obtained by time- or frequency-smoothing the (conjugate) cyclic cross-periodogram.

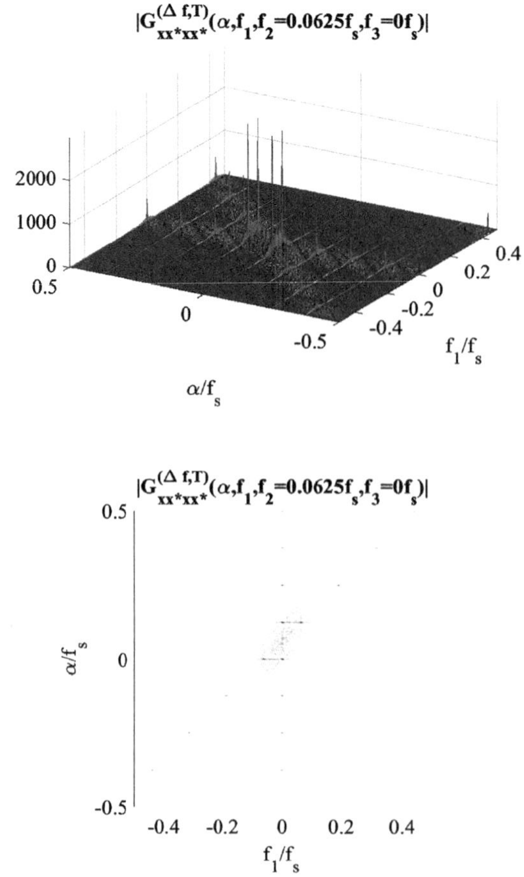

FIGURE 5.19 PAM signal. Magnitude of the time-smoothed 4th-order cyclic periodogram $G^{(\Delta f,T)}_{xx^*xx^*}(\alpha, f_1, f_2, f_3; 0)$ (5.165a) as function of α/f_s and f_1/f_s, for $f_2 = 1/(2T_p)$ and $f_3 = 0$. (Top) surface and (bottom) 2-dimensional grayscale elevation map.

Under the mild assumptions of zero mean and finite or practically finite memory of the processes, the *frequency-smoothed (conjugate) cyclic cross-periodogram* (Definition 5.29) is a mean-square consistent estimator of the (conjugate) cyclic cross-spectrum with bias $o(1/\sqrt{T\Delta f})$ and variance $\mathcal{O}((T\Delta f)^{-1})$, as $T \to \infty$ and $\Delta f \to 0$ with $T\Delta f \to \infty$, where Δf is the width of the frequency-smoothing window (Theorem 5.39). Moreover, the estimation errors –at different spectral and cycle frequencies– times $\sqrt{T\Delta f}$, are asymptotically ($T \to \infty$ and $\Delta f \to 0$ with $T\Delta f \to \infty$) jointly complex normal random variables (Theorem 5.41).

Similar results of mean-square consistency (Theorem 5.48) and asymptotic joint complex normality at different cycle and spectral frequencies (Theorem 5.49) are obtained for the *time-smoothed (conjugate) cyclic cross-periodogram* (Definition 5.42), where, in such a case, $1/\Delta f$ is the width of the short-time Fourier transform used for the cross-periodogram.

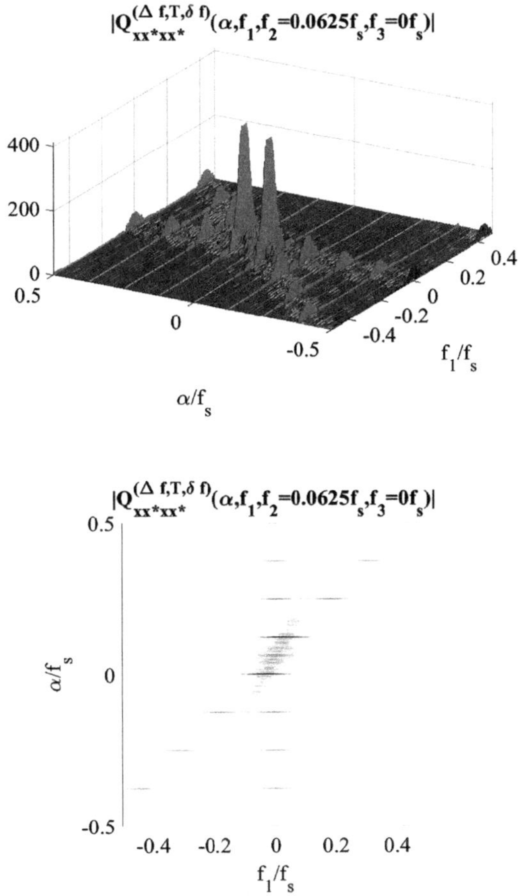

FIGURE 5.20 PAM signal. Magnitude of the median-filtered time-smoothed 4th-order cyclic periodogram $Q^{(\Delta f,T,\delta f)}_{xx^*xx^*}(\alpha, f_1, f_2, f_3)$ (5.168) as function of α/f_s and f_1/f_s, for $f_2 = 1/(2T_p)$ and $f_3 = 0$. (Top) surface and (bottom) 2-dimensional grayscale elevation map.

Both time- and frequency-smoothed (conjugate) cyclic cross-periodograms have cycle frequency resolution of the order of $1/T$ and spectral frequency resolution of the order of Δf. The finite cycle frequency resolution is responsible of the *cycle leakage* phenomenon (Section 5.3.4). It is the influence in cyclic statistic measurements at the considered cycle frequency α of the cyclic statistics at cycle frequencies different from α.

Discrete-time implementations of second-order cyclic statistic estimators are presented (Section 5.4). In particular, computationally efficient estimators for cyclic spectral analysis are described (Section 5.4.4) and Matlab/Octave code for these estimators is provided (Section 5.4.5).

If the signal contains finite-strength additive sinewave components, then the zero mean assumption required for the above consistency results is not satisfied. In such a

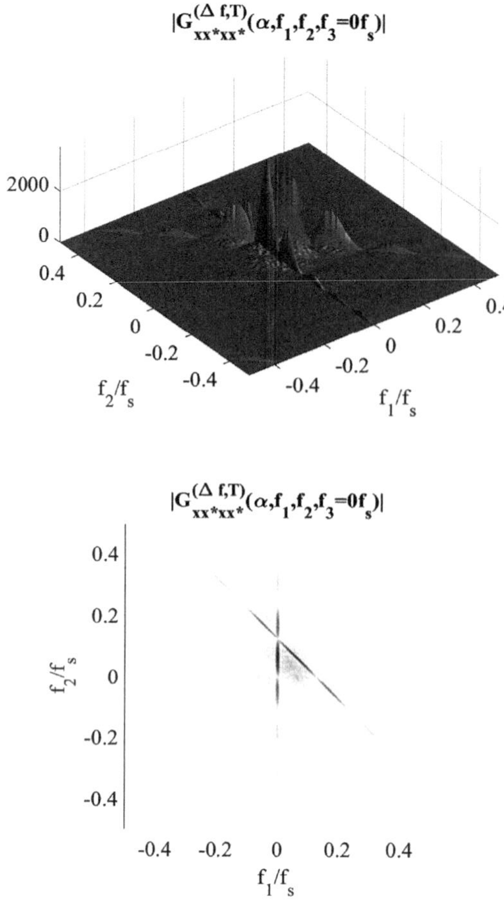

FIGURE 5.21 4PSK signal. Magnitude of the time-smoothed 4th-order cyclic periodogram $G^{(\Delta f,T)}_{xx^*xx^*}(\alpha, f_1, f_2, f_3; 0)$ (5.165a) as function of f_1/f_s and f_2/f_s, for $\alpha = 1/T_p$ and $f_3 = 0$. (Top) surface and (bottom) 2-dimensional grayscale elevation map.

case, even if the frequency of the additive sinewaves are unknown, median-filtering based smoothing techniques allow one to obtain estimates of the (conjugate) cyclic autocorrelation and cyclic spectrum of the zero-mean component of the observed signal (Section 5.2.5).

If the cycle frequency is not known, it can be estimated starting from the available data and then plugged into the cyclic statistic estimators designed for the case of known cycle frequency (cyclic correlogram and time- or frequency-smoothed cyclic periodogram). The cyclic correlogram with estimated cycle frequency is a mean-square consistent estimator of the cyclic autocorrelation, provided that the cycle frequency estimate converges almost surely to the true value with a sufficiently fast rate (Theorem 5.55). In addition, under further conditions on the speed of convergence of the cycle frequency estimates,

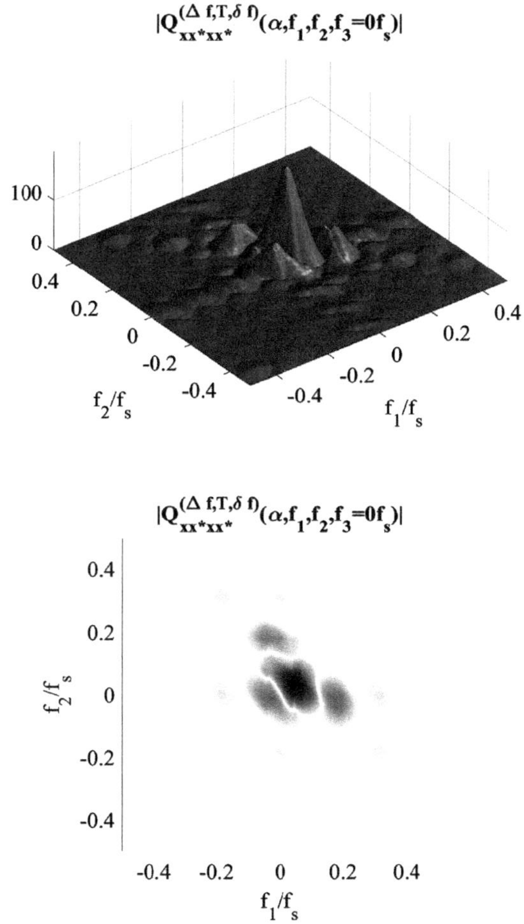

FIGURE 5.22 4PSK signal. Magnitude of the median-filtered time-smoothed 4th-order cyclic periodogram $Q_{xx^*xx^*}^{(\Delta f,T,\delta f)}(\alpha, f_1, f_2, f_3)$ (5.168) as function of f_1/f_s and f_2/f_s, for $\alpha = 1/T_p$ and $f_3 = 0$. (Top) surface and (bottom) 2-dimensional grayscale elevation map.

for the normalized estimation errors at different lags and cycle frequencies one obtains the same asymptotic joint complex normal distribution of the case of exactly known cycle frequencies (Theorem 5.58). Similarly, for the frequency-smoothed cyclic periodogram with estimated cycle frequency, the normalized estimation errors at different lags and cycle frequencies have the same asymptotic joint complex normal distribution of the case of exactly known cycle frequencies (Theorem 5.63). As a corollary, we have the mean-square consistency of the frequency-smoothed cyclic periodogram with estimated cycle frequency. The same results are obtained for the conjugate cyclic correlogram and the frequency-smoothed conjugate cyclic periodogram with estimated conjugate cycle frequency.

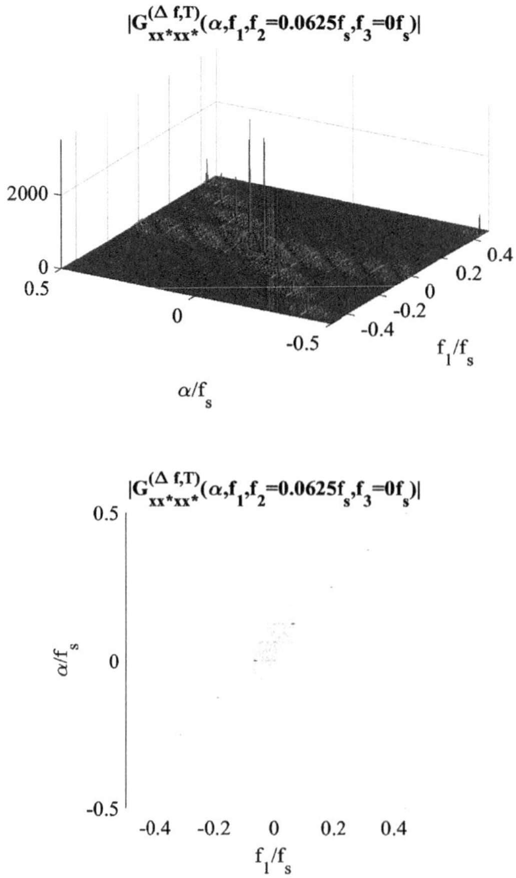

FIGURE 5.23 4PSK signal. Magnitude of the time-smoothed 4th-order cyclic periodogram $G^{(\Delta f,T)}_{xx^*xx^*}(\alpha, f_1, f_2, f_3; 0)$ (5.165a) as function of α/f_s and f_1/f_s, for $f_2 = 1/(2T_p)$ and $f_3 = 0$. (Top) surface and (bottom) 2-dimensional grayscale elevation map.

The cyclic statistic estimation problem is reinterpreted in the *fraction-of-time or functional approach* in Section 5.6. In the stochastic approach, the variability of the estimate is due to the different sample path used for the estimation. In contrast, in the fraction-of-time approach, the variability of the estimate is due to the different time segment (of the unique function of time) used for the estimate.

Estimators of temporal and spectral cyclic moments are obtained by extending to higher-order dimensions the estimators for second-order cyclic statistics (Section 5.7). Estimators of temporal and spectral cyclic cumulants are derived from those of moments accounting for the relationships between moments and cumulants and, in particular, accounting for their differences on the α-submanifold. Moreover, median filtering techniques are adopted for temporal and spectral cyclic cumulants estimates when the structure of the submanifold is unknown (Section 5.7.3).

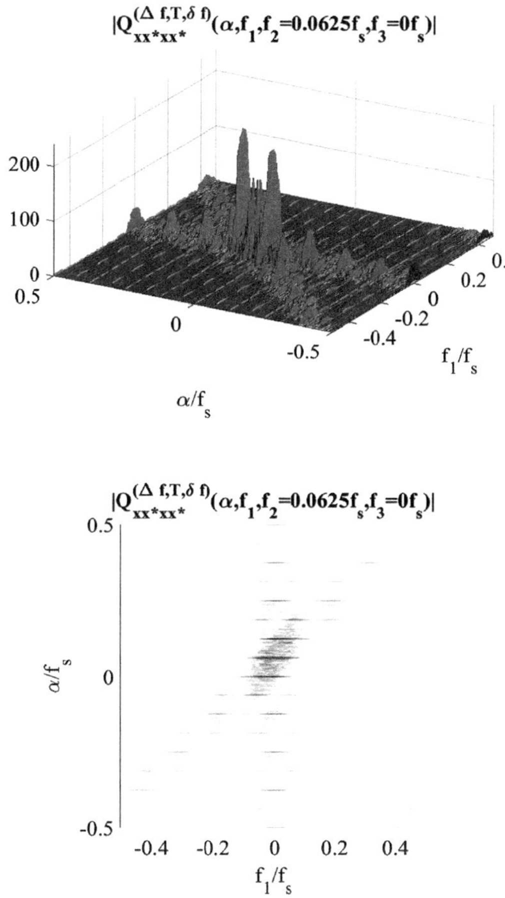

FIGURE 5.24 4PSK signal. Magnitude of the median-filtered time-smoothed 4th-order cyclic periodogram $Q_{xx^*xx^*}^{(\Delta f,T,\delta f)}(\alpha, f_1, f_2, f_3)$ (5.168) as function of α/f_s and f_1/f_s, for $f_2 = 1/(2T_p)$ and $f_3 = 0$. (Top) surface and (bottom) 2-dimensional grayscale elevation map.

For an *Nth-order cyclic temporal moment or cumulant estimate,* the asymptotic bias is $o(T^{-1})$ and the asymptotic variance is $\mathcal{O}(T^{-1})$. Moreover, the estimation errors –at different lags and cycle frequencies and possibly different orders– times $\sqrt{T}$, are asymptotically jointly complex normal random variables (Theorem 5.64). For an *Nth-order cyclic spectral moment or cumulant estimate,* the asymptotic bias is $o(1/\sqrt{T(\Delta f)^{N-1}})$ and the asymptotic variance is $\mathcal{O}(1/(T(\Delta f)^{N-1}))$. Mean-square convergence is assured provided that $T \to \infty$ and $\Delta f \to 0$, with $T(\Delta f)^{N-1} \to \infty$. Moreover, the estimation errors –at different spectral and cycle frequencies and possibly different orders– times $\sqrt{T(\Delta f)^{N-1}}$, are asymptotically jointly complex normal random variables (Theorem 5.66).

For $N > 2$, the speed of convergence to zero of bias and variance of spectral moment and cumulant estimates is slower than that at second order. Hybrid temporal-spectral mo-

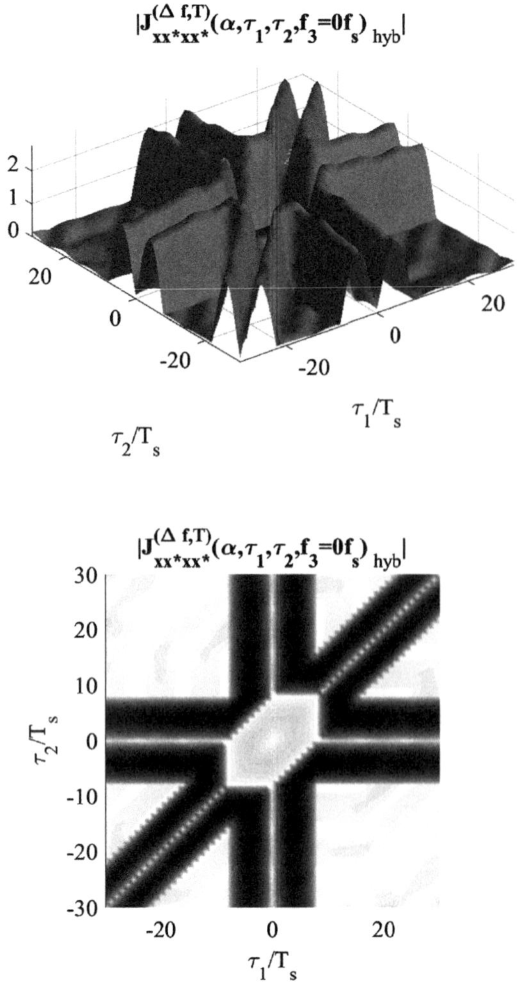

FIGURE 5.25 PAM signal. Magnitude of the estimator $J_{xx^*xx^*}^{(\Delta f,T)}(\alpha, \tau_1, \tau_2, f_3)$ (5.170b) of the 4th-order hybrid temporal-spectral cyclic cross-moment function as function of τ_1/T_s and τ_2/T_s, for $\alpha = 1/T_p$ and $f_3 = 0$. (Top) surface and (bottom) 2-dimensional grayscale elevation map.

ment and cumulant statistics allow one to exploit higher-order cyclostationarity properties by spectral functions whose estimators have bias and variance with faster speed of convergence (Section 5.7.4).

5.9 Proofs

In this section, proofs are reported for results of the present chapter. Unless otherwise specified, only formal manipulations are reported, without explicitly justifying inversion of linear operators such as time averages, infinite sums, integrals, and ensemble average

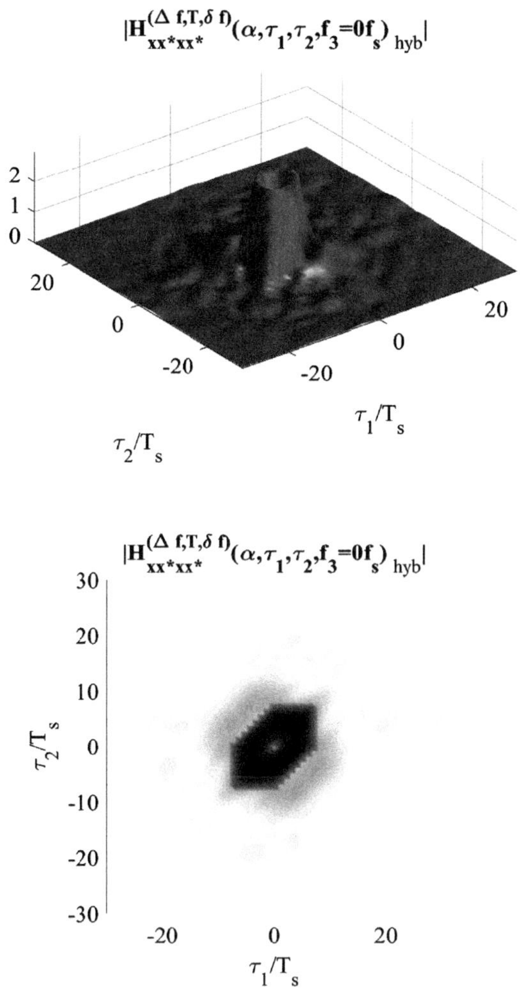

FIGURE 5.26 PAM signal. Magnitude of the estimator $H^{(\Delta f,T,\delta f)}_{xx^*xx^*}(\alpha,\tau_1,\tau_2,f_3)$ (5.171b) of the 4th-order hybrid temporal-spectral cyclic cross-cumulant function as function of τ_1/T_s and τ_2/T_s, for $\alpha = 1/T_p$ and $f_3 = 0$. (Top) surface and (bottom) 2-dimensional grayscale elevation map.

(expectation). In general, these manipulations are justified at least in the sense of generalized functions (distributions) (Champeney, 1990), (Zemanian, 1987).

The assumptions for rigorous proofs are specified for the mean-square consistency and the asymptotic complex normality of the (conjugate) cyclic cross-correlogram and the frequency-smoothed (conjugate) cyclic cross-periodogram. Results for the (conjugate) cyclic cross-correlogram of ACS processes are obtained as special cases of those for GACS processes in (Napolitano, 2012, Chapter 2). Proofs are only outlined and recall those in (Napolitano, 2012, Chapter 3). Results for the frequency-smoothed (conjugate) cyclic cross-periodogram of ACS processes are obtained as special cases of those for SC processes

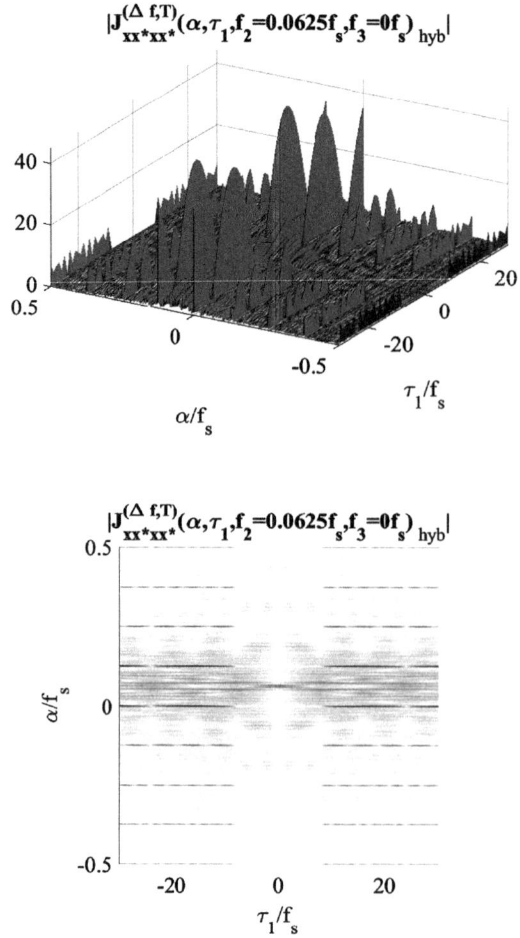

FIGURE 5.27 PAM signal. Magnitude of the estimator $J^{(\Delta f,T)}_{xx^*xx^*}(\alpha, \tau_1, f_2, f_3)$ (5.170b) of the 4th-order hybrid temporal-spectral cyclic cross-moment function as function of α/f_s and τ_1/T_s, for $f_2 = 1/(2T_p)$ and $f_3 = 0$. (Top) surface and (bottom) 2-dimensional grayscale elevation map.

in (Napolitano, 2012, Chapter 4). Proofs are only outlined and recall those in (Napolitano, 2012, Chapter 5).

Proof of Theorem 5.15 "Mean-Square Consistency of the (Conjugate) Cyclic Cross-Correlogram"

From Theorem 5.12 it follows that the (conjugate) cyclic cross-correlogram is an asymptotically unbiased estimator of the (conjugate) cyclic cross-correlation function. In addition, from Theorem 5.14 we have that its variance

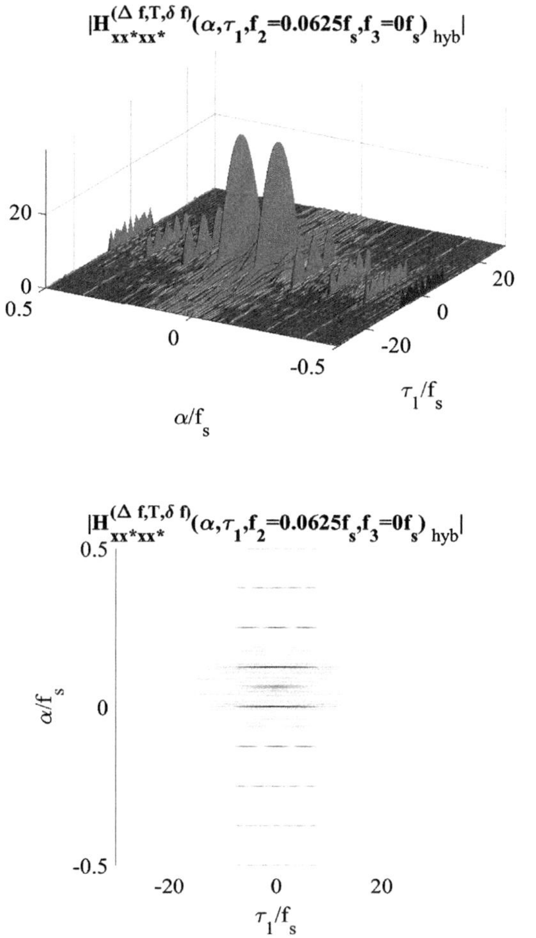

FIGURE 5.28 PAM signal. Magnitude of the estimator $H_{xx^*xx^*}^{(\Delta f,T,\delta f)}(\alpha,\tau_1,f_2,f_3)$ (5.171b) of the 4th-order hybrid temporal-spectral cyclic cross-cumulant function as function of α/f_s and τ_1/T_s, for $f_2 = 1/(2T_p)$ and $f_3 = 0$. (Top) surface and (bottom) 2-dimensional grayscale elevation map.

$$\operatorname{var}\left\{R_{yx^{(*)}}^{(T)}(\alpha,\tau;t_0)\right\} \triangleq \operatorname{cov}\left\{R_{yx^{(*)}}^{(T)}(\alpha,\tau;t_0),\, R_{yx^{(*)}}^{(T)}(\alpha,\tau;t_0)\right\} \tag{5.172}$$

is $\mathcal{O}(T^{-1})$ as $T \to \infty$. Eq. (5.33) immediately follows since

$$\begin{aligned}\mathrm{E}&\left\{\left|R_{yx^{(*)}}^{(T)}(\alpha,\tau;t_0) - R_{yx^{(*)}}^{\alpha}(\tau)\right|^2\right\}\\ &= \left|\operatorname{bias}\left\{R_{yx^{(*)}}^{(T)}(\alpha,\tau;t_0)\right\}\right|^2 + \operatorname{var}\left\{R_{yx^{(*)}}^{(T)}(\alpha,\tau;t_0)\right\}.\end{aligned} \tag{5.173}$$

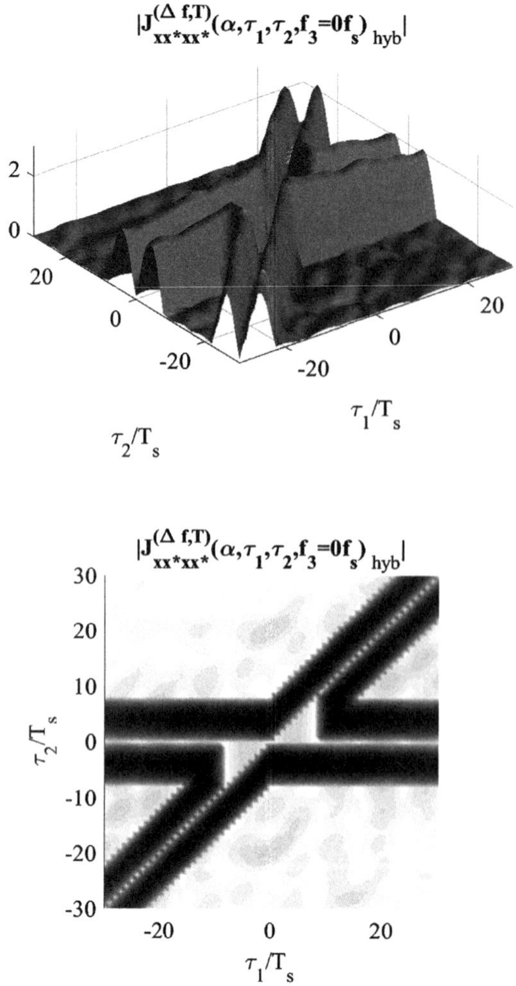

FIGURE 5.29 4PSK signal. Magnitude of the estimator $J^{(\Delta f,T)}_{xx^*xx^*}(\alpha, \tau_1, \tau_2, f_3)$ (5.170b) of the 4th-order hybrid temporal-spectral cyclic cross-moment function as function of τ_1/T_s and τ_2/T_s, for $\alpha = 1/T_p$ and $f_3 = 0$. (Top) surface and (bottom) 2-dimensional grayscale elevation map.

Proof of Theorem 5.18"Asymptotic Joint Complex Normality of (Conjugate) Cyclic Cross-Correlograms"

The proof can be obtained specializing to the case of ACS processes the proof for GACS processes (Napolitano, 2007a, Theorem 5.1), (Napolitano, 2012, Theorem 2.4.18).

The proof, only outlined here, consists in showing that, $\forall (i, j) \in \{1, \ldots, M\}^2$, as $T \to \infty$

1. $\mathrm{E}\{V_i^{(T)}\} \to 0$;

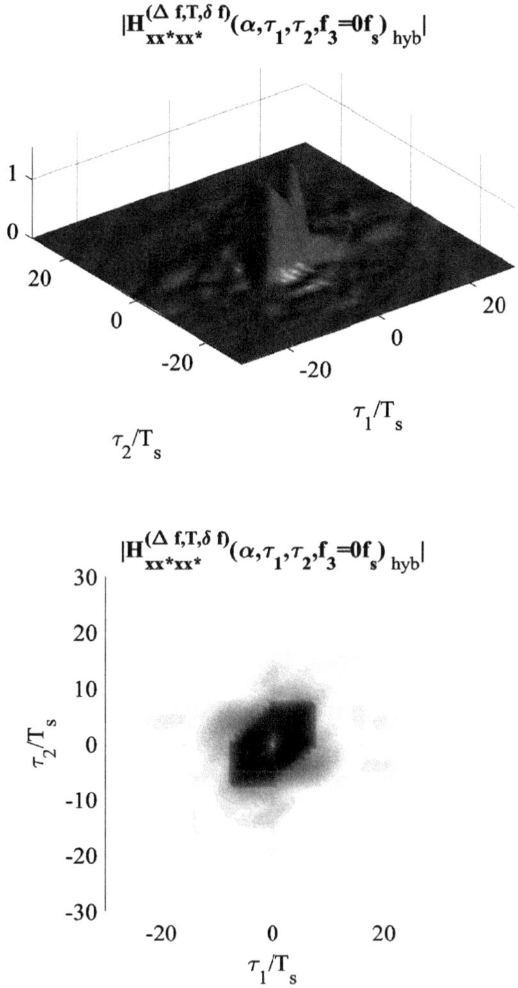

FIGURE 5.30 4PSK signal. Magnitude of the estimator $H^{(\Delta f,T,\delta f)}_{xx^*xx^*}(\alpha,\tau_1,\tau_2,f_3)$ (5.171b) of the 4th-order hybrid temporal-spectral cyclic cross-cumulant function as function of τ_1/T_s and τ_2/T_s, for $\alpha=1/T_p$ and $f_3=0$. (Top) surface and (bottom) 2-dimensional grayscale elevation map.

2a. $\text{cov}\{V_i^{(T)}, V_j^{(T)}\} \to \Sigma_{ij}$ finite;
2b. $\text{cov}\{V_i^{(T)}, V_j^{(T)*}\} \to \Sigma_{ij}^{(c)}$ finite.

In addition, starting from the asymptotic result (Napolitano, 2012, Lemma 2.4.17) (see (Napolitano, 2012, Eq. (3.126)))

$$\text{cum}\left\{V_{i_1}^{(T)[*]_1}, \ldots, V_{i_k}^{(T)[*]_k}\right\} = \mathcal{O}\left(\frac{1}{T^{k/2-1}}\right) \tag{5.174}$$

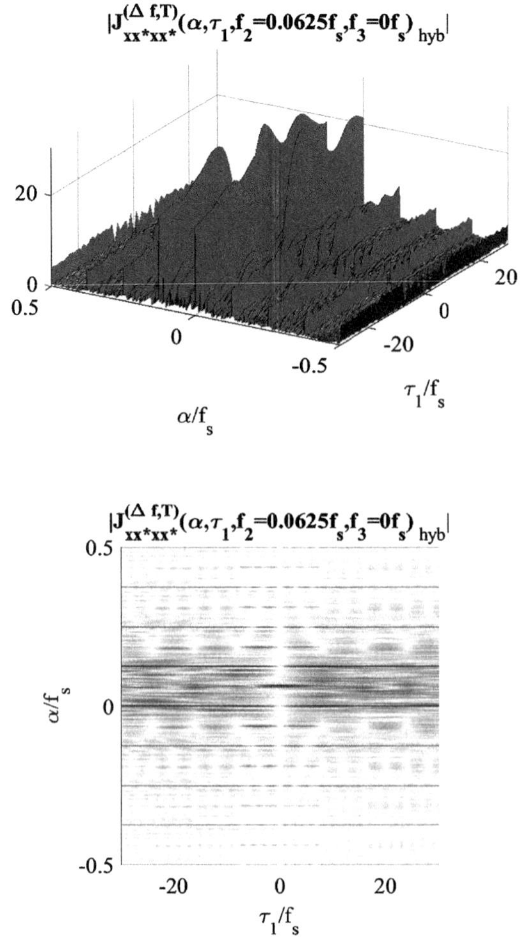

FIGURE 5.31 4PSK signal. Magnitude of the estimator $J_{xx^*xx^*}^{(\Delta f,T)}(\alpha, \tau_1, f_2, f_3)$ (5.170b) of the 4th-order hybrid temporal-spectral cyclic cross-moment function as function of α/f_s and τ_1/T_s, for $f_2 = 1/(2T_p)$ and $f_3 = 0$. (Top) surface and (bottom) 2-dimensional grayscale elevation map.

where superscripts $[*]_\ell$, $\ell = 1, \ldots, k$, denote optional complex conjugations and $(i_1, \ldots, i_k) \in \{1, \ldots, M\}^k$, we have that

3. $\operatorname{cum}\{V_{i_1}^{(T)[*]_1}, \ldots, V_{i_k}^{(T)[*]_k}\} \to 0$ for $k \geqslant 3$.

Let be $\boldsymbol{V}^{(T)} \triangleq [V_{i_1}^{(T)[*]_1}, \ldots, V_{i_k}^{(T)[*]_k}]$ and $\boldsymbol{\omega} \triangleq [\omega_1, \ldots, \omega_k]$. The cumulant of the complex random variables $V_{i_\ell}^{(T)[*]_\ell}$, $\ell = 1, \ldots, k$, defined according to (Spooner and Gardner, 1994a, App. A), (Napolitano, 2007a, App. E), (Napolitano, 2012, Sec. 1.4.2), (Section E.1.2), is

$$\operatorname{cum}\left\{V_{i_1}^{(T)[*]_1}, \ldots, V_{i_k}^{(T)[*]_k}\right\} \triangleq (-j)^k \frac{\partial^k}{\partial\omega_1 \ldots \partial\omega_k} \log \mathrm{E}\left\{e^{j\boldsymbol{\omega}^\mathrm{T}\boldsymbol{V}^{(T)}}\right\}\bigg|_{\boldsymbol{\omega}=\mathbf{0}} . \tag{5.175}$$

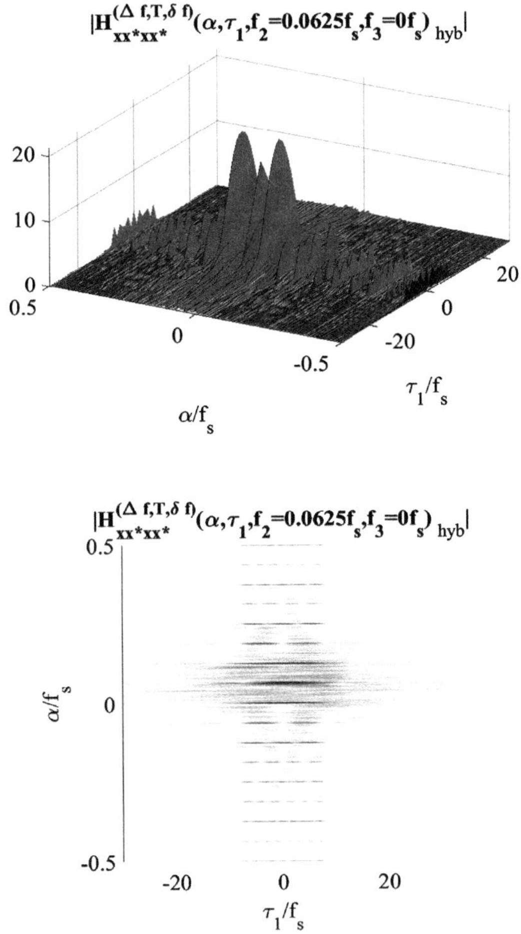

FIGURE 5.32 4PSK signal. Magnitude of the estimator $H_{xx^*xx^*}^{(\Delta f,T,\delta f)}(\alpha,\tau_1,f_2,f_3)$ (5.171b) of the 4th-order hybrid temporal-spectral cyclic cross-cumulant function as function of α/f_s and τ_1/T_s, for $f_2 = 1/(2T_p)$ and $f_3 = 0$. (Top) surface and (bottom) 2-dimensional grayscale elevation map.

Thus, accounting for items 1–3, asymptotically, the logarithm of the joint characteristic function of $V_i^{(T)[*]_i}$, $i = 1,\dots,M$ is a quadratic homogeneous polynomial, with complex coefficients, in the real variables ω_i's. That is, the (possibly conjugated) $V_i^{(T)}$ are jointly complex normal (Lemma E.1).

Proof of Theorem 5.31 "Expected Value of the Frequency-Smoothed (Conjugate) Cyclic Cross-Periodogram"

From (5.43) it follows (5.48). Thus, the assumptions are those of (Napolitano, 2007c, Theorem 3.1), (Napolitano, 2012, Theorem 4.6.3) specialized to ACS processes. Therefore, (5.67) follows from (Napolitano, 2007c, Eq. (9)), (Napolitano, 2012, Eq. (4.151)) specialized to ACS processes.

In (5.68), the width of $B_{\frac{1}{T}}(\cdot)$ is of the order of $1/T$ and the width of $A_{\Delta f}(\cdot)$ is of the order of Δf. Condition $T\Delta f \gg 1$ assures small bias (Theorems 5.36 and 5.37) and variance (Theorem 5.38). In such a case, the integrand function gives significantly nonzero contribution only for $\lambda \in [\nu - 1/2T, \nu + 1/2T]$, where $A_{\Delta f}(f-\lambda) \simeq A_{\Delta f}(f-\nu)$ and expression (5.68) of the kernel $K^{(T,\Delta f)}(\alpha, \beta, f, \nu)$ can be approximated as

$$K^{(T,\Delta f)}(\alpha, \beta, f, \nu) \simeq A_{\Delta f}(f-\nu) \quad \frac{1}{T}\int_{\mathbb{R}} B_{\frac{1}{T}}(\lambda-\nu)\, B^{(*)}_{\frac{1}{T}}((-)(\alpha-\beta+\nu-\lambda))\, \mathrm{d}\lambda\,. \tag{5.176}$$

We have

$$\begin{aligned}
&\frac{1}{T}\int_{\mathbb{R}} B_{\frac{1}{T}}(\lambda-\nu)\, B^{(*)}_{\frac{1}{T}}((-)(\alpha-\beta+\nu-\lambda))\, \mathrm{d}\lambda \\
&\quad = \int_{\mathbb{R}} W_B((\lambda-\nu)T)\, W_B^{(*)}((-)(\alpha-\beta+\nu-\lambda)T)\, T\, \mathrm{d}\lambda \\
&\quad = \int_{\mathbb{R}} W_B(\varphi)\, W_B^{(*)}((-)((\alpha-\beta)T-\varphi))\, \mathrm{d}\varphi \\
&\quad = \left[W_B(f) \underset{f}{\otimes} W_B^{(*)}((-)f) \right]_{f=(\alpha-\beta)T}
\end{aligned} \tag{5.177}$$

where in the second equality the variable change $\varphi = T(\lambda-\nu)$ is made.

Thus, by replacing (5.176) and (5.177) into (5.67) we have (5.69).

Proof of Theorem 5.39 "Mean-Square Consistency of the Frequency-Smoothed (Conjugate) Cyclic Cross-Periodogram"

From Theorem 5.36 it follows that the frequency-smoothed (conjugate) cyclic cross-periodogram is an asymptotically unbiased estimator of the (conjugate) cyclic cross-spectrum. In addition, from Theorem 5.38 we have that its variance

$$\operatorname{var}\left\{ S^{(T,\Delta f)}_{yx^{(*)}}(\alpha, f; t_0) \right\} \triangleq \operatorname{cov}\left\{ S^{(T,\Delta f)}_{yx^{(*)}}(\alpha, f; t_0),\, S^{(T,\Delta f)}_{yx^{(*)}}(\alpha, f; t_0) \right\} \tag{5.178}$$

is $\mathcal{O}((T\Delta f)^{-1})$ as $T \to \infty$ and $\Delta f \to 0$ with $T\Delta f \to \infty$. Eq. (5.79) immediately follows since

$$\begin{aligned}
&\mathrm{E}\left\{ \left| S^{(T,\Delta f)}_{yx^{(*)}}(\alpha, f; t_0) - \mathcal{E}_{(*)}\, S^{\alpha}_{yx^{(*)}}(f) \right|^2 \right\} \\
&\quad = \left| \operatorname{bias}\left\{ S^{(T,\Delta f)}_{yx^{(*)}}(\alpha, f; t_0) \right\} \right|^2 + |\mathcal{E}_{(*)}|^2 \operatorname{var}\left\{ S^{(T,\Delta f)}_{yx^{(*)}}(\alpha, f; t_0) \right\}.
\end{aligned} \tag{5.179}$$

Proof of Theorem 5.41 "Asymptotic Joint Complex Normality of the Frequency-Smoothed (Conjugate) Cyclic Cross-Periodograms"

Assumptions 5.22a (ACS statistics and Fourier series regularity), 5.24 (data-tapering window regularity), 5.30 (frequency-smoothing window regularity), 5.11 (lack of cycle frequency clusters), 5.33 (data-tapering window summability), 5.34 (frequency-smoothing window summability), and 5.35 (spectral cross-correlation density regularity), assure that Theorem 5.37 (rate of convergence of the bias of the frequency-smoothed (conjugate) cyclic cross-periodogram) holds.

Assumptions 5.24 (data-tapering window regularity), 5.30 (frequency-smoothing window regularity), and 5.40 (spectral cumulants), and the further assumption that the spectral cross-correlation densities $S^{\alpha}_{yx^{(*)}}(f)$ are continuous a.e. (a weaker condition w.r.t. that in Assumption 5.35 (spectral cross-correlation density regularity)), assure that (Napolitano, 2012, Lemma 4.7.10) (cumulants of frequency-smoothed cross-periodograms) specialized to ACS processes, holds. That is, for any $k \geqslant 2$ it results that

$$\lim_{\Delta f \to 0} \lim_{T \to \infty} (T\Delta f)^{k/2} \operatorname{cum}\left\{ S^{(T,\Delta f)}_{yx^{(*)}}(\alpha_1, f_1; t_1)^{[*]_1}, \ldots, S^{(T,\Delta f)}_{yx^{(*)}}(\alpha_k, f_k; t_k)^{[*]_k} \right\} = \begin{cases} \mathcal{O}(1) & k = 2 \\ 0 & k \geqslant 3 \end{cases} \tag{5.180}$$

where $[*]_i$ denotes ith optional complex conjugation and the order of the two limits cannot be interchanged ($T \to \infty$ and $\Delta f \to 0$ with $T\Delta f \to \infty$).

The proof of the asymptotic joint complex normality can be obtained specializing to the case of ACS processes the proof for SC processes (Napolitano, 2012, Theorem 4.7.11).

The proof, only outlined here, consists in showing that, $\forall (i, j) \in \{1, \ldots, M\}^2$, as $T \to \infty$ and $\Delta f \to 0$ with $T\Delta f \to \infty$

1. $\mathrm{E}\{U_i^{(T,\Delta f)}\} \to 0$;

2a. $\operatorname{cov}\{U_i^{(T,\Delta f)}, U_j^{(T,\Delta f)}\} \to \Sigma_{ij}$ finite;

2b. $\operatorname{cov}\{U_i^{(T,\Delta f)}, U_j^{(T,\Delta f)*}\} \to \Sigma_{ij}^{(c)}$ finite.

In addition, starting from the asymptotic result (5.180), we have that

3. $\operatorname{cum}\{U_{i_1}^{(T,\Delta f)[*]_1}, \ldots, U_{i_k}^{(T,\Delta f)[*]_k}\} \to 0$ for $k \geqslant 3$.

Let be $\boldsymbol{U}^{(T,\Delta f)} \triangleq [U_{i_1}^{(T,\Delta f)[*]_1}, \ldots, U_{i_k}^{(T,\Delta f)[*]_k}]$ and $\boldsymbol{\omega} \triangleq [\omega_1, \ldots, \omega_k]$. The cumulant of the complex random variables $U_{i_\ell}^{(T,\Delta f)[*]_\ell}$, $\ell = 1, \ldots, k$, defined according to (Spooner and Gardner, 1994a, App. A), (Napolitano, 2007a, App. E), (Napolitano, 2012, Sec. 1.4.2), is

$$\operatorname{cum}\left\{ U_{i_1}^{(T,\Delta f)[*]_1}, \ldots, U_{i_k}^{(T,\Delta f)[*]_k} \right\} \triangleq (-j)^k \frac{\partial^k}{\partial\omega_1 \ldots \partial\omega_k} \log \mathrm{E}\left\{ e^{j\boldsymbol{\omega}^{\mathsf{T}} \boldsymbol{U}^{(T,\Delta f)}} \right\} \Big|_{\boldsymbol{\omega}=\mathbf{0}} . \tag{5.181}$$

Thus, accounting for items 1–3, asymptotically, the logarithm of the joint characteristic function of $U_i^{(T,\Delta f)[*]_i}$, $i = 1, \ldots, M$ is a quadratic homogeneous polynomial, with complex coefficients, in the real variables ω_i's. That is, the (possibly conjugated) $U_i^{(T,\Delta f)}$ are jointly complex normal (Lemma E.1).

Proof of Theorem 5.44 "Expected Value of the Time-Smoothed (Conjugate) Cyclic Cross-Periodogram"

The proof of (5.89) is obtained by specializing to ACS processes the proof of (Napolitano, 2003, Lemma 4.1), (Napolitano, 2012, Lemma 4.5.5) for SC processes.

In practice, condition $T\Delta f \gg 1$ is generally verified to have small variance (Theorem 5.47).

Since $a_T(t)$ is a T-duration window, $A_{\frac{1}{T}}(f)$ has a bandwidth of the order of $1/T$. Thus, in (5.89), $A_{\frac{1}{T}}(\beta-\alpha)$ is significantly nonzero only for $|\alpha-\beta| < 1/T$. For these values of $|\alpha-\beta|$, if $\Delta f \gg 1/T$, we have $B_{\Delta f}^{(*)}((-)(\alpha-\beta-f+\nu)) \simeq B_{\Delta f}^{(*)}((-)(-f+\nu))$. Thus,

$$\begin{aligned} &\int_{\mathbb{R}} S^{\beta}_{yx^{(*)}}(\nu)\, B_{\Delta f}(f-\nu)\, B^{(*)}_{\Delta f}((-)(\alpha-\beta-f+\nu))\, \mathrm{d}\nu\, A_{\frac{1}{T}}(\beta-\alpha) \\ &\quad \simeq \int_{\mathbb{R}} S^{\beta}_{yx^{(*)}}(\nu)\, B_{\Delta f}(f-\nu)\, B^{(*)}_{\Delta f}(-(-)(f-\nu))\, \mathrm{d}\nu\, A_{\frac{1}{T}}(\beta-\alpha)\,. \end{aligned} \tag{5.182}$$

Using approximation (5.182) into (5.89) yields (5.90).

Proof of Theorem 5.52 "Strip Spectral Correlation Analyzer"

By using the definition (5.126b) for $\widetilde{Y}_M(n,\nu)$, we have

$$\begin{aligned} &\frac{1}{N}\sum_{n=-N/2}^{N/2-1} \widetilde{Y}_M(n,\nu)\, x^{(*)}(n)\, g(n)\, e^{-j2\pi\widetilde{\alpha}n} \\ &= \frac{1}{N}\sum_{n=-N/2}^{N/2-1}\left[\sum_{m=-M/2}^{M/2-1} h(m)\, e^{j2\pi\nu m}\, y(n-m)\right] x^{(*)}(n)\, g(n)\, e^{-j2\pi\widetilde{\alpha}n} \\ &= \sum_{m'=-M/2}^{M/2-1} h(-m')\, e^{-j2\pi\nu m'}\left[\frac{1}{N}\sum_{n=-N/2}^{N/2-1} y(n+m')\, x^{(*)}(n)\, g(n)\, e^{-j2\pi\widetilde{\alpha}n}\right] \\ &= \sum_{m'=-M/2}^{M/2-1} h(m')\, e^{-j2\pi\nu m'}\, R^{(N)}_{yx^{(*)}}(\widetilde{\alpha},m') = \\ &= \left[\frac{1}{N}\, Y_N(n,\nu)\, X_N^{(*)}(n,(-)(\widetilde{\alpha}-\nu))\right] \underset{\nu}{\otimes} H_{\Delta\nu}(\nu) \\ &\triangleq S^{(N,\Delta\nu)}_{yx^{(*)}}(\widetilde{\alpha},\nu) \end{aligned} \tag{5.183}$$

In the second equality, the variable change $m'=-m$ is made. In the third equality, the definition of discrete-time (conjugate) cyclic cross-correlogram with data-tapering window $g(n)$

$$R_{yx^{(*)}}^{(N)}(\widetilde{\alpha}, m) \triangleq \frac{1}{N} \sum_{n=-N/2}^{N/2-1} y(n+m)\, x^{(*)}(n)\, g(n)\, e^{-j2\pi\widetilde{\alpha}n} \tag{5.184}$$

is used. In addition, since $h(\cdot)$ is an even function, $h(m') = h(-m')$ is used.

Analogously to the continuous-time case, the (conjugate) cyclic cross-periodogram is the Fourier transform of the (conjugate) cyclic cross-correlogram (rigorously in the case of rectangular data-tapering window) (Lemma 2.39). Thus, in the fourth equality the convolution theorem for Fourier transforms is used, where $H_{\Delta\nu}(\nu)$ denotes the Fourier transform of $h(m)$. Since $h(m)$ is nonzero for $m \in \{-M/2, \ldots, M/2-1\}$, $H_{\Delta\nu}(\nu)$ has approximate bandwidth $\Delta\nu = 1/M$.

6

Quadratic time-frequency distributions

Contents

Abstract

Quadratic time-frequency distributions for almost-cyclostationary signals are linked to the second-order cyclic statistics in time and frequency domains. Moreover, a link is established with the estimators of these statistics.

Keywords

Time-frequency distribution, Wigner-Ville distribution, Spectrogram, Kernels, Cohen class

6.1 Introduction

In this chapter, results for second-order ACS signals are reinterpreted in terms of quadratic time-frequency distributions (TFDs).

Time-frequency distributions have been initially studied for finite-energy deterministic signals (Boashash, 2015, Chaps. 1–3). Definitions and properties can be extended to finite-power signals considering finite-time windowing of these signals. In addition, if signals are modeled as stochastic processes, definitions originally given for deterministic signals also include an ensemble average to remove randomness. For cyclostationary and almost-cyclostationary signals the ensemble average can be replaced by the almost-periodic component extraction operator (Definition 2.14) if the analysis is made in the functional or fraction-of-time approach.

The presence of time windowing naturally leads to establish relationships between quadratic time-frequency distributions of (windowed) ACS signals and estimators of second-order cyclic statistics (Section 5.2). An alternative approach consists in using for finite-power signals the same definitions adopted for finite-energy signals. In such a case, Dirac deltas appear in the expressions of time-frequency distributions.

Time-frequency distributions for GACS signals are derived in (Izzo and Napolitano, 2005, Sec. 5) in the fraction-of-time approach. Moments of a suitable definition of instantaneous frequency for ACS processes are derived in (Živanović, 1991).

6.2 Finite-energy signals: correlations and spectra

Let $x_i(t) \in L^2(\mathbb{R})$, $i = 1, 2$ complex-valued signals. Their *asymmetric cross-ambiguity function* is defined as

$$A_{x_1 x_2}(\nu, \tau) \triangleq \int_{\mathbb{R}} x_1(t+\tau)\, x_2(t)\, e^{-j2\pi \nu t}\, dt \tag{6.1a}$$

$$= \int_{\mathbb{R}} X_1(f)\, X_2(\nu - f)\, e^{j2\pi f \tau}\, df \tag{6.1b}$$

where $X_i(f)$ is the Fourier transform of $x_i(t)$ and the second equality is proved in Section 6.8. If $x_2(t)$ is replaced by $x_2^*(t)$, then $X_2(f)$ must be replaced by $X_2^*(-f)$.

The *spectral cross-moment* is defined as

$$E_{x_1 x_2}(\nu, f) \triangleq \int_{\mathbb{R}} A_{x_1 x_2}(\nu, \tau)\, e^{-j2\pi f \tau}\, d\tau = X_1(f)\, X_2(\nu - f)\,. \tag{6.2}$$

The *symmetric cross-ambiguity function* is defined as

$$\bar{A}_{x_1 x_2}(\nu, \tau) \triangleq \int_{\mathbb{R}} x_1(t+\tau/2)\, x_2(t-\tau/2)\, e^{-j2\pi \nu t}\, dt \tag{6.3a}$$

$$= \int_{\mathbb{R}} x_1(t'+\tau)\, x_2(t')\, e^{-j2\pi \nu (t'+\tau/2)}\, dt'$$

$$= A_{x_1 x_2}(\nu, \tau)\, e^{-j\pi \nu \tau} \tag{6.3b}$$

where the variable change $t + \tau/2 = t' + \tau$ is used.

6.3 Spectrogram

Definition (1.31) of *short-time Fourier transform (STFT)* can be generalized by considering a data-tapering window $b_T(t)$ with support $[-T/2, T/2]$. That is,

$$X_T(t, f) \triangleq \int_{\mathbb{R}} x(u)\, b_T(u - t)\, e^{-j2\pi f u}\, du \tag{6.4a}$$

$$= \int_{\mathbb{R}} x(t+s)\, b_T(s)\, e^{-j2\pi f(t+s)}\, \mathrm{d}s \tag{6.4b}$$

where in the second equality the variable change $s = u - t$ is used.

The *(conjugate) spectrogram* of the signal $x(t)$ is defined as

$$\mathsf{S}_{\boldsymbol{x}}(t, f) \triangleq X_T(t, f)\, X_T^{(*)}(t, f) \tag{6.5}$$

where subscript $\boldsymbol{x} \equiv xx^{(*)}$. If the complex conjugation is present, we have the *spectrogram*

$$\mathsf{S}_{xx^*}(t, f) \triangleq |X_T(t, f)|^2 . \tag{6.6}$$

6.3.1 Expected value of spectrogram of ACS signals

6.3.1.1 Stochastic approach

Let $x(t)$ be an ACS process. The *expected value of the (conjugate) spectrogram* is given by (Section 6.8)

$$\mathrm{E}\{\mathsf{S}_{\boldsymbol{x}}(t, f)\} = \sum_{\alpha \in A} \int_{\mathbb{R}} R_{\boldsymbol{x}}^{\alpha}(\tau)\, r_{\boldsymbol{b}}^{\kappa f - \alpha}(\tau)\, e^{-j2\pi f\tau}\, \mathrm{d}\tau\, e^{j2\pi(\alpha - \kappa f)t} \tag{6.7a}$$

$$= \sum_{\alpha \in A} \left[S_{\boldsymbol{x}}^{\alpha}(f) \underset{f}{\otimes} \left[B_{\frac{1}{T}}(f)\, B_{\frac{1}{T}}^{(*)}((-)(\beta - f)) \right] \right]_{\beta = \kappa f - \alpha} e^{j2\pi(\alpha - \kappa f)t} \tag{6.7b}$$

where $\kappa \triangleq 1 + (-)1$,

$$r_{\boldsymbol{b}}^{\beta}(\tau) \triangleq \int_{\mathbb{R}} b_T(s+\tau)\, b_T^{(*)}(s)\, e^{-j2\pi\beta s}\, \mathrm{d}s \tag{6.8}$$

and

$$B_{\frac{1}{T}}(f) \triangleq \int_{\mathbb{R}} b_T(t)\, e^{-j2\pi f t}\, \mathrm{d}t . \tag{6.9}$$

If the complex conjugation is present, one has

$$\mathrm{E}\left\{ |X_T(t, f)|^2 \right\} = \sum_{\alpha \in A} \int_{\mathbb{R}} R_{\boldsymbol{x}}^{\alpha}(\tau)\, r_{\boldsymbol{b}}^{-\alpha}(\tau)\, e^{-j2\pi f\tau}\, \mathrm{d}\tau\, e^{j2\pi\alpha t} \tag{6.10a}$$

$$= \sum_{\alpha \in A} S_{\boldsymbol{x}}^{\alpha}(f) \underset{f}{\otimes} \left[B_{\frac{1}{T}}(f)\, B_{\frac{1}{T}}^{*}(\alpha + f) \right] e^{j2\pi\alpha t} \tag{6.10b}$$

The window $b_T(t)$ has duration T. Thus, its Fourier transform $B_{\frac{1}{T}}(f)$ has approximate bandwidth $1/T$. Therefore, for $|\alpha| > 1/T$ the product $B_{\frac{1}{T}}(f)\, B_{\frac{1}{T}}^{*}(\alpha + f)$ is negligible. If T is large, then large cycle frequencies do not give significant contribution in (6.10b). If T is small, then the convolution by $B_{\frac{1}{T}}(f)\, B_{\frac{1}{T}}^{*}(\alpha + f)$ significantly modifies the shape of $S_{\boldsymbol{x}}^{\alpha}(f)$.

6.3.1.2 Fraction-of-time approach

Let us consider the *lag product decomposition* (Section 2.3.1.6)

$$x(t+s_1)\,x^{(*)}(t+s_2) = \mathrm{E}^{\{\alpha\}}\left\{x(t+s_1)\,x^{(*)}(t+s_2)\right\} + \ell_{\boldsymbol{x}}(t;s_1,s_2) \tag{6.11}$$

with $\mathrm{E}^{\{\alpha\}}\{\cdot\}$ almost-periodic component extraction operator (Definition 2.14) and $\ell_{\boldsymbol{x}}(t;s_1,s_2)$ residual term non containing any finite-strength additive sine-wave component

$$\left\langle \ell_{\boldsymbol{x}}(t;s_1,s_2)\,e^{-j2\pi\alpha t}\right\rangle_t = 0 \qquad \forall\alpha\in\mathbb{R} \tag{6.12}$$

From (6.11), and accounting for the FOT counterparts of (6.48) and (6.50), one obtains the *(conjugate) spectrogram decomposition*

$$\begin{aligned} \mathrm{S}_{\boldsymbol{x}}(t,f) &= \mathrm{E}^{\{\alpha\}}\{\mathrm{S}_{\boldsymbol{x}}(t,f)\} + \eta_{\boldsymbol{x}}(t,f) \\ &= \sum_{\alpha\in A}\int_{\mathbb{R}} R_{\boldsymbol{x}}^{\alpha}(\tau)\,r_{\boldsymbol{b}}^{\kappa f-\alpha}(\tau)\,e^{-j2\pi f\tau}\,\mathrm{d}\tau\,e^{j2\pi(\alpha-\kappa f)t} + \eta_{\boldsymbol{x}}(t,f) \qquad (6.13a)\\ &= \sum_{\alpha\in A}\left[S_{\boldsymbol{x}}^{\alpha}(f)\underset{f}{\otimes}\left[B_{\frac{1}{T}}(f)\,B_{\frac{1}{T}}^{(*)}((-)(\beta-f))\right]\right]_{\beta=\kappa f-\alpha} e^{j2\pi(\alpha-\kappa f)t} + \eta_{\boldsymbol{x}}(t,f) \qquad (6.13b) \end{aligned}$$

where

$$\eta_{\boldsymbol{x}}(t,f) \triangleq \int_{\mathbb{R}^2} \ell_{\boldsymbol{x}}(t;s_1,s_2)\,b_T(s_1)\,b_T^{(*)}(s_2)\,e^{-j2\pi f s_1}\,e^{-(-)j2\pi f s_2}\,\mathrm{d}s_1\,\mathrm{d}s_2\,e^{-j2\pi\kappa f t} \tag{6.14}$$

with

$$\begin{aligned} \left\langle \eta_{\boldsymbol{x}}(t,f)\,e^{-j2\pi\alpha t}\right\rangle_t &= \int_{\mathbb{R}^2}\left\langle \ell_{\boldsymbol{x}}(t;s_1,s_2)\,e^{-j2\pi(\alpha+\kappa f)t}\right\rangle_t \\ &\quad b_T(s_1)\,b_T^{(*)}(s_2)\,e^{-j2\pi f s_1}\,e^{-(-)j2\pi f s_2}\,\mathrm{d}s_1\,\mathrm{d}s_2 = 0 \qquad \forall\alpha\in\mathbb{R} \end{aligned} \tag{6.15}$$

The Fourier coefficients of the almost-periodic component of $\mathrm{S}_{\boldsymbol{x}}(t,f)$ are given by

$$\left\langle \mathrm{S}_{\boldsymbol{x}}(t,f)\,e^{-j2\pi(\gamma-\kappa f)t}\right\rangle_t = \begin{cases} \left[S_{\boldsymbol{x}}^{\alpha}(f)\underset{f}{\otimes}\left[B_{\frac{1}{T}}(f)\,B_{\frac{1}{T}}^{(*)}((-)(\beta-f))\right]\right]_{\beta=\kappa f-\alpha} & \gamma\in A \\ 0 & \gamma\notin A \end{cases} \tag{6.16}$$

6.4 Quadratic TFDs

Let $x(t)$ be a complex-valued continuous-time finite-power deterministic signal. For complex-valued signals, both lag-product waveforms with conjugated and not conjugated second term should be considered for a complete characterization (Section 1.3), (Schreier and Scharf, 2003b), (Schreier and Scharf, 2010), (Adali et al., 2011). In the following, only the case of conjugated second term of the lag product is considered. The case of not conjugated second term can be easily derived accounting for the guidelines in Section 1.3.

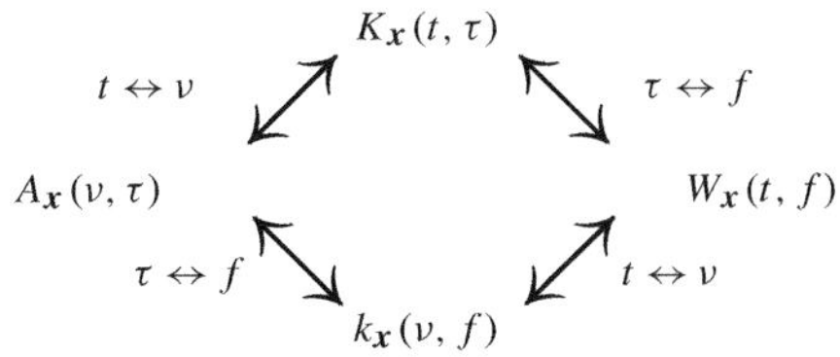

FIGURE 6.1 Quadratic TFDs.

According to the definitions given in (Boashash, 2015, Sec. 3.2.1) for finite-energy signals and those given in (Boashash, 2015, Sec. 9.4) for finite-power stochastic processes, the following functions can be defined, where Fourier transforms must be intended in the sense of distributions (generalized functions) (Gel'fand and Vilenkin, 1964, Chap. 3), (Henniger, 1970). Their mutual relationships are illustrated in Fig. 6.1.

Symmetric *second-order lag-product* or *instantaneous autocorrelation function*

$$K_x(t,\tau) \triangleq x(t+\tau/2)\, x^*(t-\tau/2) \tag{6.17}$$

Symmetric *ambiguity function*

$$A_x(\nu,\tau) \triangleq \int_{\mathbb{R}} K_x(t,\tau)\, e^{-j2\pi\nu t}\, \mathrm{d}t \tag{6.18}$$

(Symmetric) *Wigner distribution*

$$W_x(t,f) \triangleq \int_{\mathbb{R}} K_x(t,\tau)\, e^{-j2\pi f\tau}\, \mathrm{d}\tau \tag{6.19}$$

Symmetric *spectral autocorrelation function*

$$k_x(\nu,f) \triangleq \int_{\mathbb{R}} K_x(t,\tau)\, e^{-j2\pi\nu t}\, e^{-j2\pi f\tau}\, \mathrm{d}t\, \mathrm{d}\tau \tag{6.20}$$

6.4.1 Expected values of quadratic TFDs of ACS signals

Let $x(t)$ be a complex-valued continuous-time ACS stochastic process. In such a case, the quadratic TFDs defined in Section 6.4 are random functions. In the following, their expected values are expressed in terms of cyclic autocorrelation functions and cyclic spectra.

Symmetric *autocorrelation function*

$$\mathcal{R}_x(t,\tau) \triangleq \mathrm{E}\{K_x(t,\tau)\} \tag{6.21a}$$

$$= \sum_{\alpha\in A}\left[R_x^{\alpha}(\tau)\, e^{-j\pi\alpha\tau}\right] e^{j2\pi\alpha t} \tag{6.21b}$$

where

$$R_{\boldsymbol{x}}^{\alpha}(\tau)\,e^{-j\pi\alpha\tau} = \lim_{T\to\infty}\frac{1}{T}\int_{-T/2}^{T/2}\mathcal{R}_{\boldsymbol{x}}(t,\tau)\,e^{-j2\pi\alpha t}\,\mathrm{d}t \tag{6.22}$$

is the symmetric *cyclic autocorrelation function* (1.34).

Expected value of the symmetric *ambiguity function* (Napolitano, 2015b, Eq. (9.6.15))

$$\mathrm{E}\{A_{\boldsymbol{x}}(\nu,\tau)\} \triangleq \int_{\mathbb{R}}\mathcal{R}_{\boldsymbol{x}}(t,\tau)\,e^{-j2\pi\nu t}\,\mathrm{d}t \tag{6.23a}$$

$$= \sum_{\alpha\in A}\left[R_{\boldsymbol{x}}^{\alpha}(\tau)\,e^{-j\pi\alpha\tau}\right]\delta(\nu-\alpha) \tag{6.23b}$$

Expected value of the *Wigner distribution* (Martin and Flandrin, 1985), (Amin, 1992), (Flandrin, 1999), (Antoni, 2007a), (Napolitano, 2015b, Eq. (9.6.14))

$$\mathrm{E}\{W_{\boldsymbol{x}}(t,f)\} \triangleq \int_{\mathbb{R}}\mathcal{R}_{\boldsymbol{x}}(t,\tau)\,e^{-j2\pi f\tau}\,\mathrm{d}\tau \tag{6.24a}$$

$$= \sum_{\alpha\in A} S_{\boldsymbol{x}}^{\alpha}(f+\alpha/2)\,e^{j2\pi\alpha t} \tag{6.24b}$$

where $S_{\boldsymbol{x}}^{\alpha}(f+\alpha/2)$ is the symmetric *cyclic spectrum* (1.35) which is the Fourier transform of the symmetric cyclic autocorrelation function (6.22).

Spectral autocorrelation function or *rotated Loève bifrequency spectrum* (Section 6.8)

$$\mathrm{E}\{k_{\boldsymbol{x}}(\nu,f)\} \triangleq \int_{\mathbb{R}^2}\mathcal{R}_{\boldsymbol{x}}(t,\tau)\,e^{-j2\pi\nu t}\,e^{-j2\pi f\tau}\,\mathrm{d}t\,\mathrm{d}\tau \tag{6.25a}$$

$$= \mathrm{E}\Big\{X(f+\nu/2)\,X^*(f-\nu/2)\Big\} \tag{6.25b}$$

$$= \sum_{\alpha\in A} S_{\boldsymbol{x}}^{\alpha}(f+\alpha/2)\,\delta(\nu-\alpha) \tag{6.25c}$$

6.5 Filtered quadratic TFDs

6.5.1 Kernels

In this section, the finite-time counterpart of the almost-periodic component extraction operator (Definition 2.14) is considered as time-lag kernel for a given ACS signal $x(t)$. Then, the other kernels are derived by Fourier transformations. Their mutual relationships are illustrated in Fig. 6.2.

Let $\mathrm{v}(\tau)$ a unit-width lag window with Fourier transform $V(f)$ such that $\mathrm{v}(0) = \int_{\mathbb{R}} V(f)\mathrm{d}f = 1$. Thus, $\lim_{\Delta f\to 0} V(f/\Delta f)/\Delta f = \delta(f)$. For example, $\mathrm{v}(\tau) = \mathrm{rect}(\tau) \leftrightarrow V(f) = \mathrm{sinc}(f)$.

Let A be the countable set of second-order cycle frequencies of $x(t)$.

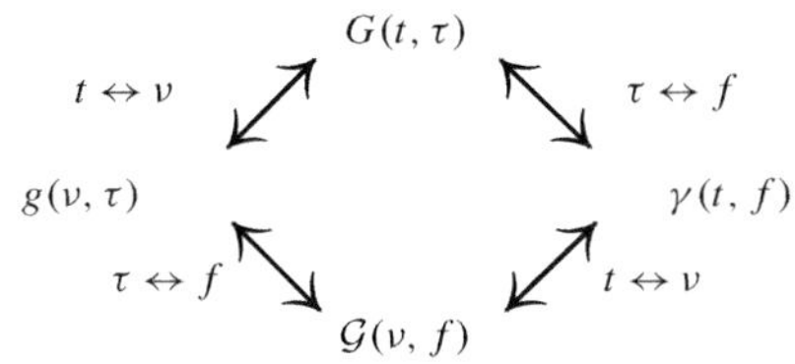

FIGURE 6.2 Kernels.

The following kernels can be defined.

Time-lag kernel

$$G(t,\tau) = \frac{1}{T}\,\mathrm{rect}\Big(\frac{t}{T}\Big)\sum_{\alpha\in A} e^{j2\pi\alpha t}\,\mathrm{v}(\Delta f\,\tau) \tag{6.26}$$

Doppler-lag kernel

$$g(\nu,\tau) \triangleq \int_{\mathbb{R}} G(t,\tau)\,e^{-j2\pi\nu t}\,\mathrm{d}t \tag{6.27a}$$

$$= \sum_{\alpha\in A} \mathrm{sinc}((\nu-\alpha)T)\,\mathrm{v}(\Delta f\,\tau) \tag{6.27b}$$

Time-frequency kernel

$$\gamma(t,f) \triangleq \int_{\mathbb{R}} G(t,\tau)\,e^{-j2\pi f\tau}\,\mathrm{d}\tau \tag{6.28a}$$

$$= \frac{1}{T}\,\mathrm{rect}\left(\frac{t}{T}\right)\sum_{\alpha\in A} e^{j2\pi\alpha t}\,\frac{1}{\Delta f}V\left(\frac{f}{\Delta f}\right) \tag{6.28b}$$

Doppler-frequency kernel

$$\mathcal{G}(\nu,f) \triangleq \int_{\mathbb{R}^2} G(t,\tau)\,e^{-j2\pi\nu t}\,e^{-j2\pi f\tau}\,\mathrm{d}t\,\mathrm{d}\tau \tag{6.29a}$$

$$= \sum_{\alpha\in A} \mathrm{sinc}((\nu-\alpha)T)\,\frac{1}{\Delta f}V\left(\frac{f}{\Delta f}\right) \tag{6.29b}$$

The presence of the window $\mathrm{v}(\Delta f\,\tau)$ is not crucial in the definitions of kernels $G(t,\tau)$ and $g(\nu,\tau)$. It will set to zero $\widetilde{K}_{\boldsymbol{x}}(t,\tau)$ and $\widetilde{A}_{\boldsymbol{x}}(\nu,\tau)$ (see Section 6.6) at values of $|\tau|$ larger than $1/2\Delta f$ since for these values of $|\tau|$ the variance of the estimators is large. In contrast, $V(f/\Delta f)/\Delta f$ is important to get consistency of estimators $\widetilde{W}_{\boldsymbol{x}}(t,f)$ and $\widetilde{k}_{\boldsymbol{x}}(\nu,f)$.

6.6 Filtered quadratic TFDs of ACS signals

Filtered quadratic TFDs for generic stochastic processes are defined according to (Boashash, 2015, Secs. 3.2.2–3.2.4). Their mutual relationships are illustrated in Fig. 6.3. In the following, their expressions for ACS processes are presented.

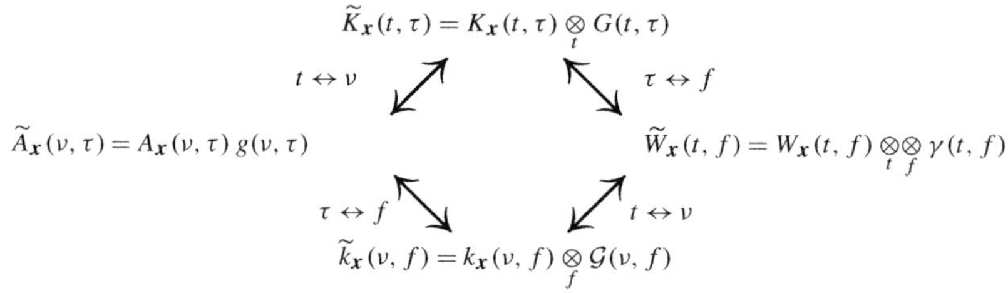

FIGURE 6.3 Filtered quadratic TFDs.

The *smoothed 2nd-order lag product* or *smoothed instantaneous autocorrelation function* is an estimator of the symmetric autocorrelation function $\mathcal{R}_{x}(t,\tau)$. For ACS processes it is given by (Napolitano, 2015b, Eq. (9.6.17)) (Section 6.8)

$$\widetilde{K}_{x}(t,\tau)\triangleq K_{x}(t,\tau)\underset{t}{\otimes}G(t,\tau) \tag{6.30a}$$

$$=\sum_{\alpha\in A}R_{x}^{(T)}(\alpha,\tau;t)\,e^{j2\pi\alpha t}\,\mathrm{v}(\Delta f\,\tau) \tag{6.30b}$$

where

$$R_{x}^{(T)}(\alpha,\tau;t)\triangleq\frac{1}{T}\int_{t-T/2}^{t+T/2}x(u+\tau/2)\,x^{*}(u-\tau/2)\,e^{-j2\pi\alpha u}\,\mathrm{d}u \tag{6.31}$$

denotes here the symmetric *cyclic correlogram* of $x(t)$. Under mild regularity assumptions on the finite or practically finite memory of $x(t)$ expressed in terms of summability of cumulants, the cyclic correlogram is a mean-square consistent (as $T\to\infty$) estimator of the cyclic autocorrelation function. Moreover, $\sqrt{T}[R_{x}^{(T)}(\alpha,\tau;t)-R_{x}^{\alpha}(\tau)]$ is zero-mean asymptotically complex normal (Section 5.2.1).

The symmetric *generalized ambiguity function* is given by

$$\widetilde{A}_{x}(\nu,\tau)=A_{x}(\nu,\tau)\,g(\nu,\tau) \tag{6.32a}$$

$$=\int_{\mathbb{R}}x(t+\tau/2)\,x^{*}(t-\tau/2)\,e^{-j2\pi\nu t}\mathrm{d}t\sum_{\alpha\in A}\mathrm{sinc}((\nu-\alpha)T)\,\mathrm{v}(\Delta f\,\tau) \tag{6.32b}$$

The *smoothed Wigner distribution* is an estimator of the Wigner distribution $W_{x}(t,f)$. For ACS processes it is given by (Napolitano, 2015b, Eq. (9.6.18))

$$\widetilde{W}_{x}(t,f)\triangleq\int_{\mathbb{R}}\widetilde{K}_{x}(t,\tau)\,e^{-j2\pi f\tau}\,\mathrm{d}\tau \tag{6.33a}$$

$$=\sum_{\alpha\in A}\left[I_{x}^{(T)}(\alpha,f;t)\underset{f}{\otimes}\frac{1}{\Delta f}V\left(\frac{f}{\Delta f}\right)\right]e^{j2\pi\alpha t} \tag{6.33b}$$

where

$$I_{x}^{(T)}(\alpha,f;t)\triangleq\frac{1}{T}X_{T}(t,f+\alpha/2)\,X_{T}^{*}(t,f-\alpha/2) \tag{6.34}$$

is the symmetric cyclic periodogram of $x(t)$ with

$$X_T(t,f) \triangleq \int_{t-T/2}^{t+T/2} x(u)\, e^{-j2\pi f u}\, \mathrm{d}u \tag{6.35}$$

short-time Fourier transform (STFT) of $x(t)$. It results (Lemma 2.39)

$$I_{\boldsymbol{x}}^{(T)}(\alpha, f; t) = \int_{\mathbb{R}} R_{\boldsymbol{x}}^{(T)}(\alpha, \tau; t)\, e^{-j2\pi f\tau}\, \mathrm{d}\tau \,. \tag{6.36}$$

Expression (6.33b) can be equivalently obtained as

$$\widetilde{W}_{\boldsymbol{x}}(t,f) = W_{\boldsymbol{x}}(t,f) \underset{t}{\otimes}\underset{f}{\otimes} \gamma(t,f)\,. \tag{6.37}$$

The cyclic periodogram $I_{\boldsymbol{x}}^{(T)}(\alpha, f; t)$ is an asymptotically (as $T \to \infty$) unbiased but not consistent estimator of $S_{\boldsymbol{x}}^{\alpha}(f+\alpha/2)$ (Section 5.2.2). The right-hand side of (6.33b) is the superposition of frequency-smoothed (conjugate) cyclic periodograms. It is a mean-square consistent estimator of the Wigner distribution of ACS signals provided that $T \to \infty$ and $\Delta f \to 0$ with $T\Delta f \to \infty$ (Section 5.2.3) and the sums over $\alpha \in A$ of biases and variances of the terms in (6.33b) are absolutely convergent.

The *smoothed spectral autocorrelation function* is given by

$$\widetilde{k}_{\boldsymbol{x}}(\nu, f) = k_{\boldsymbol{x}}(\nu, f) \underset{f}{\otimes} \mathcal{G}(\nu, f) \tag{6.38a}$$

$$= \int_{\mathbb{R}} \widetilde{W}_{\boldsymbol{x}}(t,f)\, e^{-j2\pi\nu t}\, \mathrm{d}t \tag{6.38b}$$

$$= \sum_{\alpha\in A} \int_{\mathbb{R}} \left[I_{\boldsymbol{x}}^{(T)}(\alpha, f; t) \underset{f}{\otimes} \frac{1}{\Delta f} V\left(\frac{f}{\Delta f}\right)\right] e^{-j2\pi(\nu-\alpha)t}\, \mathrm{d}t \,. \tag{6.38c}$$

6.6.1 Expected values of filtered quadratic TFDs of ACS signals

In this section, the expected values of the filtered quadratic TFDs of ACS signals are expressed in terms of cyclic autocorrelation functions and cyclic spectra. Proofs are in Section 6.8.

Smoothed autocorrelation function

$$\mathrm{E}\left\{\widetilde{K}_{\boldsymbol{x}}(t,\tau)\right\} = \mathrm{E}\{K_{\boldsymbol{x}}(t,\tau)\} \underset{t}{\otimes} G(t,\tau) \tag{6.39a}$$

$$= \sum_{\alpha\in A} \mathrm{E}\left\{R_{\boldsymbol{x}}^{(T)}(\alpha, \tau; t)\right\} e^{j2\pi\alpha t}\, \mathrm{v}(\Delta f\, \tau) \tag{6.39b}$$

$$= \mathrm{E}\left\{K_{\boldsymbol{x}}(t,\tau)\right\} \mathrm{v}(\Delta f\, \tau)$$

$$+ \sum_{\alpha\in A} \sum_{\substack{\beta\in A\\ \beta\neq\alpha}} \left[R_{\boldsymbol{x}}^{\beta}(\tau)\, e^{-j\pi\beta\tau}\right] \mathrm{sinc}(T(\beta-\alpha))\, e^{j2\pi\beta t}\, \mathrm{v}(\Delta f\, \tau) \tag{6.39c}$$

with (Theorem 5.13)

$$\mathrm{E}\Big\{R_{\boldsymbol{x}}^{(T)}(\alpha, \tau; t)\Big\} = R_{\boldsymbol{x}}^{\alpha}(\tau) + \mathcal{O}(T^{-r}) \tag{6.40}$$

where $r = \gamma$ in Theorem 5.13.

Filtered generalized ambiguity function

$$\mathrm{E}\Big\{\widetilde{A}_{\boldsymbol{x}}(\nu, \tau)\Big\} \triangleq \int_{\mathbb{R}} \mathrm{E}\Big\{\widetilde{K}_{\boldsymbol{x}}(t, \tau)\Big\}\, e^{-j2\pi\nu t}\, \mathrm{d}t \tag{6.41a}$$

$$= \mathrm{E}\Big\{A_{\boldsymbol{x}}(\nu, \tau)\Big\}\, \mathrm{v}(\Delta f\, \tau)$$

$$+ \sum_{\alpha\in A} \sum_{\substack{\beta\in A \\ \beta\neq\alpha}} R_{\boldsymbol{x}}^{\beta}(\tau)\, e^{-j\pi\beta\tau}\, \mathrm{sinc}(T(\beta-\alpha))\, \delta(\nu-\beta)\, \mathrm{v}(\Delta f\, \tau) \tag{6.41b}$$

Filtered Wigner distribution

$$\mathrm{E}\Big\{\widetilde{W}_{\boldsymbol{x}}(t, f)\Big\} = \mathrm{E}\Big\{W_{\boldsymbol{x}}(t, f)\Big\} \underset{f}{\otimes} \frac{1}{\Delta f} V\Big(\frac{f}{\Delta f}\Big)$$

$$+ \sum_{\alpha\in A} \sum_{\substack{\beta\in A \\ \beta\neq\alpha}} \left[S_{\boldsymbol{x}}^{\beta}(f+\beta/2) \underset{f}{\otimes} \frac{1}{\Delta f} V\left(\frac{f}{\Delta f}\right)\right] \mathrm{sinc}(T(\beta-\alpha))\, e^{j2\pi\beta t} \tag{6.42}$$

Filtered rotated Loéve bifrequency spectrum

$$\mathrm{E}\Big\{\widetilde{k}_{\boldsymbol{x}}(\nu, f)\Big\} = \mathrm{E}\Big\{k_{\boldsymbol{x}}(\nu, f)\Big\} \underset{f}{\otimes} \frac{1}{\Delta f} V\Big(\frac{f}{\Delta f}\Big)$$

$$+ \sum_{\alpha\in A} \sum_{\substack{\beta\in A \\ \beta\neq\alpha}} \left[S_{\boldsymbol{x}}^{\beta}(f+\beta/2) \underset{f}{\otimes} \frac{1}{\Delta f} V\left(\frac{f}{\Delta f}\right)\right] \mathrm{sinc}(T(\beta-\alpha))\, \delta(\nu-\beta) \tag{6.43}$$

6.7 Cohen's class of bilinear (or quadratic) time-frequency distributions

The bilinear or quadratic time-frequency distributions of the Cohen's class can be expressed as (Cohen, 1989), (Cohen, 1995)

$$C_{\boldsymbol{x}}(t, f) = \int_{\mathbb{R}^2} A_{\boldsymbol{x}}(\nu, \tau)\, \Phi(\nu, \tau)\, e^{j2\pi(\nu t - f\tau)}\, \mathrm{d}\nu\, \mathrm{d}\tau \tag{6.44a}$$

$$= \int_{\mathbb{R}^2} W_{\boldsymbol{x}}(\theta, \lambda)\, \Pi(t-\theta, f-\lambda)\, \mathrm{d}\theta\, \mathrm{d}\lambda \tag{6.44b}$$

where the kernels are linked by

$$\Phi(\nu, \tau) = \int_{\mathbb{R}^2} \Pi(t, f)\, e^{-j2\pi(\nu t - f\tau)}\, \mathrm{d}t\, \mathrm{d}f\, . \tag{6.45}$$

In the special case of ACS processes, the expected value of $C_x(t, f)$ is obtained by using (6.23b) into the expected value of (6.44a) or, equivalently, using (6.24b) into the expected value of (6.44b). The case of GACS signals is considered in (Izzo and Napolitano, 2005, Sec. 5) in the fraction-of-time approach.

6.8 Proofs

In this Section, proofs are reported for results of the present chapter. Unless otherwise specified, only formal manipulations are reported, without explicitly justifying inversion of linear operators such as time averages, infinite sums, integrals, and ensemble average (expectation). In general, these manipulations are justified at least in the sense of generalized functions (distributions) (Champeney, 1990), (Zemanian, 1987).

Proof of (6.1b)

Let $X_i(f)$ be the Fourier transform of $x_i(t)$ $(i = 1, 2)$. From definition (6.1a), we have

$$\begin{aligned} A_{x_1x_2}(\nu, \tau) &= \int_{\mathbb{R}} \int_{\mathbb{R}} X_1(f_1)\, e^{j2\pi f_1(t+\tau)}\, \mathrm{d}f_1 \int_{\mathbb{R}} X_2(f_2)\, e^{j2\pi f_2 t}\, \mathrm{d}f_2\, e^{-j2\pi\nu t}\, \mathrm{d}t \\ &= \int_{\mathbb{R}} \int_{\mathbb{R}} X_1(f_1)\, X_2(f_2)\, e^{j2\pi f_1\tau} \int_{\mathbb{R}} e^{j2\pi(f_1+f_2-\nu)t}\, \mathrm{d}t\, \mathrm{d}f_1\, \mathrm{d}f_2 \\ &= \int_{\mathbb{R}} \int_{\mathbb{R}} X_1(f_1)\, X_2(f_2)\, e^{j2\pi f_1\tau}\, \delta(f_1 + f_2 - \nu)\, \mathrm{d}f_1\, \mathrm{d}f_2 \\ &= \int_{\mathbb{R}} X_1(f_1)\, X_2(\nu - f_1)\, e^{j2\pi f_1\tau}\, \mathrm{d}f_1\,. \end{aligned} \tag{6.46}$$

Proof of (6.7a) and (6.7b)

Using expression (6.4b) for the STFT we have

$$\begin{aligned} &\mathsf{S}_{\boldsymbol{x}}(t, f) \\ &\quad\triangleq X_T(t, f)\, X_T^{(*)}(t, f) \\ &\quad= \int_{\mathbb{R}} x(t + s_1)\, b_T(s_1)\, e^{-j2\pi f(t+s_1)}\, \mathrm{d}s_1 \int_{\mathbb{R}} x^{(*)}(t + s_2)\, b_T^{(*)}(s_2)\, e^{-(-)j2\pi f(t+s_2)}\, \mathrm{d}s_2 \\ &\quad= \int_{\mathbb{R}^2} x(t + s_1)\, x^{(*)}(t + s_2)\, b_T(s_1)\, b_T^{(*)}(s_2)\, e^{-j2\pi f s_1}\, e^{-(-)j2\pi f s_2}\, \mathrm{d}s_1\, \mathrm{d}s_2\, e^{-j2\pi\kappa f t} \end{aligned} \tag{6.47}$$

This expression allows one, in the following, to interchange the order of expectation operator $\mathrm{E}\{\cdot\}$ and $\int_{\mathbb{R}}(\cdot)\, \mathrm{d}s_1\, \mathrm{d}s_2$ even if $\mathrm{E}\{\cdot\}$ is replaced by the almost-periodic component extraction operator $\mathrm{E}^{\{\alpha\}}\{\cdot\}$ (Definition 2.14) for the analysis the fraction-of-time (FOT) approach. In contrast, using (6.4a) for the STFT, leads to an expression where the order of $\mathrm{E}^{\{\alpha\}}\{\cdot\}$ and $\int_{\mathbb{R}}(\cdot)\, \mathrm{d}s_1\, \mathrm{d}s_2$ cannot be interchanged in the FOT approach (see also Remark 2.30).

$$
\begin{aligned}
&\mathrm{E}\{\mathrm{S}_{\boldsymbol{x}}(t,f)\} \\
&\quad\triangleq \mathrm{E}\left\{X_T(t,f)\,X_T^{(*)}(t,f)\right\} \\
&\quad= \int_{\mathbb{R}^2} \mathrm{E}\left\{x(t+s_1)\,x^{(*)}(t+s_2)\right\}\, b_T(s_1)\, b_T^{(*)}(s_2)\, e^{-j2\pi f s_1}\, e^{-(-)j2\pi f s_2}\, \mathrm{d}s_1\, \mathrm{d}s_2\, e^{-j2\pi\kappa f t} \\
&\quad= \int_{\mathbb{R}^2} \sum_{\alpha\in A} R_{\boldsymbol{x}}^{\alpha}(s_1-s_2)\, e^{j2\pi\alpha(t+s_2)}\, b_T(s_1)\, b_T^{(*)}(s_2)\, e^{-j2\pi f s_1}\, e^{-(-)j2\pi f s_2}\, \mathrm{d}s_1\, \mathrm{d}s_2\, e^{-j2\pi\kappa f t} \\
&\quad= \sum_{\alpha\in A} \int_{\mathbb{R}^2} R_{\boldsymbol{x}}^{\alpha}(\tau)\, e^{j2\pi\alpha(t+s_2)}\, b_T(s_2+\tau)\, b_T^{(*)}(s_2)\, e^{-j2\pi f(s_2+\tau)}\, e^{-(-)j2\pi f s_2}\, \mathrm{d}s_1\, \mathrm{d}s_2\, e^{-j2\pi\kappa f t} \\
&\quad= \sum_{\alpha\in A} \int_{\mathbb{R}} R_{\boldsymbol{x}}^{\alpha}(\tau)\, e^{-j2\pi f\tau} \int_{\mathbb{R}} b_T(s_2+\tau)\, b_T^{(*)}(s_2)\, e^{-j2\pi(\kappa f-\alpha)s_2} \mathrm{d}s_2\, \mathrm{d}\tau\, e^{j2\pi(\alpha-\kappa f)t} \\
&\quad= \sum_{\alpha\in A} \int_{\mathbb{R}} R_{\boldsymbol{x}}^{\alpha}(\tau)\, r_{\boldsymbol{b}}^{(\kappa f-\alpha)}(\tau)\, e^{-j2\pi f\tau} \mathrm{d}\tau\, e^{j2\pi(\alpha-\kappa f)t}
\end{aligned} \tag{6.48}
$$

where in the fourth equality the variable change $\tau = s_1 - s_2$ is made and $r_{\boldsymbol{b}}^{\beta}(\tau)$ is defined in (6.8).

Using the Fourier transform pair (3.254)

$$
r_{\boldsymbol{b}}^{\beta}(\tau) \triangleq \int_{\mathbb{R}} b_T(s+\tau)\, b_T^{(*)}(s)\, e^{-j2\pi\beta s}\, \mathrm{d}s \quad \stackrel{\mathcal{F}}{\longleftrightarrow} \quad B_{\frac{1}{T}}(f)\, B_{\frac{1}{T}}^{(*)}((-)(\beta-f)) \tag{6.49}
$$

we have

$$
\begin{aligned}
\mathrm{E}\{\mathrm{S}_{\boldsymbol{x}}(t,f)\} &= \sum_{\alpha\in A} \int_{\mathbb{R}} R_{\boldsymbol{x}}^{\alpha}(\tau)\, r_{\boldsymbol{b}}^{\beta}(\tau)\, e^{-j2\pi f\tau} \mathrm{d}\tau \Big|_{\beta=\kappa f-\alpha} e^{j2\pi(\alpha-\kappa f)t} \\
&= \sum_{\alpha\in A} \left[S_{\boldsymbol{x}}^{\alpha}(f) \underset{f}{\otimes} \left[B_{\frac{1}{T}}(f)\, B_{\frac{1}{T}}^{(*)}((-)(\beta-f)) \right] \right]_{\beta=\kappa f-\alpha} e^{j2\pi(\alpha-\kappa f)t} \\
&= \sum_{\alpha\in A} S_{\boldsymbol{x}}^{\alpha}(f) \underset{f}{\otimes} \left[B_{\frac{1}{T}}(f)\, B_{\frac{1}{T}}^{(*)}(f-(-)\alpha) \right] e^{j2\pi(\alpha-\kappa f)t}
\end{aligned} \tag{6.50}
$$

where the second equality holds provided that $B_{\frac{1}{T}}(f)$ is even (and hence $b_T(t)$ is even). In fact, $(-)(\kappa f - \alpha - f) = (-)(f + (-)f - \alpha - f) = f - (-)\alpha$ and, since $B_{\frac{1}{T}}(f)$ is even, the replacement $\beta = \kappa f - \alpha$ can be made also before doing the convolution over f.

Proof of (6.25b) and (6.25c)

Substituting (6.17) and (6.21a) into (6.25a) and making the variable change $t_1 = t + \tau/2$, $t_2 = t - \tau/2$ ($\tau = t_1 - t_2$, $t = (t_1 + t_2)/2$), we have

$$
\begin{aligned}
\mathrm{E}\{k_{\boldsymbol{x}}(\nu,f)\} &= \int_{\mathbb{R}^2} \mathrm{E}\left\{x(t+\tau/2)\,x^*(t-\tau/2)\right\}\, e^{-j2\pi f\tau}\, e^{-j2\pi\nu t}\, \mathrm{d}\tau\, \mathrm{d}t \\
&= \int_{\mathbb{R}^2} \mathrm{E}\left\{x(t_1)\,x^*(t_2)\right\}\, e^{-j2\pi f(t_1-t_2)}\, e^{-j2\pi\nu(t_1+t_2)/2}\, \mathrm{d}t_1\, \mathrm{d}t_2 \\
&= \int_{\mathbb{R}^2} \mathrm{E}\left\{x(t_1)\,x^*(t_2)\right\}\, e^{-j2\pi[(f+\nu/2)t_1-(f-\nu/2)t_2]}\, \mathrm{d}t_1\, \mathrm{d}t_2
\end{aligned}
$$

$$= \mathrm{E}\Big\{X(f+\nu/2)\, X^*(f-\nu/2)\Big\} \tag{6.51}$$

which proves (6.25b). Expression (6.25c) for ACS signals follows accounting for (1.29).

Proof of (6.30b)

$$\begin{aligned}
\mathrm{E}^{(T)}_{t,A}\{K_{\boldsymbol{x}}(t,\tau)\} &\triangleq \sum_{\alpha\in A} \frac{1}{T}\int_{t-T/2}^{t+T/2} x(u+\tau/2)\, x^*(u-\tau/2)\, e^{-j2\pi\alpha u} \mathrm{d}u\, e^{j2\pi\alpha t}\ \mathrm{v}(\Delta f\,\tau) \\
&= \sum_{\alpha\in A} \frac{1}{T}\int_{\mathbb{R}} x(u+\tau/2)\, x^*(u-\tau/2) \mathrm{rect}\Big(\frac{u-t}{T}\Big)\, e^{j2\pi\alpha(t-u)} \mathrm{d}u\ \mathrm{v}(\Delta f\,\tau) \\
&= K_{\boldsymbol{x}}(t,\tau) \underset{t}{\otimes} \Big[\frac{1}{T}\,\mathrm{rect}\Big(\frac{t}{T}\Big) \sum_{\alpha\in A} e^{j2\pi\alpha t}\Big]\, \mathrm{v}(\Delta f\,\tau) \\
&= K_{\boldsymbol{x}}(t,\tau) \underset{t}{\otimes} G(t,\tau) \\
&= \widetilde{K}_{\boldsymbol{x}}(t,\tau)
\end{aligned} \tag{6.52}$$

where (6.30a) and the fact that rect(·) is an even function have been accounted for.

Expected value of the cyclic correlogram

$$\begin{aligned}
&\mathrm{E}\Big\{R^{(T)}_{\boldsymbol{x}}(\alpha,\tau;t)\Big\}\ \mathrm{v}(\Delta f\,\tau) \\
&= \frac{1}{T}\int_{t-T/2}^{t+T/2} \mathrm{E}\Big\{x(u+\tau/2)\, x^*(u-\tau/2)\Big\}\, e^{-j2\pi\alpha u} \mathrm{d}u\ \mathrm{v}(\Delta f\,\tau) \\
&= \int_{\mathbb{R}} \mathrm{E}\Big\{K_{\boldsymbol{x}}(u,\tau)\Big\} \frac{1}{T}\,\mathrm{rect}\Big(\frac{u-t}{T}\Big)\, e^{j2\pi\alpha(t-u)} \mathrm{d}u\, e^{-j2\pi\alpha t}\ \mathrm{v}(\Delta f\,\tau) \\
&= \Big[\mathrm{E}\Big\{K_{\boldsymbol{x}}(t,\tau)\Big\} \underset{t}{\otimes} \frac{1}{T}\,\mathrm{rect}\Big(\frac{t}{T}\Big)\, e^{j2\pi\alpha t}\Big]\, e^{-j2\pi\alpha t}\ \mathrm{v}(\Delta f\,\tau) \\
\overset{t\leftrightarrow\nu}{\longleftrightarrow}\ &\Big[\mathrm{E}\{A_{\boldsymbol{x}}(\nu,\tau)\}\ \mathrm{sinc}(T(\nu-\alpha))\Big] \underset{\nu}{\otimes} \delta(\nu+\alpha)\, \mathrm{v}(\Delta f\,\tau) \\
&= \Big[\sum_{\beta\in A} R^{\beta}_{\boldsymbol{x}}(\tau)\, e^{-j\pi\beta\tau}\, \delta(\nu-\beta)\, \mathrm{sinc}(T(\nu-\alpha))\Big] \underset{\nu}{\otimes} \delta(\nu+\alpha)\, \mathrm{v}(\Delta f\,\tau) \\
&= \Big[\sum_{\beta\in A} R^{\beta}_{\boldsymbol{x}}(\tau)\, e^{-j\pi\beta\tau} \mathrm{sinc}(T(\beta-\alpha))\, \delta(\nu-\beta)\Big] \underset{\nu}{\otimes} \delta(\nu+\alpha)\, \mathrm{v}(\Delta f\,\tau) \\
&= \sum_{\beta\in A} R^{\beta}_{\boldsymbol{x}}(\tau)\, e^{-j\pi\beta\tau} \mathrm{sinc}(T(\beta-\alpha))\, \delta(\nu+\alpha-\beta)\, \mathrm{v}(\Delta f\,\tau) \\
&= R^{\alpha}_{\boldsymbol{x}}(\tau)\, e^{-j\pi\alpha\tau}\ \delta(\nu)\, \mathrm{v}(\Delta f\,\tau) \\
&\quad + \sum_{\substack{\beta\in A\\ \beta\neq\alpha}} R^{\beta}_{\boldsymbol{x}}(\tau)\, e^{-j\pi\beta\tau} \mathrm{sinc}(T(\beta-\alpha))\, \delta(\nu+\alpha-\beta)\, \mathrm{v}(\Delta f\,\tau)
\end{aligned} \tag{6.53}$$

where the last term is due to cycle leakage and becomes negligible as $T\to\infty$.

Proofs for Section 6.6.1

Accounting for (6.53) it results:

$$\mathrm{E}\left\{\widetilde{K}_{\boldsymbol{x}}(t,\tau)\right\} = \sum_{\alpha\in A} \mathrm{E}\left\{R_{\boldsymbol{x}}^{(T)}(\alpha,\tau;t)\right\} e^{j2\pi\alpha t}\, \mathrm{v}(\Delta f\,\tau)$$

$$\Big\updownarrow\; t \leftrightarrow \nu$$

$$\begin{aligned}
\mathrm{E}\left\{\widetilde{A}_{\boldsymbol{x}}(\nu,\tau)\right\} &= \sum_{\alpha\in A}\Bigg[R_{\boldsymbol{x}}^{\alpha}(\tau)\, e^{-j\pi\alpha\tau}\,\delta(\nu) \\
&\quad + \sum_{\substack{\beta\in A\\ \beta\neq\alpha}} R_{\boldsymbol{x}}^{\beta}(\tau)\, e^{-j\pi\beta\tau}\operatorname{sinc}(T(\beta-\alpha))\,\delta(\nu+\alpha-\beta)\Bigg] \underset{\nu}{\otimes} \delta(\nu-\alpha)\,\mathrm{v}(\Delta f\,\tau) \\
&= \underbrace{\sum_{\alpha\in A} R_{\boldsymbol{x}}^{\alpha}(\tau)\, e^{-j\pi\alpha\tau}\,\delta(\nu-\alpha)\;\mathrm{v}(\Delta f\,\tau)}_{\mathrm{E}\{A_{\boldsymbol{x}}(\nu,\tau)\}} \\
&\quad + \sum_{\alpha\in A}\sum_{\substack{\beta\in A\\ \beta\neq\alpha}} R_{\boldsymbol{x}}^{\beta}(\tau)\, e^{-j\pi\beta\tau}\operatorname{sinc}(T(\beta-\alpha))\,\delta(\nu-\beta)\,\mathrm{v}(\Delta f\,\tau)
\end{aligned} \tag{6.54}$$

$$\Big\updownarrow\; \tau \leftrightarrow f$$

$$\begin{aligned}
\mathrm{E}\left\{\widetilde{k}_{\boldsymbol{x}}(\nu,f)\right\} &= \mathrm{E}\{k_{\boldsymbol{x}}(\nu,f)\} \underset{f}{\otimes} \frac{1}{\Delta f} V\!\left(\frac{f}{\Delta f}\right) \\
&\quad + \sum_{\alpha\in A}\sum_{\substack{\beta\in A\\ \beta\neq\alpha}} S_{\boldsymbol{x}}^{\beta}(f+\beta/2)\operatorname{sinc}(T(\beta-\alpha))\,\delta(\nu-\beta) \underset{f}{\otimes} \frac{1}{\Delta f} V\!\left(\frac{f}{\Delta f}\right)
\end{aligned} \tag{6.55}$$

$$\Big\updownarrow\; t \leftrightarrow \nu$$

$$\begin{aligned}
\mathrm{E}\left\{\widetilde{W}_{\boldsymbol{x}}(t,f)\right\} &= \mathrm{E}\{W_{\boldsymbol{x}}(t,f)\} \underset{f}{\otimes} \frac{1}{\Delta f} V\!\left(\frac{f}{\Delta f}\right) \\
&\quad + \sum_{\alpha\in A}\sum_{\substack{\beta\in A\\ \beta\neq\alpha}} S_{\boldsymbol{x}}^{\beta}(f+\beta/2)\operatorname{sinc}(T(\beta-\alpha))\, e^{j2\pi\beta t} \underset{f}{\otimes} \frac{1}{\Delta f} V\!\left(\frac{f}{\Delta f}\right)
\end{aligned} \tag{6.56}$$

$$\Big\updownarrow\; \tau \leftrightarrow f$$

$$\begin{aligned}
\mathrm{E}\left\{\widetilde{K}_{\boldsymbol{x}}(t,\tau)\right\} &= \mathrm{E}\left\{K_{\boldsymbol{x}}(t,\tau)\right\} \mathrm{v}(\Delta f\,\tau) \\
&\quad + \sum_{\alpha\in A}\sum_{\substack{\beta\in A\\ \beta\neq\alpha}} \left[R_{\boldsymbol{x}}^{\beta}(\tau)\, e^{-j\pi\beta\tau}\right]\operatorname{sinc}(T(\beta-\alpha))\, e^{j2\pi\beta t}\,\mathrm{v}(\Delta f\,\tau)
\end{aligned} \tag{6.57}$$

7

Manufactured signals

Contents

Abstract

Cyclostationarity properties of man-made communications signals are considered. Second-order cyclic statistics in both time and frequency domains are derived for the double side-band amplitude-modulated signal, the pulse-amplitude-modulated signal, and the direct-sequence spread-spectrum signal. Higher-order cyclic spectra are derived for PAM, QAM, DSB-AM, SSB, and ASK signals.

Keywords

Double side-band amplitude modulated, Pulse-amplitude-modulated, Direct-sequence spread spectrum, Higher-order cyclic spectra

7.1 Introduction

Periodicities in the multivariate statistical functions of man-made signals used in communications, radar, sonar, and telemetry arise from the construction and/or subsequent processing of the signal, that is, from operations such as modulation, scanning, sampling, multiplexing, and coding. The corresponding cycle frequencies are related to parameters of the processes such as carrier frequency, baud rate, sampling frequency, and code rate. The shapes of the second- and higher-order cyclic statistical functions depend on the shapes of the base pulses and the correlation properties of the modulating sequences (Gardner, 1985, Chap. 12), (Gardner, 1987f), (Gardner et al., 2006, Sec. 5).

Cyclostationary Processes and Time Series. https://doi.org/10.1016/B978-0-08-102708-0.00018-2

The analytical cyclic spectral analysis of mathematical models of analog and digitally modulated signals was first carried out in (Gardner, 1985) for stochastic processes and in (Gardner, 1987e) and (Gardner et al., 1987) for nonstochastic time series. The effects of multiplexing are considered in (Bukofzer, 1990). Continuous-phase frequency-modulated signals are treated in (Chen, 1989), (Gournay and Viravau, 1998), (Viravau and Gournay, 1998), and (Napolitano and Spooner, 2001). The effects of timing jitter on the cyclostationarity properties of communications signals are addressed in (Gardner, 1987f), (Elmirghani et al., 1995), (Win, 1998).

On this general subject, see the general treatments (Gardner, 1985), (Gardner, 1987f), (Brown, 1987) (Gardner, 1990b), (Gardner, 1991c), and also see (Gardner, 1994a), (Dragan and Yavorskii, 1982b), (Leduc, 1990).

In this chapter, two fundamental examples of cyclostationary communication signals are considered and their second-order cyclic statistics are derived. Specifically, the double side-band amplitude-modulated signal and the pulse-amplitude modulated signal are considered. The derivations of the second-order cyclic statistics can be accomplished by using the results of Sections 3.2 and 3.3. Then a third example of a more sophisticated communication signal is considered. Higher-order cyclic spectra for several communications signals are derived in Section 7.5. Finally, references are provided where second- and higher-order cyclic statistics of several special cases of communications signals are derived.

Several aspects related to the second- and higher-order cyclostationarity properties of communications signals are treated in the blog (Spooner, 2015).

7.2 Double side-band amplitude-modulated signal

Let $x(t)$ be the (real-valued) double side-band amplitude-modulated (DSB-AM) signal (Fig. 7.1)

$$x(t) \triangleq s(t)\,\cos(2\pi f_0 t + \phi_0)\,. \tag{7.1}$$

The cyclic autocorrelation function and cyclic spectrum of $x(t)$ are (Gardner, 1987e)

$$\begin{aligned}R_x^\alpha(\tau) &= \frac{1}{2} R_s^\alpha(\tau)\,\cos(2\pi f_0 \tau)\\ &\quad + \frac{1}{4}\left\{R_s^{\alpha+2f_0}(\tau)\,e^{-j2\pi f_0\tau}\,e^{-j2\phi_0} + R_s^{\alpha-2f_0}(\tau)\,e^{j2\pi f_0\tau}\,e^{j2\phi_0}\right\} \end{aligned}\tag{7.2}$$

$$\begin{aligned}S_x^\alpha(f) &= \frac{1}{4}\left\{S_s^\alpha(f-f_0) + S_s^\alpha(f+f_0)\right.\\ &\quad \left. + S_s^{\alpha+2f_0}(f+f_0)\,e^{-j2\phi_0} + S_s^{\alpha-2f_0}(f-f_0)\,e^{j2\phi_0}\right\}\end{aligned}\tag{7.3}$$

respectively. If $s(t)$ is a wide-sense stationary signal then $R_s^\alpha(\tau) = R_s^0(\tau)\,\delta_\alpha$ and

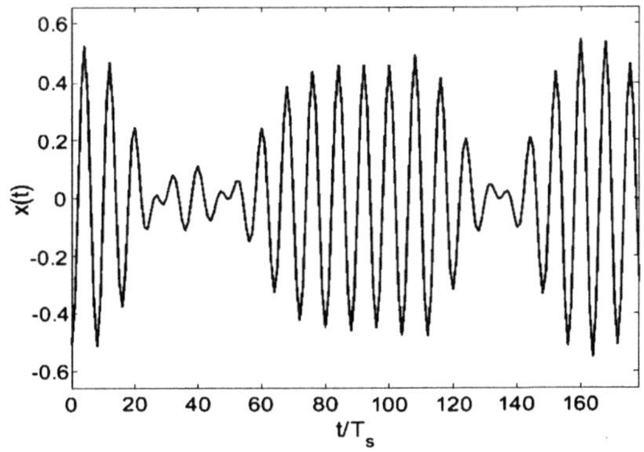

FIGURE 7.1 DSB-AM signal $x(t)$.

$$R_x^{\alpha}(\tau) = \begin{cases} \frac{1}{2} R_s^0(\tau) \cos(2\pi f_0 \tau) & \alpha = 0 \\ \frac{1}{4} R_s^0(\tau)\, e^{\pm j 2\pi f_0 \tau}\, e^{\pm j 2\phi_0} & \alpha = \pm 2 f_0 \\ 0 & \text{otherwise} \end{cases} \tag{7.4}$$

$$S_x^{\alpha}(f) = \begin{cases} \frac{1}{4} \left\{ S_s^0(f - f_0) + S_s^0(f + f_0) \right\} & \alpha = 0 \\ \frac{1}{4} S_s^0(f \mp f_0)\, e^{\pm j 2\phi_0} & \alpha = \pm 2 f_0 \\ 0 & \text{otherwise}\,. \end{cases} \tag{7.5}$$

Thus, $x(t)$ is cyclostationary with period $1/2 f_0$.

In Fig. 7.2A the magnitude of the cyclic autocorrelation function $R_x^{\alpha}(\tau)$, as a function of α and τ, and in Fig. 7.2B the magnitude of the cyclic spectrum $S_x^{\alpha}(f)$, as a function of α and f, are reported for the DSB-AM signal (7.1) with stationary modulating signal $s(t)$ having triangular autocorrelation function.

7.3 Pulse-amplitude-modulated signal

Let $x(t)$ be the complex-valued pulse-amplitude modulated (PAM) signal (Fig. 7.3)

$$x(t) \triangleq \sum_{n \in \mathbb{Z}} a_n\, q(t - n T_0)\,. \tag{7.6}$$

In (7.6), $q(t)$ is a complex-valued square integrable pulse and $\{a_n\}_{n \in \mathbb{Z}}$, $a_n \in \mathbb{C}$, is an ACS sequence whose cyclostationarity is possibly induced by framing, multiplexing, or coding (Schell, 1995b), with (conjugate) autocorrelation

$$\mathrm{E}\left\{ a_{n+m}\, a_n^{(*)} \right\} = \sum_{\widetilde{\alpha} \in \widetilde{A}} \widetilde{R}_{aa^{(*)}}^{\widetilde{\alpha}}(m)\, e^{j 2\pi \widetilde{\alpha} n} \tag{7.7}$$

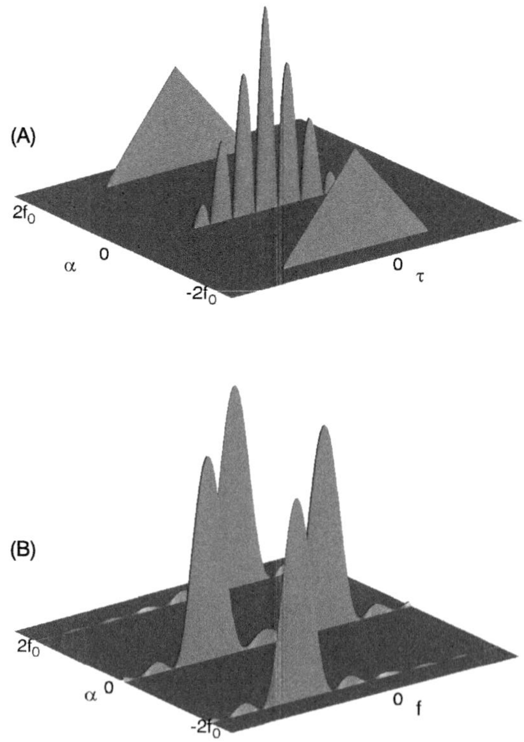

FIGURE 7.2 DSB-AM signal (7.1). (A) Magnitude of the cyclic autocorrelation function $R_x^{\alpha}(\tau)$, as a function of α and τ, and (B) magnitude of the cyclic spectrum $S_x^{\alpha}(f)$, as a function of α and f. *Source: (Gardner et al., 2006) Copyright of Elsevier.*

(conjugate) cyclic autocorrelation

$$\widetilde{R}_{aa^{(*)}}^{\widetilde{\alpha}}(m) \triangleq \lim_{N\to\infty} \frac{1}{2N+1} \sum_{n=-N}^{N} \mathrm{E}\left\{a_{n+m}\, a_n^{(*)}\right\} e^{-j2\pi\widetilde{\alpha}n} \tag{7.8}$$

and countable set of (conjugate) cycle frequencies

$$\widetilde{A} \triangleq \left\{\widetilde{\alpha} \in [-1/2, 1/2) \; : \; \widetilde{R}_{aa^{(*)}}^{\widetilde{\alpha}}(m) \neq 0\right\}. \tag{7.9}$$

The (conjugate) autocorrelation function of $x(t)$ is given by (Section 7.7)

$$\begin{aligned} &\mathrm{E}\left\{x(t+\tau)\, x^{(*)}(t)\right\} \\ &\quad = \sum_{\widetilde{\alpha}\in\widetilde{A}} \sum_{m\in\mathbb{Z}} \widetilde{R}_{aa^{(*)}}^{\widetilde{\alpha}}(m) \sum_{n\in\mathbb{Z}} e^{j2\pi\widetilde{\alpha}n}\, q(t+\tau-(n+m)T_0)\, q^{(*)}(t-nT_0)\,. \end{aligned} \tag{7.10}$$

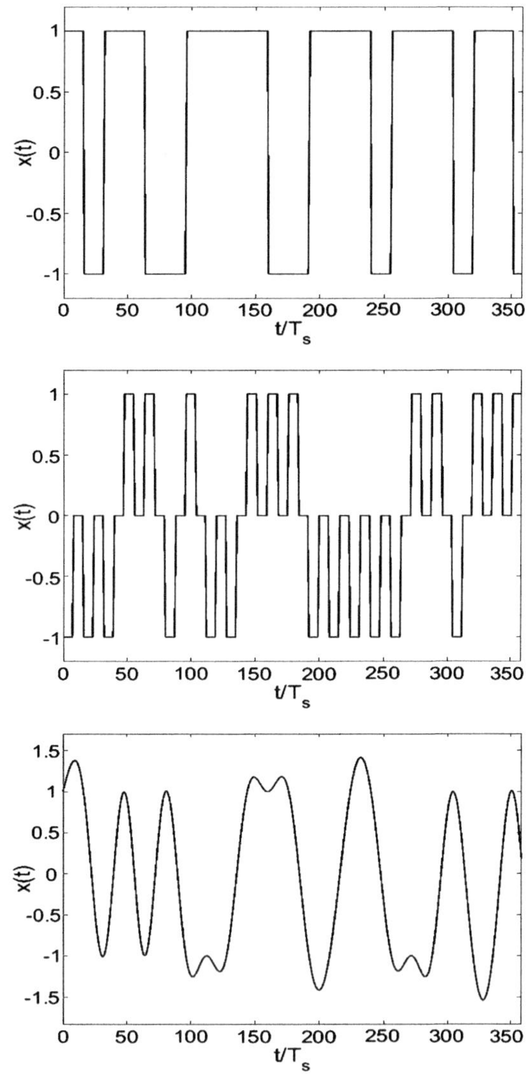

FIGURE 7.3 PAM signal $x(t)$ with stationary white binary modulating sequence $\{a_n\}$ and bit period $T_0 = 16T_s$. (Top) full duty-cycle rectangular pulse; (middle) 50% duty-cycle rectangular pulse; (bottom) raised cosine pulse.

The (conjugate) cyclic autocorrelation function and (conjugate) cyclic spectrum of $x(t)$ are (Section 7.7)

$$R^{\alpha}_{xx^{(*)}}(\tau) = \frac{1}{T_0} \sum_{m \in \mathbb{Z}} \left[\widetilde{R}^{\widetilde{\alpha}}_{aa^{(*)}}(m) \right]_{\widetilde{\alpha}=\alpha T_0} r^{\alpha}_{qq^{(*)}}(\tau - mT_0) \tag{7.11}$$

$$S^{\alpha}_{xx^{(*)}}(f) = \frac{1}{T_0} \left[\widetilde{S}^{\widetilde{\alpha}}_{aa^{(*)}}(\nu) \right]_{\nu = fT_0, \widetilde{\alpha}=\alpha T_0} Q(f)\, Q^{(*)}((-)(\alpha - f)) \tag{7.12}$$

respectively, where

$$\widetilde{S}^{\widetilde{\alpha}}_{aa^{(*)}}(\nu) \triangleq \sum_{m\in\mathbb{Z}} \widetilde{R}^{\widetilde{\alpha}}_{aa^{(*)}}(m)\, e^{-j2\pi\nu m} \tag{7.13}$$

is the (conjugate) cyclic spectrum of the sequence $\{a_n\}_{n\in\mathbb{Z}}$,

$$Q(f) \triangleq \int_{\mathbb{R}} q(t)\, e^{-j2\pi f t}\, \mathrm{d}t \tag{7.14}$$

and

$$\begin{aligned} r^{\alpha}_{qq^{(*)}}(\tau) &\triangleq q(\tau) \otimes \left[q^{(*)}(-\tau)\, e^{j2\pi\alpha\tau}\right] \\ &= \int_{\mathbb{R}} q(t+\tau)\, q^{(*)}(t)\, e^{-j2\pi\alpha t}\, \mathrm{d}t\,. \end{aligned} \tag{7.15}$$

From (7.11) and (7.12) it follows that $x(t)$ exhibits (conjugate) cyclostationarity at (conjugate) cycle frequency α only if $(\alpha T_0 \bmod 1) \in \widetilde{A}$, where mod 1 denotes the modulo 1 operation with values in $[-1/2, 1/2)$.

If the sequence $\{a_n\}_{n\in\mathbb{Z}}$ is wide-sense stationary, the (conjugate) autocorrelation (7.10) specializes into

$$\mathrm{E}\left\{x(t+\tau)\, x^{(*)}(t)\right\}$$

$$= \sum_{m\in\mathbb{Z}} \widetilde{R}^{0}_{aa^{(*)}}(m) \sum_{n\in\mathbb{Z}} q(t+\tau-(n+m)T_0)\, q^{(*)}(t-nT_0) \tag{7.16a}$$

$$= \sum_{m\in\mathbb{Z}} \widetilde{R}^{0}_{aa^{(*)}}(m)\, \mathrm{rep}_{T_0}\left[q(t+\tau-mT_0)\, q^{(*)}(t)\right] \tag{7.16b}$$

where $\mathrm{rep}_{T_0}[\cdot]$ denotes periodic replication with period T_0 (Section C.2). Thus, $x(t)$ is cyclostationary with period T_0. If the sequence $\{a_n\}_{n\in\mathbb{Z}}$ is wide-sense stationary and white, then

$$\widetilde{R}^{\widetilde{\alpha}}_{aa^{(*)}}(m) = \widetilde{R}^{0}_{aa^{(*)}}(0)\, \delta_{(\widetilde{\alpha} \bmod 1)}\, \delta_m\,. \tag{7.17}$$

In such a case, the (conjugate) autocorrelation specializes into

$$\mathrm{E}\left\{x(t+\tau)\, x^{(*)}(t)\right\} = \widetilde{R}^{0}_{aa^{(*)}}(0)\, \mathrm{rep}_{T_0}\left[q(t+\tau)\, q^{(*)}(t)\right] \tag{7.18}$$

and the (conjugate) cyclic autocorrelation and the (conjugate) cyclic spectrum of $x(t)$ become

$$R^{\alpha}_{xx^{(*)}}(\tau) = \frac{\widetilde{R}^{0}_{aa^{(*)}}(0)}{T_0}\, \delta_{(\alpha T_0 \bmod 1)}\, r^{\alpha}_{qq^{(*)}}(\tau) \tag{7.19}$$

$$S^{\alpha}_{xx^{(*)}}(f) = \frac{\widetilde{R}^{0}_{aa^{(*)}}(0)}{T_0}\, \delta_{(\alpha T_0 \bmod 1)}\, Q(f)\, Q^{(*)}((-)(\alpha-f)) \tag{7.20}$$

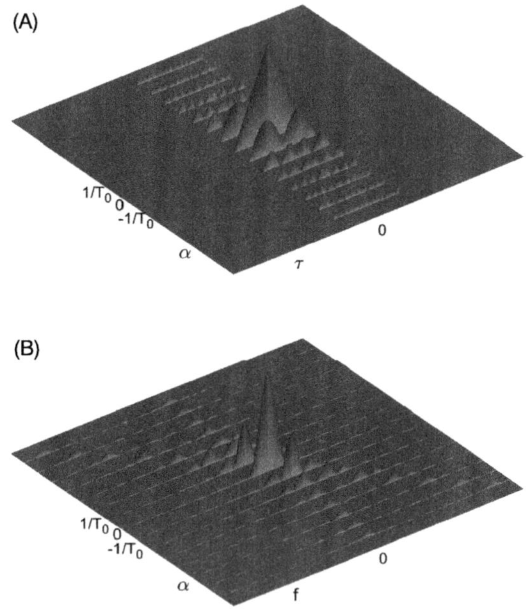

FIGURE 7.4 PAM signal (7.6) with full duty-cycle rectangular pulse (Fig. 7.3 (top)). (A) Magnitude of the cyclic autocorrelation function $R^{\alpha}_{xx^*}(\tau)$, as a function of α and τ, and (B) magnitude of the cyclic spectrum $S^{\alpha}_{xx^*}(f)$, as a function of α and f.

respectively. According with the stated cyclostationarity with period T_0, $x(t)$ exhibits cyclostationarity with cycle frequencies $\alpha = k/T_0$, $k \in \mathbb{Z}$.

The magnitude of the cyclic autocorrelation function $R^{\alpha}_{xx^*}(\tau)$, as a function of α and τ, and the magnitude of the cyclic spectrum $S^{\alpha}_{xx^*}(f)$, as a function of α and f, are reported for the PAM signal (7.6) with stationary white modulating sequence $\{a_n\}_{n\in\mathbb{Z}}$, and rectangular pulse $q(t) \triangleq \text{rect}((t - T_w/2)/T_w)$ with width T_w. In this case, $Q(f) = T_w \,\text{sinc}(f T_w)\, e^{-j\pi f T_w}$ and (7.15) reduces to

$$r^{\alpha}_{qq^*}(\tau) = e^{-j\pi\alpha(T_w-\tau)}\,\text{rect}\left(\frac{\tau}{2T_w}\right)\left(1 - \frac{|\tau|}{T_w}\right) T_w \,\text{sinc}(\alpha(T_w - |\tau|))\,. \tag{7.21}$$

In Fig. 7.4, the case of a full duty-cycle ($T_w = T_0$) pulse is represented. For $\tau = 0$ it results $x^2(t) = 1$ (constant $\forall t$). Thus $R^{\alpha}_{xx^*}(0) = 0\ \forall \alpha \neq 0$.

In Fig. 7.5, the case of a 50% duty-cycle ($T_w = T_0/2$) pulse is represented.

In Fig. 7.6A the magnitude of the cyclic autocorrelation function $R^{\alpha}_{xx^*}(\tau)$, as a function of α and τ, and in Fig. 7.6B the magnitude of the cyclic spectrum $S^{\alpha}_{xx^*}(f)$, as a function of α and f, are reported for the PAM signal (7.6) with stationary white modulating sequence

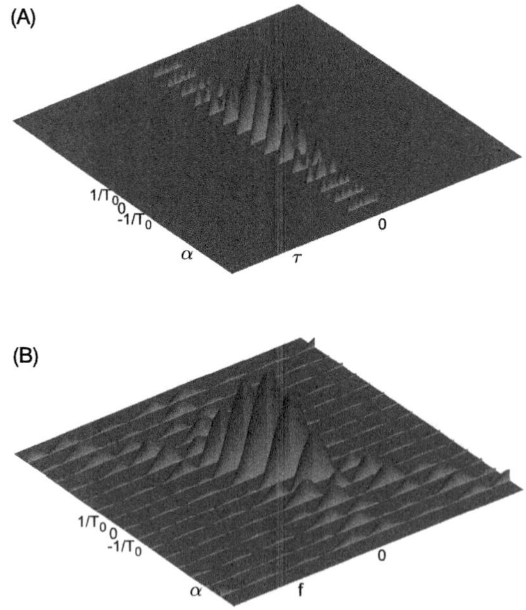

FIGURE 7.5 PAM signal (7.6) with 50% duty-cycle rectangular pulse (Fig. 7.3 (middle)). (A) Magnitude of the cyclic autocorrelation function $R^{\alpha}_{xx^*}(\tau)$, as a function of α and τ, and (B) magnitude of the cyclic spectrum $S^{\alpha}_{xx^*}(f)$, as a function of α and f.

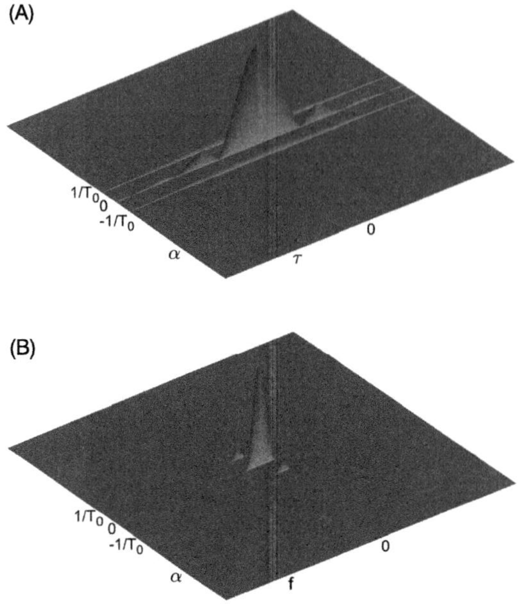

FIGURE 7.6 PAM signal (7.6) with raised cosine pulse (Fig. 7.3 (bottom)). (A) Magnitude of the cyclic autocorrelation function $R^{\alpha}_{xx^*}(\tau)$, as a function of α and τ, and (B) magnitude of the cyclic spectrum $S^{\alpha}_{xx^*}(f)$, as a function of α and f.

$\{a_n\}_{n\in\mathbb{Z}}$, and raised cosine pulse

$$q(t) \triangleq \operatorname{sinc}\left(\frac{t}{T_0}\right) \frac{\cos\left(\frac{\eta\pi t}{T_0}\right)}{1-\left(\frac{2\eta t}{T_0}\right)^2} \tag{7.22}$$

where $0 \leqslant \eta \leqslant 1$ is the roll-off factor. In this case,

$$Q(f) = \begin{cases} T_0 & |f| \leqslant \frac{1-\eta}{2T_0} \\ \frac{T_0}{2}\left[1+\cos\left(\frac{\pi T_0}{\eta}\left(|f| - \frac{1-\eta}{2T_0}\right)\right)\right] & \frac{1-\eta}{2T_0} \leqslant |f| \leqslant \frac{1+\eta}{2T_0} \\ 0 & |f| \geqslant \frac{1+\eta}{2T_0} \end{cases} \tag{7.23}$$

is strictly band limited and the cyclic autocorrelation and cyclic spectrum are nonzero only for $\alpha = 0$ and $\alpha = \pm 1/T_0$.

7.4 Direct-sequence spread-spectrum signal

Let $x(t)$ be the short-code direct-sequence spread-spectrum (DS-SS) baseband PAM signal

$$x(t) \triangleq \sum_{k\in\mathbb{Z}} a_k\, q(t-kT_0)\,, \tag{7.24}$$

where $\{a_k\}_{k\in\mathbb{Z}}$, $a_k \in \mathbb{C}$, is an ACS sequence, T_0 is the symbol period, and

$$q(t) \triangleq \sum_{n=0}^{N_c-1} c_n\, p(t-nT_c) \tag{7.25}$$

is the spreading waveform. In (7.25), $\{c_0, \ldots, c_{N_c-1}\}$ is the N_c-length spreading sequence (code) with $c_n \in \mathbb{C}$, and T_c is the chip period such that $T_0 = N_c T_c$.

The (conjugate) cyclic autocorrelation function and (conjugate) cyclic spectrum of $x(t)$ are (Napolitano and Tanda, 2001)

$$R^{\alpha}_{xx^{(*)}}(\tau) = \frac{1}{T_0} \sum_{m\in\mathbb{Z}} \left[\widetilde{R}^{\widetilde{\alpha}}_{aa^{(*)}}(m)\right]_{\widetilde{\alpha}=\alpha T_0} r^{\alpha}_{pp^{(*)}}(\tau - mT_0) \otimes \gamma^{\alpha}_{cc^{(*)}}(\tau) \tag{7.26}$$

$$\begin{aligned} S^{\alpha}_{xx^{(*)}}(f) = \frac{1}{T_0}&\left[\widetilde{S}^{\widetilde{\alpha}}_{aa^{(*)}}(\nu)\right]_{\nu=fT_0,\widetilde{\alpha}=\alpha T_0} \\ &P(f)\, P^{(*)}((-)(\alpha - f))\, \widetilde{\Gamma}^{\widetilde{\alpha}}_{cc^{(*)}}(\nu)\Big|_{\nu=fT_c,\widetilde{\alpha}=\alpha T_c} \end{aligned} \tag{7.27}$$

where

$$\gamma^{\alpha}_{cc^{(*)}}(\tau) \triangleq \sum_{n_1=0}^{N_c-1} \sum_{n_2=0}^{N_c-1} c_{n_1} c_{n_2}^{(*)} e^{-j2\pi\alpha n_2 T_c} \delta(\tau - (n_1 - n_2)T_c) \tag{7.28}$$

and

$$\widetilde{\Gamma}^{\widetilde{\alpha}}_{cc^{(*)}}(\nu) \triangleq \mathcal{C}(\nu)\, \mathcal{C}^{(*)}((-)(\widetilde{\alpha} - \nu)) \tag{7.29}$$

with

$$\mathcal{C}(\nu) \triangleq \sum_{n=0}^{N_c-1} c_n\, e^{-j2\pi\nu n} . \tag{7.30}$$

Second-order cyclic spectral analysis of more general classes of DSSS signals is addressed in (Fusco et al., 2006).

7.5 Higher-order cyclic spectra of modulated signals

In this section, results of Section 4.4 are exploited to derive higher-order cyclic cross-spectra of modulated signals. Higher-order cyclic spectra of single signals are easily obtained by considering $x_m \equiv x$, $\forall m$ in the derived relationships.

7.5.1 PAM signals

Let us consider N PAM signals

$$x_{\text{PAM}_m}(t) \triangleq \sum_{k=-\infty}^{+\infty} x_m(kT_m)\, p_m(t - kT_m) = x_{m,\delta}(t) \otimes p_m(t), \qquad m = 1, \dots, N \tag{7.31}$$

where the continuous-time signals

$$x_{m,\delta}(t) \triangleq x_m(t) \sum_{k=-\infty}^{+\infty} \delta(t - kT_m) = x_m(t)\, \frac{1}{T_m} \sum_{r=-\infty}^{+\infty} e^{j2\pi rt/T_m}, \qquad m = 1, \dots, N \tag{7.32}$$

are obtained by sampling the continuous-time signals $x_m(t)$ with rates $1/T_m$ ($m = 1, \dots, N$) and, in the second equality, the Poisson formula (C.8a) is exploited.

The RD-CSCMF of the N PAM signals, according to (4.92), can be written as

$$S^{\alpha}_{\boldsymbol{x}_{\text{PAM}}}(\boldsymbol{f}') = P_N(\alpha - \boldsymbol{f}'^{\text{T}}\mathbf{1}) \prod_{m=1}^{N-1} P_m(f_m)\, S^{\alpha}_{\boldsymbol{x}_\delta}(\boldsymbol{f}') \tag{7.33}$$

where subscripts $\boldsymbol{x}_{\text{PAM}} \equiv x_{\text{PAM}_1} \dots x_{\text{PAM}_N}$, $\boldsymbol{x}_\delta \equiv x_{1,\delta} \dots x_{N,\delta}$, and $P_m(f)$ denotes the Fourier transform of $p_m(t)$. The RD-CSCMF of the signals (7.32) can be expressed in terms of the

RD-CSCMF of the signals $x_m(t)$ (Napolitano, 1995)

$$S^{\alpha}_{\boldsymbol{x}_\delta}(\boldsymbol{f}') = \left(\prod_{m=1}^{N} \frac{1}{T_m}\right) \sum_{\boldsymbol{r} \in \mathbb{Z}^N} S^{\alpha - \boldsymbol{r}^{\mathsf{T}} \boldsymbol{f}_s}_{x_1, \ldots, x_N}(\boldsymbol{f}' - \boldsymbol{r}' \circ \boldsymbol{f}'_s) \tag{7.34}$$

where $\boldsymbol{f}_s \triangleq [1/T_1, \ldots, 1/T_N]^{\mathsf{T}}$, $\boldsymbol{r} \triangleq [r_1, \ldots, r_N]^{\mathsf{T}}$, and $\circ$ denotes the Hadamard matrix product, i.e., $\boldsymbol{r}' \circ \boldsymbol{f}'_s \triangleq [r_1/T_1, \ldots, r_{N-1}/T_{N-1}]^{\mathsf{T}}$. Accounting for the rhs of (7.32), Eq. (7.34) is derived by substituting $\sigma_m = r_m/T_m$ (with r_m any integer number) and $c_{\sigma_m} = 1/T_m$ into (4.99).

Let $x_{m,\mathrm{d}}(n) \triangleq x_m(nT_s)$, $n \in \mathbb{Z}$, be the discrete-time signals obtained by uniformly sampling with period T_s the continuous-time signals $x_m(t)$. We have that (7.34) with $T_m = T_s$, $m = 1, \ldots, N$, compared with (4.112), yields

$$\widetilde{S}^{\widetilde{\alpha}}_{\boldsymbol{x}_\mathrm{d}}(\boldsymbol{\nu}') = T_s \, S^{\widetilde{\alpha} f_s}_{\boldsymbol{x}_\delta}(\boldsymbol{f}') \,|_{\boldsymbol{f}' = \boldsymbol{\nu}' f_s} , \tag{7.35}$$

where subscript $\boldsymbol{x}_d \equiv x_{1,\mathrm{d}} \cdots x_{N,\mathrm{d}}$.

Let us consider now the N PAM signals

$$x_{\mathrm{PAM}_i}(t) = \sum_{n=-\infty}^{+\infty} x_i(n) \, p_i(t - nT_s) \qquad i = 1, \ldots, N \tag{7.36}$$

whose modulating sequences are N discrete-time signals $x_1(n), \ldots, x_N(n)$ jointly white and wide-sense stationary of order N, i.e., (Dehay et al., 2018b, Definition 4.1)

$$\widetilde{\mathcal{C}}_{\boldsymbol{x}}(n, \boldsymbol{m}) \triangleq \operatorname{cum}\{x_i(n + m_i), \; i = 1, \ldots, N\} = \widetilde{c}_{\boldsymbol{x}} \, \delta_{\boldsymbol{m}' - \mathbf{1} m_N} \tag{7.37}$$

where subscript $\boldsymbol{x} \equiv x_1 \cdots x_N$ and $\widetilde{c}_{\boldsymbol{x}}$ is the Nth-order cross-cumulant of $x_1(n), \ldots, x_N(n)$.

The CTCCF of the modulating signals is given by

$$\widetilde{\mathcal{C}}^{\widetilde{\beta}}_{\boldsymbol{x}}(\boldsymbol{m}) = \widetilde{c}_{\boldsymbol{x}} \, \delta_{\boldsymbol{m}' - \mathbf{1} m_N} \, \delta_{\widetilde{\beta} \bmod 1} \tag{7.38}$$

where

$$\delta_{a \bmod b} \triangleq \begin{cases} 1 & a/b \in \mathbb{Z}, \\ 0 & \text{otherwise.} \end{cases} \tag{7.39}$$

By Fourier transforming the discrete-time RD-CTCCF $\widetilde{C}^{\widetilde{\beta}}_{\boldsymbol{x}}(\boldsymbol{m}')$, straightforwardly derivable from (7.38), one obtains the CCP

$$\widetilde{P}^{\widetilde{\beta}}_{\boldsymbol{x}}(\boldsymbol{\nu}') = \widetilde{c}_{\boldsymbol{x}} \, \delta_{\widetilde{\beta} \bmod 1} \, . \tag{7.40}$$

Hence, by substituting (7.40) into the analogous of (7.35) in terms of CCPs and the result into the analogous of (7.33) in terms of CCPs, one has

$$P^{\beta}_{\boldsymbol{x}_{\mathrm{PAM}}}(\boldsymbol{f}') = f_s \, \widetilde{c}_{\boldsymbol{x}} \, P_N(\beta - \boldsymbol{f}'^{\mathsf{T}} \mathbf{1}) \prod_{m=1}^{N-1} P_m(f_m) \, \delta_{\beta \bmod f_s} \, . \tag{7.41}$$

Note that the relations stated in terms of CTCCFs and CCPs for PAM signals modulated by jointly white and wide-sense stationary discrete-time signals cannot be written in terms of RD-CSCMFs by simply substituting CCPs with RD-CSCMFs as in the case of continuous-time modulating signals (see also Remark 4.28). In fact, for the discrete-time TCMF, no simple relation analogous to (7.37) holds.

7.5.2 QAM signals

Let us consider N *quadrature-carrier amplitude-modulation (QAM)* time-series

$$x_{\text{QAM}_m}(t) \triangleq x_{\text{c},m}(t)\cos(2\pi f_{c,m} t + \phi_m) - x_{\text{s},m}(t)\sin(2\pi f_{c,m} t + \phi_m), \qquad m = 1, \dots, N \,. \tag{7.42}$$

These QAM signals can be considered as the outputs of a continuous-time MIMO LAPTV system excited by $\boldsymbol{x}(t) \triangleq [x_1(t), \dots, x_{2N}(t)]^{\mathrm{T}}$, where

$$x_{2m-1}(t) = x_{\text{c},m}(t) \tag{7.43}$$

$$x_{2m}(t) = x_{\text{s},m}(t) \,. \tag{7.44}$$

The input/output relationship of the system is given by (4.94), where the nonzero elements of the $N \times 2N$ matrix $\boldsymbol{C}(t)$ are given by

$$c_{m,2m-1}(t) = \cos(2\pi f_{c,m} t + \phi_m) \tag{7.45}$$

$$c_{m,2m}(t) = -\sin(2\pi f_{c,m} t + \phi_m) \,. \tag{7.46}$$

Therefore, from (4.98), letting $\sigma_{m i_m} = \pm f_{c,m}$, it follows that (Napolitano, 1995)

$$S^{\alpha}_{\boldsymbol{x}_{\text{QAM}}}(\boldsymbol{f}') = \frac{1}{2^N} \sum_{i_1=1}^{2} \cdots \sum_{i_N=2N-1}^{2N} \sum_{q_{1i_1}=\pm 1} \cdots \sum_{q_{Ni_N}=\pm 1} \left(\prod_{m=1}^{N} e^{j q_{m i_m}[\phi_m + (i_m - 2m + 1)\pi/2]} \right) \cdot S^{\alpha - \boldsymbol{f}_c^{\mathrm{T}} \boldsymbol{q}_{i_1,\dots,i_N}}_{x_{i_1},\dots,x_{i_N}}(\boldsymbol{f}' - \boldsymbol{f}'_c \circ \boldsymbol{q}'_{i_1,\dots,i_N}) \tag{7.47}$$

where $\boldsymbol{q}_{i_1,\dots,i_N} \triangleq [q_{1i_1}, \dots, q_{Ni_N}]^{\mathrm{T}}$ and $\boldsymbol{f}_c \triangleq [f_{c,1}, \dots, f_{c,N}]^{\mathrm{T}}$.

The CCP of the N QAM signals

$$x_{\text{QAM}_i}(t) \triangleq \sum_{k=-\infty}^{+\infty} x_{\text{c},i}(k)\, p_{\text{c},i}(t - kT_s)\cos(2\pi f_{c,i} t + \phi_i) - \sum_{k=-\infty}^{+\infty} x_{\text{s},i}(k)\, p_{\text{s},i}(t - kT_s)\sin(2\pi f_{c,i} t + \phi_i), \qquad i = 1, \dots, N \tag{7.48}$$

with white sequences $x_{\text{c},i}(k)$ and $x_{\text{s},i}(k)$, can be obtained accounting for (7.41), by substituting

$$P^{\beta - \boldsymbol{f}_c^{\mathsf{T}} \boldsymbol{q}_{i_1,\ldots,i_N}}_{x_{i_1},\ldots,x_{i_N}}(\boldsymbol{f}' - \boldsymbol{f}'_c \circ \boldsymbol{q}'_{i_1,\ldots,i_N})$$
$$= f_s \, \widetilde{c}_{x_{i_1},\ldots,x_{i_N}} \, P_{\mathrm{cs},i_N}(\beta - q_{N i_N} f_{\mathrm{c},N} - \boldsymbol{f}'^{\mathsf{T}} \mathbf{1}) \prod_{m=1}^{N-1} P_{\mathrm{cs},i_m}(f_m - q_{m i_m} f_{\mathrm{c},m}) \, \delta_{(\beta - \boldsymbol{f}_c^{\mathsf{T}} \boldsymbol{q}_{i_1,\ldots,i_N}) \bmod f_s} \tag{7.49}$$

into the analogous of (7.47) in terms of CCPs. In (7.49), $P_{\mathrm{cs},i_m}(f)$ is the Fourier transform of

$$p_{\mathrm{cs},i_m}(t) \triangleq p_{\mathrm{c},i_m}(t) \, \delta_{i_m,2m-1} + p_{\mathrm{s},i_m}(t) \, \delta_{i_m,2m} \tag{7.50}$$

and

$$x_{i_m}(t) = x_{\mathrm{c},i_m}(t) \text{ for } i_m = 2m-1 \qquad x_{i_m}(t) = x_{\mathrm{s},i_m}(t) \text{ for } i_m = 2m \, .$$

7.5.3 DSB-AM signals

By letting $x_{\mathrm{s},m}(t) \equiv 0$ for $m = 1, \ldots, N$ in (7.42), one obtains *double-sideband (DSB) amplitude-modulated (AM)* time-series. It results that

$$\begin{aligned} S^{\alpha}_{x_{i_1},\ldots,x_{i_N}}(\boldsymbol{f}') &= S^{\alpha}_{x_{i_1},\ldots,x_{i_N}}(\boldsymbol{f}') \, \delta_{i_1,1} \cdots \delta_{i_N,2N-1} \\ &= S^{\alpha}_{x_{\mathrm{c},1},\ldots,x_{\mathrm{c},N}}(\boldsymbol{f}') \, \delta_{i_1,1} \cdots \delta_{i_N,2N-1} \, . \end{aligned} \tag{7.51}$$

Therefore, from (7.47) we have (Napolitano, 1995)

$$S^{\alpha}_{\boldsymbol{x}_{\mathrm{DSB}}}(\boldsymbol{f}') = \frac{1}{2^N} \sum_{q_1 = \pm 1} \cdots \sum_{q_N = \pm 1} \exp\{j \boldsymbol{q}^{\mathsf{T}} \boldsymbol{\phi}\} \, S^{\alpha - \boldsymbol{q}^{\mathsf{T}} \boldsymbol{f}_c}_{x_{\mathrm{c},1},\ldots,x_{\mathrm{c},N}}(\boldsymbol{f}' - \boldsymbol{q}' \circ \boldsymbol{f}'_c) \tag{7.52}$$

where $\boldsymbol{\phi} \triangleq [\phi_1, \ldots, \phi_N]^{\mathsf{T}}$ and $\boldsymbol{q} \triangleq [q_1, \ldots, q_N]^{\mathsf{T}}$.

7.5.4 SSB signals

For *Single-Sideband (SSB)*-modulated time-series

$$x_{\mathrm{SSB}_m}(t) \triangleq x_{\mathrm{c},m}(t) \cos[2\pi f_{c_m} t + \phi_m] - \left(x_{\mathrm{c},m}(t) \otimes \frac{1}{\pi t} \right) \sin[2\pi f_{c_m} t + \phi_m] \tag{7.53}$$

one has

$$x_{i_m}(t) = x_{\mathrm{c},m}(t) \otimes \left[\delta(t) \, \delta_{i_m,2m-1} + \frac{1}{\pi t} \, \delta_{i_m,2m} \right] . \tag{7.54}$$

Then, accounting for (4.92) it follows that (Napolitano, 1995)

$$\begin{aligned} S^{\alpha}_{x_{i_1},\ldots,x_{i_N}}(\boldsymbol{f}') = {} & S^{\alpha}_{x_{\mathrm{c},1},\ldots,x_{\mathrm{c},N}}(\boldsymbol{f}') \left[\delta_{i_N,2N-1} - j \operatorname{sign}(\alpha - \boldsymbol{f}'^{\mathsf{T}} \mathbf{1}) \, \delta_{i_N,2N} \right] \\ & \cdot \prod_{m=1}^{N-1} \left[\delta_{i_m,2m-1} - j \operatorname{sign}(f_m) \, \delta_{i_m,2m} \right] \end{aligned} \tag{7.55}$$

where sign$(\cdot)$ is the signum function. Therefore, the RD-CSCMF can be obtained by substitution of (7.55) in (7.47).

7.5.5 ASK signals

N Amplitude-Shift Keyed (ASK) signals can be viewed as a special case of N QAM signals with $x_{s,i}(k) \equiv 0$ in (7.48).

Thus, their CCP is given by (Napolitano, 1995)

$$P^{\beta}_{\boldsymbol{x}_{\mathrm{ASK}}}(\boldsymbol{f}') = \frac{1}{2^N} f_s\, \widetilde{c}_{x_{c,1},\ldots,x_{c,N}} \sum_{q_1=\pm 1} \cdots \sum_{q_N=\pm 1} \exp\{j\boldsymbol{q}^{\mathsf{T}}\boldsymbol{\phi}\}$$

$$P_{c,N}(\beta - q_N f_{cN} - \boldsymbol{f}'^{\mathsf{T}}\mathbf{1}) \prod_{m=1}^{N-1} P_{c,m}(f_m - q_m f_{c,m})\, \delta_{(\beta - \boldsymbol{q}^{\mathsf{T}}\boldsymbol{f}_c) \bmod f_s} . \tag{7.56}$$

7.6 Cyclic spectral analysis of man-made signals

Second-order cyclostationarity-based spectrum sensing techniques exploit the knowledge of the (conjugate) cycle frequencies, cyclic autocorrelation functions, and cyclic spectra of the signal of interest. They have been derived for the most common analog and digitally modulated signals in (Gardner, 1987e) and (Gardner et al., 1987), respectively.

Cyclic statistics have been derived for:

- *Long-code direct-sequence spread-spectrum (DSSS) signals* (Fusco et al., 2006);
- *Continuous phase modulated (CPM) signals* (Gournay and Viravau, 1998), (Viravau and Gournay, 1998), (Cariolaro et al., 2011), (Napolitano and Spooner, 2001);
- *Digital pulse streams modulated by cyclostationary sequences* (Oner, 2009);
- *Minimum shift keying (MSK) and offset quadrature phase shift keying (QPSK)* (Vučić and Obradović, 1999);
- *Multicarrier modulated signals* (Zhang et al., 2010);
- *Ultra wide-band (UWB) signals* (Oner, 2008a), (Oner, 2008b), (Vucic and Eric, 2009);
- *Orthogonal frequency-division multiplexing (OFDM) signals* (Cui et al., 2011);
- *OFDM and single carrier linearly digitally (SCLD) modulated signals* (Punchihewa et al., 2010);
- *WiMAX and long-term evolution (LTE) OFDM signals* (Al-Habashna et al., 2012);

- *Spatial multiplexing OFDM (SM-OFDM) and Alamouti coded OFDM (AL-OFDM) signals*
 (Karami and Dobre, 2015);
- *Block transmitted SCLD (BT-SCLD) signals*
 (Zhang et al., 2013);
- *LTE single-carrier frequency-division multiple-access (SC-FDMA) signals*
 (Jerjawi et al., 2015);
- *LTE advanced signals*
 (Sutton et al., 2014);
- *Binary offset carrier (BOC) signals*
 (Lohan et al., 2014);
- *Global position system (GPS) signals*
 (Napolitano and Perna, 2014a);
- *Universal filtered multi-carrier (UFMC) signal*
 (Zerhouni et al., 2018);
- *Orthogonal frequency-division multiplexing/offset-QAM (OFDM/OQAM) signals*
 (Vučić et al., 2017).

A symbol constellation having second-order statistics with cyclostationary phase is designed in (Murphy, 2005).

7.7 Proofs

In this section, proofs are reported for results of the present chapter. Unless otherwise specified, only formal manipulations are reported, without explicitly justifying inversion of linear operators such as time averages, infinite sums, integrals, and ensemble average (expectation). In general, these manipulations are justified at least in the sense of generalized functions (distributions) (Champeney, 1990), (Zemanian, 1987).

Proof of (7.10)

$$
\begin{aligned}
&\mathrm{E}\left\{x(t+\tau)\,x^{(*)}(t)\right\} \\
&\quad = \mathrm{E}\left\{\sum_{n_1\in\mathbb{Z}} a_{n_1}\, q(t+\tau-n_1T_0) \sum_{n_2\in\mathbb{Z}} a_{n_2}^{(*)}\, q^{(*)}(t-n_2T_0)\right\} \\
&\quad = \sum_{n_1\in\mathbb{Z}}\sum_{n_2\in\mathbb{Z}} \mathrm{E}\left\{a_{n_1}\, a_{n_2}^{(*)}\right\}\, q(t+\tau-n_1T_0)\, q^{(*)}(t-n_2T_0) \\
&\quad = \sum_{n\in\mathbb{Z}}\sum_{m\in\mathbb{Z}} \mathrm{E}\left\{a_{n+m}\, a_n^{(*)}\right\}\, q(t+\tau-(n+m)T_0)\, q^{(*)}(t-nT_0) \\
&\quad = \sum_{\widetilde{\alpha}\in\widetilde{A}}\sum_{m\in\mathbb{Z}} \widetilde{R}_{aa^{(*)}}^{\widetilde{\alpha}}(m) \sum_{n\in\mathbb{Z}} e^{j2\pi\widetilde{\alpha}n}\, q(t+\tau-(n+m)T_0)\, q^{(*)}(t-nT_0)
\end{aligned}
\tag{7.57}
$$

where in the third equality the variable changes $n_1 = n+m$ and $n_2 = n$ are made, and in the fourth equality (7.7) is used.

Proof of (7.11) and (7.12)

In the right-hand side of (7.57), for every fixed $\widetilde{\alpha}$ and m, the function

$$y(t) \triangleq \sum_{n\in\mathbb{Z}} e^{j2\pi\widetilde{\alpha}n}\, q(t+\tau-(n+m)T_0)\, q^{(*)}(t-nT_0) \tag{7.58}$$

is a PAM waveform with deterministic modulating sequence $\widetilde{y}_n \triangleq e^{j2\pi\widetilde{\alpha}n}$ and pulse $p(t) \triangleq q(t+\tau-mT_0)\, q^{(*)}(t)$ whose Fourier transform is denoted by $P(f)$. The results of Section 2.4.1 can be applied to $y(t)$. In particular, we have

$$\widetilde{y}^{\widetilde{\beta}} \triangleq \left\langle e^{j2\pi\widetilde{\alpha}n}\, e^{-j2\pi\widetilde{\beta}n} \right\rangle_n = \delta_{(\widetilde{\alpha}-\widetilde{\beta}) \bmod 1} \tag{7.59}$$

where mod 1 denotes the modulo 1 operation with values in $[-1/2, 1/2)$. Thus, accounting for (2.100), the Fourier coefficient at frequency α of the almost-periodic component of $y(t)$ is given by

$$\begin{aligned} y^{\alpha} &\triangleq \left\langle y(t)\, e^{-j2\pi\alpha t} \right\rangle_t = \frac{1}{T_0}\, \widetilde{y}^{\alpha T_0}\, P(\alpha) \\ &= \frac{1}{T_0}\, \delta_{(\widetilde{\alpha}-\alpha T_0) \bmod 1} \int_{\mathbb{R}} q(t+\tau-mT_0)\, q^{(*)}(t)\, e^{-j2\pi\alpha t}\, \mathrm{d}t \end{aligned} \tag{7.60}$$

Accounting for (7.57), one has

$$\begin{aligned} R^{\alpha}_{xx^{(*)}}(\tau) &\triangleq \lim_{T\to\infty} \frac{1}{T} \int_{-T/2}^{T/2} \mathrm{E}\left\{x(t+\tau)\, x^{(*)}(t)\right\} e^{-j2\pi\alpha t}\, \mathrm{d}t \\ &= \sum_{\widetilde{\alpha}\in\widetilde{A}} \sum_{m\in\mathbb{Z}} \widetilde{R}^{\widetilde{\alpha}}_{aa^{(*)}}(m) \\ &\qquad \lim_{T\to\infty} \frac{1}{T} \int_{-T/2}^{T/2} \sum_{n\in\mathbb{Z}} e^{j2\pi\widetilde{\alpha}n}\, q(t+\tau-(n+m)T_0)\, q^{(*)}(t-nT_0)\, e^{-j2\pi\alpha t}\, \mathrm{d}t \\ &= \sum_{\widetilde{\alpha}\in\widetilde{A}} \sum_{m\in\mathbb{Z}} \widetilde{R}^{\widetilde{\alpha}}_{aa^{(*)}}(m) \\ &\qquad \frac{1}{T_0}\, \delta_{(\widetilde{\alpha}-\alpha T_0) \bmod 1} \int_{\mathbb{R}} q(t+\tau-mT_0)\, q^{(*)}(t)\, e^{-j2\pi\alpha t}\, \mathrm{d}t \\ &= \frac{1}{T_0} \sum_{m\in\mathbb{Z}} \widetilde{R}^{\alpha T_0}_{aa^{(*)}}(m)\, r^{\alpha}_{qq^{(*)}}(\tau-mT_0) \end{aligned} \tag{7.61}$$

where in the third equality (7.60) is used and the fourth equality follows since at most one value of $\widetilde{\alpha} \in \widetilde{A}$ can give nonzero contribution, provided that $(\alpha T_0 \bmod 1) \in \widetilde{A}$. The function $r^{\alpha}_{qq^{(*)}}(\tau)$ is defined in (7.15).

8

Detection and cycle frequency estimation

Contents

Abstract

Detection strategies and estimators of cycle frequencies for almost-cyclostationary signals are presented. The spectral line regeneration through homogeneous quadratic time-invariant transformation is treated. A structure for joint maximum likelihood detection and source location is considered in the weak-signal scenario. A detection strategy based on the asymptotic normality of cyclic statistical function estimators is presented for signals exhibiting second-order cyclostationarity or spectral correlation. Detectors based on subsampling or block-bootstrap techniques, robust detectors for signals with heavy tail distributions, and detectors based on higher-order cyclic statistics are considered. Several cycle frequency estimation techniques and their properties are also presented.

Cyclostationary Processes and Time Series. https://doi.org/10.1016/B978-0-08-102708-0.00019-4

Keywords

Spectral line regeneration, Detection, Asymptotic normality, Subsampling-based significance test, Robust detection, Cycle frequency estimation

8.1 Introduction

The detection of the presence of a signal-of-interest (SOI) $x(t)$ in the corrupted observation $r(t)$ (hypothesis H_1) versus its absence (hypothesis H_0), that is (in the case of additive disturbance)

$$\begin{array}{ll} \mathrm{H}_0 & r(t) = n(t) \\ \mathrm{H}_1 & r(t) = x(t) + n(t) \end{array} \qquad t \in [-T/2, T/2] \tag{8.1}$$

is a formidable problem if $n(t)$ is a non-Gaussian and possibly nonstationary disturbance signal with unknown or complicate probability density function.

If the likelihood ratio test (LRT) cannot be built, alternative (non-optimum) detectors can be derived if $x(t)$ is ACS and $n(t)$ is a disturbance signal not sharing at least one cycle frequency (say α_0) with $x(t)$. In this section, several approaches to detect the presence of an ACS SOI and to estimate its cycle frequencies are considered.

Signal detection techniques designed for cyclostationary signals take account of the periodicity or almost periodicity of the signal autocorrelation function (Gardner, 1985, Sec. 12.8.5), (Gardner, 1987f, Sec. 14.E). Single-cycle and multi-cycle detectors exploit one or multiple cycle frequencies, respectively.

The detection problem in additive Gaussian noise is addressed in (Bukofzer, 1979), (Gardner, 1982), (Gardner, 1983), (Gardner, 1987d), (Gardner and Spooner, 1988), (Gardner, 1988b), (Gardner and Spooner, 1992b), (Gardner and Spooner, 1993), (Spooner and Gardner, 1994b), (Rostaing, 1996), (Yeung and Gardner, 1996), (Rostaing et al., 1999), and in non-Gaussian noise, in (Izzo et al., 1992c), (Tanda, 1992), (Izzo and Tanda, 1995), (Gelli et al., 1996). Optimum detection of Gaussian pulsed signals is considered in (Rybakov, 1980). The problem of signal detection in cyclostationary noise is treated in (Bukofzer, 1979), (Bukofzer, 1987), (Izzo et al., 1990), (Napolitano et al., 1992). Tests for the presence of cyclostationarity are proposed in (Vecchia and Ballerini, 1991), (Dandawaté and Giannakis, 1994b), (Besson and Stoica, 1999), (Anderson and Vecchia, 1993) by exploiting the asymptotic properties of the cyclic correlogram and in (Hurd, 1982), (Hurd and Gerr, 1991) by exploiting the properties of the support of the spectral correlation function. Modulation classification techniques are proposed in (Gournay and Nicolas, 1995), (Boiteau and Le Martret, 1998), (Marchand, 1998), (Marchand et al., 1998), (Spooner, 1995). The problem of cyclostationary source separation is considered in (Ferréol and Chevalier, 2000), (Abed-Meraim et al., 2001), (Ferréol et al., 2004), (Antoni et al., 2005). Locally optimum and suboptimum detection of complex-valued cyclostationary signals is addressed in (Lunden and Koivunen, 2012). For the case of a continuous-time cyclostationary signal with known period of cyclostationarity, a generalized likelihood ratio (GLR) detector based on the Harmonic Series Representation (HSR) is presented in (Riba et al., 2014). Other cyclostationarity detectors based on coherence are proposed in (Enserink and Cochran, 1995), (Howard et al., 2018), (Sirianunpiboon et al., 2018).

On the subject of cyclostationarity-based detection see the general treatments (Gardner, 1985), (Gardner, 1990b), (Gardner and Chen, 1991), and also see (Gardner, 1988b), (Omel'chenko et al., 1988), (Sankar and Tugbay, 1988), (Spooner et al., 1989), (Živanović and Gardner, 1991), (Spooner et al., 2000), (Gelli et al., 2000), (Martone, 2000), (Spooner, 2001), (Spooner, 2015).

Patents on detection and signal recognition exploiting cyclostationarity are (Gardner, 1995), (Ferreol, 2002), (Sills and Black, 2004).

8.2 Spectral line regeneration

The presence of a signal $x(t)$ exhibiting (conjugate) second-order cyclostationarity at cycle frequency α_0 can be detected by revealing the presence of a regenerated spectral line at frequency α_0 in the homogeneous quadratic time-invariant transformation of $r(t)$

$$y(t) = \int_{\mathbb{R}^2} K(\tau_1, \tau_2)\, r(t+\tau_1)\, r^{(*)}(t+\tau_2)\, \mathrm{d}\tau_1\, \mathrm{d}\tau_2 \,. \tag{8.2}$$

In fact, if $x(t)$ exhibits second-order cyclostationarity with cycle frequency α_0, then $y(t)$ contains a finite-strength additive sine-wave component with frequency α_0. Specifically, we have

$$r(t+\tau_1)\, r^{(*)}(t+\tau_2) = \mathrm{E}\left\{r(t+\tau_1)\, r^{(*)}(t+\tau_2)\right\} + \ell_{\boldsymbol{r}}(t;\tau_1,\tau_2) \tag{8.3a}$$

$$= \sum_{\alpha\in A} R^{\alpha}_{\boldsymbol{r}}(\tau_1-\tau_2)\, e^{j2\pi\alpha(t+\tau_2)} + \ell_{\boldsymbol{r}}(t;\tau_1,\tau_2) \tag{8.3b}$$

where

$$\left\langle \ell_{\boldsymbol{r}}(t;\tau_1,\tau_2)\, e^{-j2\pi\alpha t}\right\rangle_t = 0 \qquad \forall \alpha \in \mathbb{R} \tag{8.4}$$

with (8.3a) and (8.3b) that are the stochastic and complex-valued counterparts of (2.56a) and (2.56b), respectively. Thus, it results that

$$y(t) = \sum_{\alpha\in A} \int_{\mathbb{R}^2} K(\tau_1,\tau_2)\, R^{\alpha}_{\boldsymbol{r}}(\tau_1-\tau_2)\, e^{j2\pi\alpha\tau_2}\, \mathrm{d}\tau_1\, \mathrm{d}\tau_2\, e^{j2\pi\alpha t} + \int_{\mathbb{R}^2} K(\tau_1,\tau_2)\, \ell_{\boldsymbol{r}}(t;\tau_1,\tau_2)\, \mathrm{d}\tau_1\, \mathrm{d}\tau_2 \,. \tag{8.5}$$

In additive white Gaussian noise (AWGN), the kernel $K(\tau_1, \tau_2)$ can be chosen in order to maximize the SNR defined as the ratio of the power of the regenerated spectral line at frequency α_0 and the noise power. This leads to the *single cycle detector* (Gardner, 1988b) that compares with a threshold the magnitude of the value at $t = 0$ of

$$y(t) = \frac{1}{N_0^2} \int_{\mathbb{R}} I^{(T)}_{rr^{(*)}}(\alpha_0, f; t)\, S^{\alpha_0}_{xx^{(*)}}(f)^*\, \mathrm{d}f\, e^{j2\pi\alpha_0 t} \tag{8.6a}$$

$$\simeq \frac{1}{N_0^2} \int_{-T/2}^{T/2} R_{rr^{(*)}}^{(T)}(\alpha_0, \tau; t)\, R_{xx^{(*)}}^{\alpha_0}(\tau)^* \, \mathrm{d}\tau \, e^{j2\pi\alpha_0 t} \tag{8.6b}$$

where $I_{rr^{(*)}}^{(T)}(\alpha, f; t) \triangleq \frac{1}{T} R_T(t, f) R_T^{(*)}(t, (-)(\alpha - f))$ is the (conjugate) cyclic periodogram with $R_T(t, f)$ STFT of $r(t)$ defined according to (1.31), N_0 is the noise spectral density, and $R_{rr^{(*)}}^{(T)}(\alpha, \tau; t)$ is the (conjugate) cyclic correlogram with explicit dependence on t, the central point of the observation interval $[t - T/2, t + T/2]$. The approximate equality is due to the Parseval's relation and becomes rigorous in the limit as $T \to \infty$.

If the (conjugate) cycle frequency α_0 is known, but the (conjugate) cyclic spectrum $S_{xx^{(*)}}^{\alpha_0}(f)$ or (conjugate) cyclic autocorrelation function $R_{xx^{(*)}}^{\alpha_0}(\tau)$ are not completely known, for example due to LTI propagation channel distortion, then the detector statistics (8.6a) and (8.6b) can be replaced by

$$y(t) = \frac{1}{N_0^2} \int_{\mathcal{B}} \left| I_{rr^{(*)}}^{(T)}(\alpha_0, f; t) \right|^2 \mathrm{d}f \, e^{j2\pi\alpha_0 t} \tag{8.7a}$$

$$\simeq \frac{1}{N_0^2} \int_{\mathcal{T}} \left| R_{rr^{(*)}}^{(T)}(\alpha_0, \tau; t) \right|^2 \mathrm{d}\tau \, e^{j2\pi\alpha_0 t} \tag{8.7b}$$

where $\mathcal{B}$ and $\mathcal{T}$ are sets where $S_{xx^{(*)}}^{\alpha_0}(f)$ and $R_{xx^{(*)}}^{\alpha_0}(\tau)$, respectively, are known to be nonzero (Gardner, 1988b), (Gardner and Spooner, 1992b), (Spooner and Nicholls, 2009).

In (8.6a)–(8.7b), $|y(t)|$ is practically constant with t for T sufficiently large. The detection test is

$$|y(t)| \underset{\mathrm{H}_0}{\overset{\mathrm{H}_1}{\gtrless}} \lambda \tag{8.8}$$

where the threshold λ depends on the desired probability of false alarm.

The above results can be extended to exploit higher-order cyclostationarity properties of the SOI. Specifically, the presence of the SOI $x(t)$ exhibiting Nth-order cyclostationarity at cycle frequency α_0 can be detected by detecting the presence of a regenerated spectral line at frequency α_0 in the homogeneous Nth-order time-invariant transformation of $r(t)$ (see Section 8.9).

8.3 Maximum likelihood detection and source location

The problem of jointly detecting and localizing a fixed source $x(t)$ on the basis of two *real-valued* sensor measurements is based on the model

$$\mathrm{H}_0 \quad r_k(t) = n_k(t), \quad |t| < T/2, \quad k = 1, 2 \tag{8.9a}$$

$$\mathrm{H}_1 \quad r_k(t) = x(t - d_k) + n_k(t), \quad |t| < T/2, \quad k = 1, 2 \tag{8.9b}$$

where d_k is the delay experienced by the source signal to reach the kth sensor. Under the assumption that the noise terms $n_1(t)$ and $n_2(t)$ are independent stationary AWGNs with power spectral density N_0, an equivalent statistic for the log-likelihood ratio (LLR) for low SNR is (Holt, 1987)

$$Y_{\mathrm{LLR}} = Y_{11} + 2Y_{12} + Y_{22} \tag{8.10}$$

where

$$Y_{hk} \triangleq \frac{1}{N_0^2} \int_{-T/2}^{T/2} \int_{-T/2}^{T/2} \mathrm{E}\{x(u - d_h)\, x(v - d_k)\}\, r_h(u)\, r_k(v)\, \mathrm{d}u\, \mathrm{d}v\,. \tag{8.11}$$

In the case of $x(t)$ real-valued ACS signal, by substituting the almost-periodically time-variant expression (1.14) of the autocorrelation function with $t = v - d_k$, $\tau = u - v + d_k - d_h$ into (8.11), for T sufficiently large one has (Gardner and Spooner, 1993)

$$Y_{hk} \simeq \frac{T}{N_0^2} \sum_{\alpha \in A} \int_{-T/2}^{T/2} R_{xx}^{\alpha}(\tau)\, R_{r_h r_k}^{(T)}(\alpha, \tau + d_h - d_k)^*\, \mathrm{d}\tau\, e^{-j2\pi\alpha d_k} \tag{8.12a}$$

$$\simeq \frac{T}{N_0^2} \sum_{\alpha \in A} \int_{\mathbb{R}} S_{xx}^{\alpha}(f)\, I_{r_h r_k}^{(T)}(\alpha, f)^*\, e^{-j2\pi f(d_h - d_k)} \mathrm{d}f\, e^{-j2\pi\alpha d_k} \tag{8.12b}$$

where the second equality is due to the Parseval relation (Champeney, 1990, Chap. 7) and in the (cross-)periodograms and (cross-)correlograms the time dependence is omitted since it is not relevant for T sufficiently large.[1]

If the delays d_1 and d_2 are unknown, let $d_\Delta \triangleq d_2 - d_1$ be the time-difference-of-arrival (TDOA), and $\boldsymbol{\theta}_0$ be a vector of possibly unknown deterministic parameters of the source signal $x(t)$. The maximum-likelihood (ML) estimate of the pair $(d_\Delta, \boldsymbol{\theta}_0)$ is

$$(\widehat{d}_\Delta, \widehat{\boldsymbol{\theta}}_0) = \arg\max_{(d,\boldsymbol{\theta})} Y_{\mathrm{LLR}} \tag{8.13}$$

where Y_{LLR} is given by (8.10)–(8.12b) with d_1, d_2, and $\boldsymbol{\theta}_0$ replaced by ζ_1, ζ_2, and $\boldsymbol{\theta}$, that are the variables with respect to which the maximization is made, and $d \triangleq \zeta_2 - \zeta_1$.

From (8.9b), it follows that

$$R_{r_h r_k}^{\alpha}(\tau) = R_{xx}^{\alpha}(\tau + d_k - d_h)\, e^{-j2\pi\alpha d_k} + N_0\, \delta(\tau)\, \delta_\alpha\, \delta_{h-k}\,. \tag{8.14}$$

Thus, in (8.12a), for T sufficiently large and for large SNR ($R_{xx}^0(0) \gg N_0$), we have that when $\zeta_1 \simeq d_1$ and $\zeta_2 \simeq d_2$

$$R_{r_h r_k}^{(T)}(\alpha, \tau + d_h - d_k) \simeq R_{r_h r_k}^{\alpha}(\tau + d_h - d_k) = R_{xx}^{\alpha}(\tau)\, e^{-j2\pi\alpha d_k} \quad \forall\alpha \tag{8.15}$$

where the second equality is approximate for $\alpha = 0$. Consequently, each summand in (8.12a) has zero phase and Y_{LLR} achieves its maximum value.

By replacing into Y_{LLR} given by (8.10)–(8.12a) the unknown values $(d_\Delta, \boldsymbol{\theta}_0)$ with their ML estimates $(\widehat{d}_\Delta, \widehat{\boldsymbol{\theta}}_0)$ leads to the sufficient statistic for the generalized log-likelihood ratio test (GLLRT) (Gardner and Spooner, 1993).

[1] Eq. (8.12b) corrects (Napolitano, 2016b, Eq. (6.9b)).

In the case of detection based on a single receiver, $Y_{\text{LLR}} = Y_{11}$ with (8.12a) or (8.12b) substituted into leads to the *multi-cycle detector* (Gardner and Spooner, 1992b). The performance advantages of cyclic detectors versus classical detectors and the difficulties occurring in the implementation of the multi-cycle detector that make the single cycle detector more attractive are discussed in (Gardner and Spooner, 1992b). In particular, cyclic detectors are shown to significantly outperform the energy detector (radiometer) in the case of fluctuating noise power. A non-asymptotic performance analysis of single cycle detectors is made in (da Silva and Choi, 2008). Sensitivity analysis is made in (Ojeda and Grajal, 2012).

If the signal $x(t)$ is modeled as stationary, then only the term with $\alpha = 0$ is present in the sums over $\alpha \in A$ in (8.12a) and (8.12b). The detector based on a single receiver Y_{11} reduces to the optimum radiometer (Gardner, 1988b). In the case of two receivers, only Y_{12} depends on d_1 and d_2 and the TDOA estimation reduces to a generalized cross-correlation technique (Knapp and Carter, 1976).

The results for real-valued signals can be immediately extended to the case of proper complex signals. The case of improper cyclostationary signals can be treated by a previous "properization". It consists in applying an invertible frequency-shift filter (Section 3.2.1) to the improper signal in order to obtain a proper cyclostationary signal (see (Yeo and Cho, 2012) for continuous-time signals and (Yeo and Cho, 2014) for discrete-time signals). Then, due to the theorem of reversibility (Van Trees, 1971a, pp. 289–290), any processing designed under some optimality criterion (for detection and/or estimation) can be made on the obtained proper signal retaining the same performance that would be obtained operating directly on the improper signal. A patent is described in (Lee et al., 2016).

A locally optimum (low SNR) detector for complex-valued signals in non-Gaussian noise is derived in (Lunden and Koivunen, 2012). A GLRT is derived in (Riba et al., 2014) for cyclostationary signals in fading channels.

In the case of moving source, under the narrow-band condition, on each sensor the Doppler effect can be described as a frequency-shift ν_k of the carrier (Section 11.2), (Napolitano, 2016a, Sec. 11.2). This leads, under H_1, to the model $r_k(t) = x(t - d_k)\, e^{j2\pi \nu_k t} + n_k(t)$ for the complex envelopes of the signals received on the two sensors. The LLR function for low SNR is derived in (Streight et al., 2000).

In the derivation of the locally optimum receiver (8.10), (8.11), the second-order Taylor series expansion of the exponential is exploited for small SNR. If in the derivation of the locally-optimum receiver the Nth-order expansion

$$e^x = \sum_{n=0}^{N} \frac{x^n}{n!} + o(|x|^N) \tag{8.16}$$

is used, a generalization of the locally optimum receiver (8.10), (8.11) is obtained. In such a case, not only second-order moments, but also nth-order moments up to order N are present in the locally optimum detector structure.

8.4 Detection of signals exhibiting 2nd-order cyclostationarity

In this section, the problem of detecting an ACS SOI or, more generally, a SOI exhibiting cyclostationarity (see (1.19)) is considered in the presence of additive nonstationary and/or non-Gaussian disturbance with unknown distribution. Instead of deriving the detector structure starting from the observed signal, measurements of its cyclic autocorrelation are adopted as "front-end" data. In fact, properly normalized versions of these measurements are asymptotically complex normal as the data-record length approaches infinity, provided that the SOI and disturbance are processes with finite or practically finite memory. Thus, the log-likelihood ratio test (LLRT) can be derived from these measurements (Napolitano, 2018).

Specifically, let us consider jointly ACS processes $y(t)$ and $x(t)$ or, more generally, processes $y(t)$ and $x(t)$ exhibiting joint 2nd-order (possibly generalized) cyclostationarity. Under the assumption of finite or practically finite memory of the processes expressed in terms of summability of second- and fourth-order cumulants (see Section 5.2.1 for ACS processes and (Napolitano, 2012, Sec. 2.4) for GACS processes), the (conjugate) cyclic cross-correlogram

$$R^{(T)}_{yx^{(*)}}(\alpha,\tau) \triangleq \frac{1}{T}\int_{-T/2}^{T/2} y(t+\tau)\, x^{(*)}(t)\, e^{-j2\pi\alpha t}\, \mathrm{d}t \tag{8.17}$$

is a mean-square consistent estimator of the (conjugate) cyclic cross-correlation $R^{\alpha}_{yx^{(*)}}(\tau)$. In addition, under further conditions of summability of higher-order cumulants, the normalized error

$$\sqrt{T}\,\epsilon^{(T)}_{yx^{(*)}}(\alpha,\tau) \triangleq \sqrt{T}\left[R^{(T)}_{yx^{(*)}}(\alpha,\tau) - R^{\alpha}_{yx^{(*)}}(\tau)\right] \tag{8.18}$$

is asymptotically ($T \to \infty$) zero-mean complex normal (Section 5.2.1).

The detector considered in Section 8.4.1 performs the test at specific values of cycle frequency and lag parameter of the cyclic autocorrelation function. Unlike detectors based on the spectral line regeneration (Section 8.2), (Gardner, 1987f, Sec. 14.E), (Gardner, 1988b), (Gardner and Spooner, 1992b), the considered cycle frequencies can be possibly shared with the disturbance signal, and the only requirement is that the disturbance does not exhibit cyclostationarity at the selected pairs of cycle frequencies and lags chosen for the SOI. In addition, the threshold can be analytically derived.

The detector derived in Section 8.4.1 is compared with the statistical test for presence of cyclostationarity (Section 8.4.2) (Dandawaté and Giannakis, 1994b) which is used in (Haykin et al., 2009), (Lundén et al., 2009) in the context of cognitive radio. In such a test, the null hypothesis of absence of cyclostationarity in the data is discriminated versus the alternative hypothesis of presence of cyclostationarity in the data.

The detector derived in Section 8.4.1 and the test of Section 8.4.2 are intrinsically different. In fact, in the former the observed signal is different under the two hypotheses while in the latter the observed signal is the same under both hypotheses. There is a loss in performance in adapting the test (Dandawaté and Giannakis, 1994b) to the signal detection problem as done in (Jerjawi et al., 2015), (Oner and Jondral, 2007b), and (Shen and Alsusa, 2013b). However, in such a case, the detector implementation is simplified.

Finally, note that the same detector structure derived here for a process exhibiting cyclostationarity can also be derived for a process exhibiting generalized almost cyclostationarity (see (12.27a) and (12.27b)).

8.4.1 ACS signal detection

8.4.1.1 Detector

Let us consider the binary hypothesis test

$$\begin{aligned} \mathrm{H}_0 &: \quad r(t) = n(t) \\ \mathrm{H}_1 &: \quad r(t) = x(t) + n(t) \end{aligned} \qquad t \in [-T/2, T/2] \tag{8.19}$$

where $x(t)$ and $n(t)$ are zero-mean not necessarily independent random processes.

If the joint probability density function of $x(t_i)$ and $n(t_i)$, $i = 1, \dots, N$, is unknown or complicated, then the likelihood ratio test (LRT) or the generalized likelihood ratio test (GLRT) cannot be derived.

Assumption 8.1. *For fixed α and τ, and for a given conjugation choice ($(*) = *$ or $(*)$ absent)*

1. *$x(t)$ exhibits (conjugate) cyclostationarity at cycle frequency α and lag τ, i.e.,*

$$R^{\alpha}_{xx^{(*)}}(\tau) \neq 0 \tag{8.20}$$

2. *$n(t)$ does not exhibit (conjugate) cyclostationarity at cycle frequency α and lag τ, i.e.,*

$$R^{\alpha}_{nn^{(*)}}(\tau) = 0 \tag{8.21}$$

3. *$x(t)$ and $n(t)$ do not exhibit joint (conjugate) cyclostationarity at cycle frequency α and lag τ, i.e.,*

$$R^{\alpha}_{xn^{(*)}}(\tau) = R^{\alpha}_{nx^{(*)}}(\tau) = 0\,. \tag{8.22}$$

Under Assumption 8.1, we have

$$\begin{aligned} \mathrm{H}_0 &: R^{\alpha}_{rr^{(*)}}(\tau) = R^{\alpha}_{nn^{(*)}}(\tau) = 0 \\ \mathrm{H}_1 &: R^{\alpha}_{rr^{(*)}}(\tau) = R^{\alpha}_{xx^{(*)}}(\tau) + R^{\alpha}_{xn^{(*)}}(\tau) + R^{\alpha}_{nx^{(*)}}(\tau) + R^{\alpha}_{nn^{(*)}}(\tau) \\ &\qquad = R^{\alpha}_{xx^{(*)}}(\tau)\,. \end{aligned} \tag{8.23}$$

Note that the assumptions are made for a specific pair (α, τ) and conjugation $(*)$. That is, the noise can possibly exhibit (conjugate) cyclostationarity at the same cycle frequency of the SOI, but in correspondence of values of the lag parameter that are not selected for discriminating the two hypotheses or in correspondence of a different choice for $(*)$. In addition, no statistical independence of $x(t)$ and $n(t)$ is assumed.

Eqs. (8.19) and (8.23) suggest the following ad hoc (non-optimum and not equivalent to (8.19)) binary hypothesis test to discriminate the hypotheses H_0 and H_1 when $r(t)$ is observed for $t \in [-T/2, T/2]$:

$$\begin{aligned}
\mathrm{H}_0 \;:\; & R^{(T)}_{rr^{(*)}}(\alpha,\tau) = R^{(T)}_{nn^{(*)}}(\alpha,\tau) = 0 + \epsilon_0^{(T)}(\alpha,\tau) \\
\mathrm{H}_1 \;:\; & R^{(T)}_{rr^{(*)}}(\alpha,\tau) = R^{(T)}_{xx^{(*)}}(\alpha,\tau) + R^{(T)}_{xn^{(*)}}(\alpha,\tau) + R^{(T)}_{nx^{(*)}}(\alpha,\tau) + R^{(T)}_{nn^{(*)}}(\alpha,\tau) \\
& \qquad = R^{\alpha}_{xx^{(*)}}(\tau) + \epsilon_1^{(T)}(\alpha,\tau)
\end{aligned} \tag{8.24}$$

where

$$\epsilon_0^{(T)}(\alpha,\tau) \triangleq R^{(T)}_{nn^{(*)}}(\alpha,\tau) \equiv \epsilon^{(T)}_{nn^{(*)}}(\alpha,\tau) \tag{8.25}$$

$$\epsilon_1^{(T)}(\alpha,\tau) \triangleq \epsilon^{(T)}_{xx^{(*)}}(\alpha,\tau) + R^{(T)}_{xn^{(*)}}(\alpha,\tau) + R^{(T)}_{nx^{(*)}}(\alpha,\tau) + R^{(T)}_{nn^{(*)}}(\alpha,\tau)\,. \tag{8.26}$$

In Section 5.2.1, several assumptions are made to prove the mean-square consistency and the asymptotic complex normality of the (conjugate) cyclic correlogram. These assumptions can be summarized into the lack of clusters of cycle frequencies and the following assumption of summability of cumulants.

Assumption 8.2. *Let z_k denote any of x, x^* or any of n, and n^*. For any $N \geqslant 2$ we have*

$$\begin{aligned}
\operatorname{cum}&\{z_1(t+\tau_1),\ldots,z_{N-1}(t+\tau_{N-1}), z_N(t)\} \\
&= \sum_{\beta\in\mathbb{A}_N} C^{\beta}_{z_1\cdots z_N}(\tau_1,\ldots,\tau_{N-1})\, e^{j2\pi\beta t}
\end{aligned} \tag{8.27}$$

$$\sum_{\beta\in\mathbb{A}_N} \left|C^{\beta}_{z_1\cdots z_N}(\tau_1,\ldots,\tau_{N-1})\right| \in L^1(\mathbb{R}^{N-1})\,, \tag{8.28}$$

where $C^{\beta}_{z_1\cdots z_N}(\tau_1,\ldots,\tau_{N-1})$ are the reduced-dimension cyclic temporal cross-cumulant functions (see (4.40)). □

Under very mild conditions (Assumption 8.2) and *independently of the distribution of $x(t)$ and $n(t)$* we have (Theorem 5.15), (Napolitano, 2012, Section 2.4)

$$\lim_{T\to\infty} \mathrm{E}\left\{\left|\epsilon_i^{(T)}(\alpha,\tau)\right|^2\right\} = 0 \qquad i = 0,1 \tag{8.29}$$

and the error terms $\sqrt{T}\,\epsilon_0^{(T)}(\alpha,\tau)$ and $\sqrt{T}\,\epsilon_1^{(T)}(\alpha,\tau)$, are both noncircular asymptotically $(T\to\infty)$ zero-mean complex normal with covariance and conjugate covariance matrices that are different under the two hypotheses H_0 and H_1.

Even if test (8.24) is not optimum, due to (8.29) it is expected to have asymptotically $(T\to\infty)$ good performance. Moreover, it is easily implementable due to the normal distribution of the observations under the two hypotheses H_0 and H_1.

It is worthwhile to underline that the detection problem (8.24) is different from the statistical test for presence of cyclostationarity proposed in (Dandawaté and Giannakis, 1994b) and considered in several applications in cognitive radio (see e.g., (Haykin et al., 2009), (Lundén et al., 2009)). As clarified in Section 8.4.2, in (Dandawaté and Giannakis, 1994b) the two tested hypotheses are not absence or presence of a SOI and, unlike in (8.19), the observed signal $r(t)$ is the same under both hypotheses.

In (Huang and Tugnait, 2013), the detection problem (8.19) is considered when $n(t)$ is colored Gaussian noise. The statistical characterization of the error term $\epsilon_0^{(T)}(\alpha, \tau)$ is derived in this special case.

Test (8.24) can be considered for several pairs (α_k, τ_k), and conjugation choices $(*)_k$, $k = 1, \ldots, K$, satisfying Assumption 8.1, where some of the α_ks or τ_ks can assume the same values. That is, the following test can be considered

$$\begin{aligned} \mathrm{H}_0 &: R^{(T)}_{rr^{(*)_k}}(\alpha_k, \tau_k) = 0 + \epsilon_0^{(T)}(\alpha_k, \tau_k) \\ \mathrm{H}_1 &: R^{(T)}_{rr^{(*)_k}}(\alpha_k, \tau_k) = R^{\alpha_k}_{xx^{(*)_k}}(\tau_k) + \epsilon_1^{(T)}(\alpha_k, \tau_k) \end{aligned} \qquad k = 1, \ldots, K\,. \tag{8.30}$$

Let us define the (column) random vector

$$V \triangleq \left\{\sqrt{T}\left[R^{(T)}_{rr^{(*)_k}}(\alpha_k, \tau_k) - R^{\alpha_k}_{xx^{(*)_k}}(\tau_k)\right];\ k = 1, \ldots, K\right\}. \tag{8.31}$$

It results that

$$V \mid \mathrm{H}_0 = \left\{\sqrt{T}\,\epsilon_0^{(T)}(\alpha_k, \tau_k);\ k = 1, \ldots, K\right\} \tag{8.32}$$

$$V \mid \mathrm{H}_1 = \left\{\sqrt{T}\,\epsilon_1^{(T)}(\alpha_k, \tau_k);\ k = 1, \ldots, K\right\}. \tag{8.33}$$

According to the results of Section 5.2.1, (Hurd and Leśkow, 1992b), (Dehay and Hurd, 1994), (Dandawaté and Giannakis, 1995), (Dehay and Leśkow, 1996a) for the ACS case and (Napolitano, 2007a), (Napolitano, 2009), (Napolitano, 2012, Sec. 2.4) for the GACS case, we have that under mild assumptions expressed in terms of summability of cumulants, under both hypotheses V is asymptotically zero mean ($T \to \infty$) complex normal

$$V \mid \mathrm{H}_i \sim \mathcal{N}(\mathbf{0}, \boldsymbol{\Sigma}_i, \boldsymbol{\Sigma}_i^{(c)}) \qquad i = 0, 1\,. \tag{8.34}$$

In (8.34) we have zero asymptotic mean

$$\lim_{T\to\infty} \mathrm{E}\{V \mid \mathrm{H}_i\} = \mathbf{0} \qquad i = 0, 1 \tag{8.35}$$

asymptotic covariance matrix $\boldsymbol{\Sigma}_i$ with entries

$$\begin{aligned} \Sigma_i(k_1, k_2) &= \lim_{T\to\infty} \operatorname{cov}\left\{V_{k_1}, V_{k_2} \mid \mathrm{H}_i\right\} \\ &= \lim_{T\to\infty} \operatorname{cov}\left\{\sqrt{T}\left[R^{(T)}_{rr^{(*)_{k_1}}}(\alpha_{k_1}, \tau_{k_1}) - R^{\alpha_{k_1}}_{xx^{(*)_{k_1}}}(\tau_{k_1})\right],\right. \\ &\qquad\qquad \left.\sqrt{T}\left[R^{(T)}_{rr^{(*)_{k_2}}}(\alpha_{k_2}, \tau_{k_2}) - R^{\alpha_{k_2}}_{xx^{(*)_{k_2}}}(\tau_{k_2})\right] \mid \mathrm{H}_i\right\} \end{aligned}$$

$$= \lim_{T\to\infty} T \operatorname{cov}\Big\{\epsilon_i^{(T)}(\alpha_{k_1},\tau_{k_1}),\ \epsilon_i^{(T)}(\alpha_{k_2},\tau_{k_2})\Big\} \qquad i=0,1 \tag{8.36}$$

and asymptotic conjugate covariance matrix $\boldsymbol{\Sigma}_i^{(c)}$ with entries

$$\begin{aligned}
\Sigma_i^{(c)}(k_1,k_2) &= \lim_{T\to\infty} \operatorname{cov}\Big\{V_{k_1},\, V_{k_2}^* \mid \mathrm{H}_i\Big\} \\
&= \lim_{T\to\infty} \operatorname{cov}\Big\{\sqrt{T}\Big[R^{(T)}_{rr^{(*)_{k_1}}}(\alpha_{k_1},\tau_{k_1}) - R^{\alpha_{k_1}}_{xx^{(*)_{k_1}}}(\tau_{k_1})\Big], \\
&\qquad \sqrt{T}\Big[R^{(T)}_{rr^{(*)_{k_2}}}(\alpha_{k_2},\tau_{k_2}) - R^{\alpha_{k_2}}_{xx^{(*)_{k_2}}}(\tau_{k_2})\Big]^* \mid \mathrm{H}_i\Big\} \\
&= \lim_{T\to\infty} T \operatorname{cov}\Big\{\epsilon_i^{(T)}(\alpha_{k_1},\tau_{k_1}),\ \epsilon_i^{(T)*}(\alpha_{k_2},\tau_{k_2})\Big\} \qquad i=0,1
\end{aligned} \tag{8.37}$$

Note that both asymptotic covariance matrix $\boldsymbol{\Sigma}_i$ and asymptotic conjugate covariance matrix $\boldsymbol{\Sigma}_i^{(c)}$ depend on the hypothesis H_i. In addition, the complex random vector $\boldsymbol{V}$ is in general not circular since both $\boldsymbol{\Sigma}_i$ (Napolitano, 2012, Theorem 2.4.13) and $\boldsymbol{\Sigma}_i^{(c)}$ (Napolitano, 2012, Theorem 3.7.2) are nonzero.

Let us consider now the column random vector

$$\boldsymbol{Z} \triangleq \big\{\sqrt{T}\, R^{(T)}_{rr^{(*)_k}}(\alpha_k,\tau_k);\ k=1,\dots,K\big\}\,. \tag{8.38}$$

It results that

$$\boldsymbol{Z} \mid \mathrm{H}_0 = \big\{\sqrt{T}\,\epsilon_0^{(T)}(\alpha_k,\tau_k);\ k=1,\dots,K\big\} \tag{8.39}$$

$$\boldsymbol{Z} \mid \mathrm{H}_1 = \big\{\sqrt{T}\,\epsilon_1^{(T)}(\alpha_k,\tau_k) + \sqrt{T}\, R^{\alpha_k}_{xx^{(*)_k}}(\tau_k);\ k=1,\dots,K\big\} \tag{8.40}$$

for T sufficiently large are both complex normal

$$\boldsymbol{Z} \mid \mathrm{H}_i \sim \mathcal{N}(\boldsymbol{\mu}_i, \boldsymbol{\Sigma}_i, \boldsymbol{\Sigma}_i^{(c)}) \qquad i=0,1 \tag{8.41}$$

with mean vectors

$$\boldsymbol{\mu}_0 = \mathrm{E}\{\boldsymbol{Z} \mid \mathrm{H}_0\} = \boldsymbol{0} \tag{8.42}$$

$$\boldsymbol{\mu}_1 = \mathrm{E}\{\boldsymbol{Z} \mid \mathrm{H}_1\} = \{\sqrt{T}\, R^{\alpha_k}_{xx^{(*)_k}}(\tau_k);\ k=1,\dots,K\}\,. \tag{8.43}$$

Thus, the hypothesis test (8.30) for finite T and T sufficiently large becomes

$$\begin{aligned}
\boldsymbol{Z} \mid \mathrm{H}_0 &\sim \mathcal{N}(\boldsymbol{0}, \boldsymbol{\Sigma}_0, \boldsymbol{\Sigma}_0^{(c)}) \\
\boldsymbol{Z} \mid \mathrm{H}_1 &\sim \mathcal{N}(\boldsymbol{\mu}_1, \boldsymbol{\Sigma}_1, \boldsymbol{\Sigma}_1^{(c)})
\end{aligned} \tag{8.44}$$

Let be $\boldsymbol{Z} = \mathrm{Re}[\boldsymbol{Z}] + j\mathrm{Im}[\boldsymbol{Z}]$ and $z = \mathrm{Re}[z] + j\mathrm{Im}[z] = z_{\mathrm{re}} + jz_{\mathrm{im}}$. The probability of decision error is minimized by the log-likelihood ratio test (LLRT) (Van Trees, 1971a, Sec. 2.2)

$$\ln \frac{f_{\mathrm{Re}[\boldsymbol{Z}],\mathrm{Im}[\boldsymbol{Z}] \mid \mathrm{H}_1}(z_{\mathrm{re}}, z_{\mathrm{im}})}{f_{\mathrm{Re}[\boldsymbol{Z}],\mathrm{Im}[\boldsymbol{Z}] \mid \mathrm{H}_0}(z_{\mathrm{re}}, z_{\mathrm{im}})} \underset{\mathrm{H}_0}{\overset{\mathrm{H}_1}{\gtrless}} \ln \frac{\mathrm{P}[\mathrm{H}_0]}{\mathrm{P}[\mathrm{H}_1]} \tag{8.45}$$

where $\mathrm{P}[\mathrm{H}_0]$ and $\mathrm{P}[\mathrm{H}_1]$ are the a priori probabilities of the two hypothesis.

Let us define the *augmented random vector*

$$\boldsymbol{\zeta} \triangleq \begin{bmatrix} \boldsymbol{Z} \\ \boldsymbol{Z}^* \end{bmatrix} \tag{8.46}$$

the augmented vector of the realization of $\boldsymbol{\zeta}$

$$\boldsymbol{\zeta}_a \triangleq \begin{bmatrix} \boldsymbol{z} \\ \boldsymbol{z}^* \end{bmatrix} \tag{8.47}$$

the *augmented mean vector*

$$\boldsymbol{\mu}_{\zeta\,|\mathrm{H}_i} \triangleq \begin{bmatrix} \mathrm{E}\{\boldsymbol{Z} \mid \mathrm{H}_i\} \\ \mathrm{E}\{\boldsymbol{Z}^* \mid \mathrm{H}_i\} \end{bmatrix} = \begin{bmatrix} \boldsymbol{\mu}_i \\ \boldsymbol{\mu}_i^* \end{bmatrix} \tag{8.48}$$

and the *augmented covariance matrix*

$$\boldsymbol{\Gamma}_i \triangleq \mathrm{E}\left\{(\boldsymbol{\zeta} - \boldsymbol{\mu}_{\zeta\,|\mathrm{H}_i})(\boldsymbol{\zeta} - \boldsymbol{\mu}_{\zeta\,|\mathrm{H}_i})^{\mathsf{H}} \mid \mathrm{H}_i\right\} = \begin{bmatrix} \boldsymbol{\Sigma}_i & \boldsymbol{\Sigma}_i^{(c)} \\ \boldsymbol{\Sigma}_i^{(c)\,*} & \boldsymbol{\Sigma}_i^* \end{bmatrix}. \tag{8.49}$$

Accounting for the results of Section E.1 we have (Picinbono, 1996)

$$\begin{aligned} f_{\mathrm{Re}[\boldsymbol{Z}],\mathrm{Im}[\boldsymbol{Z}]\,|\,\mathrm{H}_i}(\boldsymbol{z}_{\mathrm{re}}, \boldsymbol{z}_{\mathrm{im}}) &\equiv f_{\boldsymbol{Z},\boldsymbol{Z}^*\,|\,\mathrm{H}_i}(\boldsymbol{z}, \boldsymbol{z}^*) \\ &= \frac{1}{\pi^K |\det \boldsymbol{\Gamma}_i|^{1/2}} \exp\left[-\frac{1}{2}(\boldsymbol{\zeta}_a - \boldsymbol{\mu}_{\zeta\,|\mathrm{H}_i})^{\mathsf{H}} \boldsymbol{\Gamma}_i^{-1} (\boldsymbol{\zeta}_a - \boldsymbol{\mu}_{\zeta\,|\mathrm{H}_i})\right] \end{aligned} \tag{8.50}$$

and the test (8.45) reduces to

$$\begin{aligned} &-\frac{1}{2}(\boldsymbol{\zeta}_a - \boldsymbol{\mu}_{\zeta\,|\mathrm{H}_1})^{\mathsf{H}} \boldsymbol{\Gamma}_1^{-1} (\boldsymbol{\zeta}_a - \boldsymbol{\mu}_{\zeta\,|\mathrm{H}_1}) - \frac{1}{2} \ln |\det \boldsymbol{\Gamma}_1| \\ &\mathop{\gtrless}_{\mathrm{H}_0}^{\mathrm{H}_1} -\frac{1}{2} \boldsymbol{\zeta}_a^{\mathsf{H}} \boldsymbol{\Gamma}_0^{-1} \boldsymbol{\zeta}_a - \frac{1}{2} \ln |\det \boldsymbol{\Gamma}_0| + \ln \frac{\mathrm{P}[\mathrm{H}_0]}{\mathrm{P}[\mathrm{H}_1]} \end{aligned} \tag{8.51}$$

If the parameters $\boldsymbol{\mu}_{\zeta\,|\mathrm{H}_1}$, $\boldsymbol{\Gamma}_1$, and $\boldsymbol{\Gamma}_0$ are unknown and are replaced by their maximum likelihood (ML) estimates $\widehat{\boldsymbol{\mu}}_{\zeta\,|\mathrm{H}_1}$, $\widehat{\boldsymbol{\Gamma}}_1$, and $\widehat{\boldsymbol{\Gamma}}_0$, one obtains the generalized log-likelihood ratio test (GLLRT) (Van Trees, 1971a, Sec. 2.5)

$$\ln \frac{f_{\boldsymbol{Z},\boldsymbol{Z}^*\,|\,\mathrm{H}_1}(\boldsymbol{z}, \boldsymbol{z}^*; \widehat{\boldsymbol{\mu}}_{\zeta\,|\mathrm{H}_1}, \widehat{\boldsymbol{\Gamma}}_1)}{f_{\boldsymbol{Z},\boldsymbol{Z}^*\,|\,\mathrm{H}_0}(\boldsymbol{z}, \boldsymbol{z}^*; \widehat{\boldsymbol{\Gamma}}_0)} \mathop{\gtrless}_{\mathrm{H}_0}^{\mathrm{H}_1} \ln \frac{\mathrm{P}[\mathrm{H}_0]}{\mathrm{P}[\mathrm{H}_1]} \tag{8.52}$$

that is,

$$\begin{aligned} &-\frac{1}{2}(\boldsymbol{\zeta}_a - \widehat{\boldsymbol{\mu}}_{\zeta\,|\mathrm{H}_1})^{\mathsf{H}} \widehat{\boldsymbol{\Gamma}}_1^{-1} (\boldsymbol{\zeta}_a - \widehat{\boldsymbol{\mu}}_{\zeta\,|\mathrm{H}_1}) - \frac{1}{2} \ln |\det \widehat{\boldsymbol{\Gamma}}_1| \\ &\mathop{\gtrless}_{\mathrm{H}_0}^{\mathrm{H}_1} -\frac{1}{2} \boldsymbol{\zeta}_a^{\mathsf{H}} \widehat{\boldsymbol{\Gamma}}_0^{-1} \boldsymbol{\zeta}_a - \frac{1}{2} \ln |\det \widehat{\boldsymbol{\Gamma}}_0| + \ln \frac{\mathrm{P}[\mathrm{H}_0]}{\mathrm{P}[\mathrm{H}_1]}. \end{aligned} \tag{8.53}$$

If the estimates $\widehat{\boldsymbol{\mu}}_{\zeta|\mathrm{H}_1}$, $\widehat{\boldsymbol{\Gamma}}_1$, and $\widehat{\boldsymbol{\Gamma}}_0$, are not ML but are consistent, test (8.52), (8.53) is not GLLRT, but can be considered as ad hoc test for the considered problem (Huang and Tugnait, 2013, Sec. IV-C-2).

According to (8.43), an estimate $\widehat{\boldsymbol{\mu}}_1$ of $\boldsymbol{\mu}_1$ is given by

$$\begin{aligned}\widehat{\boldsymbol{\mu}}_1 &= \left\{\sqrt{T}\, R^{(T)}_{xx^{(*)_k}}(\alpha_k,\tau_k);\ k=1,\ldots,K\right\}\\ &= \left\{\sqrt{T}\, R^{(T)}_{rr^{(*)_k}}(\alpha_k,\tau_k);\ k=1,\ldots,K\right\}\\ &\equiv \mathbf{z}\end{aligned} \tag{8.54}$$

where the second equality holds for T sufficiently large under Assumption 8.1.

Under mild assumptions on the finite or practically finite memory of the process $r(t)$ (Assumption 8.2), (Section 5.2.1), the (conjugate) cyclic correlogram $R^{(T)}_{rr^{(*)}}(\alpha,\tau)$ is a mean-square consistent estimator of the (conjugate) cyclic autocorrelation function $R^{\alpha}_{rr^{(*)}}(\tau) = R^{\alpha}_{xx^{(*)}}(\tau)$ (Theorem 5.15), (Napolitano, 2007a), (Napolitano, 2012, Sec. 2.4.2). Let

$$\begin{aligned}R^{(b,u)}_{rr^{(*)}}(\alpha,\tau) &\triangleq \frac{1}{b}\int_{u-b/2}^{u+b/2} r(t+\tau)\, r^{(*)}(t)\, e^{-j2\pi\alpha t}\,\mathrm{d}t\\ &= \frac{1}{b}\int_{-b/2}^{b/2} r(t'+u+\tau)\, r^{(*)}(t'+u)\, e^{-j2\pi\alpha(t'+u)}\,\mathrm{d}t'\end{aligned} \tag{8.55}$$

be the (conjugate) cyclic correlogram estimated on the basis of $r(t)$ with $t \in [u-b/2, u+b/2]$, where in the notation the dependence on the central point u of the observation block is evidenced and in the second equality the variable change $t' = t-u$ is made.

A consistent estimator for the entries of covariance matrix $\boldsymbol{\Gamma}_i$ is given by the block-bootstrap or subsampling estimate built by $r(t), t \in [-T/2, T/2]$, under hypothesis H_i. For example, for the entries of the sub-matrix $\boldsymbol{\Sigma}_i$ of the augmented covariance matrix $\boldsymbol{\Gamma}_i$ (see (8.49)), we have

$$\begin{aligned}\widehat{\Sigma}_i(k_1,k_2) &= \left\langle \sqrt{b}\, R^{(b,u)}_{rr^{(*)_{k_1}}}(\alpha_{k_1},\tau_{k_1})\, \sqrt{b}\, R^{(b,u)*}_{rr^{(*)_{k_2}}}(\alpha_{k_2},\tau_{k_2}) \mid \mathrm{H}_i \right\rangle_u\\ &\quad - \left\langle \sqrt{b}\, R^{(b,u)}_{rr^{(*)_{k_1}}}(\alpha_{k_1},\tau_{k_1}) \mid \mathrm{H}_i \right\rangle_u \left\langle \sqrt{b}\, R^{(b,u)*}_{rr^{(*)_{k_2}}}(\alpha_{k_2},\tau_{k_2}) \mid \mathrm{H}_i \right\rangle_u\end{aligned} \tag{8.56}$$

In (8.56),

$$\langle \phi(u) \rangle_u \triangleq \frac{1}{T-b}\int_{-T/2+b/2}^{T/2-b/2} \phi(u)\,\mathrm{d}u \simeq \frac{1}{2L+1}\sum_{\ell=-L}^{L} \phi(\ell h) \tag{8.57}$$

where, in the approximate quality we have $2L+1 = \lfloor (T-b)/h \rfloor$ with h sampling period.

Similar estimates can be built for the entries of the other sub-matrices in (8.49). Under the mild assumption of finite or practically finite memory for the process $r(t)$, the estimate is consistent provided that $b \to \infty$ and $T \to \infty$ with $b/T \to 0$ (see Section 8.7 and (Politis, 1998), (Lenart et al., 2008), (Dehay et al., 2014) for the discrete-time case).

If the FOT counterpart of this detection problem is considered (see (Dehay et al., 2018b)), in (8.56), we have that $\mathrm{E}_u^{\{\alpha\}}\{\cdot\} \simeq \langle\cdot\rangle_u$ as $b \to \infty$ (and $T \to \infty$ with $b/T \to 0$) (Napolitano, 2012, Sec. 6.3.5).

Accounting for (8.48) and (8.54), by substituting $\widehat{\boldsymbol{\mu}}_{\zeta|\mathrm{H}_1} = [\widehat{\boldsymbol{\mu}}_1^{\mathsf{T}}, \widehat{\boldsymbol{\mu}}_1^{\mathsf{H}}]^{\mathsf{T}} = [z^{\mathsf{T}}, z^{\mathsf{H}}]^{\mathsf{T}} = \boldsymbol{\zeta}_a$ into (8.53), the quadratic form with $\widehat{\boldsymbol{\Gamma}}_1^{-1}$ disappears and the test (8.53) reduces to

$$\boldsymbol{\zeta}_a^{\mathsf{H}} \widehat{\boldsymbol{\Gamma}}_0^{-1} \boldsymbol{\zeta}_a \underset{\mathrm{H}_0}{\overset{\mathrm{H}_1}{\gtrless}} \lambda_{\mathrm{map}} \tag{8.58}$$

with

$$\lambda_{\mathrm{map}} = \ln \frac{|\det \widehat{\boldsymbol{\Gamma}}_1|}{|\det \widehat{\boldsymbol{\Gamma}}_0|} + 2 \ln \frac{\mathrm{P}[\mathrm{H}_0]}{\mathrm{P}[\mathrm{H}_1]} \tag{8.59}$$

where the estimates $\widehat{\boldsymbol{\Gamma}}_i$ are obtained according to (8.56) and the a priori probabilities $\mathrm{P}[\mathrm{H}_i]$ must be known. In cognitive radio $\mathrm{P}[\mathrm{H}_1]$ can be assumed equal to the fraction-of-time probability (Section 2.2) that the primary user is present. Note that the threshold depends on the measurements $\widehat{\boldsymbol{\Gamma}}_0$ and $\widehat{\boldsymbol{\Gamma}}_1$.

If the a priori probabilities are unknown, a Neyman-Pearson test can be performed. It is obtained maximizing the probability of detection for a given probability of false alarm (Van Trees, 1971a, Sec. 2.2). This criterion leads to the LLRT

$$\Lambda(z, z^*) \triangleq \ln \frac{f_{\mathbf{Z},\mathbf{Z}^*|\mathrm{H}_1}(z, z^*)}{f_{\mathbf{Z},\mathbf{Z}^*|\mathrm{H}_0}(z, z^*)} \underset{\mathrm{H}_0}{\overset{\mathrm{H}_1}{\gtrless}} \lambda_{\mathrm{NP}} \tag{8.60}$$

where the threshold λ_{NP} is fixed such that the probability of false alarm P_{fa} is less than the assigned value a:

$$\mathrm{P}_{\mathrm{fa}} = \mathrm{P}\big[\Lambda(\mathbf{Z}, \mathbf{Z}^*) > \lambda_{\mathrm{NP}} \mid \mathrm{H}_0\big] = \int_{\lambda_{\mathrm{NP}}}^{+\infty} f_{\Lambda|\mathrm{H}_0}(\lambda)\, \mathrm{d}\lambda \leqslant a \tag{8.61}$$

where $f_{\Lambda|\mathrm{H}_0}(\lambda)$ is the conditioned pdf of the random variable log-likelihood ratio $\Lambda(\mathbf{Z}, \mathbf{Z}^*)$ under hypothesis H_0. We also have

$$\mathrm{P}_{\mathrm{fa}} = \int_{\mathbb{D}(\lambda_{\mathrm{NP}})} f_{\mathbf{Z},\mathbf{Z}^*|\mathrm{H}_0}(z, z^*)\, \mathrm{d}z \tag{8.62}$$

where

$$\mathbb{D}(\lambda_{\mathrm{NP}}) \triangleq \left\{ z \in \mathbb{C}^K \ : \ \Lambda(z, z^*) > \lambda_{\mathrm{NP}} \right\} . \tag{8.63}$$

If the parameters $\boldsymbol{\mu}_{\zeta|\mathrm{H}_1}$, $\boldsymbol{\Gamma}_1$, and $\boldsymbol{\Gamma}_0$ are unknown and are replaced by ML estimates $\widehat{\boldsymbol{\mu}}_{\zeta|\mathrm{H}_1}$, $\widehat{\boldsymbol{\Gamma}}_1$, and $\widehat{\boldsymbol{\Gamma}}_0$, one obtains the GLLRT (Van Trees, 1971a, Sec. 2.5)

$$\Lambda(z, z^*; \widehat{\boldsymbol{\mu}}_{\zeta|\mathrm{H}_1}, \widehat{\boldsymbol{\Gamma}}_1, \widehat{\boldsymbol{\Gamma}}_0) \triangleq \ln \frac{f_{\mathbf{Z},\mathbf{Z}^*|\mathrm{H}_1}(z, z^*; \widehat{\boldsymbol{\mu}}_{\zeta|\mathrm{H}_1}, \widehat{\boldsymbol{\Gamma}}_1)}{f_{\mathbf{Z},\mathbf{Z}^*|\mathrm{H}_0}(z, z^*; \widehat{\boldsymbol{\Gamma}}_0)} \underset{\mathrm{H}_0}{\overset{\mathrm{H}_1}{\gtrless}} \lambda_{\mathrm{NP}} . \tag{8.64}$$

As already observed for test (8.52), (8.53), if the estimates $\widehat{\boldsymbol{\mu}}_{\zeta|\mathrm{H}_1}$, $\widehat{\boldsymbol{\Gamma}}_1$, and $\widehat{\boldsymbol{\Gamma}}_0$, are not ML but are consistent, test (8.64) is not GLLRT, but can be considered as ad hoc tests for the considered problem.

If the estimate (8.54) is taken for $\boldsymbol{\mu}_1$, then the numerator in (8.64) does not depend on the observation $\boldsymbol{\zeta}_a$ and test (8.64) reduces to (Napolitano, 2018)

$$\widehat{Q}(\boldsymbol{\zeta}_a) \triangleq \boldsymbol{\zeta}_a^{\mathsf{H}} \widehat{\boldsymbol{\Gamma}}_0^{-1} \boldsymbol{\zeta}_a \underset{\mathrm{H}_0}{\overset{\mathrm{H}_1}{\gtrless}} \lambda \,. \tag{8.65}$$

The threshold in (8.65), different from that in (8.58), has been renamed λ and can be analytically derived. In fact, for T large, under H_0, the quadratic form

$$Q(\boldsymbol{\zeta}) \triangleq \boldsymbol{\zeta}^{\mathsf{H}} \boldsymbol{\Gamma}_0^{-1} \boldsymbol{\zeta} \tag{8.66}$$

with $\boldsymbol{\Gamma}_0$ denoting the true augmented covariance matrix of $\boldsymbol{\zeta}$, has central χ^2 distribution with $2K$ degrees of freedom (Section E.2.3) (Anderson, 2003, Theorem 3.3.3 adapted to the complex case). Since $\widehat{\boldsymbol{\Gamma}}_0$ is a consistent estimate of $\boldsymbol{\Gamma}_0$, then $\widehat{\boldsymbol{\Gamma}}_0^{-1}$ approaches $\boldsymbol{\Gamma}_0^{-1}$ in probability as $T \to \infty$, provided that $\boldsymbol{\Gamma}_0^{-1}$ exists (Theorem E.8) (Serfling, 1980, Sec. 4.1.2, p. 140). Consequently, under H_0, $Q(\boldsymbol{\zeta})$ and $\widehat{Q}(\boldsymbol{\zeta})$ asymptotically have the same central χ^2_{2K} distribution $F_{\chi^2_{2K}}(\cdot)$ (Theorem E.10) (Serfling, 1980, Sec. 4.1.2, p. 140). Therefore, for a desired false-alarm rate

$$P_{\mathrm{fa}} = \mathrm{P}[\widehat{Q} > \lambda \mid \mathrm{H}_0] = 1 - F_{\chi^2_{2K}}(\lambda) \tag{8.67}$$

we have

$$\lambda = F^{-1}_{\chi^2_{2K}}(1 - P_{\mathrm{fa}}) \tag{8.68}$$

with $F_{\chi^2_{2K}}(\cdot)$ given in (E.38) (with k replaced by $2K$).

In particular, for $K = 1$ we have

$$\begin{aligned} P_{\mathrm{fa}} &= P\,[Q > \lambda_2 \mid H_0] \\ &= \int_{\lambda_2}^{+\infty} f_{\chi^2_2}(x)\,\mathrm{d}x = F_{\chi^2_2}(+\infty) - F_{\chi^2_2}(\lambda_2) \\ &= \left[1 - e^{-x/2}\right]_{\lambda_2}^{+\infty} = e^{-\lambda_2/2} \end{aligned} \tag{8.69}$$

where $f_{\chi^2_2}(x)$ and $F_{\chi^2_2}(x)$ are the pdf and CDF, respectively, of a χ^2_2 random variable (see (E.37) and (E.38) with $k = 2$), and for $F_{\chi^2_2}(x)$ the simple closed form (E.41) is used. Therefore, the threshold λ_2 in test (8.65) is given by

$$\lambda_2 = 2\ln\left(\frac{1}{P_{\mathrm{fa}}}\right). \tag{8.70}$$

For $K \geqslant 2$ the threshold λ_{2K} depends of the desired probability of false alarm as follows

$$P_{\mathrm{fa}} = P\,[Q > \lambda_{2K} \mid H_0]$$

$$
\begin{aligned}
&= \int_{\lambda_{2K}}^{+\infty} f_{\chi^2_{2K}}(x)\,\mathrm{d}x \\
&= F_{\chi^2_{2K}}(+\infty) - F_{\chi^2_{2K}}(\lambda_{2K}) \\
&= 1 - \frac{\gamma(K, \lambda_{2K}/2)}{\Gamma(K)} = 1 - \frac{\int_0^{\lambda_{2K}/2} t^{K-1}\, e^{-t}\,\mathrm{d}t}{\int_0^{+\infty} t^{K-1}\, e^{-t}\,\mathrm{d}t}
\end{aligned}
\tag{8.71}
$$

where $f_{\chi^2_{2K}}$ and $F_{\chi^2_{2K}}$ are the pdf and CDF, respectively, of a χ^2_{2K} random variable (see (E.37) and (E.38) with $k = 2K$).

The implementation of detector (8.65) requires auxiliary data under H_0 necessary to obtain the estimate $\widehat{\boldsymbol{\Gamma}}_0$ before performing the test. In cooperative communications, such an estimate can be obtained in time intervals during which it is known that the signal $x(t)$ is absent. In radar/sonar applications, under the assumption of sufficiently spatially homogeneous noise environment, estimate $\widehat{\boldsymbol{\Gamma}}_0$ can be obtained starting from cells neighboring the cell-of-interest.

The best choice of pairs (α_k, τ_k) and optional conjugations $(*)_k$ depends on the specific SOI $x(t)$ to be detected and on the disturbance signal $n(t)$. As a general rule, pairs (α_k, τ_k) and optional conjugations $(*)_k$ such that $|R_{\boldsymbol{x}}^{\alpha_k}(\tau_k)|$ is large are good candidates.

8.4.1.2 Discrete-time implementation

Let $r(t)$ be a continuous-time complex-valued ACS process. The discrete-time (conjugate) cyclic correlogram of $r(nT_s)$ at cycle frequency $\widetilde{\alpha} = \alpha/f_s$ and discrete lag $m = \tau/T_s$ (assumed to be integer) is defined as

$$
R_{rr^{(*)}}^{(N)}(\widetilde{\alpha}, m) \triangleq \frac{1}{2N+1} \sum_{n=-N}^{N} r((n+m)T_s)\, r^{(*)}(nT_s)\, e^{-j2\pi\widetilde{\alpha}n} \,. \tag{8.72}
$$

It is a mean-square consistent (as $N \to \infty$) estimator of the aliased (conjugate) cyclic autocorrelation function (Napolitano, 2009), (Napolitano, 2012, Sec. 2.6.2)

$$
\widetilde{R}_{rr^{(*)}}^{\widetilde{\alpha}}(m) = \sum_{p=-\infty}^{+\infty} R_{rr^{(*)}}^{(\widetilde{\alpha}+p)f_s}(mT_s) \,. \tag{8.73}
$$

The values $\widetilde{R}_{rr^{(*)}}^{\widetilde{\alpha}}(m)$ can be made as close as desired to the samples $R_{rr^{(*)}}^{\widetilde{\alpha}f_s}(mT_s)$ of the continuous-time (conjugate) cyclic autocorrelation function, provided that the sampling period T_s is sufficiently small (Napolitano, 2009), (Napolitano, 2012, Sec. 2.6.3). Moreover, under mild assumptions the random variables

$$
V_k^{(N)} \triangleq \sqrt{2N+1}\left[R_{rr^{(*)}}^{(N)}(\widetilde{\alpha}_k, m_k) - \widetilde{R}_{rr^{(*)}}^{\widetilde{\alpha}_k}(m_k) \right] \tag{8.74}
$$

are asymptotically (as $N \to \infty$) zero-mean jointly complex normal (Napolitano, 2009), (Napolitano, 2012, Sec. 2.6.2) with (finite) asymptotic covariance and conjugate covariance matrices with entries

$$\Sigma(\widetilde{\alpha}_{k_1}, m_{k_1}; \widetilde{\alpha}_{k_2}, m_{k_2}) = \lim_{N\to\infty} \operatorname{cov}\left\{V_{k_1}^{(N)}, V_{k_2}^{(N)}\right\} \tag{8.75}$$

$$\Sigma^{(c)}(\widetilde{\alpha}_{k_1}, m_{k_1}; \widetilde{\alpha}_{k_2}, m_{k_2}) = \lim_{N\to\infty} \operatorname{cov}\left\{V_{k_1}^{(N)}, V_{k_2}^{(N)*}\right\} . \tag{8.76}$$

From the mean-square convergence, we also have the convergence in probability.

Starting from (8.72)–(8.76), the discrete-time counterpart of the detection test (8.65) can be implemented.

The elements of matrix (8.75) are estimated by subsampling techniques (Dehay et al., 2014), (Lenart et al., 2008) by the discrete-time counterpart of (8.56). Specifically, let

$$R_{rr^{(*)}}^{(b,h)}(\widetilde{\alpha}, m) \triangleq \frac{1}{2b+1} \sum_{n=h-b}^{h+b} r((n+m)T_s)\, r^{(*)}(nT_s)\, e^{-j2\pi\widetilde{\alpha}n} \tag{8.77}$$

be the (conjugate) cyclic correlogram estimated on the $(2b+1)$-length data-block with $n \in \{h-b, \dots, h+b\}$. The block-bootstrap or subsampling estimate of $\Sigma(\widetilde{\alpha}_{k_1}, m_{k_1}; \widetilde{\alpha}_{k_2}, m_{k_2})$ under hypothesis H_i is given by

$$\begin{aligned}
&\widehat{\Sigma}(\widetilde{\alpha}_{k_1}, m_{k_1}; \widetilde{\alpha}_{k_2}, m_{k_2}) \\
&\quad= \left\langle \sqrt{b}\, R_{rr^{(*)}k_1}^{(b,h)}(\widetilde{\alpha}_{k_1}, m_{k_1})\, \sqrt{b}\, R_{rr^{(*)}k_2}^{(b,h)*}(\widetilde{\alpha}_{k_2}, m_{k_2}) \mid \mathrm{H}_i \right\rangle_h \\
&\quad\quad - \left\langle \sqrt{b}\, R_{rr^{(*)}k_1}^{(b,h)}(\widetilde{\alpha}_{k_1}, m_{k_1}) \mid \mathrm{H}_i \right\rangle_h \left\langle \sqrt{b}\, R_{rr^{(*)}k_2}^{(b,h)*}(\widetilde{\alpha}_{k_2}, m_{k_2}) \mid \mathrm{H}_i \right\rangle_h
\end{aligned} \tag{8.78}$$

where

$$\langle \phi(h) \rangle_h \triangleq \frac{1}{(2N+1)-(2b+1)} \sum_{h=-N+b}^{N-b} \phi(h) \tag{8.79a}$$

$$\simeq \frac{1}{2L+1} \sum_{\ell=-L}^{L} \phi(\ell K_0) . \tag{8.79b}$$

In (8.79a), $(2N+1)-(2b+1)$ data-blocks of length $2b+1$ are used to compute the average. Two adjacent blocks are temporally shifted one another by 1 sample (corresponding to a shift of T_s at continuous time) and overlap by $2b$ samples. The approximate equality (8.79b) is to reduce the computational complexity of (8.79a). In fact, in (8.79b), two adjacent blocks are temporally shifted one another by K_0 samples and overlap by $\max\{0, 2b+1-K_0\}$ samples. Only $2L+1 < (2N+1)-(2b+1)$ blocks of length $2b+1$ are considered for computing the average, where L depends on the value of K_0. If $K_0 > 2b+1$, blocks do not overlap. If blocks are sufficiently far apart to be considered independent, we have the classical estimate via Monte Carlo trials.

Under mild assumptions on the finite or practically finite memory of the process $r(nT_s)$, by following the guidelines in (Lenart et al., 2008), (Politis, 1998), the estimate (8.78) can be proved to be consistent provided that $b \to \infty$ and $N \to \infty$ with $b/N \to 0$.

An analogous result holds for the block-bootstrap or subsampling estimate of the elements of the conjugate covariance matrix (8.76).

In order to simplify the implementation, the (conjugate) covariance matrix can be computed on the available data rather than using auxiliary data under H_0. In such a case, however, there is a lack of detection performance (see Sections 8.4.2 and 8.4.3).

8.4.2 Statistical test for presence of cyclostationarity

In this section, the statistical test for presence of cyclostationarity proposed in (Dandawaté and Giannakis, 1994b) and considered in several applications in cognitive radio (see e.g., (Haykin et al., 2009), (Lundén et al., 2009)) is revisited.

The test in (Dandawaté and Giannakis, 1994b) consists in checking if the signal $r(t)$ exhibits (H_1') or not (H_0') cyclostationarity at the pair (α, τ). That is,

$$\begin{aligned} H_0' &: (\alpha, \tau) \notin \operatorname{supp}\{R^{\alpha}_{rr^{(*)}}(\tau)\} \\ H_1' &: (\alpha, \tau) \in \operatorname{supp}\{R^{\alpha}_{rr^{(*)}}(\tau)\} \end{aligned} \tag{8.80}$$

where $\operatorname{supp}\{\cdot\}$ denotes the support of the argument in the brackets. If the signal $r(t)$ is observed for $t \in [-T/2, T/2]$, test (8.80) is implemented as

$$\begin{aligned} H_0' &: R^{(T)}_{rr^{(*)}}(\alpha, \tau) = \epsilon^{(T)}(\alpha, \tau) \\ H_1' &: R^{(T)}_{rr^{(*)}}(\alpha, \tau) = R^{\alpha}_{rr^{(*)}}(\tau) + \epsilon^{(T)}(\alpha, \tau) \,. \end{aligned} \tag{8.81}$$

In (8.81), unlike (8.24), the observed signal $r(t)$ is the same under both hypotheses. Consequently, the error term $\epsilon^{(T)}(\alpha, \tau)$ is the same under both hypotheses H_0' and H_1'.

Similarly to (8.30), test (8.81) can be considered for several pairs (α_k, τ_k), $k = 1, \ldots, K$, (Lundén et al., 2009), (Shen and Alsusa, 2013a), (Shen and Alsusa, 2013b). In such a case, test (8.81) reduces to

$$\widehat{Q}'(\boldsymbol{\zeta}_a) \triangleq \boldsymbol{\zeta}_a^{\mathsf{H}} \widehat{\boldsymbol{\Gamma}}^{-1} \boldsymbol{\zeta}_a \underset{H_0'}{\overset{H_1'}{\gtrless}} \lambda' \tag{8.82}$$

which is presented here in terms of complex-valued augmented vectors and is equivalent to the test originally derived in (Dandawaté and Giannakis, 1994b) in terms of real-valued vectors of real and imaginary parts of the (conjugate) cyclic correlograms. In (8.82), $\boldsymbol{\zeta}_a$ is a realization of the random augmented vector $\boldsymbol{\zeta}$ with augmented covariance matrix $\boldsymbol{\Gamma}$ whose estimate is denoted by $\widehat{\boldsymbol{\Gamma}}$. The entries' estimates can be obtained as in (8.56). In (Dandawaté and Giannakis, 1994b), the estimates of the entries of $\boldsymbol{\Gamma}$ are obtained by frequency smoothing (conjugate) cyclic correlograms of second-order lag products of the data $r(t)$.

It is worthwhile to underline that test (8.30) and the vector-valued counterpart of test (8.81) are different. In fact, in the former, data $r(t)$ is different under the two hypotheses H_0 and H_1 and, consequently, $\epsilon_0^{(T)}$ and $\epsilon_1^{(T)}$ in (8.30) are different and have different statistical characterization. In contrast, in the latter, data $r(t)$ is the same under the two hypotheses H_0' and H_1'. Therefore, in (8.82), the estimate $\widehat{\boldsymbol{\Gamma}}$ is obtained by the available data and not from auxiliary data as in (8.65). If $r(t) = x(t) + n(t)$, in (8.81) we have

$$\begin{aligned}
\mathrm{H}_0' \; : \; R^{(T)}_{rr^{(*)}}(\alpha,\tau) &= \epsilon^{(T)}_{rr^{(*)}}(\alpha,\tau) + R^{(T)}_{nx^{(*)}}(\alpha,\tau) + R^{(T)}_{xn^{(*)}}(\alpha,\tau) + R^{(T)}_{nn^{(*)}}(\alpha,\tau) \\
\mathrm{H}_1' \; : \; R^{(T)}_{rr^{(*)}}(\alpha,\tau) &= R^{\alpha}_{xx^{(*)}}(\tau) + \epsilon^{(T)}_{rr^{(*)}}(\alpha,\tau) \\
&\quad + R^{(T)}_{nx^{(*)}}(\alpha,\tau) + R^{(T)}_{xn^{(*)}}(\alpha,\tau) + R^{(T)}_{nn^{(*)}}(\alpha,\tau)
\end{aligned} \tag{8.83}$$

The difference between the two tests should be carefully accounted for if the vector-valued version of (8.81) is used instead of (8.30) for signal detection (problem (8.19)) e.g., in the context of cognitive radio.

Test (8.82) with $K = 1$ and considered with (α, τ) ranging in a given rectangle, say $[\alpha_{\min}, \alpha_{\max}] \times [\tau_{\min}, \tau_{\max}]$, allows one to build a binary map with the support of $R^{\alpha}_{rr^{(*)}}(\tau)$ (see (8.80)) determined with the desired significance level specified by P_{fa}.

8.4.3 Performance analysis

Let us consider now, the situation in which the detection statistic $\widehat{Q}'$ of test (8.82) is adopted in the context of signal detection (see (Jerjawi et al., 2015), (Oner and Jondral, 2007b), (Shen and Alsusa, 2013b)), that is, to discriminate hypotheses H_0 and H_1 on the basis of the observation $r(t)$ as in (8.19):

$$\widehat{Q}'(\boldsymbol{\zeta}_a) \triangleq \boldsymbol{\zeta}_a^{\mathsf{H}} \widehat{\boldsymbol{\Gamma}}^{-1} \boldsymbol{\zeta}_a \underset{\mathrm{H}_0}{\overset{\mathrm{H}_1}{\gtrless}} \lambda' \tag{8.84}$$

Let as assume that the pairs (α_k, τ_k) are selected such that $x(t)$ exhibits cyclostationarity in correspondence of these pairs. We obviously have $\mathrm{H}_0 \Rightarrow \mathrm{H}_0'$. Consequently, under H_0 we have $\boldsymbol{\Gamma} = \boldsymbol{\Gamma}_0$ so that

$$Q'(\boldsymbol{\zeta}) = \boldsymbol{\zeta}^{\mathsf{H}} \boldsymbol{\Gamma}^{-1} \boldsymbol{\zeta} = \boldsymbol{\zeta}^{\mathsf{H}} \boldsymbol{\Gamma}_0^{-1} \boldsymbol{\zeta} = Q(\boldsymbol{\zeta})$$

have the same asymptotic distribution χ^2_{2K}. Thus, for a given P_{fa} we have $\lambda' = \lambda$, with λ given in (8.68).

Under H_1 we have $\boldsymbol{\Gamma} = \boldsymbol{\Gamma}_1$. Consequently,

$$Q'(\boldsymbol{\zeta}) = \boldsymbol{\zeta}^{\mathsf{H}} \boldsymbol{\Gamma}^{-1} \boldsymbol{\zeta} \quad \text{and} \quad Q(\boldsymbol{\zeta}) = \boldsymbol{\zeta}^{\mathsf{H}} \boldsymbol{\Gamma}_0^{-1} \boldsymbol{\zeta}$$

do not have the same distribution. Hence,

$$P_{\mathrm{d}}' = \mathrm{P}[Q' > \lambda \mid \mathrm{H}_1] \neq \mathrm{P}[Q > \lambda \mid \mathrm{H}_1] = P_{\mathrm{d}} \,. \tag{8.85}$$

The same results hold asymptotically, when the true covariance matrices are replaced by their consistent estimates.

If we consider the detection problem starting from "front-end" data $R^{(T)}_{rr^{(*)}}(\alpha, \tau)$ (rather than from the original data $r(t)$), we have that test (8.60) is optimum in the Neyman-

Pearson sense (among those derived from data $R^{(T)}_{rr^{(*)}}(\alpha, \tau)$). In addition,

$$\widehat{\mathbf{\Gamma}}_0 \xrightarrow{\text{p}} \mathbf{\Gamma}_0 \quad \Rightarrow \quad \widehat{\mathbf{\Gamma}}_0^{-1} \xrightarrow{\text{p}} \mathbf{\Gamma}_0^{-1}$$

provided that $\mathbf{\Gamma}_0^{-1}$ exists, and

$$\boldsymbol{\zeta} \xrightarrow{\text{d}} \boldsymbol{\zeta}_{\text{G}} \sim \mathcal{N}(\mathbf{0}, \mathbf{\Sigma}_i, \mathbf{\Sigma}_i^{(c)}) \,.$$

Thus, accounting for the Slusky theorem (Theorem E.6) (Serfling, 1980, p. 19),

$$\boldsymbol{\zeta}^{\mathsf{H}} \widehat{\mathbf{\Gamma}}_0^{-1} \boldsymbol{\zeta} \xrightarrow{\text{d}} \boldsymbol{\zeta}_{\text{G}}^{\mathsf{H}} \mathbf{\Gamma}_0^{-1} \boldsymbol{\zeta}_{\text{G}} \,.$$

Similarly,

$$\boldsymbol{\zeta}^{\mathsf{H}} \mathbf{\Gamma}_0^{-1} \boldsymbol{\zeta} \xrightarrow{\text{d}} \boldsymbol{\zeta}_{\text{G}}^{\mathsf{H}} \mathbf{\Gamma}_0^{-1} \boldsymbol{\zeta}_{\text{G}} \,.$$

Therefore, $\widehat{Q}$ approaches Q in distribution as $T \to \infty$ also under hypothesis H_1. Thus, asymptotically, test (8.64) has the same performance of the optimum test (8.60), provided that estimates are consistent. For a fixed P_{fa} the threshold (calculated under H_0) is the same and we have $P'_{\text{d}} \leqslant P_{\text{d}}$.

There is a loss in performance in using test (8.84) to discriminate between hypothesis H_0 versus H_1. However, there is the advantage that no side data is necessary for the estimation of the covariance matrix.

8.5 Detection of signals exhibiting spectral correlation

Let $x(t)$ be an ACS process. In Section 8.4, for the binary hypothesis test (8.19) a sub-optimum test (8.24) is presented, which is based on the asymptotic complex normality of the estimates of the time-domain cyclic statistics. By mimicking the procedure adopted in Section 8.4, a similar sub-optimum test can be formulated starting from the asymptotic complex normality of estimates of frequency-domain cyclic statistics.

Let us consider two jointly ACS processes $y(t)$ and $x(t)$. Under the assumption of finite or practically finite memory of the processes we have that the frequency-smoothed (conjugate) cyclic cross-periodogram (Definition 5.29)

$$S^{(T,\Delta f)}_{yx^{(*)}}(\alpha, f) \triangleq \frac{1}{\Delta f} \int_{f-\Delta f/2}^{f+\Delta f/2} \frac{1}{T} Y_T(t_0, \lambda)\, X_T^{(*)}(t_0, (-)(\alpha - \lambda))\, \mathrm{d}\lambda \tag{8.86}$$

and the time-smoothed (conjugate) cyclic cross-periodogram (Definition 5.42)

$$G^{(\Delta f,T)}_{yx^{(*)}}(\alpha, f) \triangleq \frac{1}{T} \int_{t_0-T/2}^{t_0+T/2} \Delta f\, Y_{1/\Delta f}(s, f)\, X^{(*)}_{1/\Delta f}(s, (-)(\alpha - f))\, \mathrm{d}s \tag{8.87}$$

are mean-square consistent estimators of the (conjugate) cyclic cross spectrum $S^{\alpha}_{yx^{(*)}}(f)$ as $T \to \infty$ and $\Delta f \to 0$ with $T\Delta f \to \infty$ (Sections 5.2.3 and 5.2.4). In (8.86) and (8.87), $X_Z(t_0, f)$

is the short-time Fourier transform (STFT) of $x(t)$ defined according to

$$X_Z(t_0, f) \triangleq \int_{t_0-Z/2}^{t_0+Z/2} x(t)\, e^{-j2\pi f t}\, \mathrm{d}t \,. \tag{8.88}$$

Similarly, $Y_Z(t_0, f)$ is the STFT of $y(t)$. In addition, the normalized errors

$$\sqrt{T\Delta f}\, \epsilon^{(T,\Delta f)}(\alpha, f) \triangleq \sqrt{T\Delta f}\left[S_{yx^{(*)}}^{(T,\Delta f)}(\alpha, f) - \mathcal{E}_{(*)}\, S_{yx^{(*)}}^{\alpha}(f)\right] \tag{8.89}$$

and

$$\sqrt{T\Delta f}\, \varepsilon^{(\Delta f,T)}(\alpha, f) \triangleq \sqrt{T\Delta f}\left[G_{yx^{(*)}}^{(\Delta f,T)}(\alpha, f) - k_G\, S_{yx^{(*)}}^{\alpha}(f)\right] \tag{8.90}$$

with k_G defined in (5.92), are asymptotically ($T \to \infty$ and $\Delta f \to 0$ with $T\Delta f \to \infty$) zero-mean complex normal (Theorems 5.41 and 5.49). The two estimators (8.86) and (8.87) are asymptotically equivalent in the sense specified in Section 5.3.1.

Assumption 8.3. *For fixed α and f and for a given conjugation choice ($(*) = *$ or $(*)$ absent)*

1. *$x(t)$ exhibits (conjugate) spectral correlation at cycle frequency α and spectral frequency f, i.e.,*

$$S_{xx^{(*)}}^{\alpha}(f) \neq 0 \tag{8.91}$$

2. *$n(t)$ does not exhibit (conjugate) spectral correlation at cycle frequency α and spectral frequency f, i.e.,*

$$S_{nn^{(*)}}^{\alpha}(f) = 0 \tag{8.92}$$

3. *$x(t)$ and $n(t)$ do not exhibit joint (conjugate) spectral correlation at cycle frequency α and spectral frequency f, i.e.,*

$$S_{xn^{(*)}}^{\alpha}(f) = S_{nx^{(*)}}^{\alpha}(f) = 0 \,. \tag{8.93}$$

Let us consider the binary hypothesis test (8.19). Under Assumption 8.3, we have

$$\begin{aligned} \mathrm{H}_0 &: S_{rr^{(*)}}^{\alpha}(f) = S_{nn^{(*)}}^{\alpha}(f) = 0 \\ \mathrm{H}_1 &: S_{rr^{(*)}}^{\alpha}(f) = S_{xx^{(*)}}^{\alpha}(f) + S_{xn^{(*)}}^{\alpha}(f) + S_{nx^{(*)}}^{\alpha}(f) + S_{nn^{(*)}}^{\alpha}(f) \\ &\qquad = S_{xx^{(*)}}^{\alpha}(f) \,. \end{aligned} \tag{8.94}$$

Note that assumptions are made for a specific pair (α, f) and conjugation $(*)$. That is, noise can possibly exhibit (conjugate) spectral correlation at the same cycle frequency of the SOI but in correspondence of values of the spectral frequency that are not selected for discriminating the two hypotheses or in correspondence of a different choice for $(*)$.

Eqs. (8.19) and (8.94) suggest the following ad hoc (non-optimum and not equivalent to (8.19)) binary hypothesis test to discriminate the hypotheses H_0 and H_1 when $r(t)$ is observed for $t \in [-T/2, T/2]$ and (conjugate) cyclic spectra are estimated with spectral resolution Δf by the frequency-smoothed (conjugate) cyclic periodogram:

$$
\begin{aligned}
\mathrm{H}_0 :\; & S_{rr^{(*)}}^{(T,\Delta f)}(\alpha, f) = S_{nn^{(*)}}^{(T,\Delta f)}(\alpha, f) = 0 + \epsilon_0^{(T,\Delta f)}(\alpha, f) \\
\mathrm{H}_1 :\; & S_{rr^{(*)}}^{(T,\Delta f)}(\alpha, f) = S_{xx^{(*)}}^{(T,\Delta f)}(\alpha, f) + S_{xn^{(*)}}^{(T,\Delta f)}(\alpha, f) + S_{nx^{(*)}}^{(T,\Delta f)}(\alpha, f) + S_{nn^{(*)}}^{(T,\Delta f)}(\alpha, f) \\
& \qquad = S_{xx^{(*)}}^{\alpha}(f) + \epsilon_1^{(T,\Delta f)}(\alpha, f)
\end{aligned}
\tag{8.95}
$$

where

$$
\epsilon_0^{(T,\Delta f)}(\alpha, f) \triangleq S_{nn^{(*)}}^{(T,\Delta f)}(\alpha, f) \equiv \epsilon_{nn^{(*)}}^{(T,\Delta f)}(\alpha, f) \tag{8.96}
$$

$$
\epsilon_1^{(T,\Delta f)}(\alpha, f) \triangleq \epsilon_{xx^{(*)}}^{(T,\Delta f)}(\alpha, f) + S_{xn^{(*)}}^{(T,\Delta f)}(\alpha, f) + S_{nx^{(*)}}^{(T,\Delta f)}(\alpha, f) + S_{nn^{(*)}}^{(T,\Delta f)}(\alpha, f) \,. \tag{8.97}
$$

Analogous equations are obtained if (conjugate) cyclic spectra are estimated by the time-smoothed (conjugate) cyclic periodogram.

In Section 5.2.3, assumptions are made to prove the mean-square consistency and the asymptotic complex normality of the frequency-smoothed (conjugate) cyclic periodogram. Under these assumptions and independently of the distribution of $x(t)$ and $n(t)$,

$$
\lim_{\Delta f \to 0} \lim_{T \to \infty} \mathrm{E}\left\{ \left| \epsilon_i^{(T,\Delta f)}(\alpha, f) \right|^2 \right\} = 0 \qquad i = 0, 1 \tag{8.98}
$$

and the random variables $\sqrt{T\Delta f}\, \epsilon_i^{(T,\Delta f)}(\alpha, f)$, $i = 0, 1$, are normally distributed. Specifically, the error terms $\sqrt{T\Delta f}\, \epsilon_0^{(T,\Delta f)}(\alpha, f)$ and $\sqrt{T\Delta f}\, \epsilon_1^{(T,\Delta f)}(\alpha, f)$ are both noncircular asymptotically ($T \to \infty$ and $\Delta f \to 0$ with $T\Delta f \to \infty$) zero-mean complex normal with different covariance and conjugate covariance matrices under the two hypothesis H_0 and H_1 (Section 5.2.3), (Napolitano, 2012, Sections 4.5, 4.7).

The test (8.95) can be considered for several pairs (α_k, f_k), and conjugation choices $(*)_k$, $k = 1, \dots, K$ such that (8.94) holds, where some of the α_ks or f_ks can assume the same values.

Let us define the random column vector

$$
\boldsymbol{U} \triangleq \left\{ \sqrt{T\Delta f} \left[S_{rr^{(*)_k}}^{(T,\Delta f)}(\alpha_k, f_k) - S_{xx^{(*)_k}}^{\alpha_k}(f_k) \right];\; k = 1, \dots, K \right\}. \tag{8.99}
$$

We have that

$$
\boldsymbol{U} \mid \mathrm{H}_0 = \{ \sqrt{T\Delta f}\, \epsilon_0^{(T)}(\alpha_k, f_k);\; k = 1, \dots, K \} \tag{8.100}
$$

$$
\boldsymbol{U} \mid \mathrm{H}_1 = \{ \sqrt{T\Delta f}\, \epsilon_1^{(T)}(\alpha_k, f_k);\; k = 1, \dots, K \}\,. \tag{8.101}
$$

According to the results of Section 5.2.3, (see also (Dandawaté and Giannakis, 1994a), (Dehay and Hurd, 1994), (Leskow, 1994) for the ACS case and (Napolitano, 2012, Sec. 4.7.2) for the SC case specialized to ACS signals), we have that under mild assumptions expressed in terms of summability of temporal or spectral cumulants, under both hypothesis the random vector $\boldsymbol{U}$ is asymptotically ($T \to \infty$, $\Delta f \to 0$ with $T\Delta f \to \infty$) complex normal

$$\boldsymbol{U} \mid \mathrm{H}_i \sim \mathcal{N}(\mathbf{0}, \boldsymbol{\Sigma}_i, \boldsymbol{\Sigma}_i^{(c)}) \qquad i = 1, 2 \tag{8.102}$$

with zero asymptotic mean

$$\lim_{\Delta f \to 0} \lim_{T \to \infty} \mathrm{E}\{\boldsymbol{U} \mid \mathrm{H}_i\} = \mathbf{0} \qquad i = 1, 2 \tag{8.103}$$

asymptotic covariance matrix with entries

$$\begin{aligned}\Sigma_i(k_1, k_2) = & \lim_{\Delta f \to 0} \lim_{T \to \infty} \operatorname{cov}\Big\{\sqrt{T\Delta f}\left[S_{rr^{(*)k_1}}^{(T,\Delta f)}(\alpha_{k_1}, f_{k_1}) - S_{xx^{(*)k_1}}^{\alpha_{k_1}}(f_{k_1})\right], \\ & \qquad \sqrt{T\Delta f}\left[S_{rr^{(*)k_2}}^{(T,\Delta f)}(\alpha_{k_2}, f_{k_2}) - S_{xx^{(*)k_2}}^{\alpha_{k_2}}(f_{k_2})\right] \mid \mathrm{H}_i\Big\} \\ = & \lim_{\Delta f \to 0} \lim_{T \to \infty} \operatorname{cov}\Big\{\sqrt{T\Delta f}\, \epsilon_i^{(T,\Delta f)}(\alpha_{k_1}, f_{k_1}), \sqrt{T\Delta f}\, \epsilon_i^{(T,\Delta f)}(\alpha_{k_2}, f_{k_2})\Big\}\end{aligned} \tag{8.104}$$

and asymptotic conjugate covariance matrix with entries

$$\begin{aligned}\Sigma_i^{(c)}(k_1, k_2) = & \lim_{\Delta f \to 0} \lim_{T \to \infty} \operatorname{cov}\Big\{\sqrt{T\Delta f}\left[S_{rr^{(*)k_1}}^{(T,\Delta f)}(\alpha_{k_1}, f_{k_1}) - S_{xx^{(*)k_1}}^{\alpha_{k_1}}(f_{k_1})\right], \\ & \qquad \sqrt{T\Delta f}\left[S_{rr^{(*)k_2}}^{(T,\Delta f)}(\alpha_{k_2}, f_{k_2}) - S_{xx^{(*)k_2}}^{\alpha_{k_2}}(f_{k_2})\right]^* \mid \mathrm{H}_i\Big\} \\ = & \lim_{\Delta f \to 0} \lim_{T \to \infty} \operatorname{cov}\Big\{\sqrt{T\Delta f}\, \epsilon_i^{(T,\Delta f)}(\alpha_{k_1}, f_{k_1}), \sqrt{T\Delta f}\, \epsilon_i^{(T,\Delta f)\,*}(\alpha_{k_2}, f_{k_2})\Big\}.\end{aligned} \tag{8.105}$$

Therefore, defined random column vector

$$\mathbf{Z} \triangleq \Big\{\sqrt{T\Delta f}\;\; S_{rr^{(*)k}}^{(T,\Delta f)}(\alpha_k, f_k);\; k = 1, \dots, K\Big\} \tag{8.106}$$

we have that

$$\mathbf{Z} \mid \mathrm{H}_0 = \Big\{\sqrt{T\Delta f}\, \epsilon_0^{(T,\Delta f)}(\alpha_k, f_k);\; k = 1, \dots, K\Big\} \tag{8.107}$$

$$\mathbf{Z} \mid \mathrm{H}_1 = \Big\{\sqrt{T\Delta f}\, S_{xx^{(*)k}}^{\alpha_k}(f_k) + \sqrt{T\Delta f}\, \epsilon_1^{(T,\Delta f)}(\alpha_k, f_k);\; k = 1, \dots, K\Big\} \tag{8.108}$$

for T sufficiently large, Δf sufficiently small and such that $T\Delta f \gg 1$, are both complex normal

$$\mathbf{Z} \mid \mathrm{H}_i \sim \mathcal{N}(\boldsymbol{\mu}_i, \boldsymbol{\Sigma}_i, \boldsymbol{\Sigma}_i^{(c)}) \qquad i = 1, 2 \tag{8.109}$$

with mean vectors

$$\boldsymbol{\mu}_0 \triangleq \mathrm{E}\{\mathbf{Z} \mid \mathrm{H}_0\} = \mathbf{0} \tag{8.110}$$

$$\boldsymbol{\mu}_1 \triangleq \mathrm{E}\{\mathbf{Z} \mid \mathrm{H}_1\} = \{\sqrt{T\Delta f}\; S_{xx^{(*)k}}^{\alpha_k}(f_k);\; k = 1, \dots, K\}\,. \tag{8.111}$$

Note that the complex random vector $\boldsymbol{Z}$ is in general not circular since both asymptotic covariance matrix (Napolitano, 2012, Theorem 4.7.7) and asymptotic conjugate covariance matrix (Napolitano, 2012, Theorem 5.8.3) are nonzero.

Thus, the hypothesis test (8.95) considered for all pairs (α_k, f_k), $k = 1, \dots, K$ and in vectorial form can be written as (8.44).

Starting from the frequency-domain counterpart of (8.44), the LLRT can be written as in (8.45). Therefore, (8.46)–(8.53) hold also for their frequency-domain counterpart and the detection test (8.64) also has its frequency-domain counterpart.

According to (8.111) and under mild assumptions on the finite or practically finite memory of the process $r(t)$ (Section 5.2.3), an estimate $\widehat{\boldsymbol{\mu}}_1$ of $\boldsymbol{\mu}_1$ is given by

$$\begin{aligned}\widehat{\boldsymbol{\mu}}_1 &= \left\{\sqrt{T\Delta f}\, S_{xx^{(*)}}^{(T,\Delta f)}(\alpha_k, f_k);\; k = 1, \dots, K\right\} \\ &= \left\{\sqrt{T\Delta f}\, S_{rr^{(*)}}^{(T,\Delta f)}(\alpha_k, f_k);\; k = 1, \dots, K\right\} \\ &\equiv z \end{aligned} \tag{8.112}$$

where the second equality holds asymptotically ($T \to \infty$, $\Delta f \to 0$ with $T\Delta f \to \infty$) under Assumption 8.3.

If the estimate (8.112) is taken for $\boldsymbol{\mu}_1$, then the numerator in the frequency domain counterpart of test (8.64) does not depend on the observation ζ_a and the frequency domain counterpart of (8.64) reduces to the frequency-domain counterpart of (8.65).

Let

$$S_{rr^{(*)}}^{(b,u,\Delta f)}(\alpha, f) = \frac{1}{b}\int_{u-b/2}^{u+b/2} \Delta f\, R_{1/\Delta f}(s, f)\, R_{1/\Delta f}^{(*)}(s, {(-)}(\alpha - f))\, \mathrm{d}s \tag{8.113a}$$

$$\simeq \frac{1}{\Delta f}\int_{f-\Delta f/2}^{f+\Delta f/2} \frac{1}{b}\, R_b(u, \lambda)\, R_b^{(*)}(u, {(-)}(\alpha - \lambda))\, \mathrm{d}\lambda \tag{8.113b}$$

be the (conjugate) time- or frequency-smoothed cyclic periodogram estimated on the basis of $r(t)$ with $t \in [u - b/2, u + b/2]$, with $R_Z(t_0, f)$ denoting the STFT of $r(t)$, $b\,\Delta f \gg 1$, and where in the notation the dependence on the central point u of the observation interval is evidenced. A consistent estimator for the entries of the augmented covariance matrix $\boldsymbol{\Gamma}_i$ is given by the block-bootstrap or subsampling estimate built from $r(t)$, $t \in [-T/2, T/2]$, $T \gg b$, under hypothesis H_i. For example, for the entries of the sub-matrix $\boldsymbol{\Sigma}_i$ (see (8.49)), we have

$$\begin{aligned}\widehat{\Sigma}_i(k_1, k_2) = &\left\langle \sqrt{b\Delta f}\, S_{rr^{(*)}k_1}^{(b,u,\Delta f)}(\alpha_{k_1}, f_{k_1})\, \sqrt{b\Delta f}\, S_{rr^{(*)}k_2}^{(b,u,\Delta f)*}(\alpha_{k_2}, f_{k_2}) \mid \mathrm{H}_i \right\rangle_u \\ &- \left\langle \sqrt{b\Delta f}\, S_{rr^{(*)}k_1}^{(b,u,\Delta f)}(\alpha_{k_1}, f_{k_1}) \mid \mathrm{H}_i \right\rangle_u \left\langle \sqrt{b\Delta f}\, S_{rr^{(*)}k_2}^{(b,u,\Delta f)*}(\alpha_{k_2}, f_{k_2}) \mid \mathrm{H}_i \right\rangle_u \end{aligned} \tag{8.114}$$

where

$$\langle \cdot \rangle_u \triangleq \frac{1}{T-b}\int_{-T/2+b/2}^{T/2-b/2} (\cdot)\, \mathrm{d}u\,. \tag{8.115}$$

Similar estimates can be built for the entries of the other sub-matrices in (8.49).

By following the guidelines in (Lenart et al., 2008), (Politis, 1998), under mild assumption on the finite or practically finite memory of the process $r(t)$, the estimate (8.114) can be proved to be consistent provided that $b \to \infty$, $\Delta f \to 0$, with $b\Delta f \to \infty$ and $T \to \infty$ with $b/T \to 0$ (see also Section 8.7).

8.6 Statistical test for presence of spectral coherence

The *(conjugate) spectral coherence function* of $r(t)$ at cycle frequency α is defined as (Definition 1.17), (Gardner, 1987f, Chap. 11), (Spooner and Nicholls, 2009, eqs. (18.23), (18.24))

$$\gamma_{\boldsymbol{r}}^{\alpha}(f) \triangleq \frac{S_{\boldsymbol{r}}^{\alpha}(f)}{\left[S_{rr^*}^{0}(f)\, S_{rr^*}^{0}((-)(\alpha - f))\right]^{1/2}} \tag{8.116}$$

where subscript $\boldsymbol{r} \equiv rr^{(*)}$, and it results $0 \leqslant |\gamma_{\boldsymbol{r}}^{\alpha}(f)| \leqslant 1$.

A statistical test to detect the presence of an ACS signal can be realized by detecting the presence of spectral coherence ($\boldsymbol{r} = rr^*$) along lines parallel to the main diagonal in the bifrequency plane $(f_1, f_2) \equiv (f, f - \alpha)$ (Hurd and Gerr, 1991). It generalizes the test presented in (Goodman, 1965) for stationary signals. The presence of conjugate almost-cyclostationarity can be revealed by detecting the presence of conjugate spectral coherence ($\boldsymbol{r} = rr$) along lines parallel to the secondary diagonal.

Let

$$R_N(\nu) \triangleq \sum_{n=0}^{N-1} r(nT_s)\, e^{-j2\pi\nu n} \tag{8.117}$$

be the discrete Fourier transform of $r(nT_s)$, $n = 0, 1, \ldots, N-1$, where T_s is the sampling period. The square magnitude of the sample (conjugate) spectral coherence function at cycle frequency $\alpha = \widetilde{\alpha} f_s$ and spectral frequency $f = \nu_p f_s$ is

$$|\widehat{\gamma}_M(\nu_p, \nu_q)|^2 = \frac{\left|\sum_{m=0}^{M-1} R_N(\nu_p + m/N)\, R_N^{(*)}((-)(\nu_q - m/N))\right|^2}{\sum_{m=0}^{M-1} |R_N(\nu_p + m/N)|^2 \sum_{m=0}^{M-1} |R_N((-)(\nu_q - m/N))|^2} \tag{8.118}$$

with $\nu_q = \alpha T_s - \nu_p$. The quantity M/N is the width of the frequency-smoothing window.

Let us consider the frequency sets $\mathcal{I}_1 = \{\nu_p + m/N\}_{m=0,1,\ldots,M-1}$ and $\mathcal{I}_2 = \{\nu_q + m/N\}_{m=0,1,\ldots,M-1}$ such that $\mathcal{I}_1 \cap \mathcal{I}_2 = \emptyset$. Assuming that the random variables $R_N(\nu_1)$ and $R_N(\nu_2)$ with $\nu_1 \in \mathcal{I}_1$ and $\nu_2 \in \mathcal{I}_2$ are zero mean circular complex normal with variances $\mathrm{E}\{|R_N(\nu_1)|^2\} = \sigma_1^2$ and $\mathrm{E}\{|R_N(\nu_2)|^2\} = \sigma_2^2$ constant for $\nu_1 \in \mathcal{I}_1$ and $\nu_2 \in \mathcal{I}_2$, in (Goodman, 1965) it is proved that

$$\mathrm{P}[|\widehat{\gamma}_M|^2 > |\gamma_0|^2] = (1 - |\gamma_0|^2)^{M-1}\,. \tag{8.119}$$

Thus, the threshold $|\gamma_0|^2$ (with $0 < |\gamma_0|^2 < 1$) can be computed from (8.119) in order to obtain the desired significance level and a binary map of spectral coherence with such significance level can be constructed in the (ν_p, ν_q) plane or in the $(\widetilde{\alpha}, \nu_p)$ plane (Hurd and Gerr, 1991).

8.7 Subsampling-based significance test

If the length of the observation interval is not sufficiently large to reach the asymptotic regime, then estimates of statistical functions of ACS processes cannot be assumed to be normal. Moreover, in most cases, even if the asymptotic normality condition is approximately verified, the covariance matrix is very complicated and its calculation requires the knowledge of cyclic statistics at several cycle frequencies (Theorems 5.14 and 5.28), (Dandawaté and Giannakis, 1994a), (Dandawaté and Giannakis, 1995), (Dehay and Hurd, 1994), (Napolitano, 2012, Secs. 2.4.2, 4.5, 4.7). In such cases, approximations of the unknown distribution of the estimates can be obtained by subsampling or block-bootstrap techniques. Thus, confidence intervals for these estimates can be constructed and significance tests performed.

A significance test for the (conjugate) cyclic autocorrelation estimate at given cycle frequency and lag (α, τ) is performed by subsampling-based techniques in (Lenart, 2013), (Dehay et al., 2014), (Dudek, 2015). For this purpose, the observation interval $[0, T]$ is partitioned into possibly overlapping blocks $(nh, nh + b)$ of size b, where $h \leqslant b$ is the block overlap factor, $n = 0, \ldots, N-1$, with $N = \lfloor (T-b)/h \rfloor$ number of blocks, and $\lfloor \cdot \rfloor$ denotes the floor operation (Dehay et al., 2014).

Let

$$R_{\boldsymbol{r}}^{(b,u)}(\alpha, \tau) \triangleq \frac{1}{b} \int_u^{u+b} r(t+\tau)\, r^{(*)}(t)\, e^{-j2\pi\alpha t} \mathrm{d}t \tag{8.120}$$

be the estimator of the (conjugate) cyclic autocorrelation function generated by the data in the interval $(u, u+b)$. The empirical distribution of the real part of the normalized estimation error of the (conjugate) cyclic autocorrelation function induced by the subsampling procedure is defined as

$$F_{\mathrm{Re}}^{(T,b)}(\xi) \triangleq \frac{1}{N} \sum_{n=0}^{N-1} \mathbf{1}\left\{\sqrt{b}\,\mathrm{Re}\left[R_{\boldsymbol{r}}^{(b,nh)}(\alpha, \tau) - R_{\boldsymbol{r}}^{(T)}(\alpha, \tau)\right] \leqslant \xi\right\} \tag{8.121}$$

where $\mathbf{1}\{\cdot\}$ is the indicator operator. The empirical distribution $F_{\mathrm{Im}}^{(T,b)}(\xi)$ is defined similarly for the imaginary part.

Under α-mixing conditions on the process and assuming that there is no cluster of cycle frequencies, for $T \to \infty$ and $b \to \infty$ with $b/T \to 0$, the empirical distribution $F_{\mathrm{Re}}^{(T,b)}(\xi)$ converges in the mean-squared sense to the limit zero-mean normal distribution (Section 5.2)

$$F_{\mathrm{Re}}(\xi) \triangleq \lim_{T\to\infty} \mathrm{P}\left[\sqrt{T}\,\mathrm{Re}\left[R_{\boldsymbol{r}}^{(T)}(\alpha, \tau) - R_{\boldsymbol{r}}^{\alpha}(\tau)\right] \leqslant \xi\right]. \tag{8.122}$$

That is,

$$\lim_{b\to\infty}\lim_{T\to\infty} \mathrm{E}\left\{\left|F_{\mathrm{Re}}^{(T,b)}(\xi) - F_{\mathrm{Re}}(\xi)\right|^2\right\} = 0 \tag{8.123}$$

in the continuity points of $F_{\mathrm{Re}}(\xi)$. A similar result holds for $F_{\mathrm{Im}}^{(T,b)}(\xi)$. Moreover, almost-sure convergence can also be proved (Dehay et al., 2014).

The significance test is based on the definition of *quantile* . For a significance level a, the quantile of the distribution F is defined as

$$q_a \triangleq \inf\{q \in \mathbb{R} : F(q) \geqslant a\} \tag{8.124}$$

and in the continuity points of F we have

$$a = F(q_a) \quad \Leftrightarrow \quad q_a = F^{-1}(a)\,. \tag{8.125}$$

The interval

$$[L, U] \triangleq [q_{a/2}, q_{1-a/2}] \tag{8.126}$$

is referred to as the *confidence interval* with significance level a. We have

$$\int_{q_{a/2}}^{q_{1-a/2}} \mathrm{d}F(\xi) = F(q_{1-a/2}) - F(q_{a/2}) = 1 - a\,. \tag{8.127}$$

Due to the convergence property of $F_{\mathrm{Re}}^{(T,b)}(\xi)$ and $F_{\mathrm{Im}}^{(T,b)}(\xi)$, the following significance test can be performed:

1. Fix a significance level, say a, for the estimate;
2. For the significance level a, compute the confidence interval $[L_{\mathrm{Re}}, U_{\mathrm{Re}}]$ for the estimated distribution $F_{\mathrm{Re}}^{(T,b)}(\xi)$ and the confidence interval $[L_{\mathrm{Im}}, U_{\mathrm{Im}}]$ for the estimated distribution $F_{\mathrm{Im}}^{(T,b)}(\xi)$;
3. If at least one of the two confidence intervals[2]

$$\left[\sqrt{b}\,\mathrm{Re}[R_{\boldsymbol{r}}^{(T)}(\alpha,\tau)] + L_{\mathrm{Re}},\ \sqrt{b}\,\mathrm{Re}[R_{\boldsymbol{r}}^{(T)}(\alpha,\tau)] + U_{\mathrm{Re}}\right]$$

and

$$\left[\sqrt{b}\,\mathrm{Im}[R_{\boldsymbol{r}}^{(T)}(\alpha,\tau)] + L_{\mathrm{Im}},\ \sqrt{b}\,\mathrm{Im}[R_{\boldsymbol{r}}^{(T)}(\alpha,\tau)] + U_{\mathrm{Im}}\right]$$

does not contain zero, then cyclostationarity is declared to be present at (α, τ).

For discrete-time ACS processes with finite or practically finite memory expressed in terms of α-mixing conditions, the moving block bootstrap method for the mean value and the cyclic autocorrelation functions is proposed in (Synowiecki, 2007). A bootstrap algorithm is considered in (Dudek and Leskow, 2011) and a subsampling procedure in (Lenart

[2] Corrected version of (Napolitano, 2016b, item (3), p. 394).

et al., 2008). The bootstrap of periodic random arrays with increasing period is considered in (Leskow and Synowiecki, 2010). The problem of cycle frequency identification and estimation and a subsampling procedure for the Fourier coefficients is addressed in (Lenart, 2013). A generalized seasonal block bootstrap technique is proposed in (Dudek et al., 2014b) for discrete-time cyclostationary processes with known period. An application to biological signal analysis is made in (Dudek et al., 2014c). A circular block bootstrap for ACS processes is proposed in (Dudek, 2015). It is a modification of the moving block bootstrap that consists in wrapping data on the circle with the aim of reducing edge effects. Further results are in (Dudek et al., 2014a), (Dudek and Potorski, 2014), (Gajecka-Mirek, 2014), (Dehay and Dudek, 2015), (Dudek and Uzar, 2015), (Dudek, 2018), (Dehay et al., 2018a).

8.8 Robust detectors

Consistency and asymptotic complex normality of cyclic statistic estimators are proved under the assumption of finite or practically finite memory of the process (Section 5.2). In the presence of signals with heavy-tailed (first-, second-, and higher-order) distributions, classical cyclic statistic estimators (Section 5.2) can exhibit poor performance. In such a case, one should resort to robust estimators (Lundén et al., 2010).

Robust cyclic detectors can be obtained by replacing into the detector structures of Sections 8.4 and 8.5 the classical cyclic statistic estimators with robust cyclic statistic estimators, e.g., the spatial sign (conjugate) cyclic correlogram and the spatial sign (conjugate) time-smoothed cyclic periodogram.

Let us define the *spatial sign* function of the complex number $z = |z|\, e^{j\phi_z}$ as

$$S[z] \triangleq \begin{cases} z/|z| = e^{j\phi_z} & z \neq 0 \\ 0 & z = 0 \end{cases} \tag{8.128}$$

The *spatial sign (conjugate) cyclic correlogram* of the complex-valued signal $r(t) \triangleq |r(t)|\, e^{j\phi_r(t)}$ is defined as

$$R^{(T)}_{S[r]S[r^{(*)}]}(\alpha, \tau) \triangleq \frac{1}{T} \int_{t_0-T/2}^{t_0+T/2} S[r(t+\tau)]\, S[r^{(*)}(t)]\, e^{-j2\pi\alpha t}\, \mathrm{d}t \tag{8.129a}$$

$$= \frac{1}{T} \int_{t_0-T/2}^{t_0+T/2} e^{j[\phi_r(t+\tau)+(-)\phi_r(t)-2\pi\alpha t]}\, \mathrm{d}t \tag{8.129b}$$

In (Lundén et al., 2010), a detection statistic is proposed, which is based on the discrete-time counterpart of $Z^{(T)} = \sqrt{T}\, R^{(T)}_{S[r]S[r^*]}(\alpha, \tau)$. The test performance is analytically derived in (Renard et al., 2013b).

Similarly, a robust detection test can be built starting from the *spatial sign (conjugate) time-smoothed cyclic periodogram* of $r(t)$ that is defined as (Renard et al., 2013a)

$$P^{(\Delta f,T)}_{S[R]S[R^{(*)}]}(\alpha, f) \triangleq \frac{1}{T}\int_{t_0-T/2}^{t_0+T/2} S\Big[R_{\frac{1}{\Delta f}}(t, f)\Big]\, S\Big[R^{(*)}_{\frac{1}{\Delta f}}(t, (-)(\alpha - f))\Big]\, \mathrm{d}t \tag{8.130}$$

where $R_{\frac{1}{\Delta f}}(t, f)$ is the STFT of $r(t)$ (see (1.31)).

Under mild assumptions, $R^{(T)}_{S[r]S[r^{(*)}]}(\alpha, \tau)$ and $P^{(\Delta f,T)}_{S[R]S[R^{(*)}]}(\alpha, f)$ are shown to be asymptotically complex normal with diagonal covariance matrix (Lundén et al., 2010), (Renard et al., 2013b), (Renard et al., 2013a). This simplifies the implementation of the detection test (8.65).

Other robust cyclic spectral analysis techniques are presented in (Biedka et al., 1995b), (Biedka et al., 1995a), (Sullivan and Wegman, 1995), (Lepage and Thomson, 2009), (Lunden and Koivunen, 2014), (Drake et al., 2015). The cyclic correntropy for the analysis of impulsive signals is introduced and studied in (Fontes et al., 2017b), (Fontes et al., 2017a), (Liu et al., 2018b).

8.9 Higher-order statistic based detectors

If the SOI has zero or weak second-order cyclic features or is embedded in a disturbance signal with similar second-order cyclic characteristics, by generalizing the approaches in Sections 8.2, 8.3, 8.4, 8.5, and 8.8, higher-order cyclostationarity based detectors can be designed.

Specifically, by generalizing the approach of Section 8.2, if the SOI $x(t)$ in the received (complex envelope) signal $r(t)$ exhibits Nth-order cyclostationarity at cycle frequency α_0 and the disturbance signal $n(t)$ does not exhibit Nth-order cyclostationarity at α_0, then the presence of $x(t)$ can be detected by revealing the presence of a spectral line at frequency α_0 at the output of a homogeneous Nth-order time-invariant nonlinear transformation of $r(t)$ (Spooner and Gardner, 1994a), (Spooner and Nicholls, 2009)

$$y(t) = \int_{\mathbb{R}^N} K(\tau_1, \ldots, \tau_N)\, r^{(*)_1}(t+\tau_1) \cdots r^{(*)_N}(t+\tau_N)\, \mathrm{d}\tau_1 \cdots \mathrm{d}\tau_N\,. \tag{8.131}$$

In (8.131), $(*)_i$ represents the optional complex conjugation of the ith term $r(t+\tau_i)$. The choice of conjugating or not the ith term depends on the modulation format to be detected. For each order N, different modulation formats exhibit Nth-order cyclostationarity or not depending on the number of conjugated terms (Gardner and Spooner, 1994a), (Spooner and Gardner, 1994a), (Spooner and Nicholls, 2009). Similarly to the second-order case, the kernel $K(\tau_1, \ldots, \tau_N)$ can be chosen according to an optimum criterion.

The 2nd-order cyclostationarity-based detector presented in Section 8.4, can be generalized to a higher-order cyclostationarity-based detector. In fact, under mild assumptions in terms of memory of the process, estimators of cyclic higher-order statistical functions are asymptotically complex normal (Section 5.7) (Dandawaté and Giannakis, 1994a), (Dandawaté and Giannakis, 1995).

For example, the component Z_k of $\mathbf{Z}$ of Section 8.4 can be replaced by the (normalized) Nth-order cyclic correlogram

$$Z_k \triangleq \sqrt{T}\, R_r^{(T)}(\alpha, \tau_1, \ldots, \tau_{N-1}) \tag{8.132a}$$

$$= \frac{1}{\sqrt{T}} \int_{t_0-T/2}^{t_0+T/2} r^{(*)_1}(t+\tau_1) \cdots r^{(*)_{N-1}}(t+\tau_{N-1})\, r^{(*)_N}(t)\, e^{-j2\pi\alpha t}\, \mathrm{d}t \tag{8.132b}$$

where the number of conjugated terms depends on the modulation format to be detected.

Similarly, also the techniques in Sections 8.7 and 8.8 are suitable to be generalized to higher-order statistics.

Higher-order cyclostationarity is exploited for spectrum sensing in (Renard et al., 2013a), (Renard et al., 2010).

8.10 Cycle frequency estimation

(Conjugate) cyclic autocorrelation, (conjugate) cyclic spectrum, and (conjugate) spectral coherence functions are discrete in the cycle frequency domain. That is, they are nonzero only in correspondence of a finite or at most countable set of values of α. However, in cycle frequency estimation problems, since the true values of the cycle frequencies are unknown, it is convenient to consider α varying with continuity.

Let $R_{rr^{(*)}}(\alpha, \tau)$ denote the (conjugate) cyclic autocorrelation function defined according to (1.53) with $\alpha \in \mathbb{R}$. It is coincident with $R^{\alpha}_{rr^{(*)}}(\tau)$ for $\alpha \in A$ and zero otherwise and we have

$$\lambda_{rr^{(*)}}(\alpha) \triangleq \int_{\mathbb{R}} \left| R_{rr^{(*)}}(\alpha, \tau) \right|^2 \mathrm{d}\tau \tag{8.133a}$$

$$= \sum_{\beta \in A} \int_{\mathbb{R}} \left| R^{\beta}_{rr^{(*)}}(\tau) \right|^2 \mathrm{d}\tau\ \delta_{\alpha-\beta} \tag{8.133b}$$

$$= \sum_{\beta \in A} \int_{\mathbb{R}} \left| S^{\beta}_{rr^{(*)}}(f) \right|^2 \mathrm{d}f\ \delta_{\alpha-\beta} \tag{8.133c}$$

where the third equality is a consequence of the Fourier transform relationship (1.54) and the Parseval theorem (Champeney, 1990, Chaps. 7, 8). The function $\lambda_{rr^{(*)}}(\alpha)$ is nonzero only in the countable set A. For each $\alpha \in A$ it is equal to the energy of $R^{\alpha}_{rr^{(*)}}(\tau)$ or, equivalently, of $S^{\alpha}_{rr^{(*)}}(f)$.

An estimate $\lambda^{(T)}_{rr^{(*)}}(\alpha)$ of the function $\lambda_{rr^{(*)}}(\alpha)$ can be obtained by replacing in (8.133a) the (conjugate) autocorrelation function with the (conjugate) cyclic correlogram $R^{(T)}_{rr^{(*)}}(\alpha, \tau)$ defined in (5.9) and considering the integral over a finite lag interval, say $\mathcal{T}$.

In Fig. 8.1, the function $\lambda^{(T)}_{rr^*}(\alpha)$, with $T = NT_s$ and $N = 2^{10}$ samples ($T_s = 1/f_s =$ sampling period), is reported for a PAM signal with independent and identically distributed binary symbols and full duty-cycle rectangular pulse of width $T_p = 8T_s$, in the

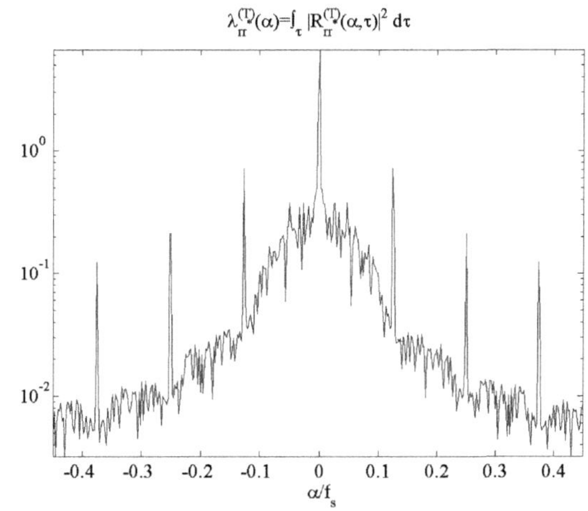

FIGURE 8.1 PAM signal: Function $\lambda_{rr^*}^{(T)}(\alpha)$ with $T = NT_s$ and $N = 2^{10}$ samples.

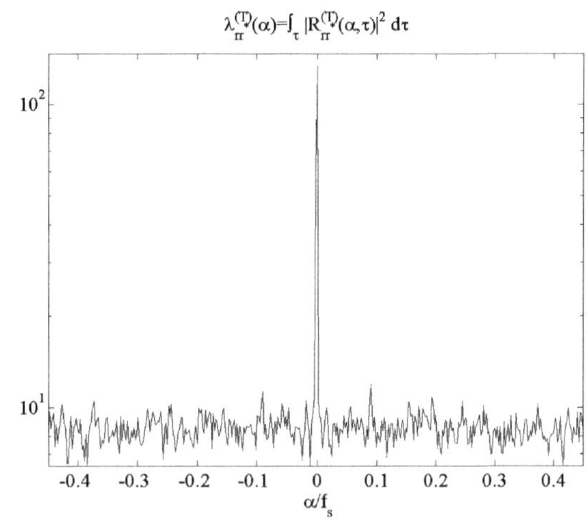

FIGURE 8.2 PAM signal in AWGN with SNR $= -10$ dB: Function $\lambda_{rr^*}^{(T)}(\alpha)$ with $T = NT_s$ and $N = 2^{10}$ samples.

absence of noise. The integration interval is $\mathcal{T} = [-2T_p, 2T_p]$. The peaks corresponding to the cycle frequencies k/T_p, k integer, are evident. In Fig. 8.2, the PAM signal is embedded in AWGN with SNR $= -10$ dB in the bandwidth $(-f_s/2, f_s/2)$ and $\lambda_{rr^*}^{(T)}(\alpha)$ is evaluated with $T = NT_s$ and $N = 2^{10}$. Due to the cycle leakage effect (Section 5.3.4) (Gardner, 1987f, Chap. 13) from $\alpha = 0$, the lines at k/T_p, $k \neq 0$, cannot be recognized. The effects of cycle leakage can be mitigated by augmenting the length of the observation interval. In Fig. 8.3, the PAM signal is embedded in AWGN with SNR $= -10$ dB in the bandwidth $(-f_s/2, f_s/2)$ and $\lambda_{rr^*}^{(T)}(\alpha)$ is evaluated with $T = NT_s$ and $N = 2^{17}$. Cycle frequencies of the PAM signal can be estimated by locating the peaks of the function $\lambda_{rr^*}^{(T)}(\alpha)$.

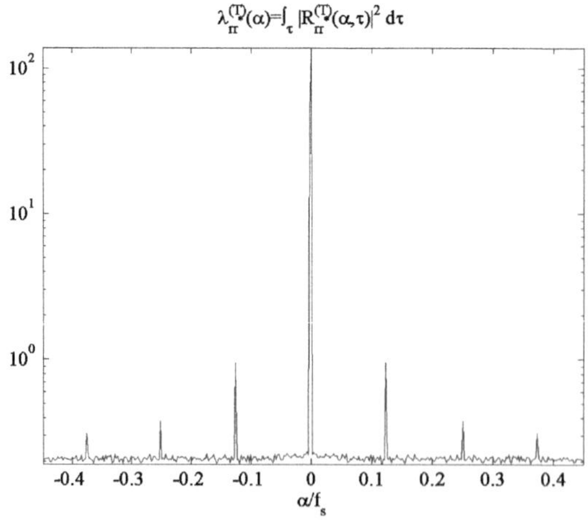

FIGURE 8.3 PAM signal in AWGN with SNR $= -10$ dB: Function $\lambda_{rr^*}^{(T)}(\alpha)$ with $T = NT_s$ and $N = 2^{17}$ samples.

A (conjugate) cycle frequency estimator based on the discrete-time counterpart of $\lambda_{rr^{(*)}}^{(T)}(\alpha)$ is proposed in (Ciblat et al., 2002a), (Ciblat et al., 2002b) where its asymptotic properties are also derived.

Cycle frequency estimator of (Ciblat et al., 2002a)

Let J be a set containing only one (conjugate) cycle frequency $\widetilde{\alpha}_0$ of the discrete-time signal $r(nT_s)$. In (Ciblat et al., 2002a), (Ciblat et al., 2002b), the cycle frequency estimator

$$\widehat{\alpha}^{(N)} = \arg\max_{\widetilde{\alpha}\in J} \lambda_{rr^{(*)}}^{(N)}(\widetilde{\alpha}) \tag{8.134}$$

is proposed, where

$$\lambda_{rr^{(*)}}^{(N)}(\widetilde{\alpha}) \triangleq \sum_{m=-M}^{M} \left| R_{rr^{(*)}}^{(N)}(\widetilde{\alpha}, m) \right|^2 . \tag{8.135}$$

In (8.135), $R_{rr^{(*)}}^{(N)}(\widetilde{\alpha}, m)$ is the discrete-time (conjugate) cyclic correlogram of $r(nT_s)$, $n \in \{-N, \ldots, N\}$, (defined according to (5.121b)) and M is such that $R_{rr^{(*)}}^{(N)}(\widetilde{\alpha}, m)$ is significantly nonzero in $\{-M, \ldots, M\}$. Under mild assumptions, the (conjugate) cycle frequency estimate $\widehat{\alpha}^{(N)}$ is mean-square consistent (as $N \to \infty$). Moreover, the estimate error is zero-mean asymptotically normal

$$(2N+1)^{3/2} \left[\widehat{\alpha}^{(N)} - \widetilde{\alpha}_0\right] \to \mathcal{N}(0, \overline{\sigma}^2) \tag{8.136}$$

with asymptotic variance

$$\overline{\sigma}^2 = \lim_{N\to\infty} (2N+1)^3 \,\mathrm{E}\left\{|\widehat{\alpha}^{(N)} - \widetilde{\alpha}_0|^2\right\} \tag{8.137}$$

and

$$\lim_{N\to\infty} (2N+1)\left[\widehat{\alpha}^{(N)} - \widetilde{\alpha}_0\right] = 0 \quad \text{almost surely.} \tag{8.138}$$

Cycle frequency domain profile

The cycle frequency estimator called *cycle frequency domain profile (CDP)* (Kim et al., 2007) consists in locating the peak of the function of α

$$\mathrm{CDP}(\alpha) \triangleq \max_f |\widehat{\gamma}_r(\alpha, f)| \tag{8.139}$$

where $\widehat{\gamma}_r(\alpha, f)$ is an estimate of the spectral coherence function defined in (8.116). If a cycle frequency interval is analyzed where more cycle frequencies are present, the function $\mathrm{CDP}(\alpha)$ exhibits more peaks and from the profile of these peaks a modulation classification technique can be designed (Kim et al., 2007).

The estimation procedure can be simplified with small performance reduction by locating the peak of the functions of α

$$\max_f |S_r^{(T,\Delta f)}(\alpha, f)| \quad \text{or} \quad \max_\tau |R_r^{(T)}(\alpha, \tau)|\,. \tag{8.140}$$

A version of the CDP technique with logarithmically compressed cycle frequency axis is proposed in (Wagstaff, 2008).

Cycle frequency estimator of (Jang, 2014)

The (conjugate) cycle frequency estimator proposed in (Jang, 2014) is an approximate computationally efficient version of the continuous-time counterpart of the estimator of (Ciblat et al., 2002a), (Ciblat et al., 2002b). It consists in locating the maximum of the function of α in the right-hand side of

$$\begin{aligned}
&\int_{\mathbb{R}} \mathrm{rect}\Big(\frac{\tau}{2\tau_{\mathrm{M}}}\Big) \left|R_r^{(T)}(\alpha, \tau)\right|^2 \mathrm{d}\tau \\
&\quad = \int_{\mathbb{R}} \Big[2\tau_{\mathrm{M}}\, \mathrm{sinc}(2\tau_{\mathrm{M}} f) \underset{f}{\otimes} I_r^{(T)}(\alpha, f) \underset{f}{\otimes} I_r^{(T)}(\alpha, -f)^*\Big] \mathrm{d}f \\
&\quad \simeq \int_{\mathbb{R}} \left|I_r^{(T)}(\alpha, f)\right|^2 \mathrm{d}f
\end{aligned} \tag{8.141}$$

where $I_r^{(T)}(\alpha, f)$ is the (conjugate) cyclic periodogram (5.41) of $r(t)$ and $\underset{f}{\otimes}$ denotes convolution with respect to f. Since the (conjugate) cyclic periodogram $I_r^{(T)}(\alpha, f)$ is the Fourier transform of the (conjugate) cyclic correlogram $R_r^{(T)}(\alpha, \tau)$ (Lemma 2.39), the first equality in (8.141) follows from Parseval theorem (Champeney, 1990, Theorems 7.5, 8.21, 8.26). The second approximate equality is aimed at reducing computational complexity and becomes rigorous as $T \to \infty$ and $\tau_{\mathrm{M}} \to \infty$, with $T/\tau_{\mathrm{M}} \to \infty$.

Cycle frequency estimator of (Dehay and Hurd, 2013)

The problem of determining a finite or countable set Λ of (radian) frequencies of an almost-periodic time series in the sense of Bohr is addressed in (Dehay and Hurd, 2013). An iterative procedure is designed. At each stage n, a finite observation $\{x_0, \ldots, x_{2n}\}$ is assumed to be available. The basic approach is to construct a finite set of frequencies Λ_n by first computing a weighted version $a_n^{(w)}(\lambda)$ of the Fourier coefficients of the observation on a uniform grid Π_n of $[0, 2\pi)$. Then, an element $\lambda_j^{(n)}$ of Π_n is put in Λ_n whenever an acceptance criterion on $a_n^{(w)}(\lambda_j^{(n)})$ is satisfied. The acceptance criterion proposed in (Dehay and Hurd, 2013) is referred to as the strong local maximum algorithm. The set of frequencies Λ_n is shown to converge to Λ as $n \to \infty$. The convergence of the algorithm is studied for noisy free almost-periodic functions, for almost-periodic functions in additive noise, and for the estimation of the second-order cycle frequencies of an ACS discrete-time process.

8.11 Detection of a moving source

In this section, the problem of detecting a moving cyclostationary source by exploiting the detector of Section 8.4.1 is addressed and illustrated with a numerical experiment (Napolitano, 2017a).

The cycle frequencies of the transmitted signal are modified at the receiver due to the relative motion between transmitter and receiver (Section 11.2). Thus, the cycle frequencies of the signal-of-interest (SOI) need to be estimated and the asymptotic complex normality of the cyclic correlogram with estimated cycle frequency (Theorem 5.58) must be invoked.

A low Earth orbit (LEO) satellite at altitude $h = 200$ km and with orbital speed $\mathrm{v}_o = 28061.5$ km h^{-1} is considered. It transmits a direct-sequence spread-spectrum (DSSS) signal with pseudo-noise (PN) modulating sequence, full duty-cycle rectangular chip pulse, chip period $T_p = 0.06$ μs, bit period $T_b = N_c T_p$, with $N_c \geqslant 2$, and carrier frequency $f_c = 2$ GHz.

The relative radial speed v_ϱ between the satellite and a fixed receiver on the Earth can be considered constant within observation intervals of interest in the applications (Napolitano and Perna, 2014b). Thus, the transmitted signal experiences a linearly time-varying delay and the complex envelope of the received signal is given by (Section 11.2, case 3), (Napolitano, 2012, Sec. 7.3), (Napolitano, 2016a, Sec. 2)

$$r(t) = a\, x(s(t - \tau_0))\, e^{j2\pi \nu t} + n(t) \tag{8.142}$$

where $x(t)$ and $n(t)$ are the zero-mean complex envelopes of the transmitted and disturbance signals, respectively. The term $n(t)$ accounts for both additive Gaussian noise and nonstationary interference and is independent of $x(t)$.

Let c be the medium propagation speed. In (8.142), the Doppler effect is modeled as a time-stretch factor $s = 1 - \mathrm{v}_\varrho / c$ in the argument of $x(\cdot)$ and a frequency shift $\nu_{\mathrm{Dop}} = (s-1) f_c$ on the carrier f_c. The overall frequency shift $\nu = \nu_{\mathrm{Dop}} + \nu_{\mathrm{osc}}$ accounts also for the possible frequency mismatch ν_{osc} between transmitter and receiver oscillators. The complex gain a accounts for attenuation and possible mismatch between the phase of the transmitter and receiver oscillators. The delay τ_0 is the propagation delay.

In the case of no motion between transmitter and receiver ($s = 1$, $\nu_{\text{Dop}} = 0$) and no frequency mismatch between transmitter and receiver oscillators ($\nu_{\text{osc}} = 0$), the detection test (8.65) (with $K = 1$) can be made at $(\alpha, \tau) = (1/T_p, T_p/2)$. For these values of cycle frequency and lag, the DSSS signal $x(t)$ with PN modulating sequence and full duty-cycle rectangular pulse exhibits a strong degree of cyclostationarity (Section 7.3). In fact, the long-code DSSS signal has the same statistical functions of a PAM signal with symbol period equal to the chip period T_p of the DSSS signal (Fusco et al., 2006).

Let α_x and β_x be a cycle frequency and a conjugate cycle frequency of $x(t)$, respectively. From (8.142) it follows that the cyclic autocorrelation function of the received signal is (Napolitano, 2012, eq. (7.265))

$$R_{rr^*}^{s\alpha_x}(\tau) = |a|^2 \, e^{j2\pi\nu\tau} \, e^{-j2\pi\alpha_x s\tau_0} \, R_{xx^*}^{\alpha_x}(s\tau) + R_{nn^*}^{s\alpha_x}(\tau) \tag{8.143}$$

and the conjugate cyclic autocorrelation function is (Napolitano, 2012, eq. (7.269))

$$R_{rr}^{s\beta_x+2\nu}(\tau) = a^2 \, e^{j2\pi\nu\tau} \, e^{-j2\pi\beta_x s\tau_0} \, R_{xx}^{\beta_x}(s\tau) + R_{nn}^{s\beta_x+2\nu}(\tau) \, . \tag{8.144}$$

That is, due to the Doppler effect, the received signal exhibits cyclostationarity at cycle frequency $\alpha_0 = s\alpha_x$ and conjugate cyclostationarity at conjugate cycle frequency $\beta_0 = s\beta_x + 2\nu$ (Section 11.2, case 3), (Napolitano, 2012, Sec. 7.7).

Since $|\mathrm{v}_\varrho| \ll c$, the value of s is close to 1 in practical applications. However, the cycle frequency $\alpha_0 = s\alpha_x$ at the receiver can be approximated to α_x only if

$$|\alpha_x - \alpha_0| = |1 - s| \, |\alpha_x| \ll 1/T \tag{8.145}$$

where $1/T$ is the cycle frequency resolution for a data-record length T (Section 5.2.1). In addition, since $R_{xx^{(*)}}^{\alpha}(s\tau) = R_{xx^{(*)}}^{\alpha}(\tau - (1-s)\tau)$, the approximation $s\tau \simeq \tau$ can be made in the argument of $R_{xx^*}^{\alpha_x}(\cdot)$ and $R_{xx}^{\beta_x}(\cdot)$, provided that

$$|1 - s| \, |\tau| \ll 1/B \tag{8.146}$$

where B is bandwidth of $x(t)$ and, hence, $1/B$ is the width of the smallest time interval where variations of $x(t)$ can be appreciated (Theorem 11.2).

In the experiment, $n(t)$ is additive white Gaussian noise (AWGN) with slowly varying power spectral density level. The average SNR ranges from -24 dB to -12 dB in the band $(-f_s/2, f_s/2)$, where $f_s = 1/T_s$ is the sampling frequency such that $1/B \simeq T_p = 4T_s$. This is the minimum sampling frequency such that the cyclic statistics of the discrete-time sampled signal are the same as those of the continuous-time signal, but for an amplitude and a frequency scale factor (Theorems 3.23 and 3.24). Due to the low values of SNR, a data record length $T = N_b T_s$, with $N_b = 2^{17}$ number of processed bits is considered in order to obtain a reliable cyclic autocorrelation estimate and, hence, a satisfactory detection performance. In the considered satellite application, the value of $|1-s|$ is of the order of 10^{-5} (Napolitano, 2014), (Napolitano and Perna, 2014b). Consequently, (8.145) is not satisfied and the cycle frequency at receiver needs to be estimated. For $\tau = T_p/2$ and $B \simeq 1/T_p$, condition (8.146) is satisfied since $|1-s|\, T_p/2 \ll 1/B \simeq T_p$.

The cycle frequency estimate $\alpha^{(T)}$ obtained by the estimator (5.136) with $\tau_{\mathrm{M}} = T_p$, is plugged into the test statistics (8.65) with $K = 1$ and $\tau = T_p/2$.

In the case of Doppler channel due to constant relative radial speed v_ϱ between transmitter and receiver, the maximum (conjugate) cycle frequency deviations $\delta\alpha$ to be considered in $D(\alpha_x, \delta\alpha)$ and $D(\beta_x, \delta\alpha)$ (see the definition after (5.136)) are

$$\delta\alpha = |\alpha_x| \frac{|\mathrm{v}_\varrho|_{\max}}{c} \qquad \text{cycle frequencies} \tag{8.147}$$

$$\delta\alpha = |\beta_x + 2f_c| \frac{|\mathrm{v}_\varrho|_{\max}}{c} \qquad \text{conjugate cycle frequencies} \tag{8.148}$$

where $|\mathrm{v}_\varrho|_{\max}$ is maximum allowable relative radial speed (in modulus).

The covariance and conjugate covariance are estimated according to the subsampling technique outlined in (8.77)–(8.79b) with block size $b = 2^9$ samples and block overlap factor $K_{\mathrm{o}} = 2^3$ samples.

In Fig. 8.4 (top), the probability of detection P_{d}, evaluated over 1000 Monte Carlo trials, for $P_{\mathrm{fa}} = 0.01$, is reported as a function of SNR. The receiver operating characteristic (ROC) (P_{d} versus P_{fa}) is shown in Fig. 8.4 (bottom) for SNR $= -18$ dB. The detector based on the test statistics (8.65) is considered in the following cases: ($\circ$) known true cycle frequency ($\alpha = \alpha_{\mathrm{true}} = s\alpha_x$); ($*$) estimated cycle frequency ($\alpha = \alpha^{(T)}$); ($\times$) neglected Doppler effect ($\alpha = \alpha_x$). For comparison purpose, the performance of the energy detector ($\Diamond$) is also considered.

From Theorem 5.58, it follows that the cyclic correlogram times $\sqrt{T}$ with estimated cycle frequency has the same complex normal distribution of the cyclic correlogram times $\sqrt{T}$ with true cycle frequency. According with such a result, the detector with estimated cycle frequencies has performance practically equivalent to that of the detector with known cycle frequency. In contrast, since (8.145) is not satisfied, the detector that neglects the Doppler effect and assumes as cycle frequency that of the transmitted signal, has poor performance. The energy detector performs poorly due to the variability of the power spectral density level that does not allow an accurate evaluation of the threshold (Gardner and Spooner, 1992b), (Tandra and Sahai, 2007).

Once the presence of the moving source has been detected, the source can be localized by estimating the parameters of signals received on two or more sensors (Section 13.6.2).

8.12 Spectrum sensing and signal classification

The signal selectivity properties of cyclostationarity-based signal processing techniques (Section 9.2) and the consistency and asymptotic complex normality of estimates of

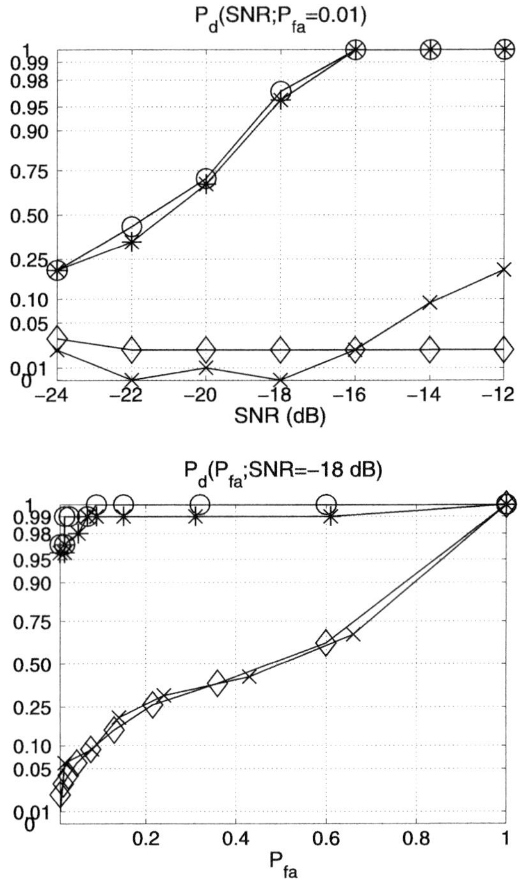

FIGURE 8.4 (Top) P_{d} as a function of SNR and (bottom) ROC for: ($\circ$) detector with the true cycle frequency ($\alpha = \alpha_{\mathrm{true}} = s\alpha_x$); ($*$) detector with the estimated cycle frequency ($\alpha = \alpha^{(T)}$); ($\times$) detector that neglects the Doppler effect ($\alpha = \alpha_x$); ($\Diamond$) energy detector.

second- and higher-order statistics of ACS signals (Chapter 5) have been suitably exploited for the purpose of signal classification in cognitive radio where a secondary user that wants to occupy frequency resources of a primary user when this user is not transmitting, must detect the presence of the primary user (Axell et al., 2012), (Dobre et al., 2007), (Haykin et al., 2009), (Ramkumar, 2009), (Spooner, 1995), (Spooner and Nicholls, 2009). In fact, even if some modulation formats exhibit identical or very similar power spectral density or even classical time-invariant higher-order statistics, they can exhibit significantly different cyclostationary features in the second- and higher-order statistics (Gardner and Spooner, 1994a), (Spooner and Gardner, 1994a), (Spooner, 1995), (Spooner et al., 2000), (Spooner, 2001).

8.12.1 Spectrum sensing

Accounting for the strength of the cyclic features exhibited by the SOI, e.g., the primary user signal (Section 8.12.2), values of cycle frequencies and lags or spectral frequencies can be selected to detect the presence of specific kinds of communications signals. Once the cyclic features are available, the classification can be made e.g., by neural networks (Fehske et al., 2005). The detection can be performed by exploiting:

1. the spectral line regeneration (Section 8.2), (Gardner, 1988b) as in (Derakhshani et al., 2011), (Sutton et al., 2008), (Wang and Chen, 2014);
2. the statistical test for presence of cyclostationarity (Section 8.4.2), (Dandawaté and Giannakis, 1994b) as in (Al-Habashna et al., 2012), (Huang and Tugnait, 2013), (Jerjawi et al., 2015), (Lundén et al., 2009), (Oner and Jondral, 2007b), (Oner and Jondral, 2007a), (Sadeghi et al., 2012b), (Sadeghi et al., 2012a);
3. the statistical test for presence of spectral coherence (Section 8.6), (Hurd and Gerr, 1991) as in (Enserink and Cochran, 1995);
4. the cycle frequency domain profile (Section 8.10), (Kim et al., 2007) as in (Li and Jayaweera, 2013), (Wagstaff, 2008);
5. higher-order cyclostationarity properties (Section 8.9) as in (Dobre et al., 2012), (Like et al., 2009), (Punchihewa et al., 2010), (Spooner, 1995), (Spooner and Nicholls, 2009).

Ad hoc receivers and processors have been derived for:

- *Blind detection of long-code code-division multiple access (CDMA) signals*
  (Li et al., 2005);
- *Quadrature amplitude modulated (QAM) signal identification*
  (Dobre et al., 2012);
- *Blind recognition, detection, and equalization of orthogonal frequency division multiplexing (OFDM) signals*
  (Bouzegzi et al., 2010), (Socheleau et al., 2011), (Xu et al., 2014), (Chopra et al., 2017); (Datta and Lin, 2018);
- *Detection of binary offset carrier (BOC) signals adopted in global navigation satellite systems (GNSSs)*
  (Sahmoudi et al., 2010);
- *Detection of other mobile satellite transmissions*
  (Dimc et al., 2015);
- *Discrimination of worldwide interoperability for microwave access (WiMax) versus ultra wide-band (UWB) signals*
  (Tani and Fantacci, 2010);
- *Blind recognition of single carrier linearly digitally modulated (SCLD) and OFDM signals*
  (Sun et al., 2014);
- *Detection of interleaved single carrier frequency division multiple access (SC-FDMA) signals*
  (Mirdamadi and Sabbaghian, 2014);

- *Wideband spectrum sensing and FM demodulation*
  (Garakyaragh et al., 2016);
- *Classification of space-time block codes*
  (Marey et al., 2012);
- *Digital television signal detection*
  (Yoo et al., 2014);
- *Design of chaos-based communications systems*
  (Kaddoum et al., 2011);
- *Specific communication signals*
  (Mesiya et al., 1978), (Roy et al., 1994), (Buzzi et al., 2001a), (Tulino and Verdù, 2001), (Sabharwal et al., 2001), (Gameiro, 2002), (Mirbagheri et al., 2006), (Nie and Mathiopoulos, 2006), (Du and Mow, 2010), (Yun and Cho, 2010), (Ahmad and Tarczynski, 2011), (Jiang et al., 2012), (Narieda and Kageyama, 2013), (Spooner et al., 2013), (Yen et al., 2013), (Astaneh and Gazor, 2014), (Rebeiz et al., 2014), (Safatly et al., 2014), (Karami and Dobre, 2015), (Yeo et al., 2015a);
- *Radar signals*
  (Bell and Grubbs, 1993), (Li and Ji, 2011).

Other addressed detection and signal classification problems are:

- *Spectrum sensing by exploiting first-order cyclostationarity*
  (Dobre et al., 2009);
- *Spectrum sensing by frequency shift filtering*
  (Chopra et al., 2014);
- *Multi antenna spectrum sensing*
  (Urriza et al., 2013), (Zhong et al., 2010), (Zhu et al., 2014);
- *Asymptotic GLRT and locally most powerful invariant test (LMPIT) for testing whether a multivariate discrete-time process is cyclostationary*
  (Ramirez et al., 2015);
- *Detection of ACS signals and cyclostationarity period estimation*
  (Horstmann et al., 2018);
- *Detection of cyclostationary signals in noise with spatio-temporal structure*
  (Pries et al., 2018);
- *Detection of cyclostationary signals by exploiting generalized coherence*
  (Howard et al., 2018), (Sirianunpiboon et al., 2018);
- *Spectrum sensing in Nakagami channels*
  (Astaneh and Gazor, 2013), (Sadeghi and Azmi, 2014);
- *Modulation recognition for cognitive radio*
  (Vučić and Obradović, 1998), (Dobre, 2015), (Spooner et al., 2017), (Cohen et al., 2018);
- *Sub-Nyquist cyclostationary detection for cognitive radio*
  (Cohen and Eldar, 2017);
- *Low-complexity modulation classification algorithm*
  (Rodriguez et al., 2017);

- *Field measurements for cognitive radio*
  (Chaudhari et al., 2016), (Chaudhari et al., 2018);
- *Wideband cyclostationary signal processing using sparse subsets of narrowband sub-channels*
  (Spooner and Mody, 2018);
- *Cyclic feature suppression for physical layer security*
  (Karabinis, 2011), (Karabinis, 2014), (Ankarali and Arslan, 2017);
- *Radio frequency (RF) steganography via linearly frequency modulated (LFM) chirp radar signals*
  (Zhang et al., 2018);
- *Effects of receiver impairments on cyclic statistics*
  (Oner and Dobre, 2011), (Rebeiz et al., 2013);
- *Effect of jitter*
  (Dehay, 2007), (Oner, 2008a), (Oner, 2009);
- *Phenomenon of "SNR wall" due to model uncertainty*
  (Tandra and Sahai, 2007);
- *Performance degradation due to presence of nonlinearities*
  (Rebeiz et al., 2015);
- *Field-programmable gate array (FPGA) realizations of cyclic detectors*
  (Kosunen et al., 2013), (Sabat et al., 2010);
- *Architectural trade-offs*
  (Noguet et al., 2010);
- *Hardware implementation of cyclostationarity-based detectors*
  (Sepidband and Entesari, 2015), (Kozłowski, 2016), (Murty and Shrestha, 2018a), (Murty and Shrestha, 2018b), (Ivanov et al., 2018);
- *Parallel algorithm for spectrum sensing*
  (Lima et al., 2018).

The properties of the continuous-time harmonic series representation of cyclostationary signals (Section 1.6.1.1) are exploited to develop the GLR detector of (Riba et al., 2014). An asymptotic GLRT for detection of cyclostationary signals based on the discrete-time harmonic series representation (Section 1.6.2.1) is presented in (Ramirez et al., 2014).

Patents on modulation format classification are described in (Jallon, 2012), (Du and Mow, 2013), (Kommate et al., 2013), (Turunen and Kosunen, 2013), (Zeng et al., 2013), (Challapali et al., 2017), (Gravely et al., 2018).

The detection of ships exploiting cyclostationary signatures is considered in (Firat and Akgul, 2018) and the passive drone surveillance in (Fu et al., 2018).

8.12.2 Cyclic spectral analysis of man-made signals

Periodicities in the multivariate statistical functions of modulated signals used in communications, radar, sonar, and telemetry arise from operations such as modulation, scanning, sampling, multiplexing, and coding. The corresponding cycle frequencies are related to parameters of the processes such as carrier frequency, baud rate, sampling frequency, and code rate. The shapes of the second- and higher-order cyclic statistical functions depend on the shapes of the base pulses and the correlation properties of the modulating sequences (Chapter 7), (Gardner, 1985, Chap. 12), (Gardner, 1987f), (Gardner et al., 2006, Sec. 5).

Second-order cyclostationarity-based spectrum sensing techniques exploit the knowledge of the (conjugate) cycle frequencies, cyclic autocorrelation functions, and cyclic spectra of the signal of interest. They have been derived for the most common analog and digitally modulated signals in (Gardner, 1987e) and (Gardner et al., 1987), respectively. In addition, cyclic statistics have been derived for many communications signals (Section 7.6).

8.12.3 Hiding the modulation format

By properly building numerical sequences that are not relatively measurable (Definition 2.2), (Leśkow and Napolitano, 2006), (Napolitano, 2012, Chap. 6), a class of signals is introduced in (Leśkow and Napolitano, 2007) such that their measured statistical functions (Section 5.2) are not convergent as the measurement interval is increased. Thus, these signals are suitable to be adopted when a secure communication channel needs to be established. In fact, an unauthorized third party cannot discover the modulation format by second- and higher-order cyclic (or stationary) statistic measurements.

Alternatively, low probability of intercept (LPI) communications signals can be designed such that they exhibit a weak degree of cyclostationarity (Karabinis, 2011), (Karabinis, 2014).

8.13 Summary

If the signal-of-interest (SOI) or the disturbance are non-Gaussian and/or non-stationary, the optimum detection structure in general is difficult to be derived or implemented. Non-optimum but easily implementable structures that exploit the almost-cyclostationarity properties of the SOI can be designed.

A *spectral line regeneration* structure (Section 8.2) detects a sinewave component at one of the SOI cycle frequencies that is regenerated by a homogeneous quadratic time-invariant transformation.

The *joint detection and localization of a fixed cyclostationary source* on the basis of two sensor measurements in the presence of additive white Gaussian noise is performed by the generalized log-likelihood ratio test (GLLRT). In the case of weak signal, the detector correlates cyclic statistic estimates with their true values for each cycle frequency of the SOI and then sums the values with appropriate phases (Section 8.3). In the case of a single sensor, the detector is dubbed *multi-cycle detector*.

Instead of deriving the detector structure starting from the received signal, measurements of its (conjugate) cyclic autocorrelation can be adopted as "front-end" data. In fact, properly normalized versions of these measurements are asymptotically complex normal as the data-record length approaches infinity, provided that the SOI and disturbance are processes with finite or practically finite memory. Thus, the log-likelihood ratio test (LLRT) can be derived starting from these measurements (Section 8.4). Similarly, a LLRT can be derived starting from a properly normalized version of the time- or frequency-smoothed cyclic periodogram since it is asymptotically complex normal (Section 8.5).

Further detection strategies are a *statistical test for spectral coherence* (Section 8.6), and a *subsampling-based significance test* (Section 8.7). In the presence of impulsive SOI or disturbance, *robust detectors* can be exploited, that are based on the asymptotic complex normality of properly normalized (conjugate) cyclic autocorrelation and spectra measurements of nonlinear transformations of the received signal (Section 8.8). In the case of signals exhibiting a low degree of second-order cyclostationarity, higher-order cyclostationarity properties can be exploited for signal detection by *higher-order spectral line regeneration* through a homogeneous higher-order time-invariant nonlinear transformation of the received signal (Section 8.9).

Several cycle frequency estimation techniques are presented in Section 8.10.

9

Communications systems: design and analysis

Contents

Abstract

The design and analysis of communications systems benefits of the exploitation of cyclostationarity properties exhibited by almost all communications signals. In particular, the signal selectivity property allows one to separate signal of interest from disturbance. The problems of minimum mean-squared error filtering, synchronization, and system identification and equalization are analyzed in detail. Other applications include parameter estimation, source location, beamforming, and source separation.

Keywords

Signal selectivity, Cyclic Wiener filtering, Synchronization, System identification, Parameter estimation, Source location

Cyclostationary Processes and Time Series. https://doi.org/10.1016/B978-0-08-102708-0.00020-0

9.1 Introduction

Cyclostationarity properties of modulated signals can be suitably exploited in the analysis and design of communications systems since the signals involved are typically ACS (Chapter 7).

A characteristic of cyclostationarity-based signal processing algorithms is the signal selectivity property when the signal of interest exhibits a cycle frequency which is not shared with any of the disturbance signals (Section 9.2). It allows one to design algorithms that are highly tolerant to noise and interference, provided that a sufficiently long observation interval is adopted for the estimation of the cyclic statistics. It has been exploited in channel identification and equalization (Section 9.5), signal parameter and waveform estimation (Section 9.6.1), source location (Section 9.6.2), beamforming (Section 9.6.3), and source separation (Section 9.6.4).

The problem of minimum mean squared-error (MMSE) linear filtering of ACS signals (cyclic Wiener filtering) leads to the design of frequency-shift (FRESH) filters. They are linear almost-periodically time-variant (LAPTV) filters whose frequencies are linked to the cycle frequencies of the almost-periodically time-variant autocorrelation and cross-correlation functions of the involved ACS signals (Section 9.3).

The spectral line regeneration property (Section 2.3.1.6) is at the basis of every synchronization algorithm (Gardner, 1986b).

The cyclostationary nature of interference in communications systems is characterized in (Campbell et al., 1983a), (Campbell et al., 1983b), (Albuquerque et al., 1984), (Smith and Potter, 1986), (Fung et al., 1989), (Petersen and Falconer, 1991), (Uzun and Haddad, 1995), (Im and Won, 2004).

On the analysis and design of communications systems, see the general treatments (Gardner, 1985), (Gardner, 1987f), (Gardner, 1990b), (Gardner and Chen, 1991), and also see (Gardner and Franks, 1975), (Gardner, 1987e), (Gardner et al., 1987), (Akinshin et al., 1988), (Janssen, 1988), (Kozlov et al., 1990), (Gardner, 1991b).

Fundamental issues relative to the exploitation of cyclostationarity in communications systems are discussed in the blog (Spooner, 2015).

Patents of inventions for the analysis and design of communications systems exploiting cyclostationarity are (Bergmans, 1994), (Golden, 1994), (Bergmans, 1995), (Nicolas and Lim, 1995), (Agee, 2000), (Thibault et al., 2001), (Agee et al., 2002), (Ding et al., 2002), (Agee, 2003), (Mandyam, 2003), (Trans, 2003), (Carroll, 2005), (Gourge and Roosen, 2005), (Vourinen et al., 2005), (Alloin, 2007).

9.2 Signal selectivity property

Cyclostationarity-based detection and estimation algorithms are signal selective. Specifically, in the case of additive disturbance signals uncorrelated with the signal-of-interest (SOI), if the SOI does not share at least one (conjugate) cycle frequency, say α_0, with the disturbance, then the (conjugate) cyclic autocorrelation function of the SOI plus noise at α_0 is coincident with the (conjugate) cyclic autocorrelation function of the SOI alone.

Let us consider the received signal

$$r(t) = \sum_{\ell=0}^{L} x_\ell(t) + n(t) \tag{9.1}$$

with $x_0(t), \ldots, x_L(t), n(t)$ zero mean and mutually uncorrelated signals.

The (conjugate) cyclic autocorrelation of $r(t)$ is given by

$$\begin{aligned} R_{\boldsymbol{r}}^{\alpha}(\tau) &= \left\langle \mathrm{E}\left\{ \left[\sum_{\ell_1=0}^{L} x_{\ell_1}(t+\tau) + n(t+\tau) \right] \left[\sum_{\ell_2=0}^{L} x_{\ell_2}(t) + n(t) \right]^{(*)} e^{-j2\pi\alpha t} \right\} \right\rangle_t \\ &= \sum_{\ell=0}^{L} R_{\boldsymbol{x}_\ell}^{\alpha}(\tau) + R_{\boldsymbol{n}}^{\alpha}(\tau) \end{aligned} \tag{9.2}$$

since all the cross-terms are zero.

Let $x_0(t)$ be the SOI, $x_\ell(t)$, $\ell = 1, \ldots, L$ be interfering signals, and $n(t)$ be stationary noise. Let us assume that $x_0(t)$ exhibits cyclostationarity at cycle frequency α_0 not shared with any other $x_\ell(t)$, $\ell = 1, \ldots, L$. That is,

$$R_{\boldsymbol{x}_0}^{\alpha_0}(\tau) \not\equiv 0 \qquad R_{\boldsymbol{x}_\ell}^{\alpha_0}(\tau) \equiv 0, \quad \ell = 1, \ldots, L\,. \tag{9.3}$$

We have that for $\alpha = \alpha_0$ (9.2) specializes into

$$R_{\boldsymbol{r}}^{\alpha_0}(\tau) = R_{\boldsymbol{x}_0}^{\alpha_0}(\tau)\,. \tag{9.4}$$

The frequency-domain counterparts of (9.2) and (9.4) are

$$S_{\boldsymbol{r}}^{\alpha}(f) = \sum_{\ell=0}^{L} S_{\boldsymbol{x}_\ell}^{\alpha}(f) + S_{\boldsymbol{n}}^{\alpha}(f) \tag{9.5}$$

and

$$S_{\boldsymbol{r}}^{\alpha_0}(f) = S_{\boldsymbol{x}_0}^{\alpha_0}(f) \tag{9.6}$$

respectively.

From (9.4) and (9.6), it follows that cyclostationarity-based detection or estimation algorithms operating at cycle frequency α_0 are intrinsically immune to the effects of noise and interference. Thus, they are highly tolerant to noise and interference in practice, provided that the estimate of the cyclic statistics is sufficiently accurate. That is, provided that a sufficiently large observation interval is adopted to estimate the cyclic statistics (Chapter 5) *regardless of the amount of temporal and spectral overlap of SOI and disturbance* (Gardner, 1985), (Gardner, 1987f). In Section 9.7, it is shown that performance is limited only by the cycle leakage phenomenon. Specifically, the larger is the data-record length adopted for the estimates, the lower is the minimum signal-to-noise ratio (SNR) and signal-to-interference ratio (SIR) for which satisfactory performance can be achieved.

9.3 Cyclic Wiener filtering

The problem of optimum linear filtering consists of designing the linear transformation of the data $x(t)$ that minimizes the mean-squared error (MSE) of the filter output relative to a desired signal, say $d(t)$. In the case of complex data, the optimum filter is obtained by processing both $x(t)$ and $x^*(t)$, leading to the *linear-conjugate-linear* (LCL) structure (Gardner, 1987f), (Brown, 1987) also referred to as *widely-linear filtering* (Picinbono and Chevalier, 1995).

If $d(t)$ and $x(t)$ are jointly ACS signals, the optimum (minimum MSE) filtering is referred to as *cyclic Wiener filtering* or *frequency-shift (FRESH) filtering* (Gardner, 1993). FRESH filtering consists of periodically or almost-periodically time-variant filtering of $x(t)$ and $x^*(t)$ and adding the results, where the frequency shifts are chosen in accordance with the cycle frequencies of $x(t)$ and $d(t)$ (e.g., cycle frequencies of the signal of interest and interference, $x(t) = d(t) + n(t)$ where $n(t)$ is the interference) (Gardner and Brown, 1989), (Gardner and Venkataraman, 1990), (Gardner, 1993). The problem of optimum LPTV or LAPTV filtering was first addressed in (Gardner, 1972a), (Gardner and Franks, 1975) and, in terms of cyclic autocorrelations and cyclic spectra in (Gardner, 1985), (Gardner, 1987f).

By exploiting the spectral correlation property of ACS signals, the optimum filter cancels spectral band of the useful signal which are corrupted by noise and then reconstructs these bands by frequency shifting and then LTI filtering the spectral components of the useful signal that are noisy-free and correlated with the canceled bands (Gardner, 1993). It provides significant performance improvement with respect to the classical Wiener filtering that assumes a stationary model for the involved signals.

For example, two complex binary-phase shift-keying (BPSK) (or two quadrature-phase shift-keying (QPSK)) signals with overlapping spectra and different baud rates, in AWGN with SNR = 20 dB, can be separated using FRESH filtering. Using 6 or more frequency shifts can lead to a reduction up to 20 dB of the MSE with respect to the case of LTI Wiener filtering (no frequency shifts) depending on the excess bandwidth of the signals (Gardner, 1993).

Let

$$\begin{aligned}\widehat{d}(t) &\triangleq y(t) + y_c(t)\\ &= \int_{\mathbb{R}} h(t,u)\, x(u)\mathrm{d}u + \int_{\mathbb{R}} h^c(t,u)\, x^*(u)\mathrm{d}u \end{aligned} \tag{9.7}$$

be the LCL estimate of the desired signal $d(t)$ obtained from the data $x(t)$. To minimize the mean-squared error

$$\mathrm{E}^{\{\alpha\}}\{|\widehat{d}(t) - d(t)|^2\} \tag{9.8}$$

a necessary and sufficient condition is that the error signal be orthogonal to the data (orthogonality condition (Gardner, 1985)); that is,

$$\mathrm{E}^{\{\alpha\}}\big\{\big[\widehat{d}(t+\tau) - d(t+\tau)\big]x^*(t)\big\} = 0 \qquad \forall t \in \mathbb{R},\ \forall \tau \in \mathbb{R} \tag{9.9a}$$

$$\mathrm{E}^{\{\alpha\}}\big\{\big[\widehat{d}(t+\tau) - d(t+\tau)\big]x(t)\big\} = 0 \qquad \forall t \in \mathbb{R},\ \forall \tau \in \mathbb{R}\,. \tag{9.9b}$$

If $x(t)$ and $d(t)$ are singularly and jointly ACS,

$$\mathrm{E}^{\{\alpha\}}\left\{x(t+\tau)\,x^{(*)}(t)\right\} = \sum_{\alpha\in A_{xx^{(*)}}} R^{\alpha}_{xx^{(*)}}(\tau)\,e^{j2\pi\alpha t} \tag{9.10}$$

$$\mathrm{E}^{\{\alpha\}}\left\{d(t+\tau)\,x^{(*)}(t)\right\} = \sum_{\gamma\in F_{dx^{(*)}}} R^{\gamma}_{dx^{(*)}}(\tau)\,e^{j2\pi\gamma t} \tag{9.11}$$

then it follows that the optimum filters are LAPTV:

$$h(t,u) = \sum_{\sigma\in G} h_\sigma(t-u)\,e^{j2\pi\sigma u} \tag{9.12}$$

$$h^c(t,u) = \sum_{\eta\in G_c} h^c_\eta(t-u)\,e^{j2\pi\eta u}\,. \tag{9.13}$$

By substituting (9.7), (9.12), and (9.13) into (9.9a) and (9.9b), we obtain the system of simultaneous filter design equations (Gardner, 1993), (Franks, 1994)

$$\sum_{\sigma\in G} e^{j2\pi\sigma\tau}\left[R^{\gamma-\sigma}_{xx^*}(\tau)\otimes h_\sigma(\tau)\right] + \sum_{\eta\in G_c} e^{j2\pi\eta\tau}\left[R^{\eta-\gamma}_{xx}(\tau)^*\otimes h^c_\eta(\tau)\right] = R^{\gamma}_{dx^*}(\tau) \quad \forall\gamma\in F_{dx^*} \tag{9.14a}$$

$$\sum_{\sigma\in G} e^{j2\pi\sigma\tau}\left[R^{\gamma-\sigma}_{xx}(\tau)\otimes h_\sigma(\tau)\right] + \sum_{\eta\in G_c} e^{j2\pi\eta\tau}\left[R^{\eta-\gamma}_{xx^*}(\tau)^*\otimes h^c_\eta(\tau)\right] = R^{\gamma}_{dx}(\tau) \quad \forall\gamma\in F_{dx} \tag{9.14b}$$

where the fact that $R^{\beta}_{x^*x^*}(\tau) = R^{-\beta}_{xx}(\tau)^*$ and $R^{\alpha}_{x^*x}(\tau) = R^{-\alpha}_{xx^*}(\tau)^*$ is used (see (1.87)). In (9.14a), for each $\gamma\in F_{dx^*}$, the sums are extended to all frequency shifts $\sigma\in G$ and $\eta\in G_c$ such that $\gamma-\sigma\in A_{xx^*}$ and $\eta-\gamma\in A_{xx}$; in (9.14b), for each $\gamma\in F_{dx}$, the sums are extended to all frequency shifts $\sigma\in G$ and $\eta\in G_c$ such that $\gamma-\sigma\in A_{xx}$ and $\eta-\gamma\in A_{xx^*}$.

Eqs. (9.14a) and (9.14b) can be re-expressed in the frequency domain:

$$\sum_{\sigma\in G} S^{\gamma-\sigma}_{xx^*}(f-\sigma)\,H_\sigma(f-\sigma) + \sum_{\eta\in G_c} S^{\eta-\gamma}_{xx}(\eta-f)^*\,H^c_\eta(f-\eta) = S^{\gamma}_{dx^*}(f) \quad \forall\gamma\in F_{dx^*} \tag{9.15a}$$

$$\sum_{\sigma\in G} S^{\gamma-\sigma}_{xx}(f-\sigma)\,H_\sigma(f-\sigma) + \sum_{\eta\in G_c} S^{\eta-\gamma}_{xx^*}(\eta-f)^*\,H^c_\eta(f-\eta) = S^{\gamma}_{dx}(f) \quad \forall\gamma\in F_{dx} \tag{9.15b}$$

where $H_\sigma(f)$ and $H^c_\eta(f)$ are the Fourier transforms of $h_\sigma(t)$ and $h^c_\eta(t)$, respectively.

Eqs. (9.15a) and (9.15b) in general involve an infinite number of frequency-shift parameters σ and η. In practice, a (best) finite set of these parameters needs to be selected, which is not an easy task. Ad hoc procedures can be designed (Gardner, 1993), (Carrick et al., 2018).

From the derivation of the optimum FRESH filtering equations (9.15a) and (9.15b) in terms of multivariate Wiener filtering, it follows that adaptive implementations of FRESH filters can be obtained from conventional multivariate adaptive filters. That is, an adaptively adjustable structure is a parallel bank of frequency-shift modulators each followed by a finite impulse-response (FIR) filter, the outputs of which are summed. The coefficients of the FIR filters can be jointly adapted by the least mean-square (LMS), recursive least square (RLS), or constant-modulus algorithms (Gardner, 1993).

Applications and problems related to FRESH filtering are the following.

- *Interference suppression or mitigation*
  (Gardner et al., 1995), (Gelli et al., 1998), (Lops et al., 1998), (Tulino, 1999), (Lops and Tulino, 2000), (Buzzi et al., 2001c), (Buzzi et al., 2001b), (Gardner, 2001), (Gardner and Reed, 2001), (Gelli and Verde, 2001), (Verde, 2001), (Gelli and Verde, 2003), (Im and Won, 2004), (Carrick et al., 2018);
- *Adaptive implementations*
  (Reed, 1987), (Reed et al., 1988), (Gardner and Brown, 1989), (Reed et al., 1989), (Gardner and Venkataraman, 1990), (Reed and Hsia, 1990), (Mendoza et al., 1991), (Ding and Fahmy, 1992), (Ding and Fahmy, 1993), (Yeung et al., 1994), (Reed et al., 1995), (Lev-Ari, 1998), (Zhang et al., 1999), (Martone, 2000), (Yu and Ueng, 2000), (Gelli and Verde, 2001), (Shlezinger et al., 2017);
- *Application to radar chirp signals*
  (Carrick and Reed, 2017);
- *Causality analysis*
  (Spurbeck and Schreier, 2007);
- *Analysis of least mean square (LMS) and normalized least mean square (NLMS) algorithms in the cyclostationary regime*
  (Gardner, 1987b), (Hajivandi and Gardner, 1990), (McLernon, 1991), (Bershad et al., 2014);
- *Periodic Levinson-Durbin algorithm for entropy maximization*
  (Boshnakov and Lambert-Lacroix, 2012);
- *Spectrum sensing*
  (Chopra et al., 2014);
- *Noise reduction in power lines*
  (Shlezinger and Dabora, 2014);
- *Other results*
  (Chopra et al., 2014), (Shlezinger and Dabora, 2014), (Chopra et al., 2016), (Sadeghi and Yu, 2016).

Cyclic Wiener filtering or FRESH filtering can be recognized to be linked to FSE's, RAKE filters, adaptive demodulators, and LMMSE despreaders for recovery of baseband symbol data from PAM, GMSK/GFSK, and DSSS signals, and for separation of baseband symbol streams in overlapped signal environments, as well as multicarrier or direct frequency diversity spread spectrum systems for adaptive transmission and combining. In particular, on blind and non-blind adaptive demodulation of PAM signals and LMMSE blind despreading of short-code DSSS/CDMA signals using FSE's and RAKE filtering structures, see (Agee, 1988), (Agee, 1989a), (Agee, 1989b), (Agee and Venkataraman, 1989), (Pateros and Saulnier, 1992), (Agee, 1993), (Fijalkow et al., 1994), (Madhow and Honig, 1994), (Rapajic and Vucetic, 1994), (Miller, 1995), (Bensley and Aazhang, 1996), (Pateros and Saulnier, 1996), (Strom et al., 1996), (Bensley and Aazhang, 1998), (Madhow, 1998), (Wang and Poor, 1998), (Madhow and Honig, 1999), (Mangalvedhe, 1999), (Strom and Miller, 1999). On direct frequency-diverse multicarrier transceivers, see (Proakis, 1993), (Stojanovic et al., 1993b), (Stojanovic et al., 1993a), (Agee et al., 1997).

FRESH filtering is exploited for global system for mobile communications (GSM) in (Gardner, 2001), (Gardner and Reed, 2001). Specifically, in (Gardner, 2001), it is shown that by FRESH filtering prior to Viterbi demodulation, GSM base-stations and/or handsets can multiply the number of cochannel signals that can be separated with M antennas by $(2M-1)/(M-1)$. That is, 1 interferer can be suppressed by one antenna, 3 interferers with a diversity antenna pair, and $2M-1$ interferers with M antennas.

Patents of inventions exploiting the FRESH filtering are (Gardner and Schell, 1998), (Arslan and Molnar, 2003).

9.4 Synchronization

Synchronization consists in generating a periodic signal synchronous with the carrier or the pulse-train that, once modulated by the random data, gave rise to the modulated communication signal (Franks, 1980). Therefore, synchronization consists in regenerating spectral lines that were hidden by the modulation process (Section 8.2).

Let $x(t)$ be a real-valued second-order wide-sense ACS time series. According to the results of Section 2.3.1.6, the second-order lag product $x(t+\tau)\,x(t)$ can be decomposed into the sum of its almost-periodic component and a residual term $\ell_x(t,\tau)$ not containing any finite-strength additive sine-wave component (see (2.56b)):

$$\begin{aligned} x(t+\tau)\,x(t) &= \mathrm{E}^{\{\alpha\}}\left\{x(t+\tau)\,x(t)\right\} + \ell_x(t,\tau) \\ &= \sum_{\alpha\in A} R_x^{\alpha}(\tau)\,e^{j2\pi\alpha t} + \ell_x(t,\tau) \end{aligned} \tag{9.16}$$

where

$$\left\langle \ell_x(t,\tau)\,e^{-j2\pi\alpha t}\right\rangle_t \equiv 0 \quad \forall\alpha\in\mathbb{R}\,. \tag{9.17}$$

For communications signals, the cycle frequencies $\alpha\in A$ are related to parameters such as sine-wave carrier frequency, pulse rate, symbol rate, frame rate, sampling frequency, etc.

(Chapter 7). Therefore, the extraction of the almost-periodic component in (9.16) leads to a signal suitable for synchronization purposes. For example, if $x(t)$ is the binary PAM signal defined in (7.6) with $q(t)$ real and duration limited to an interval strictly less than T_0, and $a_k \in \{-1, 1\}$, then it follows that

$$x^2(t) = \sum_{k \in \mathbb{Z}} q^2(t - kT_0) \tag{9.18}$$

and $\ell_x(t, 0) = 0$. Therefore, the synchronization signal $x^2(t)$ is periodic with period T_0.

More generally, by definition, ACS signals enable spectral lines to be generated by passage through a stable nonlinear time-invariant transformation (see Sections 4.2 and 9.5.4). That is, quadratic or higher-order nonlinear time-invariant transformations of an ACS signal give rise to time series containing finite-strength additive sine-wave components whose frequencies are the second or higher-order cycle frequencies of the original signal. All synchronization schemes can be recognized to exploit the second- or higher-order cyclostationarity features of signals (Gardner, 1986b).

The following synchronization issues have been addressed.

- *Spectral analysis of timing waveforms with re-generated spectral lines*
  (Franks and Bubrouski, 1974), (Bilardi and Pupolin, 1984), (Pupolin and Tomasi, 1984), (Gardner, 1986b), (Qiu and Cassara, 1989), (Blachman, 1994);
- *Phase-lock loops (PLL)*
  (Mengali and Pezzani, 1978), (Moeneclaey, 1982b), (Yoon and Lindsay, 1982), (Moeneclaey, 1983);
- *Blind or non-data-aided synchronization algorithms*
  (Bennett, 1958), (Gardner, 1987f), (Gini and Giannakis, 1998), (Ciblat, 2000), (Serpedin et al., 2000), (Bölcskei, 2001), (Napolitano and Tanda, 2001), (Wang et al., 2002), (Ciblat and Serpedin, 2003), (Ciblat and Vandendorpe, 2003), (Wang et al., 2003a), (Wang et al., 2003b), (Ciblat and Serpedin, 2004), (Napolitano and Tanda, 2004), (Napolitano et al., 2005);
- *Symbol and carrier synchronization*
  (Moeneclaey, 1982a), (Moeneclaey, 1984), (O'Reilly, 1984), (Monti and Pierobon, 1985), (Riba and Vazquez, 1994), (Scott and Olasz, 1995);
- *Symbol timing estimation*
  (Kim and Min, 2018);
- *Block timing estimation for Alamouti coded MIMO Signals*
  (Gul et al., 2017);
- *Blind recognition of space time block code in MIMO systems*
  (Buiquang and Zhang, 2018);
- *Symbol period estimation for FSK signals*
  (Wang et al., 2015), (Vito and Dobre, 2017);
- *Cycle frequency estimation*
  (Ciblat et al., 2002b), (Ciblat et al., 2002a), (Rahimzadeh et al., 2018), (Yang et al., 2018);

- *Symbol timing offset and carrier frequency offset for OFDM/OQAM systems* (Liu et al., 2018a);
- *Carrier frequency offset and channel impulse response for linear periodic channels* (Shaked et al., 2018);
- *Blind carrier frequency offset estimation* (Yang et al., 2016);
- *Blind parameter estimation of GFDM Signals* (Chang et al., 2016);
- *Jitter measurement* (Daponte et al., 2018);
- *Phase-noise measurement* (D'Arco and De Vito, 2016);
- *Steepest descent algorithm for correction of cycle frequency error* (Cai et al., 2016);
- *Soft synchronization of DSSS signals* (Agee et al., 1996);
- *Despreading* (French and Gardner, 1986).

Patents on synchronization techniques exploiting cyclostationarity are (Scott et al., 1994), (Lakkis et al., 2002), (Lakkis et al., 2004), (Zoltowski, 2012).

9.5 System identification

9.5.1 General aspects

Cyclostationarity-based techniques have been exploited for system identification and channel equalization. Linear and nonlinear systems, and time-invariant, periodically and almost-periodically time-variant systems, have been considered. Also, techniques for noisy input/output measurement, and blind adaptation algorithms, have been developed for LTI systems.

On this subject see the general treatments (Gardner, 1972a), (Gardner, 1985), (Gardner, 1987f), (Brown, 1987), (Gardner, 1990b), and also see (Isokawa, 1982), (Petersen and Falconer, 1991), (Schell and Gardner, 1994), (Napolitano, 1995), (Tsatsanis and Giannakis, 1997), (Flagiello et al., 2000), (Napolitano and Tanda, 2001).

Patents of inventions on blind system identification exploiting cyclostationarity are (Bergmans et al., 1989), (McCallister et al., 1990), (Golden, 1992b), (Golden, 1992a), (Guglielmi et al., 1998), (Celebi, 2004).

9.5.2 LTI-system identification by noisy-measurements

Cyclostationarity-based techniques can be exploited in channel identification and equalization problems in order to separate the desired and disturbance contributions in noisy input/output measurements, provided that there is at least one cycle frequency of the desired signal that is not shared with the disturbance.

Let us consider the problem of estimating the impulse-response function $h(t)$ or, equivalently, the harmonic-response function

$$H(f) \triangleq \int_{\mathbb{R}} h(t)\, e^{-j2\pi f t}\, \mathrm{d}t \tag{9.19}$$

of an LTI system with input/output relation

$$y(t) = h(t) \otimes x(t) \tag{9.20}$$

on the basis of the observed noisy signals $v(t)$ and $z(t)$

$$v(t) = x(t) + n(t) \tag{9.21}$$

$$z(t) = y(t) + m(t)\,. \tag{9.22}$$

9.5.2.1 Second-order statistic based identification

Assumption 9.1.

- *The signals $x(t)$, $n(t)$, and $m(t)$ are zero-mean;*
- *The signal $x(t)$ is uncorrelated with both $n(t)$ and $m(t)$.*

By taking in (3.17) and (3.18) $x_1 = x_2 = x$, $y_1 = y$, $y_2 = x$, $h_1 = h$ (LTI), and $h_2 = \delta$, we have

$$S^{\alpha}_{yx^{(*)}}(f) = S^{\alpha}_{xx^{(*)}}(f)\, H(f)\,. \tag{9.23}$$

Therefore, under Assumption 9.1, from the model (9.21) and (9.22) we have

$$S^{\alpha}_{vv^{(*)}}(f) = S^{\alpha}_{xx^{(*)}}(f) + S^{\alpha}_{nn^{(*)}}(f) \tag{9.24}$$

$$\begin{aligned} S^{\alpha}_{zv^{(*)}}(f) &= S^{\alpha}_{yx^{(*)}}(f) + S^{\alpha}_{mn^{(*)}}(f) \\ &= S^{\alpha}_{xx^{(*)}}(f)\, H(f) + S^{\alpha}_{mn^{(*)}}(f)\,. \end{aligned} \tag{9.25}$$

Assumption 9.2. *A (conjugate) cycle frequency α of $x(t)$ exists such that*

- *$n(t)$ does not exhibit (conjugate) cyclostationarity at cycle frequency α. That is,*

$$S^{\alpha}_{nn^{(*)}}(f) \equiv 0 \tag{9.26}$$

- *$n(t)$ and $m(t)$ do not exhibit joint (conjugate) cyclostationarity at cycle frequency α. That is,*

$$S^{\alpha}_{mn^{(*)}}(f) \equiv 0 \tag{9.27}$$

Eqs. (9.24) and (9.25) reveal the ability of cyclostationarity-based algorithms to be signal selective.

Theorem 9.3. *Under Assumptions 9.1 and 9.2, the harmonic-response function for the LTI system (9.20) is given by*

$$H(f) = \frac{S^{\alpha}_{yx^{(*)}}(f)}{S^{\alpha}_{xx^{(*)}}(f)} = \frac{S^{\alpha}_{zv^{(*)}}(f)}{S^{\alpha}_{vv^{(*)}}(f)} \tag{9.28}$$

for all values of frequency f such that $S^{\alpha}_{xx^{(*)}}(f) \neq 0$. □

That is, $H(f)$ can be expressed in terms of the (conjugate) cyclic spectra of the noisy input and output signals. Therefore, (9.28) provides a system identification formula that is intrinsically immune to the effects of noise and interference (Gardner, 1987f), (Gardner, 1990a). Thus, this identification method is highly tolerant to disturbances in practice, provided that a sufficiently long integration time is used for the (conjugate) cyclic spectral estimates (Sections 5.3.4 and 9.7). For $\alpha = 0$, (9.28) reduces to the classical Wiener (noncausal) identification formula for LTI systems (Papoulis, 1991, Sec. 14-1).

The identification of LTI systems based on noisy input/output measurements is treated in (Gardner, 1985), (Gardner, 1987f), (Gardner and Chen, 1988b), (Gardner and Chen, 1988a), (Gardner, 1990a), (Gardner and Chen, 1992), (Chen and Gardner, 1992), (Izzo et al., 1992a), (Gelli et al., 1993), (Izzo et al., 1994a), (Napolitano, 1994), (Giannakis, 1995), (Antoni et al., 2004b).

9.5.2.2 Higher-order statistic based identification

The LTI identification formula (9.28) based on second-order (conjugate) cyclic spectra cannot be exploited in the following cases:

1. The input signal $x(t)$ does not exhibit a significant degree of second-order cyclostationarity;
2. The LTI system is very narrow-band so that the output signal does not exhibit a significant degree of second-order cyclostationarity (see Section 4.1, Example 4.12);
3. There is no second-order (conjugate) cycle frequency such that Assumption 9.2 is verified.

In these cases, higher-order cyclostationarity based identification formulas can be considered.

Let us assume that for a given conjugation configuration $\boldsymbol{x} \equiv x\, x^{(*)_2} \ldots x^{(*)_N} \equiv x\, \boldsymbol{x}_0$, a cycle frequency β exists such that the Nth-order CCP $P^{\beta}_{\boldsymbol{x}}(\boldsymbol{f}')$ is not identically zero.

By taking in (4.93) $H_1(f_1) = H(f_1)$, $H_m(f_m) = 1$ for $m = 2, \ldots, M$, we have

$$P^{\beta}_{y\boldsymbol{x}_0}(\boldsymbol{f}') = P^{\beta}_{\boldsymbol{x}}(\boldsymbol{f}')\, H(f_1) \tag{9.29}$$

from which we obtain the LTI system identification formula

$$H(f_1) = \frac{P^{\beta}_{y\boldsymbol{x}_0}(\boldsymbol{f}')}{P^{\beta}_{\boldsymbol{x}}(\boldsymbol{f}')} \tag{9.30}$$

for all values of $\boldsymbol{f}'$ such that $P^{\beta}_{\boldsymbol{x}}(\boldsymbol{f}') \neq 0$.

In the case of noisy measurements of the input and output signals, we have model (9.21) and (9.22).

By assuming $x(t)$ independent of both $n(t)$ and $m(t)$, and accounting for the additivity property of the cumulant of the sum of independent signals (Section 4.2.2), we have

$$\frac{P^{\beta}_{z\boldsymbol{v}_0}(\boldsymbol{f}')}{P^{\beta}_{\boldsymbol{v}}(\boldsymbol{f}')} = \frac{P^{\beta}_{y\boldsymbol{x}_0}(\boldsymbol{f}') + P^{\beta}_{m\boldsymbol{n}_0}(\boldsymbol{f}')}{P^{\beta}_{\boldsymbol{x}}(\boldsymbol{f}') + P^{\beta}_{\boldsymbol{n}}(\boldsymbol{f}')} \tag{9.31}$$

where $\boldsymbol{n} \equiv n\, n^{(*)_2} \dots n^{(*)_N} \equiv n\, \boldsymbol{n}_0$ and $\boldsymbol{v} \equiv v\, v^{(*)_2} \dots v^{(*)_N} \equiv v\, \boldsymbol{v}_0$.

The higher-order counterpart of Assumption 9.2 is the following.

Assumption 9.4.

- *$n(t)$ does not exhibit Nth-order cyclostationarity at cycle frequency β for the considered conjugation configuration $\boldsymbol{n} \equiv n\, n^{(*)_2} \dots n^{(*)_N}$. That is,*

$$P^{\beta}_{\boldsymbol{n}}(\boldsymbol{f}') \equiv 0 \tag{9.32}$$

- *$n(t)$ and $m(t)$ do not exhibit joint Nth-order cyclostationarity at cycle frequency β for the considered conjugation configuration $m\boldsymbol{n}_0 \equiv m\, n^{(*)_2} \dots n^{(*)_N}$. That is,*

$$P^{\beta}_{m\boldsymbol{n}_0}(\boldsymbol{f}') \equiv 0 \tag{9.33}$$

Accounting for (9.30) and (9.31), we have the following LTI system identification formula based on the CCPs of noisy versions of the input and output signals (Napolitano, 1995).

Theorem 9.5. (Napolitano, 1995). *If $x(t)$ is independent of both $n(t)$ and $m(t)$, under Assumption 9.4 we have*

$$\frac{P^{\beta}_{z\boldsymbol{v}_0}(\boldsymbol{f}')}{P^{\beta}_{\boldsymbol{v}}(\boldsymbol{f}')} = \frac{P^{\beta}_{y\boldsymbol{x}_0}(\boldsymbol{f}')}{P^{\beta}_{\boldsymbol{x}}(\boldsymbol{f}')} = H(f_1)\,. \tag{9.34}$$

□

Consequently, the LTI system can be identified by estimating both $P^{\beta}_{z\boldsymbol{v}_0}(\boldsymbol{f}')$ and $P^{\beta}_{\boldsymbol{v}}(\boldsymbol{f}')$ involved in (9.34). Estimation methods for the CCP are presented in Section 5.7.

In the noise free case, an identification formula analogous to (9.30) can be established in therms of RD-CSCMFs. In fact, by taking in (4.92) $H_1(f_1) = H(f_1)$, $H_m(f_m) = 1$ for $m = 2, \dots, M$, we have

$$S^{\alpha}_{y\boldsymbol{x}_0}(\boldsymbol{f}') = S^{\alpha}_{\boldsymbol{x}}(\boldsymbol{f}')\, H(f_1) \tag{9.35}$$

from which we obtain the LTI system identification formula

$$H(f_1) = \frac{S^{\alpha}_{y\boldsymbol{x}_0}(\boldsymbol{f}')}{S^{\alpha}_{\boldsymbol{x}}(\boldsymbol{f}')} \tag{9.36}$$

for all values of $\boldsymbol{f}'$ such that $S^{\alpha}_{\boldsymbol{x}}(\boldsymbol{f}') \neq 0$.

The identification formula in terms of RD-CSCMFs (9.36), however, cannot be immediately extended to the case of noisy measurements of the input and output signals. In fact, unlike cumulants, moments of the sum of independent signals do not factorize into the sum of the moments of each signal. In addition, RD-CSCMFs in general contain impulses (Section 4.2.3.1) so that the ratio in (9.36) can be not well defined.

Assumption 9.6. *N is the lowest-order of cyclostationarity of $x(t)$ for the conjugation configuration $\boldsymbol{x} \equiv x\, x^{(*)_2} \dots x^{(*)_N} \equiv x\, \boldsymbol{x}_0$. That is,*

$$S_{\boldsymbol{x}}^{\gamma}(\boldsymbol{f}')_k \equiv 0, \qquad \gamma \neq 0, \quad k = 1, \dots, N-1 \tag{9.37}$$

$$S_{\boldsymbol{x}}^{\gamma}(\boldsymbol{f}')_N \not\equiv 0 \tag{9.38}$$

where subscript k is introduced in notation to explicitly denote the order of the RD-CSCMFs. □

Assumption 9.7. *$n(t)$ and $m(t)$ do not exhibit cyclostationarity and joint cyclostationarity for the following orders and conjugation configurations:*

$$S_{\boldsymbol{n}}^{\alpha}(\boldsymbol{f}')_k \equiv 0, \quad k = 1, \dots, N \tag{9.39}$$

$$S_{m\boldsymbol{n}_0}^{\alpha}(\boldsymbol{f}')_k \equiv 0, \quad k = 1, \dots, N\,. \tag{9.40}$$

□

Theorem 9.8. (Napolitano, 1995), (Flagiello et al., 2000). *Under Assumptions 9.6 and 9.7, we have the following equalities among Nth-order RD-CSCMF and CCPs*

$$S_{y\boldsymbol{x}_0}^{\alpha}(\boldsymbol{f}') = P_{y\boldsymbol{x}_0}^{\alpha}(\boldsymbol{f}') = P_{z\boldsymbol{v}_0}^{\alpha}(\boldsymbol{f}') = S_{z\boldsymbol{v}_0}^{\alpha}(\boldsymbol{f}') \tag{9.41}$$

$$S_{\boldsymbol{x}}^{\alpha}(\boldsymbol{f}') = P_{\boldsymbol{x}}^{\alpha}(\boldsymbol{f}') = P_{\boldsymbol{v}}^{\alpha}(\boldsymbol{f}') = S_{\boldsymbol{v}}^{\alpha}(\boldsymbol{f}') \tag{9.42}$$

and the LTI system identification formula

$$\frac{S_{z\boldsymbol{v}_0}^{\alpha}(\boldsymbol{f}')}{S_{\boldsymbol{v}}^{\alpha}(\boldsymbol{f}')} = \frac{S_{y\boldsymbol{x}_0}^{\alpha}(\boldsymbol{f}')}{S_{\boldsymbol{x}}^{\alpha}(\boldsymbol{f}')} = H(f_1) \tag{9.43}$$

holds for all values of f_1 such that $S_{\boldsymbol{x}}^{\alpha}(\boldsymbol{f}') \neq 0$.

Proof. See Section 9.9. □

It is worthwhile to underline that (9.34) and (9.43) are interesting identification formulas in those cases where the degree of second-order cyclostationarity of $x(t)$ is low or zero, whereas the degree of HOCS is substantial. Moreover, no problem arises from Assumption 9.6 since the cyclostationarity of lowest order exhibited by $x(t)$ is usually exploited to avoid unnecessary computational complexity. Furthermore, note that Assumption 9.7 on noise and interfering signals is more restrictive than Assumption 9.4. Finally, note that under Assumptions 9.6 and 9.7, the RD-CTCMFs in (9.43) are coincident with the corresponding CCPs so that the ratios in (9.43) are well defined.

Formulas analogous to (9.34) and (9.43) can be straightforwardly stated for discrete-time LTI system identification.

9.5.2.3 Higher-order hybrid temporal-spectral cyclic statistic based identification

Although the application of the identification methods based on (9.34) or (9.43) requires only the estimation of a slice of the involved (conjugate) cyclic spectra, the computational cost grows with N and could result prohibitive for high orders of cyclostationarity and/or large collect times. Thus, to overcome this problem an identification method is proposed in (Flagiello et al., 2000), which is based on higher-order hybrid temporal-spectral cyclic statistic (Section 4.7). The method exploits the higher-order cyclostationarity properties of the involved signals, but exhibits a computational cost comparable to that of the second-order cyclostationarity based method (9.28).

The $(N-2)$-fold inverse Fourier transform of the RD-CSCMF $S^{\alpha}_{y\boldsymbol{x}_0}(\boldsymbol{f}')$ with respect to the $(N-2)$-dimensional vector $\boldsymbol{f}'' \triangleq [f_2, \cdots, f_{N-1}]^{\mathrm{T}}$ is referred to as *hybrid temporal-spectral cyclic moment* (Definition 4.38). It can be expressed as

$$
\begin{aligned}
\int_{\mathbb{R}^{N-2}} & S^{\alpha}_{y\boldsymbol{x}_0}(\boldsymbol{f}')\, e^{j2\pi \boldsymbol{f}''^{\mathrm{T}}\boldsymbol{\tau}''}\, \mathrm{d}\boldsymbol{f}'' \\
&= \int_{\mathbb{R}^{N-2}} \int_{\mathbb{R}^{N-1}} R^{\alpha}_{y\boldsymbol{x}_0}(\boldsymbol{u}')\, e^{-j2\pi \boldsymbol{f}'^{\mathrm{T}}\boldsymbol{u}'}\, \mathrm{d}\boldsymbol{u}'\, e^{j2\pi \boldsymbol{f}''^{\mathrm{T}}\boldsymbol{\tau}''}\, \mathrm{d}\boldsymbol{f}'' \\
&= \int_{\mathbb{R}} R^{\alpha}_{y\boldsymbol{x}_0}(\boldsymbol{\tau}')\, e^{-j2\pi f_1\tau_1}\, \mathrm{d}\tau_1 \\
&= \int_{\mathbb{R}} R^{\alpha}_{y\chi}(\tau_1; \boldsymbol{\tau}'')\, e^{-j2\pi f_1\tau_1}\, \mathrm{d}\tau_1 \\
&\equiv S^{\alpha}_{y\chi}(f_1; \boldsymbol{\tau}'') \qquad (N > 2)
\end{aligned} \tag{9.44}
$$

where $\boldsymbol{\tau}' \triangleq [\tau_1, \boldsymbol{\tau}''^{\mathrm{T}}]^{\mathrm{T}} \triangleq [\tau_1, \ldots, \tau_{N-1}]^{\mathrm{T}}$ and

$$
\chi(t; \boldsymbol{\tau}'') \triangleq x^{(*)_N}(t) \prod_{k=2}^{N-1} x^{(*)_k}(t+\tau_k)\,. \tag{9.45}
$$

That is, $S^{\alpha}_{y\chi}(f_1; \boldsymbol{\tau}'')$ is the second-order conjugate cyclic cross-spectrum of $y(t)$ and the lag-product wave form $\chi(t; \boldsymbol{\tau}'')$, for a fixed $\boldsymbol{\tau}''$, at conjugate cycle frequency α.

Therefore, by considering the $(N-2)$-fold inverse Fourier transform of both sides of (9.35) one has

$$
H(f_1) = \frac{S^{\alpha}_{y\chi}(f_1; \boldsymbol{\tau}'')}{S^{\alpha}_{x\chi}(f_1; \boldsymbol{\tau}'')}\,. \tag{9.46}
$$

Analogously, by considering the $(N-2)$-fold inverse Fourier transform of the reduced-dimension CSCMFs of the noisy input and output signals, that is $S^{\alpha}_{z\boldsymbol{v}_0}(\boldsymbol{f}')$ and $S^{\alpha}_{\boldsymbol{v}}(\boldsymbol{f}')$, from (9.43) one obtains an identification formula based on noisy input and output signal measurements.

Theorem 9.9. (Flagiello et al., 2000). *Under Assumptions 9.6 and 9.7, it results that*

$$
\frac{S^{\alpha}_{z\varrho}(f_1; \boldsymbol{\tau}'')}{S^{\alpha}_{v\varrho}(f_1; \boldsymbol{\tau}'')} = \frac{S^{\alpha}_{y\chi}(f_1; \boldsymbol{\tau}'')}{S^{\alpha}_{x\chi}(f_1; \boldsymbol{\tau}'')} = H(f_1) \tag{9.47}
$$

where

$$\varrho(t;\boldsymbol{\tau}'') \triangleq v^{(*)_N}(t) \prod_{k=2}^{N-1} v^{(*)_k}(t+\tau_k) \tag{9.48}$$

and $S^{\alpha}_{z\varrho}(f_1;\boldsymbol{\tau}'')$ and $S^{\alpha}_{v\varrho}(f_1;\boldsymbol{\tau}'')$ are defined according to (9.44). □

It is worth to observe that, since $S^{\alpha}_{\boldsymbol{v}}(\boldsymbol{f}') = P^{\alpha}_{\boldsymbol{v}}(\boldsymbol{f}')$ and $S^{\alpha}_{z\boldsymbol{v}_0}(\boldsymbol{f}') = P^{\alpha}_{z\boldsymbol{v}_0}(\boldsymbol{f}')$, accounting for (9.44), neither $S^{\alpha}_{z\varrho}(f_1;\boldsymbol{\tau}'')$ nor $S^{\alpha}_{v\varrho}(f_1;\boldsymbol{\tau}'')$ contain impulsive terms, that is, their α-submanifolds are empty and, hence, they are coincident with the corresponding second-order cyclic cross-poly-spectra. Thus, the ratio (9.47) is well defined for all values of f_1 and $\boldsymbol{\tau}''$ such that the denominator is not zero. Moreover, although only second-order cyclic spectra are involved in the identification formula (9.47), the method exploits the Nth-order cyclostationarity features of the input signal, since these features are extracted by considering the cyclic correlations of the noisy input and output signals $v(t)$ and $z(t)$ with the $(N-1)$th-order lag product waveform (9.48). In addition, since (according to (9.44)) $S^{\alpha}_{v\varrho}(f_1;\boldsymbol{\tau}'')$ is the $(N-2)$-fold inverse Fourier transform of $S^{\alpha}_{\boldsymbol{v}}(\boldsymbol{f}')$, both these functions exhibit the same set of cycle frequencies, i.e., the same values of α for which they are not identically zero. As the computational complexity of the identification method (9.47), it is practically independent of the order N and moreover is just that of methods based on second-order (cyclic) statistics.

Another significant advantage of the identification formula (9.47) with respect to (9.43) is the better performance in terms of bias and variance of the estimator.

In both formulas (9.43) and (9.47) the harmonic response estimate is expressed as a ratio of complex variables. Thus, the approximate results derived in Section E.3 can be exploited for the expected value and the variance of the estimator.

The estimator of the harmonic response function $H(f_1)$ based on the identification formula (9.43) is

$$\widehat{H}_{\text{hocs}}(f_1) = \frac{S^{(T,\Delta f)}_{z\boldsymbol{v}_0}(\alpha,\boldsymbol{f}')}{S^{(T,\Delta f)}_{\boldsymbol{v}}(\alpha,\boldsymbol{f}')}, \tag{9.49}$$

where the frequency-smoothed Nth-order cyclic periodogram (Section 5.7) $S^{(T,\Delta f)}_{\boldsymbol{v}}(\alpha,\boldsymbol{f}')$ is taken as estimator of the Nth-order RD-CTCMF $S^{\alpha}_{\boldsymbol{v}}(\boldsymbol{f}')$ and the frequency-smoothed Nth-order cyclic cross-periodogram $S^{(T,\Delta f)}_{z\boldsymbol{v}_0}(\alpha,\boldsymbol{f}')$ is taken as estimator of the Nth-order RD-CTCMF $S^{\alpha}_{z\boldsymbol{v}_0}(\boldsymbol{f}')$.

Under mild conditions expressed in terms of summability of cumulants of $v(t)$ and $z(t)$, we have the mean-square convergence of the higher-order cyclic spectra estimators to the true values outside the α-submanifold (Section 5.7), (Dandawaté and Giannakis, 1994a). Thus we also have the convergence in probability. Consequently, accounting for (9.43), the ratio of these estimators in the right-hand-side of (9.49) converges in probability to $H(f_1)$ (Theorem E.5).

With reference to the notation introduced in Section E.3, let us consider $\mathsf{Z} = g(\mathsf{X},\mathsf{Y}) = \mathsf{Y}/\mathsf{X}$ with Y and X given by the numerator and the denominator in the right-hand side of

(9.49), respectively. The expressions of the mean value (E.76) and of the variance (E.77) of Z specialize to

$$
\begin{aligned}
\mathrm{E}\left\{\widehat{H}_{\mathrm{hocs}}(f_1)\right\} &= \frac{\mathrm{E}\left\{S_{z\boldsymbol{v}_0}^{(T,\Delta f)}(\alpha, \boldsymbol{f}')\right\}}{\mathrm{E}\left\{S_{\boldsymbol{v}}^{(T,\Delta f)}(\alpha, \boldsymbol{f}')\right\}} \\
&\quad + \text{terms proportional to the conjugate variance and covariance} \\
&\quad\ \text{of } S_{z\boldsymbol{v}_0}^{(T,\Delta f)}(\alpha, \boldsymbol{f}') \text{ and } S_{\boldsymbol{v}}^{(T,\Delta f)}(\alpha, \boldsymbol{f}') \\
&= \frac{\mathrm{E}\left\{S_{z\boldsymbol{v}_0}^{(T,\Delta f)}(\alpha, \boldsymbol{f}')\right\}}{\mathrm{E}\left\{S_{\boldsymbol{v}}^{(T,\Delta f)}(\alpha, \boldsymbol{f}')\right\}} + \mathcal{O}\left(\frac{1}{T(\Delta f)^{N-1}}\right)
\end{aligned}
\tag{9.50}
$$

$$
\begin{aligned}
\operatorname{var}\left\{\widehat{H}_{\mathrm{hocs}}(f_1)\right\} &= \text{terms proportional to the variance and covariance} \\
&\quad\ \text{of } S_{z\boldsymbol{v}_0}^{(T,\Delta f)}(\alpha, \boldsymbol{f}') \text{ and } S_{\boldsymbol{v}}^{(T,\Delta f)}(\alpha, \boldsymbol{f}') \\
&= \mathcal{O}\left(\frac{1}{T(\Delta f)^{N-1}}\right)
\end{aligned}
\tag{9.51}
$$

where the asymptotic behavior of variance and covariance are consequence of Theorem 5.66.

The estimator of the harmonic response function $H(f_1)$ based on the identification formula (9.47) is

$$
\widehat{H}_{\mathrm{hybrid}}(f_1) = \frac{S_{zQ}^{(T,\Delta f)}(\alpha, f_1; \boldsymbol{\tau}'')}{S_{vQ}^{(T,\Delta f)}(\alpha, f_1; \boldsymbol{\tau}'')}
\tag{9.52}
$$

where the frequency-smoothed 2nd-order cyclic cross-periodogram (Section 5.2.3) $S_{vQ}^{(T,\Delta f)}(\alpha, f_1; \boldsymbol{\tau}'')$ is taken as estimator for the 2nd-order cyclic cross-spectrum $S_{vQ}^{\alpha}(f_1; \boldsymbol{\tau}'')$ (with $\boldsymbol{\tau}''$ fixed parameter) and the frequency-smoothed 2nd-order cyclic cross-periodogram $S_{zQ}^{(T,\Delta f)}(\alpha, f_1; \boldsymbol{\tau}'')$ is taken as estimator for the 2nd-order cyclic cross-spectrum $S_{zQ}^{\alpha}(f_1; \boldsymbol{\tau}'')$.

Under mild conditions expressed in terms of summability of cumulants of $v(t)$ and $z(t)$, we have the mean-square convergence of the 2nd-order cyclic cross-spectra estimators to the true values (Theorem 5.39), (Section 5.7.4). Thus we also have the convergence in probability. Consequently, accounting for (9.47), the ratio of these estimators in the right-hand-side of (9.52) converges in probability to $H(f_1)$ (Theorem E.5).

In such a case, the expressions of the mean value (E.76) and of the variance (E.77) specialize to

$$
\begin{aligned}
\mathrm{E}\left\{\widehat{H}_{\mathrm{hybrid}}(f_1)\right\} &= \frac{\mathrm{E}\left\{S_{zQ}^{(T,\Delta f)}(\alpha, f_1; \boldsymbol{\tau}'')\right\}}{\mathrm{E}\left\{S_{vQ}^{(T,\Delta f)}(\alpha, f_1; \boldsymbol{\tau}'')\right\}} \\
&\quad + \text{terms depending on the conjugate variance and covariance}
\end{aligned}
$$

$$
\begin{aligned}
&\text{of } S_{zQ}^{(T,\Delta f)}(\alpha, f_1; \boldsymbol{\tau}'') \text{ and } S_{vQ}^{(T,\Delta f)}(\alpha, f_1; \boldsymbol{\tau}'') \\
&= \frac{\mathrm{E}\left\{S_{zQ}^{(T,\Delta f)}(\alpha, f_1; \boldsymbol{\tau}'')\right\}}{\mathrm{E}\left\{S_{vQ}^{(T,\Delta f)}(\alpha, f_1; \boldsymbol{\tau}'')\right\}} + \mathcal{O}\left(\frac{1}{T\Delta f}\right)
\end{aligned} \tag{9.53}
$$

$$
\begin{aligned}
\operatorname{var}\left\{\widehat{H}_{\mathrm{hocs}}(f_1)\right\} &= \text{terms depending on the variance and covariance} \\
&\quad \text{of } S_{zQ}^{(T,\Delta f)}(\alpha, f_1; \boldsymbol{\tau}'') \text{ and } S_{vQ}^{(T,\Delta f)}(\alpha, f_1; \boldsymbol{\tau}'') \\
&= \mathcal{O}\left(\frac{1}{T\Delta f}\right)
\end{aligned} \tag{9.54}
$$

where the asymptotic behavior of variance and covariance are consequence of Theorem 5.38.

When the data record length $T \to \infty$ and the spectral frequency resolution $\Delta f \to 0$, the term $\mathcal{O}(1/(T(\Delta f)^{N-1}))$ in (9.50) and (9.51) approaches zero provided that $T(\Delta f)^{N-1} \to \infty$. In contrast, the term $\mathcal{O}(1/(T\Delta f))$ in (9.53) and (9.54) approaches zero provided that $T\Delta f \to \infty$.

Thus, for a given order of magnitude of bias and variance, a finer spectral resolution Δf can be obtained for the identification method (9.47) based on higher-order hybrid temporal-spectral cyclic moments with respect to method (9.43) based on RD-CSCFMs. Alternatively, for a given spectral resolution, a smaller bias and variance can be obtained for the identification method (9.47) with respect to (9.43).

9.5.3 Blind LTI-system identification and equalization

From (3.22) we have

$$
S_{yy^*}^{\alpha}(f) = S_{xx^*}^{\alpha}(f)\, H(f)\, H^*(f-\alpha) \tag{9.55}
$$

which, for $\alpha = 0$, reduces to the input/output relationship in terms of power spectra:

$$
S_{yy^*}^{0}(f) = S_{xx^*}^{0}(f)\, |H(f)|^2 \,. \tag{9.56}
$$

From (9.55) and (9.56) it follows that input/output relationships for LTI systems in terms of cyclic statistics, unlike those in terms of autocorrelation functions and power spectra, preserve phase information of the harmonic-response function $H(f)$. Thus, cyclostationarity properties of the output signal are suitable to be exploited for recovering both phase and magnitude of the system harmonic-response function as first observed in (Gardner, 1991b).

Cyclostationarity properties are exploited for blind identification (without measurements of the system input) of linear systems and for blind equalization techniques in (Cerrato and Eisenstein, 1977), (Gardner, 1991b), (Duel-Hallen, 1992), (Chen and Nikias, 1994), (Ding, 1994), (Hatzinakos, 1994), (Li and Ding, 1994), (Schell and Gardner, 1994), (Tong et al., 1994b), (Tong et al., 1994a), (Tong et al., 1995), (Tugnait, 1995), (Liu et al., 1996), (Tugnait, 1996), (Wong and Chambers, 1996), (Chevreuil and Loubaton, 1997),

(Chevreuil, 1997), (Giannakis and Halford, 1997), (Tsatsanis and Giannakis, 1997), (Zeng and Tong, 1997), (Giannakis and Serpedin, 1998), (Mazzenga, 1998), (Serpedin and Giannakis, 1998), (Chevreuil and Loubaton, 1999), (Heat and Giannakis, 1999), (Chevreuil et al., 2000), (Mazzenga, 2000b), (Bölcskei et al., 2001), (Bölcskei et al., 2002), (Gelli and Verde, 2002), (Tugnait and Luo, 2002), (Bradaric et al., 2003), (Houcke et al., 2003), (Narasimhan et al., 2005).

9.5.4 Nonlinear-system identification

Let $y(t)$ the output signal of a Volterra system excited by the input signal $x(t)$:

$$y(t) = \sum_{n=1}^{+\infty} \int_{\mathbb{R}^n} k_n(\tau_1, \ldots, \tau_n)\, x(t+\tau_1) \cdots x(t+\tau_n) \mathrm{d}\tau_1 \cdots \mathrm{d}\tau_n \,. \tag{9.57}$$

Accounting for the results of Sections 2.3 and 4.2, the input lag-product waveform can be decomposed into the sum of an almost-periodic component, referred to as the temporal moment function, and a residual term not containing any finite-strength additive sine-wave component (see (4.4a)):

$$x(t+\tau_1) \cdots x(t+\tau_n) = \sum_{\alpha \in A_{\boldsymbol{x}}} \mathcal{R}_{\boldsymbol{x}}^{\alpha}(\boldsymbol{\tau})\, e^{j2\pi\alpha t} + \ell_{\boldsymbol{x}}(t, \boldsymbol{\tau}) \,. \tag{9.58}$$

Thus, identification and equalization techniques for Volterra systems excited by ACS signals make use of higher-order cyclostationarity properties. In fact, there are potentially substantial advantages to using cyclostationary input signals, relative to stationary input signals, for purposes of Volterra system modeling and identification (Gardner and Archer, 1993).

Both time-invariant and almost-periodically time-variant nonlinear systems are treated in (Archer and Gardner, 1990), (Gardner and Archer, 1991), (Gardner and Archer, 1993), (Izzo et al., 1994b), (Gardner and Paura, 1995), (Prakriya and Hatzinakos, 1997), (Mattera and Paura, 1998), (Mattera, 1999).

9.6 Applications

9.6.1 Signal parameter estimation

Cyclostationarity properties can be exploited to design signal selective algorithms for signal parameter and waveform estimation. In fact, if the desired and interfering signals have different cyclic parameters such as carrier frequency or baud rate, then they exhibit cyclostationarity at different cycle frequencies and, consequently, parameters of the desired signal can be extracted by estimating cyclic statistics of the received data, consisting of desired signal plus interfering signal, at a cycle frequency exhibited by the desired signal but not by the interference (Section 9.2). This signal selectivity by exploitation of cyclostationarity was first suggested in (Gardner, 1980). Parameters that can be estimated in the presence of interference and/or high noise include carrier frequency and phase, pulse rate and phase, signal power level, modulation indices, bandwidths, time- and frequency-difference of arrival, direction of arrival, and so on.

The following applications have been considered:

- *Array processing problems, including spatial filtering and direction finding*
  (Agee et al., 1987), (Agee et al., 1988), (Gardner, 1988c), (Schell and Agee, 1988), (Agee, 1989a), (Schell et al., 1989), (Schell and Gardner, 1989), (Xu and Kailath, 1989), (Agee et al., 1990), (Agee and Young, 1990), (Schell, 1990), (Schell and Gardner, 1990d), (Schell and Gardner, 1990c), (Schell and Gardner, 1990b), (Schell and Gardner, 1990a), (Biedka and Agee, 1991), (Gelli et al., 1991a), (Izzo et al., 1992b), (Schell and Gardner, 1992), (Xu and Kailath, 1992), (Biedka, 1993), (Castedo et al., 1993), (Castedo and Tseng, 1993), (Schell and Gardner, 1993b), (Schell and Gardner, 1993a), (Biedka and Kahn, 1994), (Castedo et al., 1994), (Gelli, 1994), (Kahn et al., 1994), (Schell, 1994a), (Schell and Gardner, 1994), (Schell, 1994b), (Shamsunder and Giannakis, 1994), (Wu et al., 1994), (Biedka, 1995), (Castedo and Figueiras-Vidal, 1995), (Gelli, 1995), (Schell and Gardner, 1995), (Wu and Wong, 1996), (Gelli and Izzo, 1997), (Lee and Lee, 1999), (Martone, 1999), (Lee et al., 2000), (Biedka, 2001), (Lee and Lee, 2001a), (Lee and Lee, 2001b), (Huang et al., 2002a), (Chargé et al., 2003), (Gelli and Izzo, 2003), (Lee and Lee, 2003), (Gelli et al., 2005), (Yan and Fan, 2005c);
- *Polynomial-phase signal parameter estimation*
  (Shamsunder et al., 1995), (Zhou, 1995), (Gini and Giannakis, 1999), (Morelande and Zoubir, 2002);
- *Harmonics in additive and multiplicative noise*
  (Giannakis and Zhou, 1994), (Giannakis and Zhou, 1995), (Zhou and Giannakis, 1995b), (Zhou and Giannakis, 1995a), (Ghogo et al., 1999b), (Ghogo et al., 1999a);
- *Modulation index estimation*
  (Bianchi et al., 2004).

But, there are difficulties in application of some of this work–difficulties associated with lack of ergodic properties of stochastic-process models adopted for polynomial-phase signals and signals with coupled harmonics.

On this subject see the general treatments (Gardner, 1972a), (Gardner and Franks, 1975), (Gardner, 1985), (Gardner, 1987f), (Brown, 1987), (Gardner, 1990b), and also see (Gardner, 1973), (Gardner, 1976), (Bilyik, 1977a), (Graef, 1983), (Gardner and Brown, 1989), (Chen, 1989), (Gardner and Venkataraman, 1990), (Mendoza et al., 1991), (Živanović and Gardner, 1991), (Allen and Hobbs, 1992), (Napolitano and Tanda, 2001), (Anderson and Meerschaert, 2005).

Patents of inventions on signal parameter and waveform estimation exploiting cyclostationarity are (Friedman et al., 1989), (Gardner and Agee, 1993), (Gardner and Schell, 1993), (Gardner et al., 1994), (Sullivan, 1999), (Krikorian and Rosen, 2003).

9.6.2 Source location

Signal selective and interference tolerant algorithms for source location can be designed by exploiting the cyclostationarity properties of the source signal (Agee et al., 1990).

Direction of arrival (DOA) estimation algorithms are proposed in (Yan and Fan, 2005a), (Yan and Fan, 2007), (Liu et al., 2017). The presence of cycle frequency error is considered in (Zhang et al., 2005). In (Reuven and Weiss, 2009), a technique referred to as cyclic direct position determination is presented. Transmitter localization in the presence of interference is considered in (Chaudhari and Cabric, 2016).

Source location by estimating the TDOA of the cyclostationary signals impinging on two sensors is addressed in (Gardner and Chen, 1988b), (Gardner and Chen, 1988a), (Spooner et al., 1989), (Xu and Kailath, 1990), (Gelli et al., 1991b), (Gardner and Chen, 1992), (Chen and Gardner, 1992), (Gardner and Spooner, 1992a), (Gardner and Spooner, 1993), (Gerr and Allen, 1994b), (Izzo et al., 1994c), (Mazzenga and Vatalaro, 1996), (Hongyi and Zhang, 1998), (Mazzenga, 2000a), (Teplitsky and Yeredor, 2009), (Wen and Wan, 2013). Specifically, in (Chen and Gardner, 1992), (Gardner and Chen, 1992), under the assumption that a cycle frequency of the SOI exists which is not shared with the disturbance signals, relationships in terms of cyclic autocorrelation and cross-correlation functions and cyclic spectra and cross-spectra are derived to be exploited for the TDOA estimation problem (Example 3.8). The performance of estimation techniques based on these relations is only limited by the cycle leakage phenomenon (Section 9.7). The presence of impulsive noise is considered in (Liu et al., 2012), (Liu et al., 2018c). TDOA estimation relaxing the assumption of the existence of a cycle frequency of the SOI not shared with the disturbance is made in (Izzo et al., 1994c). On the subject of TDOA estimation see also (Efimov et al., 2018).

In the case of relative motion with constant relative radial speed between the transmitting source and the receiving sensors, the Doppler effect can be modeled as a frequency shift of the carrier provided that the so called narrow-band condition is satisfied (Section 11.2), (Van Trees, 1971b, pp. 339–340), (Napolitano, 2012, Sec. 7.5.1), (Napolitano, 2016a, Sec. 2). Joint TDOA and frequency-shift estimation are made in (Streight et al., 2000), (Huang et al., 2002c), (Huang and Zhou, 2006), (Huang et al., 2008), (Liu et al., 2015).

9.6.3 Beamforming

Beamforming techniques are addressed in (Hongsheng et al., 2005), (Gelli et al., 2005), (Tong et al., 2007), (Huang et al., 2011), (Li et al., 2013a). The presence of cycle frequency error is treated in (Lee and Huang, 2010), (Lee and Huang, 2012). The cyclic multiple signal classification (MUSIC) algorithm and its variants are considered in (Charge and Wang, 2005), (Liu et al., 2009), (Yan and Fan, 2005b). Adaptive beamforming of cyclostationary sources is treated in (Meng et al., 2017). The case of impulsive noise is considered in (Jin et al., 2019).

9.6.4 Source separation

Cyclostationary source separation algorithms are presented in (Antoni et al., 2005), (Jallon et al., 2010), (Capdessus and Nandi, 2011), (Florian et al., 2012). They are based on the assumption that the source signal to be extracted does not share at least a cycle frequency (assumed to be known) with the other signals of the mixture.

The case of blind source extraction from convolutional mixtures is treated in (Wang et al., 2004), (Xiang et al., 2017).

9.6.5 Miscellaneous

Further applications are in the following subjects:

- *Capacity evaluation of second-order cyclostationary complex Gaussian noise channels* (Han and Cho, 2012);
- *Capacity evaluation of orthogonal overlay channels* (Yeo et al., 2015b);
- *Capacity evaluation of power line communication channels* (Shlezinger and Dabora, 2015);
- *Distortion rate function evaluation for cyclostationary processes* (Kipnis et al., 2018);
- *Timing and signal parameter estimation* (Lopez-Salcedo and Vazquez, 2006), (Jallon and Chevreuil, 2008), (Mosquera et al., 2008), (Shi et al., 2009), (Punchihewa et al., 2011), (Gonzalez et al., 2013), (Sun et al., 2013), (Liu et al., 2013), (Güner and Kaya, 2014), (El Waled, 2015), (Majhi and Ho, 2015);
- *Frequency-domain equalizer (FDE) design* (Yoo and Cho, 2010);
- *Blind multiple-input multiple-output (MIMO) system identification* (Sabri et al., 2009);
- *Signal power, SNR, and signal-to-interference-and-noise ratio (SINR) estimation* (Angrisani et al., 2009), (Riba et al., 2010), (Hong et al., 2012);
- *Doppler spread estimation* (Zhang and Abdi, 2009);
- *Antenna processing* (Ferreol et al., 2008);
- *Doppler effect analysis and micro-Doppler estimation* (Li et al., 2010), (Zhang and Abdi, 2012), (Zhao and Su, 2018), (Heydari et al., 2018);
- *Noise modeling in power lines* (Katayama et al., 2006), (Gianaroli et al., 2012), (Shlezinger and Dabora, 2014);
- *Interference mitigation or self-interference cancellation* (Tani et al., 2016), (Carrick et al., 2018), (Soriano-Irigaray et al., 2018), (Elgenedy et al., 2018);
- *Single-carrier frequency-domain equalizer (SC-FDE)* (Kim et al., 2016);

- *Filter bank analysis*
  (Ohno and Sakai, 1996), (Lall et al., 1999);
- *Subband coding*
  (Francis and Dasgupta, 1999), (Pandharipande and Dasgupta, 1999);
- *Radio frequency interference (RFI) mitigation in radio astronomy*
  (Hellbourg et al., 2012).

9.7 Performance of cyclostationarity-based algorithms

As observed in Chapter 5 and Section 9.2, good performance of cyclostationarity-based signal processing algorithms can be achieved, provided that the data-record length T for the cyclic statistic estimation is sufficiently large. In such a case, unlike classical algorithms based on the WSS signal model, cyclostationarity-based detection or estimation algorithms operating at a cycle frequency of the SOI not shared with the disturbance signals are potentially immune to the effects of disturbance, *regardless of the amount of temporal and spectral overlap of SOI and disturbance* (Section 9.2), (Gardner, 1985), (Gardner, 1987f), (Gardner et al., 2006).

The performance degradation is only due to the *cycle leakage* effect (Section 5.3.4) (Gardner, 1987f, Chap. 13). It is the influence, in cyclic statistic measurements at the considered cycle frequency α_0, of the cyclic statistics of the disturbance signals and the SOI itself at cycle frequencies different from α_0. For example, let us consider the additive model

$$r(t) = x(t) + n(t) \tag{9.59}$$

where $x(t)$ is the SOI and $n(t)$, uncorrelated with $x(t)$, is zero-mean white noise with power spectral density $N_0/2$ in the bandwidth $(-B, B)$, that is, with autocorrelation function

$$\mathrm{E}\{n(t+\tau)\, n^*(t)\} = R^0_{nn^*}(\tau) = \frac{N_0}{2}\, 2B\, \mathrm{sinc}(2B\tau)\,. \tag{9.60}$$

For every $\alpha \in \mathbb{R}$ it results that

$$\begin{aligned} R^\alpha_{rr^*}(\tau) &= R^\alpha_{xx^*}(\tau) + R^\alpha_{nn^*}(\tau) \\ &= R^\alpha_{xx^*}(\tau) + R^0_{nn^*}(\tau)\, \delta_\alpha \end{aligned} \tag{9.61}$$

and for every finite T we have

$$R^{(T)}_{rr^*}(\alpha,\tau) = R^{(T)}_{xx^*}(\alpha,\tau) + R^{(T)}_{xn^*}(\alpha,\tau) + R^{(T)}_{nx^*}(\alpha,\tau) + R^{(T)}_{nn^*}(\alpha,\tau)\,. \tag{9.62}$$

In the case of rectangular data-tapering window $w_T(t) = \mathrm{rect}(t/T)/T$, the expected value of the cyclic correlogram of $n(t)$ can be obtained by specializing (5.21) to the case $y \equiv x \equiv n$ with only the zero cycle frequency and $W_{\frac{1}{T}}(f) = \mathrm{sinc}(fT)$. We have

$$\mathrm{E}\left\{R^{(T)}_{nn^*}(\alpha,\tau)\right\} = R^0_{nn^*}(\tau)\, W_{\frac{1}{T}}(\alpha) \tag{9.63a}$$

$$= N_0 B \operatorname{sinc}(2B\tau) \operatorname{sinc}(\alpha T) . \tag{9.63b}$$

In addition, since $x(t)$ and $n(t)$ are uncorrelated,

$$\mathrm{E}\left\{R_{xn^*}^{(T)}(\alpha,\tau)\right\} = \mathrm{E}\left\{R_{nx^*}^{(T)}(\alpha,\tau)\right\} = 0 . \tag{9.64}$$

Accounting for (9.63b) and (9.64), the amount of cycle leakage in (9.62) arising from the stationary noise at each point of the (α, τ) plane can be roughly estimated by replacing random quantities with their expected values. However, for better estimates, also variances should be accounted for since they are $\mathcal{O}(1/T)$ (Theorem 5.14). For $\alpha \neq 0$, the smaller is the value of T or the larger is the noise power $N_0 B$, the larger is the cycle leakage phenomenon. In contrast, at $\alpha = 0$, $R_{nn^*}^{(T)}(0, \tau)$ cannot be made arbitrarily small by taking T sufficiently large. A similar result can be obtained by considering the cycle leakage at a given cycle frequency arising from the cyclic statistics of the SOI itself at cycle frequencies different from the one considered.

From the above considerations it follows that the larger is the adopted data-record length, the lower is the minimum SNR and SIR for which satisfactory performance can be achieved.

However, note that the accuracy on the knowledge of the cycle frequency α puts an upper bound to the maximum data-record length T. Specifically, the cycle frequency resolution of the cyclic statistic estimates, which is of the order of $1/T$, cannot be smaller than the error ϵ on the cycle frequency value, in order to avoid significant bias in the cyclic statistic estimates. In fact, if $\alpha = \alpha_0 + \epsilon$, where α_0 is the nominal cycle frequency value, (5.21) specializes into

$$\begin{aligned}
\mathrm{E}\left\{R_{yx^{(*)}}^{(T)}(\alpha_0+\epsilon,\tau;t_0)\right\} &= \sum_{\beta \in A_{yx^{(*)}}} R_{yx^{(*)}}^{\beta}(\tau)\, A((\alpha_0+\epsilon-\beta)T)\, e^{-j2\pi[\alpha_0+\epsilon-\beta]t_0} \\
&= R_{yx^{(*)}}^{\alpha_0}(\tau)\, A(\epsilon\, T)\, e^{-j2\pi\epsilon t_0} \\
&\quad + \sum_{\beta \neq \alpha_0} R_{yx^{(*)}}^{\beta}(\tau)\, A((\alpha_0+\epsilon-\beta)T)\, e^{-j2\pi[\alpha_0+\epsilon-\beta]t_0} .
\end{aligned} \tag{9.65}$$

Since $W_{\frac{1}{T}}(f) = A(f\ T)$ has bandwidth of the order of $1/T$, the first term in (9.65) approaches zero when $1/T$ becomes smaller than $|\epsilon|$. See also the discussions following Theorems 5.7, 5.31, and 5.47.

In addition, a further limitation to increase without bound the data-record length T is the possible lack of validity of the ACS model for the SOI $x(t)$ for large T (Section 11.3).

9.8 Summary

Cyclostationarity properties of modulated signals can be suitably exploited in the analysis and design of communications systems since the signals involved are typically almost cyclostationary (ACS). Most techniques are based on the *signal selectivity property* of cyclostationarity-based detection and estimation algorithms (Section 9.2). Specifically, in the case of additive disturbance signals uncorrelated with the signal-of-interest (SOI), if the SOI does not share at least one (conjugate) cycle frequency α with the disturbance, then the (conjugate) cyclic autocorrelation function of the SOI plus noise at (conjugate) cycle frequency α is coincident with the (conjugate) cyclic autocorrelation function of the SOI alone. Consequently, cyclostationarity-based detection or estimation algorithms are potentially immune to the effects of noise and interference, provided that a sufficiently large observation interval is adopted to accurately estimate the cyclic statistics.

Cyclic Wiener filtering is the optimum (in the minimum mean-squared error (MMSE) sense) linear filtering of almost-cyclostationary signals. It leads to the design of *frequency-shift (FRESH) filters* (Section 9.3). By exploiting the spectral correlation property of ACS signals, the optimum filter cancels spectral band of the SOI which are corrupted by noise and then reconstructs these bands by frequency shifting and then by linear time-invariant filtering the spectral components of the SOI that are noisy-free and correlated with the canceled bands. Significant performance improvement is obtained with respect to the classical Wiener filtering that assumes a stationary model for the involved signals.

Quadratic or higher-order nonlinear time-invariant transformations of an ACS signal give rise to time series containing finite-strength additive sine-wave components whose frequencies are the second or higher-order cycle frequencies of the ACS signal. All synchronization schemes can be recognized to exploit the second- or higher-order cyclostationarity properties of ACS signals (Section 9.4).

By exploiting the signal selectivity properties of ACS signals, *linear time-invariant (LTI) system identification algorithms* are built starting from noisy measurements of the system input and output signals. The identification formula for the system harmonic response based on second-order cyclic spectra measurements extends to the cyclostationary case the Wiener identification formula (Section 9.5.2.1). If the input signal exhibits a low degree of second-order cyclostationarity, then identification formulas based higher-order cyclic statistic measurements can be exploited (Section 9.5.2.2). These formulas, however, give rise to higher bias and variance of the system harmonic response estimate. Bias and variance can be made comparable to those of second-order statistic based methods by resorting to identification formulas based on higher-order hybrid temporal-spectral cyclic statistic estimates (Section 9.5.2.3).

Blind system identification techniques are based on the fact that input/output relations for LTI systems in terms of cyclic statistics preserve phase information (Section 9.5.3). *Nonlinear system identification techniques* are based on higher-order cyclic statistic measurements (Section 9.5.4).

In communications system analysis and design, cyclostationarity has also been exploited for signal parameter estimation (Section 9.6.1), source location (Section 9.6.2), beamforming (Section 9.6.3), source separation (Section 9.6.4). In most cases, the signal selectivity property of cyclostationarity-based algorithms is exploited and the performance is only limited by the *cycle leakage* phenomenon (Section 9.7).

9.9 Proofs

Proof of Theorem 9.8

Let us first reformulate Assumptions 9.6 and 9.7 in the time domain.

Assumption 9.10. (Assumption 9.6). *N is the lowest-order of cyclostationarity of $x(t)$ for the conjugation configuration $\boldsymbol{x} \equiv x\, x^{(*)_2} \ldots x^{(*)_N} \equiv x\, \boldsymbol{x}_0$. That is,*

$$\text{a.} \qquad R_{\boldsymbol{x}}^{\gamma}(\boldsymbol{\tau}')_k \equiv 0, \qquad \gamma \neq 0, \quad k = 1, \ldots, N-1 \tag{9.66}$$

$$\text{b.} \qquad R_{\boldsymbol{x}}^{\gamma}(\boldsymbol{\tau}')_N \not\equiv 0 \tag{9.67}$$

where subscript k is introduced in notation to explicitly denote the order of the RD-CTCMFs. □

Assumption 9.11. (Assumption 9.7). *$n(t)$ and $m(t)$ do not exhibit cyclostationarity and joint cyclostationarity for the following orders and conjugation configurations:*

$$\text{a.} \qquad R_{\boldsymbol{n}}^{\alpha}(\boldsymbol{\tau}')_k \equiv 0, \quad k = 1, \ldots, N \tag{9.68}$$

$$\text{b.} \qquad R_{m\boldsymbol{n}_0}^{\alpha}(\boldsymbol{\tau}')_k \equiv 0, \quad k = 1, \ldots, N\,. \tag{9.69}$$

□

Note that from Assumptions 9.10 and 9.11 it follows that both $n(t)$ and $m(t)$ are zero mean. In particular, they do not contain any finite-strength additive sine-wave component.

Let us consider the Nth-order CTCMF between the noisy input and output $z(t)$ and $v(t)$ which, accounting for (9.21) and (9.22), is given by

$$\begin{aligned}
\mathcal{R}_{z\boldsymbol{v}_0}^{\alpha}(\boldsymbol{\tau})_N &\triangleq \left\langle z(t+\tau_1) \prod_{k=2}^{N} v^{(*)_k}(t+\tau_k)\, e^{-j2\pi\alpha t} \right\rangle_t \\
&= \left\langle \left[x(u) \otimes h(u)|_{u=t+\tau_1} + m(t+\tau_1) \right] \prod_{k=2}^{N} [x(t+\tau_k) + n(t+\tau_k)]^{(*)_k}\, e^{-j2\pi\alpha t} \right\rangle_t \\
&= \sum_{\mu_x,\mu_n} \left\langle \left[x(u) \otimes h(u)|_{u=t+\tau_1} + m(t+\tau_1) \right] \right. \\
&\qquad \left. \prod_{k\in\mu_x} x^{(*)_k}(t+\tau_k) \prod_{i\in\mu_n} n^{(*)_i}(t+\tau_i)\, e^{-j2\pi\alpha t} \right\rangle_t
\end{aligned} \tag{9.70}$$

where the sum is extended to all partitions $\{\mu_x, \mu_n\}$ of $\{2, \cdots, N\}$ (hence, for each partition $\mu_x \cap \mu_n = \emptyset$ and $\mu_x \cup \mu_n = \{2, \cdots, N\}$).

Under Assumptions 9.10 and 9.11 all but one terms in (9.70) are identically zero. In fact, the terms with $x(\cdot) \otimes h(\cdot)$ and $\mu_n \neq \emptyset$ (i.e., $|\mu_n| \leqslant N-1$ and $|\mu_x| \leqslant N-2$, where $|\mu_n|$ denotes the number of elements in μ_n and similarly $|\mu_x|$), say T_1, can be expressed as

$$T_1 \triangleq \left\langle x(u) \otimes h(u)|_{u=t+\tau_1} \prod_{k\in\mu_x} x^{(*)_k}(t+\tau_k) \prod_{i\in\mu_n} n^{(*)_i}(t+\tau_i)\, e^{-j2\pi\alpha t} \right\rangle_t$$
$$= \left\langle \sum_{\alpha_{\mu'_x}} \mathcal{R}_{\boldsymbol{x}}^{\alpha_{\mu'_x}}(\boldsymbol{\tau}_{\mu'_x})_{|\mu'_x|}\, e^{j2\pi\alpha_{\mu'_x} t} \sum_{\alpha_{\mu_n}} \mathcal{R}_{\boldsymbol{n}}^{\alpha_{\mu_n}}(\boldsymbol{\tau}_{\mu_n})_{|\mu_n|}\, e^{j2\pi\alpha_{\mu_n} t}\, e^{-j2\pi\alpha t} \right\rangle_t \underset{\tau_1}{\otimes} h(\tau_1) \tag{9.71}$$

where $\mu'_x \triangleq \mu_x \cup \{1\}$, $\boldsymbol{\tau}_{\mu'_x}$ is the $|\mu'_x|$-dimensional vector whose components are those of $\boldsymbol{\tau}$ having indices in μ'_x, and $\alpha_{\mu'_x}$ and α_{μ_n} denote $|\mu'_x|$th-order and $|\mu_n|$th-order cycle frequencies of $x(t)$ and $n(t)$, respectively, for the considered conjugation configuration. Then, from Assumption 9.10a it follows that

$$T_1 = \mathcal{R}_{\boldsymbol{x}}^{0}(\boldsymbol{\tau}_{\mu'_x})_{|\mu'_x|} \underset{\tau_1}{\otimes} h(\tau_1) \sum_{\alpha_{\mu_n}} \mathcal{R}_{\boldsymbol{n}}^{\alpha_{\mu_n}}(\boldsymbol{\tau}_{\mu_n})_{|\mu_n|} \left\langle e^{j2\pi\alpha_{\mu_n} t}\, e^{-j2\pi\alpha t} \right\rangle_t$$
$$= 0 \tag{9.72}$$

where, in the last equality, Assumption 9.11a has been accounted for. Moreover, the terms in (9.70) with $m(t+\tau_1)$ and $\mu_n \neq \emptyset$, say T_2, can be written as

$$T_2 \triangleq \left\langle m(t+\tau_1) \prod_{k\in\mu_x} x^{(*)_k}(t+\tau_k) \prod_{i\in\mu_n} n^{(*)_i}(t+\tau_i)\, e^{-j2\pi\alpha t} \right\rangle_t$$
$$= \mathcal{R}_{\boldsymbol{x}}^{0}(\boldsymbol{\tau}_{\mu_x})_{|\mu_x|} \sum_{\alpha_{m\mu_n}} \mathcal{R}_{m\boldsymbol{n}}^{\alpha_{m\mu_n}}([\tau_1, \boldsymbol{\tau}_{\mu_n}^{\mathsf{T}}]^{\mathsf{T}})_{|\mu_n|+1} \left\langle e^{j2\pi\alpha_{m\mu_n} t}\, e^{-j2\pi\alpha t} \right\rangle_t$$
$$= 0 \tag{9.73}$$

where $\alpha_{m\mu_n}$ are the cycle frequencies of the joint $(|\mu_n|+1)$th-order cyclostationarity exhibited by $m(t)$ and $\{n^{(*)_i}(t),\ i \in \mu_n\}$. In the penultimate equality, Assumption 9.10a has been exploited and in the last equality Assumption 9.11b has been accounted for. Finally, the term in (9.70) with $\mu_n = \emptyset$, say T_3, is given by

$$T_3 \triangleq \left\langle m(t+\tau_1) \prod_{k=2}^{N} x^{(*)_k}(t+\tau_k)\, e^{-j2\pi\alpha t} \right\rangle_t$$
$$= \mathcal{R}_{\boldsymbol{x}}^{0}(\boldsymbol{\tau}')_{N-1} \sum_{\alpha_m} \mathcal{R}_{m}^{\alpha_m}(\tau_1)_1 \left\langle e^{j2\pi\alpha_m t}\, e^{-j2\pi\alpha t} \right\rangle_t$$
$$= 0 \tag{9.74}$$

where α_m are the frequencies of the additive sine-wave components contained in $m(t)$, in the penultimate equality Assumption 9.10a and in the last equality Assumption 9.11b (with $k = 1$) have been accounted for. Therefore, under Assumptions 9.10–9.11, the only nonzero term in (9.70) is the one with $x(\cdot) \otimes h(\cdot)$ and $\mu_n = \emptyset$, that is,

$$\begin{aligned}
\mathcal{R}^{\alpha}_{z\boldsymbol{v}_0}(\boldsymbol{\tau})_N &= \mathcal{R}^{\alpha}_{\boldsymbol{x}}(\boldsymbol{\tau})_N \underset{\tau_1}{\otimes} h(\tau_1) \\
&= \mathcal{C}^{\alpha}_{\boldsymbol{x}}(\boldsymbol{\tau})_N \underset{\tau_1}{\otimes} h(\tau_1) \\
&= \mathcal{C}^{\alpha}_{y\boldsymbol{x}_0}(\boldsymbol{\tau})_N \, .
\end{aligned} \tag{9.75}$$

Let us now consider the Nth-order CTMF of the noisy input $v(t)$. It is given by

$$\begin{aligned}
\mathcal{R}^{\alpha}_{\boldsymbol{v}}(\boldsymbol{\tau})_N &\triangleq \left\langle \prod_{k=1}^{N} [x(t+\tau_k) + n(t+\tau_k)]^{(*)_k} \, e^{-j2\pi\alpha t} \right\rangle_t \\
&= \sum_{\mu_x, \mu_n} \left\langle \prod_{k\in\mu_x} x^{(*)_k}(t+\tau_k) \prod_{i\in\mu_n} n^{(*)_i}(t+\tau_i) \, e^{-j2\pi\alpha t} \right\rangle_t
\end{aligned} \tag{9.76}$$

where the sum is extended now to all partitions $\{\mu_x, \mu_n\}$ of $\{1, \cdots, N\}$ and the optional conjugation $(*)_1$ is absent.

Under Assumptions 9.10a and 9.11a all but one terms in (9.76) are identically zero. In fact, the terms with $\mu_n \neq \emptyset$, say T_4, can be written as

$$\begin{aligned}
T_4 &\triangleq \left\langle \prod_{k\in\mu_x} x^{(*)_k}(t+\tau_k) \prod_{i\in\mu_n} n^{(*)_i}(t+\tau_i) \, e^{-j2\pi\alpha t} \right\rangle_t \\
&= \mathcal{R}^{0}_{\boldsymbol{x}}(\boldsymbol{\tau}_{\mu_x})_{|\mu_x|} \sum_{\alpha_{\mu_n}} \mathcal{R}^{\alpha_{\mu_n}}_{\boldsymbol{n}}(\boldsymbol{\tau}_{\mu_n})_{|\mu_n|} \left\langle e^{j2\pi\alpha_{\mu_n} t} \, e^{-j2\pi\alpha t} \right\rangle_t \\
&= 0
\end{aligned} \tag{9.77}$$

where in the last but one equality Assumption 9.10a and in the last equality Assumption 9.11a have been accounted for. Therefore, under Assumptions 9.10a and 9.11a, the only nonzero term in (9.76) is that with $\mu_n = \emptyset$, that is, it results that

$$\mathcal{R}^{\alpha}_{\boldsymbol{v}}(\boldsymbol{\tau})_N = \mathcal{R}^{\alpha}_{\boldsymbol{x}}(\boldsymbol{\tau})_N = \mathcal{C}^{\alpha}_{\boldsymbol{x}}(\boldsymbol{\tau})_N = \mathcal{C}^{\alpha}_{\boldsymbol{v}}(\boldsymbol{\tau})_N \, . \tag{9.78}$$

Finally, let us note that (9.75) and (9.78) are just equivalent to (9.41) and (9.42). Eq. (9.43) follows immediately accounting for (9.29) and (9.35).

10

Selected topics and applications

Contents

Abstract

Periodic autoregressive moving average systems, compressive sensing, random fields, and level crossing analysis are briefly reviewed and relevant references cited. Applications in several fields are summarized and relevant references classified. They include systems, circuits and control, acoustics and mechanics, econometrics, biology, climatology, hydrology, optics, and astrophysics.

Keywords

PARMA systems, Compressive sensing, Random fields, Control, Acoustics, Mechanics, Econometrics, Biology

10.1 PARMA systems

Autoregressive (AR) and autoregressive moving average (ARMA) (discrete-time) systems with periodic coefficients and/or order, are referred to as periodic AR (PAR) and periodic ARMA (PARMA) systems, respectively.

PAR and PARMA systems are characterized by input/output relationships described by difference equations with periodically time-varying coefficients and system orders (Bittanti et al., 1994), (Miamee, 1994), (Hurd and Miamee, 2007, Secs. 2.1.7, 2.1.8)

$$\sum_{k=0}^{P(n)} a_k(n)\, y(n-k) = \sum_{m=0}^{Q(n)} b_m(n)\, x(n-m) \tag{10.1}$$

where $x(n)$ and $y(n)$ are the input and output signals, respectively, and the coefficients $a_k(n)$ and $b_m(n)$ and the orders $P(n)$ and $Q(n)$ are periodic functions with the same period N_0.

Thus, periodic ARMA systems are a special case of discrete-time periodically time-varying systems. When they are excited by a stationary or cyclostationary input signal $x(n)$, they give rise to a cyclostationary output signal $y(n)$. When they are excited by an almost-cyclostationary input signal, they give rise to an almost-cyclostationary output signal. The problem of fitting an AR model to data $y(n)$ is equivalent to the problem of solving for a linear predictor for $y(n)$, where $x(n)$ is the model-fitting-error time series.

The following topics and problems have been addressed:

- *Modeling and prediction*
  (Brelsford, 1967), (Granger, 1978), (Pagano, 1978), (Troutman, 1979), (Miamee and Salehi, 1980), (Salas and Smith, 1980), (Tiao and Grupe, 1980), (Shell et al., 1981), (Sakai, 1982), (Andel, 1985b), (Andel, 1985a), (Cipra, 1985), (Thompstone et al., 1985), (Vecchia, 1985b), (Bittanti, 1986), (Obeysekera and Salas, 1986), (McLeod et al., 1987), (Bartolini et al., 1988), (Li and Hui, 1988), (Andel, 1989), (Kato and Yamaguchi, 1990), (Ula, 1990), (Bittanti et al., 1991a), (Sakai, 1991), (Bentarzi and Hallin, 1993), (Bittanti and De Nicolao, 1993), (Miamee, 1993), (Ula, 1993), (Bittanti et al., 1994), (Miamee, 1994), (Makagon et al., 1994), (Dandawaté and Giannakis, 1996), (Lund and Basawa, 1997), (Makagon and Miamee, 1997), (Bentarzi and Hallin, 1998), (Anderson et al., 1999), (Makagon, 1999b), (Hurd et al., 2002), (Shao and Lund, 2004), (Shishebor et al., 2006), (Makagon and Miamee, 2014), (Mohammadpour and Soltani, 2014), (Aknouche, 2015), and (Garcia-Vega et al., 2015);
- *PAR system analysis*
  (Andel and Rubio, 1986), (Dehay, 2015a), (Wylomanska et al., 2015);
- *PARMA system analysis*
  (Anderson et al., 2013), (Dudek et al., 2015);
- *Periodically autoregressive Hilbertian processes*
  (Ruiz-Medina, 2011), (Soltani and Hashemi, 2011a), (Soltani and Hashemi, 2011b);
- *Parameter estimation*
  (Kapustinskas, 1977), (Kapustinskas, 1978), (Newton, 1982), (Salas et al., 1982), (Vecchia, 1983), (Vecchia, 1985a), (Cipra and Tusty, 1987), (Jacob, 1990), (Sakai, 1990), (Sakai and Sakaguchi, 1990), (Anderson and Vecchia, 1993), (Bittanti et al., 1994), (Adams and Goodwin, 1995), (Boshnakov, 1995), (Bentarzi and Hallin, 1996), (Lambert-Lacroix, 2000), (Lund and Basawa, 2000), (Basawa and Lund, 2001), (Bibi and Franq, 2003);
- *Estimation of PAR time series*
  (Manouchehri and Nematollahi, 2019);
- *Modeling of atmospheric and hydrologic signals*
  (Jones and Brelsford, 1967), (Hasselmann and Barnett, 1981), (Barnett et al., 1984), and (Johnson et al., 1985);
- *Modeling helicopter noise*
  (Hardin et al., 1991);
- *Applications to econometrics*
  (Parzen and Pagano, 1979), (Osborn and Smith, 1989), (Franses and Paap, 1994), and (Franses, 1995).

On the subject of periodic AR and ARMA modeling and prediction, see also the general treatments (Gardner, 1985), (Gardner, 1987f), (Gardner, 1990b), (Gardner, 1991c).

10.2 Compressive sensing

Compressive sensing techniques have been proposed for parsimonious representation and computation of cyclic statistics, mainly for applications in spectrum sensing.

In (Lim and Wakin, 2012), (Lim and Wakin, 2015), it is observed that an ACS communication signal $x(t)$ not containing any pilot tone is not compressible in the frequency domain. In fact its Fourier spectrum does not present any significant peak. In contrast, the second- or higher-order lag product waveform

$$y(t; \boldsymbol{\tau}) \triangleq \prod_{i=1}^{k} x^{(*)_i}(t + \tau_i) \tag{10.2}$$

where $\boldsymbol{\tau} \triangleq [\tau_1, \ldots, \tau_k]$, can be decomposed into the sum of an almost-periodic component and a residual term not containing any finite-strength additive sine-wave component (Sections 2.3.1.6 and 4.2.1). The Fourier coefficients of the almost-periodic component of $y(t; \boldsymbol{\tau})$ are the cyclic temporal moment functions

$$\mathcal{R}_{\boldsymbol{x}}^{\alpha}(\boldsymbol{\tau}) = \lim_{T \to \infty} \frac{1}{T} \int_{-T/2}^{T/2} y(t; \boldsymbol{\tau})\, e^{-j2\pi\alpha t}\, \mathrm{d}t \,. \tag{10.3}$$

Under mild regularity assumptions, $\mathcal{R}_{\boldsymbol{x}}^{\alpha}(\boldsymbol{\tau})$ is coincident with probability 1 with its stochastic counterpart, that is, the Fourier coefficient of the almost-periodic function $\mathrm{E}\{y(t; \boldsymbol{\tau})\}$.

The almost-periodic component is compressible in the frequency domain. That is, nonuniform samples of the lag product $y(t; \boldsymbol{\tau})$ taken at a rate lower than the Nyquist rate allow one to reconstruct the cyclic temporal moment functions. Cyclic temporal moment functions for several orders k, several values of the cycle frequency α, several numbers of conjugated terms, and several lag vectors $\boldsymbol{\tau}$ are features that can be used to infer the modulation format of a communication signal (Section 8.12).

In (Lim and Wakin, 2015), instead of nonuniformly sampling the lag product $y(t; \boldsymbol{\tau})$, low-rate nonuniform samples of the signal $x(t)$ and its delayed versions $x(t + \tau_i)$ are taken and then a nonlinear transformation of these samples is considered. Then, the task of inferring the modulation format is realized directly in this compressed domain without requiring signal reconstruction.

The MMSE reconstruction for undersampled cyclostationary processes is addressed in (Prendergast and Nguyen, 2006). In (Bollig and Mathar, 2013), a method for estimating the cyclic autocorrelation function using a short observation interval is presented.

The problem of sub-Nyquist sampling for cyclic spectrum reconstruction has been addressed resorting to compressive sampling. The case of a cyclostationary signal with known cyclostationarity period is addressed in (Ariananda and Leus, 2014), (Leus and Tian,

2011), and (Tian et al., 2012). Specifically, in (Ariananda and Leus, 2014), nonuniform sampling is adopted for the reconstruction of the correlation matrix of a cyclostationary (not bandpass) signal, in (Leus and Tian, 2011) random linear projections are adopted for cyclic spectrum reconstruction, and in (Tian et al., 2012) a convex ℓ_1-norm minimization problem is solved for the recovery of the 2D cyclic spectrum. The case of uniform sampling is considered in (Cohen et al., 2013) and random sampling in (Zhu et al., 2013). Sub-Nyquist sampling of bandpass PAM signals is presented in (Cohen and Eldar, 2017) by exploiting the structure of the spectral correlation matrix for this specific modulation format.

The sparsity of the cyclic autocorrelation function in the cycle frequency domain is exploited in (Khalaf et al., 2011a), (Khalaf et al., 2011b) for efficient spectrum sensing techniques. The sparsity of the Loève bifrequency spectrum is exploited in (Rebeiz et al., 2012), (Tian, 2011), (Tian et al., 2012). An application to the automatic modulation classification problem is considered in (Zhou and Man, 2013). Compressed sensing for cyclic spectrum estimation is exploited in (Zhang et al., 2015).

10.3 Random fields

A periodically correlated or cyclostationary random field is a second-order random field whose mean and correlation have periodic structure (Bilyik, 1979), (Alekseev, 1991a), (Hurd, 1995), (Li et al., 1998), (Dehay and Hurd, 2002), (Hurd et al., 2004). Specifically, a random field $X(n_1, n_2)$ indexed on $\mathbb{Z}^2$ is called strongly periodically correlated with period (N_{01}, N_{02}) if and only if there exists no smaller $N_{01} > 0$ and $N_{02} > 0$ for which the mean and correlation satisfy

$$\mathrm{E}\{X(n_1 + N_{01}, n_2 + N_{02})\} = \mathrm{E}\{X(n_1, n_2)\} \tag{10.4}$$

$$\begin{aligned} &\mathrm{E}\left\{X(n_1 + N_{01}, n_2 + N_{02})\, X^*(n_1' + N_{01}, n_2' + N_{02})\right\} \\ &\quad = \mathrm{E}\left\{X(n_1, n_2)\, X^*(n_1', n_2')\right\} \end{aligned} \tag{10.5}$$

for all $n_1, n_2, n_1', n_2' \in \mathbb{Z}$.

Bi-dimensional cyclostationarity has been exploited in the following subjects:

- *Statistical test for the presence of almost cyclostationarity in images* (Vazquez-Padin et al., 2010);
- *Sea clutter texture estimation* (Gini and Greco, 2002), (Abdelaziz et al., 2013);
- *Image forensic* (Mahdian and Saic, 2009), (Mahdian et al., 2015);
- *Filtering and prediction* (Dubovetska et al., 2014), (Dubovetska et al., 2015);
- *Ocean waves modeling* (Dragan and Yavorskii, 1982a).

10.4 Level crossing

Level crossings of cyclostationary signals have been characterized in (Markelov, 1966c), (Markelov, 1966a), (Dragan, 1972b), (Bilyik, 1973), (Bilyik, 1974b), (Bilyik, 1974a), (Bilyik, 1976), (Bilyik, 1977b), (Bilyik, 1980), (Martin, 1999).

10.5 Applications to systems, circuits, and control

Applications of cyclostationary theory to systems, circuits, and control are addressed in the following fields:

- *Modeling noise in circuit theory*
  (Bruyevich, 1968), (Voloshin, 1971), (Zalud and Kulesov, 1980), (Vokurka, 1982), (Dragan and Yavorskii, 1986), (Leonov et al., 1987), (Goette and Gobet, 1989), (Okumura et al., 1993), (Roychowdhury et al., 1998), (Bonani et al., 2002), (Shiktorov et al., 2003), (Cho, 2004), (Graffeuil et al., 2010), (Ngoya, 2011), (Wu et al., 2015);
- *Periodically or almost-periodically time variant systems in system theory (Section 3.2)*
  (Renger, 1971), (Dragan and Yavorskii, 1974a), (Rootenberg and Ghozati, 1977), (Strom and Signell, 1977), (Silaiev and Yakimov, 1979), (Zaytsev and Pechinkin, 1979), (Semenov and Smirnov, 1983), (Ferrara, 1985), (Bittanti and De Nicolao, 1993), (Hurd and Jones, 1994);
- *Control theory*
  (Rozenvasser, 1973), (Bittanti and Guardabassi, 1982), (Williamson and Kadiman, 1985), (Kadiman and Williamson, 1987), (Bittanti et al., 1991b), (Williamson, 1991), (Bittanti and Colaneri, 2008);
- *Analysis of least mean-square (LMS) and normalized least mean square (NLMS) algorithms and their variants with cyclostationary inputs*
  (McLernon, 1991), (Bershad et al., 2014), (Eweda and Bershad, 2017), (Wang and Zhao, 2017), (Wang and Zhao, 2018), (Bershad et al., 2018), (Eweda et al., 2018), (Shlezinger and Todros, 2019);
- *Adaptive filters with cyclostationary inputs*
  (Bershad et al., 2016), (Yellepeddi and Preisig, 2018);
- *Tracking performance of stochastic gradient algorithms*
  (Hajivandi, 1982).

On this subject, also see (Braun and Seth, 1980), (Ferrara and Widrow, 1981), (Gardner, 1990a), (Todd, 1990), (Schell and Gardner, 1994), (Gardner and Spooner, 1994b), (Giannakis, 1995), (Flagiello et al., 2000).

Patents on cyclostationarity-based methods for system and circuit analysis and design are (Roychowdhury et al., 2000), (Gullapalli, 2003), (Roychowdhury, 2004).

10.6 Applications to acoustics and mechanics

Cyclostationarity has been exploited in acoustics and mechanics for modeling road traffic noise (Yamaguchi and Kato, 1989), (Kato and Yamaguchi, 1990), for analyzing music signals (Black and Donohue, 2000b), and for describing the vibration signals in mechanical systems (Javorskyj, 2013). In mechanical systems with moving parts, such as engines, if some parameter such as speed, temperature, and load torque can be assumed to be constant, then the dynamic physical processes generate vibrations that can be modeled as originating from periodic mechanisms such as rotation and reciprocation of gears, belts, chains, shafts, propellers, pistons, and so on (Gardner, 1985). Consequently, vibration signals exhibit periodic behavior of one type or another with periods related to the engine cycle.

Mechanical systems with moving parts generate vibro-acoustic signals that contain both an additive almost-periodic component and a zero-mean ACS component (see (1.63)), (Braun and Seth, 1980), (McCormick and Nandi, 1998), (Capdessus et al., 2000), (Antoniadis and Glossiotis, 2001), (Antoni et al., 2002a), (Raad et al., 2002), (Antoni et al., 2004a), (Antoni, 2007b), (Gardner et al., 2006, Sec. 15), (Randall, 2011). The almost-periodic component in the signal generates an almost-periodic term in the (conjugate) cyclic autocorrelation function (see (1.66)) and, hence, Dirac impulses in the (conjugate) cyclic spectrum (see (1.67)). In general, the zero-mean ACS component has a power smaller than that of the almost-periodic one. This almost periodic component must be removed before estimating the second-order cyclic statistics of the zero-mean ACS component (Section 5.2), (Antoni, 2007a), (Javorskyj et al., 2007), (Antoni, 2009), (Javorskyj et al., 2010).

Even if the almost-cyclostationary component has a power smaller than the power of the almost-periodic component, it can be useful; for example it can be successfully exploited in early diagnosis of gear faults (Koenig and Boehme, 1994), (Capdessus et al., 2000), (Dalpiaz et al., 2000), (Bouillaut and Sidahmed, 2001), (Randall et al., 2001), (Antoni et al., 2002a), (Antoni et al., 2002b), (Antoni and Randall, 2002), (Antoni and Randall, 2003), (Li and Qu, 2003), (Raad, 2003), (Zhu et al., 2005).

The zero-mean ACS component has been exploited for the following problems:

- *Early diagnostic of gear faults*
  (Bouguerriou et al., 2005), (Boustany and Antoni, 2005), (Raad et al., 2008), (Feng et al., 2011), (Kilundu et al., 2011), (Ming et al., 2011), (Borghesani et al., 2013), (Cioch et al., 2013), (Botero et al., 2014), (Chaari and Haddar, 2014), (Dabrowski et al., 2014), (Lamraoui et al., 2014), (Peng and Bo, 2014), (Vicuña and Acuña, 2014), (Wylomanska et al., 2014);
- *Acoustic blind source separation*
  (Antoni et al., 2005), (Bonnardot et al., 2005), (Bouguerriou et al., 2005), (Boustany and Antoni, 2005), (Boustany and Antoni, 2008), (Delvecchio and Antoni, 2012), (Antoni et al., 2013), (Ming et al., 2014);
- *Acoustic source location*
  (Chen et al., 2015);

- *Acoustic emission analysis*
  (Quezada Acuña and Molina Vicuña, 2015).

More general cyclostationarity-based techniques are adopted in (Urbanek et al., 2013), (Antoni et al., 2014), (Antoni, 2016), (Borghesani, 2015), (Delvecchio et al., 2015), (Léonard, 2015), (Pennacchi et al., 2015).

Further applications of cyclostationarity in acoustics are:

- *Detection of surface ships*
  (Antoni and Hanson, 2012);
- *Measurement and prediction of wind turbine swishing noise*
  (Cheong and Joseph, 2014);
- *Fault detection in helicopters*
  (Estupiñan et al., 2007), (Girondin et al., 2013);
- *Modal analysis of MIMO systems*
  (Hanson et al., 2007a), (Hanson et al., 2007b);
- *Acoustic imaging*
  (Lafon et al., 2011);
- *Acoustic holography*
  (Zhang et al., 2008);
- *Cyclic spectral analysis of rotating machines with fluctuating rotation speed*
  (Leclere and Hamzaoui, 2014);
- *Analysis of forced oscillations of cracked beam*
  (Matsko et al., 2018);
- *Analysis of ultrasonic modulation components*
  (Liu and Sohn, 2017);
- *Analysis of sound field of rotor blades*
  (Yu et al., 2019b);
- *Bearing fault detection*
  (Teng et al., 2017);
- *Rolling element bearings diagnosis*
  (Moshrefzadeh and Fasana, 2018);
- *Fault identification*
  (Buzzoni et al., 2018);
- *Local damage detection in gears/bearings*
  (Zak et al., 2019);
- *Sound source reconstruction*
  (Yu et al., 2019a);
- *Condition monitoring of rotating machinery under varying operating conditions*
  (Gryllias et al., 2017);
- *Analysis of vibration signals from centrifugal pumps*
  (Li et al., 2019);

- *Rolling element bearing pseudo-fault analysis*
  (Chi et al., 2019);
- *Fan broadband noise prediction*
  (Wohlbrandt et al., 2018);
- *Centrifugal pump fault characterization*
  (Sun et al., 2018);
- *Multiple local damage detection in gearbox*
  (Kruczek et al., 2019).

Additional applications in the field of acoustics and mechanics can be found in (Hardin and Miamee, 1990), (Bittanti et al., 1991b), (Yamaguchi and Kato, 1991), (Gerr and Allen, 1994a), (Sadler, 1995).

Patents on applications of cyclostationarity to engine diagnosis are (Dusserre-Telmon et al., 2001), (Dusserre-Telmon et al., 2002).

10.7 Applications to econometrics

In high-frequency financial time series, such as asset return, the repetitive patterns of openings and closures of markets, the number of active markets throughout the day, seasonally varying preferences, and so forth, are sources of periodic variations in financial-market volatility and other statistical parameters. Autoregressive models with periodically varying parameters provide appropriate descriptions of seasonally varying economic time series (Parzen and Pagano, 1979), (Osborn and Smith, 1989), (Todd, 1990), (Osborn, 1991), (Franses and Paap, 1994), (Boswijk and Franses, 1995), (Franses, 1995), (Bollerslev and Ghysels, 1996), (Dehay and Leśkow, 1996b), (Ghysels et al., 1996), (Franses and McAleer, 1998), (Ghysels and Osborne, 2001), (Broszkiewicz-Suwaj et al., 2004), (Li and Jayaweera, 2017). Furthermore, neglecting the periodic behavior gives rise to a loss in forecast efficiency (Gardner, 1994a), (Bollerslev and Ghysels, 1996), (Ghysels et al., 1996).

On this subject see also (Bentarzi and Hallin, 1996), (Iqelan, 2011), (Bibi and Lescheb, 2014).

10.8 Applications to biology

Applications of the concept of spectral redundancy (spectral correlation or cyclostationarity) have been proposed in medical image signal processing (Varghese and Donohue, 1994b), (Varghese and Donohue, 1995a), (Varghese et al., 1995), and nondestructive evaluation (Donohue et al., 1993a). Methods of averaging were developed for estimating the generalized spectrum that allow for the meaningful characterization of phase information when classic assumptions of stationarity do not hold. This has led to many significant performance improvements in ultrasonic tissue characterization, particularly for liver and breast tissues (Varghese and Donohue, 1994a), (Varghese and Donohue, 1995b), (Donohue et al., 1998), (Donohue et al., 1999a), (Donohue et al., 1999b), (Huang et al.,

2000), (Donohue et al., 2001), (Donohue and Huang, 2001), (Donohue et al., 2003), (Gefen et al., 2003).

Applications are in the following subjects:

- *Analysis of genome signals*
  (Arora et al., 2008);
- *Molecular dynamics modeling*
  (Ho and Hamelberg, 2016);
- *Neuroscience*
  (Aru et al., 2015);
- *Ballistocardiogram analysis*
  (Ghaderi et al., 2010);
- *Heart and lung sound separation*
  (Ghaderi et al., 2009);
- *Analysis of the embolic blood Doppler signal*
  (Girault et al., 2006);
- *Foetal PQRST extraction from electrocardiogram (ECG) recordings*
  (Haritopoulos et al., 2010);
- *Heart and respiration rates monitoring*
  (Finelli and Jenkins, 1991), (Kazemi et al., 2014);
- *Heart sound signal selection*
  (Li et al., 2013b);
- *Heart sound cancellation from lung sound*
  (Li et al., 2013c);
- *Detection and characterization of a runner's fatigue*
  (Boungou et al., 2015), (Maiz et al., 2014);
- *Analysis of electromyographic signals*
  (Roussel et al., 2013), (Roussel et al., 2017);
- *Modeling and analysis of ground reaction force signals*
  (Dudek et al., 2014c), (Sabri et al., 2010);
- *Analysis of myoelectric signals*
  (Agostini and Knaflitz, 2012);
- *Heart and respiration monitoring by Doppler radar systems*
  (Kazemi et al., 2015);
- *ECG modeling*
  (Haritopoulos et al., 2014), (Mihandoost and Amirani, 2017);
- *Cholesterol detection*
  (Savic, 1994);
- *Other applications*
  (Johnson, 1981).

10.9 Other applications

Further applications of cyclostationarity are in the following fields:

- *Linear extrapolation*
  (Dubovetska and Moklyachuk, 2013), (Dubovetska and Moklyachuk, 2014);
- *Operations management*
  (Kochel, 1982), (Saghafian et al., 2014);
- *Queuing theory in computer networks*
  (Anderson et al., 1979), (Clare and Rubin, 1983), (Kaplan, 1983), (Ackroyd, 1984), (Habibi and Lewis, 1996);
- *Queuing theory in car traffic*
  (Nakatsuka, 1986);
- *Network analysis*
  (Maciá-Fernández et al., 2018);
- *Ecosystem analysis*
  (Wouters et al., 2016);
- *Climatology*
  (Monin, 1963), (Barnett, 1983), (Dragan et al., 1984b), (Johnson et al., 1985), (Ortiz and Ruiz de Elvira, 1985), (Dragan et al., 1987), (Kim and North, 1992), (Bloomfield et al., 1994), (Lund et al., 1995), (Huang and North, 1996), (Kim and North, 1997b), (Kim and North, 1997a), (Kim and Wu, 1999), (Kim, 2000), (Kim, 2002), (Kim and Kim, 2004), (Stefanakos, 2008), (Gaucherel, 2010), (Hamlington et al., 2011), (Dowell et al., 2013), (Pierini, 2014), (Schkoda et al., 2014), (Kim et al., 2015);
- *Hydrology*
  (Bhuiya, 1971), (Fernandez and Salas, 1986), (Obeysekera and Salas, 1986), (McLeod et al., 1987), (Kacimov et al., 1999);
- *Optics and spectroscopy*
  (Davis, 2007), (Fernández-Pousa, 2013), (Schoonover et al., 2008), (Schoonover et al., 2009b), (Schoonover et al., 2009a), (Schoonover et al., 2011), (Serena and Bononi, 2015); (Ionescu et al., 2015), (Zhao et al., 2018);
- *Astrophysics*
  (Demorest, 2011), (Edlund et al., 2005), (Li and Ke, 2010), (Li et al., 2012);
- *Voice speech analysis*
  (Chorna, 2003), (Jalili et al., 2018);
- *Geophysics*
  (Javors'kyj et al., 2015);
- *Image processing*
  (Chen et al., 1996).

Miscellaneous results can be found in (Voychishin et al., 1972), (Dubois and Besserides, 1976), (Dragan et al., 1980a), (Rocha, 1980), (Gitlin and Weinstein, 1981), (Pelkowitz, 1981), (Obrezkov and Sizyakov, 1985), (Bergmans and Janssen, 1987), (Joshi and Falconer, 1988), (Joshi and Falconer, 1990), (Wilbur and Bono, 1991), (Abdulrahman and Falconer, 1992),

(Wilbur and McDonald, 1992), (Donohue et al., 1993b), (Donohue and Cheah, 1995), (Andrieu et al., 1996), (Anderson and Meerschaert, 1997), (Cui et al., 1998), (Subotic et al., 1998), (Arciplani et al., 1979), (Arciplani et al., 1980), (Black and Donohue, 2000a), (Xin and Sano, 2001), (Lee and Tung, 2002), (Tardu, 2003), (Ghozzi et al., 2006).

PART 2

Generalizations

11

Limits of the ACS model

Contents

Abstract

Limits of the applicability of the almost-cyclostationary model to the received signal in the presence of relative motion between transmitter and receiver are analyzed when the transmitted signal is almost-cyclostationary. Conditions are derived in terms of length of the observation interval, signal bandwidth, and motion parameters under which the almost-cyclostationary model or one of its generalizations can be adopted for the received signal.

Keywords

Doppler, Narrow-band condition, Generalizations of cyclostationarity, Irregular statistical cyclicity, Model limits and mismatch

11.1 Introduction

Cyclostationary signal analysis has been successfully exploited in several fields of applications (Chapters 8, 9, and 10). In communications, radar/sonar, and telemetry, cyclostationarity-based signal processing algorithms have been shown to significantly outperform classical algorithms that model signals as stationary and neglect the periodic or almost-periodic time variability in the probabilistic functions which is characteristic of cyclostationary or almost-cyclostationary processes.

In communications, radar/sonar, and telemetry applications, the effect of relative motion between transmitter, receiver, surrounding scatterers, and targets is to introduce a time-varying delay in the received signal with respect to the transmitted one. Due to this time-varying delay, the nonstationarity kind of the transmitted signal, in particular its ACS nature, changes at the receiver, unless some assumptions are made in terms of signal bandwidth, observation interval, and motion parameters (Napolitano, 2012, Sec. 7.1).

In the case of constant relative radial speed between transmitter and receiver, most of the cyclostationarity-based signal processing algorithms assume that the Doppler effect can be modeled just as a frequency shift of the carrier and the transmitted and received signals are jointly ACS. This assumption is practically verified provided that the product of signal bandwidth and observation interval is much smaller than the ratio of the medium propagation speed and the relative radial speed (Van Trees, 1971b, pp. 339–340), (Napolitano, 2012, Sec. 7.5.1). This condition, called narrow-band condition, puts a limit on the maximum signal bandwidth and/or length of the observation interval and/or on the mobility scenario. Consequently, this condition puts a limit to the best achievable performance of most of the cyclostationarity-based techniques whose performance can be improved by increasing the data-record length. Thus, the narrow-band condition puts a limit to the minimum signal-to-noise ratio (SNR) and signal-to-interference ratio (SIR) for which satisfactory performance can be obtained.

The constraint on the data-record length and, hence, on the minimum acceptable values of SNR and SIR, can be relaxed if Doppler-channel and signal models can be handled such that the narrow-band condition does not need to be satisfied. For this purpose, the new classes of the *generalized almost-cyclostationary (GACS)* processes (Izzo and Napolitano, 1998a), (Izzo and Napolitano, 2002a), (Izzo and Napolitano, 2002b), (Izzo and Napolitano, 2003), (Napolitano, 2007a), (Napolitano, 2009), (Napolitano and Tesauro, 2011), (Napolitano, 2012, Chap. 2), *spectrally correlated processes (SC)* (Napolitano, 2003), (Napolitano, 2010b), (Napolitano, 2011), (Napolitano, 2012, Chap. 4), and *oscillatory almost-cyclostationary (OACS)* processes (Napolitano, 2016a, Sec. 6), (Napolitano, 2017b), (Napolitano, 2017c) have been introduced.

Models for signals that exhibit irregular cyclicity in the statistical functions, unlike the cyclostationary and almost-cyclostationary signals, have been recently proposed in (Napolitano and Gardner, 2016), (Napolitano, 2017c), (Gardner, 2018b). These models are appropriate not only to account for the modification of regular cyclicity in the statistical functions due, for example, to relative motion between transmitter and receiver, but also to describe signals where the pace is naturally irregular as in the case of biological signals. For example, in the analysis of the electrocardiogram (ECG) signal, the ACS model turns out to be valid only if the data-record length does not exceed few seconds. In contrast, the adoption of the time-warped ACS model (a special case of OACS signal), allows one to model the ECG signal in much longer observation intervals opening new perspectives in diagnostic and biometrics.

11.2 Doppler effect on ACS signals

In this section, the Doppler effect on ACS signals is analyzed and limits of cyclostationary signal processing in terms of length of the observation interval, signal bandwidth, motion parameters, and SNR and SIR are discussed.

In the case of relative motion between transmitter (TX) and receiver (RX), and under the assumption of free-space propagation and wide-band TX and RX antennas, a band-pass transmitted signal $z_x(t)$ is received as

$$z_y(t) = A(t)\, z_x(t - D(t))\,. \tag{11.1}$$

That is, the relative motion between TX and RX produces at receiver a replica of the transmitted signal with time-varying gain $A(t)$ and time-varying delay $D(t)$. The received signal $z_y(t)$ is an amplitude-modulated and time-warped version of the transmitted signal $z_x(t)$ (Section 14.3). It results that

$$c\,D(t) = \varrho(t) \tag{11.2}$$

where c is the medium propagation speed and $\varrho(t)$ is the time-varying distance between RX at time t and TX at time $t - D(t)$ (Napolitano, 2012, Sec. 7.1.3). Note that $\varrho(t)$ should not be confused with the time-varying distance $R(t)$ between TX and RX at the same time instant t. The distance $\varrho(t)$, and not $R(t)$, is responsible of the Doppler effect (Napolitano, 2012, Sec. 7.1.3).

Under very mild assumptions, the time-varying amplitude can be approximated as a constant within observation intervals of interest in the applications (Napolitano, 2012, pp. 396, 399). The time-varying delay introduces nonstationarity to stationary signals and modifies the nonstationarity of nonstationary signals. In particular, the almost cyclostationarity exhibited by almost all modulated signals adopted in communications, radar, sonar, and telemetry, can be modified into other kinds of nonstationarity. Conditions under which the ACS model is valid for the received signal involve both signal and motion parameters.

Let us consider a transmitted band-pass signal

$$z_x(t) = \mathrm{Re}\left[x(t)\,e^{j2\pi f_c t}\right] \tag{11.3}$$

where Re[·] denotes real part, f_c is the carrier frequency, and $x(t)$ is the transmitted-signal complex envelope (Section D.1). The band-pass received signal is

$$z_y(t) = \mathrm{Re}\left[y(t)\,e^{j2\pi f_c t}\right] \tag{11.4}$$

where, according to (11.1), the received-signal complex envelope is given by

$$y(t) = A\,x(t - D(t))\,e^{-j2\pi f_c D(t)}\,. \tag{11.5}$$

Let B denote the bandwidth of $x(t)$. Thus, $1/B$ represents the width of the minimum time interval in which variations of $x(t)$ can be appreciated (Napolitano, 2012, Sec. 7.5). That is, a delay Δt can be neglected in the argument of $x(\cdot)$, namely $x(t+\Delta t) \simeq x(t)$, provided that $|\Delta t| \ll 1/B$. Specifically, we have the following results.

Theorem 11.1. Narrow-Band Condition – Deterministic Signals (Napolitano, 2012, Theorem 7.5.1). *Let $x(t)$ be a differentiable deterministic signal with Fourier transform $X(f)$ such that $X(f) = \mathcal{O}(|f|^{-\gamma})$ with $\gamma > 2$ for $|f| > B$ (that is, $|X(f)| \leqslant K|f|^{-\gamma}$ for $|f| > B$). It results*

$$x(t+\Delta t) = x(t) + \mathcal{O}(B|\Delta t|) + \mathcal{O}(B^{-\gamma+2}|\Delta t|) \tag{11.6}$$

Proof. See (Napolitano, 2012, Sec. 7.5). □

Theorem 11.2. Narrow-Band Condition – Stochastic Processes (Napolitano, 2012, Theorem 7.5.2). *Let $x(t)$ be a finite-power stochastic process with a.e. differentiable time-averaged autocorrelation function $R_{xx^*}(\tau)$ (Section A.5.1) and with power spectrum $S_{xx^*}(f)$ such that $S_{xx^*}(f) = \mathcal{O}(|f|^{-\gamma})$ with $\gamma > 2$ for $|f| > B$ (that is, $S_{xx^*}(f) \leqslant K|f|^{-\gamma}$ for $|f| > B$). For all values of Δt such that $R_{xx^*}(\tau)$ is differentiable in $(0, |\Delta t|)$, it results*

$$\left\langle \mathrm{E}\left\{|x(t+\Delta t) - x(t)|^2\right\}\right\rangle_t = \mathcal{O}(B|\Delta t|) + \mathcal{O}(B^{-\gamma+2}|\Delta t|) \tag{11.7}$$

Proof. See (Napolitano, 2012, Sec. 7.5). □

Fact 11.3. *A delay Δt can be neglected in the complex carrier with frequency f_c provided that $2\pi f_c|\Delta t| \ll 1$. That is,*

$$e^{j2\pi f_c \Delta t} \simeq 1\,. \tag{11.8}$$

□

Let

$$\varrho_2(t) \triangleq \varrho(0) + \left.\frac{\mathrm{d}\varrho}{\mathrm{d}t}\right|_{t=0} t + \frac{1}{2}\left.\frac{\mathrm{d}^2\varrho}{\mathrm{d}t^2}\right|_{t=0} t^2 \tag{11.9}$$

be the quadratic part (constant acceleration) of $\varrho(t)$.

Assumption 11.4. *The quadratic part $\varrho_2(t)$ satisfies the condition*

$$\sup_{t\in[0,T]} |\varrho(t) - \varrho_2(t)|\,\frac{1}{c} \ll \frac{1}{B}\,. \tag{11.10}$$

□

Under Assumption 11.4, we have (Theorems 11.1 and 11.2)

$$x\left(t - \frac{\varrho(t)}{c}\right) \simeq x\left(t - \frac{\varrho_2(t)}{c}\right) \qquad \forall t \in [0, T]\,. \tag{11.11}$$

Assumption 11.5. *The quadratic part $\varrho_2(t)$ satisfies the condition*

$$\sup_{t\in[0,T]} 2\pi f_c\ |\varrho(t) - \varrho_2(t)|\,\frac{1}{c} \ll 1\,. \tag{11.12}$$

□

Under Assumption 11.5, we have (Fact 11.3)

$$e^{j2\pi f_c(t-\varrho(t)/c)} \simeq e^{j2\pi f_c(t-\varrho_2(t)/c)} \qquad \forall t \in [0, T]\,. \tag{11.13}$$

Since in general $f_c \gg B$, we have that if (11.12) is satisfied, then also (11.10) holds.

Lemma 11.6. *Under Assumptions 11.4 and 11.5, the relative radial acceleration* $\mathrm{d}^2\varrho/\mathrm{d}t^2$ *can be considered constant within the observation interval* $[0, T]$ *and we have*

$$D(t) = \frac{\varrho(t)}{c} = \frac{\varrho_0}{c} + \frac{\mathrm{v}_\varrho}{c}\, t + \frac{a_\varrho}{2c}\, t^2 \tag{11.14}$$

where $\varrho_0 = \varrho(0)$, $\mathrm{v}_\varrho \triangleq (\mathrm{d}\varrho/\mathrm{d}t)|_{t=0}$ *is the relative radial speed at* $t = 0$, *and* $a_\varrho \triangleq (\mathrm{d}^2\varrho/\mathrm{d}t^2)|_{t=0}$ *is the constant relative radial acceleration.* □

In the following, conditions a)–d) on signal and motion parameters are considered. The validity of different combinations of these conditions leads to the simplifications 1)–6) for $y(t)$ in (11.5) when $t \in [0, T]$.

a. The quadratically time-varying component of $D(t)$, that is $(a_\varrho/2c)\, t^2$, can be neglected in the argument of $x(\cdot)$ provided that $|a_\varrho| T^2/(2c) \ll 1/B$, that is,

$$BT^2 \ll \frac{2c}{|a_\varrho|} \tag{11.15}$$

b. The linearly time-varying component of $D(t)$, that is $(\mathrm{v}_\varrho/c)\, t$, can be neglected in the argument of $x(\cdot)$ provided that $|\mathrm{v}_\varrho| T/c \ll 1/B$, that is,

$$BT \ll \frac{c}{|\mathrm{v}_\varrho|} \tag{11.16}$$

Condition (11.16), albeit it involves the product BT, is referred to as the "narrow-band condition" (Van Trees, 1971b, pp. 339–340), (Napolitano, 2012, Sec. 7.5.1).

c. The quadratically time-varying component of $D(t)$ can be neglected in the complex exponential provided that $2\pi f_c |a_\varrho| T^2/(2c) \ll 1$, that is,

$$f_c T^2 \ll \frac{c}{\pi\, |a_\varrho|} \tag{11.17}$$

d. The linearly time-varying component of $D(t)$ can be neglected in the complex exponential provided that $2\pi f_c |\mathrm{v}_\varrho| T/c \ll 1$, that is,

$$f_c T \ll \frac{c}{2\pi\, |\mathrm{v}_\varrho|}\, . \tag{11.18}$$

Note that in general $B \ll f_c$. Thus, c) ⇒ a) and d) ⇒ b).

The following models for the received complex envelope $y(t)$ are of interest:

1. Assumptions 11.4 and 11.5 hold. All conditions a)–d) are satisfied. Thus, the time-varying component of the delay can be neglected in both the argument of $x(\cdot)$ and the argument of the complex exponential. The channel is linear time-invariant (LTI). It introduces just a constant delay $d = \varrho_0/c$ and there is no Doppler effect. That is,

$$y(t) = a\, x(t - d) \tag{11.19}$$

where $a = Ae^{-j2\pi f_c d}$ is a complex gain.

In this case TX and RX can be considered fixed with respect to each other. Note that the opposite implication, in general, is not true. In fact, even if TX and RX are firm with respect to each other, we can have Doppler effect if they move with generic motion law along parallel lines and with constant reciprocal distance (Napolitano, 2012, Sec. 7.1.3).

2. Assumptions 11.4 and 11.5 hold. Conditions a)–c) are satisfied and condition d) is not satisfied. The linear time-variant delay is relevant only in the argument of the complex exponential. The channel is linear periodically time-variant (LPTV). The Doppler effect is modeled just as a frequency shift ν of the carrier. That is,

$$y(t) = a\,x(t-d)\,e^{j2\pi\nu t} \tag{11.20}$$

where a and d have the same expression of case 1) and $\nu = -(\mathrm{v}_\varrho/c) f_c$.
If $x(t)$ is ACS with cycle frequencies $\{\alpha_x\}$ and conjugate cycle frequencies $\{\beta_x\}$, $y(t)$ is ACS with cycle frequencies $\{\alpha_x\}$ and conjugate cycle frequencies $\{\beta_x + 2\nu\}$. Moreover, $x(t)$ and $y(t)$ are jointly ACS (Napolitano, 2012, Sec. 1.3.3.1).
Since both a) and c) are verified, $D(t)$ is modeled as a linear function. That is, TX and RX are in relative motion with constant relative radial speed v_ϱ.

3. Assumptions 11.4 and 11.5 hold. Conditions a) and c) are satisfied and condition b) is not satisfied. A fortiori, condition d) is not satisfied. The linear time-variant delay is relevant in both the arguments of $x(\cdot)$ and of the complex exponential. The channel is LTV. It belongs to the class of the FOT-deterministic linear systems (Izzo and Napolitano, 2002a), (Napolitano, 2012, Sec. 6.3.8). The Doppler effect is modeled as a time-scale factor s in the complex envelope and a frequency shift ν of the carrier. That is,

$$y(t) = a\,x(st-d)\,e^{j2\pi\nu t} \tag{11.21}$$

where $s = 1 - \mathrm{v}_\varrho/c$, a and d have the same expression of case 1) and ν of case 2).
If $x(t)$ is ACS with cycle frequencies $\{\alpha_x\}$ and conjugate cycle frequencies $\{\beta_x\}$, $y(t)$ is ACS with cycle frequencies $\{s\alpha_x\}$ and conjugate cycle frequencies $\{s\beta_x + 2\nu\}$ (Section 8.11). However, $x(t)$ and $y(t)$ are not jointly ACS but, rather, jointly SC (Sections 13.6.1 and 13.6.2) (Napolitano, 2012, Sec. 7.7.1).
As in case 2), TX and RX are in relative motion with constant relative radial speed v_ϱ. Even if the ratio $c/|\mathrm{v}_\varrho|$ is in general very high, the BT product could not satisfy condition (11.16) if the bandwidth B and/or the data-record length T is large. The narrow-band condition (11.16) is not verified in several propagation conditions in radar/sonar (Van Trees, 1971b, pp. 339–340), wide-band and ultra wide-band (UWB) communications (Jin et al., 1995), (Schlotz, 2002), space communications (Oberg, 2004), indoor applications of global navigation satellite systems (GNSSs) (Napolitano and Perna, 2014b), (Seco-Granados et al., 2012), and underwater acoustics (Munk et al., 1995).

4a. Assumptions 11.4 and 11.5 hold. Conditions a) and b) are satisfied and conditions c) and d) are not satisfied. Both linear and quadratic components of the time-varying delay must be considered in the argument of the complex exponential and can be neglected in the argument of $x(\cdot)$. The channel is LTV and the Doppler effect consists in a frequency shift ν and a chirp modulation with chirp rate $\gamma = -f_c a_Q/c$ (Napolitano, 2012, Sec. 7.5.2). That is,

$$y(t) = a\,x(t-d)\,e^{j2\pi\nu t}\,e^{j\pi\gamma t^2}\,. \tag{11.22}$$

If $x(t)$ is ACS, then $y(t)$ is GACS (Section 12.7.1) (Napolitano, 2012, Sec. 7.8.1.1). This model finds application in radar/sonar in the presence of accelerating targets (Kelly, 1961), (Kelly and Wishner, 1965).

4b. Assumptions 11.4 and 11.5 hold. Condition a) is satisfied and conditions b), c), and d) are not satisfied. Both linear and quadratic components of the time-varying delay must be considered in the argument of the complex exponential. The quadratic component of the time-varying delay can be neglected in the argument of $x(\cdot)$. The channel is LTV and we have

$$y(t) = a\,x(st-d)\,e^{j2\pi\nu t}\,e^{j\pi\gamma t^2}\,. \tag{11.23}$$

If $x(t)$ is ACS, then $y(t)$ is GACS.

5. Assumptions 11.4 and 11.5 hold. All conditions a)–d) are not satisfied. All time-varying components must be considered in the arguments of both $x(\cdot)$ and complex exponential. That is,

$$y(t) = a\,x(rt^2+st-d)\,e^{j2\pi\nu t}\,e^{j\pi\gamma t^2} \tag{11.24}$$

with $r = -a_Q/2c$. If $x(t)$ is ACS, $y(t)$ is a degenerate case of GACS signal with cyclic autocorrelation function constituted by a single Kronecker delta in the origin of the (α, τ) plane (Napolitano, 2012, Sec. 7.8.1.2).

6. Assumptions 11.4 and 11.5 do not hold. The received signal model is (11.5). If $x(t)$ is ACS, then $y(t)$ is an amplitude-modulated time-warped (AM-TW) signal (14.3)

$$y(t) = a(t)\,x(\psi(t)) \tag{11.25}$$

with $a(t) = A\,e^{-j2\pi f_c D(t)}$ and $\psi(t) = t - D(t)$. It is a special case of OACS signal (Section 14.3.1).

The results of cases 1)–6) are summarized in Table 11.1.

For multipath propagation, we have

$$z_y(t) = \sum_{k=1}^{K} A_k\, z_x(t - D_k(t)) \tag{11.26a}$$

$$= \sum_{k=1}^{K} A_k\, \mathrm{Re}\Big[x(t-D_k(t))\,e^{j2\pi f_c(t-D_k(t))}\Big]\,. \tag{11.26b}$$

In such a case, even if the relative radial speed is constant for each path of the channel, the ACS nature of the transmitted signal is lost at receiver. The output signal is SC (Section 13.6.1) (Napolitano, 2012, Secs. 4.2.4.1 and 7.7.2.1). However, if (11.16) holds for the

Table 11.1 Doppler effect on the ASC process $x(t)$. Received signal $y(t)$ in the cases 1)–6).

	1	**2**	**3**	**4a**	**4b**	**5**	**6**
Assumption 11.4 Assumption 11.5	✓	✓	✓	✓	✓	✓	
a.	✓	✓	✓	✓	✓		
b.	✓	✓		✓			
c.	✓	✓	✓				
d.	✓						
Doppler channel	**LTI**	**LPTV**	**FOT-D LTV**	**LTV**	**LTV**	**LTV**	**LTV**
$y(t)$	$y_1(t)$ ACS	$y_2(t)$ ACS	$y_3(t)$ ACS	$y_{4a}(t)$ GACS	$y_{4b}(t)$ GACS	$y_5(t)$ D-GACS	$y_6(t)$ OACS
$y(t)$ & $x(t)$	jointly ACS	jointly ACS	jointly SC	jointly GACS			jointly OACS

1. $y_1(t) = a\,x(t-d)$
2. $y_2(t) = a\,x(t-d)\,e^{j2\pi\nu t}$
3. $y_3(t) = a\,x(st-d)\,e^{j2\pi\nu t}$

4a. $y_{4a}(t) = a\,x(t-d)\,e^{j2\pi\nu t}\,e^{j\pi\gamma t^2}$

4b. $y_{4b}(t) = a\,x(st-d)\,e^{j2\pi\nu t}\,e^{j\pi\gamma t^2}$

5. $y_5(t) = a\,x(rt^2+st-d)\,e^{j2\pi\nu t}\,e^{j\pi\gamma t^2}$
6. $y_6(t) = a\,x(t-D(t))\,e^{-j2\pi f_c D(t)}$

a. $BT^2 \ll \dfrac{2c}{|a_Q|}$

b. $BT \ll \dfrac{c}{|v_Q|}$

c. $f_c T^2 \ll \dfrac{c}{\pi|a_Q|}$

d. $f_c T \ll \dfrac{c}{2\pi|v_Q|}$

radial speed of each path k, then the Doppler effect for each path can be modeled as a frequency shift of the carrier and the multipath channel is linear almost-periodically time variant (LAPTV). Thus, if $x(t)$ is ACS, then also $y(t)$ is ACS even if with cyclic features different from those of $x(t)$. Moreover, $x(t)$ and $y(t)$ are jointly ACS. Consequently, cyclostationarity-based algorithms can be suitably exploited for signal detection and estimation, noise and interference suppression, blind identification and equalization, synchronization, and signal classification purposes (Sections 9.4–9.6).

11.3 Mismatch to the ACS model

A suitable exploitation of almost-cyclostationarity properties of signals requires the adoption of large or moderately large values of the data-record T. In fact, estimators of cyclic features ((conjugate) cyclic autocorrelation functions and spectra) are consistent as $T \to$

∞ (Section 5.2) (Dandawaté and Giannakis, 1994a), (Dehay and Hurd, 1994), (Hurd and Leśkow, 1992b), (Hurd and Miamee, 2007). That is, reliable estimates can be obtained in practice provided that T is sufficiently large. In severe noise and interference environments, the value of T must be large enough to reduce the effects of the cycle leakage phenomenon in the estimates due to strong cyclic (or stationary) features of the disturbance that are close to the cyclic features of the signal of interest (Sections 5.3.4 and 9.7), (Gardner, 1987f, Chap. 13), (Napolitano, 2016b, Sec. 9).

If $x(t)$ is ACS, conditions (11.15), (11.16), and (11.17) determine an upper bound on the maximum value T such that, within the observation interval, $y(t)$ can be modeled as ACS and $x(t)$ and $y(t)$ as jointly ACS. Such an upper bound on T puts a limit to the minimum SNR and SIR for which satisfactory performance can be achieved with cyclostationarity-based signal processing algorithms.

The constraints on bandwidth, data-record length, and motion parameters imposed by conditions (11.15), (11.16), and (11.17) are widely adopted in the applications and arise from the necessity of modeling the propagation channel as LAPTV in order to preserve the joint almost cyclostationarity between transmitted and received signal. Such constraints can be relaxed by a more accurate description of the Doppler effect and the adoption of higher fidelity models for the received signal and for the joint characterization of transmitted and received signals.

The adoption of the SC, GACS, and OACS models allows the use of data-record lengths that are significantly larger than those compatible with techniques based on WSS or ACS models for the received signals. Consequently, good performance can be achieved by techniques based on the new models also when SNR and SIR are such that classical methods fail (Napolitano, 2013), (Napolitano, 2014), (Napolitano and Perna, 2014a), (Napolitano and Perna, 2014b), (Napolitano, 2017c). Moreover, in communications, even at SNR and SIR for which classical techniques have good performance, the possibility of using larger observation intervals will allow one to increase the pay-load of packets.

11.4 Irregular statistical cyclicity

Cyclostationary processes are originated by the combination of random phenomena with periodic phenomena. This combination produces processes whose statistical functions are periodic functions of time. In the case of more periodicities with incommensurate periods, one obtains statistical functions that are almost-periodic functions of time and the resulting processes are said almost cyclostationary.

In cyclostationary and almost-cyclostationary processes, the random phenomena do not alter the clocks or pacemakers of the underlying periodic phenomena. For example in an AM-DSB signal (Section 7.2), the modulation of the carrier does not modify the zero crossings of the carrier. Similarly, in a PAM signal (Section 7.3), the modulating sequence does not modify the location of each pulse in the pulse train.

More generally, the random phenomenon can interact with the periodic phenomenon in such a way that the resulting process exhibits statistical functions that are not perfectly periodic. Furthermore, the timing process itself could be a not perfectly periodic signal as in the case of biological signals. In such situations, statistical functions exhibit *irregular statistical cyclicity* unlike the case of cyclostationary and almost-cyclostationary signal that exhibit *regular statistical cyclicity* (Napolitano and Gardner, 2016), (Napolitano, 2017c), (Gardner, 2018b).

The OACS processes, and in particular the amplitude-modulated time-warped almost-cyclostationary (AM-TW ACS) signals, constitute a suitable model for processes that exhibit irregular statistical cyclicity.

12

Generalized almost-cyclostationary signals

Contents

Abstract

Generalized almost-cyclostationary signals have autocorrelation function almost periodic in time with generalized Fourier series expansion having both coefficients and frequencies depending on the lag parameter. This class of signals generalizes that of the almost-cyclostationary signals for which frequencies do not depend on the lag parameter. The second-order characterization is provided in the time domain. The spectral characterization is obtained by resorting to the mu functional. Uniform sampling and issues on discrete-time representation are treated. Mean-square consistency and asymptotic normality of the estimator of the cyclic cross-correlation function are addressed. This signal model is adopted to describe the received signal when the transmitted one is almost-cyclostationary and there is relative motion between transmitter and receiver with constant relative radial acceleration.

Keywords

Generalized almost cyclostationary, Mu functional, Cyclic autocorrelation, Lag dependent cycle frequency, Doppler, Chirp

Cyclostationary Processes and Time Series. https://doi.org/10.1016/B978-0-08-102708-0.00024-8

12.1 Introduction

Second-order ACS processes have autocorrelation function almost-periodic in time whose (generalized) Fourier series expansion has coefficients depending on the lag parameter and frequencies not depending on the lag parameter (Theorem 1.9). A generalization of this class of processes consists in considering processes for which also the frequencies depend on the lag parameter.

The generalized almost-cyclostationary (GACS) processes have multivariate statistical functions almost periodic with respect to time with (generalized) Fourier series expansion having both coefficients and frequencies depending on the lag parameters of the time-shifted versions of the process (Izzo and Napolitano, 1998a), (Izzo and Napolitano, 2002a), (Izzo and Napolitano, 2002b), (Izzo and Napolitano, 2003), (Izzo and Napolitano, 2005), (Napolitano, 2007a), (Napolitano, 2009), (Napolitano, 2012, Chap. 2), (Napolitano, 2016a, Sec. 4).

GACS processes turn out to be an appropriate model for the received signal under the narrow-band condition (11.15), (11.16) when constant relative radial acceleration exists between transmitter and receiver (in communications), between target and transmitter/receiver (in active radar/sonar), or between source and receiver (in passive radar/sonar) (Napolitano, 2012, Secs. 2.2.6.1, 7.5.2, 7.8).

The GACS processes are characterized at second order in the time domain by the cyclic autocorrelation function, as function of the two parameters cycle frequency and lag. The knowledge of this function for all values of cycle frequency and lag completely determines the almost-periodically time varying autocorrelation function.

GACS processes are not harmonizable in the sense of Loève. They are characterized in the frequency domain by the power spectral density by resorting to the μ functional, a generalized function obtained as the limit of a unit area rectangular window when the width approaches infinite and the height approaches zero.

Under mild regularity assumptions on the finite or practically finite memory of the process, the cyclic correlogram is a mean-square consistent estimator of the cyclic autocorrelation function. However, the estimation is complicated in practice by the lack of a discrete-time counterpart of the continuous-time GACS processes. In addition, uniformly sampling a continuous-time GACS process leads to a discrete-time ACS process. Starting from this discrete-time ACS process, the GACS or ACS nature of the underlying continuous-time process can only be conjectured.

The results on the probabilistic characterization of the GACS signals addressed in this chapter have a twofold interpretation:

1. In the classical stochastic approach, where signals are modeled as stochastic processes and the expectation operator is the ensemble average operator;
2. In the functional or fraction-of-time approach, where signals are modeled as single functions of time and the expectation operator is the almost-periodic component extraction operator (Definition 2.14). For GACS signals, however, unlike the ACS signals,

the almost-periodic component extraction operator extracts frequencies belonging to a set that depends on the lag parameters of the time-shifted version of the time series.

For a comprehensive treatment on the characterization of GACS processes and on the problem of their statistical function estimation, see (Napolitano, 2012, Chaps. 2, 3).

12.2 Strict-sense characterization

Definition 12.1. The real-valued signal $\{x(t),\ t \in \mathbb{R}\}$, is said to *exhibit almost-periodic structure in the strict sense* with lags $\tau_1, \ldots, \tau_{N-1}$ and parameters $\xi_1, \ldots, \xi_N$, if the Nth-order distribution function

$$\begin{aligned} & F_{x(t+\tau_1)\cdots x(t+\tau_{N-1})x(t)}(\xi_1, \ldots, \xi_{N-1}, \xi_N) \\ & \quad \triangleq P\{x(t+\tau_1) \leqslant \xi_1, \ldots, x(t+\tau_{N-1}) \leqslant \xi_{N-1}, x(t) \leqslant \xi_N\} \\ & \quad \equiv \mathrm{E}\left\{\mathbf{1}_{\{x(t+\tau_1)\leqslant\xi_1\}} \cdots \mathbf{1}_{\{x(t+\tau_{N-1})\leqslant\xi_{N-1}\}}\,\mathbf{1}_{\{x(t)\leqslant\xi_N\}}\right\} \end{aligned} \tag{12.1}$$

is uniformly almost-periodic in t (Section B.2) for the considered $\tau_1, \ldots, \tau_{N-1}$ and $\xi_1, \ldots, \xi_N$. □

If the signal is complex valued, then the $2N$th-order joint distribution of the real and imaginary parts must be considered (see (A.2b) for the case $N = 2$). In addition, almost-periodicity can also be considered in one of the generalized senses of Section B.4.

Theorem 12.2. *If the series of the Fourier coefficients*

$$F_{\boldsymbol{x}}^{\gamma}(\boldsymbol{\xi}; \boldsymbol{\tau}) \triangleq \lim_{T\to\infty} \frac{1}{T} \int_{-T/2}^{T/2} F_{x(t+\tau_1)\cdots x(t+\tau_{N-1})x(t)}(\xi_1, \ldots, \xi_{N-1}, \xi_N)\, e^{-j2\pi\gamma t}\,\mathrm{d}t \tag{12.2}$$

is absolutely convergent

$$\sum_{\gamma\in\Gamma_{\boldsymbol{\tau},\boldsymbol{\xi}}} \left|F_{\boldsymbol{x}}^{\gamma}(\boldsymbol{\xi}; \boldsymbol{\tau})\right| < \infty \tag{12.3}$$

where $\Gamma_{\boldsymbol{\tau},\boldsymbol{\xi}}$ is a countable set depending on $\boldsymbol{\tau} \triangleq [\tau_1, \ldots, \tau_{N-1}]$ and $\boldsymbol{\xi} \triangleq [\xi_1, \ldots, \xi_N]$, then the Nth-order distribution function is the sum of the uniformly convergent w.r.t. time t (generalized) Fourier series (Theorem B.2)

$$F_{x(t+\tau_1)\cdots x(t+\tau_{N-1})x(t)}(\xi_1, \ldots, \xi_{N-1}, \xi_N) = \sum_{\gamma\in\Gamma_{\boldsymbol{\tau},\boldsymbol{\xi}}} F_{\boldsymbol{x}}^{\gamma}(\boldsymbol{\xi}; \boldsymbol{\tau})\, e^{j2\pi\gamma t} \tag{12.4}$$

□

Definition 12.3. The real-valued signal $\{x(t),\ t \in \mathbb{R}\}$, is said to be *$N$th-order generalized almost-cyclostationary in the strict sense* if for every $\boldsymbol{\tau} \in \mathbb{R}^{N-1}$ the set

$$\Gamma_{\boldsymbol{\tau}} \triangleq \bigcup_{\boldsymbol{\xi}\in\mathbb{R}^N} \Gamma_{\boldsymbol{\tau},\boldsymbol{\xi}} \tag{12.5}$$

is countable. See (Izzo and Napolitano, 2005) for the definition in the functional (or fraction-of-time) approach. □

According with the results of Section 1.2.1.1, we have the following definition.

Definition 12.4. The real-valued signal $\{x(t),\ t \in \mathbb{R}\}$, is said to be *Nth-order almost-cyclostationary in the strict sense* if the set

$$\Gamma \triangleq \bigcup_{\boldsymbol{\tau} \in \mathbb{R}^{N-1}} \Gamma_{\boldsymbol{\tau}} \tag{12.6}$$

is countable. □

Theorem 12.5. (Napolitano, 2012, Theorems 2.2.4 and 2.2.5). *For a Nth-order GACS signal in the strict sense, the Nth-order distribution function and the Nth-order moment are almost-periodic functions of time that can be expressed as the sum of (generalized) Fourier series with both coefficient and frequencies depending on the lag parameters.* □

12.3 Second-order characterization

12.3.1 Time domain

Definition 12.6. A finite-power complex-valued continuous-time stochastic process $x(t)$ is said *second-order GACS in the wide sense* if its mean value is an almost-periodic function and its autocorrelation and conjugate autocorrelation functions

$$\mathcal{R}_{xx^{(*)}}(t, \tau) \triangleq \mathrm{E}\left\{x(t+\tau)\, x^{(*)}(t)\right\} \tag{12.7}$$

for each $\tau \in \mathbb{R}$ are almost-periodic in t in the sense of Bohr (Bohr, 1933, par. 84–92) or, equivalently, uniformly almost periodic in sense of Besicovitch (Besicovitch, 1932, Chap. 1), (Corduneanu, 1989). □

Theorem 12.7. *Let $x(t)$ be a second-order GACS process in the wide-sense and let us consider the Fourier coefficients*

$$R_{xx^{(*)}}(\alpha, \tau) \triangleq \lim_{T \to \infty} \frac{1}{T} \int_{t_0 - T/2}^{t_0 + T/2} \mathcal{R}_{xx^{(*)}}(t, \tau)\, e^{-j2\pi\alpha t}\, \mathrm{d}t \tag{12.8}$$

where the limit exists and is independent of t_0. If the series of the Fourier coefficients is absolutely convergent for each τ

$$\sum_{\alpha \in A_\tau} \left| R_{xx^{(*)}}(\alpha, \tau) \right| < \infty \tag{12.9}$$

then the (conjugate) autocorrelation is the sum of the uniformly convergent (generalized) Fourier series

$$\mathcal{R}_{xx^{(*)}}(t, \tau) = \sum_{\alpha \in A_\tau} R_{xx^{(*)}}(\alpha, \tau)\, e^{j2\pi\alpha t}\ . \tag{12.10}$$

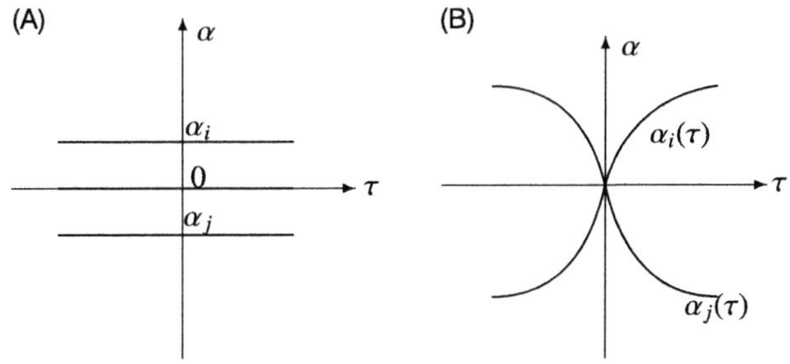

FIGURE 12.1 Support in the (α, τ)-plane of the cyclic autocorrelation function $R_{xx^*}(\alpha, \tau)$ of (A) an ACS process and (B) a GACS (not ACS) process.

In (12.8)–(12.10), the real numbers α *and the complex-valued functions* $R_{xx^{(*)}}(\alpha, \tau)$, *are referred to as* (conjugate) cycle frequencies *and* (conjugate) cyclic autocorrelation functions. *The set*

$$A_\tau \triangleq \left\{\alpha \in \mathbb{R} \,:\, R_{xx^{(*)}}(\alpha, \tau) \neq 0\right\} \tag{12.11}$$

is a countable set of possibly incommensurate (conjugate) cycle frequencies which, in general, depends on $(*)$ *and* τ. □

Note that, even if the set A_τ is countable, the set

$$A \triangleq \bigcup_{\tau \in \mathbb{R}} A_\tau \tag{12.12}$$

is not necessarily countable. Thus, the class of the second-order wide-sense GACS processes extends that of the wide-sense ACS which are obtained as a special case of GACS processes when the set A is countable (Theorem 1.11) (Dehay and Hurd, 1994), (Hanin and Schreiber, 1998), (Hurd, 1991).

Fact 12.8. *From (12.11) it immediately follows that the set* A_τ *can be expressed as*

$$A_\tau = \bigcup_{k \in \mathbb{I}} \left\{\alpha \in \mathbb{R} \,:\, \alpha = \alpha^{(k)}_{xx^{(*)}}(\tau)\right\} \equiv \left\{\alpha^{(k)}_{xx^{(*)}}(\tau)\right\}_{k \in \mathbb{I}} \tag{12.13}$$

where $\mathbb{I}$ *is a countable set and the functions* $\alpha^{(k)}_{xx^{(*)}}(\tau)$, *referred to as* lag-dependent (conjugate) cycle frequencies *are such that, for each* α *and* τ, *there exists at most one* $k \in \mathbb{I}$ *such that* $\alpha = \alpha^{(k)}_{xx^{(*)}}(\tau)$. □

Since the set A_τ is countable for each τ, the support in the (α, τ)-plane of the (conjugate) cyclic autocorrelation function $R_{xx^{(*)}}(\alpha, \tau)$ is constituted by the closure of the set of curves defined by the explicit equations $\alpha = \alpha^{(k)}_{xx^{(*)}}(\tau)$, $k \in \mathbb{I}$ (Fig. 12.1)

$$\begin{aligned} \operatorname{supp}\left\{R_{xx^{(*)}}(\alpha, \tau)\right\} &\triangleq \operatorname{cl}\left\{(\alpha, \tau) \in A_\tau \times \mathbb{R} \,:\, R_{xx^{(*)}}(\alpha, \tau) \neq 0\right\} \\ &= \operatorname{cl} \bigcup_{k \in \mathbb{I}} \left\{(\alpha, \tau) \in \mathbb{R} \times \mathcal{T}^{(k)} \,:\, \alpha = \alpha^{(k)}_{xx^{(*)}}(\tau)\right\} \end{aligned} \tag{12.14}$$

where

$$\mathcal{T}^{(k)} \triangleq \left\{\tau \in \mathbb{R} : \alpha^{(k)}_{xx^{(*)}}(\tau) \text{ is defined}\right\} . \tag{12.15}$$

The closure of a countable set of curves can be the whole plane $\mathbb{R}^2$ if infinities of clusters of curves exist. In such a case, the more appropriate concept of "being concentrated on" (Dehay, 1994) can be used to describe the region of the (α, τ)-plane where $R_{xx^{(*)}}(\alpha, \tau) \neq 0$.

A very mild regularity assumption is that the functions $\tau \mapsto \alpha^{(k)}_{xx^{(*)}}(\tau)$ are continuous.

Starting from (12.13), the following result (Izzo and Napolitano, 1998a), (Izzo and Napolitano, 2002a), (Napolitano, 2012, Sec. 2.2.2) provides an alternative representation for the (conjugate) autocorrelation function of GACS processes.

Theorem 12.9. (Napolitano, 2012, Theorem 2.2.7). *Let $x(t)$ be a second-order GACS process in the wide sense and let us consider the Fourier coefficients, referred to as* generalized (conjugate) cyclic autocorrelation functions,

$$R^{(k)}_{xx^{(*)}}(\tau) \triangleq \begin{cases} \displaystyle\lim_{T\to\infty} \frac{1}{T} \int_{t_0-T/2}^{t_0+T/2} \mathcal{R}_{xx^{(*)}}(t, \tau)\, e^{-j2\pi\alpha^{(k)}_{xx^{(*)}}(\tau)t}\, \mathrm{d}t & \tau \in \mathcal{T}^{(k)} \\ 0 & \tau \in \mathbb{R} - \mathcal{T}^{(k)} \end{cases} \tag{12.16}$$

where the limit exists and is independent of t_0. If (12.9) holds, then we also have that

$$\sum_{k\in\mathbb{I}} \left|R^{(k)}_{xx^{(*)}}(\tau)\right| < \infty \tag{12.17}$$

and the (conjugate) autocorrelation function can be expressed by the uniformly convergent w.r.t. time t (generalized) Fourier series

$$\mathcal{R}_{xx^{(*)}}(t, \tau) = \sum_{k\in\mathbb{I}} R^{(k)}_{xx^{(*)}}(\tau)\, e^{j2\pi\alpha^{(k)}_{xx^{(*)}}(\tau)t} . \tag{12.18}$$

□

In the following, in order to simplify notation, we use $\boldsymbol{x} \equiv xx^{(*)}$ in subscripts, that is,

$$R_{\boldsymbol{x}}(\alpha, \tau) \equiv R_{xx^{(*)}}(\alpha, \tau) \qquad \alpha^{(k)}_{\boldsymbol{x}}(\tau) \equiv \alpha^{(k)}_{xx^{(*)}}(\tau) \qquad R^{(k)}_{\boldsymbol{x}}(\tau) \equiv R^{(k)}_{xx^{(*)}}(\tau) .$$

If the generalized (conjugate) cyclic autocorrelation functions $R^{(k)}_{\boldsymbol{x}}(\tau)$ are defined by continuity for those τ where two or more support curves $\alpha = \alpha_k(\tau)$ intercept each other then we have (Napolitano, 2012, Fact 2.2.8)

$$R_{\boldsymbol{x}}(\alpha, \tau) = \sum_{k\in\mathbb{I}} R^{(k)}_{\boldsymbol{x}}(\tau)\, \delta_{\alpha-\alpha^{(k)}_{\boldsymbol{x}}(\tau)} \tag{12.19}$$

where δ_γ denotes Kronecker delta, that is, $\delta_\gamma = 1$ for $\gamma = 0$ and $\delta_\gamma = 0$ for $\gamma \neq 0$. Furthermore, for all values of τ such that two different support curves $\alpha = \alpha^{(k)}_{\boldsymbol{x}}(\tau)$ and $\alpha = \alpha^{(k')}_{\boldsymbol{x}}(\tau)$, $k \neq k'$, do not intersect, it results that (Izzo and Napolitano, 2002a), (Napolitano, 2007a)

$$R^{(k)}_{\boldsymbol{x}}(\tau) = R_{\boldsymbol{x}}(\alpha, \tau)\big|_{\alpha=\alpha^{(k)}_{\boldsymbol{x}}(\tau)} . \tag{12.20}$$

Note that a different notation is used for the (conjugate) cyclic autocorrelation of ACS and GACS processes (compare (1.12) and (12.8)). In fact, for ACS processes, the values of (conjugate) cycle frequencies such that the (conjugate) cyclic autocorrelation function is nonzero range in a countable set, while for GACS processes these values range in an uncountable set.

Fact 12.10. *The wide-sense ACS processes are obtained as a special case of GACS processes when the lag-dependent (conjugate) cycle frequencies are constant with respect to τ and, hence, are coincident with the cycle frequencies (Izzo and Napolitano, 1998a). That is*

$$\alpha^{(k)}_{xx^{(*)}}(\tau) = \alpha_k \qquad k \in \mathbb{I}\,. \tag{12.21}$$

We also have

$$A = \{\alpha_k\}_{k\in\mathbb{I}} \tag{12.22}$$

$$\mathcal{R}_{xx^{(*)}}(t,\tau) = \sum_{k\in\mathbb{I}} R^{\alpha_k}_{xx^{(*)}}(\tau)\, e^{j2\pi\alpha_k t} \tag{12.23}$$

$$R_{xx^{(*)}}(\alpha,\tau) = \begin{cases} R^{\alpha_k}_{xx^{(*)}}(\tau) & \alpha = \alpha_k \in A \\ 0 & \textit{otherwise} \end{cases} \tag{12.24}$$

Moreover, for ACS processes only one term is present in the sum in (12.19) and, consequently, the generalized (conjugate) cyclic autocorrelation functions are coincident with the (conjugate) cyclic autocorrelation functions. Furthermore, the (conjugate) autocorrelation function $\mathcal{R}_{xx^{(*)}}(t,\tau)$ depends uniformly on the parameter τ (Corduneanu, 1989, Chap. 2) and is uniformly continuous and the (conjugate) cyclic autocorrelation functions $R_{xx^{(*)}}(\alpha,\tau)$ are continuous in τ for each $\alpha \in A$ (which is countable) (Hurd, 1991), (Dehay and Hurd, 1994). □

Example 12.11. As an example of GACS process, consider the chirp-modulated signal

$$x(t) = s(t)\, e^{j2\pi\nu t}\, e^{j\pi\gamma t^2} \tag{12.25}$$

where $s(t)$ is a (complex envelope) PAM signal with raised-cosine pulse and binary white modulating sequence. The signal $s(t)$ is cyclostationary with cycle frequencies $\alpha_k = k/T_p$, $k \in \{0, \pm 1\}$, where T_p is the symbol period (Section 7.3). The signal $x(t)$ models the output of the Doppler channel existing between transmitter and receiver with constant relative radial acceleration (Sections 11.2 and 12.7.1).

The lag-dependent cycle frequencies of $x(t)$ are linear functions of τ and the generalized cyclic autocorrelation functions have the same magnitude of the cyclic autocorrelation functions of $s(t)$. Specifically, by computing the autocorrelation of $x(t)$ and accounting for (12.19) and (12.20) we have (Napolitano, 2007a), (Napolitano, 2012, Sec. 7.8)

$$\alpha^{(k)}_{xx^*}(\tau) = k/T_p + \gamma\tau \qquad R^{(k)}_{xx^*}(\tau) = R^{k/T_p}_{ss^*}(\tau)\, e^{j2\pi\nu\tau}\, e^{j\pi\gamma\tau^2} \qquad k \in \{0, \pm 1\}\,. \tag{12.26}$$

In Fig. 12.2, the magnitude of the cyclic autocorrelation function, as a function of α and τ, for both $s(t)$ and $x(t)$ is reported. The support of the cyclic autocorrelation function

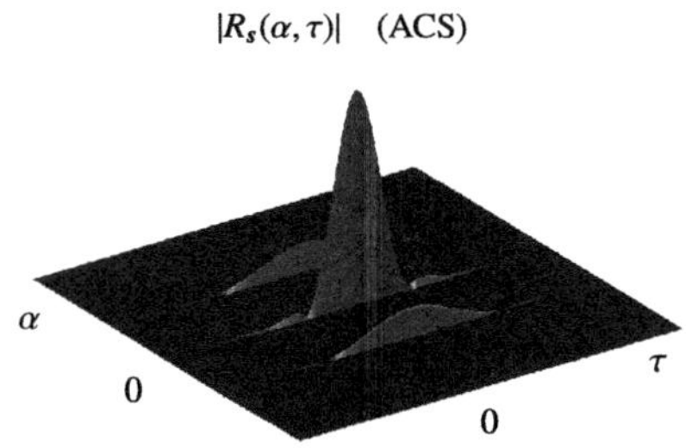

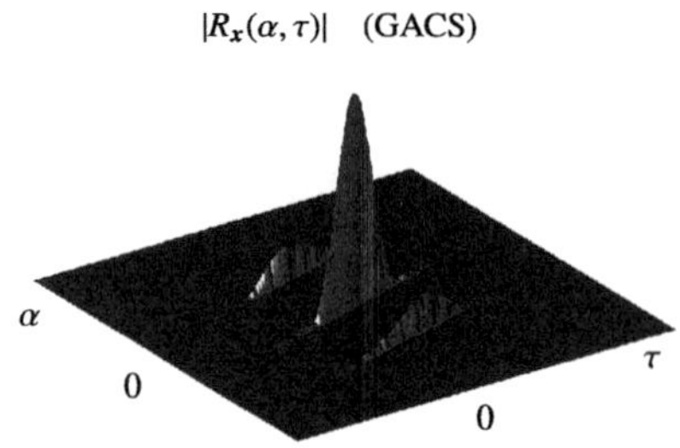

FIGURE 12.2 ACS input and GACS output of the Doppler channel existing between transmitter and receiver with constant relative radial acceleration. Magnitude of the cyclic autocorrelation function (top) of the input ACS signal $s(t)$ and (bottom) of the output GACS signal $x(t)$ as functions of α and τ.

of the ACS signal $s(t)$ is constituted by three lines parallel to the τ axis (Fig. 12.2 (top)). In contrast, $x(t)$ is GACS with cyclic autocorrelation function whose support are three parallel lines with slope γ (Fig. 12.2 (bottom)). □

More general definition of GACS processes are given in (Napolitano, 2012, Sec. 2.2.2) by considering the almost-periodicity property in a generalized sense (Section B.4), (Besicovitch, 1932, Chap. 2), (Napolitano, 2012, Sec. 1.2), (Napolitano and Tesauro, 2011). In these cases, the function $\mathcal{R}_x(t,\tau)$ is constituted by a uniformly almost-periodic function plus a residual term not containing any additive finite-strength sine-wave component. Thus, (12.10) and (12.18) generalize to

$$\mathcal{R}_x(t,\tau) = \sum_{\alpha \in A_\tau} R_x(\alpha,\tau)\, e^{j2\pi\alpha t} + r_x(t,\tau) \tag{12.27a}$$

$$= \sum_{k \in \mathbb{I}} R_x^{(k)}(\tau)\, e^{j2\pi\alpha_x^{(k)}(\tau)t} + r_x(t,\tau) \tag{12.27b}$$

with

$$\lim_{T\to\infty} \frac{1}{T} \int_{-T/2}^{T/2} r_x(t,\tau)\, e^{-j2\pi\alpha t} \mathrm{d}t = 0 \quad \forall \alpha \in \mathbb{R}\,. \tag{12.28}$$

An example of GACS process with autocorrelation function (12.27a), (12.27b) is provided in Section 12.7.1.

12.3.2 Frequency domain

The characterization of GACS processes in the frequency domain is a challenging problem. From the lack of continuity w.r.t. τ of the cyclic autocorrelation function for at least one value of α (see (12.19)) it can be proved that the time-averaged autocorrelation function $R_{xx^*}(0,\tau)$ is discontinuous in $\tau = 0$ (Napolitano, 2012, p. 57). In fact, for every finite power process $x(t)$, $R_{xx^*}(0,\tau)$ is continuous in $\tau = 0$ (and hence for every τ) if and only if the cross-correlation function $R_{xy^*}(0,\tau)$ (defined by replacing x^* with y^* in the definition of $R_{xx^*}(0,\tau)$) is continuous for any $\tau \in \mathbb{R}$ and for any finite-power process $y(t)$ (Lee, 1967, pp. 74–78). By taking $y_\alpha(t) = x(t)\, e^{j2\pi\alpha t}$ one obtains $R_{xy_\alpha^*}(0,\tau) = R_{xx^*}(\alpha,\tau)$ which is not continuous in τ due to the presence of the Kronecker deltas (see (12.19)). Thus, in order to respect the symmetry property $R_{xx^*}(0,\tau) = R_{xx^*}(0,-\tau)^*$ for every $\tau > 0$, $R_{xx^*}(0,\tau)$ must admit the decomposition

$$R_{xx^*}(0,\tau) = \overline{x_0^2}\, \delta_\tau + b(\tau) \tag{12.29}$$

with δ_τ Kronecker delta and $b(\tau)$ definite nonnegative, continuous in $\tau = 0$, possibly discontinuous in some $\tau \neq 0$, and such that $b(\tau) = b^*(-\tau)$ (Napolitano, 2012, Sec. 2.2.3). Every positive definite function can be decomposed into the sum of a continuous function $b_0(\tau)$ and a function zero almost everywhere (Krein, 1978). Thus, the positive definite function $R_{xx^*}(0,\tau)$ admits such a decomposition. Consequently, the possible discontinuities of $b(\tau)$ in $\tau \neq 0$ are Kronecker deltas. That is,

$$b(\tau) = b_0(\tau) + \sum_\ell \beta_\ell\, \delta_{\tau-\tau_\ell} \qquad \tau_\ell \neq 0\,. \tag{12.30}$$

From the discontinuous behavior (12.29) it follows that GACS processes are not second-order continuous (Napolitano, 2012, Theorem 2.2.13) and, hence, are not harmonizable in the sense of Loève (Loève, 1963, p. 474). That is, the autocorrelation function (12.10), (12.18) cannot be expressed by a Fourier-Stieljes integral (A.7) with spectral correlation function $\gamma_{\boldsymbol{x}}(f_1, f_2)$ of bounded variation.

A spectral characterization of GACS processes can be obtained by exploiting the μ *functional*. It is defined through the following relation on the test function $\Phi(f)$

$$\mu : \Phi(f) \mapsto \lim_{W\to\infty} \frac{1}{W} \int_{-W/2}^{W/2} \Phi(f)\, \mathrm{d}f \triangleq \int_{\mathbb{R}} \Phi(f)\mu(f)\mathrm{d}f \tag{12.31}$$

and can be formally expressed as

$$\mu(f) \triangleq \lim_{W\to\infty} \frac{1}{W} \operatorname{rect}\left(\frac{f}{W}\right). \tag{12.32}$$

That is, the μ functional provides the continuous component (i.e., the average over the whole real line) of the test function $\Phi(f)$. Formally, $\mu(f)$ is obtained as the limit of a very tiny and large rectangular window with unit area (Fig. 12.3 (right)) and can be managed

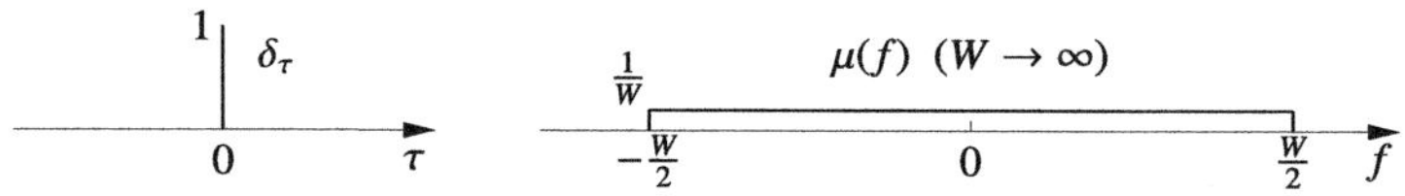

FIGURE 12.3 Formal Fourier transform pair $\delta_\tau \leftrightarrow \mu(f)$.

similarly to the formal manipulations of the Dirac delta (Napolitano, 2012, Sec. 3.2.1). In particular, we have

$$\int_{\mathbb{R}} \mu(f)\,\mathrm{d}f = 1\,. \tag{12.33}$$

Accounting for the Fourier transform relationship

$$\operatorname{sinc}(W\tau) = \frac{\sin(\pi W\tau)}{\pi W\tau} \stackrel{\mathcal{F}}{\longleftrightarrow} \frac{1}{W}\operatorname{rect}\Big(\frac{f}{W}\Big) \tag{12.34}$$

in the limit as $W \to \infty$ one obtains the formal Fourier transform (Napolitano, 2012, Sec. 3.2.1) (Fig. 12.3)

$$\delta_\tau = \lim_{W\to\infty} \operatorname{sinc}(W\tau) \stackrel{\mathcal{F}}{\longleftrightarrow} \mu(f) = \lim_{W\to\infty} \frac{1}{W}\operatorname{rect}\Big(\frac{f}{W}\Big)\,. \tag{12.35}$$

The formal Fourier transform pair in (12.35) is meaningless in the sense of ordinary Fourier transforms in $L^p(\mathbb{R})$, $p \geqslant 1$ (Champeney, 1990, Chap. 7.8). In fact, δ_τ is a function zero almost everywhere and, hence, its ordinary Fourier transform is zero. In addition, (12.35) is meaningless in the sense of Schwartz or tempered distributions (Champeney, 1990, Chap. 12–14), (Zemanian, 1987, Chap. 1, 4). In fact, the test function $\Phi(f)$ in (12.31) is not with compact support or of rapid descent. Test functions for the μ functional are persistent functions (Napolitano, 2012, Sec. 3.2.1). A sufficient condition for the existence of the limit in (12.31) is that $\Phi(f)$ is bounded and relatively measurable (Leśkow and Napolitano, 2006), (Napolitano, 2012, Sec. 6.2.2). By using (12.35), the Fourier transform pair for a translated Kronecker delta is formally given by

$$\delta_{\tau-\tau_\ell} \stackrel{\mathcal{F}}{\longleftrightarrow} \mu(f)\, e^{-j2\pi f\tau_\ell}\,. \tag{12.36}$$

By replacing (12.30) into (12.29) one has the following expression for the time-averaged autocorrelation function

$$R_{xx^*}(0,\tau) = \overline{x_0^2}\,\delta_\tau + b_0(\tau) + \sum_\ell \beta_\ell\, \delta_{\tau-\tau_\ell} \tag{12.37}$$

By Fourier transforming both sides of (12.37) and using (12.36), the power spectrum of a GACS process can be formally expressed as

$$S_{xx^*}(f) = \overline{x_0^2}\,\mu(f) + B_0(f) + \sum_\ell \beta_\ell\, \mu(f)\, e^{-j2\pi f\tau_\ell} \tag{12.38}$$

where $B_0(f)$ is the ordinary Fourier transform of $b_0(\tau)$.

The terms $\beta_\ell\,\mu(f)\,e^{-j2\pi f\tau_\ell}$ with $\tau_\ell \neq 0$ do not give contribution to the GACS signal power. In fact, for $\tau_\ell \neq 0$ we have

$$\int_{\mathbb{R}} \beta_\ell\,\mu(f)\,e^{-j2\pi f\tau_\ell}\,\mathrm{d}f = \beta_\ell \lim_{F\to\infty}\frac{1}{F}\int_{-F/2}^{F/2} e^{-j2\pi f\tau_\ell}\,\mathrm{d}f = 0\,. \tag{12.39}$$

Therefore, the signal power is given by

$$P_x \triangleq \int_{\mathbb{R}} S_{xx^*}(f)\,\mathrm{d}f = \overline{x_0^2}\int_{\mathbb{R}}\mu(f)\,\mathrm{d}f + \int_{\mathbb{R}} B_0(f)\,\mathrm{d}f = \overline{x_0^2} + b_0(0)\,. \tag{12.40}$$

The term $\overline{x_0^2}\,\mu(f)$ in the power spectrum describes a quota $\overline{x_0^2}$ of the total power $P_x = R_{xx^*}(0,0) = \overline{x_0^2} + b_0(0)$ which is uniformly spread over an infinite bandwidth. That is, a GACS process is not strictly band limited or essentially band limited. The power contribution $\overline{x_0^2}$ is canceled by low-pass or band-pass filtering with finite or essentially finite bandwidth (Izzo and Napolitano, 2002b), (Napolitano, 2012, Sec. 2.7.7).

Note that the presence of the term $\overline{x_0^2}\,\mu(f)$ in the power spectrum (12.38) corresponding to the discontinuous autocorrelation (12.29) agrees with the Wiener's observation in (Wiener, 1949, pp. 39–40) that if the time-averaged autocorrelation function is discontinuous in the origin, then "there is a portion of the energy which does not belong to any finite frequencies, and which in a certain sense we must associate with infinite frequencies."

12.4 Discrete-time processes

Continuous-time GACS processes, unlike WSS and ACS processes, do not have a discrete-time counterpart (Izzo and Napolitano, 2003), (Napolitano, 2009), (Napolitano, 2012, Sec. 2.5). Specifically, any discrete-time (conjugate) autocorrelation function $\mathrm{E}\{x_\mathrm{d}(n+m)x_\mathrm{d}^{(*)}(n)\}$ which is almost-periodic in n can be expressed by a (generalized) Fourier series where neither frequencies nor their set depend on m. In fact, let us consider the discrete-time counterpart of (12.10)

$$\mathrm{E}\Big\{x_\mathrm{d}(n+m)x_\mathrm{d}^{(*)}(n)\Big\} = \sum_{\widetilde{\alpha}\in\widetilde{A}_m} R_{\boldsymbol{x}_d}^{\widetilde{\alpha}}(m)\,e^{j2\pi\widetilde{\alpha}n} \tag{12.41}$$

where, for each $m \in \mathbb{Z}$, the (conjugate) cyclic autocorrelation function

$$R_{\boldsymbol{x}_d}^{\widetilde{\alpha}}(m) \triangleq \lim_{N\to\infty}\frac{1}{2N+1}\mathrm{E}\Big\{x_\mathrm{d}(n+m)x_\mathrm{d}^{(*)}(n)\Big\}\,e^{-j2\pi\widetilde{\alpha}n} \tag{12.42}$$

is nonzero for $\widetilde{\alpha}$ ranging in the countable set $\widetilde{A}_m$ depending on m. Unlike the case of the continuous-time processes where the set A defined in (12.12) is uncountable for GACS processes that do not reduce to ACS processes, the set

$$\widetilde{A} \triangleq \bigcup_{m\in\mathbb{Z}} \widetilde{A}_m \tag{12.43}$$

is always countable. Consequently, the sum in (12.41) can be equivalently extended over $\widetilde{A}$ since $R^{\widetilde{\alpha}}_{\boldsymbol{x}_{\rm d}}(m) = 0$ for $\widetilde{\alpha} \notin \widetilde{A}_m$. Therefore, all discrete-time processes with almost-periodic (conjugate) autocorrelation are ACS and discrete-time GACS processes do not exist.

Uniformly sampling a continuous-time GACS process leads to a discrete-time ACS process (Izzo and Napolitano, 2003), (Napolitano, 2012, Sec. 2.5). Moreover, starting from a discrete-time ACS process, the possible GACS or ACS nature of the underlying continuous-time process can only be conjectured (Napolitano, 2012, Sec. 2.7.6).

The conjecture can be made accounting for the following considerations. Let $x_{\rm d}(n) = x(t)|_{t=nT_s}$ be the discrete-time ACS process obtained by uniformly sampling with period $T_s = 1/f_s$ the continuous-time GACS process $x(t)$ with (conjugate) cyclic autocorrelation function having support curves $\alpha = \alpha^{(k)}_{\boldsymbol{x}}(\tau)$, $k \in \mathbb{I}$, in the (α, τ) plane. The (conjugate) cyclic autocorrelation function of the discrete-time ACS signal $x_{\rm d}(n)$ can be expressed in terms of the (conjugate) cyclic autocorrelation functions of the continuous-time GACS signal $x(t)$ by the same aliasing formula (3.191) valid if $x(t)$ is ACS (Napolitano, 2012, Theorem 2.5.3). By substituting (12.19) into (3.191), it results in

$$\widetilde{R}^{\widetilde{\alpha}}_{\boldsymbol{x}_{\rm d}}(m) = \sum_{k \in \mathbb{I}} R^{(k)}_{\boldsymbol{x}}(mT_s)\, \delta_{[\widetilde{\alpha} - \alpha^{(k)}_{\boldsymbol{x}}(mT_s)T_s] \bmod 1} \tag{12.44}$$

where mod 1 denotes modulo 1 operation with values in $[-1/2, 1/2)$. From (12.44) it follows that the support of the (conjugate) cyclic autocorrelation function of $x_{\rm d}(n)$ (in the main domain $(\widetilde{\alpha}, m) \in [-1/2, 1/2) \times \mathbb{Z}$) is contained in the countable set of points

$$\widetilde{S} \triangleq \bigcup_{m \in \mathbb{Z}} \bigcup_{k \in \mathbb{I}} \left\{ \widetilde{\alpha} \in [-1/2, 1/2) \, : \, \widetilde{\alpha} = \alpha^{(k)}_{\boldsymbol{x}}(mT_s) T_s \mod 1 \right\} \tag{12.45}$$

Let the (conjugate) cyclic correlogram be estimated with cycle frequency step $\Delta\widetilde{\alpha} = 1/(NP)$, where N is the number of points for the discrete Fourier transform (DFT) and P is the zero-padding factor. Let us assume that the cycle frequency variations within a sampling period are smaller than twice the (normalized) cycle frequency step, that is, let us assume that $\forall k \in \mathbb{I}$ it results

$$\left| \alpha^{(k)}_{\boldsymbol{x}}(mT_s) - \alpha^{(k)}_{\boldsymbol{x}}((m+1)T_s) \right| \leqslant 2\Delta\widetilde{\alpha}/T_s \tag{12.46}$$

for any m such that $R^{(k)}_{\boldsymbol{x}}(mT_s)$ and $R^{(k)}_{\boldsymbol{x}}((m+1)T_s)$ are significantly nonzero. Then, due to the interpolation made by the graphic software, the support will appear to be continuous in correspondence of the continuous tracts of the functions $\alpha^{(k)}_{\boldsymbol{x}}(\tau)$. In the case of piecewise constant functions $\alpha^{(k)}_{\boldsymbol{x}}(mT_s)$, the continuous-time signal should be conjectured to be ACS. In contrast, when the $\alpha^{(k)}_{\boldsymbol{x}}(mT_s)$ are not piecewise constant, the continuous-time GACS signal should be conjectured to be not ACS (Napolitano, 2012, Sec. 2.7.6).

12.5 Jointly GACS processes

Two finite-power complex-valued processes $x(t)$ and $y(t)$ are said *jointly GACS* in the wide sense if their *(conjugate) cross-correlation* function can be expressed in one of the two equivalent forms

$$\begin{aligned}\mathcal{R}_{yx^{(*)}}(t,\tau) &\triangleq \mathrm{E}\left\{y(t+\tau)\,x^{(*)}(t)\right\} \\ &= \sum_{\alpha\in A_\tau} R_{yx^{(*)}}(\alpha,\tau)\,e^{j2\pi\alpha t} &\text{(12.47a)}\\ &= \sum_{k\in\mathbb{I}} R^{(k)}_{yx^{(*)}}(\tau)\,e^{j2\pi\alpha^{(k)}_{yx^{(*)}}(\tau)t}\,. &\text{(12.47b)}\end{aligned}$$

In the Fourier series (12.47a), A_τ is a countable set (depending on $(*)$) of possibly incommensurate (conjugate) cycle frequencies α. In the Fourier series (12.47b), $\mathbb{I}$ is a countable set and the frequencies $\alpha^{(k)}_{yx^{(*)}}(\tau)$ are referred to as *lag-dependent (conjugate) cycle frequencies.*

The *(conjugate) cyclic cross-correlation* function

$$R_{yx^{(*)}}(\alpha,\tau) \triangleq \lim_{T\to\infty}\frac{1}{T}\int_{t_0-T/2}^{t_0+T/2} \mathcal{R}_{yx^{(*)}}(t,\tau)\,e^{-j2\pi\alpha t}\,\mathrm{d}t \tag{12.48}$$

$$= \sum_{k\in\mathbb{I}} R^{(k)}_{yx^{(*)}}(\tau)\,\delta_{\alpha-\alpha^{(k)}_{yx^{(*)}}(\tau)}\,, \tag{12.49}$$

describes the Fourier coefficients in (12.47a) and the coefficients $R^{(k)}_{yx^{(*)}}(\tau)$ in (12.47b) are referred to as *generalized (conjugate) cyclic cross-correlation* functions. The equations $\alpha=\alpha^{(k)}_{yx^{(*)}}(\tau)$, $k=\mathbb{I}$, describe the support curves of the (conjugate) cyclic cross-correlation function.

For all values of τ such that two different support curves do not intersect, we have (Izzo and Napolitano, 2002a), (Napolitano, 2007a)

$$R^{(k)}_{yx^{(*)}}(\tau) = R_{yx^{(*)}}(\alpha,\tau)\Big|_{\alpha=\alpha^{(k)}_{yx^{(*)}}(\tau)}\,. \tag{12.50}$$

12.6 Estimation of the cyclic cross-correlation function

12.6.1 Continuous time

From (12.47a), it follows that the (conjugate) cross-correlation function is completely known if the (conjugate) cyclic cross-correlation function $R_{yx^{(*)}}(\alpha,\tau)$ is known $\forall(\alpha,\tau)\in\mathbb{R}^2$. Thus, the estimation of the (conjugate) cyclic cross-correlation function, for all values of α and τ, is fundamental for the second-order analysis of jointly GACS processes.

A comprehensive treatment on the estimation of the (conjugate) cyclic cross-correlation function of jointly GACS processes is made in (Napolitano, 2012, Sec. 2.4). In this section, results are summarized without proof and refer to (Napolitano, 2012, Sec. 2.4).

Let us consider two jointly GACS processes $x(t)$ and $y(t)$ satisfying the following assumptions:

- the second and fourth-order statistics of $x(t)$ and $y(t)$ can be expressed by (generalized) Fourier series satisfying some regularity assumptions (Napolitano, 2012, Assumptions 2.4.2–2.4.4);
- $x(t)$ and $y(t)$ have finite or practically-finite memory, that is, they satisfy mixing assumptions expressed in terms of summability of second- and higher-order cumulants (Napolitano, 2012, Assumptions 2.4.8, 2.4.9, 2.4.15);
- there are no clusters of support curves (Napolitano, 2012, Assumption 2.4.10).

Under the above assumptions, in (Napolitano, 2012, Sec. 2.4) the following results are derived:

- the (conjugate) cyclic cross-correlogram is a mean-square consistent estimator of the cyclic autocorrelation function (Napolitano, 2012, Theorems 2.4.12 and 2.4.13);
- the properly normalized (conjugate) cyclic cross-correlogram is asymptotically complex normal (Napolitano, 2012, Theorem 2.4.18).

In the following, the main results of (Napolitano, 2012, Sec. 2.4) are presented and discussed.

Let $w_T(t) \triangleq a(t/T)/T$ be a data-tapering window with $a(t)$ with unit-area and nonzero in $(-1/2, 1/2)$ (Napolitano, 2012, Assumption 2.4.5). The *(conjugate) cyclic cross-correlogram*

$$R^{(T)}_{yx^{(*)}}(\alpha,\tau) \triangleq \int_{\mathbb{R}} w_T(t-t_0)\, y(t+\tau)\, x^{(*)}(t)\, e^{-j2\pi\alpha t}\,\mathrm{d}t \tag{12.51}$$

is the natural estimator of the (conjugate) cyclic cross-correlation function of the jointly GACS signals $y(t)$ and $x(t)$ on the basis of the observation for $t \in [t_0 - T/2, t_0 + T/2]$.

By taking the expected value of both sides of (12.51) and using (12.47b), one obtains the expected value of the (conjugate) cyclic cross-correlogram (Napolitano, 2007a, Theorem 3.1), (Napolitano, 2012, Theorem 2.4.6)

$$\mathrm{E}\left\{R^{(T)}_{yx^{(*)}}(\alpha,\tau)\right\} = \sum_{k\in\mathbb{I}} R^{(k)}_{yx^{(*)}}(\tau)\, W_{\frac{1}{T}}\left(\alpha - \alpha^{(k)}_{yx^{(*)}}(\tau)\right) e^{-j2\pi[\alpha-\alpha^{(k)}_{yx^{(*)}}(\tau)]t_0} \tag{12.52}$$

where $W_{\frac{1}{T}}(f)$ is the Fourier transform of $w_T(t)$.

The function $w_T(t)$ has duration T and, hence, $W_{\frac{1}{T}}(f)$ has bandwidth of the order of $1/T$. Consequently, from (12.52), it follows that the expected value of the (conjugate) cyclic cross-correlogram is significantly different from zero within strips of width $1/T$ around the support curves $\alpha = \alpha^{(k)}_{yx^{(*)}}(\tau)$, $k \in \mathbb{I}$, of the (conjugate) cyclic cross-correlation function. Thus, $\mathrm{E}\{R^{(T)}_{yx^{(*)}}(\alpha,\tau)\}$ approaches the (conjugate) cyclic cross-correlation function $R_{yx^{(*)}}(\alpha,\tau)$ (12.47b) as $T \to \infty$, that is, the estimator (12.51) is asymptotically unbiased. In addition, if there is no cluster of support curves $\alpha = \alpha^{(k)}_{yx^{(*)}}(\tau)$ (Napolitano, 2012, Assumption 2.4.10) we have (Napolitano, 2012, Theorem 2.4.12)

$$\text{bias}\left\{R^{(T)}_{yx^{(*)}}(\alpha,\tau)\right\} = \mathcal{O}(T^{-\gamma}) \tag{12.53}$$

where $\gamma \geqslant 1$ is the rate of decay to zero of the Fourier transform $A(f)$ of $a(t)$, that is $A(f) = \mathcal{O}(|f|^{-\gamma})$ as $|f| \to \infty$. Note that, according to (12.52), the expected value of the (conjugate) cyclic cross-correlogram along a given lag-dependent (conjugate) cycle-frequency curve $\alpha = \alpha^{(k)}_{yx^{(*)}}(\tau)$ is influenced not only by $R^{(k)}_{yx^{(*)}}(\tau)$ but also by the values of the generalized (conjugate) cyclic cross-correlation functions $R^{(h)}_{yx^{(*)}}(\tau)$ corresponding to all the other lag-dependent (conjugate) cycle-frequency curves $\alpha = \alpha^{(h)}_{yx^{(*)}}(\tau)$, $h \neq k$, the influence being stronger from curves $\alpha = \alpha^{(h)}_{yx^{(*)}}(\tau)$ closer to $\alpha = \alpha^{(k)}_{yx^{(*)}}(\tau)$ and with larger $|R^{(h)}_{yx^{(*)}}(\tau)|$. Such a phenomenon is a generalization of the cycle leakage occurring in cyclic statistic estimates of ACS processes (Section 5.3.4) (Gardner, 1987f, Chap. 13).

For a zero-mean process with summable second- and fourth-order cumulants, the (conjugate) covariance of cyclic correlograms at (possibly) different lags and cycle frequencies is such that (Napolitano, 2004), (Napolitano, 2007a, Theorem 4.3), (Napolitano, 2012, Theorems 2.4.13 and 3.7.2)

$$\text{cov}\left\{R^{(T)}_{yx^{(*)}}(\alpha_i,\tau_i),\, R^{(T)\,[*]}_{yx^{(*)}}(\alpha_j,\tau_j)\right\} = \mathcal{O}(T^{-1}) \tag{12.54}$$

as $T \to \infty$, where $[*]$ is an optional complex conjugation. Thus, the variance of the (conjugate) cyclic cross-correlogram is such that $\text{var}\{R^{(T)}_{yx^{(*)}}(\alpha,\tau)\} = \mathcal{O}(T^{-1})$.

From (12.53) and (12.54) it immediately follows that the (conjugate) cyclic cross-correlogram is a mean-square consistent estimator of the (conjugate) cyclic cross-correlation function:

$$\lim_{T\to\infty} \mathrm{E}\left\{\left|R^{(T)}_{yx^{(*)}}(\alpha,\tau) - R_{yx^{(*)}}(\alpha,\tau)\right|^2\right\} = 0\,. \tag{12.55}$$

In addition, under further summability assumptions on kth-order cumulants, the complex random variables

$$V^{(T)}_i \triangleq \sqrt{T}\left[R^{(T)}_{yx^{(*)}}(\alpha_i,\tau_i) - R_{yx^{(*)}}(\alpha_i,\tau_i)\right] \qquad (i = 1, 2, \ldots) \tag{12.56}$$

are such that, as $T \to \infty$,

1. $\mathrm{E}\left\{V^{(T)}_i\right\} \to 0$;

2a. $\text{cov}\left\{V^{(T)}_i, V^{(T)}_j\right\} \to \Sigma_{ij}$ finite;

2b. $\text{cov}\left\{V^{(T)}_i, V^{(T)*}_j\right\} \to \Sigma^{(c)}_{ij}$ finite;

3. $\text{cum}\left\{V^{(T)[*]_1}_1, \ldots, V^{(T)[*]_k}_k\right\} \to 0$ for $k \geqslant 3$;

with superscript $[*]_i$ denoting ith optional complex conjugation. Thus, asymptotically, the logarithm of the joint characteristic function of $V_i^{(T)}$, $i = 1, \ldots, k$, is a quadratic form and, hence, the $V_i^{(T)}$ are asymptotically ($T \to \infty$) zero-mean jointly complex normal (Lemma E.1), (Napolitano, 2005). The expressions of the covariance Σ_{ij} and conjugate covariance $\Sigma_{ij}^{(c)}$ are given in (Napolitano, 2007a), (Napolitano, 2012, Theorems 2.4.13, 3.7.2).

Finally, note that well-known consistency and asymptotic normality results for the (conjugate) cyclic cross-correlogram of ACS processes (Section 5.2.1), (Gardner, 1987f), (Hurd and Leśkow, 1992a), (Dehay and Hurd, 1994), (Giannakis, 1998), (Hurd and Miamee, 2007), can be obtained by specializing the results for GACS processes.

12.6.2 Discrete time

Let $x_{\rm d}(n) = x(t)|_{t=nT_s}$ and $y_{\rm d}(n) = y(t)|_{t=nT_s}$ be the discrete-time ACS processes obtained by uniformly sampling with period $T_s = 1/f_s$ the continuous-time jointly GACS processes $x(t)$ and $y(t)$ (see Section 12.5).

Under mild conditions, the *discrete-time (conjugate) cyclic cross-correlogram*

$$\widetilde{R}^{(N)}_{y_{\rm d}x_{\rm d}^{(*)}}(\widetilde{\alpha}, m) \triangleq \sum_{n=-\infty}^{+\infty} v_N(n-n_0)\, y_{\rm d}(n+m)\, x_{\rm d}^{(*)}(n)\, e^{-j2\pi\widetilde{\alpha}n} \tag{12.57}$$

where $v_N(n)$ is a unit-sum data-tapering window nonzero in $\{-N, \ldots, N\}$, in (Napolitano, 2009), (Napolitano, 2012, Sec. 2.6.2) is shown to be a mean-square consistent and asymptotically ($N \to \infty$) complex normal estimator of the aliased (conjugate) cyclic cross-correlation function of $x(t)$ and $y(t)$

$$\widetilde{R}^{\widetilde{\alpha}}_{y_{\rm d}x_{\rm d}^{(*)}}(m) = \sum_{p\in\mathbb{Z}} R^{(\widetilde{\alpha}+p)f_s}_{yx^{(*)}}(mT_s) \tag{12.58}$$

whose expression is derived similarly to (3.191).

Since GACS processes cannot be strictly bandlimited (Section 12.3.2), (Izzo and Napolitano, 1998a), (Napolitano, 2007a), (Napolitano, 2009), (Napolitano, 2012, Secs. 2.2.3, 2.7.7), the sampling period T_s should approach zero in order to reduce the aliasing effects. In (Napolitano, 2009), it is shown that the mean squared error can be made arbitrarily small by taking the sampling period T_s sufficiently small. The asymptotic analysis of the discrete-time (conjugate) cyclic cross-correlogram (12.57) as $T_s \to 0$ has the drawback that $\alpha = \widetilde{\alpha} f_s \to \infty$ and $\tau = mT_s \to 0$. For this reason, in (Napolitano, 2009), (Napolitano, 2012, Sec. 2.6.3) asymptotic results as $N \to \infty$ and $T_s \to 0$ with $NT_s \to \infty$ are carried out resorting to the *hybrid (conjugate) cyclic cross-correlogram*

$$\varrho^{(N,T_s)}_{y_{\rm d}x_{\rm d}^{(*)}}(\alpha, \tau) \triangleq \sum_{n=-\infty}^{+\infty} v_N(n-n_0)\, y(nT_s+\tau)\, x^{(*)}(nT_s)\, e^{-j2\pi\alpha nT_s} \tag{12.59}$$

with α and τ not depending on T_s.

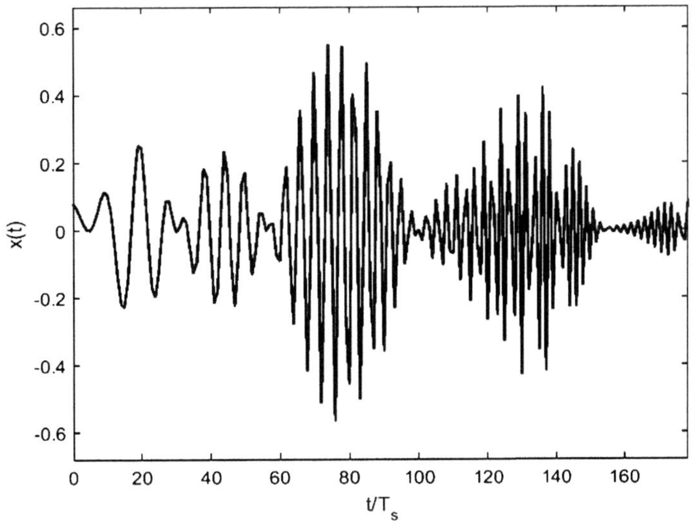

FIGURE 12.4 Chirp-modulated DSB-AM signal $x(t)$.

12.7 Examples and applications

12.7.1 Doppler channel due to constant relative radial acceleration between transmitter and receiver

In this section, the GACS model is shown to be useful in describing the received signal in the presence of relative radial acceleration between transmitter and receiver when the transmitted signal is ACS. Let

$$x_0(t) = s_0(t)\ \cos(2\pi f_c t) \tag{12.60}$$

be the double-sideband amplitude-modulated (DSB-AM) transmitted signal (Section 7.2), (Fig. 7.1).

The Doppler channel due to relative motion between transmitter and receiver with constant relative radial acceleration within the observation interval, introduces a quadratically time-variant delay on the transmitted signal (Section 11.2), (Napolitano, 2012, Sec. 7.4). If the narrow-band conditions (11.15) and (11.16) are satisfied and (11.17) and (11.18) do not hold, then time variations of the delay can be neglected in the complex envelope of the received signal but not in the carrier and the Doppler channel introduces a frequency shift and a chirp modulation on the transmitted signal (Section 11.2, case 4) (Napolitano, 2012, Sec. 7.5.2). Thus, but for an attenuation factor, the (noise-free) received signal can be modeled as (Fig. 12.4)

$$x(t) = s(t)\ \cos(2\pi f_v(t)\, t + \phi_0) \tag{12.61}$$

where $s(t) = s_0(t - d_0)$ is the modulating process with d_0 the propagation delay, and

$$f_v(t) = f_c + f_D + \frac{\gamma}{2} t \,. \tag{12.62}$$

In (12.62), f_D and γ are the frequency shift and the chirp rate due to the Doppler effect (Section 11.2, case 4).

Define $f_0 \triangleq f_c + f_D$ and, for the sake of simplicity, assume that the modulating process $s(t)$ is WSS with autocorrelation function

$$\mathrm{E}\left\{s(t+\tau)\, s^*(t)\right\} = R_{ss^*}(\tau) \,. \tag{12.63}$$

The autocorrelation function of $x(t)$ is

$$\begin{aligned}
\mathrm{E}\left\{x(t+\tau)\, x^*(t)\right\} = \frac{1}{4}\, &\mathrm{E}\left\{s(t+\tau)\, s^*(t)\right\} \\
&\cdot \Big[e^{j2\pi[f_0\tau + \gamma\tau^2/2]}\, e^{j2\pi\,(\gamma\tau)\,t} + e^{-j2\pi[f_0\tau + \gamma\tau^2/2]}\, e^{-j2\pi\,(\gamma\tau)\,t} \\
&+ e^{j2\pi[f_0\tau + \gamma\tau^2/2]}\, e^{j2\phi_0}\, e^{j2\pi\, 2f_0\, t}\, e^{j2\pi\,\gamma\, t^2} \\
&+ e^{-j2\pi[f_0\tau + \gamma\tau^2/2]}\, e^{-j2\phi_0}\, e^{-j2\pi\, 2f_0\, t}\, e^{-j2\pi\,\gamma\, t^2} \Big] \,.
\end{aligned} \tag{12.64}$$

If also condition (11.17) is satisfied, then $\gamma = 0$ can be assumed in the model of the received signal and $x(t)$ can be modeled as double side-band amplitude-modulated (DSB-AM) signal. It is cyclostationary with period $1/(2f_0)$ and cyclic autocorrelation function (see (7.4)), (Fig. 12.5)

$$\begin{aligned}
R_{xx^*}(\alpha, \tau) = \; &\frac{1}{2}\, R_{ss^*}(\tau)\, \cos(2\pi f_0 \tau)\, \delta_\alpha \\
&+ \frac{1}{4}\, R_{ss^*}(\tau)\, e^{j2\phi_0}\, e^{j2\pi f_0 \tau}\, \delta_{[\alpha - 2f_0]} \\
&+ \frac{1}{4}\, R_{ss^*}(\tau)\, e^{-j2\phi_0}\, e^{-j2\pi f_0 \tau}\, \delta_{[\alpha + 2f_0]} \,.
\end{aligned} \tag{12.65}$$

The signal $x(t)$ exhibits cyclostationarity at cycle frequencies $\alpha = 0$ and $\alpha = \pm 2 f_0$.

If condition (11.17) does not hold, then $\gamma \neq 0$ must be considered in the model of $x(t)$. The autocorrelation function in (12.64) is constituted by the sum of a uniformly almost-periodic component in the sense of Besicovitch with cycle frequencies depending on the lag parameter τ (first and second term in the right-hand side of (12.64)) and a residual term not containing any finite-strength additive sine-wave component (third and fourth term in the right-hand side of (12.64)). In fact, the following result holds

$$\lim_{T\to\infty} \frac{1}{T} \int_{-T/2}^{T/2} e^{jat}\, e^{jbt^2}\, \mathrm{d}t = \delta_a\, \delta_b = \begin{cases} 1 & a = b = 0 \\ 0 & \text{otherwise} \end{cases} \tag{12.66}$$

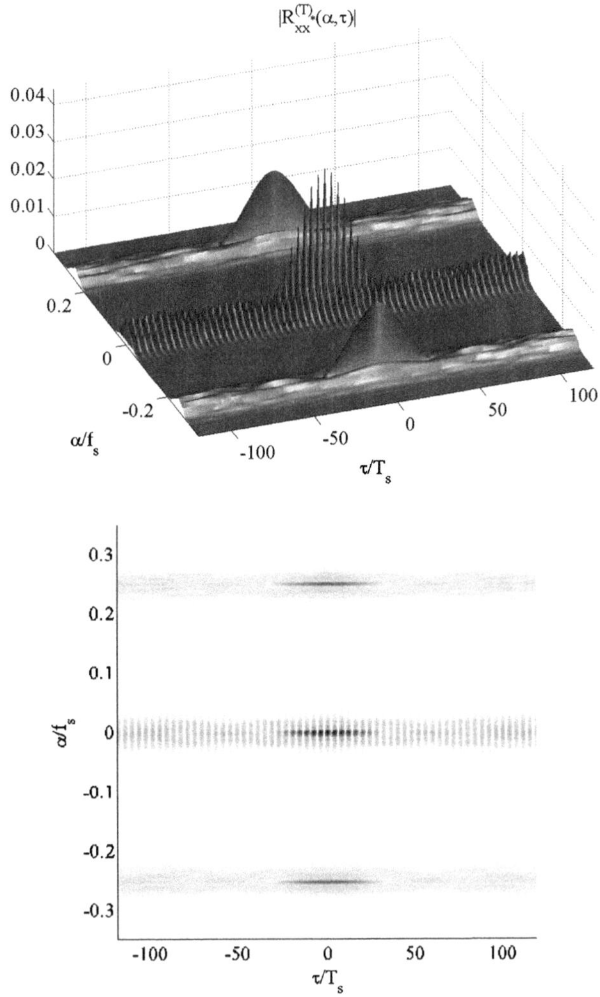

FIGURE 12.5 (Top) magnitude and (bottom) 2-dimensional gray-scale elevation map (or "checkerboard" plot) of the cyclic correlogram of the DSB-AM signal $x(t)$. *Source: (Napolitano, 2016a) Copyright of Elsevier.*

Thus, $x(t)$ is GACS according to the model (12.27a)–(12.28). The cyclic autocorrelation function is (Fig. 12.6)

$$R_{xx^*}(\alpha, \tau) = \frac{1}{4} R_{ss^*}(\tau)\, e^{j2\pi[f_0\tau + \gamma\tau^2/2]}\, \delta_{[\alpha - \gamma\tau]} + \frac{1}{4} R_{ss^*}(\tau)\, e^{-j2\pi[f_0\tau + \gamma\tau^2/2]}\, \delta_{[\alpha + \gamma\tau]} \,. \tag{12.67}$$

The signal $x(t)$ has the lag-dependent cycle frequencies and corresponding generalized cyclic autocorrelation functions given by

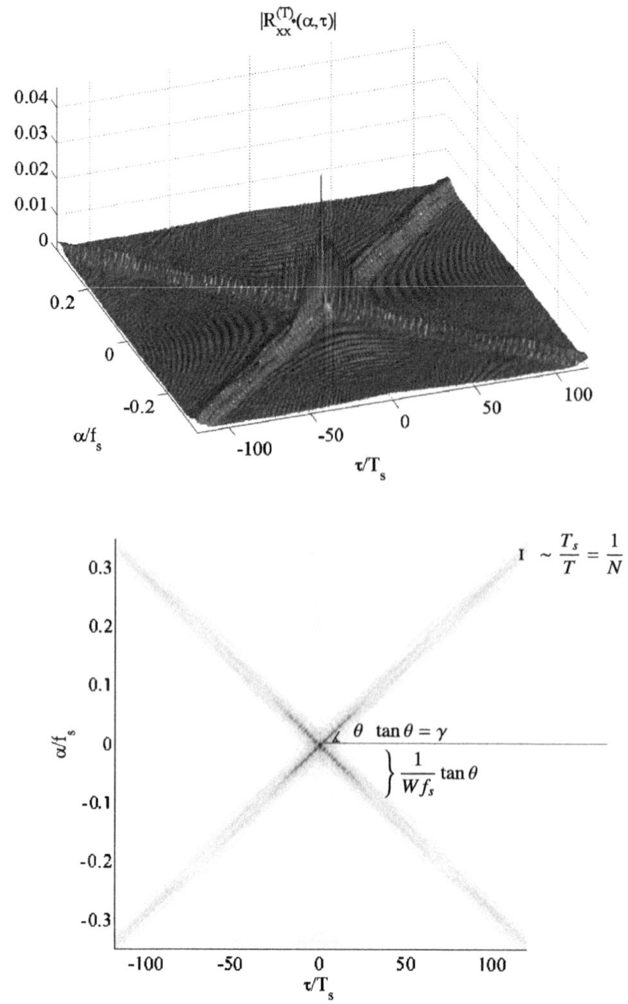

FIGURE 12.6 (Top) magnitude and (bottom) 2-dimensional gray-scale elevation map (or "checkerboard" plot) of the cyclic correlogram of the GACS signal $x(t)$. *Source: (Napolitano, 2016a) Copyright of Elsevier.*

$$\alpha^{(1)}_{xx^*}(\tau) = \gamma\tau \qquad R^{(1)}_{xx^*}(\tau) = \frac{1}{4}\, R_{ss^*}(\tau)\, e^{j2\pi[f_0\tau+\gamma\tau^2/2]} \tag{12.68}$$

$$\alpha^{(2)}_{xx^*}(\tau) = -\gamma\tau \qquad R^{(2)}_{xx^*}(\tau) = \frac{1}{4}\, R_{ss^*}(\tau)\, e^{-j2\pi[f_0\tau+\gamma\tau^2/2]}\,. \tag{12.69}$$

The support curves $\alpha = \gamma\tau$ and $\alpha = -\gamma\tau$ intercept each other in the point $(\alpha,\tau) = (0,0)$ where $R_{xx^*}(\alpha,\tau)$ presents a Kronecker delta.

Let W the bandwidth of $s(t)$, that is, the width of its power spectrum. Then, the width of the autocorrelation $R_{ss^*}(\tau)$ is approximatively equal to $1/W$. The cross pattern in Fig. 12.6

(bottom) can be recognized provided that

$$\frac{1}{W f_s} \tan\theta = \frac{1}{W f_s} \gamma > \frac{2T_s}{T} = \frac{2}{N} \tag{12.70}$$

where $f_s = 1/T_s$ is the sampling frequency and $T = NT_s$ the data-record length, with N the number of samples. In (12.70), the left-hand side is the maximum separation between the τ axis and one of the two segments where $R^{(T)}_{xx^*}(\alpha, \tau)$ is significantly nonzero and the right-hand side is the cycle frequency resolution achievable using N samples for the cyclic correlogram computation, that is, the width of the strips around $\alpha = \pm\gamma\tau$ (Section 12.6), (Napolitano, 2012, Secs. 2.4.1, 2.7.6). Note that the dimensions of γ are s^{-2}.

If the cross pattern can be recognized, then $x(t)$ can be inferred to be GACS with linear lag-dependent cycle frequencies. The chirp parameter γ is the slope of the linear lag-dependent cycle frequency $\alpha^{(1)}_{xx^*}(\tau) = \gamma\tau$. It can be estimated by locating, for each τ, the value of α corresponding to the peak of $|R^{(T)}_{xx^*}(\alpha, \tau)|$ and then making a linear least-square fitting. The search must be made in an a priori known thin rectangular strip containing only the support line $\alpha = \gamma\tau$ and for values of $\tau > 0$ less than the width of the cyclic autocorrelation function. Once γ, and hence the support line, is estimated, the generalized cyclic autocorrelation function $R^{(1)}_{xx^*}(\tau)$ can be estimated along the estimated support line according to (12.20) (Napolitano, 2013).

If the GACS nature of the received signal is neglected and the ACS model is adopted, increasing the observation interval in order to counteract noise and interference in cyclostationarity-based algorithms does not produce the expected beneficial effects but, rather, it worsens performance. In fact, the cyclic autocorrelation function estimate turns out to be biased. In particular, cyclic features at $\alpha = \pm 2 f_0$ disappear (Fig. 12.6). An algorithm which is based on measurements of cyclic autocorrelation functions can improve performance only if the estimates of the cyclic autocorrelation functions are replaced by estimates of the generalized cyclic autocorrelation functions (Section 12.7.2) (Napolitano, 2013).

If the bandwidth W is increased so that condition (12.70) is not satisfied, then the estimate of $R_{xx^*}(\alpha, \tau)$ reduces to a single spike in $(\alpha, \tau) = (0, 0)$ (Fig. 12.7). Such a behavior is in agreement with the result in (Napolitano, 2012, Sec. 7.8.1.2) where it is shown that increasing the bandwidth W so that the narrow-band conditions (11.15) and (11.16) (with $B = W$) are not satisfied, leads to a process with $R_{xx^*}(\alpha, \tau) = \overline{x^2}\, \delta_\alpha\, \delta_\tau$.

Finally, note that if the complex envelope $\widetilde{x}(t) = s(t)\, e^{j[2\pi f_v(t) t + \phi_0]}$ is considered, we have

$$\mathrm{E}\{\widetilde{x}(t+\tau)\, \widetilde{x}^*(t)\} = R_{ss^*}(\tau)\, e^{j2\pi[f_0\tau + \gamma\tau^2/2]}\, e^{j2\pi(\gamma\tau)t} \,. \tag{12.71}$$

That is, the autocorrelation function does not contain any residual term and model (12.10), (12.18) must be considered. See also the example in Fig. 12.8 and (Napolitano, 2007a), (Napolitano, 2012, Sec. 2.2.6.1).

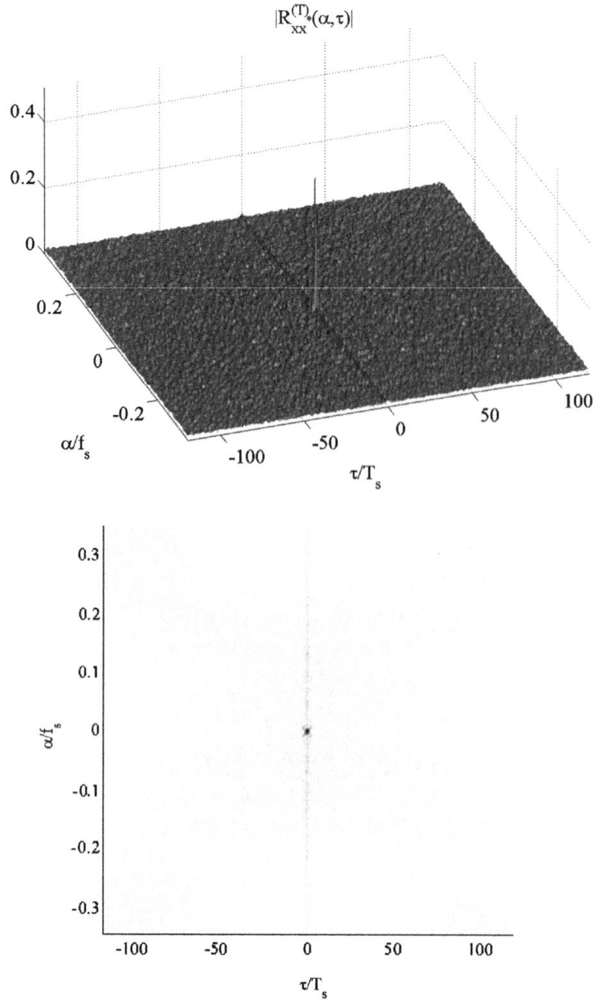

FIGURE 12.7 (Top) magnitude and (bottom) 2-dimensional gray-scale elevation map (or "checkerboard" plot) of the cyclic correlogram of the GACS signal $x(t)$ under the wide-band condition. *Source: (Napolitano, 2016a) Copyright of Elsevier.*

12.7.2 Performance degradation due to ACS model mismatch

Several signal-selective algorithms for source location, channel identification and equalization, signal detection, estimation, and classification, are based on the estimation of the cyclic autocorrelation function of the received ACS signal at given cycle frequencies (Agee et al., 1990), (Gardner and Chen, 1992), (Gardner, 1993), (Gardner and Spooner, 1993), (Streight et al., 2000), (Gardner et al., 2006), (Dobre et al., 2007), (Lundén et al., 2009), (Spooner and Nicholls, 2009), (Teplitsky and Yeredor, 2009). Good performance of the signal processing algorithm is achieved when the estimate of the cyclic autocorrelation function is accurate and the accuracy can be improved by increasing the data-record length T (Section 5.2.1). In particular, increasing T allows one to counteract the cycle leakage coming from stationary noise, interference, and the useful signal itself (Section 5.3.4)

In the case of relative radial acceleration between transmitter and receiver, if the data-record length is increased too much, the GACS model for the received signal must be considered since the cyclic autocorrelation estimate that assumes the ACS model becomes biased. Specifically, modeling a GACS signal as ACS leads to performance degradation in the cyclic autocorrelation estimate when the data-record length is sufficiently large so that the variability with τ of the lag dependent cycle frequencies cannot be neglected. Consequently, performance degradation is expected for cyclostationarity-based algorithms that exploit cyclic autocorrelation measurements.

In the case of Doppler channel due to transmitter and receiver with constant relative radial acceleration, the transmitted signal experiences a quadratically time-variant delay. If the narrow-band conditions (11.15), (11.16) are verified and the phase conditions (11.17), (11.18) are not verified, then the received signal $y(t)$ is a chirp-modulated version of the transmitted signal $x(t)$. That is (Section 11.2, case 4a)

$$y(t) = b\, x(t-d_0)\, e^{j2\pi \nu t}\, e^{j\pi\gamma t^2} \tag{12.72}$$

where b is the complex gain, d_0 the delay, ν the frequency shift, and γ the chirp rate.

Let us assume that $x(t)$ is ACS with cycle frequencies α_k and cyclic autocorrelation functions $R^{\alpha_k}_{xx^*}(\tau)$. Thus, $y(t)$ is GACS with linear lag-dependent cycle frequencies and generalized cyclic autocorrelation functions given by (Section 12.7.1) (Napolitano, 2007a), (Napolitano, 2012, Sec. 7.8)

$$\alpha^{(k)}_{yy^*}(\tau) = \alpha_k + \gamma\tau \qquad R^{(k)}_{yy^*}(\tau) = |b|^2\, R^{\alpha_k}_{xx^*}(\tau)\, e^{j2\pi\nu\tau}\, e^{j\pi\gamma\tau^2}\, e^{-j2\pi\alpha_k d_0} \tag{12.73}$$

If $x(t)$ is a PAM signal with full duty-cycle rectangular pulse and binary white modulating sequence, its cycle frequencies are $\alpha_k = k/T_p$, $k \in \mathbb{Z}$, where T_p is the symbol period (Section 7.3). In Fig. 12.8, (top) magnitude and (bottom) 2-dimensional gray-scale elevation map (or "checkerboard" plot) of the cyclic autocorrelation function of $y(t)$ are reported. According to (12.73), the generalized cyclic autocorrelation functions of $y(t)$ have the same shapes of the cyclic autocorrelation functions of $x(t)$.

From (12.73) we have that the chirp rate γ and the cycle frequency α_k are the slope and the intercept of the linear lag-dependent cycle frequency $\alpha^{(k)}_{yy^*}(\tau) = \alpha_k + \gamma\tau$. They can be estimated by first locating, for each τ, the value of α corresponding to the peak of the magnitude of the cyclic autocorrelation function and then making a linear least-square fitting. The search should be made in an a priori known thin (along α) rectangle in the (α, τ)-plane containing only one support line $\alpha = \alpha_k + \gamma\tau$ and for values of τ less than the width of the signal autocorrelation function. Then, the generalized cyclic autocorrelation function can be estimated along the estimated support line.

Let $x(t)$ be a transmitted PAM signal with full duty-cycle rectangular pulse of width $T_p = 64T_s$ and stationary-white binary modulating sequence. The width of all cyclic au-

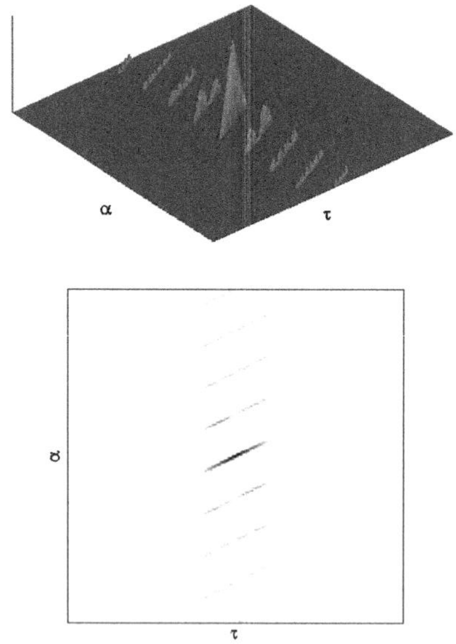

FIGURE 12.8 (Top) magnitude and (bottom) 2-dimensional gray-scale elevation map (or "checkerboard" plot) of the cyclic autocorrelation function of the (complex envelope) chirp modulated PAM signal as function of α and τ.

tocorrelation functions is $2T_p$ (Fig. 7.4A). The signal $x(t)$ passes through a Doppler channel characterized by $\nu = 0.02/T_s$ and $\gamma = 10^{-5}/T_s^2$, where T_s is the sampling period.

If the received signal $y(t)$ is observed over an interval of length $T = NT_s$, then the cycle frequency resolution is $1/T$. That is, variations of the cycle frequency smaller than $1/T$ cannot be appreciated. Consequently, for $|\gamma|T_p < 1/T$, $y(t)$ can be modeled as ACS. In contrast, if this condition is not satisfied, $y(t)$ should be modeled as GACS (Fig. 12.8).

The cyclic autocorrelation function $R_{yy^*}^{\alpha_k}(\tau)$ at the nominal cycle frequency $\alpha_k = 1/T_p$ is estimated by the cyclic correlogram by $N = 2^{10}$ samples (Fig. 12.9 (top)) and $N = 2^{13}$ samples (Fig. 12.10 (top)). Increasing the data-record length does not improve performance. In fact, for $N = 2^{10}$ it results $|\gamma|T_p < 1/T$ while for $N = 2^{13}$ we have $|\gamma|T_p > 1/T$ and the estimate is biased. In contrast, the estimate of the generalized cyclic autocorrelation function $R_{yy^*}^{(k)}(\tau) = R_{yy^*}(\alpha_k + \gamma\tau, \tau)$ improves passing from $N = 2^{10}$ (Fig. 12.9 (bottom)) to $N = 2^{13}$ (Fig. 12.10 (bottom)).

According to (12.50), the estimate of the generalized cyclic autocorrelation function is obtained by the cyclic correlogram $R_{yy^*}^{(T)}(\alpha_k, \tau)$ by plugging the estimate $\alpha_{yy^*}^{(k),(T)}(\tau)$ of the lag-dependent cycle frequency in place of the cycle frequency α_k. The consistency of such an estimator can be proved by following the guidelines for ACS processes in Section 5.5.1 for the cyclic correlogram with estimated cycle frequency.

12.7.3 Signal detection

The random variables $V_i^{(T)}$ defined in (12.56) with $y(t) \equiv x(t)$ are asymptotically $(T \to \infty)$ zero-mean jointly complex normal. Thus, a detection test at cycle frequency α and lag τ

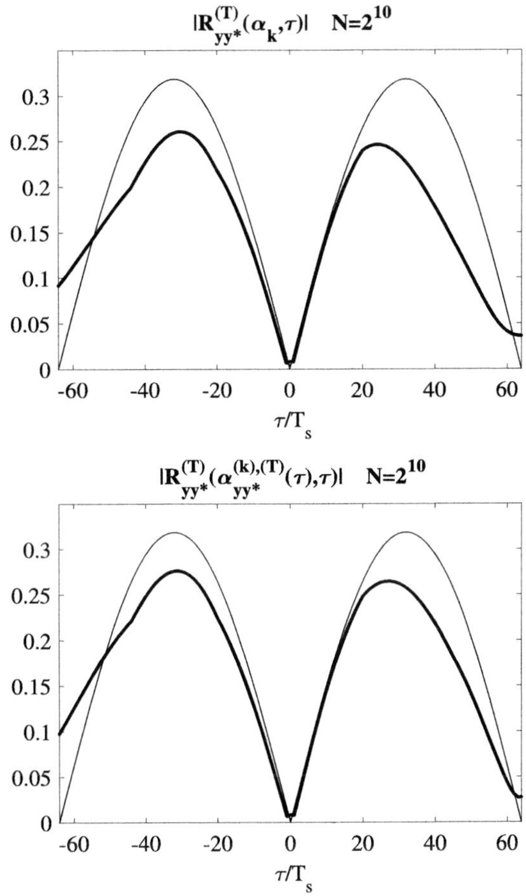

FIGURE 12.9 Chirp modulated PAM signal. (Top) Magnitude of the cyclic correlogram at cycle frequency $\alpha = 1/T_p$ estimated by $N = 2^{10}$ samples (thick line) and of the numerically evaluated cyclic autocorrelation without chirp modulation (thin line). (Bottom) Magnitude of the estimate of the generalized cyclic autocorrelation obtained by $N = 2^{10}$ samples (thick line) and of the numerically evaluated generalized cyclic autocorrelation (thin line).

can be designed similarly to the case of ACS processes (Section 8.4). In particular, spectrum sensing techniques based on this test and developed in the case of fixed transmitter and receiver (Section 8.12) can be extended to the case of relative motion between transmitter and receiver with constant relative radial acceleration (Sections 11.2, 12.7.1).

12.8 Summary

Generalized almost-cyclostationary (GACS) signals have multivariate statistical functions that are almost-periodic functions of time whose Fourier series has both frequencies and coefficients depending on the lag parameters. For each fixed value of the lag pa-

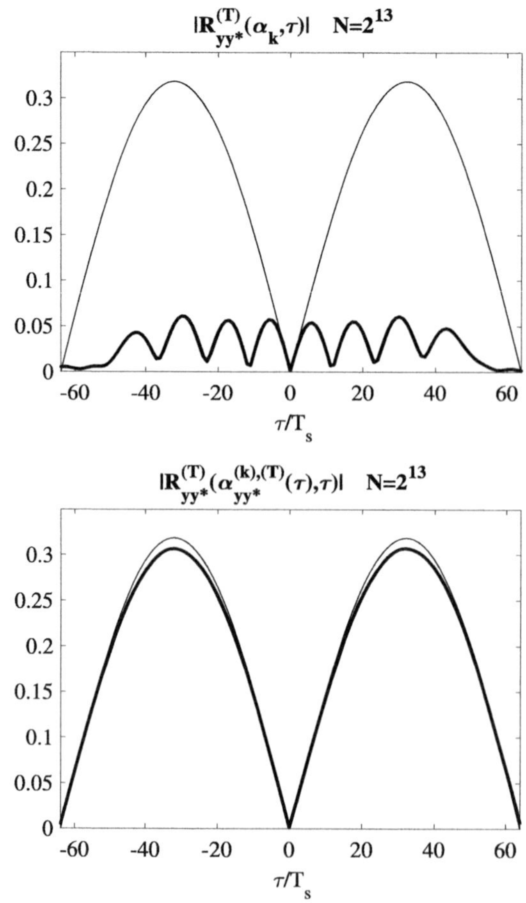

FIGURE 12.10 Chirp modulated PAM signal. (Top) Magnitude of the cyclic correlogram at cycle frequency $\alpha = 1/T_p$ estimated by $N = 2^{13}$ samples (thick line) and of the numerically evaluated cyclic autocorrelation without chirp modulation (thin line). (Bottom) Magnitude of the estimate of the generalized cyclic autocorrelation obtained by $N = 2^{13}$ samples (thick line) and of the numerically evaluated generalized cyclic autocorrelation (thin line).

rameter, the autocorrelation function can be expressed as the superposition of complex sinewaves whose frequencies range in a countable set depending on the lag parameter (Theorem 12.7). Equivalently, the autocorrelation can be expressed as a Fourier series whose frequencies are named *lag-dependent cycle frequencies* and whose coefficients are called *generalized cyclic autocorrelation functions* (Theorem 12.9). Similar representations hold for the conjugate autocorrelation. Almost-cyclostationary (ACS) signals are obtained as special case of GACS signals when the lag-dependent cycle frequencies are constant with respect to lag.

The time-averaged autocorrelation of GACS processes contains a Kronecker delta at zero lag. Consequently, the process is not harmonizable in the sense of Loève and is not mean-square continuous. At least a portion of the signal power is spread on the whole frequency axis and the power spectrum is described in terms of the μ distribution (Section 12.3.2).

Continuous-time GACS processes, unlike wide-sense stationary and ACS processes, do not have a discrete-time counterpart. Uniformly sampling a continuous-time GACS process leads to a discrete-time ACS process. Moreover, starting from a discrete-time ACS process, the possible GACS or ACS nature of the underlying continuous-time process can only be conjectured (Section 12.4).

Under mild regularity assumptions expressed in terms of summability of cumulants, the (conjugate) cyclic cross-correlogram is a mean-square consistent estimator of the cyclic autocorrelation function with bias at least $\mathcal{O}(T^{-1})$ and variance $\mathcal{O}(T^{-1})$, where T is the data-record length. In addition, the estimation errors times $\sqrt{T}$ at different lags and cycle frequencies are asymptotically ($T \to \infty$) jointly complex normal (Section 12.6). A leakage phenomenon similar to that for cyclic statistic estimates of ACS processes exists.

The Doppler channel existing between transmitter and receiver in relative motion with constant relative radial acceleration transforms a transmitted ACS signal into a GACS signal with lag-dependent cycle frequencies linearly depending on the lag parameter (Section 12.7.1). The GACS nature of the received signal can be appreciated provided that the data-record length is sufficiently large. Thus, in such a case, cyclostationarity-based algorithms that in principle are expected to improve performance by increasing the data-record length, provide biased estimates when the data-record length is such that the ACS model is not appropriate and the GACS model should be considered (Section 12.7.2).

13

Spectrally correlated signals

Contents

Abstract

Spectrally correlated signals have Loève bifrequency spectrum with spectral masses concentrated on a countable set of support curves. This class of signals generalizes that of the almost-cyclostationary signals for which spectral masses are concentrated on lines with unit slope. The class of the spectrally correlated signals is characterized at second-order in the time and frequency domains. The estimation of the density of Loève bifrequency spectrum is addressed in both cases of known and unknown support curves. Applications to multipath Doppler channel analysis, moving source location, signal detection, fractional Brownian motion spectral characterization, and spectral analysis with nonuniform frequency spacing are considered.

Keywords

Spectrally correlated, Loève bifrequency spectrum, Frequency warping, Fractional Brownian motion, Multipath Doppler channel, Support curves

13.1 Introduction

Wide-sense stationary processes are characterized by the lack of correlation between distinct spectral components (Section A.5). The presence of correlation between distinct spectral components is an indicator of nonstationarity, that is, of time dependence in the autocorrelation (Loève, 1963), (Gardner, 1987f), (Hurd and Gerr, 1991), (Genossar, 1992), (Dehay and Hurd, 1994), (Napolitano, 2003), (Dmochowski et al., 2009), (Napolitano, 2012, Chap. 4). If the frequency separation between correlated spectral components belongs to a countable set of values, then the process is almost-cyclostationary (ACS). The set of these values is the set of the cycle frequencies, that is, the frequencies of the Fourier series expansion of the almost-periodically time varying autocorrelation function. Therefore, for ACS processes, the Loève bifrequency spectrum has spectral masses concentrated, in the bifrequency plane, on a countable set of lines with unit slope and whose intercepts are the cycle frequencies.

Spectrally correlated (SC) processes have Loève bifrequency spectrum whose spectral masses are concentrated on a countable set of support curves in the bifrequency plane (Napolitano, 2001), (Napolitano, 2003), (Soltani and Parvardeh, 2006), (Napolitano, 2007c), (Napolitano, 2011), (Napolitano, 2012, Chap. 4), (Mahmoudi et al., 2015). SC processes are referred to as simple processes in (Soltani and Parvardeh, 2006), (Mahmoudi et al., 2015), (Soltani et al., 2019). The special case of support constituted by lines with not necessarily unit slope is treated in (Allen and Hobbs, 1992), (Lii and Rosenblatt, 2002).

SC processes are a suitable model for signals encountered in several applications.

In communications, SC processes are an appropriate model for the received signal in high mobility scenarios and in the presence of wide-band transmitted ACS signals. In fact, if the so called narrow-band condition is not satisfied, that is, if the product of bandwidth and data-record length is not much smaller than the ratio of the medium propagation speed and the radial speed between transmitter and receiver, then the Doppler effect on the received signal cannot be described just by a frequency shift of the carrier and also a nonunit time-scale factor in the complex envelope must be accounted for (Section 11.2), (Van Trees, 1971b, pp. 339–340), (Napolitano, 2012, Sec. 7.5.1). If the transmitted signal is ACS and a multipath Doppler channel introduces different gain, time-scale factor, frequency shift, and delay for each path, the received signal is SC (Section 13.6.1), (Napolitano, 2003), (Napolitano, 2012, Sec. 4.2.4.1).

In spectral analysis with nonuniform frequency spacing (Oppenheim et al., 1971), (Oppenheim and Johnson, 1972), (Braccini and Oppenheim, 1974), the SC process model with nonlinear support curves in the bifrequency plane must be accounted for if the analyzed signal is ACS (Section 13.6.6) (Napolitano, 2012, Sec. 4.13). Similar considerations hold for signal processing algorithms exploiting frequency warping techniques (Makur and Mitra, 2001), (Franz et al., 2003) when applied to ACS processes.

Jointly SC processes occur in the presence of linear time-variant (LTV) systems that are referred to as FOT-deterministic (Izzo and Napolitano, 2002a), (Napolitano, 2012, Sec. 4.3.2). Special cases of these systems are considered in (Franaszek, 1967), (Franaszek and Liu, 1967), (Liu and Franaszek, 1969), (Claasen and Mecklenbräuker, 1982).

In (Øigård et al., 2006) it is shown that fractional Brownian motion (fBm) processes have Loève bifrequency spectrum with spectral masses concentrated on the axes and the main diagonal of the bifrequency plane. Finally, in discrete-time, ACS processes and their multirate processed versions are jointly SC (Napolitano, 2010a), (Napolitano, 2012, Sec. 4.10).

The problem of estimating the bifrequency spectral correlation density of the Loève bifrequency spectrum has different peculiarities in the two cases of unknown and known support curves. If the location of the support curves is unknown, then the time-smoothed cross-periodogram provides a reliable (small bias and variance) estimate of the bifrequency spectral correlation density only if the departure of the nonstationarity from the almost-cyclostationarity is not too large (Section 13.5.1), (Napolitano, 2001), (Napolitano, 2003), (Napolitano, 2012, Sec. 4.5). In contrast, if the location of the support curves is known, then the cross-periodogram frequency smoothed along a support curve, provides a mean-square consistent estimator of the spectral correlation density on this curve. Moreover, a properly normalized version of the error estimate is asymptotically complex normal (Section 13.5.2), (Napolitano, 2007c), (Napolitano, 2012, Sec. 4.5).

The lack of consistency of the spectral density estimators when the kind of nonstationarity is not perfectly known, i.e., when the location of the support curves is unknown, is in agreement with the general result in (Gardner, 1988a) where it is shown that in order to consistently estimate an autocorrelation function, the nonstationarity must be of periodic or almost periodic type or with known structure. This result is also in agreement with the fact that a fraction-of-time or non-stochastic characterization for SC signals is not available.

For a comprehensive treatment on the characterization of SC processes and on the problem of their statistical function estimation, see (Napolitano, 2012, Chaps. 4, 5).

13.2 Second-order characterization

Definition 13.1. Let $x(t)$ be a finite-power continuous-time complex-valued second-order harmonizable stochastic process not containing any finite-strength additive sine-wave component. The process $x(t)$ is said to be *spectrally correlated* (Napolitano, 2003), (Napolitano, 2012, Sec. 4.2), if its Loève bifrequency spectrum (A.19) has spectral masses concentrated on a countable set of support curves. That is, it can be expressed in the two equivalent forms

$$\mathcal{S}_{\boldsymbol{x}}(f_1, f_2) \triangleq \mathrm{E}\left\{X(f_1)\, X^{(*)}(f_2)\right\}$$

$$= \sum_{k \in \mathbb{I}} S_{\boldsymbol{x}}^{(k)}(f_1)\, \delta\left(f_2 - \Psi_{\boldsymbol{x}}^{(k)}(f_1)\right) \tag{13.1a}$$

$$= \sum_{k \in \mathbb{I}} G_{\boldsymbol{x}}^{(k)}(f_2)\, \delta\left(f_1 - \Phi_{\boldsymbol{x}}^{(k)}(f_2)\right). \tag{13.1b}$$

In (13.1a) and (13.1b), subscript $\boldsymbol{x} \equiv x x^{(*)}$, $\delta(\cdot)$ is Dirac delta, $\mathbb{I}$ is a countable set, and $X(f)$ denotes the Fourier transform of $x(t)$ in the sense of distributions (Section A.3), (Gel'fand and Vilenkin, 1964, Chap. 3), (Henniger, 1970), (Napolitano, 2012, Secs. 1.1.2, 4.2.1). □

In (13.1a) and (13.1b), the curves $f_2 = \Psi_{\boldsymbol{x}}^{(k)}(f_1)$ or, equivalently, $f_1 = \Phi_{\boldsymbol{x}}^{(k)}(f_2)$, describe the support of $\mathcal{S}_{\boldsymbol{x}}(f_1, f_2)$, and the complex-valued functions $S_{\boldsymbol{x}}^{(k)}(f_1)$ and $G_{\boldsymbol{x}}^{(k)}(f_2)$, called *spectral correlation density functions*, represent the density of the Loève spectrum on its support curves.

The functions $S_{\boldsymbol{x}}^{(k)}(f_1)$ and $G_{\boldsymbol{x}}^{(k)}(f_2)$ are not unambiguously determined. They can be chosen with supports such that the functions $\Psi_{\boldsymbol{x}}^{(k)}(\cdot)$ and $\Phi_{\boldsymbol{x}}^{(k)}(\cdot)$ are invertible and the curves $f_2 = \Psi_{\boldsymbol{x}}^{(k)}(f_1)$ and $f_2 = \Psi_{\boldsymbol{x}}^{(k')}(f_1)$ for $k \neq k'$ intercept at most in a countable set of points.

Assumption 13.2. *The functions $\Psi_{\boldsymbol{x}}^{(k)}(\cdot)$ and $\Phi_{\boldsymbol{x}}^{(k)}(\cdot)$ are assumed to be invertible and differentiable.* □

Theorem 13.3. *Under Assumption 13.2, the functions $\Psi_{\boldsymbol{x}}^{(k)}(\cdot)$ and $\Phi_{\boldsymbol{x}}^{(k)}(\cdot)$, for each k, are inverse one another:*

$$\Phi_{\boldsymbol{x}}^{(k)}(\cdot) = \left[\Psi_{\boldsymbol{x}}^{(k)}(\cdot)\right]^{-1}. \tag{13.2}$$

In addition, from the variable change formula in the argument of Dirac delta (Zemanian, 1987, Sec. 1.7) it follows that

$$S_{\boldsymbol{x}}^{(k)}(f_1) = \left|\Psi_{\boldsymbol{x}}^{(k)\prime}(f_1)\right| G_{\boldsymbol{x}}^{(k)}\left(\Psi_{\boldsymbol{x}}^{(k)}(f_1)\right) \tag{13.3a}$$

$$G_{\boldsymbol{x}}^{(k)}(f_2) = \left|\Phi_{\boldsymbol{x}}^{(k)\prime}(f_2)\right| S_{\boldsymbol{x}}^{(k)}\left(\Phi_{\boldsymbol{x}}^{(k)}(f_2)\right) \tag{13.3b}$$

where superscript ′ denotes derivative. □

Accounting for (A.19)–(A.20b), we can formally write

$$\mathrm{d}\gamma_{\boldsymbol{x}}(f_1, f_2) = \mathrm{E}\left\{X(f_1)\, X^{(*)}(f_2)\right\}\, \mathrm{d}f_1\, \mathrm{d}f_2\,. \tag{13.4}$$

Thus, the bounded variation condition (A.8), which is consequence of the harmonizability assumption, for SC processes can be written as (Napolitano, 2003)

$$\int_{\mathbb{R}^2} |\mathrm{d}\gamma_{\boldsymbol{x}}(f_1, f_2)| = \sum_{k\in\mathbb{I}} \int_{\mathbb{R}} \left|S_{\boldsymbol{x}}^{(k)}(f_1)\right|\, \mathrm{d}f_1 = \sum_{k\in\mathbb{I}} \int_{\mathbb{R}} \left|G_{\boldsymbol{x}}^{(k)}(f_2)\right|\, \mathrm{d}f_2 < \infty\,. \tag{13.5}$$

Theorem 13.4. *The* (conjugate) autocorrelation *and the* time-varying spectrum *(see (A.22)) of SC processes are given by*

$$\begin{aligned}\mathcal{R}_{\boldsymbol{x}}(t, \tau) &\triangleq \mathrm{E}\left\{x(t+\tau)\, x^{(*)}(t)\right\} \\ &= \int_{\mathbb{R}^2} e^{j2\pi[f_1(t+\tau)+(-)f_2 t]}\, \mathrm{d}\gamma_{\boldsymbol{x}}(f_1, f_2) \\ &= \sum_{k\in\mathbb{I}} \int_{\mathbb{R}} S_{\boldsymbol{x}}^{(k)}(f)\, e^{j2\pi f(t+\tau)}\, e^{(-)j2\pi \Psi_{\boldsymbol{x}}^{(k)}(f)t}\, \mathrm{d}f\,,\end{aligned} \tag{13.6}$$

and

$$\mathcal{S}_{\boldsymbol{x}}(t,f) \triangleq \int_{\mathbb{R}} \mathcal{R}_{\boldsymbol{x}}(t,\tau)\, e^{-j2\pi f \tau} \mathrm{d}\tau$$
$$= \sum_{k\in\mathbb{I}} S_{\boldsymbol{x}}^{(k)}(f)\, e^{j2\pi[f+(-)\Psi_{\boldsymbol{x}}^{(k)}(f)]t}\,, \tag{13.7}$$

respectively.

Proof. Eq. (13.6) is obtained by substituting (13.1a) and (A.20b) into (A.7).
Eq. (13.7) is obtained by Fourier transforming with respect to τ both sides of (13.6). □

From (13.7) we have that the time-varying spectrum of SC processes is an almost-periodic function of t whose generalized Fourier series expansion has both coefficients and frequencies depending on f. An equivalent result is obtained for the Rihaczek distribution (A.23). Analogously, the Wigner spectrum (A.24a), (A.24b) of SC processes is an almost-periodic function of t' with both coefficient and frequencies depending on f (Napolitano, 2012, Sec. 4.2).

Fact 13.5. *ACS processes are obtained as special case of SC processes when the support curves of the Loève bifrequency spectrum are* $\Psi_{\boldsymbol{x}}^{(k)}(f) = (-)(\alpha - f)$*, where k is an index in one-to-one correspondence with the cycle frequencies* α*. That is, the support curves are lines with slopes* ± 1 *(see (1.56)). In addition, the time-varying spectrum is an almost-periodic function of t whose generalized Fourier series expansion has coefficients depending on f and constant frequencies coincident with the cycle frequencies (see (1.57)).* □

Example 13.6. As an example of SC process $x(t)$, consider the frequency warped version of an ACS signal $z(t)$, that is,

$$x(t) = \int_{\mathbb{R}} Z(\psi(f))\, e^{j2\pi f t} \mathrm{d}f \tag{13.8}$$

where $\psi(f)$ is the frequency warping function and $Z(f)$ is the Fourier transform of $z(t)$. The Loève bifrequency spectrum of $x(t)$ is given by

$$\begin{aligned}
&\mathrm{E}\Big\{X(f_1)\, X^{(*)}(f_2)\Big\} \\
&\quad = \mathrm{E}\Big\{Z\Big(\psi(f_1)\Big)\, Z^{(*)}\Big(\psi(f_2)\Big)\Big\} \\
&\quad = \sum_{\alpha\in A} S_z^{\alpha}\Big(\psi(f_1)\Big)\, \delta\Big(\psi(f_2) - (-)(\alpha - \psi(f_1))\Big) \\
&\quad = \sum_{\alpha\in A} S_z^{\alpha}\Big(\psi(f_1)\Big)\, \Big|\phi'((-)(\alpha - \psi(f_1)))\Big|\, \delta\Big(f_2 - \phi((-)(\alpha - \psi(f_1)))\Big)
\end{aligned} \tag{13.9}$$

where $\phi(\cdot)$ is the inverse function of $\psi(\cdot)$.

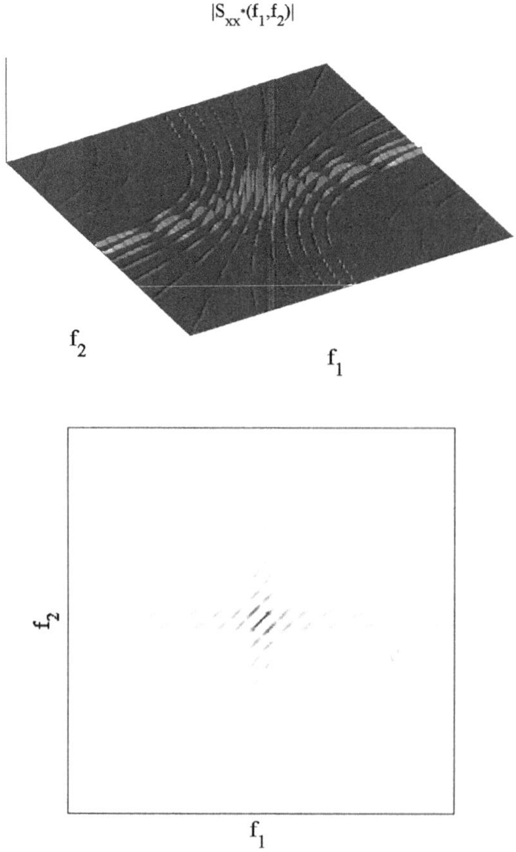

FIGURE 13.1 (Top) magnitude and (bottom) 2-dimensional gray-scale elevation map (or "checkerboard" plot) of the density of Loève bifrequency spectrum of the frequency warped PAM signal as function of f_1 and f_2. *Source: (Napolitano, 2016a) Copyright of Elsevier.*

For a PAM signal $z(t)$ and a frequency-warping function $\psi(f) = B_m \tan^{-1}(f/B_s)$, in Fig. 13.1, (top) magnitude and (bottom) 2-dimensional gray-scale elevation map ("checkerboard" plot) of the density of the Loève bifrequency spectrum (with $(*) = *$) of the frequency warped PAM signal $x(t)$ are reported. The support curves are described by the equations

$$f_2 = \Psi_{\boldsymbol{x}}^{(k)}(f_1) = \phi(\psi(f_1) - k/T_p) \quad k \in \mathbb{Z} \tag{13.10}$$

where and T_p is the symbol period of $z(t)$ and k/T_p, $k \in \mathbb{Z}$, are the cycle frequencies (Section 7.3). The densities

$$S_{\boldsymbol{x}}^{(k)}(f_1) = S_{\boldsymbol{z}}^{k/T_p}\Big(\psi(f_1)\Big) \left|\phi'(\psi(f_1) - k/T_p)\right| \quad k \in \mathbb{Z} \tag{13.11}$$

are warped versions of the cyclic spectra of $z(t)$.

Note that if $z(t)$ in (13.8) is WSS, then only the term with $\alpha = 0$ is present in (13.9). Thus, since $\phi(\psi(f_1)) = f_1$ and $\phi(-\psi(f_1)) = -f_1$, also the frequency-warped process $x(t)$ is WSS, according with the results in (Franaszek, 1967), (Franaszek and Liu, 1967), (Liu and Franaszek, 1969), (Claasen and Mecklenbräuker, 1982). □

The SC processes are a generalization of the ACS processes since the spectral masses of the Loève bifrequency spectrum are concentrated on a countable set of curves rather than only on lines parallel to the main diagonal (compare Fig. 13.1 (top) and Fig. 1.2 (bottom)).

In Definition 13.1, the process $x(t)$ is assumed to not contain any finite-strength additive sine-wave component. That is, in decomposition (1.63) the almost-periodic term is identically zero.

Let $x(t)$ be given by the superposition of a zero-mean SC process and a deterministic or random almost-periodic component (Corduneanu, 1989, Sect. II.3), (Napolitano, 2012, Sec. 1.2.7), that is,

$$x(t) = \xi(t) + \sum_{\ell} x_\ell \, e^{j2\pi\lambda_\ell t} \tag{13.12}$$

with x_ℓ complex-valued deterministic coefficients or random variables, λ_ℓ deterministic frequencies, and $\xi(t)$ zero-mean and SC (Definition 13.1) with Loève bifrequency spectrum

$$\mathrm{E}\Big\{\Xi(f_1)\,\Xi^{(*)}(f_2)\Big\} = \sum_{k\in\mathbb{I}} S^{(k)}_{\xi\xi^{(*)}}(f_1)\,\delta\Big(f_2 - \Psi^{(k)}_{\xi\xi^{(*)}}(f_1)\Big) \tag{13.13}$$

where $\Xi(f)$ denotes the Fourier transform in the sense of distributions of $\xi(t)$. The Loève bifrequency spectrum of $x(t)$ is given by

$$\begin{aligned}
\mathcal{S}_{xx^{(*)}}(f_1, f_2) &= \mathrm{E}\Big\{\Big[\Xi(f_1) + \sum_{\ell_1} x_{\ell_1}\,\delta(f_1 - \lambda_{\ell_1})\Big]\Big[\Xi(f_2) + \sum_{\ell_2} x_{\ell_2}\,\delta(f_2 - \lambda_{\ell_2})\Big]^{(*)}\Big\} \\
&= \sum_{k\in\mathbb{I}} S^{(k)}_{\xi\xi^{(*)}}(f_1)\,\delta\Big(f_2 - \Psi^{(k)}_{\xi\xi^{(*)}}(f_1)\Big) \\
&\quad + \sum_{\ell_1} \mathrm{E}\Big\{x_{\ell_1}\,\Xi^{(*)}(f_2)\Big\}\,\delta(f_1 - \lambda_{\ell_1}) \\
&\quad + \sum_{\ell_2} \mathrm{E}\Big\{\Xi(f_1)\,x^{(*)}_{\ell_2}\Big\}\,\delta(f_2 - \lambda_{\ell_2}) \\
&\quad + \sum_{\ell_1}\sum_{\ell_2} \mathrm{E}\Big\{x_{\ell_1}\,x^{(*)}_{\ell_2}\Big\}\,\delta(f_1 - \lambda_{\ell_1})\,\delta(f_2 - \lambda_{\ell_2})\,.
\end{aligned} \tag{13.14}$$

That is, spectral masses are concentrated on curves, lines parallel to the axes, and isolated points in the bifrequency plane.

13.3 Discrete-time processes

Let $x_{\mathrm{d}}(n)$ be a finite-power harmonizable second-order discrete-time process with Fourier transform (defined in a generalized sense (Section A.3)) $X_{\mathrm{d}}(\nu)$ and not containing any finite-strength additive sine-wave component. The process $x_{\mathrm{d}}(n)$ is said to be SC if its Loève bifrequency spectrum (A.38) is constituted by spectral masses concentrated on a countable set of support curves (Napolitano, 2012, Sec. 4.8). That is,

$$\mathrm{E}\left\{X_{\mathrm{d}}(\nu_1)\, X_{\mathrm{d}}^{(*)}(\nu_2)\right\} = \sum_{k\in\mathbb{I}} \widetilde{S}_{\boldsymbol{x}_{\mathrm{d}}}^{(k)}(\nu_1)\, \widetilde{\delta}\left(\nu_2 - \widetilde{\Psi}_{\boldsymbol{x}_{\mathrm{d}}}^{(k)}(\nu_1)\right) \tag{13.15}$$

where $\mathbb{I}$ is a countable set and $\widetilde{\delta}(\nu) \triangleq \sum_{p\in\mathbb{Z}} \delta(\nu - p)$. For each $k \in \mathbb{I}$, the complex-valued function $\widetilde{S}_{\boldsymbol{x}_{\mathrm{d}}}^{(k)}(\nu_1)$ describes the density of spectral correlation along the support curve $\nu_2 = \widetilde{\Psi}_{\boldsymbol{x}_{\mathrm{d}}}^{(k)}(\nu_1) \bmod 1$, where mod 1 is the modulo 1 operation with values in $[-1/2, 1/2)$. Both $\widetilde{S}_{\boldsymbol{x}_{\mathrm{d}}}^{(k)}(\nu_1)$ and $\widetilde{\Psi}_{\boldsymbol{x}_{\mathrm{d}}}^{(k)}(\nu_1)$ are periodic functions of ν_1 with period 1. The spectral mass distribution is periodic with period 1 in both frequency variables ν_1 and ν_2.

By replacing (13.1a) into the right-hand side of (A.40), allows one to prove that uniformly sampling a continuous-time SC process leads to a discrete-time SC process (Napolitano, 2011), (Napolitano, 2012, Sec. 4.9). Moreover, if the continuous-time SC process is strictly band-limited and the sampling frequency is sufficiently high, then, in the main frequency domain $[-1/2, 1/2)^2$, the discrete-time SC process has spectral functions (the Loève bifrequency spectrum and the spectral correlation density functions) proportional to those of the continuous-time process but for a scale factor on the frequency axes (Napolitano, 2011), (Napolitano, 2012, Sec. 4.9).

Several aspects related to the sampling problem and the statistical function estimation in the special case of linear support curves (with non-necessarily unit slope) are treated in (Allen and Hobbs, 1992), (Lii and Rosenblatt, 2002), (Napolitano, 2010b), (Napolitano, 2012, Secs. 7.7.5, 7.7.6).

13.4 Jointly SC processes

Let $x(t)$ and $y(t)$ be finite-power continuous-time complex-valued second-order jointly harmonizable stochastic processes not containing any finite-strength additive sine-wave component. The processes are said *jointly spectrally correlated* if their *Loève bifrequency cross-spectrum* can be expressed in one of the two equivalent forms

$$\begin{aligned}\mathcal{S}_{yx^{(*)}}(f_1, f_2) &\triangleq \mathrm{E}\left\{Y(f_1)\, X^{(*)}(f_2)\right\} \\ &= \sum_{k\in\mathbb{I}} S_{yx^{(*)}}^{(k)}(f_1)\, \delta\left(f_2 - \Psi_{yx^{(*)}}^{(k)}(f_1)\right) \qquad (13.16\text{a}) \\ &= \sum_{k\in\mathbb{I}} G_{yx^{(*)}}^{(k)}(f_2)\, \delta\left(f_1 - \Phi_{yx^{(*)}}^{(k)}(f_2)\right) \qquad (13.16\text{b})\end{aligned}$$

In (13.16a) and (13.16b), the curves $f_2 = \Psi_{yx^{(*)}}^{(k)}(f_1)$ or, equivalently, $f_1 = \Phi_{yx^{(*)}}^{(k)}(f_2)$, describe the support of $\mathcal{S}_{yx^{(*)}}(f_1, f_2)$, and the complex-valued functions $S_{yx^{(*)}}^{(k)}(f_1)$ and

$G^{(k)}_{yx^{(*)}}(f_2)$, called *spectral cross-correlation density functions*, represent the density of the Loève bifrequency cross-spectrum on its support curves.

Similarly to case of the Loève bifrequency spectrum of a single process, the functions $\Psi^{(k)}_{yx^{(*)}}(f_1)$ and $\Phi^{(k)}_{yx^{(*)}}(f_2)$ can be one the inverse of the other and the functions $S^{(k)}_{yx^{(*)}}(f_1)$ and $G^{(k)}_{yx^{(*)}}(f_2)$ are not unambiguously determined and are related each other.

13.5 Estimation of the spectral cross-correlation density function

The amount of spectral cross-correlation existing between separate spectral components of two jointly SC process is characterized by the *bifrequency spectral cross-correlation density function* which is the density of the Loève bifrequency cross-spectrum on its support curves.

Two different bifrequency densities can be considered by replacing Dirac deltas with Kronecker deltas in (13.16a) or in (13.16b). Specifically, the densities corresponding to the (13.16a) and (13.16b) are given

$$\bar{\mathbb{S}}_{yx^{(*)}}(f_1, f_2) \triangleq \sum_{k \in \mathbb{I}} S^{(k)}_{yx^{(*)}}(f_1)\, \delta_{f_2 - \Psi^{(k)}_{yx^{(*)}}(f_1)} \tag{13.17a}$$

$$\bar{\mathbb{G}}_{yx^{(*)}}(f_1, f_2) \triangleq \sum_{k \in \mathbb{I}} G^{(k)}_{yx^{(*)}}(f_2)\, \delta_{f_1 - \Phi^{(k)}_{yx^{(*)}}(f_2)} \tag{13.17b}$$

In the following, the problem of estimating the bifrequency density $\bar{\mathbb{S}}_{yx^{(*)}}(f_1, f_2)$ is considered. Results for $\bar{\mathbb{G}}_{yx^{(*)}}(f_1, f_2)$ are similar. A comprehensive treatment is made in (Napolitano, 2012, Secs. 4.4–4.7). In this section, results are summarized without proof and refer to (Napolitano, 2012, Chap. 4). Results for a single process are simply obtained by considering $y(t) \equiv x(t)$.

The *(conjugate) cross-periodogram* of the two processes $x(t)$ and $y(t)$ is defined as

$$I^{(1/\Delta f)}_{yx^{(*)}}(f_1, f_2; t) \triangleq \Delta f\, Y_{\frac{1}{\Delta f}}(t, f_1)\, X^{(*)}_{\frac{1}{\Delta f}}(t, f_2) \tag{13.18}$$

where the STFTs are defined according to

$$Z_{\frac{1}{\Delta f}}(t, f) \triangleq \int_{\mathbb{R}} z(u)\, b_{\frac{1}{\Delta f}}(u - t)\, e^{-j2\pi f u}\, \mathrm{d}u \tag{13.19}$$

with $b_{\frac{1}{\Delta f}}(u)$ being a $(1/\Delta f)$-duration data-tapering window satisfying some regularity conditions (Napolitano, 2012, Assumption 4.4.5). Note that, with a small abuse of notation, the same symbol is used for the (conjugate) cyclic cross-periodogram (5.41) which is a function of $(\alpha, f; t)$ and the (conjugate) cross-periodogram (13.18) which is a function of $(f_1, f_2; t)$.

Let us consider two jointly SC processes $x(t)$ and $y(t)$ satisfying the following assumptions:

Assumptions AS1-SC

- The second-order harmonizable stochastic processes $x(t)$ and $y(t)$ are singularly and jointly spectrally correlated with function series of the densities satisfying some regularity assumptions (Napolitano, 2012, Assumptions 4.4.2a and 4.4.3a);
- The fourth-order spectral cross-cumulant of $x(t)$ and $y(t)$ is concentrated on a countable set of 3-dimensional manifolds with densities satisfying some regularity assumptions (Napolitano, 2012, Assumptions 4.4.2b and 4.4.3b);
- The functions describing the support curves are a.e. differentiable with a.e. continuous derivatives (Napolitano, 2012, Assumption 4.4.4).

Under Assumptions AS1-SC, $I_{yx^{(*)}}^{1\Delta f}(f_1, f_2; t)$ is an asymptotically ($\Delta f \to 0$) unbiased estimator of the spectral cross-correlation density but for a multiplicative factor depending on f_1 and, for each support curve, on the shape of the support curve (Napolitano, 2003, Theorem 3.1), (Napolitano, 2012, Theorem 4.4.8). The estimator, however, has an asymptotic finite variance and, hence, is not consistent (Napolitano, 2003, Theorem 3.2), (Napolitano, 2012, Theorem 4.4.9).

13.5.1 Unknown support curves

If the location of the spectral masses is unknown, time- or frequency-smoothed versions of the cross-periodogram do not provide estimates of the bifrequency spectral cross-correlation density function asymptotically unbiased and with zero asymptotic variance (Napolitano, 2003), (Napolitano, 2012, Sec. 4.5).

However, if the departure of the nonstationarity from the almost-cyclostationarity is not too large, the bifrequency spectral cross-correlation density function (13.17a) of jointly SC processes that are not jointly ACS can be estimated with some degree of reliability by the *bifrequency time-smoothed cross-periodogram*

$$S_{yx^{(*)}}^{(\Delta f,T)}(t; f_1, f_2) \triangleq I_{yx^{(*)}}^{(1/\Delta f)}(f_1, f_2; t) \underset{t}{\otimes} a_T(t) \tag{13.20}$$

where $\underset{t}{\otimes}$ denotes convolution w.r.t. t and $a_T(t)$ is a T-duration time-smoothing window satisfying some regularity conditions (Napolitano, 2012, Assumption 4.5.2).

When $(*)$ is present, the bifrequency time-smoothed cross-periodogram (13.20) represents the temporal cross-correlation (with zero lag) of two spectral components with bandwidth Δf at frequencies f_1 and f_2. It is the generalization to jointly SC processes of the time-smoothed cyclic cross-periodogram for jointly ACS processes (see (5.85)).

Under Assumptions AS1-SC, the expected value of the time-smoothed bifrequency cross-periodogram (13.20) is given by (Napolitano, 2003, Lemma 4.1), (Napolitano, 2012, Lemma 4.5.5)

$$\begin{aligned} \mathrm{E}\left\{S_{yx^{(*)}}^{(\Delta f,T)}(t; f_1, f_2)\right\} &= \Delta f \sum_{k\in\mathbb{I}} \int_{\mathbb{R}} S_{yx^{(*)}}^{(k)}(\nu_1)\, \mathcal{B}_{\Delta f}(f_1-\nu_1; t)\, \mathcal{B}_{\Delta f}^{(*)}\Big(f_2 - \Psi_{yx^{(*)}}^{(k)}(\nu_1); t\Big) \\ &\quad A_{\frac{1}{T}}\Big(\nu_1 - f_1 +_{(-)}(\Psi_{yx^{(*)}}^{(k)}(\nu_1) - f_2)\Big)\, \mathrm{d}\nu_1 \end{aligned} \tag{13.21}$$

where $A_{\frac{1}{T}}(f)$ is the Fourier transform of $a_T(t)$ and $\mathcal{B}_{\Delta f}(f;t) \triangleq B_{\Delta f}(f)\, e^{-j2\pi f t}$, with $B_{\Delta f}(f)$ Fourier transform of $b_{\frac{1}{\Delta f}}(t)$.

Let us consider two jointly SC processes $x(t)$ and $y(t)$ satisfying the following assumptions:

Assumptions AS2-SC

- There is no cluster of support curves (Napolitano, 2012, Assumption 4.5.3);
- The functions describing the support curves have a.e. bounded second-order derivatives (Napolitano, 2012, Assumption 4.5.4).

Under Assumptions AS1-SC and AS2-SC, the time-smoothed cross-periodogram has asymptotic ($T \to \infty$, $\Delta f \to 0$, with $T\Delta f \to \infty$) variance $\mathcal{O}((T\Delta f)^{-1})$ (Napolitano, 2003, Lemma 4.2), (Napolitano, 2012, Lemma 4.5.6). However, it is asymptotically biased. That is, the time-smoothed cross-periodogram (13.20) is asymptotically zero in every point of the support where the slope of the support curve is not unit in a neighborhood of the point (Napolitano, 2003, Theorem 4.1), (Napolitano, 2012, Theorem 4.5.7).

A reliable estimate (small bias and variance) can be obtained in those points where the spectral cross-correlation density is significant, provided that the observation interval is not too large and the slope of the support curve is close to one (Napolitano, 2013, Corollary 4.1), (Napolitano, 2012, Corollary 4.5.8). Specifically, let (f_1, f_2) be a point of the bifrequency plane satisfying both conditions

$$\left| f_2 - \Psi^{(k)}_{yx^{(*)}}(f_1) \right| \ll \frac{1}{T} \qquad \left| 1 +_{(-)} \Psi^{(k)}_{yx^{(*)}}{}'(f_1) \right| \ll \frac{1}{T\Delta f} \tag{13.22}$$

that is, sufficiently close to a support curve in a point where the slope of the curve is not too far from unity. In (13.22), $\Psi^{(k)}_{yx^{(*)}}{}'(\cdot)$ is the first-order derivative of $\Psi^{(k)}_{yx^{(*)}}(\cdot)$. Accounting for the variance behavior with $T\Delta f$ and the bias expression (see (13.21)), in those points (f_1, f_2) such that (13.22) is satisfied, the smoothing product $T\Delta f$ can be chosen large enough to obtain a small variance and maintaining, at the same time, small bias.

For jointly WSS and ACS signals, $\Psi^{(k)}_{yx^{(*)}}{}'(f) = -_{(-)}1$. Consequently, the second condition in (13.22) is verified and the bifrequency time-smoothed cross-periodogram (13.20) is a consistent estimator of the bifrequency spectral cross-correlation density function (13.17a) as $T \to \infty$, $\Delta f \to 0$, and $T\Delta f \to \infty$ according to the results of Section 5.2.4 (see also (Dandawaté and Giannakis, 1994a), (Dehay and Hurd, 1994), (Hurd and Miamee, 2007)).

Let

$$D^{(k)}_{\epsilon} \triangleq \{ f_1 \in \mathbb{R} : |S^{(k)}_{yx^{(*)}}(f_1)| > \epsilon \} \tag{13.23}$$

be the support of $S^{(k)}_{yx^{(*)}}(f_1)$ with approximation ϵ. For jointly SC processes, the smaller is the number

$$d_{\epsilon} \triangleq \sup_{k \in \mathbb{I}} \sup_{f_1 \in D^{(k)}_{\epsilon}} \left| 1 +_{(-)} \Psi^{(k)}_{yx^{(*)}}{}'(f_1) \right| \tag{13.24}$$

the smaller is the departure of the nonstationarity from the almost-cyclostationarity. For jointly SC processes that are not jointly ACS, the smaller is d_ϵ the smaller is the left-hand-side of the second condition in (13.22) for every f_1 and, hence, the larger is the allowed value of $T\Delta f$. That is, for jointly SC processes that are not jointly ACS, if the location of the support curves is unknown, a trade-off exists between the departure of the nonstationarity from the almost-cyclostationarity and the reliability of spectral cross-correlation measurements obtainable by a single sample path for each process.

13.5.2 Known support curves

Let us consider now the case of known support curves. The cross-periodogram frequency smoothed along the support curve $f_2 = \Psi^{(k)}_{yx^{(*)}}(f_1)$ is defined as

$$S^{(k),(T,\Delta f)}_{yx^{(*)}}(t; f_1) \triangleq \left[I^{(T)}_{yx^{(*)}}(f_1, f_2; t)\Big|_{f_2=\Psi^{(k)}_{yx^{(*)}}(f_1)} \right] \underset{f_1}{\otimes} A_{\Delta f}(f_1) \tag{13.25}$$

where $A_{\Delta f}(f_1)$ is a Δf-bandwidth unit-area frequency-smoothing window satisfying some regularity assumptions (Napolitano, 2012, Assumption 4.6.2). It is the generalization to jointly SC processes of the frequency-smoothed (conjugate) cyclic cross-periodogram for jointly ACS processes (5.63).

Under Assumptions AS1-SC and regularity assumptions on the data-tapering and frequency-smoothing windows, it results (Napolitano, 2007c), (Napolitano, 2012, Theorem 4.6.3)

$$\mathrm{E}\left\{ S^{(k),(T,\Delta f)}_{yx^{(*)}}(t; f) \right\} = \sum_{h \in \mathbb{I}} \int_{\mathbb{R}} S^{(h)}_{yx^{(*)}}(\nu)\, K^{(T,\Delta f)}(f, \nu; k, h; t)\, \mathrm{d}\nu \tag{13.26}$$

where

$$K^{(T,\Delta f)}(f, \nu; k, h; t) \triangleq \frac{1}{T} \int_{\mathbb{R}} \mathcal{B}_{\frac{1}{T}}(\lambda - \nu; t)\, \mathcal{B}^{(*)}_{\frac{1}{T}} \left(\Psi^{(k)}_{yx^{(*)}}(\lambda) - \Psi^{(h)}_{yx^{(*)}}(\nu); t \right) A_{\Delta f}(f - \lambda)\, \mathrm{d}\lambda\,. \tag{13.27}$$

The expected value (13.26) of the frequency-smoothed cross-periodogram along a given support curve $f_2 = \Psi^{(k)}_{yx^{(*)}}(f_1)$ is influenced not only by $S^{(k)}_{yx^{(*)}}(f_1)$, but also by the spectral cross-correlation densities $S^{(h)}_{yx^{(*)}}(f_1)$ relative to all the others support curves $f_2 = \Psi^{(h)}_{yx^{(*)}}(f_1)$, $h \neq k$, the influence being stronger from support curves $f_2 = \Psi^{(h)}_{yx^{(*)}}(f_1)$ closer to $f_2 = \Psi^{(k)}_{yx^{(*)}}(f_1)$ and with larger $|S^{(h)}_{yx^{(*)}}(f_1)|$. Such a phenomenon is similar to the cycle leakage phenomenon occurring in cyclic spectra estimates of ACS processes (Section 5.3.4) (Gardner, 1986a), (Gardner, 1987f, Chap. 13) and the leakage phenomenon occurring in the estimation of cyclic statistics of GACS processes (Section 12.6), (Napolitano, 2007a).

Moreover, accounting for (13.27), in the sum over h in (13.26), all terms such that $\left| \Psi^{(h)}_{yx^{(*)}}(f) - \Psi^{(k)}_{yx^{(*)}}(f) \right| \leqslant 1/T$, give a significantly nonzero contribution. Thus, in correspondence of the at most countable set of values of f in which $\Psi^{(h)}_{yx^{(*)}}(f) = \Psi^{(k)}_{yx^{(*)}}(f)$ for $h \neq k$, we have spikes in the spectral density estimate. In the special case of linear support curves, this behavior was observed in (Lii and Rosenblatt, 2002).

Under Assumptions AS1-SC and regularity assumptions on the data-tapering and frequency-smoothing windows, we have that (Napolitano, 2012, Theorem 4.7.5)

$$\lim_{\Delta f \to 0} \lim_{T \to \infty} \mathrm{E}\left\{ S^{(k),(T,\Delta f)}_{yx^{(*)}}(t; f) \right\} = S^{(k)}_{yx^{(*)}}(f)\, \mathcal{E}^{(k)}(f) \tag{13.28}$$

for all values of f such that $\Psi^{(k)}_{yx^{(*)}}(f) \neq \Psi^{(k')}_{yx^{(*)}}(f)$ for $k \neq k'$. In (13.28), the order of the two limits cannot be interchanged and $\mathcal{E}^{(k)}(f)$ is a multiplicative bias term depending on the data-tapering window and the first-order derivative of $\Psi^{(k)}_{yx^{(*)}}(f)$. The variance of the estimator is such that (Napolitano, 2012, Theorem 4.7.7)

$$\mathrm{var}\left\{ S^{(k),(T,\Delta f)}_{yx^{(*)}}(t; f) \right\} = \mathcal{O}((T\Delta f)^{-1}) \tag{13.29}$$

as $T \to \infty$, $\Delta f \to 0$, with $T\Delta f \to \infty$. From (13.28) and (13.29) it follows that as the observation interval T approaches infinity and the spectral resolution Δf approaches zero, the properly normalized (by $\mathcal{E}^{(k)}(f)$) frequency-smoothed cross-periodogram is a mean-square consistent estimator of the spectral correlation density. In addition, the expression (13.26) of the expected value of the frequency-smoothed cross-periodogram on a given support curve reveals that for finite T and Δf, leakage exists from the neighboring support curves, the influence being stronger from closer support curves and with larger spectral density.

Assumptions AS3-SC

- The spectral cross-correlation density functions are a.e. derivable with uniformly bounded derivative (Napolitano, 2012, Assumption 4.7.4);
- The processes $x(t)$ and $y(t)$ are kth-order jointly spectrally correlated with kth-order spectral cross-cumulants satisfying regularity properties (Napolitano, 2012, Assumption 4.7.8);
- The support functions $\Psi^{(k)}_{yx^{(*)}}(f)$ are invertible. The functions $\Psi^{(k)}_{yx^{(*)}}(f)$ and their inverse $\Phi^{(k)}_{yx^{(*)}}(f)$ are differentiable with uniformly bounded derivatives.

Under Assumptions AS1-SC, AS2-SC, and AS3-SC and under regularity assumptions on the data-tapering and frequency-smoothing windows, the random variables

$$U_i^{(T,\Delta f)} \triangleq \sqrt{T\Delta f}\left[S^{(k_i),(T,\Delta f)}_{yx^{(*)}}(t_i, f_i) - \mathcal{E}^{(k_i)}(f_i)\, S^{(k_i)}_{yx^{(*)}}(f_i) \right] \qquad (i = 1, 2, \ldots) \tag{13.30}$$

are asymptotically ($T \to \infty$ and $\Delta f \to 0$ with $T\Delta f \to \infty$) zero-mean jointly complex normal (Napolitano, 2012, Theorem 4.7.11).

Consistency results for cyclic spectrum estimates of ACS processes (Section 5.2), (Dandawaté and Giannakis, 1994a), (Dehay and Hurd, 1994), (Hurd and Miamee, 2007), can be obtained by specializing the results for SC processes. In particular, for ACS processes $\mathcal{E}^{(k)}(f)$ does not depend on f.

13.6 Examples and applications

In this section, examples of the usefulness of the SC model are presented. They include the modeling of communications signals in high mobility scenarios when the constraints on bandwidth, data-record length, and source mobility parameters that generally allow one to use the (joint) ACS model are not satisfied. In addition, the spectral analysis of fractional Brownian motion, the input-output characterization of multirate systems, and the spectral analysis with nonuniform frequency spacing are considered.

13.6.1 Multipath Doppler channel

A wide-band transmitted signal $x(t)$ passing through a multipath Doppler channel such that, for each path, the relative radial speed v_k between transmitter and receiver and/or reflecting moving objects can be considered constant within the observation interval, gives rise to the output signal (Section 11.2), (Napolitano, 2012, Secs. 7.3, 7.6), (Van Trees, 1971b, pp. 339–340)

$$y(t) = \sum_{k=1}^{K} a_k \, x(s_k t - d_k) \, e^{j2\pi \nu_k t} \,. \tag{13.31}$$

In (13.31), for each path k, a_k is the complex gain, d_k the delay, s_k the time-scale factor, and ν_k the frequency shift. If $x(t)$ is ACS with Loève bifrequency spectrum (1.29), by substituting the Fourier transform (in a generalized sense) of (13.31) into the definition of Loève bifrequency spectrum and using (1.29) leads to (Napolitano, 2003), (Napolitano, 2012, Secs. 7.7.2)

$$\begin{aligned} \mathrm{E}\{Y(f_1)\, Y^*(f_2)\} = \sum_{k_1=1}^{K} \sum_{k_2=1}^{K} \sum_{\alpha} \frac{a_{k_1} a_{k_2}^*}{|s_{k_1}|} \, e^{-j2\pi(f_1-\nu_{k_1}) d_{k_1}/s_{k_1}} \, e^{j2\pi(f_1/s_{k_1}-\alpha-\nu_{k_1}/s_{k_1}) d_{k_2}} \\ S_{\boldsymbol{x}}^{\alpha}\left(\frac{f_1-\nu_{k_1}}{s_{k_1}}\right) \delta\left(f_2 - \frac{s_{k_2}}{s_{k_1}} f_1 - \nu_{k_2} + s_{k_2}\alpha + \frac{s_{k_2}}{s_{k_1}} \nu_{k_1}\right) . \end{aligned} \tag{13.32}$$

That is, $y(t)$ is SC with Loève bifrequency spectrum having spectral masses concentrated on lines with slopes s_{k_2}/s_{k_1}. More generally, in mobile multi-input multi-output (MIMO) multipath Doppler communications channels, if the transmitted signals are jointly ACS, the received signals are jointly SC with Loève bifrequency cross-spectra having spectral masses concentrated on lines with nonunit slopes (Napolitano, 2012, Sec. 4.2.4.2).

In Fig. 13.2, the magnitude of the bifrequency spectral correlation density, as a function of f_1 and f_2, for both transmitted and received signals $x(t)$ and $y(t)$ is reported in the case of a multipath Doppler channel with $K = 2$. The case of $x(t)$ PAM signal with Nyquist-shaped pulse and stationary-white binary modulating sequence is considered (Section 7.3). Since

(A) $|\bar{\mathbb{S}}_{\boldsymbol{x}}(f_1, f_2)|$ (ACS)

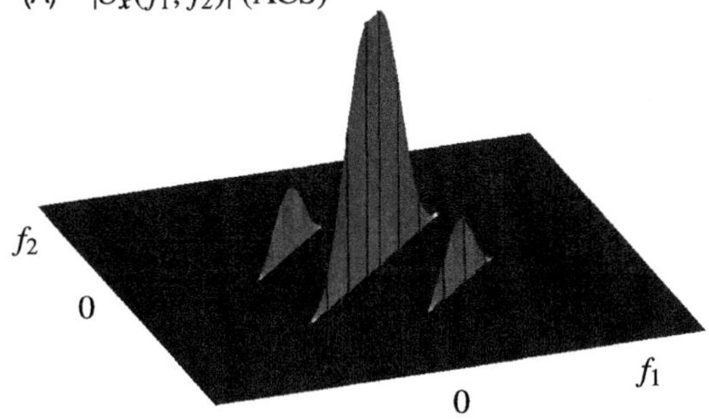

(B) $|\bar{\mathbb{S}}_{\boldsymbol{y}}(f_1, f_2)|$ (SC)

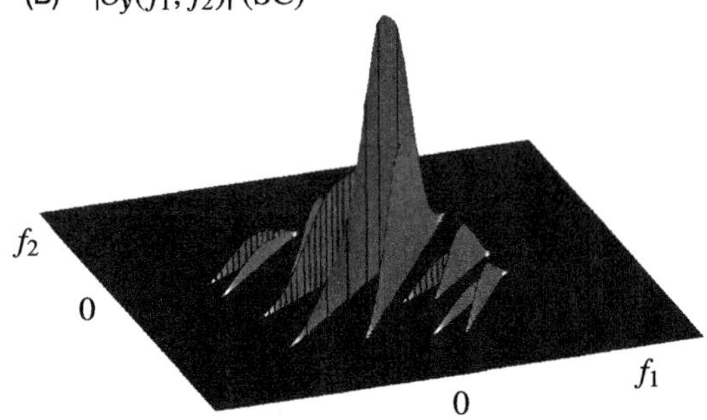

FIGURE 13.2 ACS input and SC output of the multipath Doppler channel existing between transmitter and receiver with constant relative radial speeds for each path. Magnitude of the bifrequency spectral correlation density (A) of the input ACS signal $x(t)$ and (B) of the output SC signal $y(t)$ as functions of f_1 and f_2.

$x(t)$ is cyclostationary with cycle frequencies $\beta_h = h/T_p$, $h \in \{0, \pm 1\}$, where T_p is the symbol period, according to (1.29), the support of its bifrequency spectral correlation density is constituted by three lines parallel to the main diagonal (Fig. 13.2A). In contrast, according to (13.32), the signal $y(t)$ is SC with bifrequency spectral correlation density whose support are lines with nonunit slopes (Fig. 13.2B).

13.6.2 Moving source location

Let $x(t)$ be the complex envelope of a signal transmitted by a source in relative motion with respect to two sensors and let $r_1(t)$ and $r_2(t)$ denote the noisy received complex-envelope signals. If the relative radial speeds of source and sensors can be assumed constant within the observation interval and the narrow-band condition (11.16) is not verified (but (11.15) holds), and, in addition, (11.17) holds and (11.18) does not hold, then the received signals are given by (Section 11.2, case 3)

$$\begin{aligned} r_k(t) &= y_k(t) + n_k(t) \\ &= a_k\, x(s_k t - d_k)\, e^{j2\pi \nu_k t} + n_k(t) \qquad (k = 1, 2) \end{aligned} \tag{13.33}$$

where d_k, ν_k, and s_k are the delay, the frequency shift, and the time-scale factor for the kth sensor, respectively, and $n_k(t)$ are the disturbance signals independent of $x(t)$.

If the source signal $x(t)$ is ACS, the received signal $y_k(t)$ on each sensor is in turn ACS but with different cyclic features. Specifically, if α_x and β_x are a cycle frequency and a conjugate cycle frequency of $x(t)$, the received signal $y_k(t)$ on the kth sensor has a cycle frequency $s_k\alpha_x$ and a conjugate cycle frequency $s_k\beta_x + 2\nu_k$ (Section 8.11). In contrast, $y_1(t)$ and $y_2(t)$ are not jointly ACS but, rather, jointly SC. For every cycle frequency α_x of $x(t)$, the Loève bifrequency cross-spectrum (with $(*)$ present) of $y_1(t)$ and $y_2(t)$ has spectral masses concentrated on the line $f_2 = -(s_1\alpha_x - (f_1 - \nu_\Delta)/s_\Delta)$ of the bifrequency plane (f_1, f_2), where $s_\Delta = s_2/s_1$ is the time-scale ratio (TSR), and $\nu_\Delta = \nu_2 - s_\Delta \nu_1$ is the frequency-difference-of-arrival (FDOA) (Napolitano, 2014).

A procedure referred to as wide-band spectral coherence alignment (WB-SPECCOA) is proposed in (Napolitano, 2014) to estimate TSR, FDOA, and time-difference-of-arrival (TDOA) $\tau_\Delta = (d_2 - d_1)/s_2$. Specifically, estimates of parameters s_k, ν_k (and, hence of TSR and FDOA) are obtained starting from (conjugate) cycle frequency estimates (Section 8.10). The estimation of the TDOA is based on the characterization of $r_1(t)$ and $r_2(t)$ as jointly SC and the assumption that the noise terms $n_1(t)$ and $n_2(t)$ do not exhibit spectral cross-correlation on the same lines of the bifrequency plane where $y_1(t)$ and $y_2(t)$ exhibit spectral cross-correlation. The density of spectral cross-correlation along $f_2 = -(s_1\alpha_x - (f_1 - \nu_\Delta)/s_\Delta)$ can be expressed in terms of the cyclic spectrum of $r_1(t)$ as

$$S^{(k)}_{r_2 r_1^*}(f_1) = \frac{a_\Delta}{|s_\Delta|}\, e^{-j2\pi(f_1-\nu_\Delta)\tau_\Delta}\, S^{s_1\alpha_x}_{r_1 r_1^*}\left(\frac{f_1 - \nu_\Delta}{s_\Delta}\right) \tag{13.34}$$

where $a_\Delta \triangleq (a_2/a_1)\, e^{j2\pi\nu_1(s_\Delta\tau_2 - \tau_1)}$ is the complex gain ratio (CGR). The TDOA estimate is obtained by minimizing the L^2-norm of the difference between the estimates of the left-hand-side and right-hand-side of (13.34). From TSR, FDOA, and TDOA estimates, the angle position and the relative radial speeds of the source w.r.t. the sensors can be derived.

The algorithm performance is tested in the case of a direct-sequence spread-spectrum (DSSS) signal with bit period T_p transmitted by a low Earth orbit (LEO) satellite and received by two widely separated sensors in the presence of SNR $= -3$ dB and a strong interfering signal impinging on both sensors with SIR $= -5$ dB. The DSSS signal has a bandwidth $B \simeq 1/T_c = 16.5$ MHz, where T_c is the chip period and $T_p = N_c T_c$, with $N_c = 64$ number of chips per bit. The carrier frequency is $f_c = 2$ GHz.

Due to the relatively low altitude, LEO satellites have orbital speeds of the order of tens of thousands kilometers per hour. Consequently, a strong Doppler effect is present on signals received on the Earth. In particular, the narrow-band condition is not verified even for relatively small data-record lengths. In such a case, in (13.33) the time-scale factors s_k cannot be considered unity in the argument of the complex envelope $x(\cdot)$ (Napolitano and Perna, 2014b). In particular, in the considered scenario, $|1 - s_\Delta|$ is of the order of 10^{-5} and ν_Δ is of the order of 20 kHz.

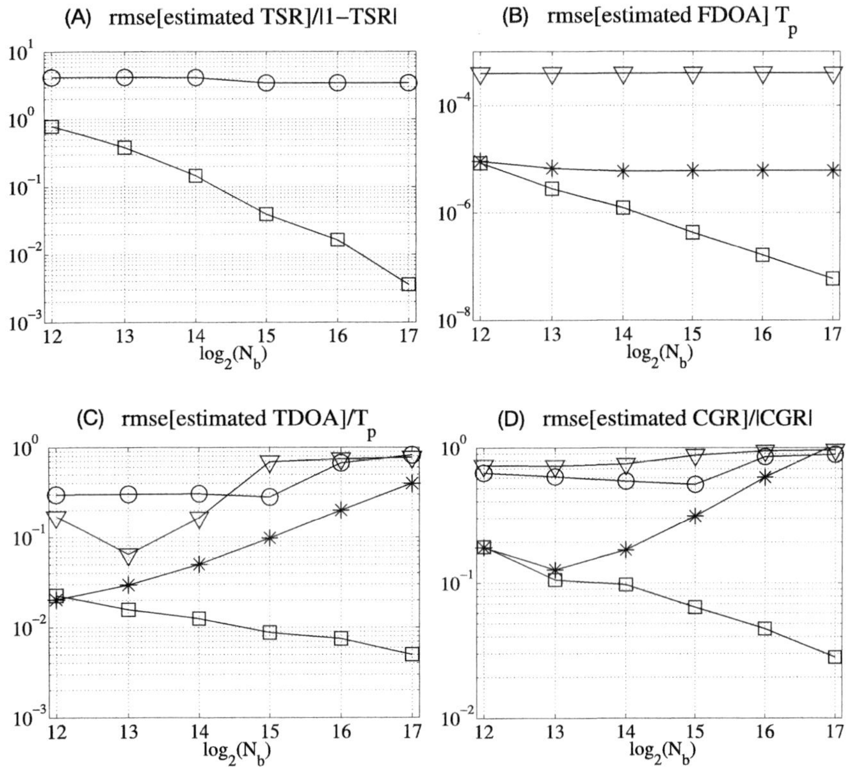

FIGURE 13.3 Normalized sample root mean squared error (rmse) of estimated parameters as function of the number $N_b = T/T_p$ of processed bits. (□) proposed WB-SPECCOA; (∗) NB-SPECCOA; (○) WB-CAF; (▽) NB-CAF. (A) rmse of normalized TSR estimate; (B) rmse of normalized FDOA estimate; (C) rmse of normalized TDOA estimate; (D) rmse of normalized CGR estimate. *Source: (Napolitano, 2016a) Copyright of Elsevier.*

For the considered DSSS signal, $T = N_b N_c T_c \simeq N_b N_c / B$ and the narrow-band condition (11.16) for the TSR reduces to $BT = N_b N_c \ll |\mathrm{v}_1 - \mathrm{v}_2|/c = 1/|1 - s_\Delta|$, that is, $N_b \ll 10^5/64 \simeq 2^{10.6}$. v_k is the relative radial speed of the satellite with respect to the kth sensor.

The WB-SPECCOA method is compared with the SPECCOA method with compensated frequency shift (Huang et al., 2002c), here referred to as narrow-band SPECCOA (NB-SPECCOA), that *assumes* the narrow-band condition (11.16) be verified and models the received signals $y_1(t)$ and $y_2(t)$ as jointly ACS ($s_1 = s_2 = 1$ in (13.33)). Moreover, the performance of the estimators based on the narrow-band cross-ambiguity function (NB-CAF) (Van Trees, 1971b, Chap. 10) and the wide-band cross-ambiguity function (WB-CAF) (Weiss, 1994), is evaluated for comparison purposes (Fig. 13.3).

The WB-SPECCOA method that properly models the received signals as jointly SC improves its performance as the number of processed bits $N_b = T/T_p$ is increased. In fact, the sample normalized root mean-squared error (rmse), evaluated by 100 Monte Carlo runs, decreases with N_b for all the estimated parameters. In contrast, the NB-SPECCOA method that models the received signals as jointly ACS performs worse and worse as the data record increases. In fact, for the considered values of N_b, the narrow-band condition (11.16) is not verified. The bad performance of the NB-CAF and WB-CAF based methods is due to the presence of correlated interference on the two sensors.

13.6.3 Signal detection

By following the guidelines for detecting a signal exhibiting second-order spectral correlation (Section 8.5), a test to detect the presence of spectral correlation at (f_1, f_2), with $f_2 = \Psi_{\boldsymbol{x}}^{(k)}(f_1)$, can be established starting from the asymptotic complex normality of the cross-periodogram frequency-smoothed along the known support curve $f_2 = \Psi_{\boldsymbol{x}}^{(k)}(f_1)$ (see (13.30)). Moreover, the test can be extended to check the presence of spectral correlation in more points of the bifrequency plane. Thus, spectrum sensing techniques based on the test for presence of spectral correlation for cyclostationary signals (Section 8.5) can be extended to the case of received SC signal (e.g., at the output of a multipath Doppler channel between transmitter and receiver with constant relative radial speed for each path (Sections 11.2 and 13.6.1)).

13.6.4 Fractional Brownian motion

The fractional Brownian motion (fBm) process $B_H(t)$ is defined by (Mandelbrot and Van Ness, 1968)

$$B_H(t) - B_H(0) = \frac{1}{\Gamma(H+1/2)} \left[\int_{-\infty}^{0} [|t-s|^{H-1/2} - |s|^{H-1/2}] \, \mathrm{d}B(s) + \int_0^t |t-s|^{H-1/2} \, \mathrm{d}B(s) \right] \tag{13.35}$$

where $B(t)$ is a zero-mean standard Brownian motion or Wiener process, $\Gamma(\cdot)$ is the Gamma function, and $H \in (0, 1)$ is the Hurst parameter. The autocorrelation function of $B_H(t)$ is (Flandrin, 1989)

$$\mathrm{E}\left\{B_H(t_1)\, B_H^*(t_2)\right\} = \Gamma(1-2H)\, \frac{\cos(\pi H)}{2\pi H} \left[|t_1|^{2H} + |t_2|^{2H} - |t_1 - t_2|^{2H} \right]. \tag{13.36}$$

Accounting for (A.21) and using the Fourier transform (to be intended in the sense of distributions) (Champeney, 1990, pp. 137-138)

$$\int_{\mathbb{R}} |t|^{\lambda-1} e^{-j2\pi f t} \, \mathrm{d}t = \frac{2\Gamma(\lambda)\, \cos(\lambda\pi/2)}{(2\pi)^\lambda} |f|^{-\lambda} \qquad \lambda > 0, \quad \lambda \notin \mathbb{N} \tag{13.37}$$

with $\lambda = 2H + 1$, the bifrequency spectrum of fBm is obtained in a generalized sense since fBm is not harmonizable in the sense of Loève (Øigård et al., 2006), (Napolitano, 2012, Sec. 4.2.4.4):

$$\mathcal{S}_{B_H B_H^*}(f_1, f_2) = |2\pi f_1|^{-(2H+1)}\, \delta(f_1 - f_2)$$

$$- |2\pi f_2|^{-(2H+1)} \, \delta(f_1) - |2\pi f_1|^{-(2H+1)} \, \delta(f_2) \, . \tag{13.38}$$

That is, spectral masses are concentrated on the main diagonal and the two frequency axes.

In general, if spectral masses of the bifrequency spectrum are concentrated on a countable set of curves that possibly contains segments parallel to the frequency axes, a representation of the bifrequency spectrum that accounts for both expressions (13.1a) and (13.1b) should be considered. See for example (13.14) and (Napolitano, 2012, Eq. (4.23)).

The density of spectral correlation of the bifrequency spectrum of fBm cannot be estimated by directly exploiting the results of Section 13.5. In fact, the density $|2\pi f|^{-(2H+1)}$ does not fulfill the required regularity conditions (Napolitano, 2012, Secs. 4.4–4.7). An artifice is exploited in the following to get density estimates by the time-smoothed cross-periodogram with two different measurements.

A fBm with $H = 1/2$ (standard Brownian motion) is considered. An estimate of a portion of the power spectrum (that is, the density along the main diagonal $f_2 = f_1$) is obtained by considering a filtered version of the fBm where the spectral components close to the zero frequency are removed (Fig. 13.4 (top)). In such a case, the contribution to the bifrequency spectrum along the two frequency axes $f_1 = 0$ and $f_2 = 0$ disappears. In fact, the density along $f_1 = 0$ and $f_2 = 0$ is representative of correlation existing between the spectral component at zero frequency and spectral components at every other frequency. An attempt to obtain an estimate of the spectral densities along the axes can be made by considering the time-smoothed cross-periodogram (13.20) with a very small amount of smoothing (see Fig. 13.4 (bottom), where high values of the densities close to $(f_1, f_2) = (0, 0)$ are clipped for representation purpose). In such a case, the contribution to the bifrequency spectrum along the main diagonal is clouded.

Once the density of spectral correlation is estimated, the Hurst parameter H can be estimated by fitting the rate of decay to zero of the spectral correlation density (Øigård et al., 2006).

13.6.5 Multirate processing

Input/output relations for several 2-input/2-output interconnections of expansors (upsamplers), decimators (downsamplers), and LTI systems are derived in (Napolitano, 2012, Sec. 4.10) when the input signals are jointly SC. The class of the jointly SC processes is shown to be closed under 2-input/2-output multirate processing.

The special case of jointly ACS input signals is analyzed in Section 3.7. It is shown that expansion and/or decimation of the same WSS or ACS process with two different rates generates jointly SC processes.

13.6.6 Spectral analysis with nonuniform frequency spacing of ACS signals

Nonlinear frequency warping is exploited for spectral analysis with nonuniform (or unequal) frequency spacing (Claasen and Mecklenbräuker, 1982), (Makur and Mitra, 2001),

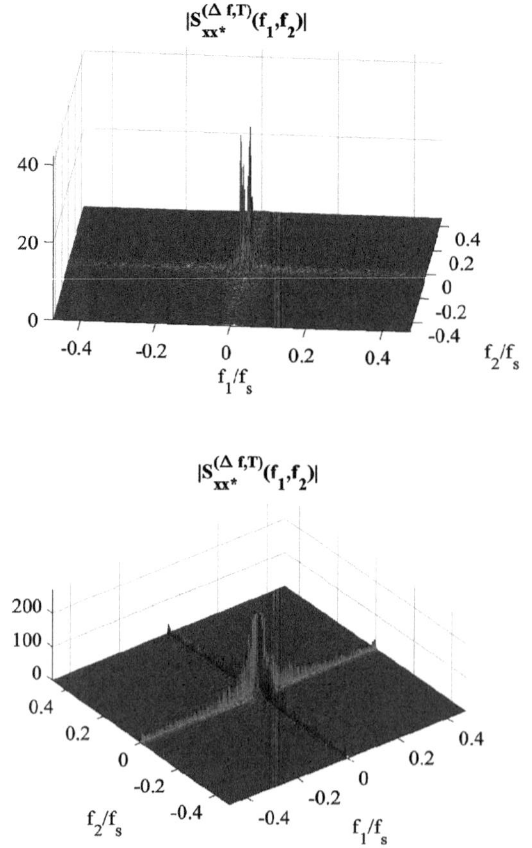

FIGURE 13.4 fBm: Estimated density of the bifrequency spectrum. (Top) power spectral density evidenced, (bottom) density along $f_1 = 0$ and $f_2 = 0$ evidenced. *Source: (Napolitano, 2016a) Copyright of Elsevier.*

(Oppenheim et al., 1971). Let $X(\nu)$ be the Fourier transform of the sequence $x(n)$, and let $\psi(\nu)$, $|\nu| \leqslant 1/2$, be a (nonlinear) invertible frequency-warping function with values in $[-1/2, 1/2]$, with inverse $\phi(\nu)$, $|\nu| \leqslant 1/2$. The sequence

$$y(n) = \int_{-1/2}^{1/2} X(\psi(\nu))\, e^{j2\pi\nu n}\, \mathrm{d}\nu \tag{13.39}$$

is a frequency-warped version of $x(n)$. Spectral analysis of $y(n)$ with uniform frequency spacing is equivalent to spectral analysis of $x(n)$ with nonuniform frequency spacing.

If the analyzed signal is WSS, then its frequency warped version is still WSS (Example 13.6), (Claasen and Mecklenbräuker, 1982). However if the analyzed signal is ACS, then its frequency warped version is SC. Specifically, let $\widetilde{\psi}(\nu)$ and $\widetilde{\phi}(\nu)$ the periodic replications with period 1 of $\psi(\nu)$ and $\phi(\nu)$, respectively. The sequence $y(n)$ is SC with Loève bifrequency spectrum $\mathrm{E}\{Y(\nu_1)\,Y^*(\nu_2)\}$ concentrated on the support curves

$$\nu_2 = \widetilde{\phi}(\widetilde{\psi}(\nu_1) - \widetilde{\alpha}) \quad \bmod 1 \tag{13.40}$$

where $\widetilde{\alpha}$ are the cycle frequencies of the discrete-time ACS signal $x(n)$. Moreover, the processes $x(n)$ and $y(n)$ are jointly SC (Napolitano, 2012, Sec. 4.13). Similarly, in passing from a continuous-time signal to a discrete-time one through a bilinear transformation, if the original continuous-time signal is ACS, the obtained discrete-time signal is SC.

In the following example, the estimation with nonuniform frequency spacing of the cyclic spectrum of an ACS process $x(n)$ at cycle frequency $\widetilde{\alpha}$ is made by estimating with uniform frequency spacing the spectral correlation density of the frequency-warped SC process $y(n)$ along the support curve given by (13.40).

Let $x(n)$ be the sequence obtained by uniformly sampling with period $T_s = 1/f_s$ a continuous-time PAM signal with full-duty-cycle rectangular pulse, binary white modulating sequence, and bit period $T_p = 8T_s$ (Section 7.3). The frequency-smoothed cyclic periodogram $S_{xx^*}^{(N,\Delta\nu)}(\widetilde{\alpha}, \nu_1)$ of the discrete-time signal is the discrete-time counterpart of the definition for a continuous-time signal (see (5.124)). The aliasing in cycle and spectral frequencies (see (3.192)) is not significant. Thus, the frequency-smoothed cyclic periodogram of the discrete-time signal is approximately proportional to that of the continuous-time signal, but for a scale factor on the frequency axis. Let $\psi(\nu_1) = \tan^{-1}(\nu_1)/2$, $|\nu_1| \leqslant 1/2$, and $\phi(\nu_2) = \tan(2\nu_2)$, $|\nu_2| \leqslant 0.22$, its inverse. In Fig. 13.5, (top) magnitude and (bottom) phase of the frequency-smoothed cyclic periodogram of $x(n)$ at cycle frequency $\widetilde{\alpha} = \widetilde{\alpha}_p = T_s/T_p$ as function of $\nu_1 = f_1 T_s$, are reported. The estimate is obtained by $N = 2^9$ samples (2^6 independent symbols) and with a frequency-smoothing window-width $\Delta\nu = 1/2^5$. The actual value is drawn by a thin line. Sample mean (thick line) and standard deviation (shaded area) are evaluated by 100 Monte Carlo trials. Although it would be correct to represent mean plus/minus standard deviation for real and imaginary parts of the frequency-smoothed cyclic periodogram, in Fig. 13.5, mean plus/minus standard deviation of magnitude and phase are reported to show the accuracy of the phase estimate when the magnitude is small.

For the discrete-time signal, the cross-periodogram frequency smoothed along a given support curve has a definition similar to that in (13.25) for continuous-time signals with the obvious changes. In Fig. 13.6, the magnitude of the cross-periodogram of the SC signal $y(n)$ frequency smoothed along the support curve $\nu_2 = \widetilde{\phi}(\widetilde{\psi}(\nu_1) - \widetilde{\alpha}_p) \bmod 1$, with $\widetilde{\alpha}_p = T_s/T_p$, say $S_{yy^*}^{(1),(N,\Delta\nu)}(\nu_1)$, is reported. The main three lobes of the cyclic spectrum of $x(n)$, due to the frequency warping, are spread over a larger bandwidth with no significant increase of the estimate variance. Due to the consistency of the frequency-smoothed cross-periodogram (Section 13.5) bias and variance can be reduced by increasing the length of the observation interval (see Fig. 13.7, where $N = 2^{11}$ samples (2^8 independent symbols) are used).

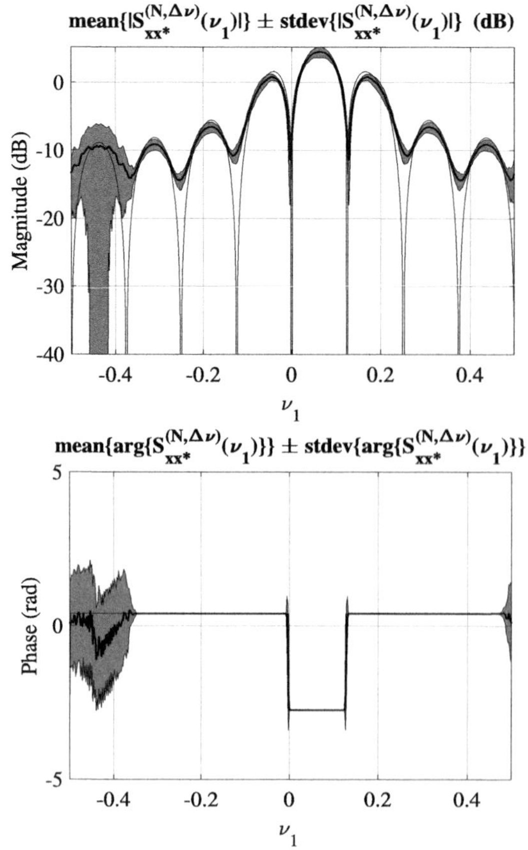

FIGURE 13.5 PAM signal $x(n)$: (Top) magnitude and (bottom) phase of the frequency smoothed cyclic periodogram at cycle frequency $\widetilde{\alpha} = \widetilde{\alpha}_p = T_s/T_p$, as a function of $\nu_1 = f_1 T_s$, estimated by $N = 2^9$ samples. Thin line: Actual value; Thick line: Sample mean; Shaded area: Region within one sample standard deviation. *Source: (Napolitano, 2016a) Copyright of Elsevier.*

13.7 Summary

Spectrally correlated (SC) processes have Loève bifrequency spectrum with spectral masses concentrated on a countable set of support curves in the bifrequency plane (Definition 13.1). The almost-cyclostationary processes are obtained as special case when the support curves are lines with unit slope.

The amount of spectral cross-correlation existing between separate spectral components of two jointly SC process is characterized by the *bifrequency spectral cross-correlation density* function which is the density of the Loève bifrequency cross-spectrum on its support curves.

If the location of the spectral masses is unknown (unknown support curves), time- or frequency-smoothed versions of the cross-periodogram do not provide estimates of the

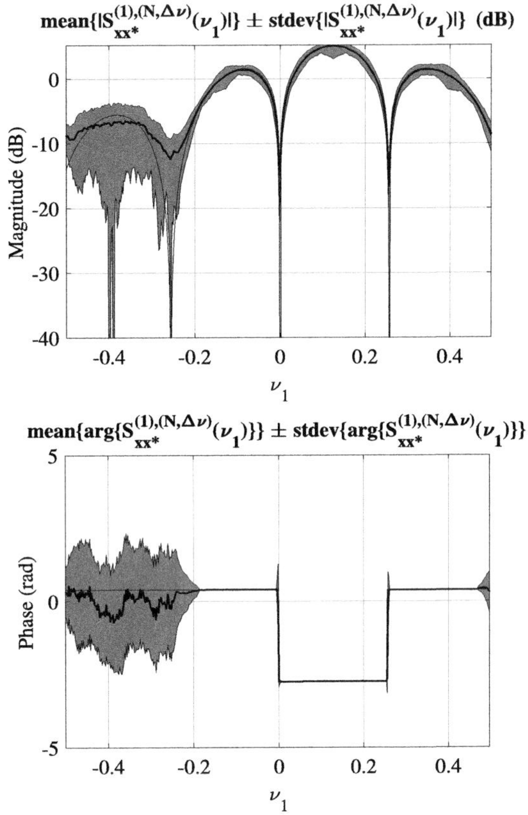

FIGURE 13.6 Frequency-warped PAM signal $y(n)$: (Top) magnitude and (bottom) phase of the cross-periodogram frequency smoothed along the known support curve $\nu_2 = \widetilde{\phi}(\widetilde{\psi}(\nu_1) - \widetilde{\alpha}_p) \mod 1$ with $\widetilde{\alpha}_p = T_s/T_p$, as a function of $\nu_1 = f_1 T_s$, estimated by $N = 2^9$ samples. Thin line: Actual value; Thick line: Sample mean; Shaded area: Region within one sample standard deviation. *Source: (Napolitano, 2016a) Copyright of Elsevier.*

bifrequency spectral cross-correlation density function asymptotically unbiased and with zero asymptotic variance. However, reliable estimates (small bias and variance) can be obtained if the departure of the nonstationarity from the almost-cyclostationarity is not too large (Section 13.5.1).

If the location of the spectral masses is known (known support curves), under mild regularity assumptions, the cross-periodogram frequency smoothed along each known curve is a mean-square consistent estimator of the spectral density along the considered curve with bias $o(1/\sqrt{T\Delta f})$ and variance $\mathcal{O}((T\Delta f)^{-1})$. In addition, the estimation errors –at different frequencies and for different support curves– times $\sqrt{T\Delta f}$ are asymptotically ($T \to \infty$ and $\Delta f \to 0$ with $T\Delta f \to \infty$) jointly complex normal random variables (Section 13.5.2).

Spectrally correlated processes can be suitably exploited to model the output signal of a *multipath Doppler channel* when the input is almost-cyclostationary (Section 13.6.1), in

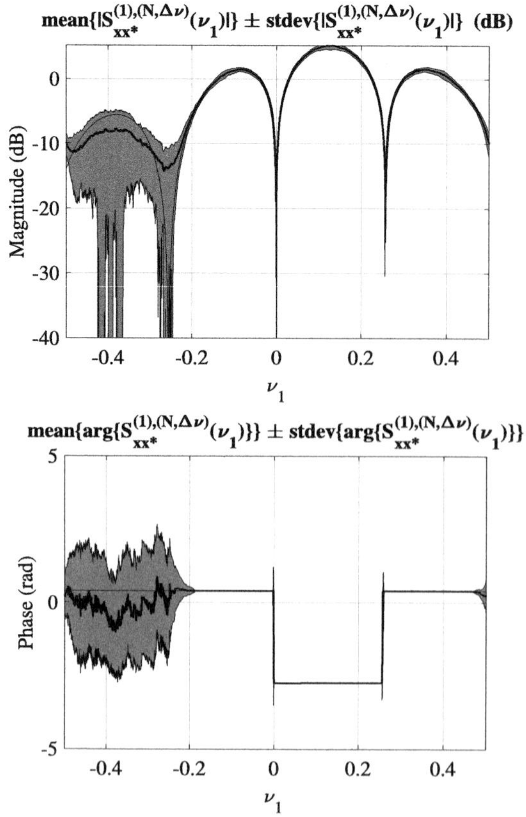

FIGURE 13.7 Frequency-warped PAM signal $y(n)$: (Top) magnitude and (bottom) phase of the cross-periodogram frequency smoothed along the known support curve $\nu_2 = \widetilde{\phi}(\widetilde{\psi}(\nu_1) - \widetilde{\alpha}_p) \mod 1$ with $\widetilde{\alpha}_p = T_s/T_p$, as a function of $\nu_1 = f_1 T_s$, estimated by $N = 2^{11}$ samples. Thin line: Actual value; Thick line: Sample mean; Shaded area: Region within one sample standard deviation. *Source: (Napolitano, 2016a) Copyright of Elsevier.*

source location problems when the narrow-band condition is not satisfied (Section 13.6.2), in characterizing fractional Brownian motion in the frequency domain (Section 13.6.4), in the analysis with nonuniform frequency spacing of almost-cyclostationary signals (Section 13.6.6).

14

Oscillatory almost-cyclostationary signals

Contents

Abstract

Second-order oscillatory almost-cyclostationary signals have autocorrelation function which is the superposition of amplitude- and angle-modulated complex sinewaves, where the modulating functions, referred to as evolutionary cyclic autocorrelation functions, depend on both time and lag parameter. This class of signals generalizes that of the almost-cyclostationary signals for which complex sinewaves are not modulated. The problem of statistical function measurements is addressed for the special case of amplitude-modulated time-warped almost-cyclostationary signals. They constitute a suitable model for the electrocardiogram signal.

Keywords

Oscillatory almost-cyclostationary, Evolutionary cyclic autocorrelation, Time warping, Irregular cyclicity, Monolateral processes, Electrocardiogram

14.1 Introduction

The way in which a random and a periodic phenomenon mix in order to generate a cyclostationary process or time series is such that the resulting process is not periodic, but its statistical functions are exactly periodic functions of time. In such a case, the mix with random phenomenon does not modify the timing of the periodic one. For example, the DSB-AM signal (Fig. 7.1) preserves the zero crossings of the carrier. Analogously, the PAM signal in Fig. 7.3 (middle) has the same zero time intervals of the periodic pulse train.

Similarly, in the case of almost-cyclostationary processes, each sine wave of the autocorrelation function is ideal and its frequency is related to one of the timing parameters of the signal.

The cyclostationary or almost-cyclostationary models are appropriate in the case of exact pace in the periodic generation mechanisms (even if with possibly incommensurate periodicities). However, this is not the case of most of the biological signals that are more suitably described by processes exhibiting irregular cyclicity (Napolitano, 2017c), (Gardner, 2018b). In communications, the relative motion –with arbitrary motion law– between transmitter and receiver and, in radar/sonar, the relative motion between sensors and targets introduces, in the received signal, a time-varying delay that results in a time-warping that modifies the timing of the transmitted signal. In mechanics, the presence of a variable load in rotating machinery gives rise to vibro-acoustic signals whose timing is not regular.

For these cases, a suitable model is obtained by resorting to oscillatory processes in the sense of Priestley. A process of this class is expressed as the integral, made with respect to the integrated complex spectrum of a WSS process, of an oscillatory function that modulates a complex exponential (Priestley, 1965).

The oscillatory almost cyclostationary (OACS) processes introduced in (Napolitano, 2016a, Sec. 6), extend the Priestley's class by considering the integral with respect to the integrated complex spectrum of an ACS process, which is referred to as the underlying ACS process. For an OACS process, the autocorrelation function is the superposition of amplitude and angle modulated sine waves whose frequencies are the cycle frequencies of the underlying ACS process (Napolitano, 2016a, Sec. 6), (Napolitano, 2017b), (Napolitano, 2017c).

The amplitude-modulated time-warped ACS processes are a notable example of OACS processes that finds application in several fields. They model the received signal in the presence of relative motion between transmitter and receiver with general motion law, when the transmitted signal is ACS (Section 11.2). In addition, they provide a suitable model for the electrocardiogram signal (Section 14.6). Monolateral ACS processes (Section 14.5) are a special case of amplitude-modulated ACS processes.

14.2 Second-order characterization

The class of the zero-mean *Priestley's oscillatory processes* admitting the representation

$$y(t) = \int_{\mathbb{R}} A_t(f)\, e^{j2\pi f t}\, \mathrm{d}Z(f) \tag{14.1}$$

is introduced in (Priestley, 1965). In (14.1), $Z(f)$ is an orthogonal increment process, that is,

$$\mathrm{E}\left\{\mathrm{d}Z(f_1)\,\mathrm{d}Z^*(f_2)\right\} = \delta(f_2 - f_1)\,\mathrm{d}\mu_{xx^*}(f_1)\,\mathrm{d}f_2 \tag{14.2}$$

with $\mu_{xx^*}(f_1)$ bounded measure on $\mathbb{R}$. In other words, $Z(f)$ is the integrated spectrum of a WSS process $x(t)$ (see (A.26)). In (14.1), the modulating functions $A_t(f)$, as functions of t, are low-pass functions (Priestley, 1965), (Napolitano, 2012, Sec. 1.1.5). From (14.1) and (14.2), and using (A.7) and (A.20b), it follows that

$$\mathrm{E}\left\{x(t_1)\,x^*(t_2)\right\} = \int_{\mathbb{R}} A_{t_1}(f)\,A^*_{t_2}(f)\,e^{j2\pi f(t_1-t_2)}\,\mathrm{d}\mu_{xx^*}(f) \tag{14.3}$$

and, hence

$$\mathrm{E}\left\{|x(t)|^2\right\} = \int_{\mathbb{R}} |A_t(f)|^2\,\mathrm{d}\mu_{xx^*}(f) \tag{14.4}$$

so that the function $\mathrm{d}F_t(f) \triangleq |A_t(f)|^2\,\mathrm{d}\mu_{xx^*}(f)$ is referred to as *evolutionary spectrum*.

The class of the oscillatory processes with representation (14.1) can be generalized by considering as $Z(f)$ the integrated complex spectrum of an ACS process.

Definition 14.1. A second-order harmonizable stochastic process $y(t)$ is said to be *oscillatory almost-cyclostationary (OACS)* with respect to the family of oscillatory functions $\{A_t(f)\,e^{j2\pi ft}\}$ if it admits the representation (14.1) where $Z(f)$ is the integrated complex spectrum of an ACS process $x(t)$, that is (see (1.24) and (1.56))

$$\mathrm{E}\left\{\mathrm{d}Z(f_1)\,\mathrm{d}Z^{(*)}(f_2)\right\} = \sum_{\alpha\in A} \delta(f_2 - _{(-)}(\alpha - f_1))\,\mathrm{d}\mu^{\alpha}_{\boldsymbol{x}}(f_1)\,\mathrm{d}f_2 \tag{14.5}$$

and the modulating functions $A_t(f)$, as functions of t, are in general low-pass functions (Napolitano, 2016a, Sec. 6). In (14.5), subscript $\boldsymbol{x} \equiv xx^{(*)}$ and $\mu^{\alpha}_{\boldsymbol{x}}(f)$, $\alpha \in A$, is a family of complex measures. The process $x(t)$ is referred to as the *underlying ACS process* for the OACS process $y(t)$. □

If $\mu^{\alpha}_{\boldsymbol{x}}(f)$ does not contain the singular component (Hurd and Miamee, 2007, p. 197), (Cramér, 1940), then $\mu^{\alpha}_{\boldsymbol{x}}(f) = S^{\alpha}_{\boldsymbol{x}}(f)\,\mathrm{d}f$, where $S^{\alpha}_{\boldsymbol{x}}(f)$ is the (conjugate) cyclic spectrum of $x(t)$. $S^{\alpha}_{\boldsymbol{x}}(f)$ contains Dirac deltas in correspondence of the jumps of $\mu^{\alpha}_{\boldsymbol{x}}(f)$.

If $Z(f)$ is an orthogonal-increment process (the set A contains only the element $\alpha = 0$, see (14.2)) then $y(t)$ is an oscillatory process in the sense of Priestley (Priestley, 1965), (Napolitano, 2012, Sec. 1.1.5).

Theorem 14.2. *Let $y(t)$ be an OACS process with representation (14.1) with $Z(f)$ satisfying (14.5). Its (conjugate) autocorrelation is given by (Napolitano, 2016a, Sec. 6)*

$$\mathrm{E}\left\{y(t+\tau)\,y^{(*)}(t)\right\} = \sum_{\alpha\in A} \rho^{\alpha}_{\boldsymbol{y}}(t,\tau)\,e^{j2\pi\alpha t} \tag{14.6}$$

where the functions

$$\rho_{y}^{\alpha}(t,\tau) \triangleq \int_{\mathbb{R}} A_{t+\tau}(f)\, A_t^{(*)}((-)(\alpha - f))\, e^{j2\pi f\tau}\, \mathrm{d}\mu_{x}^{\alpha}(f) \tag{14.7}$$

are referred to as evolutionary (conjugate) cyclic autocorrelation functions.

Proof. See Section 14.8. □

For ACS processes the (conjugate) autocorrelation function is the superposition of complex sine waves (Theorem 1.9) whose frequencies are the (conjugate) cycle frequencies and whose complex amplitudes are the (conjugate) cyclic autocorrelation functions (that depend only on the lag parameter τ).

For OACS processes, by letting

$$\rho_{y}^{\alpha}(t,\tau) = r_{y}^{\alpha}(t,\tau)\, e^{j\phi_{y}^{\alpha}(t,\tau)} \tag{14.8}$$

with $r_{y}^{\alpha}(t,\tau) \triangleq |\rho_{y}^{\alpha}(t,\tau)|$ and $\phi_{y}^{\alpha}(t,\tau) \triangleq \arg[\rho_{y}^{\alpha}(t,\tau)]$, it follows that the (conjugate) autocorrelation function is the superposition of amplitude- and angle-modulated complex sine waves (see (14.6)) whose frequencies are the (conjugate) cycle frequencies of the underlying ACS process (see (14.5)) and the amplitude and angle modulating functions are the magnitude and phase, respectively, of the evolutionary (conjugate) cyclic autocorrelation functions. That is

$$\mathrm{E}\left\{y(t+\tau)\, y^{(*)}(t)\right\} = \sum_{\alpha \in A} r_{y}^{\alpha}(t,\tau)\, e^{j[2\pi\alpha t + \phi_{y}^{\alpha}(t,\tau)]}\,. \tag{14.9}$$

The function

$$\mathrm{d}\sigma_t^{\alpha}(f) = A_t(f)\, A_t^{(*)}((-)(\alpha - f))\, \mathrm{d}\mu_{x}^{\alpha}(f) \tag{14.10}$$

is referred to as *evolutionary (conjugate) cyclic spectrum*. It results

$$\mathrm{E}\left\{y(t)\, y^{(*)}(t)\right\} = \sum_{\alpha \in A} \int_{\mathbb{R}} \mathrm{d}\sigma_t^{\alpha}(f)\, e^{j2\pi\alpha t}\,. \tag{14.11}$$

By Fourier transforming w.r.t. τ both sides of (14.6), one obtains the *time-variant spectrum*

$$\mathcal{S}_{y}(t,f) = \sum_{\alpha \in A} S_{y}^{\alpha}(t,f)\, e^{j2\pi\alpha t} \tag{14.12}$$

where $S_{y}^{\alpha}(t,f)$ are the Fourier transforms (w.r.t. τ) of the evolutionary (conjugate) cyclic autocorrelation functions $\rho_{y}^{\alpha}(t,\tau)$.

The ACS processes are a special case of OACS processes when $A_t(f) = 1$. In such case, $y(t) \equiv x(t)$ and the evolutionary (conjugate) cyclic autocorrelation functions $\rho_{y}^{\alpha}(t,\tau)$ are coincident with the (conjugate) cyclic autocorrelation functions $R_{x}^{\alpha}(\tau)$.

Theorem 14.3. *Let $y(t)$ be a harmonizable OACS process with modulating functions $A_t(f)$ and underlying ACS process $x(t)$ having (conjugate) cyclic spectra $S_{\boldsymbol{x}}^{\alpha}(f)$. The* Loève bifrequency spectrum *of $y(t)$ is given by*

$$\begin{aligned}\mathcal{S}_{\boldsymbol{y}}(f_1, f_2) &\triangleq \mathrm{E}\left\{Y(f_1)\, Y^{(*)}(f_2)\right\} \\ &= \sum_{\alpha \in A} \int_{\mathbb{R}} S_{\boldsymbol{x}}^{\alpha}(\nu)\, \mathcal{A}(f_1 - \nu; \nu)\, \mathcal{A}^{(*)}(f_2 - {\scriptstyle(-)}(\alpha - \nu); {\scriptstyle(-)}(\alpha - \nu))\, \mathrm{d}\nu \end{aligned} \tag{14.13}$$

where

$$\mathcal{A}(f; \nu) \triangleq \int_{\mathbb{R}} A_t(\nu)\, e^{-j2\pi f t}\, \mathrm{d}t\,. \tag{14.14}$$

Proof. See Section 14.8. □

If $A_t(f) = 1$, we have $\mathcal{A}(f; \nu) = \delta(f)$, $y(t) \equiv x(t)$, and the Loève bifrequency spectrum of $y(t)$ is coincident with that of $x(t)$ (see (14.5) and (1.29))

$$\mathrm{E}\left\{Y(f_1)\, Y^{(*)}(f_2)\right\} = \sum_{\alpha \in A} S_{\boldsymbol{x}}^{\alpha}(f_1)\, \delta(f_2 - {\scriptstyle(-)}(\alpha - f_1))\,. \tag{14.15}$$

Therefore, the Loève bifrequency spectrum (14.13) of an OACS process can be seen as obtained by the Loève bifrequency spectrum of the underlying ACS process $x(t)$ by spreading around the support lines $f_2 = {\scriptstyle(-)}(\alpha - f_1)$, $\alpha \in A$, the (conjugate) spectral correlation densities $S_{\boldsymbol{x}}^{\alpha}(f_1)$, $\alpha \in A$.

By making the variable change $f_1 = f$, $f_2 = {\scriptstyle(-)}(\delta f - f)$ (that is, $f = f_1$, $\delta f = f_1 + {\scriptstyle(-)} f_2$) into the expression of the Loève bifrequency spectrum, one obtains the *rotated Loève bifrequency spectrum*

$$\mathcal{S}_{\boldsymbol{y}}(f, \delta f) \triangleq \left. \mathcal{S}_{\boldsymbol{y}}(f_1, f_2)\right|_{f_1 = f,\ f_2 = {\scriptstyle(-)}(\delta f - f)} \tag{14.16a}$$

$$= \sum_{\alpha \in A} \int_{\mathbb{R}} S_{\boldsymbol{x}}^{\alpha}(\nu)\, \mathcal{A}(f - \nu; \nu)\, \mathcal{A}^{(*)}({\scriptstyle(-)}(\delta f - \alpha + f - \nu); {\scriptstyle(-)}(\alpha - \nu))\, \mathrm{d}\nu \tag{14.16b}$$

where, with a slight abuse of notation, we use the same symbol $\mathcal{S}$ to denote both normal and rotated Loève bifrequency spectrum.

In Fig. 14.1 for an ACS Process $x(t)$, the density of the Loève bifrequency spectrum $\mathcal{S}_{xx^*}(f_1, f_2)$ is represented as function of f_1/f_s and f_2/f_s. The cyclic spectrum $S_{xx^*}^{\alpha}(f_1)$ is the density of $\mathcal{S}_{xx^*}(f_1, f_2)$ along the slice $f_1 - f_2 = \alpha$. In Fig. 14.2, for an OACS process $y(t)$ with underlying ACS process $x(t)$, the Loève bifrequency spectrum $\mathcal{S}_{yy^*}(f_1, f_2)$ is represented as function of f_1/f_s and f_2/f_s.

In Fig. 14.3 for an ACS Process $x(t)$, the density of the rotated Loève bifrequency spectrum $\mathcal{S}_{xx^*}(f_1, \delta f)$ is represented as function of f_1/f_s and $\delta f/f_s$. The cyclic spectrum $S_{xx^*}^{\alpha}(f_1)$ is the density of $\mathcal{S}_{xx^*}(f_1, \delta f)$ along the slice $\delta f = \alpha$. In Fig. 14.4, for an OACS process $y(t)$

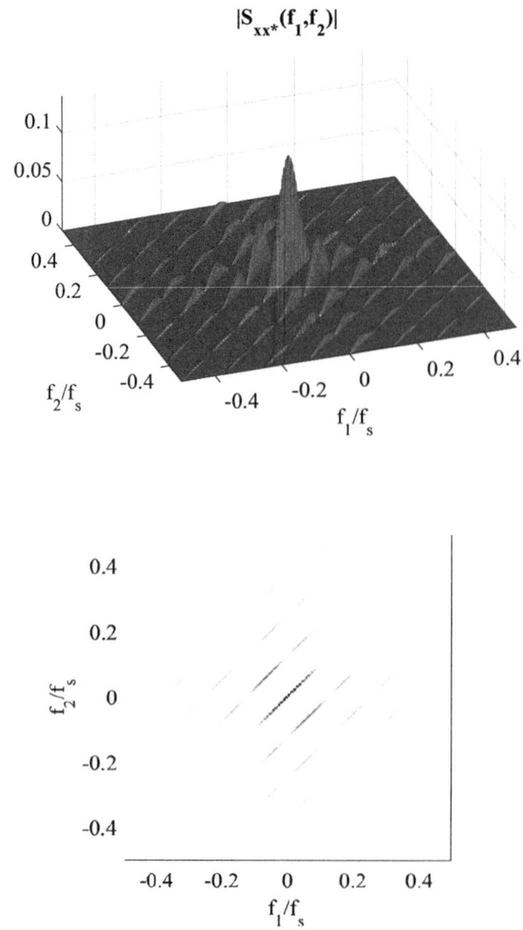

FIGURE 14.1 ACS process $x(t)$. Density of the Loève bifrequency spectrum $\mathcal{S}_{xx^*}(f_1, f_2)$ as function of f_1/f_s and f_2/f_s. (Top) magnitude and (bottom) 2-dimensional grayscale elevation map.

with underlying ACS process $x(t)$, the rotated Loève bifrequency spectrum $\mathcal{S}_{yy^*}(f_1, \delta f)$ is represented as function of f_1/f_s and $\delta f/f_s$.

Both normal and rotated Loève bifrequency spectrum of the OACS signal are a smeared version of that of the underlying ACS signal. Other examples of processes with continuous Loève bifrequency spectrum are analyzed in (Soedjack, 2002), (Wisdom et al., 2014).

14.2.1 LTV filtering of ACS processes

Every process $y(t)$ obtained by LTV filtering an ACS process

$$x(t) = \int_{\mathbb{R}} e^{j2\pi f t}\, \mathrm{d}Z(f) \tag{14.17}$$

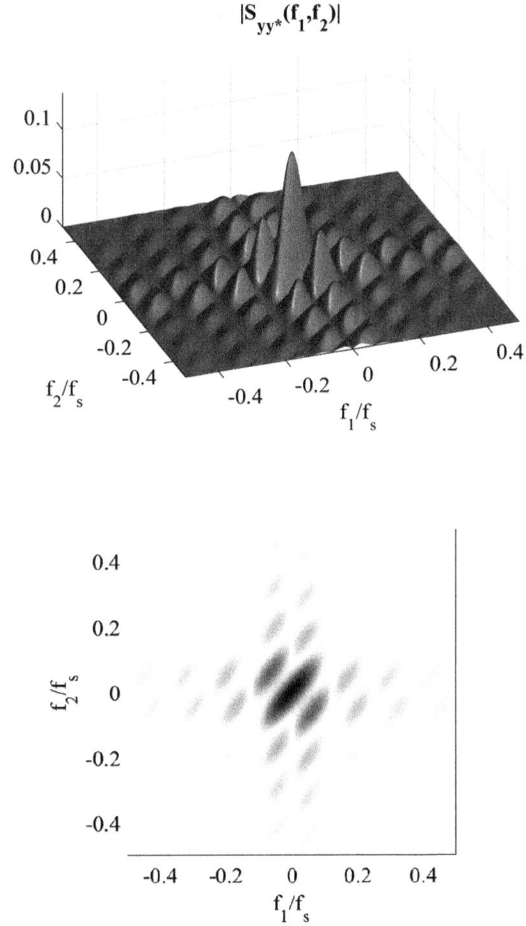

FIGURE 14.2 OACS process $y(t)$. Loève bifrequency spectrum $\mathcal{S}_{yy^*}(f_1, f_2)$ as function of f_1/f_s and f_2/f_s. (Top) magnitude and (bottom) 2-dimensional grayscale elevation map.

with $Z(f)$ satisfying (14.5) can be modeled as OACS process. In fact, let be

$$y(t) = \int_{\mathbb{R}} h(t, u)\, x(u)\, \mathrm{d}u \tag{14.18}$$

where $h(t, u)$ is the impulse response function of the LTV system. By replacing (14.17) into (14.18) and interchanging the order the of integrals, $y(t)$ can be written in the form (14.1) with

$$A_t(f) = \int_{\mathbb{R}} h(t, u)\, e^{j2\pi f(u-t)}\, \mathrm{d}u\,. \tag{14.19}$$

If the system is LAPTV (Definition 3.1) then $A_t(f)$ is an almost-periodic function of t and $y(t)$ is ACS.

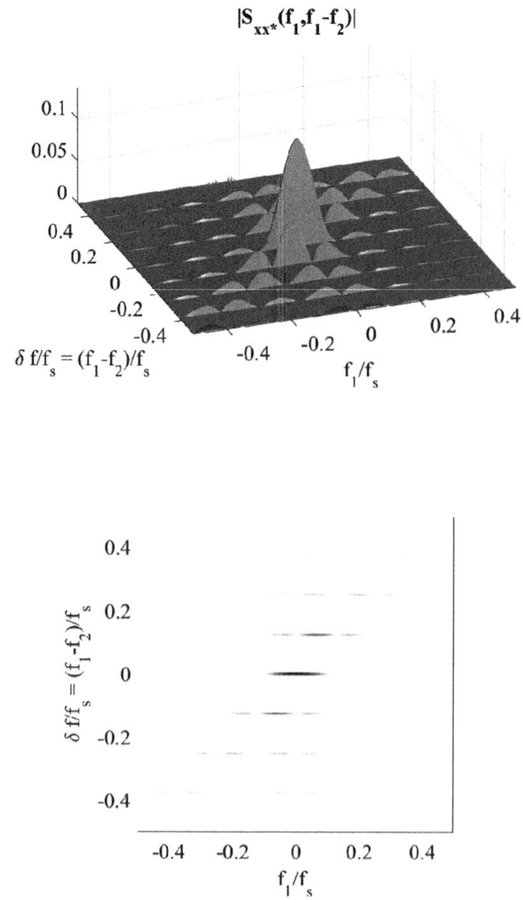

FIGURE 14.3 ACS process $x(t)$. Density of the rotated Loève bifrequency spectrum $\mathcal{S}_{xx^*}(f_1, \delta f)$ as function of f_1/f_s and $\delta f/f_s \triangleq (f_1 - f_2)/f_s$. (Top) magnitude and (bottom) 2-dimensional grayscale elevation map.

14.2.2 Modulated cyclical processes

Let us consider an OACS process with separable evolutionary (conjugate) cyclic autocorrelation function. That is,

$$\rho_y^{\alpha}(t,\tau) = r_y^{\alpha}(\tau)\, m_y^{\alpha}(t)\,. \tag{14.20}$$

In such a case, the (conjugate) autocorrelation function (14.6) specializes into

$$\mathrm{E}\left\{y(t+\tau)\, y^{(*)}(t)\right\} = \sum_{\alpha \in A} r_y^{\alpha}(\tau)\, m_y^{\alpha}(t)\, e^{j2\pi\alpha t} \tag{14.21}$$

and the process is referred to as *modulated cyclical* (MC) process, according to the terminology of (Olhede and Ombao, 2013), where the discrete-time counterpart of these

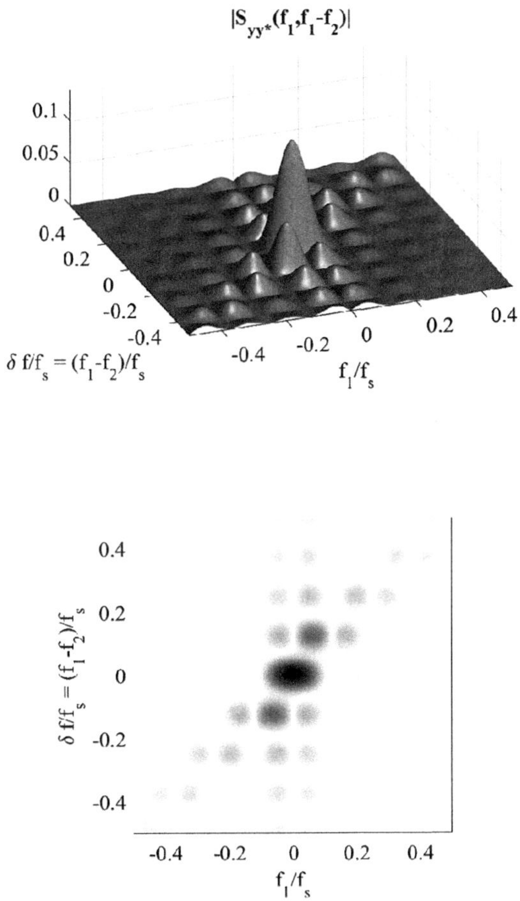

FIGURE 14.4 OACS process $y(t)$. Rotated Loève bifrequency spectrum $\mathcal{S}_{yy^*}(f_1, \delta f)$ as function of f_1/f_s and $\delta f/f_s \triangleq (f_1 - f_2)/f_s$. (Top) magnitude and (bottom) 2-dimensional grayscale elevation map.

processes is introduced when the modulating functions $m_y^\alpha(t)$ are low-pass narrow-band functions and the set A contains frequencies α all multiple of a fundamental one.

The Loève bifrequency spectrum corresponding to the (conjugate) autocorrelation function (14.21) is

$$\mathcal{S}_y(f_1, f_2) = \sum_{\alpha \in A} s_y^\alpha(f_1)\, M_y^\alpha(f_1 + (-) f_2 - \alpha) \tag{14.22}$$

where the functions $s_y^\alpha(f_1)$ and $M_y^\alpha(f)$ are the Fourier transforms of $r_y^\alpha(\tau)$ and $m_y^\alpha(t)$, respectively.

If the $m_y^\alpha(t)$ are narrow-band, the considered class of processes exhibit significant spectral correlation within thin strips around the parallel lines $f_2 = (-)(\alpha - f_1)$ of the bifrequency plane. The widths of these strips are equal to the bandwidths of the functions $m_y^\alpha(t)$. Unlike the case of the ACS and SC processes, the Loève bifrequency spectrum in (14.22) is a smooth function, provided that the functions $M_y^\alpha(f)$ are smooth. An estimator of the Loève bifrequency spectrum is proposed in (Olhede and Ombao, 2013) using a multitaper method (Thomson, 1982).

The rotated Loève bifrequency spectrum is obtained by making the variable change $f = f_1$ and $\delta f = f_1 +_{(-)} f_2$ into (14.22). With the usual slight abuse of notation we have

$$\begin{aligned} \mathcal{S}_y(f, \delta f) &= \mathcal{S}_y(f_1, f_2)\big|_{f_1=f, f_2=_{(-)}(\delta f - f)} \\ &= \sum_{\alpha \in A} s_y^\alpha(f)\, M_y^\alpha(\delta f - \alpha)\,. \end{aligned} \tag{14.23}$$

By Fourier transforming w.r.t. τ the right-hand side of (14.21) one obtains the time-variant spectrum

$$\mathcal{S}_y(t, f) = \sum_{\alpha \in A} s_y^\alpha(f)\, m_y^\alpha(t)\, e^{j2\pi\alpha t}\,. \tag{14.24}$$

An example of a process with (conjugate) autocorrelation (14.21) is given by $y(t) = a(t)\, x(t)$, where $x(t)$ is ACS with cyclic autocorrelation functions $R_x^\alpha(\tau)$, $\alpha \in A$, and $a(t)$ is a narrow-band (compared to $x(t)$) deterministic signal such that $a(t+\tau) \simeq a(t)$ for all values of τ such that $|R_x^\alpha(\tau)|$ is significantly nonzero. It results

$$\begin{aligned} \mathrm{E}\left\{y(t+\tau)\, y^{(*)}(t)\right\} &= a(t+\tau)\, a^{(*)}(t)\, \mathrm{E}\left\{x(t+\tau)\, x^{(*)}(t)\right\} \\ &\simeq a(t)\, a^{(*)}(t)\, \mathrm{E}\left\{x(t+\tau)\, x^{(*)}(t)\right\}\,. \end{aligned} \tag{14.25}$$

Thus, in (14.21) we have $r_y^\alpha(\tau) = R_x^\alpha(\tau)$, $\alpha \in A$, and $m_y^\alpha(t) = a(t)\, a^{(*)}(t)$ independent of α. In addition, in (14.22), (14.23), and (14.24) we have $s_y^\alpha(f) = S_x^\alpha(f)$.

Model (14.25) allows one to describe the presence of a time-varying gain in the propagation channel between transmitter and receiver in relative motion (Napolitano, 2012, Sec. 7.1.3).

If $a(t) = \mathrm{rect}(t/T)$, then $x(t)$ is a truncated ACS signal. Approximation in (14.25) holds provided that $|\tau| \ll T$ for all τ such that $|R_x^\alpha(\tau)|$ is significantly nonzero. In such a case, $m_y^\alpha(t) = \mathrm{rect}(t/T)$ and $M_y^\alpha(f) = T\,\mathrm{sinc}(fT)\ \forall\alpha \in A$.

ACS signals are obtained as a special case when $m_y^\alpha(t) = 1\ \forall\alpha \in A$ (and, hence $M_y^\alpha(f) = \delta(f)$).

14.3 Amplitude-modulated time-warped almost-cyclostationary processes

14.3.1 Probabilistic characterization

An *amplitude-modulation and time-warping transformation* is a linear time variant (LTV) transformation that modifies time scale and amplitude of the input signal $x(t)$. That is,

$$y(t) = a(t)\, x(\psi(t)) = \int_{\mathbb{R}} h(t, u)\, x(u)\, \mathrm{d}u \tag{14.26}$$

In (14.26),

$$h(t,u) = a(t)\,\delta(u - \psi(t)) \tag{14.27}$$

is the impulse-response function, with $\delta(\cdot)$ Dirac delta. The function $a(t)$ is the (possibly complex-valued) deterministic amplitude-modulation function. The function $\psi(t)$ is the deterministic, real-valued, asymptotically unbounded, and nondecreasing time-warping function. It should not be confused with the timing jitter which, in general, is modeled as a discrete-time random process.

If $x(t)$ is ACS with (conjugate) autocorrelation (1.55), then $y(t)$ is referred to as *amplitude-modulated (AM) time-warped (TW) ACS process*. Its (conjugate) autocorrelation is

$$\begin{aligned}
&\mathrm{E}\left\{y(t+\tau)\,y^{(*)}(t)\right\} \\
&\quad = a(t+\tau)\,a^{(*)}(t)\,\mathrm{E}\left\{x(\psi(t+\tau))\,x^{(*)}(\psi(t))\right\} && (14.28a)\\
&\quad = a(t+\tau)\,a^{(*)}(t)\sum_{\alpha\in A} R_{\boldsymbol{x}}^{\alpha}\big(\psi(t+\tau)-\psi(t)\big)\,e^{j2\pi\alpha\psi(t)}\,. && (14.28b)
\end{aligned}$$

The signal $y(t)$ is an LTV transformation of an ACS signal. Thus, $y(t)$ can be modeled as OACS signal (Section 14.2.1). By exploiting the sampling property of the Dirac delta, one has that modulating functions, evolutionary (conjugate) cyclic autocorrelation functions, and evolutionary (conjugate) cyclic spectra of $y(t)$ specialize as follows.

Modulating functions:

$$A_t(f) = a(t)\,e^{j2\pi f(\psi(t)-t)} \tag{14.29}$$

Evolutionary (conjugate) cyclic autocorrelation functions (14.7):

$$\rho_{\boldsymbol{y}}^{\alpha}(t,\tau) = a(t+\tau)\,a^{(*)}(t)\,e^{j2\pi\alpha(\psi(t)-t)}\,R_{\boldsymbol{x}}^{\alpha}\big(\psi(t+\tau)-\psi(t)\big) \tag{14.30}$$

Evolutionary (conjugate) cyclic spectra (14.10):

$$\mathrm{d}\sigma_t^{\alpha}(f) = a(t)\,a^{(*)}(t)\,e^{j2\pi\alpha(\psi(t)-t)}\,\mathrm{d}\mu_{\boldsymbol{x}}^{\alpha}(f) \tag{14.31}$$

14.3.2 Time warping

In this section, different kinds of time warping are considered and their effects on the probabilistic characterization of the AM-TW ACS process (14.26) is discussed (Napolitano, 2017c).

14.3.2.1 Strong time warping

Assumption 14.4. Strong Time Warping. *The time warping function $\psi(t)$ is such that, for $\tau \neq 0$, the increment $|\psi(t+\tau)-\psi(t)|$ is unbounded as $|t|\to\infty$, that is,*

$$\lim_{|t|\to\infty} |\psi(t+\tau)-\psi(t)| = \infty \qquad \tau \neq 0\,. \tag{14.32}$$

□

If $\psi(t)$ satisfies Assumption 14.4, it means that $\psi(t)$ is rapidly varying. In particular, if $\psi(t)$ is differentiable with first-order derivative $\dot{\psi}(t)$, from the first-order Taylor series expansion for small $|\tau|$

$$\psi(t+\tau) - \psi(t) = \dot{\psi}(t)\,\tau + \mathcal{O}(\tau^2) \tag{14.33}$$

with $\mathcal{O}$ denoting the "big oh" Landau symbol, we have that Assumption 14.4 is satisfied, provided that $|\dot{\psi}(t)|$ is unbounded as $|t| \to \infty$.

Lemma 14.5. (Napolitano, 2017c, Lemma 3.2). *Let $R_x^\alpha(\tau)$ satisfy $R_x^\alpha(\tau) \in L^1(\mathbb{R})$ or at least $\lim_{|\tau|\to\infty} R_x^\alpha(\tau) = 0$. Under Assumption 14.4 we have*

$$\lim_{|t|\to\infty} R_x^\alpha\big(\psi(t+\tau) - \psi(t)\big) = 0 \qquad \tau \neq 0 \tag{14.34}$$

from which it follows that

$$\left\langle R_x^\alpha\big(\psi(t+\tau) - \psi(t)\big)\, e^{-j2\pi\beta t} \right\rangle_t = R_x^\alpha(0)\, \delta_\tau\, \delta_\beta\,. \tag{14.35}$$

□

Let $[\theta]$ denote the integer part (floor) of a real number θ and $\bar{\theta} = \theta - [\theta]$ its fractional part.

Definition 14.6. The function $x(t)$ is said to be *uniformly distributed modulo 1* or *uniformly dense in* $(0, 1)$ if

$$\mu_R\Big(\{t \in \mathbb{R} : \bar{x}(t) \leqslant \xi\}\Big) \triangleq \big\langle \mathbf{1}_{\{\bar{x}(t)\leqslant\xi\}} \big\rangle_t = \begin{cases} 0 & \xi < 0 \\ \xi & 0 \leqslant \xi \leqslant 1 \\ 1 & \xi > 1 \end{cases} \tag{14.36}$$

where $\mu_R(A)$ denotes the relative measure of the set A (Definition 2.1) and $\mathbf{1}_S$ is the indicator function of the set S (Leśkow and Napolitano, 2006, Definition 6.7). That is, the FOT density of $\bar{x}(t)$ is the uniform-in-$(0, 1)$ density. □

Lemma 14.7. (Napolitano, 2017c, Lemma 3.3). *Let $\psi(t)$ satisfy Assumption 14.4 and $\beta\psi(t)$ be uniformly distributed modulo 1 for $\beta \neq 0$. Let $R_x^\alpha(\tau)$ satisfy $R_x^\alpha(\tau) \in L^1(\mathbb{R})$ or at least $\lim_{|\tau|\to\infty} R_x^\alpha(\tau) = 0$. It results that*

$$\left\langle R_x^\alpha\big(\psi(t+\tau) - \psi(t)\big)\, e^{-j2\pi\beta\psi(t)} \right\rangle_t = R_x^\alpha(0)\, \delta_\tau\, \delta_\beta \tag{14.37}$$

See (Napolitano, 2012, eq. (7.369)) for an example.

Proof. See (Napolitano, 2017c). □

Theorem 14.8. (Napolitano, 2017c, Theorem 3.5). *Let $y(t)$ be the AM-TW ACS signal (14.26). Under Assumption 14.4, using (14.37), the time-averaged (conjugate) autocorrelation function of $y(t)$ is given by*

$$R_y(\tau) \triangleq \left\langle \mathrm{E}\{y(t+\tau)\, y^{(*)}(t)\} \right\rangle_t = R_a^0(0)\, R_x^0(0)\, \delta_\tau \tag{14.38}$$

Consequently, the power spectrum of the TW-ACS signal $y(t)$ can be formally expressed accounting for the formal Fourier transform pair

$$R_{yy^*}(\tau) = R^0_{aa^*}(0)\, R^0_{xx^*}(0)\, \delta_\tau \stackrel{\mathcal{F}}{\longleftrightarrow} S_{yy^*}(f) = R^0_{aa^*}(0)\, R^0_{xx^*}(0)\, \mu(f) \tag{14.39}$$

where $\mu(f)$ denotes the μ functional (Section 12.3.2) (Napolitano, 2012, Sec. 3.2.1), (Napolitano, 2016a, Sec. 4.1). □

From (14.39) it follows that the finite power $R_{yy^*}(0) = R^0_{aa^*}(0)\, R^0_{xx^*}(0)$ of $y(t)$ is uniformly spread over the whole frequency axis. That is, $y(t)$ is a low-probability-of-intercept (LPI) signal.

Note that a process with time averaged autocorrelation and power spectrum as in (14.39) is not harmonizable in the sense of Loève (Napolitano, 2012, Sec. 2.2.3). Thus, no Loève bifrequency spectrum is derived in the case of strong time warping.

14.3.2.2 Moderate time warping

Assumption 14.9. Moderate Time Warping. *The time-warping function $\psi(t)$ is such that $\psi(t) = t + \epsilon(t)$ with $\epsilon(t)$ differentiable and slowly varying with respect to t (whose derivative is 1), that is,*

$$\sup_t \left|\dot{\epsilon}(t)\right| \ll 1 \tag{14.40}$$

where $\dot{\epsilon}(t)$ is the first-order derivative of $\epsilon(t)$. □

Assumption 14.10. *The deterministic function $a(t)$ is slowly varying with respect to the stochastic process $x(t)$. That is*

$$a(t+\tau) \simeq a(t) \tag{14.41}$$

for all τ such that $R^\alpha_{\boldsymbol{x}}(\tau)$ is significantly nonzero.

Under Assumptions 14.9 and 14.10, the AM-TW ACS model characterized by (14.29), (14.31), and (14.30) reduces to an MC model (Section 14.2.2), (Napolitano, 2016a, Sec. 6.2.2), (Olhede and Ombao, 2013). We have the following result.

Theorem 14.11. *Under Assumptions 14.9 and 14.10, the process $y(t)$ is MC and it results that*

1. *The evolutionary (conjugate) cyclic autocorrelation functions of $y(t)$ are given by*

$$\rho^\alpha_{\boldsymbol{y}}(t,\tau) = m^\alpha_{\boldsymbol{y}}(t)\, R^\alpha_{\boldsymbol{x}}(\tau) \tag{14.42}$$

with

$$m^\alpha_{\boldsymbol{y}}(t) \triangleq a(t)\, a^{(*)}(t)\, e^{j2\pi\alpha\epsilon(t)} \tag{14.43}$$

(independent of $x(t)$).

2. *The (conjugate) autocorrelation (14.28b) of $y(t)$ simplifies into*

$$\mathrm{E}\left\{y(t+\tau)\, y^{(*)}(t)\right\} = a(t)\, a^{(*)}(t) \sum_{\alpha \in A} e^{j2\pi\alpha\epsilon(t)}\, R_{\boldsymbol{x}}^{\alpha}(\tau)\, e^{j2\pi\alpha t}\,. \tag{14.44}$$

3. *The Loève bifrequency spectrum of $y(t)$ is*

$$\begin{aligned} \mathcal{S}_{\boldsymbol{y}}(f_1, f_2) &\triangleq \mathrm{E}\left\{Y(f_1)\, Y^{(*)}(f_2)\right\} \\ &= \sum_{\alpha \in A} S_{\boldsymbol{x}}^{\alpha}(f_1)\, M_{\boldsymbol{a}\epsilon}^{\alpha}(f_1 + (-) f_2 - \alpha) \end{aligned} \tag{14.45}$$

where

$$\begin{aligned} M_{\boldsymbol{a}\epsilon}^{\alpha}(f) &\triangleq \int_{\mathbb{R}} m_{\boldsymbol{y}}^{\alpha}(t)\, e^{-j2\pi f t}\, \mathrm{d}t \\ &= \int_{\mathbb{R}} a(t)\, a^{(*)}(t)\, e^{j2\pi\alpha\epsilon(t)}\, e^{-j2\pi f t}\, \mathrm{d}t \\ &= \mathcal{A}(f) \otimes \mathcal{A}^{(*)}((-) f) \otimes M^{\alpha}(f) \end{aligned} \tag{14.46}$$

with $\mathcal{A}(f)$ and $M^{\alpha}(f)$ Fourier transforms of $a(t)$ and $e^{j2\pi\alpha\epsilon(t)}$, respectively, and subscript $\boldsymbol{a} \triangleq aa^{(*)}$.

4. *The rotated Loève bifrequency spectrum of $y(t)$ is (with the usual slight abuse of notation)*

$$\begin{aligned} \mathcal{S}_{\boldsymbol{y}}(f, \delta f) &= \mathcal{S}_{\boldsymbol{y}}(f_1, f_2)\big|_{f_1 = f,\, f_2 = (-)(\delta f - f)} \\ &= \sum_{\alpha \in A} S_{\boldsymbol{x}}^{\alpha}(f)\, M_{\boldsymbol{a}\epsilon}^{\alpha}(\delta f - \alpha)\,. \end{aligned} \tag{14.47}$$

Proof. See Section 14.8. □

The Loève bifrequency spectrum (14.45) can be seen as obtained by the Loève bifrequency spectrum (1.56) of the ACS signal $x(t)$ by replacing the Dirac deltas with the functions $M_{\boldsymbol{a}\epsilon}^{\alpha}(\cdot)$. That is, the spectral correlation densities of the ACS signal $x(t)$ that are concentrated on the lines $f_2 = (-)(\alpha - f_1)$, $\alpha \in A$, due to the amplitude modulation and the time warping, spread on strips around these lines. The widths of the strips are given by the widths of the functions $M_{\boldsymbol{a}\epsilon}^{\alpha}(f)$, $\alpha \in A$, that is, the bandwidths of the functions $m_{\boldsymbol{y}}^{\alpha}(t) = a(t)\, a^{(*)}(t)\, e^{j2\pi\alpha\epsilon(t)}$, $\alpha \in A$.

Note that the shape of $M_{\boldsymbol{a}\epsilon}^{\alpha}(f)$ depends on α. Thus, both location in the (f_1, f_2) plane and shape of $M_{\boldsymbol{a}\epsilon}^{\alpha}(f_1 + (-) f_2 - \alpha)$ in (14.45) depend on α. The bigger is $|\alpha|$, the larger is the bandwidth of $m_{\boldsymbol{y}}^{\alpha}(t)$, that is, the width of $M_{\boldsymbol{a}\epsilon}^{\alpha}(f)$. In the case of absence of amplitude modulation ($a(t) =$ constant) we have that $M_{\boldsymbol{a}\epsilon}^{0}(f) = \delta(f)$, that is, the power spectral density (PSD) on the main diagonal $f_2 = f_1$ is not spread on a strip.

14.3.2.3 Weak time warping

Let be $\psi(t) = t + \epsilon(t)$ and let us define

$$\epsilon_0(t) \triangleq \epsilon(t) - \langle \epsilon(t) \rangle_{t \in [-T/2, T/2]} \ . \tag{14.48}$$

Assumption 14.12. Weak Time Warping. *The function $\epsilon_0(t)$ is such that*

$$2\pi |\alpha| \sup_{t \in [-T/2, T/2]} |\epsilon_0(t)| \ll 1 \tag{14.49}$$

for all α such that $|R_x^{\alpha}(\tau)|$ is significantly different from zero. □

Using (14.48), each angle-modulation term in (14.44) can be decomposed as $e^{j2\pi\alpha\epsilon(t)} = e^{j2\pi\alpha\epsilon_0(t)}\, e^{j2\pi\alpha\langle\epsilon\rangle}$. Thus, for each fixed value of α, the time variation of $\epsilon(t)$ in $e^{j2\pi\alpha\epsilon(t)}$ can be neglected within the observation interval $t \in [-T/2, T/2]$ provided that (14.49) holds.

It results

$$\begin{aligned} \left| \frac{\mathrm{d}}{\mathrm{d}t} \left[a(t)\, a^{(*)}(t) \right] \right| &= \left| \dot{a}(t)\, a^{(*)}(t) + a(t)\, \dot{a}^{(*)}(t) \right| \\ &\leqslant \left| \dot{a}(t) \right| \left| a^{(*)}(t) \right| + |a(t)| \left| \dot{a}^{(*)}(t) \right| \\ &= 2|a(t)| \left| \dot{a}(t) \right| \end{aligned} \tag{14.50}$$

Assumption 14.13. *The complex gain $a(t)$ is such that*

$$\sup_{t \in [-T/2, T/2]} |a(t)| \left| \dot{a}(t) \right| \ll 1 \tag{14.51}$$

□

Accounting for (14.50), under Assumption 14.13 the complex gain $a(t)$ can be considered constant within the observation interval. That is,

$$a(t) \simeq a_0 \qquad \forall t \in [-T/2, T/2] \ . \tag{14.52}$$

The following result can be immediately derived.

Theorem 14.14. *Under Assumptions 14.12 and 14.13, the right-hand side of (14.44) reduces to the right-hand side of (1.56) multiplied by $a_0\, a_0^{(*)}$. That is, $y(t)$ can be modeled as ACS for $t \in [-T/2, T/2]$ and its (conjugate) autocorrelation function is coincident, but for a multiplicative constant, with that of $x(t)$.* □

Note that (14.49) holds only for those values of α such that $|S_x^{\alpha}(f)|$ or $|R_x^{\alpha}(\tau)|$ is significant. Thus, large values of $|\alpha|$ should not be considered in (14.49) if the corresponding $|S_x^{\alpha}(f)|$ or $|R_x^{\alpha}(\tau)|$ are small.

Finally, note that the larger is the left-hand side of (14.49), the better the time-warping operation destroys the cyclostationarity at cycle frequency α of the underlying ACS signal $x(t)$.

14.3.3 Estimation of the time-warping and amplitude-modulation functions

Let us consider the AM-TW ACS signal defined in (14.26) in the case of moderate time warping (Section 14.3.2.2) and let α_0 be a (conjugate) cycle frequency of $x(t)$.

Let α_0 be unknown, and assume that its coarse estimate, say $\widetilde{\alpha}_0$ is available. Such an estimate can be obtained, for example, by locating one of the peaks of the PSD of $y(t + \tau)\, y^{(*)}(t)$. In fact, from (14.44) it follows that spectral components of $\mathrm{E}\{y(t+\tau)\, y^{(*)}(t)\}$ are concentrated around the cycle frequencies $\alpha \in A$. The value of τ such that $|R_{x}^{\alpha}(\tau)|$ peaks should be chosen. In most cases $\tau = 0$ is the best choice for all α.

Let $h_W(t)$ be the impulse response function of an ideal low-pass filter with monolateral bandwidth W and let us define

$$z^{(W)}(\widetilde{\alpha}_0; t, \tau) \triangleq \left[y(t+\tau)\, y^{(*)}(t)\, e^{-j2\pi\widetilde{\alpha}_0 t} \right] \otimes h_W(t) \,. \tag{14.53}$$

Let Assumptions 14.9 (moderate time warping) and 14.10 ($a(t)$ slowly varying) be satisfied and let $a(t)$ be real and $a(t) > 0$.

The signal $y(t)$ is MC (Section 14.2.2). Accounting for (14.44), we have

$$\begin{aligned} \mathrm{E}\left\{ z^{(W)}(\widetilde{\alpha}_0; t, \tau) \right\} &= \left[a^2(t)\, e^{j2\pi\alpha_0\epsilon(t)}\, R_{x}^{\alpha_0}(\tau)\, e^{j2\pi(\alpha_0-\widetilde{\alpha}_0)t} \right] \otimes h_W(t) \\ &\quad + \left[a^2(t) \sum_{\substack{\alpha\in A \\ \alpha\neq\alpha_0}} e^{j2\pi\alpha\epsilon(t)}\, R_{x}^{\alpha}(\tau)\, e^{j2\pi(\alpha-\widetilde{\alpha}_0)t} \right] \otimes h_W(t) \\ &\simeq a^2(t)\, R_{x}^{\alpha_0}(\tau)\, e^{j2\pi\alpha_0\epsilon(t)}\, e^{j2\pi(\alpha_0-\widetilde{\alpha}_0)t} \end{aligned} \tag{14.54}$$

provided that (see Section 14.8)

$$|\alpha_0 - \widetilde{\alpha}_0| + B^{\alpha_0} < W < \inf_{\substack{\alpha\in A \\ \alpha\neq\alpha_0}} \left(|\alpha - \alpha_0| - B^{\alpha} \right) - |\alpha_0 - \widetilde{\alpha}_0| \tag{14.55}$$

where B^{α} is the monolateral bandwidth of $a^2(t)\, e^{j2\pi\alpha\epsilon(t)}$. In fact, each term

$$a^2(t)\, e^{j2\pi\alpha\epsilon(t)}\, R_{x}^{\alpha}(\tau)\, e^{j2\pi(\alpha-\widetilde{\alpha}_0)t}$$

in (14.54) has most of the spectral components concentrated in the frequency interval $(\alpha - \widetilde{\alpha}_0 - B^{\alpha}, \alpha - \widetilde{\alpha}_0 + B^{\alpha})$. Therefore, in (14.55), the first inequality assures that $a^2(t)\, e^{j2\pi\alpha_0\epsilon(t)}\, e^{j2\pi(\alpha_0-\widetilde{\alpha}_0)t}$ passes unaltered through $h_W(t)$ and the second inequality assures that the terms $a^2(t)\, e^{j2\pi\alpha\epsilon(t)}\, e^{j2\pi(\alpha-\widetilde{\alpha}_0)t}$ with $\alpha \neq \alpha_0$ are filtered out. Eq. (14.55) is the corrected version of (Napolitano, 2017b, Eq. (31)) and (Napolitano, 2017c, Eq. (5.13)).

Inequalities in (14.55) put a lower limit to the separation between α_0 and the other cycle frequencies and an upper limit to the bandwidth of the angle modulated terms $e^{j2\pi\alpha\epsilon(t)}$, that is, to the strength of the time warping. Moreover, the worse is the approximation

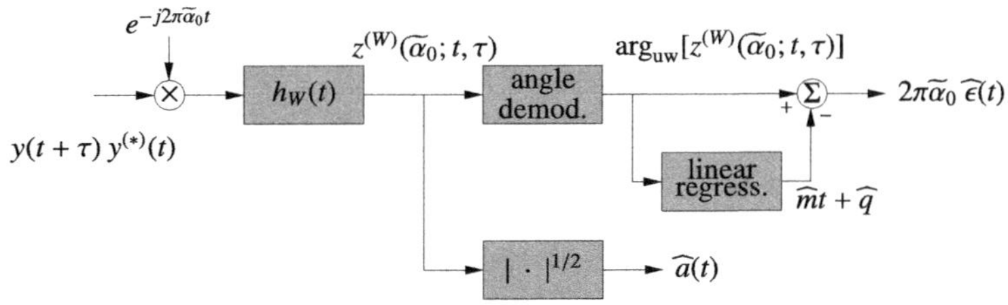

FIGURE 14.5 Block diagram for amplitude estimation and angle demodulation. The least-square linear regression is made off-line on the available data ($t = nT_s$, $n =, 0, 1, \ldots, N-1$).

$\alpha_0 \simeq \widetilde{\alpha}_0$, that is the larger is $|\alpha_0 - \widetilde{\alpha}_0|$, the more stringent are the constraints in (14.55). Condition (14.55) can be relaxed by considering the infimum only for those values of α such that $|R_{\boldsymbol{x}}^{\alpha}(\tau)|$ is large.

Thus, under condition (14.55),

$$\left|\mathrm{E}\left\{z^{(W)}(\widetilde{\alpha}_0; t, \tau)\right\}\right| \simeq a^2(t)\,\left|R_{\boldsymbol{x}}^{\alpha_0}(\tau)\right| \tag{14.56}$$

and $a(t)$ can be estimated, but for a constant (with respect to t) multiplicative factor, as

$$\widehat{a}(t) = \left|z^{(W)}(\widetilde{\alpha}_0; t, \tau)\right|^{1/2}. \tag{14.57}$$

Since $a^2(t) > 0$, the estimation of $\epsilon(t)$ is made starting from the relationship (see (14.54))

$$\begin{aligned} &\arg\left[\mathrm{E}\left\{z^{(W)}(\widetilde{\alpha}_0; t, \tau)\right\}\right] \\ &\quad \simeq \arg\left[R_{\boldsymbol{x}}^{\alpha_0}(\tau)\right] + 2\pi\alpha_0\epsilon(t) + 2\pi(\alpha_0 - \widetilde{\alpha}_0)t \mod 2\pi \end{aligned} \tag{14.58}$$

where $\arg[\cdot]$ denotes the argument of the complex number in the brackets and mod 2π is the modulo 2π operation.

The affine term in (14.58), say $mt + q$, depends on the cycle frequency error. Estimates $\widehat{m}$ and $\widehat{q}$ of the coefficients m and q can be obtained by least-squares linear regression on the available data $\arg_{uw}[z^{(W)}(\widetilde{\alpha}_0; t, \tau)]$, $t = nT_s$, $n =, 0, 1, \ldots, N-1$, where $T = NT_s$ is the sample size, T_s the sampling period, and $\arg_{uw}[\cdot]$ denotes the unwrapped phase.

Therefore, $\epsilon(t)$ can be estimated as (Fig. 14.5)

$$\widehat{\epsilon}(t) = \frac{1}{2\pi\widetilde{\alpha}_0}\left[\arg_{uw}\left[z^{(W)}(\widetilde{\alpha}_0; t, \tau)\right] - (\widehat{m}t + \widehat{q})\right]. \tag{14.59}$$

Note that, since $R_{\boldsymbol{x}}^{\alpha}(\tau)$ is generally unknown, the estimate of $\epsilon(t)$ is determined to within an unknown constant representing a fixed time delay.

By letting $\widehat{m} = 2\pi(\widehat{\alpha}_0 - \widetilde{\alpha}_0)$ (see (14.58)), the following estimator for the cycle frequency α_0 is obtained:

$$\widehat{\alpha}_0 = \frac{\widehat{m}}{2\pi} + \widetilde{\alpha}_0\,. \tag{14.60}$$

Finally, note that in order to improve performance, a two step procedure can be carried out by repeating twice the estimation procedure based on (14.53)–(14.59). Specifically, once the estimate $\widehat{\alpha}_0$ (14.60) is available after the first step, the procedure based on (14.53)–(14.59) is repeated by replacing $\widetilde{\alpha}_0$ with the estimate $\widehat{\alpha}_0$ obtained in the previous step and possibly reducing the bandwidth W of the low-pass filter $h_W(t)$.

Alternative methods for estimating $\epsilon(t)$ are proposed in (Gardner, 2018b) and (Napolitano, 2017c). For a comparison of these methods see (Napolitano and Gardner, 2016).

14.3.4 De-warping and amplitude-modulation compensation

If time warping never reverse time, then $\psi(t)$ is invertible (Gardner, 2018b). Denoted by $\psi^{-1}(t)$ its inverse, $x(t)$ can be obtained by de-warping $y(t)$, that is, by compensating the warping effect in $y(t)$. In addition, the amplitude modulation can be easily compensated. From (14.26) it follows that

$$x(t) = y(\psi^{-1}(t))/a(\psi^{-1}(t)) \tag{14.61}$$

provided that $a(\psi^{-1}(t)) \neq 0$. Thus, once the amplitude modulation function $a(t)$ and the warping function $\psi(t)$ are estimated, an estimate $\widehat{x}(t)$ of the underlying ACS signal $x(t)$ is given by

$$\widehat{x}(t) = y(\widehat{\psi}^{-1}(t))/\widehat{a}(\widehat{\psi}^{-1}(t)) \tag{14.62}$$

where $\widehat{a}(t)$ is an estimate of $a(t)$ and $\widehat{\psi}^{-1}(t)$ is the inverse function of $\widehat{\psi}(t)$ which is used as estimate of $\psi^{-1}(t)$.

Then, $\widehat{x}(t)$ can be processed with all the available tools for ACS signals (e.g., FRESH filtering).

Let us consider $\psi(t) = t + \epsilon(t)$ with $\epsilon(t)$ slowly varying. Let $\widehat{a}(t)$ and $\widehat{\epsilon}(t)$ be the estimates of $a(t)$ and $\epsilon(t)$ obtained by the procedure described in Section 14.3.3. For the purpose of de-warping, a suitable estimate of $\psi^{-1}(t)$ is given by $t - \widehat{\epsilon}(t)$ (Section 14.8). Thus, since $y(t) = x(\psi(t))/a(\psi(t)) = x(t + \epsilon(t))/a(t + \epsilon(t))$, an estimate of the signal $x(t)$ is obtained by $y(t)$ as (Section 14.8)

$$\widehat{x}(t) = \frac{y(t - \widehat{\epsilon}(t))}{\widehat{a}(t - \widehat{\epsilon}(t))} \tag{14.63}$$

provided that

1. $a(t)/\widehat{a}(t) \simeq 1$, that is, the estimate $\widehat{a}(t)$ is sufficiently accurate
2. the estimation error on $\epsilon(t)$ is sufficiently small in the sense that

$$\sup_t |\epsilon(t) - \widehat{\epsilon}(t)| \ll 1/B \tag{14.64}$$

with B bandwidth of $x(t)$.

Note that in the case of slowly varying amplitude modulation (Assumption 14.13), $a(t)$ is almost constant. If $\widehat{a}(t)$ is a close estimate of $a(t)$, no regularization is needed for the ratio in (14.63).

14.4 Cyclostationarity restoral

An alternative approach (Gardner, 2018b) for the analysis of time-warped cyclostationary signals consists of directly estimating the time-warping function $\psi(t)$ (or its inverse $\psi^{-1}(t)$) in a manner that transforms the data $y(t)$ having an autocorrelation reflecting irregular cyclicity into data having regular cyclicity. In such an approach, signals are modeled as single functions of time (FOT approach).

Let $x(t)$ exhibit cyclostationarity with at least one cycle frequency α_0. Estimates $\widehat{\psi}$ or $\varphi = \widehat{\psi}^{-1}$ of ψ or ψ^{-1} are determined such that, for the recovered signal $x_\varphi(t) = y(\varphi(t))$, the amplitude of the complex sine wave at frequency α_0 contained in the second-order lag-product $x_\varphi(t+\tau)\, x_\varphi^{(*)}(t)$ is maximized.

Let

$$\{c_k(t)\}_{k=1,\ldots,K} \tag{14.65}$$

be a set of (not necessarily orthonormal) functions. Two procedures are proposed in (Gardner, 2018b):

a. Consider the expansion

$$\varphi(t) = \widehat{\psi}^{-1}(t) = \mathbf{a}^{\mathsf{T}}\mathbf{c}(t) \tag{14.66}$$

where $\mathbf{c}(t) = [c_1(t), \ldots, c_K(t)]^{\mathsf{T}}$ and $\mathbf{a} = [a_1, \ldots, a_K]^{\mathsf{T}}$ and maximize with respect to $\mathbf{a}$ the objective function

$$J_a(\mathbf{a}) = \left|\widehat{R}_{x_\varphi}^{\alpha_0}(\tau)\right|^2 \tag{14.67}$$

with

$$\widehat{R}_{x_\varphi}^{\alpha_0}(\tau) \triangleq \frac{1}{T}\int_{t_0}^{t_0+T} x_\varphi(t+\tau)\, x_\varphi^{(*)}(t)\, e^{-j2\pi\alpha_0 t}\, \mathrm{d}t \tag{14.68}$$

$$= \frac{1}{T}\int_{t_0}^{t_0+T} y(\mathbf{a}^{\mathsf{T}}\mathbf{c}(t+\tau))\, y^{(*)}(\mathbf{a}^{\mathsf{T}}\mathbf{c}(t))\, e^{-j2\pi\alpha_0 t}\, \mathrm{d}t \tag{14.69}$$

for some specific value of lag τ. (Generalizations of $J_a(\mathbf{a})$ that include sums over more cycle frequencies α and/or lags τ are possible.)

b. Consider the expansion

$$\widehat{\psi}(t) = \mathbf{b}^{\mathsf{T}}\mathbf{c}(t) \tag{14.70}$$

where $\mathbf{b} = [b_1, \ldots, b_K]^{\mathsf{T}}$ and maximize with respect to $\mathbf{b}$ the objective function

$$J_b(\mathbf{b}) = \left|\widehat{R}_{x_\varphi}^{\alpha_0}(\tau)\right|^2 \tag{14.71}$$

with

$$\widehat{R}^{\alpha_0}_{x_\varphi}(\tau) = \frac{1}{T}\int_{\varphi(t_0)}^{\varphi(t_0+T)} y(u + \Delta^\tau_\varphi[\varphi^{-1}(u)])\, y^{(*)}(u)\, e^{-j2\pi\alpha_0\varphi^{-1}(u)}\, \dot{\varphi}^{-1}(u)\, \mathrm{d}u \tag{14.72}$$

$$\simeq \frac{1}{T}\int_{t_0}^{t_0+T} y(u + \tau/\mathbf{b}^{\mathsf{T}}\dot{\mathbf{c}}(u))\, y^{(*)}(u)\, e^{-j2\pi\alpha_0\mathbf{b}^{\mathsf{T}}\mathbf{c}(u)}\, \mathbf{b}^{\mathsf{T}}\dot{\mathbf{c}}(u)\, \mathrm{d}u \tag{14.73}$$

where (14.72) is obtained from (14.68) by the variable change $u = \varphi(t)$ and

$$\Delta^\tau_\varphi[\varphi^{-1}(u)] \triangleq \varphi[\varphi^{-1}(u) + \tau] - \varphi[\varphi^{-1}(u)] \simeq \tau[1/\dot{\varphi}^{-1}(u)] = \tau/\mathbf{b}^{\mathsf{T}}\dot{\mathbf{c}}(u) \tag{14.74}$$

The value of the vector $\mathbf{a}$ or $\mathbf{b}$ that maximizes the corresponding objective function is taken as an estimate of the coefficient vector for the expansion of $\varphi(t) = \widehat{\psi}^{-1}(t)$ or $\widehat{\psi}(t)$. The maximization can be performed by a gradient-ascent algorithm, with starting points throughout a sufficiently fine grid. With reference to the objective function in Procedure a), we have

$$\widehat{\mathbf{a}}^{(i+1)} = \widehat{\mathbf{a}}^{(i)} + \mu(i)\, \nabla J(\widehat{\mathbf{a}}^{(i)}) \tag{14.75}$$

where $\widehat{\mathbf{a}}^{(i)}$ is the estimate of $\mathbf{a}$ at iteration i and $\mu(i)$ is the step size at iteration i. The Barzilai-Borwein step size sequence (Barzilai and Borwein, 1988)

$$\mu(i) = -\frac{(\widehat{\mathbf{a}}^{(i)} - \widehat{\mathbf{a}}^{(i-1)})^{\mathsf{T}}(\nabla J(\widehat{\mathbf{a}}^{(i)}) - \nabla J(\widehat{\mathbf{a}}^{(i-1)}))}{\|\nabla J(\widehat{\mathbf{a}}^{(i)}) - \nabla J(\widehat{\mathbf{a}}^{(i-1)})\|_2^2} \tag{14.76}$$

is used since it provides fast convergence for many kinds of objective functions. The case of the objective function in Procedure b) is treated in an analogous manner.

In both cases a) and b), the gradient of the objective function must be computed. We have

$$\nabla\left|\widehat{R}^{\alpha_0}_{x_\varphi}(\tau)\right|^2 = 2\,\mathrm{Re}\left\{\widehat{R}^{\alpha_0}_{x_\varphi}(\tau)^*\, \nabla\widehat{R}^{\alpha_0}_{x_\varphi}(\tau)\right\} \tag{14.77}$$

and in the two cases the gradient expression is

Procedure a)

$$\nabla\widehat{R}^{\alpha_0}_{x_\varphi}(\tau) = \frac{1}{T}\int_{t_0}^{t_0+T} \dot{y}(\mathbf{a}^{\mathsf{T}}\mathbf{c}(t+\tau))\, y^{(*)}(\mathbf{a}^{\mathsf{T}}\mathbf{c}(t))\, e^{-j2\pi\alpha_0 t}\, \mathbf{c}(t+\tau)\, \mathrm{d}t + \frac{1}{T}\int_{t_0}^{t_0+T} y(\mathbf{a}^{\mathsf{T}}\mathbf{c}(t+\tau))\, \dot{y}^{(*)}(\mathbf{a}^{\mathsf{T}}\mathbf{c}(t))\, e^{-j2\pi\alpha_0 t}\, \mathbf{c}(t)\, \mathrm{d}t \tag{14.78}$$

Procedure b)

$$\nabla\widehat{R}^{\alpha_0}_{x_\varphi}(\tau) \simeq \frac{1}{T}\int_{t_0}^{t_0+T} \dot{y}(u + \tau/\mathbf{b}^{\mathsf{T}}\dot{\mathbf{c}}(u))\, y^{(*)}(u)\, e^{-j2\pi\alpha_0\mathbf{b}^{\mathsf{T}}\mathbf{c}(u)} \left[-\dot{\mathbf{c}}(u)\tau/\mathbf{b}^{\mathsf{T}}\dot{\mathbf{c}}(u)\right] \mathrm{d}u$$

$$+\frac{1}{T}\int_{t_0}^{t_0+T} y(u+\tau/\mathbf{b}^\mathsf{T}\dot{\mathbf{c}}(u))\, y^{(*)}(u)\, e^{-j2\pi\alpha_0\mathbf{b}^\mathsf{T}\mathbf{c}(u)} \left[\dot{\mathbf{c}}(u) - j2\pi\alpha_0\mathbf{b}^\mathsf{T}\dot{\mathbf{c}}(u)\,\mathbf{c}(u)\right] \mathrm{d}u \tag{14.79}$$

In both approaches a) and b) data must be re-interpolated in order to evaluate the gradient of the objective function at every iteration of the iterative gradient-ascent search method. This is very time-consuming. However, for $\tau = 0$, method b) (i.e., (14.79)) does not require data interpolation, and its implementation can be made very efficient. There are several important design parameters, which are discussed in (Gardner, 2018b).

14.5 Monolateral ACS signals

In this section, monolateral ACS processes are defined and characterized as special case of OACS processes.

Let $x(t)$ be an ACS process defined for $t \in \mathbb{R}$ with (conjugate) autocorrelation function (1.55) and Loève bifrequency spectrum (1.56).

A *monolateral ACS* process is defined as

$$y(t) \triangleq x(t)\,\mathrm{u}(t-t_0) \tag{14.80}$$

where

$$\mathrm{u}(t-t_0) \triangleq \begin{cases} 1 & t \geqslant t_0 \\ 0 & t < t_0 \end{cases} \tag{14.81}$$

is the unit step function centered in t_0.

The (conjugate) autocorrelation function of $y(t)$ is given by

$$\mathrm{E}\left\{y(t+\tau)\,y^{(*)}(t)\right\} = \mathrm{u}(t+\tau-t_0)\,\mathrm{u}(t-t_0)\,\mathrm{E}\left\{x(t+\tau)\,x^{(*)}(t)\right\} \tag{14.82a}$$

$$= \sum_{\alpha\in A} \mathrm{u}(t+\tau-t_0)\,\mathrm{u}(t-t_0)\,R_{\boldsymbol{x}}^{\alpha}(\tau)\,e^{j2\pi\alpha t}\,. \tag{14.82b}$$

By comparing (14.82b) with (14.6), it follows that $y(t)$ can be modeled as OACS process with evolutionary (conjugate) cyclic autocorrelation functions

$$\rho_{\boldsymbol{y}}^{\alpha}(t,\tau) = \mathrm{u}(t+\tau-t_0)\,\mathrm{u}(t-t_0)\,R_{\boldsymbol{x}}^{\alpha}(\tau)\,. \tag{14.83}$$

In addition, by comparing (14.80) with (14.1), it follows that the modulating functions are independent of f and are given by

$$A_t(f) = \mathrm{u}(t-t_0) \tag{14.84}$$

and

$$\mathcal{A}(f;\nu) \triangleq \int_{\mathbb{R}} A_t(\nu)\,e^{-j2\pi f t}\,\mathrm{d}t$$

$$= \int_{\mathbb{R}} \mathrm{u}(t - t_0)\, e^{-j2\pi f t}\, \mathrm{d}t$$
$$= \frac{1}{2}\left[\delta(f) + \frac{1}{j\pi f}\right] e^{-j2\pi f t_0} \tag{14.85}$$

Thus, by replacing (14.85) into the expression (14.13) of the Loève bifrequency spectrum of OACS processes, one has (Section 14.8)

$$\begin{aligned}
\mathcal{S}_y(f_1, f_2) &\triangleq \mathrm{E}\left\{Y(f_1)\, Y^{(*)}(f_2)\right\} \\
&= \frac{1}{4}\sum_{\alpha \in A} e^{-j2\pi(f_1 + (-)f_2 - \alpha)t_0} \Bigg[S_{\boldsymbol{x}}^{\alpha}(f_1)\, \delta(f_2 - (-)(\alpha - f_1)) \\
&\quad + S_{\boldsymbol{x}}^{\alpha}(\alpha - (-)f_2)\, \frac{1}{(-)j\pi(f_2 - (-)(\alpha - f_1))} \\
&\quad + S_{\boldsymbol{x}}^{\alpha}(f_1)\, \frac{1}{(-)j\pi(f_2 - (-)(\alpha - f_1))} \\
&\quad + \int_{\mathbb{R}} S_{\boldsymbol{x}}^{\alpha}(\nu)\, \frac{1}{j\pi(f_1 - \nu)}\, \frac{1}{(-)j\pi(f_2 - (-)(\alpha - \nu))}\, \mathrm{d}\nu \Bigg]
\end{aligned} \tag{14.86}$$

Albeit the characterization of the monolateral ACS process $y(t)$ in the time domain is coincident with that of the bilateral ACS process $x(t)$ for $t > \max\{t_0, t_0 - \tau\}$, their frequency-domain characterization is different. In fact, the Loève bifrequency spectrum of $y(t)$ has spectral masses concentrated on the lines $f_2 = (-)(\alpha - f_1)$, $\alpha \in A$, of the bifrequency plane but, unlike that of $x(t)$, it also contains continuous components that are spread around these lines.

14.6 Electrocardiogram

Several attempts have been made to model the electrocardiogram (ECG) signal as cyclostationary (Wang et al., 2004), (Haritopoulos et al., 2014), (Zhang et al., 2016), (Mihandoost and Amirani, 2017). In all these works, in order to consider the heart rate practically constant as requested by the cyclostationary model, the data-record length never exceeds few seconds. On larger data-record lengths the ECG signal $y(t)$ must be described by the higher fidelity model AM-TW ACS (14.26) where $x(t)$ is cyclostationary.

For the ECG signal, the AM-TW ACS model is conjectured and then validated by measurements made on several patients taken from the data base `https://physionet.org` (Goldberger et al., 2000). In the following, the signal stored in file `m001.dat` of the CEBS database is considered (Fig. 14.6). It corresponds to a healthy 30 year male and provides a typical example.

The continuous-time signal is modeled as

$$y(t) = a(t)\, x(\psi(t)) \tag{14.87}$$

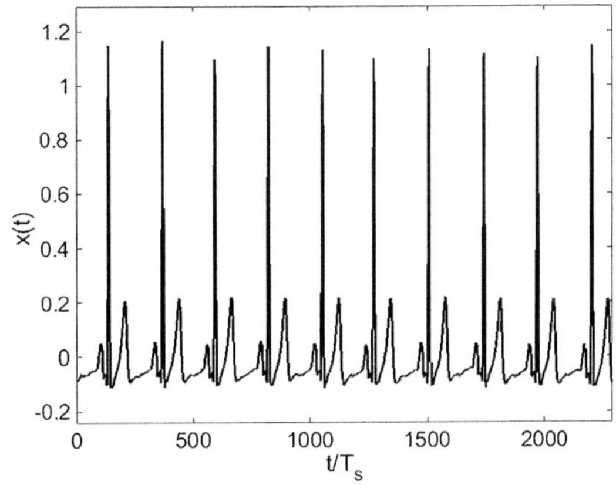

FIGURE 14.6 ECG signal $y(t)$ (first 2300 samples).

where $x(t)$ is a real-valued cyclostationary process that can be decomposed as

$$x(t) = x_r(t) + \sum_{h=-\infty}^{+\infty} x_{h/T_0}\, e^{j2\pi(h/T_0)t} \tag{14.88}$$

with $x_r(t)$ real-valued zero-mean cyclostationary process with period T_0 and x_{h/T_0} deterministic coefficients. T_0 represents the average cardiac cycle, that is, the reciprocal of the average heart rate α_0. In (14.87), the time-warping function $\psi(t) = t + \epsilon(t)$ describes the slow time variability (i.e., satisfying Assumption 14.9) of the heart rate and $a(t)$ is a slowly time-varying amplitude (i.e., satisfying Assumption 14.10). Both $\psi(t)$ and $a(t)$ account for variation in the propagation of the electrical wave throughout the hearth and also describe possible artifacts.

In order to corroborate the conjectured model (14.87), (14.88), the time-warping and amplitude-modulation functions are estimated by the procedure described in Section 14.3.3. Then, the underlying ACS signal is recovered as explained in Section 14.3.4 and its cyclic spectrum and second-order cyclic polyspectrum estimated. The validity of Assumptions 14.9 and 14.10 to obtain the autocorrelation (14.44) are verified a posteriori.

From (14.87) and (14.88) and the slow variability of $\epsilon(t)$ and $a(t)$, accounting for the results of Section 14.3.2.2, the evolutionary cyclic autocorrelation functions of $y(t)$ are given by

$$\rho_{\boldsymbol{y}}^{\alpha}(t,\tau) = a^2(t)\, e^{j2\pi\alpha\epsilon(t)}\, R_{\boldsymbol{x}}^{\alpha}(\tau) \qquad \alpha = \frac{k}{T_0},\ k \in \mathbb{Z} \tag{14.89}$$

with (see (1.66) with deterministic Fourier coefficients x_{h/T_0})

$$R_{\boldsymbol{x}}^{k/T_0}(\tau) = R_{\boldsymbol{x}_r}^{k/T_0}(\tau) + \sum_{h=-\infty}^{+\infty} x_{h/T_0}\, x_{(h-k)/T_0}^{*}\, e^{j2\pi(h/T_0)\tau}\,. \tag{14.90}$$

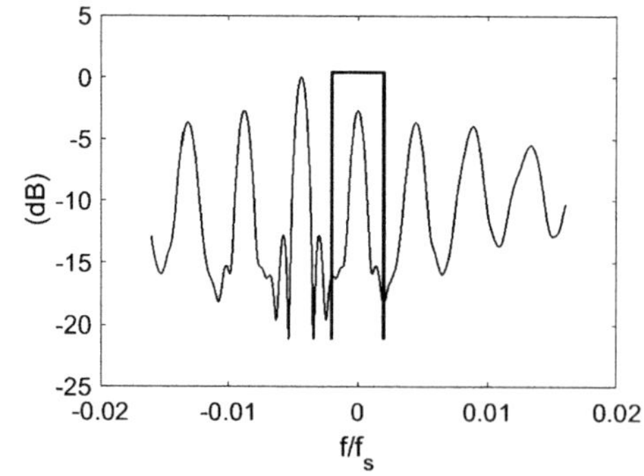

FIGURE 14.7 ECG signal $y(t)$. Zoom around $f = 0$ of the estimate of the PSD of the frequency-shifted lag product $y(t+\tau)\, y^*(t)\, e^{-j2\pi\widetilde{\alpha}_0 t}$, with $\tau = 0$. The spectral band of the low-pass filter $h_W(t)$ that extracts $z^{(W)}(\widetilde{\alpha}_0; t, \tau)$ is evidenced.

The signal $y(t) - \langle y(t) \rangle_t$ is analyzed, where $\langle y(t) \rangle_t$ is the temporal mean of $y(t)$. The digitalized signal $y(t)|_{t=nT_s}$ is obtained with sampling frequency $f_s = 1/T_s = 250$ Hz (sampling period $T_s = 4$ ms). The data is converted into Matlab®/Octave format by the `wfdb` toolbox (Silva and Moody, 2014). In the experiment, $N = 2^{15}$ samples are taken. This corresponds to a data-record length $T = NT_s \simeq 131$ s which is significantly larger than that usually considered when the cyclostationary model is adopted (Wang et al., 2004), (Haritopoulos et al., 2014), (Zhang et al., 2016), (Mihandoost and Amirani, 2017).

In Fig. 14.9 (top), the magnitude of the frequency-smoothed cyclic periodogram (Definition 5.29) $S_{yy^*}^{(T,\Delta f)}(\alpha, f)$ of the ECG signal $y(t)$ is reported. According to Theorem 14.11, cyclic features of the underlying cyclostationary signal $x(t)$ are spread around the cycle frequencies k/T_0 (see (14.47)). Since $a(t) \simeq 1$, for $\alpha = 0$ we have $M_{a\epsilon}^{\alpha}(f) = \delta(f)$ in (14.46) and there is no spread of the power spectral density (PSD) around the line $\alpha = 0$.

A coarse estimate of the average heart rate $\alpha_0 \triangleq 1/T_0$ starting from the observation interval $[0, T]$ is derived as $\widetilde{\alpha}_0 = n(T)/T$, where $n(T)$ is the number of spikes in $y(t)$ within the observation interval. We obtain $\widetilde{\alpha}_0 = 0.004385 f_s$. Thus, the average cardiac cycle is $\widetilde{T}_0 = 1/\widetilde{\alpha}_0 \simeq 228\, T_s \simeq 0.91$ s. This corresponds to $60/\widetilde{T}_0 \simeq 66$ beats per minute. The low-pass filter $h_W(t)$ has monolateral bandwidth $W = 0.002\, f_s$ whose value is obtained by visual inspection of the PSD of the frequency-shifted lag product $y(t+\tau)\, y(t)\, e^{-j2\pi\widetilde{\alpha}_0 t}$ (Fig. 14.7). The value $\tau = 0$ is used. The time-warping and amplitude modulation functions are estimated by the procedure outlined in Section 14.3.3. The estimates $\widehat{\epsilon}(t)$ and $\widehat{a}(t)$ are reported in Fig. 14.8. From Fig. 14.8 (top), we see that $\sup_t |\epsilon(t)| \simeq 250\, T_s$ which is greater than $\widetilde{T}_0$. The cycle frequency estimate $\widehat{\alpha}_0 = 0.00438\, f_s$ (see (14.60)) is obtained. Finally, an estimate $\widehat{x}(t)$ of the underlying cyclostationary signal $x(t)$ is obtained by (14.63).

In Fig. 14.9 (middle), the magnitude of the frequency-smoothed cyclic periodogram $S_{\widehat{x}\widehat{x}^*}^{(T,\Delta f)}(\alpha, f)$ of the de-warped and amplitude-modulation compensated signal $\widehat{x}(t)$ is reported. According with the conjectured model, $x(t)$ is cyclostationary since cyclic features are evident in correspondence of cycle frequencies k/T_0. Moreover, spikes whose shape is that of the frequency-smoothing window are present. They correspond to the Dirac

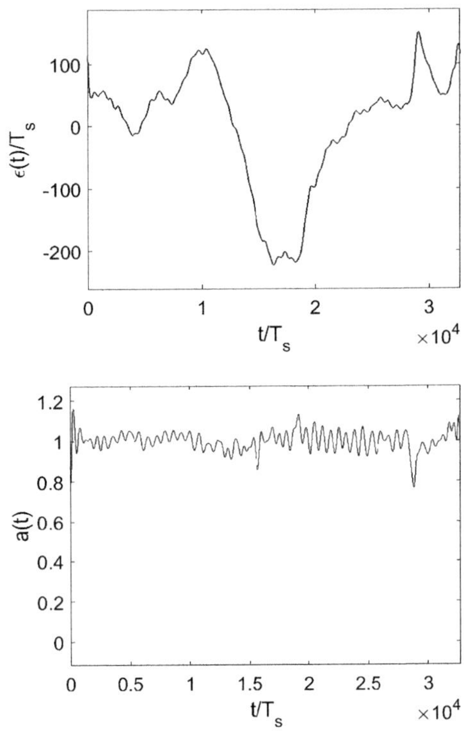

FIGURE 14.8 ECG signal $y(t)$. Top: Estimate $\widehat{\epsilon}(t)$ by (14.59). Bottom: Estimate $\widehat{a}(t)$ by (14.57).

deltas in the cyclic spectrum as a consequence of the additive periodic component in $x(t)$ (see (1.67)). These spikes are removed by median filtering the cyclic periodogram (Section 5.2.5). In Fig. 14.9 (bottom), the magnitude of the median-frequency-smoothed cyclic periodogram $P_{\widehat{x}\widehat{x}^*}^{(T,\Delta f,\delta f)}(\alpha, f)$ of $\widehat{x}(t)$ is reported. It is an estimate of the 2nd-order cyclic polyspectrum, that is, the inverse Fourier transform of the cyclic covariance which is the cyclic autocorrelation function of $x_r(t)$, the zero-mean ACS component of $x(t)$ (Section 1.4).

The measurement results reported in Fig. 14.9, corroborate the conjectured model for the ECG signal. The same behavior for the estimated cyclic statistics is found using ECGs in the database measured on different subjects.

14.7 Summary

The Priestley's oscillatory processes have spectral representation expressed as a slowly varying frequency-dependent modulating function times a complex exponential integrated with respect to the differential of the integrated spectrum of a wide-sense stationary process (Eqs. (14.1) and (14.2)). This class is extended to that of the oscillatory almost-

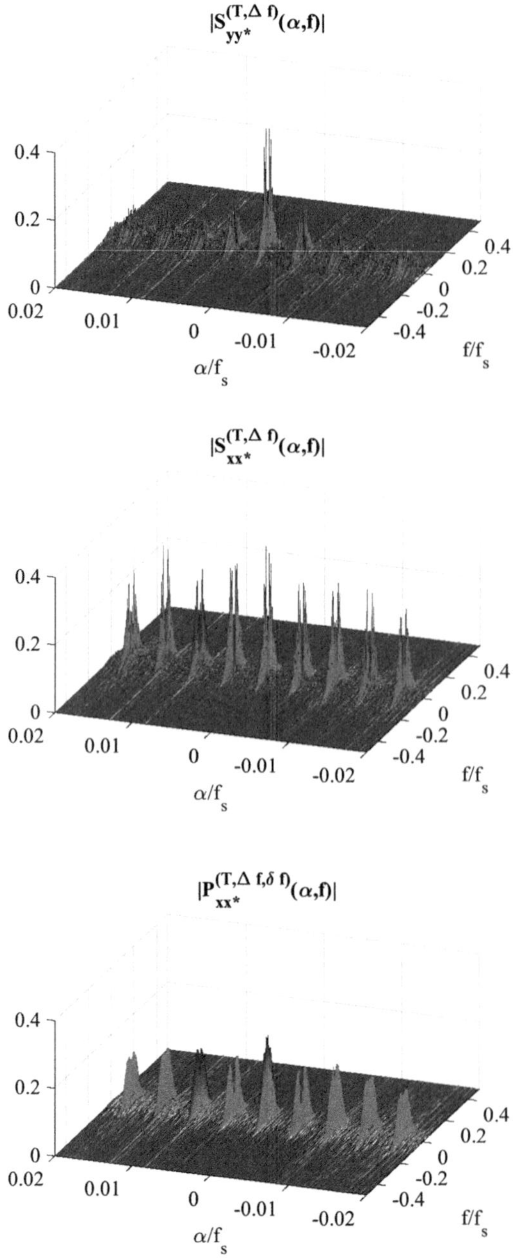

FIGURE 14.9 Top: Frequency-smoothed cyclic periodogram $S_{yy^*}^{(T,\Delta f)}(\alpha, f)$ for the ECG signal $y(t)$. Middle: Frequency-smoothed cyclic periodogram $S_{\widehat{x}\widehat{x}^*}^{(T,\Delta f)}(\alpha, f)$ for the de-warped signal $\widehat{x}(t)$. Bottom: Median-frequency-smoothed cyclic periodogram $P_{\widehat{x}\widehat{x}^*}^{(T,\Delta f,\delta f)}(\alpha, f)$ for the de-warped signal $\widehat{x}(t)$. All estimates are functions of α/f_s and f/f_s.

cyclostationary (OACS) processes by considering the differential of the integrated spectrum of an almost-cyclostationary (ACS) process which is referred to as the underlying ACS process (Definition 14.1).

OACS processes have autocorrelation function given by the superposition of amplitude- and angle-modulated sine waves whose frequencies are the cycle frequencies of the underlying ACS process (Theorem 14.2). The Loève bifrequency spectrum is obtained by the Loève bifrequency spectrum of the underlying ACS process by spreading around the support lines the (conjugate) spectral correlation densities (Theorem 14.3).

A notable class of OACS processes is the class of the amplitude-modulated time-warped (AM-TW) ACS process (Section 14.3). A deterministic amplitude modulation function multiplies an ACS process whose time scale is modified by a deterministic time-warping function. In the case of moderate time warping and slowly varying amplitude modulating function, amplitude-modulation and time-warping functions are estimated by amplitude and angle demodulation of one of the modulated sine waves of the autocorrelation, provided that this sine wave can be filtered out from the other components (Section 14.3.3). In such a case, the underlying ACS signal can be recovered by time dewarping and amplitude-modulation compensation (Section 14.3.4).

An alternative approach for the analysis of time-warped ACS signals consists of directly estimating the time-warping function (or its inverse) in a manner that transforms the data having an autocorrelation reflecting irregular cyclicity into data having regular cyclicity. The time-warping function (or its inverse) is expressed as an expansion in terms of basis functions and then the expansion coefficients are estimated such that the magnitude of the cyclic autocorrelation of the ACS signal for a given lag and cycle frequency is maximized (Section 14.4).

The electrocardiogram (ECG) signal is shown to fit the AM-TW ACS model (Section 14.6).

14.8 Proofs

In this Section, proofs are reported for results of the present chapter. Unless otherwise specified, only formal manipulations are reported, without explicitly justifying inversion of linear operators such as time averages, infinite sums, integrals, and ensemble average (expectation). In general, these manipulations are justified at least in the sense of generalized functions (distributions) (Champeney, 1990), (Zemanian, 1987).

Proof of Theorem 14.2

Using (14.1) and (14.5), one has

$$\begin{aligned}
&\mathrm{E}\left\{y(t+\tau)\, y^{(*)}(t)\right\} \\
&\quad = \mathrm{E}\left\{\int_{\mathbb{R}} A_{t+\tau}(f_1)\, e^{j2\pi f_1(t+\tau)}\, \mathrm{d}Z(f_1) \int_{\mathbb{R}} A_t^{(*)}(f_2)\, e^{(-)j2\pi f_2 t}\, \mathrm{d}Z^{(*)}(f_2)\right\}
\end{aligned}$$

$$
\begin{aligned}
&= \int_{\mathbb{R}} \int_{\mathbb{R}} A_{t+\tau}(f_1)\, A_t^{(*)}(f_2)\, e^{j2\pi f_1(t+\tau)}\, e^{(-)j2\pi f_2 t}\, \mathrm{E}\left\{\mathrm{d}Z(f_1)\, \mathrm{d}Z^{(*)}(f_2)\right\} \\
&= \int_{\mathbb{R}} \int_{\mathbb{R}} A_{t+\tau}(f_1)\, A_t^{(*)}(f_2)\, e^{j2\pi f_1(t+\tau)}\, e^{(-)j2\pi f_2 t} \sum_{\alpha \in A} \delta(f_2 - (-)(\alpha - f_1))\, \mathrm{d}\mu_{\boldsymbol{x}}^{\alpha}(f_1)\, \mathrm{d}f_2 \\
&= \sum_{\alpha \in A} \int_{\mathbb{R}} A_{t+\tau}(f_1)\, A_t^{(*)}((-)(\alpha - f_1))\, e^{j2\pi f_1(t+\tau)}\, e^{(-)j2\pi(-)(\alpha - f_1)t}\, \mathrm{d}\mu_{\boldsymbol{x}}^{\alpha}(f_1) \\
&= \sum_{\alpha \in A} \int_{\mathbb{R}} A_{t+\tau}(f_1)\, A_t^{(*)}((-)(\alpha - f_1))\, e^{j2\pi f_1 \tau}\, \mathrm{d}\mu_{\boldsymbol{x}}^{\alpha}(f_1)\, e^{j2\pi \alpha t}
\end{aligned}
\tag{14.91}
$$

Proof of Theorem 14.3

The Fourier transform of $y(t)$ in (14.1), defined in a distributional sense (Section A.3), (Napolitano, 2012, Secs. 1.1.2, 4.2.1), (Napolitano, 2016a, Sec. 3), can be formally expressed as

$$
\begin{aligned}
Y(f) &\triangleq \int_{\mathbb{R}} y(t)\, e^{-j2\pi f t}\, \mathrm{d}t \\
&= \int_{\mathbb{R}} \int_{\mathbb{R}} A_t(\nu)\, e^{j2\pi \nu t} \mathrm{d}Z(\nu)\, e^{-j2\pi f t}\, \mathrm{d}t \\
&= \int_{\mathbb{R}} \int_{\mathbb{R}} A_t(\nu)\, e^{-j2\pi(f-\nu)t} \mathrm{d}t\, \mathrm{d}Z(\nu) \\
&= \int_{\mathbb{R}} \mathcal{A}(f - \nu; \nu)\, \mathrm{d}Z(\nu)
\end{aligned}
\tag{14.92}
$$

where $\mathcal{A}(f; \nu)$ is defined in (14.14).

Using (14.92) and (14.5), we have

$$
\begin{aligned}
&\mathrm{E}\left\{Y(f_1)\, Y^{(*)}(f_2)\right\} \\
&\quad = \mathrm{E}\left\{\int_{\mathbb{R}} \mathcal{A}(f_1 - \nu_1; \nu_1)\, \mathrm{d}Z(\nu_1) \int_{\mathbb{R}} \mathcal{A}^{(*)}(f_2 - \nu_2; \nu_2)\, \mathrm{d}Z^{(*)}(\nu_2)\right\} \\
&\quad = \int_{\mathbb{R}} \int_{\mathbb{R}} \mathcal{A}(f_1 - \nu_1; \nu_1)\, \mathcal{A}^{(*)}(f_2 - \nu_2; \nu_2)\, \mathrm{E}\left\{\mathrm{d}Z(\nu_1)\, \mathrm{d}Z^{(*)}(\nu_2)\right\} \\
&\quad = \int_{\mathbb{R}} \int_{\mathbb{R}} \mathcal{A}(f_1 - \nu_1; \nu_1)\, \mathcal{A}^{(*)}(f_2 - \nu_2; \nu_2) \sum_{\alpha \in A} \delta(\nu_2 - (-)(\alpha - \nu_1))\, \mathrm{d}\mu_{\boldsymbol{x}}^{\alpha}(\nu_1)\, \mathrm{d}\nu_2 \\
&\quad = \sum_{\alpha \in A} \int_{\mathbb{R}} \mathcal{A}(f_1 - \nu_1; \nu_1)\, \mathcal{A}^{(*)}(f_2 - (-)(\alpha - \nu_1); (-)(\alpha - \nu_1))\, \mathrm{d}\mu_{\boldsymbol{x}}^{\alpha}(\nu_1)
\end{aligned}
\tag{14.93}
$$

Eq. (14.13) follows using $\mathrm{d}\mu_{\boldsymbol{x}}^{\alpha}(\nu_1) = S_{\boldsymbol{x}}^{\alpha}(\nu_1)\, \mathrm{d}\nu_1$ into (14.93), under the assumption that $\mu_{\boldsymbol{x}}^{\alpha}(\nu_1)$ does not contain singular component for every $\alpha \in A$.

Proof of Theorem 14.11

1. Since $\epsilon(t)$ is differentiable, for every fixed τ there exists $\bar{t}_\tau$ (with $\bar{t}_\tau \in (t, t+\tau)$ if $\tau > 0$ and $\bar{t}_\tau \in (t+\tau, t)$ if $\tau < 0$) such that

$$\epsilon(t+\tau) - \epsilon(t) = \dot{\epsilon}(\bar{t}_\tau)\,\tau\,. \tag{14.94}$$

From (14.94) it follows that the time-varying terms in the argument of $R_{x}^{\alpha}(\cdot)$ in (14.28b) can be neglected provided that

$$\sup_t \left|\dot{\epsilon}(t)\right| |\tau| \ll \frac{1}{B} \tag{14.95}$$

In (14.95), B denotes the bandwidth of $x(t)$ and $1/B$ is the approximate width of $R_{x}^{0}(\cdot)$. For $|\tau| < |\tau|_{\max} = 1/B$, condition (14.95) is satisfied provided that condition (14.40)

$$\sup_t \left|\dot{\epsilon}(t)\right| \ll 1 \tag{14.96}$$

of $\epsilon(t)$ slowly varying (with respect to t, whose derivative is 1) holds.
Thus, if condition (14.96) is satisfied,

$$R_{x}^{\alpha}\big(\tau + \epsilon(t+\tau) - \epsilon(t)\big) = R_{x}^{\alpha}\big(\tau + \dot{\epsilon}(\bar{t}_\tau)\,\tau\big) \simeq R_{x}^{\alpha}(\tau)\,. \tag{14.97}$$

In addition, if also condition (14.41) is satisfied, the (conjugate) autocorrelation can be written as (14.44).
In the case of affine time-warping function $\psi(t) = st - d$, approximation (14.97) specializes into $R_{x}^{\alpha}(s\tau) \simeq R_{x}^{\alpha}(\tau)$ for all values of τ such that $|R_{x}^{\alpha}(\tau)|$ is significantly nonzero (see also Section 8.11).
2. By doubly Fourier transforming the (conjugate) autocorrelation $\mathrm{E}\{y(t_1)\,y^{(*)}(t_2)\}$ one obtains (14.45).
3. Eq. (14.47) is obtained by making the variable change $f = f_1$ and $\delta f = f_1 + (-) f_2$ into (14.45).

Proof of (14.55)

Each term $a^2(t)\,e^{j2\pi\alpha\epsilon(t)}\,R_{x}^{\alpha}(\tau)\,e^{j2\pi(\alpha-\widetilde{\alpha}_0)t}$ in (14.54) has most of the spectral components concentrated in the frequency interval $(\alpha - \widetilde{\alpha}_0 - B^{\alpha}, \alpha - \widetilde{\alpha}_0 + B^{\alpha})$, where B^{α} is the monolateral bandwidth of $a^2(t)\,e^{j2\pi\alpha\epsilon(t)}$.

The term $a^2(t)\,e^{j2\pi\alpha_0\epsilon(t)}\,e^{j2\pi(\alpha_0-\widetilde{\alpha}_0)t}$ passes unaltered through $h_W(t)$ provided that both conditions

$$\alpha_0 - \widetilde{\alpha}_0 + B^{\alpha_0} < W \tag{14.98a}$$

$$\alpha_0 - \widetilde{\alpha}_0 - B^{\alpha_0} > -W \tag{14.98b}$$

are satisfied. Since we have that

$$|\alpha_0 - \widetilde{\alpha}_0| + B^{\alpha_0} < W \qquad \Rightarrow \qquad \alpha_0 - \widetilde{\alpha}_0 + B^{\alpha_0} < W \tag{14.99}$$

$$-\,|\alpha_0 - \widetilde{\alpha}_0| - B^{\alpha_0} > -W \qquad \Rightarrow \qquad \alpha_0 - \widetilde{\alpha}_0 - B^{\alpha_0} > -W \tag{14.100}$$

then a sufficient condition such that both (14.98a) and (14.98b) are satisfied is

$$|\alpha_0 - \widetilde{\alpha}_0| + B^{\alpha_0} < W \tag{14.101}$$

The terms $a^2(t)\, e^{j2\pi\alpha\epsilon(t)}\, e^{j2\pi(\alpha-\widetilde{\alpha}_0)t}$ with $\alpha \neq \alpha_0$ in (14.54) are filtered out provided that

$$\inf_{\substack{\alpha\in A\\ \alpha>\alpha_0}} \left(\alpha - \widetilde{\alpha}_0 - B^{\alpha}\right) > W \tag{14.102a}$$

$$\sup_{\substack{\alpha\in A\\ \alpha<\alpha_0}} \left(\alpha - \widetilde{\alpha}_0 + B^{\alpha}\right) < -W\,. \tag{14.102b}$$

Since (14.102b) is equivalent to

$$\inf_{\substack{\alpha\in A\\ \alpha<\alpha_0}} \left(-\alpha + \widetilde{\alpha}_0 - B^{\alpha}\right) > W \tag{14.103}$$

we have that a sufficient condition such that both (14.102a) and (14.102b) are satisfied is

$$\inf_{\substack{\alpha\in A\\ \alpha\neq\alpha_0}} \left(|\alpha - \widetilde{\alpha}_0| - B^{\alpha}\right) > W\,. \tag{14.104}$$

It results that

$$|\alpha - \widetilde{\alpha}_0| = |\alpha - \alpha_0 + \alpha_0 - \widetilde{\alpha}_0| \geqslant |\alpha - \alpha_0| - |\alpha_0 - \widetilde{\alpha}_0| \tag{14.105}$$

Thus a sufficient condition such that (14.104) holds is

$$\inf_{\substack{\alpha\in A\\ \alpha\neq\alpha_0}} \left(|\alpha - \alpha_0| - B^{\alpha}\right) - |\alpha_0 - \widetilde{\alpha}_0| > W\,. \tag{14.106}$$

Putting (14.101) and (14.106) together gives

$$|\alpha_0 - \widetilde{\alpha}_0| + B^{\alpha_0} < W < \inf_{\substack{\alpha\in A\\ \alpha\neq\alpha_0}} \left(|\alpha - \alpha_0| - B^{\alpha}\right) - |\alpha_0 - \widetilde{\alpha}_0|\,. \tag{14.107}$$

Proof of (14.63)

If $\epsilon(t)$ is slowly varying, then we can use the 1st-order Taylor series approximation

$$\epsilon(t - \widehat{\epsilon}(t)) \simeq \epsilon(t) - (\mathrm{d}\epsilon/\mathrm{d}t)\widehat{\epsilon}(t) \tag{14.108}$$

provided that $\widehat{\epsilon}(t)$ (and hence $\epsilon(t)$) is small $\forall t \in [-T/2, T/2]$.

We have for all t such that $\widehat{a}(t - \widehat{\epsilon}(t)) \neq 0$

$$\widehat{x}(t) \triangleq y(t - \widehat{\epsilon}(t))/\widehat{a}(t - \widehat{\epsilon}(t))$$

$$
\begin{aligned}
&= \frac{a(t-\widehat{\epsilon}(t))}{\widehat{a}(t-\widehat{\epsilon}(t))}\, x(t-\widehat{\epsilon}(t)+\epsilon(t-\widehat{\epsilon}(t))) \\
&\simeq \frac{a(t-\widehat{\epsilon}(t))}{\widehat{a}(t-\widehat{\epsilon}(t))}\, x(t-\widehat{\epsilon}(t)+\epsilon(t)-(\mathrm{d}\epsilon/\mathrm{d}t)\widehat{\epsilon}(t)) \\
&= \frac{a(t-\widehat{\epsilon}(t))}{\widehat{a}(t-\widehat{\epsilon}(t))}\, x(t-[1+(\mathrm{d}\epsilon/\mathrm{d}t)]\widehat{\epsilon}(t)+\epsilon(t)) \\
&\simeq x(t)
\end{aligned} \tag{14.109}
$$

where the last approximate equality holds since $1+(\mathrm{d}\epsilon/\mathrm{d}t)\simeq 1$ (see (14.40)) and provided that $a(t)/\widehat{a}(t)\simeq 1$, that is, the estimate $\widehat{a}(t)$ is sufficiently accurate, and the estimation error on $\epsilon(t)$ is sufficiently small in the sense that (14.64) holds.

Condition (14.64) assures that the time-varying delay $\epsilon(t)-\widehat{\epsilon}(t)$ can be neglected in the argument of $x(\cdot)$.

Proof of (14.86)

By replacing (14.85) into the expression (14.13) of the Loève bifrequency spectrum of OACS processes, one has

$$
\begin{aligned}
\mathcal{S}_{\boldsymbol{y}}(f_1,f_2) &\triangleq \mathrm{E}\left\{Y(f_1)\,Y^{(*)}(f_2)\right\} \\
&= \sum_{\alpha\in A}\int_{\mathbb{R}} S_{\boldsymbol{x}}^{\alpha}(\nu)\,\mathcal{A}(f_1-\nu;\nu)\,\mathcal{A}^{(*)}(f_2-{\scriptstyle(-)}(\alpha-\nu);{\scriptstyle(-)}(\alpha-\nu))\,\mathrm{d}\nu \\
&= \sum_{\alpha\in A}\int_{\mathbb{R}} S_{\boldsymbol{x}}^{\alpha}(\nu)\,\frac{1}{2}\left[\delta(f_1-\nu)+\frac{1}{j\pi(f_1-\nu)}\right]e^{-j2\pi(f_1-\nu)t_0} \\
&\qquad \frac{1}{2}\left[\delta(f_2-{\scriptstyle(-)}(\alpha-\nu))+\frac{1}{j\pi(f_2-{\scriptstyle(-)}(\alpha-\nu))}\right]^{(*)} e^{-(-)j2\pi(f_2-(-)(\alpha-\nu))t_0}\,\mathrm{d}\nu \\
&= \frac{1}{4}\sum_{\alpha\in A} e^{-j2\pi(f_1+(-)f_2-\alpha)t_0}\int_{\mathbb{R}} S_{\boldsymbol{x}}^{\alpha}(\nu)\left[\delta(f_1-\nu)\,\delta(f_2-{\scriptstyle(-)}(\alpha-\nu))\right. \\
&\qquad +\frac{1}{j\pi(f_1-\nu)}\,\delta(f_2-{\scriptstyle(-)}(\alpha-\nu)) \\
&\qquad +\delta(f_1-\nu)\,\frac{1}{{\scriptstyle(-)}j\pi(f_2-{\scriptstyle(-)}(\alpha-\nu))} \\
&\qquad \left.+\frac{1}{j\pi(f_1-\nu)}\,\frac{1}{{\scriptstyle(-)}j\pi(f_2-{\scriptstyle(-)}(\alpha-\nu))}\right]\mathrm{d}\nu
\end{aligned} \tag{14.110}
$$

from which (14.86) follows accounting for the sampling property of the Dirac delta.

15

The big picture

Contents

Abstract

The new class of the oscillatory spectrally correlated processes is introduced. It includes, as special cases, all the previously considered classes of nonstationary signals, namely the almost-cyclostationary, generalized almost-cyclostationary, spectrally correlated, and oscillatory almost-cyclostationary signals. Inclusion relationships among these classes and those of the wide-sense stationary and of the Priestley oscillatory signals are enlightened.

Keywords

Oscillatory spectrally correlated, Relationships among process classes, Nonstationary processes, Autocorrelation, Loève bifrequency spectrum, Time-variant spectrum

15.1 Introduction

In this chapter, relationships among the classes of nonstationary signals considered in the previous chapters are enlightened. Inclusion relationships among these classes are presented and the structure of their statistical functions in time and frequency domains are compared.

The new class of the oscillatory spectrally correlated processes is introduced and its second-order characterization in the time and frequency domains is provided. This class of processes is shown to contain, as special cases, all the other classes of processes considered in the previous chapters.

15.2 Oscillatory spectrally correlated processes

Let us consider a second-order harmonizable stochastic process $y(t)$ that admits the representation (14.1)

Cyclostationary Processes and Time Series. https://doi.org/10.1016/B978-0-08-102708-0.00027-3

$$y(t) = \int_{\mathbb{R}} A_t(f)\, e^{j2\pi f t}\, \mathrm{d}Z(f)\,. \tag{15.1}$$

The process $y(t)$ is said to be *oscillatory spectrally correlated (OSC)* with respect to the family of oscillatory functions $\{A_t(f)\, e^{j2\pi f t}\}$ if $Z(f)$ is the integrated complex spectrum of an SC process $x(t)$, that is (see (13.1a))

$$\mathrm{E}\left\{\mathrm{d}Z(f_1)\, \mathrm{d}Z^{(*)}(f_2)\right\} = \sum_{k \in \mathbb{I}} \delta\left(f_2 - \Psi_{\boldsymbol{x}}^{(k)}(f_1)\right) \mathrm{d}\mu_{\boldsymbol{x}}^{(k)}(f_1)\, \mathrm{d}f_2\,, \tag{15.2}$$

where $\mathbb{I}$ is a countable set. The modulating functions $A_t(f)$, as functions of t, are in general low-pass functions. In (15.2), $\mu_{\boldsymbol{x}}^{(k)}(f)$, $k \in \mathbb{I}$, is a family of complex measures. The process $x(t)$ is referred to as the underlying SC process.

If $\mu_{\boldsymbol{x}}^{(k)}(f)$ does not contain the singular component (Hurd and Miamee, 2007, p. 197), (Cramér, 1940), then $\mu_{\boldsymbol{x}}^{(k)}(f) = S_{\boldsymbol{x}}^{(k)}(f)\, \mathrm{d}f$, where $S_{\boldsymbol{x}}^{(k)}(f)$ is the density of spectral correlation of the SC process $x(t)$ (see (13.1a)). $S_{\boldsymbol{x}}^{(k)}(f)$ contains Dirac deltas in correspondence of the jumps of $\mu_{\boldsymbol{x}}^{(k)}(f)$.

The (conjugate) autocorrelation function of $y(t)$ is given by

$$\begin{aligned}
&\mathrm{E}\left\{y(t+\tau)\, y^{(*)}(t)\right\} \\
&= \sum_{k \in \mathbb{I}} \int_{\mathbb{R}} A_{t+\tau}(f_1)\, A_t^{(*)}\left(\Psi_{\boldsymbol{x}}^{(k)}(f_1)\right) e^{j2\pi[f_1 + (-)\Psi_{\boldsymbol{x}}^{(k)}(f_1)]t}\, e^{j2\pi f_1 \tau}\, \mathrm{d}\mu_{\boldsymbol{x}}^{(k)}(f_1)
\end{aligned} \tag{15.3}$$

The *Loève bifrequency spectrum* of $y(t)$ is given by

$$\begin{aligned}
\mathcal{S}_{\boldsymbol{y}}(f_1, f_2) &\triangleq \mathrm{E}\left\{Y(f_1)\, Y^{(*)}(f_2)\right\} \\
&= \sum_{k \in \mathbb{I}} \int_{\mathbb{R}} \mathcal{A}(f_1 - \nu; \nu)\, \mathcal{A}^{(*)}\left(f_2 - \Psi_{\boldsymbol{x}}^{(k)}(\nu); \Psi_{\boldsymbol{x}}^{(k)}(\nu)\right) \mathrm{d}\mu_{\boldsymbol{x}}^{(k)}(\nu)
\end{aligned} \tag{15.4}$$

where

$$\mathcal{A}(f; \nu) \triangleq \int_{\mathbb{R}} A_t(\nu)\, e^{-j2\pi f t}\, \mathrm{d}t\,. \tag{15.5}$$

From (15.4) and (15.2) it follows that the Loève bifrequency spectrum of an OSC process $y(t)$ can be seen as obtained by the Loève bifrequency spectrum of the underlying SC process $x(t)$ by spreading around the support curves $f_2 = \Psi_{\boldsymbol{x}}^{(k)}(f_1)$, $k \in \mathbb{I}$, the spectral correlation densities $S_{\boldsymbol{x}}^{(k)}(f)$.

15.3 Relationships among classes of nonstationary processes

15.3.1 ACS, CS, and WSS processes

ACS processes in the wide sense have the autocorrelation function which is an almost-periodic function of time with coefficients depending on the lag parameter τ and frequencies, referred to as cycle frequencies, not depending on τ. Cycle frequencies are possibly incommensurate, that is, they are not necessarily harmonically related. Poly-cyclostationary

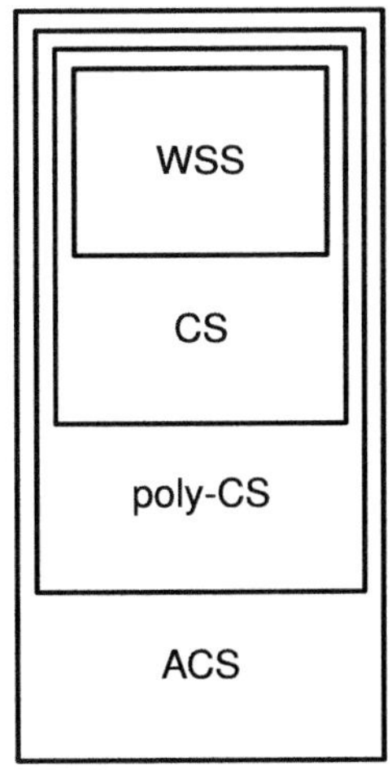

FIGURE 15.1 Inclusion relationships for ACS, poly-CS, CS, and WSS process classes.

(poly-CS) processes are obtained as special case of ACS processes when the cycle frequencies are integer multiples of the reciprocals of two or more incommensurate periods (Remark 1.10). Cyclostationary (CS) processes in the wide sense are obtained as special case of poly-CS processes (and hence of ACS processes) when the cycle frequencies are all integer multiples of a fundamental one. Finally, WSS processes are obtained as further specialization when the only cycle frequency is zero. Therefore, the class of the ACS processes includes that of the poly-CS processes that includes that of the CS processes which in turn includes that of the WSS processes (Fig. 15.1).

The same inclusion relationships are obtained if almost-cyclostationarity, cyclostationarity, and stationarity are considered in the strict sense for a fixed order or if the almost-periodicity or periodicity of the statistical functions are defined in one of the generalized senses considered in Section B.4.

Note that a different inclusion relationship among the ACS, poly-CS, CS, and WSS process classes is obtained if the classification is made in the FOT approach on the basis of the existence of FOT probabilistic models. In fact, the existence of an almost-periodic distribution requires more stringent conditions on the time series with respect to those for the existence of the stationary distribution (Section 2.3.3). Therefore, time series for which a periodic, poly-periodic, or almost-periodic FOT probabilistic model exists are a subclass of the time series for which a stationary model exists.

15.3.2 GACS, SC, and ACS processes

SC processes have the Loève bifrequency spectrum with spectral masses concentrated on a countable set of curves in the bifrequency plane. ACS processes are the special case when the support curves are lines parallel to the main diagonal (compare Fig. 13.1 (top) and Fig. 1.2 (bottom)).

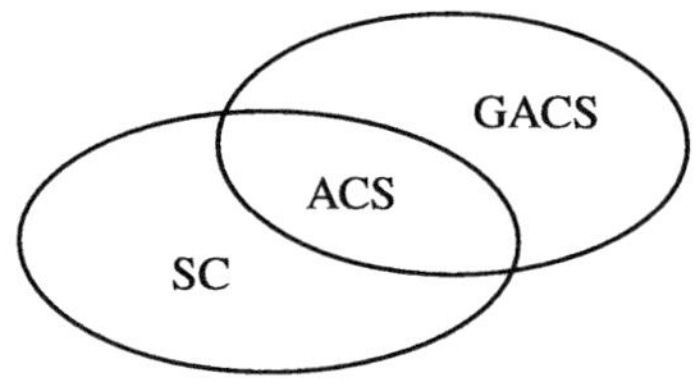

FIGURE 15.2 GACS, SC, and ACS process classes.

Table 15.1 (Conjugate) autocorrelation $\mathcal{R}_x(t,\tau)$, Loève bifrequency spectrum $\mathcal{S}_x(f_1,f_2)$, and time-variant spectrum $\mathcal{S}_x(t,f)$ of WSS, ACS, GACS, and SC processes. Void boxes correspond to not easily interpretable expressions.

	$\mathcal{R}_x(t,\tau) = \mathrm{E}\{x(t+\tau)\,x^{(*)}(t)\}$ **(Conjugate) Autocorrelation**	$\mathcal{S}_x(f_1,f_2) = \mathrm{E}\{X(f_1)\,X^{(*)}(f_2)\}$ **Loève Bifrequency Spectrum**	$\mathcal{S}_x(t,f) = \mathcal{F}_{\tau\to f}[\mathcal{R}_x(t,\tau)]$ **Time-Variant Spectrum**
WSS	$R_x(\tau)$	$S_x(f_1)\,\delta(f_2+(-)f_1)$	$S_x(f)$
ACS	$\sum_{\alpha\in A} R_x^\alpha(\tau)\,e^{j2\pi\alpha t}$	$\sum_{\alpha\in A} S_x^\alpha(f_1)\delta(f_2-(-)(\alpha-f_1))$	$\sum_{\alpha\in A} S_x^\alpha(f)\,e^{j2\pi\alpha t}$
GACS	$\sum_{k\in\mathbb{I}} R_x^{(k)}(\tau)\,e^{j2\pi\alpha_x^{(k)}(\tau)t}$		
SC	†	$\sum_{k\in\mathbb{I}} S_x^{(k)}(f_1)\delta\left(f_2-\Psi_x^{(k)}(f_1)\right)$	$\sum_{k\in\mathbb{I}} S_x^{(k)}(f)\,e^{j2\pi[f+(-)\Psi_x^{(k)}(f)]t}$

† $\sum_{k\in\mathbb{I}} \int_{\mathbb{R}} S_x^{(k)}(f)\,e^{j2\pi f(t+\tau)}\,e^{(-)j2\pi\Psi_x^{(k)}(f)t}\,df$

GACS processes have an almost-periodically time-variant autocorrelation function whose (generalized) Fourier series expansion has coefficients and frequencies (cycle frequencies) depending on the lag parameter (Chapter 12). ACS processes are obtained as special case when the frequencies are constant with respect to the lag parameter. This corresponds to a Loève bifrequency spectrum with spectral masses concentrated along lines with unit slope. Thus, the class of the ACS processes is the intersection of the classes of the GACS and SC processes (Fig. 15.2), (Table 15.1), (Napolitano, 2012, Sec. 4.2.2).

15.3.3 OSC, OACS, SC, and ACS processes

OSC and OACS processes admit the representation (15.1). For OSC processes, $Z(f)$ is the integrated complex spectrum of an SC process while for ACS processes $Z(f)$ is the integrated complex spectrum of an ACS process. Since ACS processes are a subclass of the SC process class, OACS processes are obtained as special case of OSC processes (Fig. 15.3).

If $A_t(f)=1$ in (15.1), OSC processes reduce to SC processes (Fig. 15.3). Specifically, the Loève bifrequency spectrum of an OSC process has spectral density significantly different from zero in strips around the support curves of the underlying SC process (Table 15.2). If $A_t(f)=1$, then $\mathcal{A}(f;\nu)=\delta(f)$ independent of ν and the Loève bifrequency spectrum (15.4) reduces to that of an SC process.

If $A_t(f)=1$ in (15.1), OACS processes reduce to ACS processes (Fig. 15.3). Specifically, the Loève bifrequency spectrum of an OACS process has spectral density significantly

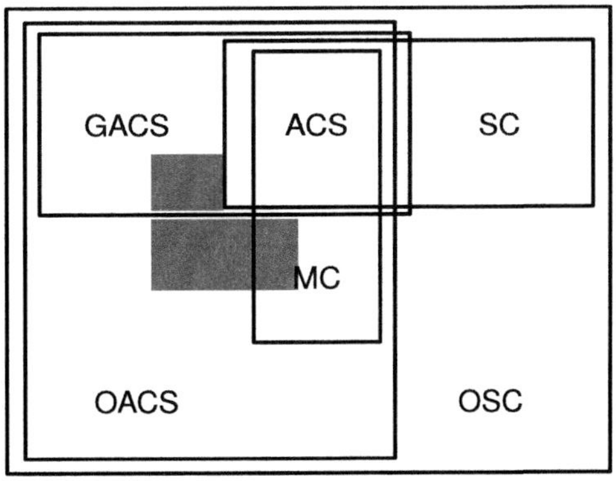

FIGURE 15.3 Inclusion relationships for GACS, SC, ACS, MC, OACS, and OSC process classes. Gray area represents AM-TW ACS processes that do not reduce to ACS processes.

Table 15.2 (Conjugate) autocorrelation $\mathcal{R}_{\boldsymbol{y}}(t,\tau)$, Loève bifrequency spectrum $\mathcal{S}_{\boldsymbol{y}}(f_1,f_2)$, and time-variant spectrum $\mathscr{S}_{\boldsymbol{y}}(t,f)$ of MC, OACS and OSC processes $y(t)$. Void boxes correspond to not easily interpretable expressions. For an OACS process $y(t)$, $x(t)$ denotes the underlying ACS process. For an OSC process $y(t)$, $x(t)$ denotes the underlying SC process.

	$\mathcal{R}_{\boldsymbol{y}}(t,\tau) = \mathrm{E}\{y(t+\tau)\, y^{(*)}(t)\}$ **(Conjugate) Autocorrelation**	$\mathcal{S}_{\boldsymbol{y}}(f_1,f_2) = \mathrm{E}\{Y(f_1)\, Y^{(*)}(f_2)\}$ **Loève Bifrequency Spectrum**	$\mathscr{S}_{\boldsymbol{y}}(t,f) = \mathcal{F}_{\tau\to f}[\mathcal{R}_{\boldsymbol{y}}(t,\tau)]$ **Time-Variant Spectrum**
MC	$\sum_{\alpha\in A} r_{\boldsymbol{y}}^{\alpha}(\tau)\, m_{\boldsymbol{y}}^{\alpha}(t)\, e^{j2\pi\alpha t}$	$\sum_{\alpha\in A} s_{\boldsymbol{y}}^{\alpha}(f_1) M_{\boldsymbol{y}}^{\alpha}(f_1 + (-)f_2 - \alpha)$	$\sum_{\alpha\in A} s_{\boldsymbol{y}}^{\alpha}(f)\, m_{\boldsymbol{y}}^{\alpha}(t)\, e^{j2\pi\alpha t}$
OACS	$\sum_{\alpha\in A} \rho_{\boldsymbol{y}}^{\alpha}(t,\tau)\, e^{j2\pi\alpha t}$	†	$\sum_{\alpha\in A} S_{\boldsymbol{y}}^{\alpha}(t,f)\, e^{j2\pi\alpha t}$
OSC	††	†††	

† $\displaystyle\sum_{\alpha\in A} \int_{\mathbb{R}} \mathcal{A}(f_1-\nu;\nu)\, \mathcal{A}^{(*)}\Big(f_2-(-)(\alpha-\nu);(-)(\alpha-\nu)\Big)\, \mathrm{d}\mu_{\boldsymbol{x}}^{\alpha}(\nu)$

†† $\displaystyle\sum_{k\in\mathbb{I}} \int_{\mathbb{R}} A_{t+\tau}(f_1)\, A_t^{(*)}\Big(\Psi_{\boldsymbol{x}}^{(k)}(f_1)\Big)\, e^{j2\pi[f_1+(-)\Psi_{\boldsymbol{x}}^{(k)}(f_1)]t}\, e^{j2\pi f_1\tau}\, \mathrm{d}\mu_{\boldsymbol{x}}^{(k)}(f_1)$

††† $\displaystyle\sum_{k\in\mathbb{I}} \int_{\mathbb{R}} \mathcal{A}(f_1-\nu;\nu)\, \mathcal{A}^{(*)}\Big(f_2-\Psi_{\boldsymbol{x}}^{(k)}(\nu);\Psi_{\boldsymbol{x}}^{(k)}(\nu)\Big)\, \mathrm{d}\mu_{\boldsymbol{x}}^{(k)}(\nu)$

different from zero in strips around the support lines of the underlying ACS process. It reduces to the Loève bifrequency spectrum of an ACS process when $A_t(f)=1$. Equivalently, the autocorrelation function of an OACS process is constituted by the superposition of amplitude- and angle-modulated sine waves. When $A_t(f)=1$, the modulation is not present and one obtains the autocorrelation of an ACS process which is given by the superposition of (non-modulated) sine waves (Tables 15.1 and 15.2).

MC processes are obtained as subclass of the OACS processes when each evolutionary cyclic autocorrelation function factorizes into the product of a function of t and a function of τ (see (14.20)). If the function of t is constant, then we obtain the ACS processes. Thus

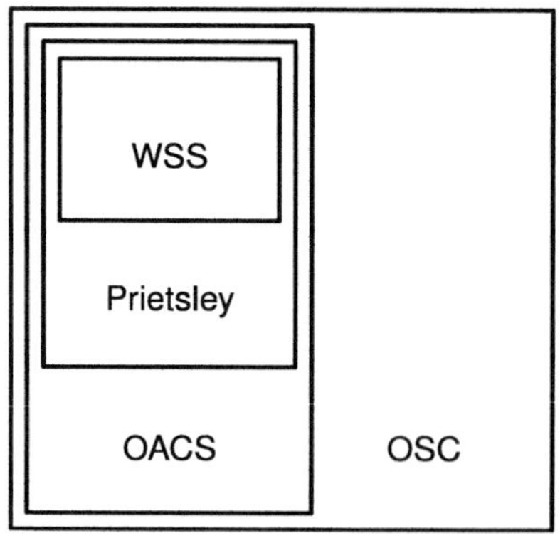

FIGURE 15.4 Inclusion relationships for WSS, Priestley's, OACS, and OSC process classes.

the class of the OACS processes contains that of the MC processes that in turn contains that of the ACS processes (Fig. 15.3).

A GACS process with generalized (conjugate) cyclic autocorrelation functions $R^{(k)}_{yy^{(*)}}(\tau)$ and lag-dependent (conjugate) cycle frequencies $\alpha^{(k)}_{yy^{(*)}}(\tau)$ is a special case of OACS process (Fig. 15.3) with evolutionary (conjugate) cyclic autocorrelation functions $\rho_{yy^{(*)}}(t,\tau) = R^{(k)}_{yy^{(*)}}(\tau)\, e^{j2\pi\epsilon^{(k)}_{yy^{(*)}}(\tau)\,t}$, where $\alpha^{(k)}_{yy^{(*)}}(\tau) \triangleq \alpha_k + \epsilon^{(k)}_{yy^{(*)}}(\tau)$ and α_k are the (conjugate) cycle frequencies of the underlying ACS process for the OACS process. As already observed in the general case of OACS processes, the case of slowly varying evolutionary (conjugate) cyclic autocorrelation functions is of interest. This corresponds to the condition $\sup_\tau |\epsilon^{(k)}_{yy^{(*)}}(\tau)| \ll |\alpha_k|$ for $\alpha_k \neq 0$. Note that in (Napolitano, 2016a, Fig. 13), only OACS processes with slowly varying evolutionary cyclic autocorrelation functions are considered.

AM-TW ACS processes are a special case of OACS processes (Section 14.3). In the absence of amplitude modulation and time warping AM-TW ACS processes reduce to ACS processes. If the evolutionary (conjugate) cyclic autocorrelation factorizes, AM-TW ACS processes are also MC processes. In some cases they are GACS (e.g., if the time-warping function is quadratic). In Fig. 15.3, the gray area represents only those AM-TW ACS processes that do not reduce to ACS processes.

15.3.4 Oscillatory processes

OACS processes admit the representation (15.1) with $Z(f)$ integrated complex spectrum of an ACS process. Thus, OACS processes are obtained as special case of OSC processes when the support curves in (15.2) reduces to the support lines with unit slope in (14.5). If $Z(f)$ is an orthogonal-increment process, the only support line in the bifrequency plane is the main diagonal and one obtains the oscillatory processes in the sense of Priestley. The Priestley's oscillatory processes reduce to the WSS processes when $A_t(f) = 1$ (Fig. 15.4).

PART 3

Appendices

A

Nonstationary signal analysis

Contents

A.1 Introduction

In this appendix, the second-order characterization of nonstationary stochastic processes is reviewed in both time and frequency domains and for both continuous- and discrete-time cases (Kampé de Fériet and Frenkiel, 1962), (Priestley, 1965), (Martin and Flandrin, 1985), (Hurd and Miamee, 2007). For the higher-order characterization, see (Hanssen and Scharf, 2003).

A.2 Second-order processes

Let us consider a continuous-time complex-valued stochastic process $\{x(t;\omega), t \in \mathbb{R}, \omega \in \Omega\}$, with abbreviated notation $x(t)$ when this does not create ambiguity, defined on a probability space $(\Omega, \mathcal{F}, P)$, where Ω is the sample space, equipped with the σ-field $\mathcal{F}$, and P is a probability measure defined on the elements of $\mathcal{F}$.

If $x(t)$ is complex valued, that is

$$x(t) = x_{\mathrm{r}}(t) + j\, x_{\mathrm{i}}(t) \tag{A.1}$$

then the first-order probabilistic characterization is made by the second-order joint probability distribution function of $x_{\mathrm{r}}(t)$ and $x_{\mathrm{i}}(t)$.

The second-order probabilistic characterization is made by the fourth-order joint probability distribution function of $x_{\mathrm{r}}(t+\tau)$, $x_{\mathrm{i}}(t+\tau)$, $x_{\mathrm{r}}(t)$, and $x_{\mathrm{i}}(t)$, that is

$$F_{x_{\mathrm{r}}(t+\tau)\, x_{\mathrm{i}}(t+\tau)\, x_{\mathrm{r}}(t)\, x_{\mathrm{i}}(t)}(\xi_{1,\mathrm{r}},\ \xi_{1,\mathrm{i}},\ \xi_{2,\mathrm{r}},\ \xi_{2,\mathrm{i}})$$

$$\triangleq P\left[x_{\mathrm{r}}(t+\tau) \leqslant \xi_{1,\mathrm{r}},\ x_{\mathrm{i}}(t+\tau) \leqslant \xi_{1,\mathrm{i}},\ x_{\mathrm{r}}(t) \leqslant \xi_{2,\mathrm{r}},\ x_{\mathrm{i}}(t) \leqslant \xi_{2,\mathrm{i}}\right] \tag{A.2a}$$

$$= \mathrm{E}\left\{\mathbf{1}_{\{x_{\mathrm{r}}(t+\tau) \leqslant \xi_{1,\mathrm{r}}\}}\, \mathbf{1}_{\{x_{\mathrm{i}}(t+\tau) \leqslant \xi_{1,\mathrm{i}}\}}\, \mathbf{1}_{\{x_{\mathrm{r}}(t) \leqslant \xi_{2,\mathrm{r}}\}}\, \mathbf{1}_{\{x_{\mathrm{i}}(t) \leqslant \xi_{2,\mathrm{i}}\}}\right\} \tag{A.2b}$$

Cyclostationary Processes and Time Series. https://doi.org/10.1016/B978-0-08-102708-0.00028-5

where $t \mapsto \mathbf{1}_A$ is the indicator function of the set A, that is, $\mathbf{1}_A = 1$ if $t \in A$ and $\mathbf{1}_A = 0$ if $t \notin A$. The fourth-order joint probability density function of $x_r(t+\tau)$, $x_i(t+\tau)$, $x_r(t)$, and $x_i(t)$, is

$$f_{x_r(t+\tau)\, x_i(t+\tau)\, x_r(t)\, x_i(t)}(\xi_{1,r},\ \xi_{1,i},\ \xi_{2,r},\ \xi_{2,i}) \triangleq \frac{\partial^4}{\partial\xi_{1,r}\,\partial\xi_{1,i}\,\partial\xi_{2,r}\,\partial\xi_{2,i}} F_{x_r(t+\tau)\, x_i(t+\tau)\, x_r(t)\, x_i(t)}(\xi_{1,r},\ \xi_{1,i},\ \xi_{2,r},\ \xi_{2,i}) \tag{A.3}$$

provided that the derivative exists. The second-order moments are

$$\mathcal{R}_{\boldsymbol{x}}(t,\tau) \triangleq \mathrm{E}\left\{x(t+\tau)\, x^{(*)}(t)\right\} \tag{A.4a}$$

$$= \int_\Omega x(t+\tau;\omega)\, x^{(*)}(t;\omega)\, \mathrm{d}P(\omega) \tag{A.4b}$$

$$= \int_{\mathbb{R}^4} (\xi_{1,r} + j\xi_{1,i})\,(\xi_{2,r} + j\xi_{2,i})^{(*)}\, \mathrm{d}F_{x_r(t+\tau)\, x_i(t+\tau)\, x_r(t)\, x_i(t)}(\xi_{1,r}, \xi_{1,i}, \xi_{2,r}, \xi_{2,i}) \tag{A.4c}$$

$$= \int_{\mathbb{R}^4} (\xi_{1,r} + j\xi_{1,i})\,(\xi_{2,r} + j\xi_{2,i})^{(*)} f_{x_r(t+\tau)\, x_i(t+\tau)\, x_r(t)\, x_i(t)}(\xi_{1,r}, \xi_{1,i}, \xi_{2,r}, \xi_{2,i})\, \mathrm{d}\xi_{1,r}\, \mathrm{d}\xi_{1,i}\, \mathrm{d}\xi_{2,r}\, \mathrm{d}\xi_{2,i} \tag{A.4d}$$

where the fourth equality holds provided that the joint probability density function exists. In (A.4a), superscript $(*)$ denotes an optional complex conjugation and subscript $\boldsymbol{x} \equiv x x^{(*)}$. If conjugation is present, then the function $\mathcal{R}_{xx^*}(t,\tau)$ in (A.4a) is the *autocorrelation function*. If conjugation is absent, then the function $\mathcal{R}_{xx}(t,\tau)$ in (A.4a) is the *conjugate autocorrelation function* also called *relation function* (Picinbono and Bondon, 1997) or *complementary correlation* (Schreier and Scharf, 2003b).

Definition A.1. The process $x(t)$ is said to be a *second order process* if $\mathrm{E}\{|x(t)|^2\}$ is finite $\forall t$. □

Fact A.2. *For every complex-valued random variables* X *and* Y *it results*

$$|\mathrm{E}\{|\mathsf{X}\mathsf{Y}^{(*)}|\}|^2 \leqslant \mathrm{E}\{|\mathsf{X}|^2\}\, \mathrm{E}\{|\mathsf{Y}|^2\}\,. \tag{A.5}$$

Thus, for a second-order process $x(t)$ the second-order moments

$$\mathcal{R}_{\boldsymbol{x}}(t,\tau) \triangleq \mathrm{E}\left\{x(t+\tau)\, x^{(*)}(t)\right\} \tag{A.6}$$

exist $\forall t$ and $\forall \tau$. □

Fact A.3. *By taking* $\mathsf{Y} = 1$ *in (A.5) we have that for a second-order process $x(t)$ $\mathrm{E}\{|x(t)|\}$ is finite $\forall t$.* □

A.3 Harmonizable processes

Definition A.4. Let $x(t)$ be a second-order stochastic process $x(t)$. The (conjugate) covariance (or autocorrelation) function $\mathrm{E}\{x(t_1)\, x^{(*)}(t_2)\}$ is said to be *harmonizable* if it can be expressed by the Fourier-Stieltjes integral

$$\mathrm{E}\{x(t_1)\, x^{(*)}(t_2)\} = \int_{\mathbb{R}^2} e^{j2\pi[f_1 t_1 + (-) f_2 t_2]} \mathrm{d}\gamma_{\boldsymbol{x}}(f_1, f_2) \tag{A.7}$$

where $(-)$ is an optional minus sign that is linked to $(*)$ and $\gamma_{\boldsymbol{x}}(f_1, f_2)$ is a (conjugate) spectral correlation function of bounded variation (Loève, 1963)

$$\int_{\mathbb{R}^2} \left|\mathrm{d}\gamma_{\boldsymbol{x}}(f_1, f_2)\right| < \infty \,. \tag{A.8}$$

□

Definition A.5. A second-order stochastic process $x(t)$ is said to be harmonizable if there exists a second-order stochastic process $\chi(f)$ with covariance function $\mathrm{E}\{\chi(f_1)\, \chi^*(f_2)\} = \gamma_{xx^*}(f_1, f_2)$ of bounded variation on $\mathbb{R} \times \mathbb{R}$ such that

$$x(t) = \int_{\mathbb{R}} e^{j2\pi f t} \mathrm{d}\chi(f) \tag{A.9}$$

with probability one. □

Theorem A.6. Cramér decomposition (Cramér, 1940). *A second-order harmonizable process $x(t)$ can be expressed by the Cramér representation (Cramér, 1940) (A.9). The random function $\chi(f)$ can be written by the* Cramér decomposition

$$\chi(f) = \chi_1(f) + \chi_2(f) + \chi_3(f) \,. \tag{A.10}$$

In (A.10),

1. $\chi_1(f)$ *is the* absolutely continuous component,

$$\mathrm{d}\chi_1(f) = X_1(f)\, \mathrm{d}f \tag{A.11}$$

with $X_1(f)$ the ordinary first-order derivative of $\chi_1(f)$;

2. $\chi_2(f)$ *is the component containing only jumps referred to as* discrete component,

$$\chi_2(f) = \sum_k x_k\, \mathrm{u}(f - f_k) \tag{A.12}$$

with x_k possibly correlated random variables. In the sense of distributions (generalized functions), it results

$$\mathrm{d}\chi_2(f) = \sum_k x_k\, \delta(f - f_k)\, \mathrm{d}f \,; \tag{A.13}$$

3. $\chi_3(f)$ *is the singular component (i.e., with derivative zero almost everywhere) (Doob, 1953, Sec. XI.5), (Hurd and Miamee, 2007, p. 197).*

The singular component can in turn be decomposed into a continuous component and a discrete component (Hurd and Miamee, 2007, p. 197). □

Theorem A.7. (Loève, 1963). *Assume that the mean function of the stochastic process $x(t)$ is the Fourier transform of a measure. A necessary condition for $x(t)$ to be harmonizable is that it is second-order continuous. Moreover, $x(t)$ is harmonizable if and only if its covariance is harmonizable.* □

For a WSS harmonizable process, $\chi(f)$ has orthogonal increments (Doob, 1953, Sec. XI.4–XI.5), (Hurd and Miamee, 2007, Sec. 4.1.2). In contrast, if the process is nonstationary, then the increments are not orthogonal. In particular, if the harmonizable process is ACS, then $\mathrm{d}\chi(f_1)$ and $\mathrm{d}\chi(f_2)$ are correlated if and only if $f_2 = f_1 - \alpha$, with α cycle frequency (see (1.29)) (Dehay and Hurd, 1994, Sec. 3), (Hurd and Miamee, 2007, Sec. 6.2).

If $\chi(f)$ does not contain singular components with probability 1 (w.p.1) (Cramér, 1940), (Hurd and Miamee, 2007, p. 197), we can formally write $\mathrm{d}\chi(f) = X(f)\,\mathrm{d}f$ (Gardner, 1985, Chap. 10.1.2), where

$$X(f) = \int_{\mathbb{R}} x(t)\, e^{-j2\pi f t}\, \mathrm{d}t \tag{A.14}$$

is the Fourier transform of $x(t)$. For *finite-power processes*, that is such that the time-averaged power

$$P_x \triangleq \lim_{T\to\infty} \frac{1}{T} \int_{-T/2}^{T/2} \mathrm{E}\left\{|x(t)|^2\right\}\, \mathrm{d}t \tag{A.15}$$

exists and is finite, relation (A.14) is intended in the sense of distributions (Gel'fand and Vilenkin, 1964, Chap. 3), (Henniger, 1970), (Napolitano, 2012, Secs. 1.1.2, 4.2.1), and $X(f)$ contains Dirac deltas in correspondence of the jumps of $\chi(f)$. The function $\chi(f)$ is also called the *integrated spectrum* of $x(t)$ and we have

$$\chi(f) = \int_{-\infty}^{f} X(\lambda)\, \mathrm{d}\lambda\,. \tag{A.16}$$

The structure of the support the spectral measure of harmonizable sequences is analyzed in (Makagon and Wylomanska, 2008).

Definition A.8. The *Loève bifrequency spectrum* of the process $x(t)$ is defined as

$$\mathcal{S}_{\boldsymbol{x}}(f_1, f_2) \triangleq \lim_{T\to\infty} \left\langle \mathrm{E}\left\{X_T(t, f_1)\, X_T^{(*)}(t, f_2)\right\}\right\rangle_t \tag{A.17}$$

where

$$X_T(t, f) \triangleq \int_{t-T/2}^{t+T/2} x(u)\, e^{-j2\pi f u}\, \mathrm{d}u \tag{A.18}$$

is the short-time Fourier transform (STFT) of $x(t)$ and the limit is in the sense of generalized functions (distributions) (Zemanian, 1987). □

Proposition A.9. *The* Loève bifrequency spectrum *(A.17) can be formally expressed as (Loève, 1963), (Gerr and Allen, 1994a)*

$$\mathcal{S}_{\boldsymbol{x}}(f_1, f_2) = \mathrm{E}\left\{X(f_1)\, X^{(*)}(f_2)\right\} . \tag{A.19}$$

□

We can formally write

$$\mathrm{d}\gamma_{\boldsymbol{x}}(f_1, f_2) = \mathrm{E}\left\{\mathrm{d}\chi(f_1)\, \mathrm{d}\chi^{(*)}(f_2)\right\} \tag{A.20a}$$

$$= \mathcal{S}_{\boldsymbol{x}}(f_1, f_2)\, \mathrm{d}f_1\, \mathrm{d}f_2 \tag{A.20b}$$

provided that $\gamma_{\boldsymbol{x}}(f_1, f_2)$ does not contain singular components and $\chi(f)$ does not contain singular components w.p.1 (Napolitano, 2012, Sec. 1.1.2).

According to (A.7) and (A.14), we formally have

$$\mathcal{S}_{\boldsymbol{x}}(f_1, f_2) = \int_{\mathbb{R}^2} \mathrm{E}\{x(t_1)\, x^{(*)}(t_2)\}\, e^{-j2\pi[f_1 t_1 + (-) f_2 t_2]}\, \mathrm{d}t_1\, \mathrm{d}t_2 . \tag{A.21}$$

A.4 Time-frequency representations

A time-frequency second-order characterization is given by the *time-variant spectrum* (Boashash, 2003, Sec. 9.4) which is defined as the Fourier transform of the second-order moment (A.4a) with respect to (w.r.t.) the lag parameter τ

$$\mathcal{S}_{\boldsymbol{x}}(t, f) \triangleq \int_{\mathbb{R}} \mathcal{R}_{\boldsymbol{x}}(t, \tau)\, e^{-j2\pi f \tau}\, \mathrm{d}\tau \tag{A.22}$$

and it results that (Section A.7)

$$\mathcal{S}_{\boldsymbol{x}}(t, f)\, \mathrm{d}f = \mathrm{E}\left\{\mathrm{d}\chi(f)\, x^{(*)}(t)\right\}\, e^{j2\pi f t} \tag{A.23}$$

where the right-hand-side is referred to as the *(conjugate) Rihaczek distribution* of $x(t)$ (Scharf et al., 2005). After the variable change $t' = t + \tau/2$ in (A.4a) and the Fourier transform w.r.t. τ, we obtain the *Wigner spectrum* for stochastic processes (Boashash, 2003, Sec. 9.4), (Martin and Flandrin, 1985)

$$\mathcal{W}_{\boldsymbol{x}}(t', f) \triangleq \int_{\mathbb{R}} \mathrm{E}\left\{x(t' + \tau/2)\, x^{(*)}(t' - \tau/2)\right\}\, e^{-j2\pi f \tau}\, \mathrm{d}\tau \tag{A.24a}$$

$$= \int_{\mathbb{R}} \mathrm{E}\left\{X(f + \nu/2)\, X^{(*)}(f - \nu/2)\right\}\, e^{j2\pi \nu t'}\, \mathrm{d}\nu . \tag{A.24b}$$

If $x(t)$ is an analytic signal, then $\mathcal{W}_{\boldsymbol{x}}(t', f)$ is referred to as the *Wigner-Ville distribution*.

Tables 15.1 and 15.2 summarize the behavior of (conjugate) autocorrelation, Loève bifrequency spectrum, and time-variant spectrum of WSS, ACS, GACS, SC, MC, OACS, and OSC processes.

A.5 Wide-sense stationary processes

Second-order wide-sense stationary (WSS) processes have (conjugate) autocorrelation function not depending on t:

$$\mathcal{R}_{\boldsymbol{x}}(t,\tau) = R_{\boldsymbol{x}}(\tau) . \tag{A.25}$$

For a harmonizable process, the independence on t of the (conjugate) autocorrelation function corresponds, in the spectral domain, to the property that

$$\mathrm{E}\left\{\mathrm{d}\chi(f_1)\,\mathrm{d}\chi^{(*)}(f_2)\right\} = \delta(f_2 +_{(-)} f_1)\,\mathrm{d}\mu_{\boldsymbol{x}}(f_1)\,\mathrm{d}f_2 \tag{A.26}$$

where $\delta(\cdot)$ denotes Dirac delta and $\mu_{\boldsymbol{x}}(f_1)$ is a spectral measure. In fact,

$$\begin{aligned}\mathrm{E}\left\{x(t_1)\,x^{(*)}(t_2)\right\} &= \int_{\mathbb{R}^2} e^{j2\pi[f_1t_1+_{(-)}f_2t_2]}\,\mathrm{E}\left\{\mathrm{d}\chi(f_1)\,\mathrm{d}\chi^{(*)}(f_2)\right\} \\ &= \int_{\mathbb{R}^2} e^{j2\pi[f_1t_1+_{(-)}f_2t_2]}\,\delta(f_2 +_{(-)} f_1)\,\mathrm{d}\mu_{\boldsymbol{x}}(f_1)\,\mathrm{d}f_2 \\ &= \int_{\mathbb{R}} e^{j2\pi f_1(t_1-t_2)}\,\mathrm{d}\mu_{\boldsymbol{x}}(f_1)\end{aligned} \tag{A.27}$$

Thus, one obtains the *Wiener-Khinchin relation*

$$R_{\boldsymbol{x}}(\tau) = \int_{\mathbb{R}} e^{j2\pi f\tau}\,\mathrm{d}\mu_{\boldsymbol{x}}(f) \tag{A.28a}$$

$$= \int_{\mathbb{R}} e^{j2\pi f\tau}\,S_{\boldsymbol{x}}(f)\,\mathrm{d}f \tag{A.28b}$$

where the second equality holds if $\mu_{\boldsymbol{x}}(f_1)$ does not contain singular component and, hence, $\mathrm{d}\mu_{\boldsymbol{x}}(f_1) = S_{\boldsymbol{x}}(f_1)\,\mathrm{d}f_1$ with $S_{\boldsymbol{x}}(f_1)$ denoting the *power spectral density* of $x(t)$.

If $\mu_{\boldsymbol{x}}(f_1)$ does not contain singular component, from (A.26) it follows that the Loève bifrequency spectrum (A.19) is nonzero only on the diagonal $f_2 = -_{(-)} f_1$. That is,

$$\mathcal{S}_{\boldsymbol{x}}(f_1, f_2) = S_{\boldsymbol{x}}(f_1)\,\delta(f_2 +_{(-)} f_1) . \tag{A.29}$$

Thus, WSS processes have distinct spectral component that are uncorrelated. Equivalently, the process $\chi(f)$ has orthogonal increments (Doob, 1953, Sec. XI.4–XI.5), (Hurd and Miamee, 2007, Sec. 4.1.2) (see (A.26)). The presence of spectral correlation between distinct spectral components is an indicator of nonstationarity.

Consequently, also the time-variant spectrum does not depend on t and is coincident with the (conjugate) power spectrum

$$\mathcal{S}_{\boldsymbol{x}}(t, f) = S_{\boldsymbol{x}}(f) \tag{A.30}$$

and for $(*)$ present, the Fourier-transform (A.22) specializes into the Wiener-Khinchin relationship (A.28b). In addition, for WSS processes the Wigner spectrum is independent of t' and is coincident with the power spectrum. That is, $\mathcal{W}_{\boldsymbol{x}}(t', f) = S_{\boldsymbol{x}}(f)$.

A.5.1 Time-averaged autocorrelation

The Wiener-Khinchin relation for WSS processes can be extended to a wide class of nonstationary processes, provided that the time-averaged (conjugate) autocorrelation function exists and is Fourier transformable (Wu and Lev-Ari, 1997).

Let $x(t)$ be a finite-power *nonstationary* process such that its *time-averaged autocorrelation function*

$$R_{xx^*}(\tau) \triangleq \left\langle \mathrm{E}\left\{x(t+\tau)\, x^*(t)\right\}\right\rangle_t \tag{A.31}$$

exists for all $\tau \in \mathbb{R}$.

The *power spectrum* or *power spectral density* is defined as

$$S_{xx^*}(f) \triangleq \lim_{T\to\infty} \frac{1}{T} \mathrm{E}\left\{|X_T(f)|^2\right\} \tag{A.32}$$

provided that the limit exists. In (A.32), $X_T(f)$ denotes the Fourier transform of $x(t)\,\mathrm{rect}(t/T)$.

If $R_{xx^*}(\tau) \in L^1(\mathbb{R})$ we have

$$S_{xx^*}(f) = \int_{\mathbb{R}} R_{xx^*}(\tau)\, e^{-j2\pi f\tau}\, \mathrm{d}\tau\ . \tag{A.33}$$

Similarly, the (conjugate) autocorrelation and (conjugate) power spectral density can be defined. We have the Fourier transform pair

$$R_{xx^{(*)}}(\tau) \triangleq \left\langle \mathrm{E}\left\{x(t+\tau)\, x^{(*)}(t)\right\}\right\rangle_t \stackrel{\mathcal{F}}{\longleftrightarrow} S_{xx^{(*)}}(f) \triangleq \lim_{T\to\infty} \frac{1}{T} \mathrm{E}\left\{X_T(f)\, X_T^{(*)}(f)\right\} \tag{A.34}$$

provided that all the involved functions are well defined.

A.6 Discrete-time nonstationary stochastic processes

Discrete-time nonstationary stochastic processes can be characterized similarly to the continuous-time processes with obvious modifications.

The covariance (or autocorrelation) of the discrete-time process $x_{\mathrm{d}}(n)$ is harmonizable if

$$\mathrm{E}\left\{x_{\mathrm{d}}(n+m)\, x_{\mathrm{d}}^{(*)}(n)\right\} = \int_{[-1/2,1/2]^2} e^{j2\pi[\nu_1(n+m)+(-)\nu_2 n]} \mathrm{d}\widetilde{\gamma}_{x_{\mathrm{d}}}(\nu_1,\nu_2) \tag{A.35}$$

with $\widetilde{\gamma}_{x_{\mathrm{d}}}(\nu_1,\nu_2)$ spectral correlation function of bounded variation when $(\nu_1,\nu_2) \in [-1/2, 1/2]^2$. The process $x_{\mathrm{d}}(n)$ is harmonizable if it can be expressed as

$$x_{\mathrm{d}}(n) = \int_{-1/2}^{1/2} e^{j2\pi\nu n} \mathrm{d}\chi_{\mathrm{d}}(\nu) \tag{A.36}$$

where $\mathrm{E}\left\{\mathrm{d}\chi_{\mathrm{d}}(\nu_1)\, \mathrm{d}\chi_{\mathrm{d}}^{(*)}(\nu_2)\right\}$ is of bounded variation when $(\nu_1,\nu_2) \in [-1/2, 1/2]^2$. In addition, if $\chi_{\mathrm{d}}(\nu)$ does not contain singular components w.p.1, one can formally write $\mathrm{d}\chi_{\mathrm{d}}(\nu) =$

$X_{\mathrm{d}}(\nu)\mathrm{d}\nu$ (Gardner, 1985, Chap. 10.1.2) where

$$X_{\mathrm{d}}(\nu) = \sum_{n \in \mathbb{Z}} x_{\mathrm{d}}(n)\, e^{-j2\pi \nu n} \tag{A.37}$$

is the Fourier transform of $x_{\mathrm{d}}(n)$ defined in a generalized sense (Gel'fand and Vilenkin, 1964, Chap. 3), (Henniger, 1970) that accounts for the presence of periodic delta trains with period 1 in correspondence of the jumps of the periodic function $\chi_{\mathrm{d}}(\nu)$. The Loève bifrequency spectrum of $x_{\mathrm{d}}(n)$ is defined as (Loève, 1963)

$$\widetilde{\mathbb{S}}_{\boldsymbol{x}_{\mathrm{d}}}(\nu_1, \nu_2) \triangleq \mathrm{E}\left\{X_{\mathrm{d}}(\nu_1)\, X_{\mathrm{d}}^{(*)}(\nu_2)\right\} . \tag{A.38}$$

Let $x_{\mathrm{d}}(n) \triangleq x(t)|_{t=nT_s}$ be the discrete-time process obtained by uniformly sampling with period $T_s = 1/f_s$ the continuous-time process $x(t)$. The (conjugate) autocorrelation function and the Loève bifrequency spectrum of $x_{\mathrm{d}}(n)$ are linked to those of $x(t)$ by

$$\mathrm{E}\left\{x_{\mathrm{d}}(n+m)\, x_{\mathrm{d}}^{(*)}(n)\right\} = \mathrm{E}\left\{x(t+\tau)\, x^{(*)}(t)\right\}\Big|_{t=nT_s,\tau=mT_s} \tag{A.39}$$

and

$$\widetilde{\mathbb{S}}_{\boldsymbol{x}_{\mathrm{d}}}(\nu_1, \nu_2) = f_s^2 \sum_{p_1 \in \mathbb{Z}} \sum_{p_2 \in \mathbb{Z}} \mathbb{S}_{\boldsymbol{x}}\Big((\nu_1 - p_1) f_s, (\nu_2 - p_2) f_s\Big) , \tag{A.40}$$

respectively.

A.7 Proofs

Proof of (A.23)

In this section, a formal proof of (A.23) is provided for the case where $\chi(f)$ does not contain singular component.

Using (A.22), we have

$$\begin{aligned}
\mathcal{S}_{\boldsymbol{x}}(t, f)\, \mathrm{d}f &= \int_{\mathbb{R}} \mathrm{E}\left\{x(t+\tau)\, x^{(*)}(t)\right\} e^{-j2\pi f\tau}\, \mathrm{d}\tau\, \mathrm{d}f \\
&= \mathrm{E}\left\{\int_{\mathbb{R}} x(t+\tau)\, e^{-j2\pi f\tau}\, \mathrm{d}\tau\, x^{(*)}(t)\right\} \mathrm{d}f \\
&= \mathrm{E}\left\{X(f)\, e^{j2\pi ft}\, x^{(*)}(t)\right\} \mathrm{d}f \\
&= \mathrm{E}\left\{\mathrm{d}\chi(f)\, x^{(*)}(t)\right\} e^{j2\pi ft}
\end{aligned} \tag{A.41}$$

B Almost-periodic functions

Contents

B.1 Almost-periodic functions

In this appendix, definitions and main results on almost-periodic (AP) functions and their generalizations are presented for both continuous- and discrete-time cases. For extensive treatments on almost-periodic functions see (Besicovitch, 1932), (Bohr, 1933), (Corduneanu, 1989) for continuous-time and (Corduneanu, 1989, Chap. VII), (Jessen and Tornehave, 1945), (von Neumann, 1934) for discrete-time. A review on AP functions and their generalizations is in (Napolitano, 2012, Sec. 1.2).

B.2 Uniformly almost-periodic functions

Definition B.1. *(Besicovitch, 1932, Chap. 1).* A complex-valued function $z(t)$, $t \in \mathbb{R}$, is said to be *uniformly almost-periodic* if $\forall \epsilon > 0 \; \exists \; \ell_\epsilon > 0$ such that for any interval $I_\epsilon = (t_0, t_0 + \ell_\epsilon)$ $\exists \; \tau_\epsilon \in I_\epsilon$ such that

$$\sup_{t \in \mathbb{R}} |z(t + \tau_\epsilon) - z(t)| < \epsilon \, . \tag{B.1}$$

The quantity τ_ϵ is said translation number of $z(t)$ corresponding to ϵ. □

A set $D \subseteq \mathbb{R}$ is said to be *relatively dense* in $\mathbb{R}$ if $\exists \; \ell > 0$ such that $\forall \; I = (t_0, t_0 + \ell)$ it results $D \cap I \neq \emptyset$.

Thus, defined the set of the translation numbers of $z(t)$ corresponding to ϵ

$$E(z, \epsilon) \triangleq \left\{ \tau \in \mathbb{R} \, : \, \sup_{t \in \mathbb{R}} |z(t + \tau) - z(t)| < \epsilon \right\} \tag{B.2}$$

according to Definition B.1, the function $z(t)$ is uniformly almost periodic if and only if $\forall \epsilon > 0$ the set $E(z, \epsilon)$ is relatively dense in $\mathbb{R}$. That is, there are many translation numbers of $z(t)$ corresponding to ϵ.

Cyclostationary Processes and Time Series. https://doi.org/10.1016/B978-0-08-102708-0.00029-7

Theorem B.2. (Besicovitch, 1932, Chap. 1). *Any uniformly AP function is the limit of a uniformly convergent sequence of trigonometric polynomials in t. Thus, $z(t)$ is bounded and uniformly continuous.*

Let us define the Fourier coefficient

$$z_\alpha \triangleq \lim_{T\to\infty} \frac{1}{T} \int_{t_0-T/2}^{t_0+T/2} z(t)\, e^{-j2\pi\alpha t}\, \mathrm{d}t \tag{B.3}$$

where the limit exists and is independent of t_0. It results that $z_\alpha \neq 0$ for $\alpha \in A$, where A is a countable set of possibly incommensurate reals and possibly containing cluster points.

Under the assumption

$$\sum_{\alpha\in A} |z_\alpha| < \infty \tag{B.4}$$

we have the uniform pointwise convergence of the generalized Fourier series:

$$z(t) = \sum_{\alpha\in A} z_\alpha\, e^{j2\pi\alpha t}\,. \tag{B.5}$$

□

Theorem B.3. (Besicovitch, 1932, Chap. 1), (Corduneanu, 1989, Theorem 1.18). *For any uniformly AP function the following Parseval's equality holds*

$$\lim_{T\to\infty} \frac{1}{T} \int_{-T/2}^{T/2} |z(t)|^2 \mathrm{d}t = \sum_{\alpha\in A} |z_\alpha|^2\,. \tag{B.6}$$

□

Theorem B.4. (Corduneanu, 1989, Theorem 1.19). *Two distinct almost-periodic functions have distinct Fourier series.*

Definition B.5. A function $z(t)$ is *periodic* with period $T_0 > 0$ if T_0 is the smallest nonzero value such that

$$z(t) = z(t + T_0) \qquad \forall t \in \mathbb{R}\,. \tag{B.7}$$

□

Thus, uniformly continuous periodic functions are obtained as special case of almost-periodic functions with $\tau_\epsilon = kT_0$ independent of ϵ, $k \in \mathbb{Z}$. In such a case, the frequencies of the set A are all multiple of a fundamental frequency $1/T_0$, that is, $A = \{k/T_0\}_{k\in\mathbb{Z}}$ and (B.5) is the ordinary Fourier series expansion of the periodic function $z(t)$.

An example of uniformly AP function which is not periodic is

$$z(t) = \cos(2\pi t/T_0) + \cos(2\pi t/(\sqrt{2}T_0))\,. \tag{B.8}$$

Both cosines are periodic functions with periods T_0 and $\sqrt{2}T_0$, respectively. However, their sum is not periodic since the two period are incommensurate. In fact, the ratio of the two periods T_0 and $\sqrt{2}T_0$ is the irrational number $1/\sqrt{2}$.

The functions defined in Definition B.1 and characterized in Theorem B.2 are called almost-periodic in the sense of Bohr (Bohr, 1933, par. 84–92) or, equivalently, *uniformly almost periodic* in sense the of Besicovitch (Besicovitch, 1932, Chap. 1), or, equivalently, *almost-periodic with respect to the sup norm*. More general classes of almost-periodic functions, including possibly discontinuous functions, are treated in (Besicovitch, 1932, Chap. 2) and Section B.4.

Definition B.6. A uniformly almost-periodic function is said to be *poly-periodic* (Gardner, 1987f, Chap. 15), (Gardner and Brown, 1991), (Gardner, 1994a) if the set A of frequencies can be written as

$$A = \bigcup_{p=1}^{M} \left\{ \alpha = \frac{\ell}{T_p},\ \ell \in \mathbb{Z} - \{0\} \right\} \bigcup \{0\} \tag{B.9}$$

where the fundamental periods T_p, $p = 1, \ldots, M$ are incommensurate. In such a case, under assumption (B.4) we have

$$z(t) = \sum_{\substack{\ell_1=-\infty \\ \ell_1 \neq 0}}^{+\infty} z_{\ell_1/T_1}\, e^{j2\pi(\ell_1/T_1)t} + \cdots + \sum_{\substack{\ell_M=-\infty \\ \ell_M \neq 0}}^{+\infty} z_{\ell_M/T_M}\, e^{j2\pi(\ell_M/T_M)t} + z_0\,. \tag{B.10}$$

□

Definition B.7. The trigonometric series of Bohl's class (Besicovitch, 1932, p. xi) are almost-periodic functions that can be expressed as

$$z(t) = \sum_{\boldsymbol{\ell} \in \mathbb{Z}^M} z_{\boldsymbol{\ell}}\, e^{j2\pi[(\ell_1/T_1)+\cdots+(\ell_M/T_M)]t} \tag{B.11}$$

where $\boldsymbol{\ell} \triangleq [\ell_1, \ldots, \ell_M]$, provided that the series is absolutely convergent. □

A function belonging to the Bohl class (Definition B.7) can be obtained by the product of periodic functions with incommensurate periods:

$$z(t) = z_{T_1}(t) \cdots z_{T_M}(t) \tag{B.12}$$

with

$$z_{T_m}(t) = \sum_{\ell_m \in \mathbb{Z}} z_{\ell_m/T_m}\, e^{j2\pi(\ell_m/T_m)t}\,. \tag{B.13}$$

In (B.11) one has $z_{\boldsymbol{\ell}} = z_{\ell_1/T_1} \cdots z_{\ell_M/T_M}$.

Note that the frequencies of the poly-periodic functions are integer multiples of the fundamental frequencies $1/T_1, \ldots, 1/T_M$ (see (B.10)) while the frequencies of the Bohl functions are linear combinations of these fundamental frequencies (see (B.11)).

Theorem B.8. (Besicovitch, 1932, Chap. I, Par. 1), (Corduneanu, 1989, Theorem 1.4). *Let $z(t)$ be a uniformly almost periodic function. For every real constant a and complex constant c, $z^*(t)$, $c\, z(t)$, $z(t+a)$, and $z(at)$ are uniformly almost periodic.* □

Theorem B.9. (Besicovitch, 1932, Chap. I, Par. 1), (Corduneanu, 1989, Theorem 1.5). *Let $z_1(t)$ and $z_2(t)$ be uniformly almost periodic functions. The function $z_1(t) + z_2(t)$ and $z_1(t)\, z_2(t)$ are uniformly almost periodic.* □

Theorem B.10. (Besicovitch, 1932, Chap. I, Par. 1), (Corduneanu, 1989, Theorem 1.7). *Let $z_1(t), \ldots, z_M(t)$ be complex-valued uniformly almost periodic functions and the function $z \in \mathbb{C}^M \mapsto \Phi(z) \in \mathbb{C}$ uniformly continuous. The function $\Phi(z_1(t), \ldots, z_M(t))$ is uniformly almost periodic.* □

Corollary B.11. (Besicovitch, 1932, Chap. I, Par. 1), (Corduneanu, 1989, Coroll. p. 13). *Let $z_1(t)$ and $z_2(t)$ be uniformly almost periodic functions with $0 < m \leqslant |z_2(t)|$. The function $z_1(t)/z_2(t)$ is uniformly almost periodic.* □

Theorem B.12. (Besicovitch, 1932, Chap. I, Par. 1), (Corduneanu, 1989, Theorem 1.8). *If $z(t)$ is a complex-valued uniformly almost periodic function and $\mathrm{d}z(t)/\mathrm{d}t$ is uniformly continuous on the real line, then $\mathrm{d}z(t)/\mathrm{d}t$ is uniformly almost periodic.* □

B.3 Almost-periodic sequences

Definition B.13. *Discrete-Time Almost-Periodic Functions (Corduneanu, 1989, Chap. VII), (Jessen and Tornehave, 1945), (von Neumann, 1934).* A sequence $z(n)$, $n \in \mathbb{Z}$, is said to be *almost-periodic* if $\forall \epsilon > 0\ \exists\ \ell_\epsilon \in \mathbb{N}$ such that for any set $I_\epsilon \triangleq \{n_0, n_0 + 1, \ldots, n_0 + \ell_\epsilon\}\ \exists\ m_\epsilon \in I_\epsilon$ such that

$$\sup_{n \in \mathbb{Z}} |z(n + m_\epsilon) - z(n)| < \epsilon \tag{B.14}$$

The integer m_ϵ is said translation number of $z(n)$ corresponding to ϵ. □

Theorem B.14. (Corduneanu, 1989, Chap. VII), (von Neumann, 1934). *Every AP sequence $\{z(n)\}_{n \in \mathbb{Z}}$ is the limit of a sequence of trigonometric polynomials in n. Thus $z(n)$ is bounded.*

Let us define the Fourier coefficient

$$z_{\widetilde{\alpha}} \triangleq \lim_{N \to \infty} \frac{1}{2N+1} \sum_{n=n_0-N}^{n_0+N} z(n)\, e^{-j2\pi\widetilde{\alpha}n} \tag{B.15}$$

where the limit exists and is independent of n_0. It results that $z_{\widetilde{\alpha}} \neq 0$ for $\widetilde{\alpha} \in \widetilde{A}$, where $\widetilde{A}$ is a countable set of possibly incommensurate real numbers in $[-1/2, 1/2)$ and possibly containing cluster points.

Under the assumption

$$\sum_{\widetilde{\alpha} \in \widetilde{A}} |z_{\widetilde{\alpha}}| < \infty \tag{B.16}$$

we have the uniform convergence of the generalized Fourier series:

$$z(n) = \sum_{\widetilde{\alpha} \in \widetilde{A}} z_{\widetilde{\alpha}}\, e^{j2\pi\widetilde{\alpha}n}\,. \tag{B.17}$$

□

A sequence $z(n)$ is *periodic* with period N_0 if N_0 is the smallest nonzero integer such that

$$z(n) = z(n + N_0) \qquad \forall n \in \mathbb{Z}\,. \tag{B.18}$$

Thus, periodic sequences are obtained as special case of almost-periodic sequences with $m_\epsilon = kN_0$ independent of ϵ, $k \in \mathbb{Z}$. In such a case, $\widetilde{A}$ is the finite set $\{0, 1/N_0, \dots, (N_0 - 1)/N_0\}$ or any equivalent set $k_0 + A \triangleq \{k_0, k_0 + 1/N_0, \dots, k_0 + (N_0 - 1)/N_0\}$ with k_0 integer, and (B.17) is the discrete Fourier series (DFS) of $z(n)$.

In continuous-time, the complex sine wave $z(t) = e^{j2\pi f_0 t}$ is periodic with period $T_0 = 1/f_0$ and the polynomial phase signal $z(t) = e^{j2\pi f_0 t^\gamma}$, $\gamma > 1$, is not almost periodic. Complex sine waves and polynomial phase sequences require more attention. The sequence $z(n) = e^{j2\pi \nu_0 n}$ with $\nu_0 = p/q \in \mathbb{Q}$, p, q relative prime integers or co-prime (that is, they have no common positive divisor other than 1 or, equivalently, their greatest common divisor is 1 or, equivalently, p/q is an irreducible fraction), is periodic with period q. In contrast, the sequence $z(n) = e^{j2\pi \nu_0 n}$ with $\nu_0 \notin \mathbb{Q}$, is almost periodic (not periodic). For every positive integer $L \geqslant 2$, the sequence $z(n) = e^{j2\pi \nu_0 n^L}$ with $\nu_0 = p/q \in \mathbb{Q}$, p, q relative prime integers, is periodic with period q whereas it is not almost periodic for $\nu_0 \notin \mathbb{Q}$.

B.4 Generalizations of AP functions

B.4.1 AP functions in the sense of Stepanov, Weyl, and Besicovitch

The almost-periodicity property can be defined with respect to the following norms or seminorms, with $p \geqslant 1$, (Besicovitch, 1932, Chap. 2):

1. *Stepanov S_T^p-norm:*

$$\|z\|_{S_T^p} \triangleq \left[\sup_{a\in\mathbb{R}} \frac{1}{T}\int_{a-T/2}^{a+T/2} |z(t)|^p \,\mathrm{d}t\right]^{1/p} \tag{B.19}$$

2. *Weyl W^p-norm:*

$$\|z\|_{W^p} \triangleq \lim_{T\to\infty} \|z\|_{S_T^p} = \left[\lim_{T\to\infty} \sup_{a\in\mathbb{R}} \frac{1}{T}\int_{a-T/2}^{a+T/2} |z(t)|^p \,\mathrm{d}t\right]^{1/p} \tag{B.20}$$

3. *Besicovitch B^p-seminorm:*

$$\|z\|_{B^p} \triangleq \left[\limsup_{T\to\infty} \frac{1}{T}\int_{-T/2}^{T/2} |z(t)|^p \,\mathrm{d}t\right]^{1/p}. \tag{B.21}$$

Functions belonging to $L^p_{\text{loc}}(\mathbb{R})$ with finite Besicovitch B^p-seminorm form a seminormed space called Marcinkiewicz space $\mathcal{M}^p$.

Let $\|\cdot\|_{G^p}$ denote any of the above defined (semi)norms, that is, $\|\cdot\|_{S_T^p}$, $\|\cdot\|_{W^p}$, or $\|\cdot\|_{B^p}$. For each (semi)norm, a definition of almost-periodicity can be given.

Definition B.15. *S_T^p, W^p, and B^p Almost Periodicity (Besicovitch, 1932, Chap. 2).* A function $z(t)$ is said to be *G^p almost-periodic (G^p-AP),* $p \geqslant 1$, if $\forall \epsilon > 0 \; \exists \; \ell_\epsilon > 0$ such that for any interval $I_\epsilon = (t_0, t_0 + \ell_\epsilon) \; \exists \; \tau_\epsilon \in I_\epsilon$ such that

$$\|z(t + \tau_\epsilon) - z(t)\|_{G^p} < \epsilon \,. \tag{B.22}$$

Specifically, if $G^p = S_T^p$, then $z(t)$ is said S_T^p-AP; If $G^p = W^p$, then $z(t)$ is said W^p-AP; If $G^p = B^p$, then $z(t)$ is said B^p-AP. □

Theorem B.16. (Besicovitch, 1932, Chap. 2). *Under assumption (B.4), any G^p-AP function is G^p-bounded ($\|z\|_{G^p} < \infty$) and is the G^p-limit of a sequence of trigonometric polynomials in t:*

$$\lim_n \left\| z(t) - \sum_{\alpha \in A_n} z_\alpha \, e^{j2\pi\alpha t} \right\|_{G^p} = 0 \tag{B.23}$$

where A_n is an increasing sequence of countable sets such that $m < n \Rightarrow A_m \subseteq A_n$ and $\lim_n A_n \triangleq \cup_{n \in \mathbb{N}} A_n = A$. The coefficients z_α of the generalized Fourier series are given by (B.3). □

In (Besicovitch, 1932, p. 74) it is shown that: If $\|z_1(t) - z_2(t)\|_{S_T^p} = 0$ then $z_1(t) = z_2(t)$ a.e.; If $\|z_1(t) - z_2(t)\|_{W^p} = 0$ or $\|z_1(t) - z_2(t)\|_{B^p} = 0$, then $z_1(t)$ and $z_2(t)$ may differ at a set of points of finite and even of infinite measure. In addition, for $p \geqslant 1$ it results that (Besicovitch, 1932, p. 73) $\sup_{t \in \mathbb{R}} |z(t)| \geqslant \|z(t)\|_{S_T^p} \geqslant \|z(t)\|_{W^p} \geqslant \|z(t)\|_{B^p}$.

Theorem B.17. (Besicovitch, 1932, Chap. 2). *For any G^2-AP function the following Parseval's equality holds*

$$\|z(t)\|_{G^2}^2 = \sum_{\alpha \in A} |z_\alpha|^2 \,. \tag{B.24}$$

□

B.4.2 Other generalizations of AP functions

Further generalizations of almost-periodic functions can be found in (Besicovitch, 1932, Chap. 2), (Bohr, 1933, Par. 94–102), (Corduneanu, 1989, Chap. VI). Furthermore, the following generalizations have been considered.

- Weakly AP Functions in the Sense of Eberlein (Eberlein, 1949), (Eberlein, 1956), (Napolitano, 2012, Sec. 1.2.3).
  The weakly almost-periodic functions include as special cases the uniformly almost periodic functions (in the sense of Definition B.1), the positive definite functions (hence Fourier-Stieltjes transforms), and functions vanishing at infinity (Eberlein, 1949, Theorems 11.1 and 11.2). A weakly almost-periodic function is uniformly continuous (Eberlein, 1949, Theorem 13.1).
  Other definitions of weakly almost-periodic functions different from that in (Eberlein, 1949), (Eberlein, 1956), are given in (Amerio and Prouse, 1971, Chap. 3), (Corduneanu, 1989, Sect. VI.5), (Zhang and Liu, 2010).

- Pseudo AP Functions (Zhang, 1994), (Zhang, 1995), (Ait Dads and Arino, 1996), (Napolitano, 2012, Sec. 1.2.4).
- Asymptotically AP Functions (Leśkow and Napolitano, 2006, Section 6.2).
- AP Functions in the Sense of Hartman and Ryll-Nardzewski (Kahane, 1961), (Urbanik, 1962), (Kahane, 1962), (Andreas et al., 2006, Def. 7.1), (Napolitano, 2012, Sec. 1.2.5).
- AP functions and their generalizations on groups (Corduneanu, 1989, Chap. VII), (von Neumann, 1934), (Casinovi, 2009).
- AP functions with values in Hilbert spaces (Phong, 2007).
- AP functions with values in Banach spaces and surveys of their properties (Andreas et al., 2006), (Chérif, 2011a), (Chérif, 2011b).
- AP Random Functions in Probability (Corduneanu, 1989, Sect. II.3), (Napolitano, 2012, Sec. 1.2.7).
- AP Random Sequences in Probability (Han and Hong, 2007), (Napolitano, 2012, Sec. 1.2.9).

In most cases, a function $z(t)$ which is AP in one of the generalized senses considered above can be expressed in the form

$$z(t) = \sum_{\alpha \in A} z_\alpha \, e^{j2\pi\alpha t} + z_r(t) \tag{B.25}$$

where the possibly finite-power residual term $z_r(t)$ does not contain any finite-strength additive sine-wave component

$$\lim_{T\to\infty} \frac{1}{T} \int_{t_0-T/2}^{t_0+T/2} z_r(t)\, e^{-j2\pi\alpha t}\, \mathrm{d}t \equiv 0 \qquad \forall \alpha \in \mathbb{R}\,. \tag{B.26}$$

B.5 Synchronized averaging for (poly-)periodic functions

Assumption B.18. *Let us assume that the periodic function $z_{T_0}(t)$ has convergent Fourier series*

$$z_{T_0}(t) = \sum_{k=-\infty}^{+\infty} z_{k/T_0}\, e^{j2\pi(k/T_0)t} \tag{B.27}$$

with Fourier coefficients

$$z_{k/T_0} = \frac{1}{T_0} \int_{-T_0/2}^{T_0/2} z_{T_0}(t)\, e^{-j2\pi(k/T_0)t}\, \mathrm{d}t \tag{B.28a}$$

$$= \lim_{T\to\infty} \frac{1}{T} \int_{-T/2}^{T/2} z(t)\, e^{-j2\pi(k/T_0)t}\, \mathrm{d}t\,. \tag{B.28b}$$

□

Theorem B.19. Synchronized Averaging Theorem (Gardner, 1985), (Gardner, 1987f). *Let $z(t)$ be a time series containing a periodic component $z_{T_0}(t)$ with period T_0 (see Proposi-*

tion 2.11) satisfying Assumption B.18 (see (B.33) and (B.34)). It results that

$$\sum_{k=-\infty}^{+\infty} z_{k/T_0}\, e^{j2\pi(k/T_0)t} = \lim_{N\to\infty} \frac{1}{2N+1} \sum_{n=-N}^{N} z(t - nT_0)\,. \tag{B.29}$$

Proof. See Section B.6. □

Theorem B.20. (Gardner and Brown, 1991) *Let $z(t)$ be a time series containing a poly-periodic component with periods $T_1, \ldots, T_P$ (see Proposition 2.13). Let*

$$\mathrm{E}^{\{T_p\}}\{z(t)\} \triangleq \lim_{N\to\infty} \frac{1}{2N+1} \sum_{n=-N}^{N} z(t - nT_p) \tag{B.30}$$

denote the periodic component with period T_p. Thus, the poly-periodic component of $z(t)$ with periods $T_1, \ldots, T_P$ can be expressed as

$$\mathrm{E}^{\{T_1,\ldots,T_P\}}\{z(t)\} = \sum_{T_p} \Big[\mathrm{E}^{\{T_p\}}\{z(t)\} - \langle z(t)\rangle_t \Big] + \langle z(t)\rangle_t \tag{B.31}$$

□

If two periods are commensurate, then the two periodic functions can be merged into a unique periodic function.

The sum of discrete-time periodic functions can be expressed as a unique periodic function whose period is at most equal to the least common multiple (lcm) of the integer periods of the individual periodic functions (Restrepo and Chacon, 1998). Thus, no discrete-time counterpart of continuous-time poly-periodic functions exists.

B.6 Proofs

Proof of Theorem B.19

Let be

$$z(t) = \sum_{p=0}^{P} \sum_{\substack{k_p=-\infty \\ k_p\neq 0}}^{+\infty} z_{k_p/T_p}\, e^{j2\pi(k_p/T_p)t} + z_0 + z_r(t) \tag{B.32}$$

with $T_0, T_1, \ldots, T_P$ incommensurate periods,

$$\sum_{k=-\infty}^{+\infty} \left| z_{k/T_p} \right| < \infty \qquad p = 0, \ldots, P \tag{B.33}$$

and

$$\left\langle z_r(t)\, e^{-j2\pi\eta t} \right\rangle_t = 0 \qquad \forall \eta \in \mathbb{R}\,. \tag{B.34}$$

With reference to the notation in (2.17), it results

$$z_{\varrho}(t) = \sum_{p=1}^{P} \sum_{\substack{k_p=-\infty \\ k_p \neq 0}}^{+\infty} z_{k_p/T_p} \, e^{j2\pi(k_p/T_p)t} + z_r(t) \,. \tag{B.35}$$

We have that

$$\begin{aligned}
&\lim_{N\to\infty} \frac{1}{2N+1} \sum_{n=-N}^{N} z(t-nT_0) \\
&\quad = \lim_{N\to\infty} \frac{1}{2N+1} \sum_{n=-N}^{N} \sum_{p=0}^{P} \sum_{\substack{k_p=-\infty \\ k_p \neq 0}}^{+\infty} z_{k_p/T_p} \, e^{j2\pi(k_p/T_p)(t-nT_0)} + z_0 \\
&\qquad + \lim_{N\to\infty} \frac{1}{2N+1} \sum_{n=-N}^{N} z_r(t-nT_0) \\
&\quad = \sum_{\substack{k_0=-\infty \\ k_0 \neq 0}}^{+\infty} z_{k_0/T_0} \, e^{j2\pi(k_0/T_0)t} \lim_{N\to\infty} \frac{1}{2N+1} \sum_{n=-N}^{N} e^{-j2\pi k_0 n} \\
&\qquad + \sum_{p=1}^{P} \sum_{\substack{k_p=-\infty \\ k_p \neq 0}}^{+\infty} z_{k_p/T_p} \, e^{j2\pi(k_p/T_p)t} \lim_{N\to\infty} \frac{1}{2N+1} \sum_{n=-N}^{N} e^{-j2\pi k_p n(T_0/T_p)} + z_0
\end{aligned} \tag{B.36}$$

where the first limit ($T_p = T_0$) is equal to 1 and the second limit ($T_p \neq T_0$) is zero since $T_0/T_p \notin \mathbb{Q}$ for $p = 1, \ldots, P$ (the periods are incommensurate) (Bass, 1974a). In addition,

$$\lim_{N\to\infty} \frac{1}{2N+1} \sum_{n=-N}^{N} z_r(t-nT_0) = 0 \,. \tag{B.37}$$

In fact, let be

$$y_N(t) \triangleq \frac{1}{2N+1} \sum_{n=-N}^{N} z_r(t-nT_0) \,. \tag{B.38}$$

For every finite N, accounting for (B.34), we have

$$\left\langle y_N(t) \, e^{-j2\pi\eta t} \right\rangle_t = \frac{1}{2N+1} \sum_{n=-N}^{N} \left\langle z_r(t-nT_0) \, e^{-j2\pi\eta t} \right\rangle_t = 0 \quad \forall \eta \in \mathbb{R} \,. \tag{B.39}$$

Therefore, also in the limit as $N \to \infty$ it results

$$\lim_{N\to\infty} \left\langle y_N(t) \, e^{-j2\pi\eta t} \right\rangle_t = 0 \quad \forall \eta \in \mathbb{R} \,. \tag{B.40}$$

Let $y(t)$ be the temporal mean-squared sense (t.m.s.s.) limit

$$\lim_{N\to\infty} y_N(t) = y(t) \qquad \text{t.m.s.s.} \tag{B.41}$$

that is

$$\lim_{N\to\infty} \left\langle |y_N(t) - y(t)|^2 \right\rangle_t = 0\,. \tag{B.42}$$

The function $y(t)$ is periodic with period T_0:

$$\begin{aligned} y(t+T_0) &= \lim_{N\to\infty} \frac{1}{2N+1} \sum_{n=-N}^{N} z_r(t+T_0-nT_0) \\ &= \lim_{N\to\infty} \frac{1}{2N+1} \sum_{n'=-N-1}^{N-1} z_r(t-n'T_0) \\ &= y(t) \end{aligned} \tag{B.43}$$

where in the second equality we set $n' = n - 1$ and equalities should be intended in the t.m.s.s.

Using the Cauchy-Schwarz inequality for the integrand functions $y_N(t) - y(t)$ and $e^{-j2\pi\eta t}$, it results

$$\begin{aligned} &\left| \frac{1}{T} \int_{-T/2}^{T/2} y_N(t)\, e^{-j2\pi\eta t}\, \mathrm{d}t - \frac{1}{T} \int_{-T/2}^{T/2} y(t)\, e^{-j2\pi\eta t}\, \mathrm{d}t \right|^2 \\ &\quad = \left| \frac{1}{T} \int_{-T/2}^{T/2} \left[y_N(t) - y(t) \right] e^{-j2\pi\eta t}\, \mathrm{d}t \right|^2 \\ &\quad \leqslant \frac{1}{T} \int_{-T/2}^{T/2} |y_N(t) - y(t)|^2\, \mathrm{d}t \end{aligned} \tag{B.44}$$

Thus, in the limit as $T \to \infty$ it results

$$\left| \left\langle y_N(t)\, e^{-j2\pi\eta t} \right\rangle_t - \left\langle y(t)\, e^{-j2\pi\eta t} \right\rangle_t \right|^2 \leqslant \left\langle |y_N(t) - y(t)|^2 \right\rangle_t \tag{B.45}$$

and, hence in the limit as $N \to \infty$, accounting for (B.42), we have

$$\left\langle y(t)\, e^{-j2\pi\eta t} \right\rangle_t = \lim_{N\to\infty} \left\langle y_N(t)\, e^{-j2\pi\eta t} \right\rangle_t = 0 \quad \forall \eta \in \mathbb{R} \tag{B.46}$$

where the equality to zero is due to (B.40). Thus the periodic function $y(t)$ has all the Fourier coefficients equal to zero. That is, it is zero in the t.m.s.s.

Therefore, from (B.36) it follows that

$$\mathrm{E}^{\{T_0\}}\{z(t)\} \triangleq \lim_{N\to\infty} \frac{1}{2N+1} \sum_{n=-N}^{N} z(t-nT_0) = \sum_{k=-\infty}^{+\infty} z_{k/T_0}\, e^{j2\pi(k/T_0)t} \tag{B.47}$$

C

Sampling and replication

Contents

C.1 Sampling in time

Fourier transform of a continuous-time signal $x_a(t)$, $t \in \mathbb{R}$

$$x_a(t) \stackrel{\mathcal{F}}{\longleftrightarrow} X_a(f) \quad \Leftrightarrow \quad \begin{cases} X_a(f) = \displaystyle\int_{\mathbb{R}} x_a(t)\, e^{-j2\pi f t}\, \mathrm{d}t \\ x_a(t) = \displaystyle\int_{\mathbb{R}} X_a(f)\, e^{j2\pi f t}\, \mathrm{d}f \end{cases} \tag{C.1}$$

Let be $x_a(t) \in L^1(\mathbb{R})$. Then the Fourier transform $X_a(f)$ exists in the sense of summability (Champeney, 1990). If $x_a(t) \in L^2(\mathbb{R})$, then $X_a(f) \in L^2(\mathbb{R})$ and the convergence of the integral to $X_a(f)$ is in the mean-square sense (Plancherel Theorem (Wiener, 1933, pp. 69–70)).

Fourier transform of a discrete-time signal $x(n)$, $n \in \mathbb{Z}$

$$x(n) \stackrel{\mathcal{F}_{\mathrm{D}}}{\longleftrightarrow} X(\nu) \quad \Leftrightarrow \quad \begin{cases} X(\nu) = \displaystyle\sum_{n \in \mathbb{Z}} x(n)\, e^{-j2\pi \nu n} \\ x(n) = \displaystyle\int_{-1/2}^{1/2} X(\nu)\, e^{j2\pi \nu n}\, \mathrm{d}\nu \end{cases} \tag{C.2}$$

Aliasing formula (sampling period $T_s = 1/f_s$):

$$x(n) \triangleq x_a(t)|_{t=nT_s} \stackrel{\mathcal{F}_{\mathrm{D}}}{\longleftrightarrow} X(\nu) = \frac{1}{T_s} \sum_{k=-\infty}^{+\infty} X_a((\nu - k) f_s) \tag{C.3}$$

Assumptions:

1. $x_a(t)$ strictly band-limited, i.e., $X_a(f) = 0$ for $|f| > B$.
2. $f_s \geqslant 2B$ (sufficient condition assuring the replicas in (C.3) do not overlap).

Cyclostationary Processes and Time Series. https://doi.org/10.1016/B978-0-08-102708-0.00030-3

Shannon Sampling Theorem. Under Assumptions 1 and 2 it results:

$$X(\nu) = \frac{1}{T_s} X_a(f)|_{f=\nu f_s} \qquad |\nu| \leqslant \frac{1}{2} \tag{C.4}$$

$$X_a(f) = \begin{cases} T_s X(\nu)|_{\nu=f/f_s} & |f| \leqslant \frac{f_s}{2} \\ 0 & \text{otherwise} \end{cases} \tag{C.5}$$

Shannon Interpolation Formula. Under Assumptions 1 and 2 it results:

$$x_a(t) = \sum_{n\in\mathbb{Z}} x(n) \operatorname{sinc}\left(\frac{t-nT_s}{T_s}\right) \qquad \forall t \in \mathbb{R} \tag{C.6}$$

where $\operatorname{sinc}(u) \triangleq \sin(\pi u)/(\pi u)$.

Since $\operatorname{sinc}(t/T_s) \leftrightarrow T_s \operatorname{rect}(fT_s)$, by Fourier transforming both sides of (C.6), it follows that

$$X_a(f) = \sum_{n\in\mathbb{Z}} x(n)\, e^{-j2\pi f n T_s}\, T_s \operatorname{rect}(fT_s)\,. \tag{C.7}$$

That is, the strictly band-limited Fourier transform $X_a(f)$ can be expressed as truncation to $[-1/(2T_s), 1/(2T_s)]$ of a Fourier series with "period in f" equal to $1/T_s$ and coefficients $x(n)T_s$.

C.2 Sampling in frequency

A signal $x_a(t) \in L^1(\mathbb{R})$ or $\in L^2(\mathbb{R})$ can be used as generator for a periodic signal $\widetilde{x}_a(t)$. That is, by using the Poisson formula (Zemanian, 1987, p. 189) and its Fourier transform counterpart

$$\sum_{n=-\infty}^{+\infty} \delta(t-nT_0) = \frac{1}{T_0} \sum_{k=-\infty}^{+\infty} e^{j2\pi(k/T_0)t} \tag{C.8a}$$

$$\overset{\mathcal{F}}{\longleftrightarrow} \quad \frac{1}{T_0} \sum_{k=-\infty}^{+\infty} \delta\left(f - \frac{k}{T_0}\right) = \sum_{n=-\infty}^{+\infty} e^{-j2\pi f n T_0} \tag{C.8b}$$

one has that the periodic signal $\widetilde{x}_a(t)$ can be expressed as periodic replication with period T_0 of the generator $x_a(t)$. It results that

$$\begin{aligned} \widetilde{x}_a(t) \triangleq \operatorname{rep}_{T_0}[x_a(t)] &\triangleq \sum_{n=-\infty}^{+\infty} x_a(t-nT_0) \\ &= x_a(t) \otimes \sum_{n=-\infty}^{+\infty} \delta(t-nT_0) \end{aligned}$$

$$= x_a(t) \otimes \frac{1}{T_0} \sum_{k=-\infty}^{+\infty} e^{j2\pi(k/T_0)t} \tag{C.9a}$$

$$\overset{\mathcal{F}}{\longleftrightarrow} \quad \widetilde{X}_a(f) = X_a(f) \frac{1}{T_0} \sum_{k=-\infty}^{+\infty} \delta\left(f - \frac{k}{T_0}\right)$$

$$= \sum_{k=-\infty}^{+\infty} \frac{1}{T_0} X_a\left(\frac{k}{T_0}\right) \delta\left(f - \frac{k}{T_0}\right) \tag{C.9b}$$

$$\overset{\mathcal{F}}{\longleftrightarrow} \quad \widetilde{x}_a(t) = \sum_{k=-\infty}^{+\infty} \frac{1}{T_0} X_a\left(\frac{k}{T_0}\right) e^{j2\pi(k/T_0)t} \tag{C.9c}$$

that is, $\frac{1}{T_0} X_a\left(\frac{k}{T_0}\right)$ are the coefficients of the Fourier series expansion of the periodic signal $\widetilde{x}_a(t)$.

In general, there is no one-to-one correspondence between a periodic signal $\widetilde{x}_a(t)$ and its generator $x_a(t)$. Two different generators $x_{a1}(t)$ and $x_{a2}(t)$ of the same T_0-periodic signal $\widetilde{x}_a(t)$ have Fourier transforms $X_{a1}(f)$ and $X_{a2}(f)$ such that $X_{a1}(k/T_0) = X_{a2}(k/T_0)$, $\forall k \in \mathbb{Z}$, since the uniqueness of the coefficients of a Fourier series. However, the generator $x_a(t) \triangleq \widetilde{x}_a(t)\,\mathrm{rect}(t/T_0)$, that is, the generator with finite support in $[-T_0/2, T_0/2]$ is unique. We obtain a dual of the Shannon sampling theorem for strictly band-limited signals.

$$x_a(t) \triangleq \widetilde{x}_a(t)\,\mathrm{rect}(t/T_0) \tag{C.10a}$$

$$\overset{\mathcal{F}}{\longleftrightarrow} \quad X_a(f) = \widetilde{X}_a(f) \otimes T_0\,\mathrm{sinc}(f T_0)$$

$$= \left[\sum_{k=-\infty}^{+\infty} \frac{1}{T_0} X_a\left(\frac{k}{T_0}\right) \delta\left(f - \frac{k}{T_0}\right)\right] \otimes T_0\,\mathrm{sinc}(f T_0)$$

$$= \sum_{k=-\infty}^{+\infty} X_a\left(\frac{k}{T_0}\right) \mathrm{sinc}(f T_0 - k) \tag{C.10b}$$

$$\overset{\mathcal{F}}{\longleftrightarrow} \quad x_a(t) = \sum_{k=-\infty}^{+\infty} \frac{1}{T_0} X_a\left(\frac{k}{T_0}\right) e^{j2\pi(k/T_0)t}\,\mathrm{rect}\left(\frac{t}{T_0}\right) \tag{C.10c}$$

C.3 Poisson's summation formulas

From the *Poisson's Summation Formula* (C.9c)

$$\sum_{n=-\infty}^{+\infty} x_a(t - nT_0) = \sum_{k=-\infty}^{+\infty} \frac{1}{T_0} X_a\left(\frac{k}{T_0}\right) e^{j2\pi(k/T_0)t} \tag{C.11}$$

by using the duality theorem for Fourier transforms (if $x(t) \leftrightarrow X(f)$, then $X(t) \leftrightarrow x(-f)$) and making the variable replacements $t \curvearrowright f$, $(1/T_0) \curvearrowright T$, and $h = -k$, we get the *Second*

Poisson's Summation Formula

$$\sum_{n=-\infty}^{+\infty} X_a\left(f - \frac{n}{T}\right) = T \sum_{h=-\infty}^{+\infty} x_a(hT)\, e^{-j2\pi f h T} \tag{C.12}$$

Alternatively, (C.12) can be proved as follows.

$$x_a(t) \sum_{n=-\infty}^{+\infty} \delta(t - nT) = \sum_{n=-\infty}^{+\infty} x_a(nT)\, \delta(t - nT) \tag{C.13a}$$

$$\overset{\mathcal{F}}{\longleftrightarrow} \quad X_a(f) \otimes \frac{1}{T} \sum_{k=-\infty}^{+\infty} \delta\left(f - \frac{k}{T}\right) = \sum_{n=-\infty}^{+\infty} x_a(nT)\, e^{-j2\pi f n T} \tag{C.13b}$$

$$\Leftrightarrow \quad \frac{1}{T} \sum_{k=-\infty}^{+\infty} X_a\left(f - \frac{k}{T}\right) = \sum_{n=-\infty}^{+\infty} x_a(nT)\, e^{-j2\pi f n T} \tag{C.13c}$$

C.4 LTI filtering of continuous-time periodic signals

$$\widetilde{y}_a(t) = \widetilde{x}_a(t) \otimes h(t) \tag{C.14a}$$

$$\overset{\mathcal{F}}{\longleftrightarrow} \quad \widetilde{Y}_a(f) = H(f) \sum_{k=-\infty}^{+\infty} \frac{1}{T_0} X_a\left(\frac{k}{T_0}\right) \delta\left(f - \frac{k}{T_0}\right)$$

$$= \sum_{k=-\infty}^{+\infty} \frac{1}{T_0} H\left(\frac{k}{T_0}\right) X_a\left(\frac{k}{T_0}\right) \delta\left(f - \frac{k}{T_0}\right) \tag{C.14b}$$

$$\overset{\mathcal{F}}{\longleftrightarrow} \quad \widetilde{y}_a(t) = \sum_{k=-\infty}^{+\infty} \frac{1}{T_0} H\left(\frac{k}{T_0}\right) X_a\left(\frac{k}{T_0}\right) e^{j2\pi (k/T_0) t} \tag{C.14c}$$

$$\widetilde{y}_a(t) = \widetilde{x}_a(t) \otimes h(t) \tag{C.15a}$$

$$= \left[\sum_{n=-\infty}^{+\infty} x_a(t - nT_0)\right] \otimes h(t)$$

$$= \sum_{n=-\infty}^{+\infty} \left[x_a(t) \otimes h(t) \otimes \delta(t - nT_0)\right] \tag{C.15b}$$

$$= \sum_{n=-\infty}^{+\infty} y_a(t - nT_0) \tag{C.15c}$$

where

$$y_a(t) = x_a(t) \otimes h(t) \tag{C.16}$$

C.5 Sampling of discrete-time signals

The *discrete-time sample train with sampling period M* and its Fourier transform are given by

$$\delta_{n \bmod M} \triangleq \sum_{\ell=-\infty}^{+\infty} \delta_{n-\ell M} = \frac{1}{M}\sum_{p=0}^{M-1} e^{j2\pi(p/M)n} \overset{\mathcal{F}_D}{\longleftrightarrow} \frac{1}{M}\sum_{p=0}^{M-1} \widetilde{\delta}\left(\nu - \frac{p}{M}\right) \tag{C.17}$$

where

$$\widetilde{\delta}(\nu) \triangleq \sum_{k=-\infty}^{+\infty} \delta(\nu - k) \tag{C.18}$$

is the *periodic delta train* with period 1.

The periodic delta train $\widetilde{\delta}(\nu)$ with period 1 defined in (C.18) and the unit element with respect to convolution of complex envelopes $\widetilde{\delta}(t)$ defined in (D.52) are denoted by the same symbol. Which one of the two generalized functions is considered is understood from the context and the variable used in the argument.

Accounting for (C.17), we have the following Fourier transform for the *sampled discrete-time signal*

$$x_\delta(n) \triangleq x(n) \sum_{\ell=-\infty}^{+\infty} \delta_{n-\ell M} = \sum_{\ell=-\infty}^{+\infty} x(\ell M)\, \delta_{n-\ell M} \tag{C.19a}$$

$$\overset{\mathcal{F}_D}{\longleftrightarrow} \quad X_\delta(\nu) = X(\nu) \otimes \frac{1}{M}\sum_{p=0}^{M-1} \widetilde{\delta}\left(\nu - \frac{p}{M}\right) = \frac{1}{M}\sum_{p=0}^{M-1} X\left(\nu - \frac{p}{M}\right) \tag{C.19b}$$

The *rectangular window of length N* is defined as

$$\mathcal{R}_N(n) \triangleq \begin{cases} 1 & n \in \{0, 1, \ldots, N-1\} \\ 0 & \text{otherwise} \end{cases} \tag{C.20a}$$

$$\overset{\mathcal{F}_D}{\longleftrightarrow} \quad W_{\frac{1}{N}}(\nu) = \sum_{n=-\infty}^{+\infty} \mathcal{R}_N(n)\, e^{-j2\pi\nu n} = e^{-j\pi\nu(N-1)} \frac{\sin(\pi\nu N)}{\sin(\pi\nu)} \tag{C.20b}$$

If the sequence $x(n)$ is defined only for $n \in \{0, 1, \ldots, N-1\}$ (and is assumed to be zero otherwise), then its Fourier transform $X(\nu)$ is completely determined in term of the samples $X(k/N)$, $k = 0, 1, \ldots, N-1$. In fact,

$$\widetilde{x}(n) = \text{rep}_N[x(n)] = \sum_{\ell=-\infty}^{+\infty} x(n - \ell N)$$

$$= x(n) \otimes \sum_{\ell=-\infty}^{+\infty} \delta_{n-\ell N} = x(n) \otimes \frac{1}{N}\sum_{k=0}^{N-1} e^{j2\pi(k/N)n} \tag{C.21a}$$

$$\overset{\mathcal{F}_{\mathrm{D}}}{\longleftrightarrow} \quad \widetilde{X}(\nu) = X(\nu)\,\frac{1}{N}\sum_{k=0}^{N-1}\widetilde{\delta}\left(\nu - \frac{k}{N}\right) = \sum_{k=0}^{N-1}\frac{1}{N}\,X\left(\frac{k}{N}\right)\widetilde{\delta}\left(\nu - \frac{k}{N}\right) \tag{C.21b}$$

$$\overset{\mathcal{F}_{\mathrm{D}}}{\longleftrightarrow} \quad \widetilde{x}(n) = \sum_{k=0}^{N-1}\frac{1}{N}\,X\left(\frac{k}{N}\right)e^{j2\pi(k/N)n} \tag{C.21c}$$

$$x(n) = \mathcal{R}_N(n)\,\widetilde{x}(n) \tag{C.22a}$$

$$\begin{aligned}\overset{\mathcal{F}_{\mathrm{D}}}{\longleftrightarrow} \quad X(\nu) &= W_{\frac{1}{N}}(\nu)\otimes\widetilde{X}(\nu)\\ &= W_{\frac{1}{N}}(\nu)\otimes\sum_{k=0}^{N-1}\frac{1}{N}\,X\left(\frac{k}{N}\right)\widetilde{\delta}\left(\nu-\frac{k}{N}\right)\\ &= \sum_{k=0}^{N-1}\frac{1}{N}\,X\left(\frac{k}{N}\right)W_{\frac{1}{N}}\left(\nu-\frac{k}{N}\right)\end{aligned} \tag{C.22b}$$

$$\overset{\mathcal{F}_{\mathrm{D}}}{\longleftrightarrow} \quad x(n) = \mathcal{R}_N(n)\sum_{k=0}^{N-1}\frac{1}{N}\,X\left(\frac{k}{N}\right)e^{j2\pi(k/N)n} \tag{C.22c}$$

$$X\left(\frac{k}{N}\right) = \mathrm{DFT}[x(n)] \triangleq \sum_{n=0}^{N-1}x(n)\,e^{-j2\pi(k/N)n} \qquad k = 0, 1, \ldots, N-1 \tag{C.23a}$$

$$x(n) = \mathrm{IDFT}\left[X\left(\frac{k}{N}\right)\right] \triangleq \frac{1}{N}\sum_{n=0}^{N-1}X\left(\frac{k}{N}\right)e^{j2\pi(k/N)n} \qquad n = 0, 1, \ldots, N-1 \tag{C.23b}$$

$$x(n) = \mathcal{R}_N(n)\,\widetilde{x}(n) \tag{C.24a}$$

$$\widetilde{x}(n) = \sum_{p=0}^{N-1}x(p)\,\delta_{[(n+p) \bmod N]} \tag{C.24b}$$

C.6 LTI filtering of discrete-time periodic signals

$$\widetilde{y}(n) = \widetilde{x}(n)\otimes h(n) \tag{C.25a}$$

$$\overset{\mathcal{F}_{\mathrm{D}}}{\longleftrightarrow} \quad \widetilde{Y}(\nu) = H(\nu)\,\widetilde{X}(\nu) = \sum_{k=0}^{N-1}\frac{1}{N}\,H\left(\frac{k}{N}\right)\widetilde{X}\left(\frac{k}{N}\right)\widetilde{\delta}\left(\nu-\frac{k}{N}\right) \tag{C.25b}$$

$$\overset{\mathcal{F}_{\mathrm{D}}}{\longleftrightarrow} \quad \widetilde{y}(n) = \sum_{k=0}^{N-1}\frac{1}{N}\,H\left(\frac{k}{N}\right)\widetilde{X}\left(\frac{k}{N}\right)e^{j2\pi(k/N)n} \tag{C.25c}$$

$$\widetilde{y}(n) = \widetilde{x}(n) \otimes h(n) \tag{C.26a}$$

$$= \left[\sum_{\ell=-\infty}^{+\infty} x(n - \ell N) \right] \otimes h(n)$$

$$= \sum_{\ell=-\infty}^{+\infty} \left[x(n) \otimes h(n) \otimes \delta(n - \ell N) \right] \tag{C.26b}$$

$$= \sum_{\ell=-\infty}^{+\infty} y(n - \ell N) \tag{C.26c}$$

where

$$y(n) = x(n) \otimes h(n) \tag{C.27}$$

D Hilbert transform, analytic signal, and complex envelope

Contents

Let $x(t)$ be a *real-valued* signal. In this appendix, main results on signals involved in Rice's representation of $x(t)$ are briefly reviewed.

In the following, k is a normalizing factor which is chosen differently from different authors: $k = \sqrt{2}$ in (Van Trees, 1971b), $k = 1$ in (Papoulis, 1983), (Gardner, 1987c), (Gardner, 1987f, Chap. 11, Sec. D-2), (Papoulis, 1991), and $k = 2$ in (Izzo and Napolitano, 1997).

Let p.v. denote the Cauchy principal value. It will be omitted in the sequel to simplify notation. The following Fourier-transform pairs hold in the sense of distributions (generalized functions) (Champeney, 1990), (Zemanian, 1987):

$$\text{p.v.}\ \frac{1}{\pi t} \stackrel{\mathcal{F}}{\longleftrightarrow} -j\ \text{sign}(f) \tag{D.1}$$

where $\text{sign}(\cdot)$ is the *signum function*

$$\text{sign}(f) \triangleq \begin{cases} 1 & f > 0 \\ 0 & f = 0 \\ -1 & f < 0 \end{cases} \tag{D.2}$$

$$\left[\text{p.v.}\ \frac{1}{\pi t}\right] \otimes \left[\text{p.v.}\ \frac{1}{\pi t}\right] = -\delta(t) \stackrel{\mathcal{F}}{\longleftrightarrow} \left[-j\ \text{sign}(f)\right] \left[-j\ \text{sign}(f)\right] = -1 \tag{D.3}$$

$$\mathring{\delta}(t) \triangleq \frac{1}{k}\left[\delta(t) + j\frac{1}{\pi t}\right] \stackrel{\mathcal{F}}{\longleftrightarrow} \frac{2}{k}\,\text{u}(f) \tag{D.4}$$

where $\text{u}(\cdot)$ is the *unit step function*

Cyclostationary Processes and Time Series. https://doi.org/10.1016/B978-0-08-102708-0.00031-5

$$\mathrm{u}(f) \triangleq \begin{cases} 1 & f \geqslant 0 \\ 0 & f < 0 \end{cases} \tag{D.5}$$

$$\overset{\circ}{\delta}{}^{*}(t) = \frac{1}{k}\left[\delta(t) - j\frac{1}{\pi t}\right] \overset{\mathcal{F}}{\longleftrightarrow} \frac{2}{k}\,\mathrm{u}(-f) \tag{D.6}$$

D.1 Rice's representation

Hilbert transform

$$\widehat{x}(t) \triangleq \text{p.v.}\, x(t) \otimes \frac{1}{\pi t} = \text{p.v.} \int_{-\infty}^{+\infty} x(\tau)\,\frac{1}{\pi(t-\tau)}\,\mathrm{d}\tau \tag{D.7}$$

$$\overset{\mathcal{F}}{\longleftrightarrow} \quad -j\,\mathrm{sign}(f)\,X(f) \tag{D.8}$$

Analytic signal

$$\overset{\circ}{x}(t) \triangleq x(t) \otimes \overset{\circ}{\delta}(t) = x(t) \otimes \frac{1}{k}\left[\delta(t) + j\frac{1}{\pi t}\right] = \frac{1}{k}\left[x(t) + j\widehat{x}(t)\right] \tag{D.9}$$

$$\overset{\mathcal{F}}{\longleftrightarrow} \quad \frac{2}{k}\,\mathrm{u}(f)\,X(f) \tag{D.10}$$

Complex envelope

$$\widetilde{x}(t) \triangleq \overset{\circ}{x}(t)\,e^{-j2\pi f_0 t} \triangleq \frac{1}{k}\left[x_{\mathrm{c}}(t) + j x_{\mathrm{s}}(t)\right] \tag{D.11}$$

$$\overset{\mathcal{F}}{\longleftrightarrow} \quad \frac{2}{k}\,\mathrm{u}(f+f_0)\,X(f+f_0) \tag{D.12}$$

where f_0 is an arbitrary real number.

The Fourier transform of a real-valued signal $x(t)$ has Hermitian symmetry, that is $X(f) = X^*(-f)$. We have the following Fourier-transform pairs:

$$\overset{\circ}{x}{}^{*}(t) \triangleq x(t) \otimes \overset{\circ}{\delta}{}^{*}(t) \overset{\mathcal{F}}{\longleftrightarrow} \frac{2}{k}\,\mathrm{u}(-f)\,X(f) \tag{D.13}$$

$$\widetilde{x}^{*}(t) \triangleq \overset{\circ}{x}{}^{*}(t)\,e^{j2\pi f_0 t} \overset{\mathcal{F}}{\longleftrightarrow} \frac{2}{k}\,\mathrm{u}(-f+f_0)\,X(f-f_0) \tag{D.14}$$

In-phase and in-quadrature components

$$x_{\mathrm{c}}(t) \triangleq k\,\mathrm{Re}\left[\widetilde{x}(t)\right] = x(t)\cos(2\pi f_0 t) + \widehat{x}(t)\sin(2\pi f_0 t) \tag{D.15}$$

$$x_{\mathrm{s}}(t) \triangleq k\,\mathrm{Im}\left[\widetilde{x}(t)\right] = -x(t)\sin(2\pi f_0 t) + \widehat{x}(t)\cos(2\pi f_0 t) \tag{D.16}$$

$$x_{\mathrm{c}}(t) \triangleq \frac{k}{2}\left[\widetilde{x}(t) + \widetilde{x}^{*}(t)\right]$$

$$\overset{\mathcal{F}}{\longleftrightarrow} \quad u(f+f_0)\,X(f+f_0) + u(-f+f_0)\,X(f-f_0) \tag{D.17}$$

$$x_s(t) \triangleq -j\frac{k}{2}\left[\widetilde{x}(t) - \widetilde{x}^*(t)\right]$$

$$\overset{\mathcal{F}}{\longleftrightarrow} \quad -j\left[u(f+f_0)\,X(f+f_0) - u(-f+f_0)\,X(f-f_0)\right] \tag{D.18}$$

Rice's representation

For every real-valued signal $x(t)$ and arbitrary f_0, the following representation holds:

$$x(t) = k\,\mathrm{Re}\left[\mathring{x}(t)\right] \tag{D.19a}$$

$$= k\,\mathrm{Re}\left[\widetilde{x}(t)\,e^{j2\pi f_0 t}\right] \tag{D.19b}$$

$$= x_c(t)\,\cos(2\pi f_0 t) - x_s(t)\,\sin(2\pi f_0 t) \tag{D.19c}$$

Coherent demodulation

Let $x(t)$ be a signal strictly band-limited around f_0. That is

$$X(f) \neq 0 \quad \text{only if} \quad \big||f| - f_0\big| \leqslant B, \quad f_0 > B\,. \tag{D.20}$$

The complex envelope and the in-phase and in-quadrature components can be extracted from the real signal by coherent demodulation:

$$x_c(t) = \left[2\,x(t)\,\cos(2\pi f_0 t)\right] \otimes h_{lp}(t) \tag{D.21}$$

$$x_s(t) = -\left[2\,x(t)\,\sin(2\pi f_0 t)\right] \otimes h_{lp}(t) \tag{D.22}$$

$$\widetilde{x}(t) = \left[x(t)\,e^{-j2\pi f_0 t}\right] \otimes h_{lp}(t) \tag{D.23}$$

where

$$h_{lp}(t) = 2B\,\mathrm{sinc}(2Bt) \overset{\mathcal{F}}{\longleftrightarrow} H_{lp}(f) = \mathrm{rect}\left(\frac{f}{2B}\right) \tag{D.24}$$

D.2 Polar representation

Complex envelope

$$\widetilde{x}(t) \triangleq \mathring{x}(t)\,e^{-j2\pi f_0 t} \triangleq \frac{1}{k}\left[x_c(t) + jx_s(t)\right] \triangleq \frac{1}{k}\left[r(t)\,e^{j\varphi(t)}\right] \tag{D.25}$$

where

$$r(t) = \left[x_c^2(t) + x_s^2(t)\right]^{1/2} \tag{D.26}$$

$$\mathrm{tg}\varphi(t) = \frac{x_s(t)}{x_c(t)} \tag{D.27}$$

$$\varphi(t) = \tan^{-1}_{(4)}\left(\frac{x_S(t)}{x_C(t)}\right) \triangleq \tan^{-1}\left(\frac{x_S(t)}{x_C(t)}\right) + \pi\, \mathrm{u}(-x_C(t))\, \mathrm{sign}(x_S(t)) \tag{D.28}$$

$$x_C(t) = r(t)\, \cos\varphi(t) \tag{D.29}$$

$$x_S(t) = r(t)\, \sin\varphi(t) \tag{D.30}$$

with $\tan^{-1}_{(4)}[\cdot]$ denoting the arctangent with values in 4 quadrants.

Rice's representation

$$x(t) = x_C(t)\, \cos(2\pi f_0 t) - x_S(t)\, \sin(2\pi f_0 t) \tag{D.31a}$$

$$= r(t)\, \cos(2\pi f_0 t + \varphi(t)) \tag{D.31b}$$

Time derivatives

$$\begin{aligned}\dot{\widetilde{x}}(t) &= \frac{1}{k}\frac{\mathrm{d}}{\mathrm{d}t}\left[r(t)\, e^{j\varphi(t)}\right] \\ &= \frac{1}{k}\left[\dot{r}(t)\, e^{j\varphi(t)} + r(t)\, j\dot{\varphi}(t)\, e^{j\varphi(t)}\right] \\ &= \frac{\dot{r}(t)}{r(t)}\, \widetilde{x}(t) + j\dot{\varphi}(t)\, \widetilde{x}(t)\end{aligned} \tag{D.32}$$

$$\frac{\dot{\widetilde{x}}(t)}{\widetilde{x}(t)} = \frac{\dot{r}(t)}{r(t)} + j\dot{\varphi}(t) \tag{D.33}$$

$$\frac{\mathrm{d}}{\mathrm{d}t}\, \mathrm{tg}\varphi(t) = \dot{\varphi}(t)\left[1 + \mathrm{tg}^2\varphi(t)\right] = \dot{\varphi}(t)\left[1 + \frac{x_S^2(t)}{x_C^2(t)}\right] \tag{D.34}$$

$$\frac{\mathrm{d}}{\mathrm{d}t}\frac{x_S(t)}{x_C(t)} = \frac{\dot{x}_S(t)\, x_C(t) - \dot{x}_C(t)\, x_S(t)}{x_C^2(t)} \tag{D.35}$$

$$\dot{\varphi}(t) = \frac{\dot{x}_S(t)\, x_C(t) - \dot{x}_C(t)\, x_S(t)}{x_C^2(t)\left[1 + \frac{x_S^2(t)}{x_C^2(t)}\right]} \tag{D.36}$$

Instantaneous frequency

$$f_i(t) \triangleq \frac{1}{2\pi}\, \dot{\varphi}(t) = \frac{1}{2\pi}\, \frac{\dot{x}_S(t)\, x_C(t) - \dot{x}_C(t)\, x_S(t)}{x_C^2(t) + x_S^2(t)} \tag{D.37}$$

D.3 Non-uniqueness of QAM representation

For a give real-valued signal $x(t)$, the representation in terms of in-phase and in-quadrature components, also referred to as quadrature amplitude modulation (QAM) representation, is not unique. In fact, let be

$$f_0' \triangleq f_0 + \Delta f \tag{D.38}$$

we have

$$\begin{aligned}
x(t) &= x_c(t)\ \cos(2\pi f_0 t) - x_s(t)\ \sin(2\pi f_0 t) \\
&= r(t)\ \cos(2\pi f_0 t + \varphi(t)) \\
&= r(t)\ \cos(2\pi[f_0 + \Delta f]t + \varphi(t) - \Delta f\ t) \\
&= r(t)\left[\cos(2\pi[f_0 + \Delta f]t)\ \cos(\varphi(t) - \Delta f\ t) - \sin(2\pi[f_0 + \Delta f]t)\ \sin(\varphi(t) - \Delta f\ t)\right] \\
&= r(t)\ \cos(\varphi(t) - \Delta f\ t)\cos(2\pi[f_0 + \Delta f]t) - r(t)\ \sin(\varphi(t) - \Delta f\ t)\sin(2\pi[f_0 + \Delta f]t) \\
&= x'_c(t)\ \cos(2\pi f'_0 t) - x'_s(t)\ \sin(2\pi f'_0 t)
\end{aligned} \tag{D.39}$$

where

$$x'_c(t) \triangleq r(t)\ \cos(\varphi(t) - \Delta f\ t) \tag{D.40}$$

$$x'_s(t) \triangleq r(t)\ \sin(\varphi(t) - \Delta f\ t) \tag{D.41}$$

Let us consider the generic QAM representation of $x(t)$

$$x(t) = \bar{x}_c(t)\ \cos(2\pi f_0 t) - \bar{x}_s(t)\ \sin(2\pi f_0 t) \tag{D.42}$$

and its *dual signal*

$$y(t) = \bar{x}_s(t)\ \cos(2\pi f_0 t) + \bar{x}_c(t)\ \sin(2\pi f_0 t)\,. \tag{D.43}$$

In (Papoulis, 1991, pp. 367–368), it is shown that if f_0 is the center of gravity of $S_{\overset{\circ}{x}\overset{\circ}{x}^*}(f)$, that is,

$$f_0 \triangleq \frac{\displaystyle\int_0^{+\infty} f\ S_{\overset{\circ}{x}\overset{\circ}{x}^*}(f)\mathrm{d}f}{\displaystyle\int_0^{+\infty} S_{\overset{\circ}{x}\overset{\circ}{x}^*}(f)\mathrm{d}f} \tag{D.44}$$

and, moreover, $y(t) = \widehat{x}(t)$, that is,

$$\bar{x}_c(t) \equiv x_c(t) = x(t)\ \cos(2\pi f_0 t) + \widehat{x}(t)\ \sin(2\pi f_0 t) \tag{D.45}$$

$$\bar{x}_s(t) \equiv x_s(t) = -x(t)\ \sin(2\pi f_0 t) + \widehat{x}(t)\ \cos(2\pi f_0 t) \tag{D.46}$$

then the functional

$$J \triangleq \left\langle \mathrm{E}\left\{\left|\frac{\mathrm{d}}{\mathrm{d}t}\widetilde{x}(t)\right|^2\right\}\right\rangle_t \tag{D.47}$$

is minimum.

D.4 Linear time-invariant systems

Let $y(t)$ obtained by linear time-invariant (LTI) filtering of $x(t)$. That is

$$y(t) = x(t) \otimes h(t) \ \overset{\mathcal{F}}{\longleftrightarrow}\ Y(f) = X(f)\ H(f) \tag{D.48}$$

Analytic signal

$$\begin{aligned}\mathring{y}(t) &= y(t) \otimes \mathring{\delta}(t) = x(t) \otimes h(t) \otimes \mathring{\delta}(t) \\ &\stackrel{\mathcal{F}}{\longleftrightarrow} \quad X(f)\, H(f)\, \frac{2}{k}\, \mathrm{u}(f) = \frac{k}{2} \left[X(f)\, \frac{2}{k}\, \mathrm{u}(f) \right] \left[H(f)\, \frac{2}{k}\, \mathrm{u}(f) \right] \\ &\stackrel{\mathcal{F}}{\longleftrightarrow} \quad \frac{k}{2}\, \mathring{x}(t) \otimes \mathring{h}(t) \end{aligned} \tag{D.49}$$

Complex envelope

$$\begin{aligned}\widetilde{y}(t) &= \mathring{y}(t)\, e^{-j2\pi f_0 t} \\ &\stackrel{\mathcal{F}}{\longleftrightarrow} \quad \frac{k}{2} \left[X(f+f_0)\, \frac{2}{k}\, \mathrm{u}(f+f_0) \right] \left[H(f+f_0)\, \frac{2}{k}\, \mathrm{u}(f+f_0) \right] \\ &\stackrel{\mathcal{F}}{\longleftrightarrow} \quad \frac{k}{2}\, \widetilde{x}(t) \otimes \widetilde{h}(t) \end{aligned} \tag{D.50}$$

$$y(t) = k\, \mathrm{Re}\left[\widetilde{y}(t)\, e^{j2\pi f_0 t} \right] = \frac{k^2}{4}\, \mathrm{Re}\left[\left(\widetilde{x}(t) \otimes \widetilde{h}(t) \right) e^{j2\pi f_0 t} \right] \tag{D.51}$$

The generalized function $\widetilde{\delta}(t)$ is the unit element with respect to convolution of complex envelopes:

$$\widetilde{\delta}(t) \triangleq \mathring{\delta}(t)\, e^{-j2\pi f_0 t} = \frac{1}{k} \left[\delta(t) + j \frac{1}{\pi t} \right] e^{-j2\pi f_0 t} \stackrel{\mathcal{F}}{\longleftrightarrow} \frac{2}{k} \mathrm{u}(f+f_0) \tag{D.52}$$

$$\begin{aligned}\frac{k}{2}\, \widetilde{x}(t) \otimes \widetilde{\delta}(t) \stackrel{\mathcal{F}}{\longleftrightarrow} &\frac{k}{2} \left[X(f+f_0)\, \frac{2}{k}\, \mathrm{u}(f+f_0) \right] \left[\frac{2}{k}\, \mathrm{u}(f+f_0) \right] \\ &= X(f+f_0)\, \frac{2}{k}\, \mathrm{u}(f+f_0) \\ \stackrel{\mathcal{F}}{\longleftrightarrow} &\ \widetilde{x}(t) \end{aligned} \tag{D.53}$$

The unit element with respect to convolution of complex envelopes $\widetilde{\delta}(t)$ defined in (D.52) and the periodic delta train $\widetilde{\delta}(\nu)$ with period 1 defined in (C.18) are denoted by the same symbol. Which one of the two generalized functions is considered is understood from the context and the variable used in the argument.

D.4.1 Incoherent channel with amplitude fading

The incoherent channel with amplitude fading is a random linear time-invariant system characterized by the following input/output relationship in terms of complex envelopes:

$$\widetilde{y}(t) = A\, e^{j\Theta}\, \widetilde{x}(t)\,. \tag{D.54}$$

A is generally modeled as a Rayleigh random variable and Θ as a uniform in $[0, 2\pi)$ random variable.

Input/output relationship in terms of real signals:

$$\begin{aligned} y(t) &= k\,\mathrm{Re}\big[\widetilde{y}(t)\, e^{j2\pi f_0 t}\big] \\ &= k\,\mathrm{Re}\big[A\, e^{j\Theta}\, \widetilde{x}(t)\, e^{j2\pi f_0 t}\big] \\ &= k\,\mathrm{Re}\Big[A\, \frac{1}{k}[x_c(t) + j x_s(t)]\,[\cos(2\pi f_0 t + \Theta) + j \sin(2\pi f_0 t + \Theta)]\Big] \\ &= A\, x_c(t)\, \cos(2\pi f_0 t + \Theta) - A\, x_s(t)\, \sin(2\pi f_0 t + \Theta) \end{aligned} \tag{D.55}$$

Impulse-response function for the complex envelopes:

$$\widetilde{h}(t) = A\, e^{j\Theta}\, \widetilde{\delta}(t) = A\, e^{j\Theta}\, \frac{1}{k}\Big[\delta(t) + j\frac{1}{\pi t}\Big]\, e^{-j2\pi f_0 t} \overset{\mathcal{F}}{\longleftrightarrow} A\, e^{j\Theta}\, \frac{2}{k} \mathrm{u}(f + f_0) \tag{D.56}$$

Input/output relationship in terms of complex envelopes:

$$\begin{aligned} \widetilde{y}(t) &= \frac{k}{2}\, \widetilde{x}(t) \otimes \widetilde{h}(t) \\ &\overset{\mathcal{F}}{\longleftrightarrow} \quad \frac{k}{2}\Big[X(f + f_0)\, \frac{2}{k}\, \mathrm{u}(f + f_0)\Big]\Big[\frac{2}{k} A\, e^{j\Theta}\, \mathrm{u}(f + f_0)\Big] \\ &\qquad = A\, e^{j\Theta}\, \frac{2}{k}\, X(f + f_0)\, \mathrm{u}(f + f_0) \\ &\overset{\mathcal{F}}{\longleftrightarrow} \quad A\, e^{j\Theta}\, \widetilde{x}(t) \end{aligned} \tag{D.57}$$

Impulse-response function for the real signals:

$$\begin{aligned} h(t) &= k\,\mathrm{Re}\big[\widetilde{h}(t)\, e^{j2\pi f_0 t}\big] \\ &= k\,\mathrm{Re}\big[A\, e^{j\Theta}\, \widetilde{\delta}(t)\, e^{j2\pi f_0 t}\big] \\ &= k\,\mathrm{Re}\Big[A\, e^{j\Theta}\, \frac{1}{k}\Big[\delta(t) + j\frac{1}{\pi t}\Big]\, e^{-j2\pi f_0 t}\, e^{j2\pi f_0 t}\Big] \\ &= \mathrm{Re}\Big[A\, (\cos\Theta + j \sin\Theta)\, \Big[\delta(t) + j\frac{1}{\pi t}\Big]\Big] \\ &= A\Big[\delta(t)\, \cos\Theta - \frac{1}{\pi t}\, \sin\Theta\Big] \end{aligned} \tag{D.58}$$

For linear time-variant systems see (Aiken, 1967), (Yoo, 2008).

D.5 Inner product

Let $x(t)$ and $y(t)$ be finite energy signals.

Inner product of the complex envelopes:

$$\begin{aligned} \big(\widetilde{x}(t), \widetilde{y}(t)\big) &\triangleq \int_{-\infty}^{+\infty} \widetilde{x}(t)\, \widetilde{y}^*(t)\, \mathrm{d}t \\ &= \int_{-\infty}^{+\infty} \mathring{x}(t)\, e^{-j2\pi f_0 t}\, \mathring{y}^*(t)\, e^{j2\pi f_0 t}\, \mathrm{d}t = \int_{-\infty}^{+\infty} \mathring{x}(t)\, \mathring{y}^*(t)\, \mathrm{d}t \end{aligned}$$

$$
\begin{aligned}
&= \int_{-\infty}^{+\infty} \left[\frac{2}{k} X(f)\,\mathrm{u}(f)\right] \left[\frac{2}{k} Y(f)\,\mathrm{u}(f)\right]^* \mathrm{d}f \\
&= \frac{4}{k^2} \int_0^{+\infty} X(f)\,Y^*(f)\,\mathrm{d}f
\end{aligned} \tag{D.59}
$$

Inner product of the real signals:

$$
\begin{aligned}
\big(x(t), y(t)\big) &\triangleq \int_{-\infty}^{+\infty} x(t)\,y(t)\,\mathrm{d}t \\
&= \int_{-\infty}^{+\infty} X(f)\,Y^*(f)\,\mathrm{d}f \\
&= \int_{-\infty}^{+\infty} |X(f)|\,|Y(f)|\,e^{j[\arg X(f) - \arg Y(f)]}\,\mathrm{d}f \\
&= \int_{-\infty}^{+\infty} |X(f)|\,|Y(f)|\,\cos\big[\arg X(f) - \arg Y(f)\big]\,\mathrm{d}f \\
&\qquad + j \int_{-\infty}^{+\infty} |X(f)|\,|Y(f)|\,\sin\big[\arg X(f) - \arg Y(f)\big]\,\mathrm{d}f \\
&= 2 \int_0^{+\infty} |X(f)|\,|Y(f)|\,\cos\big[\arg X(f) - \arg Y(f)\big]\,\mathrm{d}f \\
&= 2\mathrm{Re} \int_0^{+\infty} X(f)\,Y^*(f)\,\mathrm{d}f
\end{aligned} \tag{D.60}
$$

where the Parseval theorem for Fourier transforms and the fact that for a real-valued signal $v(t)$ the Fourier transform $V(f) = |V(f)|^{j\arg V(f)}$ has $|V(f)|$ even and $\arg V(f)$ odd are used.

By comparing (D.59) and (D.60) it follows that

$$
\int_{-\infty}^{+\infty} x(t)\,y(t)\,\mathrm{d}t = \frac{k^2}{2} \mathrm{Re} \int_{-\infty}^{+\infty} \widetilde{x}(t)\,\widetilde{y}^*(t)\,\mathrm{d}t \tag{D.61}
$$

In particular, for $y(t) \equiv x(t)$ it results

$$
\int_{-\infty}^{+\infty} x^2(t)\,\mathrm{d}t = \frac{k^2}{2} \int_{-\infty}^{+\infty} |\widetilde{x}(t)|^2\,\mathrm{d}t\,. \tag{D.62}
$$

D.6 Miscellaneous results

Let

$$
\mathcal{H}[x(t)] \equiv \widehat{x}(t) \triangleq x(t) \otimes \frac{1}{\pi t} \tag{D.63}
$$

1. Inverse Hilbert transform

$$
\mathcal{H}\big[\mathcal{H}[x(t)]\big] = x(t) \otimes \frac{1}{\pi t} \otimes \frac{1}{\pi t}
$$

$$\begin{aligned} &\overset{\mathcal{F}}{\longleftrightarrow} \quad X(f)\left[-j\,\text{sign}(f)\right]\left[-j\,\text{sign}(f)\right] = -X(f) \\ &\overset{\mathcal{F}}{\longleftrightarrow} \quad -x(t) \end{aligned} \tag{D.64}$$

provided that Assumption 3.14 holds.

2. Time-scale change

$$\begin{aligned} \mathcal{H}[x(st)] &= x(st) \otimes \frac{1}{\pi t} \\ &\overset{\mathcal{F}}{\longleftrightarrow} \quad -j\,\text{sign}(f)\,\frac{1}{|s|}X\left(\frac{f}{s}\right) = -j\,\text{sign}\left(\frac{f}{s}\right)\text{sign}(s)\,\frac{1}{|s|}X\left(\frac{f}{s}\right) \\ &\overset{\mathcal{F}}{\longleftrightarrow} \quad \text{sign}(s)\,\widehat{x}(st) \end{aligned} \tag{D.65}$$

3. Orthogonality of $x(t)$ and $\widehat{x}(t)$ when $x(t)$ is real valued.

$$\begin{aligned} \int_{\mathbb{R}} x(t)\,\widehat{x}(t)\,\mathrm{d}t &= \int_{\mathbb{R}} X(f)\left[-j\,\text{sign}(f)\,X(f)\right]^*\,\mathrm{d}f \\ &= j\int_{\mathbb{R}} \text{sign}(f)\,|X(f)|^2\,\mathrm{d}f = 0 \end{aligned} \tag{D.66}$$

where, in the last equality, the fact that $\text{sign}(f)$ is an odd function and $|X(f)|$ is even for real-valued signals is used.

4. Time derivative

$$\mathcal{H}\left[\frac{\mathrm{d}}{\mathrm{d}t}x(t)\right] = \frac{\mathrm{d}}{\mathrm{d}t}\mathcal{H}[x(t)] \tag{D.67}$$

5. Convolution

$$\begin{aligned} \mathcal{H}\big[x(t) \otimes y(t)\big] &= x(t) \otimes y(t) \otimes \frac{1}{\pi t} \\ &= \widehat{x}(t) \otimes y(t) = x(t) \otimes \widehat{y}(t) \end{aligned} \tag{D.68}$$

6. Correlation of finite-energy real-valued signals

Let

$$r_{yx}(\tau) \triangleq \int_{\mathbb{R}} y(t+\tau)\,x(t)\mathrm{d}t = x(\tau) \otimes y(-\tau) \tag{D.69}$$

It results that

$$\begin{aligned} \mathcal{H}[r_{yx}(\tau)] &= x(\tau) \otimes y(-\tau) \otimes \frac{1}{\pi\tau} \\ &= \begin{cases} \widehat{x}(\tau) \otimes y(-\tau) = r_{y\widehat{x}}(\tau) \\ x(\tau) \otimes \mathcal{H}[y(-\tau)] = -x(\tau) \otimes \widehat{y}(-\tau) = r_{\widehat{y}x}(\tau) \end{cases} \end{aligned} \tag{D.70}$$

7. Hilbert transform of strictly band-limited modulated signals (Bedrosian's theorem)
Let $x(t)$ strictly band-limited: $X(t) = 0$ for $|f| > B$ with $|f_0| > B$ ($f_0 \gtrless 0$)

$$
\begin{aligned}
\mathcal{H}\left[x(t)\, e^{j2\pi f_0 t}\right] &= \left[x(t)\, e^{j2\pi f_0 t}\right] \otimes \frac{1}{\pi t} \\
&\overset{\mathcal{F}}{\longleftrightarrow} \quad -j\,\text{sign}(f)\, X(f - f_0) = \begin{cases} -j\, X(f - f_0) & f_0 > 0 \\ j\, X(f - f_0) & f_0 < 0 \end{cases} \\
&= \quad X(f - f_0)\, e^{-j\,\text{sign}(f_0)\pi/2} \\
&\overset{\mathcal{F}}{\longleftrightarrow} \quad x(t)\, e^{j[2\pi f_0 t - \text{sign}(f_0)\pi/2]} = x(t)\, e^{j\,\text{sign}(f_0)[2\pi|f_0|t - \pi/2]}
\end{aligned}
\tag{D.71}
$$

For $f_0 > 0$ it results

$$
\begin{aligned}
\mathcal{H}\left[x(t)\, \cos(2\pi f_0 t + \phi_0)\right] &= \left[x(t)\, \cos(2\pi f_0 t + \phi_0)\right] \otimes \frac{1}{\pi t} \\
&\overset{\mathcal{F}}{\longleftrightarrow} \quad -j\,\text{sign}(f) \left[\frac{1}{2}\left(X(f - f_0)\, e^{j\phi_0} + X(f + f_0)\, e^{-j\phi_0}\right)\right] \\
&\qquad - j\frac{1}{2} X(f - f_0)\, e^{j\phi_0} + j\frac{1}{2} X(f + f_0)\, e^{-j\phi_0} \\
&\overset{\mathcal{F}}{\longleftrightarrow} \quad x(t)\, \sin(2\pi f_0 t + \phi_0)
\end{aligned}
\tag{D.72}
$$

E

Complex random vectors, quadratic forms, and chi squared distribution

Contents

E.1 Complex (normal) random variables and vectors

In this section, results concerning complex-valued random variables and vectors are reported. The case of complex normal variables and vectors is considered in detail. For extensive treatments see (Brandwood, 1983), (van den Bos, 1994), (van den Bos, 1995), (Picinbono, 1996), (Picinbono and Bondon, 1997), (Schreier and Scharf, 2003a), (Schreier and Scharf, 2003b), (Schreier and Scharf, 2010), (Adali et al., 2011).

E.1.1 Complex random variables

Let $Z = X + jY$ be a complex-valued random variable with X and Y real-valued random variables. We have

$$\begin{aligned}
\sigma_Z^2 &\triangleq \mathrm{E}\left\{|Z - \mathrm{E}\{Z\}|^2\right\} \\
&= \mathrm{E}\left\{[Z - \mathrm{E}\{Z\}]\,[Z - \mathrm{E}\{Z\}]^*\right\} \\
&= \mathrm{E}\left\{|Z|^2\right\} - |\mathrm{E}\{Z\}|^2 \\
&= \mathrm{E}\left\{|X + jY|^2\right\} - |\mathrm{E}\{X\} + j\mathrm{E}\{Y\}|^2 \\
&= \mathrm{E}\left\{X^2\right\} + \mathrm{E}\left\{Y^2\right\} - [\mathrm{E}\{X\}]^2 - [\mathrm{E}\{Y\}]^2
\end{aligned}$$

Cyclostationary Processes and Time Series. https://doi.org/10.1016/B978-0-08-102708-0.00032-7

$$
\begin{aligned}
&= \sigma_X^2 + \sigma_Y^2 && \text{(E.1)} \\
\gamma_Z &\triangleq \mathrm{E}\left\{[Z - \mathrm{E}\{Z\}]^2\right\} \\
&= \mathrm{E}\{[Z - \mathrm{E}\{Z\}]\,[Z - \mathrm{E}\{Z\}]\} \\
&= \mathrm{E}\left\{Z^2\right\} - [\mathrm{E}\{Z\}]^2 \\
&= \mathrm{E}\left\{[X + jY]^2\right\} - \left[\mathrm{E}\{X\} + j\mathrm{E}\{Y\}\right]^2 \\
&= \mathrm{E}\left\{X^2\right\} - [\mathrm{E}\{X\}]^2 - \mathrm{E}\left\{Y^2\right\} + [\mathrm{E}\{Y\}]^2 + 2j\,[\mathrm{E}\{XY\} - \mathrm{E}\{X\}\mathrm{E}\{Y\}] \\
&= \sigma_X^2 - \sigma_Y^2 + 2j\,\mathrm{cov}(X,Y) && \text{(E.2)} \\
\mathrm{cov}(X,Y) &\triangleq \mathrm{E}\{[X - \mathrm{E}\{X\}]\,[Y - \mathrm{E}\{Y\}]\} \\
&= \mathrm{E}\{XY\} - \mathrm{E}\{X\}\mathrm{E}\{Y\} \\
&= \frac{1}{2}\mathrm{Im}\left[\gamma_Z\right] && \text{(E.3)} \\
\sigma_X^2 &= \frac{1}{2}\mathrm{Re}\left[\sigma_Z^2 + \gamma_Z\right] && \text{(E.4)} \\
\sigma_Y^2 &= \frac{1}{2}\mathrm{Im}\left[\sigma_Z^2 - \gamma_Z\right] && \text{(E.5)}
\end{aligned}
$$

E.1.2 Complex random vectors

Let $\boldsymbol{V} = [V_1, \ldots, V_M, V_{M+1}, \ldots, V_{2M}]^{\mathsf{T}} \triangleq [\boldsymbol{X}^{\mathsf{T}}, \boldsymbol{Y}^{\mathsf{T}}]^{\mathsf{T}}$ be the $2M$-dimensional column vector of real-valued random variables obtained by stacking the M-dimensional column vectors $\boldsymbol{X} \triangleq [X_1, \ldots, X_M]^{\mathsf{T}}$, and $\boldsymbol{Y} \triangleq [Y_1, \ldots, Y_M]^{\mathsf{T}}$ of real-valued random variables. It is characterized by the $2M$th-order joint probability density function (pdf)

$$
f_{\boldsymbol{V}}(\boldsymbol{v}) = f_{V_1\cdots V_{2M}}(v_1, \ldots, v_{2M}) = f_{X_1\cdots X_M Y_1\cdots Y_M}(x_1, \ldots, x_M, y_1, \ldots, y_M)\,. \tag{E.6}
$$

Its *moment generating function* is the $2M$-dimensional Laplace transform

$$
\Phi_{\boldsymbol{V}}(\boldsymbol{s}) \triangleq \mathrm{E}\left\{e^{\boldsymbol{s}^{\mathsf{T}}\boldsymbol{V}}\right\} = \int_{\mathbb{R}^{2M}} f_{\boldsymbol{V}}(\boldsymbol{v})\, e^{\boldsymbol{s}^{\mathsf{T}}\boldsymbol{v}} \mathrm{d}\boldsymbol{v} \tag{E.7}
$$

with $\boldsymbol{s} \triangleq [\boldsymbol{s}_X^{\mathsf{T}}, \boldsymbol{s}_Y^{\mathsf{T}}]^{\mathsf{T}} \in \mathbb{C}^{2M}$, which is analytic in the region of convergence of the integral. The mean vector of $\boldsymbol{V}$ is

$$
\boldsymbol{\mu}_V \triangleq \mathrm{E}\{\boldsymbol{V}\} = [\boldsymbol{\mu}_X^{\mathsf{T}}, \boldsymbol{\mu}_Y^{\mathsf{T}}]^{\mathsf{T}} \tag{E.8}
$$

where $\boldsymbol{\mu}_X$ and $\boldsymbol{\mu}_Y$ are the mean vectors of $\boldsymbol{X}$ and $\boldsymbol{Y}$, respectively. The covariance matrix is

$$
\boldsymbol{C}_{VV} \triangleq \mathrm{E}\{(\boldsymbol{V} - \mathrm{E}\{\boldsymbol{V}\})(\boldsymbol{V} - \mathrm{E}\{\boldsymbol{V}\})^{\mathsf{T}}\} = \begin{bmatrix} \boldsymbol{C}_{XX} & \boldsymbol{C}_{XY} \\ \boldsymbol{C}_{YX} & \boldsymbol{C}_{YY} \end{bmatrix} \tag{E.9}
$$

where $\boldsymbol{C}_{XX}$ and $\boldsymbol{C}_{YY}$ are the covariance matrices of $\boldsymbol{X}$ and $\boldsymbol{Y}$, respectively, and

$$\boldsymbol{C}_{XY} \triangleq \mathrm{E}\{(\boldsymbol{X} - \mathrm{E}\{\boldsymbol{X}\})(\boldsymbol{Y} - \mathrm{E}\{\boldsymbol{Y}\})^{\mathsf{T}}\} = \boldsymbol{C}_{YX}^{\mathsf{T}} \tag{E.10}$$

is their cross-covariance matrix.

The $2M$th-order cumulant of the real random variables $V_1, \ldots, V_{2M}$ is defined as

$$\begin{aligned} \mathrm{cum}\{V_1, \ldots, V_{2M}\} \triangleq (-j)^{2M} & \frac{\partial^{2M}}{\partial \omega_{X1} \cdots \partial \omega_{XM} \partial \omega_{Y1} \cdots \partial \omega_{YM}} \\ & \log_e \mathrm{E}\left\{\exp[j(\boldsymbol{\omega}_X^{\mathsf{T}} \boldsymbol{X} + \boldsymbol{\omega}_Y^{\mathsf{T}} \boldsymbol{Y})]\right\}\Big|_{\boldsymbol{\omega}_X = \boldsymbol{\omega}_Y = \boldsymbol{0}} \end{aligned} \tag{E.11}$$

where $\boldsymbol{\omega}_X \triangleq [\omega_{X1}, \ldots, \omega_{XM}]^{\mathsf{T}}$ and $\boldsymbol{\omega}_Y \triangleq [\omega_{Y1}, \ldots, \omega_{YM}]^{\mathsf{T}}$.

Let us consider, now, the M-dimensional complex-valued column vector $\boldsymbol{Z} = [Z_1, \ldots, Z_M]^{\mathsf{T}} \triangleq \boldsymbol{X} + j\boldsymbol{Y}$. It is characterized by the same $2M$th-order joint pdf $f_{\boldsymbol{V}}(\boldsymbol{v})$ of the $2M$-dimensional real-valued vector $\boldsymbol{V}$ defined above.

Instead of considering the vector $\boldsymbol{V}$, the following complex *augmented-dimension vector*

$$\boldsymbol{\zeta} \triangleq \begin{bmatrix} \boldsymbol{Z} \\ \boldsymbol{Z}^* \end{bmatrix} \tag{E.12}$$

can be considered. Its mean vector and covariance matrix, referred to as *augmented mean vector* and *augmented covariance matrix*, respectively, are given by

$$\boldsymbol{\mu}_{\zeta} \triangleq \begin{bmatrix} \mathrm{E}\{\boldsymbol{Z}\} \\ \mathrm{E}\{\boldsymbol{Z}^*\} \end{bmatrix} \tag{E.13}$$

$$\boldsymbol{\Gamma} \triangleq \mathrm{E}\left\{(\boldsymbol{\zeta} - \boldsymbol{\mu}_{\zeta})(\boldsymbol{\zeta} - \boldsymbol{\mu}_{\zeta})^{\mathsf{H}}\right\} = \begin{bmatrix} \boldsymbol{C}_{ZZ^*} & \boldsymbol{C}_{ZZ} \\ \boldsymbol{C}_{ZZ}^* & \boldsymbol{C}_{ZZ^*}^* \end{bmatrix} \tag{E.14}$$

where

$$\begin{aligned} \boldsymbol{C}_{ZZ^*} &\triangleq \mathrm{cov}\{\boldsymbol{Z}, \boldsymbol{Z}\} = \mathrm{E}\left\{(\boldsymbol{Z} - \mathrm{E}\{\boldsymbol{Z}\})(\boldsymbol{Z} - \mathrm{E}\{\boldsymbol{Z}\})^{\mathsf{H}}\right\} \\ &= \boldsymbol{C}_{XX} + \boldsymbol{C}_{YY} - j\boldsymbol{C}_{XY} + j\boldsymbol{C}_{YX} \end{aligned} \tag{E.15}$$

$$\begin{aligned} \boldsymbol{C}_{ZZ} &\triangleq \mathrm{cov}\left\{\boldsymbol{Z}, \boldsymbol{Z}^*\right\} = \mathrm{E}\left\{(\boldsymbol{Z} - \mathrm{E}\{\boldsymbol{Z}\})(\boldsymbol{Z} - \mathrm{E}\{\boldsymbol{Z}\})^{\mathsf{T}}\right\} \\ &= \boldsymbol{C}_{XX} - \boldsymbol{C}_{YY} + j\boldsymbol{C}_{YX} + j\boldsymbol{C}_{XY} \end{aligned} \tag{E.16}$$

$$\boldsymbol{C}_{XX} = \frac{1}{2}\mathrm{Re}\{\boldsymbol{C}_{ZZ^*} + \boldsymbol{C}_{ZZ}\} \tag{E.17}$$

$$\boldsymbol{C}_{YY} = \frac{1}{2}\mathrm{Re}\{\boldsymbol{C}_{ZZ^*} - \boldsymbol{C}_{ZZ}\} \tag{E.18}$$

$$\boldsymbol{C}_{XY} = -\frac{1}{2}\mathrm{Im}\{\boldsymbol{C}_{ZZ^*} - \boldsymbol{C}_{ZZ}\} \tag{E.19}$$

$$\boldsymbol{C}_{YX} = \frac{1}{2}\mathrm{Im}\{\boldsymbol{C}_{ZZ^*} + \boldsymbol{C}_{ZZ}\} \tag{E.20}$$

with $\boldsymbol{C}_{ZZ^*}$ and $\boldsymbol{C}_{ZZ}$ referred to as *covariance matrix* and *conjugate covariance matrix*, respectively.

The Mth-order cumulant of the complex random variables $Z_1, \ldots, Z_M$ can be defined as (Spooner and Gardner, 1994a, App. A), (Napolitano, 2007a)

$$\text{cum}\{Z_1, \ldots, Z_M\} \triangleq (-j)^M \frac{\partial^M}{\partial\omega_1 \cdots \partial\omega_M} \log_e \mathrm{E}\left\{\exp[j\boldsymbol{\omega}^{\mathsf{T}}(\boldsymbol{X} + j\boldsymbol{Y})]\right\}\Big|_{\boldsymbol{\omega}=\mathbf{0}} \tag{E.21a}$$

$$= \sum_{\mathsf{P}} (-1)^{p-1}(p-1)! \prod_{i=1}^{p} \mathrm{E}\left\{\prod_{\ell \in \mu_i} Z_\ell\right\} \tag{E.21b}$$

where $\boldsymbol{\omega} \triangleq [\omega_1, \ldots, \omega_M]^{\mathsf{T}}$ and P is the set of distinct partitions of $\{1, \ldots, M\}$ each constituted by the subsets $\{\mu_i,\ i = 1, \ldots, p\}$. Expression (E.21b) is the first of Leonov and Shiryaev formulas (Leonov and Shiryaev, 1959) that links moments and cumulants (see (4.45)). Note that since each complex variable Z_k is arbitrary, it can also be the complex conjugate of another complex variable. Such a definition turns out to be useful when applied to complex-valued stochastic processes or time series as shown in (Spooner and Gardner, 1994a, App. A), (Izzo and Napolitano, 1998a), (Izzo and Napolitano, 2002a). In particular, it is useful since it allows to preserve the same relationship between moments and cumulants of real random variables (Leonov and Shiryaev, 1959), (Brillinger, 1965), (Brillinger and Rosenblatt, 1967), (Brillinger, 1981), being such a relationship purely algebric. In addition, the characteristic function $\mathrm{E}\left\{\exp[j\boldsymbol{\omega}^{\mathsf{T}}(\boldsymbol{X} + j\boldsymbol{Y})]\right\}$ in (E.21a), when it exists, is an analytic function of the complex variables $X_k + jY_k$, $k = 1, \ldots, M$. In contrast, the characteristic function $\mathrm{E}\left\{\exp[j(\boldsymbol{\omega}_X^{\mathsf{T}}\boldsymbol{X} + \boldsymbol{\omega}_Y^{\mathsf{T}}\boldsymbol{Y})]\right\}$ in (E.11), even if is always well defined, depends separately on X_k and Y_k and hence, in general, is not an analytic function of the complex variables $X_k + jY_k$. Finally, note that, for deterministic liner time-variant systems, input/output relationships in terms of cumulants defined as in (E.21a) have the same form as that in terms of moments (Theorem 4.27 and Remark 4.28), (Napolitano, 1995), (Izzo and Napolitano, 2002a).

E.1.3 Multivariate complex normal distribution

Let $\boldsymbol{Z}$ be a M-dimensional column vector of jointly complex normal random variables. That is, $\boldsymbol{V}$ is a $2M$-dimensional column vector of jointly normal real-valued random variables:

$$f_{\boldsymbol{V}}(\boldsymbol{v}) = \frac{1}{(2\pi)^M |\det \boldsymbol{C}_{VV}|^{1/2}} \exp\left[-\frac{1}{2}(\boldsymbol{v} - \boldsymbol{\mu}_V)^{\mathsf{T}} \boldsymbol{C}_{VV}^{-1} (\boldsymbol{v} - \boldsymbol{\mu}_V)\right] \tag{E.22}$$

that can also be written in the complex form (Picinbono, 1996), (van den Bos, 1995)

$$f_{\boldsymbol{Z},\boldsymbol{Z}^*}(\boldsymbol{z}, \boldsymbol{z}^*) = \frac{1}{\pi^M |\det \boldsymbol{\Gamma}|^{1/2}} \exp\left[-\frac{1}{2}(\boldsymbol{\zeta}_a - \boldsymbol{\mu}_\zeta)^{\mathsf{H}} \boldsymbol{\Gamma}^{-1} (\boldsymbol{\zeta}_a - \boldsymbol{\mu}_\zeta)\right] \tag{E.23}$$

where $\boldsymbol{\zeta}_a \triangleq [\boldsymbol{z}^{\mathsf{T}}, \boldsymbol{z}^{\mathsf{H}}]^{\mathsf{T}}$. We write $\boldsymbol{Z} \sim \mathcal{N}(\boldsymbol{\mu}_\zeta, \boldsymbol{C}_{ZZ^*}, \boldsymbol{C}_{ZZ})$.

In (Picinbono, 1996), the Cholesky block factorization or block LU decomposition is provided for the square matrix $\mathbf{\Gamma}$ defined in (E.14) with $\boldsymbol{C}_{ZZ^*}$ invertible.

E.1.4 Cumulants of complex normal vectors

The following lemma extends to complex-valued jointly normal random variables, a well known result for real-valued jointly normal random variables, provided that the cumulant of complex-valued random variables is properly defined.

Lemma E.1. Cumulants of Jointly Normal Complex Random Variables (Napolitano, 2017a, Lemma 23). *For complex-valued jointly normal random variables, the kth-order cumulant defined according to (E.21a), (Napolitano, 2007a, App. E), (Napolitano, 2012, Sec. 1.4.2), (Spooner and Gardner, 1994a, App. A) is zero for $k \geqslant 3$.*

Proof. Characteristic Function: The moment generating function (E.7) of $f_V(\boldsymbol{v})$ is the $2M$-dimensional Laplace transform

$$\Phi_V(\boldsymbol{s}) = \exp[\boldsymbol{s}^{\mathsf{T}} \boldsymbol{\mu}_V] \exp\left[\frac{1}{2} \boldsymbol{s}^{\mathsf{T}} \boldsymbol{C}_{VV} \boldsymbol{s}\right] \tag{E.24}$$

whose region of convergence is the whole complex space $\mathbb{C}^{2M}$. Therefore, both characteristic functions involved in the cumulant definitions (E.11) and (E.21a) can be expressed as slices of the moment generating function $\Phi_V(\boldsymbol{s})$ given in (E.24). Specifically, in (E.11) we have

$$\begin{aligned}
&\mathrm{E}\left\{\exp[j(\boldsymbol{\omega}_X^{\mathsf{T}} \boldsymbol{X} + \boldsymbol{\omega}_Y^{\mathsf{T}} \boldsymbol{Y})]\right\} \\
&\quad = \Phi_V(\boldsymbol{s})|_{\boldsymbol{s}_X = j\boldsymbol{\omega}_X, \boldsymbol{s}_Y = j\boldsymbol{\omega}_Y} \\
&\quad = \exp[j(\boldsymbol{\omega}_X^{\mathsf{T}} \boldsymbol{\mu}_X + \boldsymbol{\omega}_Y^{\mathsf{T}} \boldsymbol{\mu}_Y)] \\
&\qquad \exp\left[-\frac{1}{2}\left(\boldsymbol{\omega}_X^{\mathsf{T}} \boldsymbol{C}_{XX} \boldsymbol{\omega}_X + \boldsymbol{\omega}_X^{\mathsf{T}} \boldsymbol{C}_{XY} \boldsymbol{\omega}_Y + \boldsymbol{\omega}_Y^{\mathsf{T}} \boldsymbol{C}_{YX} \boldsymbol{\omega}_X + \boldsymbol{\omega}_Y^{\mathsf{T}} \boldsymbol{C}_{YY} \boldsymbol{\omega}_Y\right)\right]
\end{aligned} \tag{E.25}$$

from which it follows that $\log_e \mathrm{E}\left\{\exp[j(\boldsymbol{\omega}_X^{\mathsf{T}} \boldsymbol{X} + \boldsymbol{\omega}_Y^{\mathsf{T}} \boldsymbol{Y})]\right\}$ is a quadratic homogeneous polynomial in the real variables $\omega_{X1}, \ldots, \omega_{XM}, \omega_{Y1}, \ldots, \omega_{YM}$. As a consequence, we have the well known result that jointly real-valued normal random variables have kth-order cumulants of order $k \geqslant 3$ equal to zero. In addition, in (E.21a) we have

$$\begin{aligned}
&\mathrm{E}\left\{\exp[j\boldsymbol{\omega}^{\mathsf{T}}(\boldsymbol{X} + j\boldsymbol{Y})]\right\} \\
&\quad = \Phi_V(\boldsymbol{s})|_{\boldsymbol{s}_X = j\boldsymbol{\omega}, \boldsymbol{s}_Y = -\boldsymbol{\omega}} \\
&\quad = \exp[j\boldsymbol{\omega}^{\mathsf{T}} \boldsymbol{\mu}_X - \boldsymbol{\omega}^{\mathsf{T}} \boldsymbol{\mu}_Y] \\
&\qquad \exp\left[-\frac{1}{2} \boldsymbol{\omega}^{\mathsf{T}} (\boldsymbol{C}_{XX} - \boldsymbol{C}_{YY}) \boldsymbol{\omega} - j \frac{1}{2} \boldsymbol{\omega}^{\mathsf{T}} (\boldsymbol{C}_{XY} + \boldsymbol{C}_{YX}) \boldsymbol{\omega}\right].
\end{aligned} \tag{E.26}$$

Hence, $\log_e \mathrm{E}\left\{\exp[j\boldsymbol{\omega}^{\mathsf{T}}(\boldsymbol{X} + j\boldsymbol{Y})]\right\}$ is a quadratic homogeneous polynomial, with complex coefficients, in the real variables ω_i, $i = 1, \ldots, M$.

Cumulants: For a function ψ of a real variable, n times differentiable, the following identity holds

$$\frac{\partial^n}{\partial\omega_1 \dots \partial\omega_n} \psi(\omega_1 + \dots + \omega_n) = \frac{\partial^n}{\partial\omega^n} \psi(\omega)|_{\omega=\omega_1+\dots+\omega_n} . \tag{E.27}$$

Similarly, for a function ψ of M variables, differentiable a sufficient number of times, it results

$$\begin{aligned}
&\frac{\partial^{n_1+\dots+n_M}}{\partial\omega_{11}\dots\partial\omega_{1n_1}\dots\partial\omega_{M1}\dots\partial\omega_{Mn_M}} \psi(\omega_{11}+\dots+\omega_{1n_1},\dots,\omega_{M1}+\dots+\omega_{Mn_M}) \\
&\quad = \frac{\partial^{n_1+\dots+n_M}}{\partial\omega_1^{n_1}\dots\partial\omega_M^{n_M}} \psi(\omega_1,\dots,\omega_M)\bigg|_{\substack{\omega_1=\omega_{11}+\dots+\omega_{1n_1} \\ \dots \\ \omega_M=\omega_{M1}+\dots+\omega_{Mn_M}}}
\end{aligned} \tag{E.28}$$

Let

$$\Phi_{\boldsymbol{Z}}(\boldsymbol{\omega}) \triangleq \mathrm{E}\left\{e^{j[\omega_1 Z_1+\dots+\omega_M Z_M]}\right\} \tag{E.29}$$

be the characteristic function of the M complex-valued random variables $Z_1,\dots,Z_M$, where $\boldsymbol{Z} \triangleq [Z_1,\dots,Z_M]^{\mathrm{T}}$ and $\boldsymbol{\omega} \triangleq [\omega_1,\dots,\omega_M]^{\mathrm{T}}$. The M-variate Taylor series expansion around $\boldsymbol{\omega} = \mathbf{0}$ of the logarithm of $\Phi_{\boldsymbol{Z}}(\boldsymbol{\omega})$ is

$$\log\Phi_{\boldsymbol{Z}}(\boldsymbol{\omega}) = \sum_{n_1=0}^{+\infty}\dots\sum_{n_M=0}^{+\infty}\left[\frac{\partial^{n_1+\dots+n_M}}{\partial\omega_1^{n_1}\dots\partial\omega_M^{n_M}}\log\Phi_{\boldsymbol{Z}}(\boldsymbol{\omega})\right]_{\boldsymbol{\omega}=\mathbf{0}} \frac{\omega_1^{n_1}}{n_1!}\dots\frac{\omega_M^{n_M}}{n_M!} . \tag{E.30}$$

Thus, $\log\Phi_{\boldsymbol{Z}}(\boldsymbol{\omega})$ is completely characterized by the cumulants

$$\mathrm{cum}\left\{Z_1,n_1;\dots;Z_M,n_M\right\}$$

$$\triangleq (-j)^k\left[\frac{\partial^{n_1+\dots+n_M}}{\partial\omega_1^{n_1}\dots\partial\omega_M^{n_M}}\log\Phi_{\boldsymbol{Z}}(\boldsymbol{\omega})\right]_{\boldsymbol{\omega}=\mathbf{0}} \tag{E.31a}$$

$$= (-j)^k\left[\frac{\partial^{n_1+\dots+n_M}}{\partial\omega_{11}\dots\partial\omega_{1n_1}\dots\partial\omega_{M1}\dots\partial\omega_{Mn_M}}\log_e\Phi_{\boldsymbol{Z}'}(\boldsymbol{\omega}')\right]_{\boldsymbol{\omega}'=\mathbf{0}} \tag{E.31b}$$

$$\triangleq \mathrm{cum}\left\{\underbrace{Z_1,\dots,Z_1}_{n_1},\dots,\underbrace{Z_M,\dots,Z_M}_{n_M}\right\}$$

where $k = n_1 + \dots + n_M$ is the order,

$$\boldsymbol{Z}' \triangleq [\underbrace{Z_1,\dots,Z_1}_{n_1},\dots,\underbrace{Z_M,\dots,Z_M}_{n_M}]^{\mathrm{T}} \tag{E.32}$$

$$\boldsymbol{\omega}' \triangleq [\omega_{11},\dots,\omega_{1n_1},\dots,\omega_{M1},\dots,\omega_{Mn_M}]^{\mathrm{T}} \tag{E.33}$$

$$\Phi_{\boldsymbol{Z}'}(\boldsymbol{\omega}') \triangleq \mathrm{E}\left\{e^{j[(\omega_{11}+\dots+\omega_{1n_1})Z_1+\dots+(\omega_{M1}+\dots+\omega_{Mn_M})Z_M]}\right\} \tag{E.34}$$

and the equality between (E.31a) and (E.31b) is a consequence of identity (E.28). The cumulant defined in (E.31b) is referred to as "simple" since each $\omega_{\ell m}$ variable is derived only once. The cumulant defined in (E.31a) is not simple if at least one $n_i > 1$. Therefore, the nonsimple cumulant defined in (E.31a) can be expressed as the simple cumulant (E.31b) where each variable Z_i is repeated n_i times.

If the complex-valued random variables $Z_1, \ldots, Z_M$ are jointly complex normal, then their log characteristic function in (E.21a) is a quadratic homogeneous polynomial, with complex coefficients, in the real variables $\omega_i, i = 1, \ldots, M$ (see (E.26)). Therefore, terms with derivatives of order greater than or equal to 3 in (E.31a) and (E.31b) are zero. Consequently, jointly complex normal random variables $Z_i, i = 1, \ldots, M$, are characterized by the fact that their simple cumulants

$$\operatorname{cum}\left\{Z_{i_1}, \ldots, Z_{i_k}\right\} \qquad (i_1, \ldots, i_k) \in \{1, \ldots, M\}^k \tag{E.35}$$

of order $k \geqslant 3$ are equal to zero. □

E.2 Chi squared distribution and complex normality

In this section, results on chi squared distribution are briefly summarized. In addition, the link with complex normal random variables is considered.

E.2.1 Chi squared distribution

Let $X_1, \ldots, X_k$ be independent real-valued standard normal random variables, that is $X_i \sim \mathcal{N}(0, 1), i = 1, \ldots, k$. The random variable

$$Q = X_1^2 + \cdots + X_k^2 \tag{E.36}$$

has *central chi squared distribution with k degrees of freedom*: $Q \sim \chi_k^2$. The probability density function (pdf) and cumulative distribution functions (CDF) are

$$f_{\chi_k^2}(x) = \frac{x^{(k/2)-1}\, e^{-x/2}}{2^{k/2}\, \Gamma\left(\frac{k}{2}\right)}\, \mathrm{u}(x) \tag{E.37}$$

$$F_{\chi_k^2}(x) = \frac{\gamma\left(\frac{k}{2}, \frac{x}{2}\right)}{\Gamma\left(\frac{k}{2}\right)}\, \mathrm{u}(x) = P\left(\frac{k}{2}, \frac{x}{2}\right)\, \mathrm{u}(x) \tag{E.38}$$

respectively, where $\mathrm{u}(\cdot)$ is the unit step function, $\Gamma(\cdot)$ denotes the Gamma function (NIST, 2010, Sec. 5.2, Eq. 5.2.1)

$$\Gamma(z) = \int_0^{+\infty} t^{z-1}\, e^{-t}\, \mathrm{d}t \qquad \operatorname{Re}[z] > 0 \tag{E.39}$$

$\gamma(\cdot, \cdot)$ is the lower incomplete Gamma function (NIST, 2010, Sec. 8.2, Eq. 8.2.1)

$$\gamma(a, z) = \int_0^{z} t^{a-1}\, e^{-t}\, \mathrm{d}t \qquad \operatorname{Re}[a] > 0 \tag{E.40}$$

and $P(\cdot,\cdot)$ is the normalized or regularized incomplete Gamma function (NIST, 2010, Sec. 8.2, Eq. 8.2.4) $P(a,z) = \gamma(a,z)/\Gamma(a)$ (this function is called incomplete gamma function in Matlab® (MathWorks, 1994)/Octave (GNU Octave, 1998)). In the special case $k = 2$, the CDF has the simple form

$$F_{\chi_2^2}(x) = (1 - e^{-x/2})\, \mathrm{u}(x) \tag{E.41}$$

and the pdf is

$$f_{\chi_2^2}(x) = \frac{\mathrm{d}}{\mathrm{d}x} F_{\chi_2^2}(x) = \frac{1}{2}\, e^{-x/2}\, \mathrm{u}(x)\,. \tag{E.42}$$

Let $X_1, \ldots, X_k$ be independent real-valued normal random variables, that is $X_i \sim \mathcal{N}(\mu_i, \sigma_i^2)$, $i = 1, \ldots, k$. The random variable

$$Q = \frac{X_1^2}{\sigma_1^2} + \cdots + \frac{X_k^2}{\sigma_k^2} \tag{E.43}$$

has *noncentral chi squared distribution with k degrees of freedom* and noncentrality parameter

$$\lambda = \frac{\mu_1^2}{\sigma_1^2} + \cdots + \frac{\mu_k^2}{\sigma_k^2} \tag{E.44}$$

and is denoted as $Q \sim \chi_k^2(\lambda)$. Its pdf is

$$f(x;k,\lambda) = \frac{1}{2}\, e^{-(x+\lambda)/2} \left(\frac{x}{\lambda}\right)^{k/4-1/2} I_{k/2-1}(\sqrt{\lambda x})\, \mathrm{u}(x) \tag{E.45}$$

where $I_\nu(z)$ is the modified Bessel function of first kind (NIST, 2010, Sec. 10.25, Eq. 10.25.2)

$$I_\nu(z) = \left(\frac{1}{2}z\right)^{\nu} \sum_{k=0}^{+\infty} \frac{(\frac{1}{4}z^2)^k}{k!\,\Gamma(\nu+k+1)}\,. \tag{E.46}$$

The random variable

$$Q = \frac{(X_1-\mu_1)^2}{\sigma_1^2} + \cdots + \frac{(X_k-\mu_k)^2}{\sigma_k^2} \tag{E.47}$$

has central chi squared distribution with k degrees of freedom: $Q \sim \chi_k^2$.

Let $\boldsymbol{X} = (X_1, \ldots, X_k)^{\mathsf{T}}$ be $\mathcal{N}(\boldsymbol{\mu}, \boldsymbol{I}_k)$, where $\boldsymbol{I}_k$ is the identity matrix. That is, X_i are independent normal random variables with mean μ_i and unit variance. Let $\boldsymbol{C}$ a symmetric $k \times k$ matrix. Then, $\boldsymbol{X}^{\mathsf{T}}\boldsymbol{C}\boldsymbol{X}$ has (possibly noncentral) chi squared distribution if and only if $\boldsymbol{C}$ is idempotent, that is, $\boldsymbol{C}^2 = \boldsymbol{C}$. In such a case, the chi squared distribution has $\mathrm{rank}(\boldsymbol{C}) = \mathrm{trace}(\boldsymbol{C})$ degrees of freedom and noncentrality parameter $\boldsymbol{\mu}^{\mathsf{T}}\boldsymbol{C}\boldsymbol{\mu}$ (Lemma in (Serfling, 1980, Sec. 3.5, p. 128)).

E.2.2 Quadratic form of real-valued normal random vector

Theorem E.2. (Anderson, 2003, Theorem 3.3.3). *If the k-component real random vector $\boldsymbol{X}$ is distributed according to $\mathcal{N}(\boldsymbol{\mu}, \boldsymbol{\Sigma})$ (nonsingular distribution), then $\boldsymbol{X}^{\mathsf{T}}\boldsymbol{\Sigma}^{-1}\boldsymbol{X}$ is distributed according to the noncentral χ^2 distribution with k degrees of freedom and noncentrality parameter $\boldsymbol{\mu}^{\mathsf{T}}\boldsymbol{\Sigma}^{-1}\boldsymbol{\mu}$. That is*

$$\boldsymbol{X}^{\mathsf{T}}\boldsymbol{\Sigma}^{-1}\boldsymbol{X} \sim \chi_k^2(\boldsymbol{\mu}^{\mathsf{T}}\boldsymbol{\Sigma}^{-1}\boldsymbol{\mu}) . \tag{E.48}$$

If $\boldsymbol{\mu} = \mathbf{0}$, the distribution is the central χ^2 distribution with k degrees of freedom. □

Corollary E.3. *Let $\boldsymbol{X}$ be a real-valued $k \times 1$ random vector $\boldsymbol{X} \sim \mathcal{N}(\boldsymbol{\mu}, \boldsymbol{\Sigma})$, with $\boldsymbol{\Sigma}$ definite positive covariance matrix. Then*

$$Q(\boldsymbol{X}) = (\boldsymbol{X} - \boldsymbol{\mu})^{\mathsf{T}}\boldsymbol{\Sigma}^{-1}(\boldsymbol{X} - \boldsymbol{\mu}) \sim \chi_k^2 . \tag{E.49}$$

□

E.2.3 Quadratic form of complex-valued normal random vector

Let be $\boldsymbol{V} = [X, Y]^{\mathsf{T}}$, $Z = X + jY$, $\boldsymbol{\zeta} = [Z, Z^*]^{\mathsf{T}}$, with Z complex-valued (non-circular) normal random variable. By using notation and results of Section E.1, it results that

$$(\boldsymbol{\zeta} - \boldsymbol{\mu}_\zeta)^{\mathsf{H}}\boldsymbol{\Gamma}^{-1}(\boldsymbol{\zeta} - \boldsymbol{\mu}_\zeta) = (\boldsymbol{V} - \boldsymbol{\mu}_V)^{\mathsf{T}}\boldsymbol{C}_{VV}^{-1}(\boldsymbol{V} - \boldsymbol{\mu}_V) \sim \chi_2^2 . \tag{E.50}$$

Let be $\boldsymbol{Z} \in \mathbb{C}^k$, $\boldsymbol{Z} \sim \mathcal{N}(\boldsymbol{\mu}, \boldsymbol{C}_{ZZ^*}, \boldsymbol{C}_{ZZ})$ complex-valued normal column vector, $\boldsymbol{\zeta} = [\boldsymbol{Z}^{\mathsf{T}}, \boldsymbol{Z}^{\mathsf{H}}]^{\mathsf{T}}$ its augmented-dimension vector, and $\boldsymbol{\Gamma}$ the augmented covariance matrix defined in (E.14). If $\boldsymbol{\Gamma}$ has rank $q \leqslant 2k$, denoting by $\boldsymbol{\Gamma}^\dagger$ the Moore-Penrose inverse of $\boldsymbol{\Gamma}$, it results that (Anderson, 2003, Chap. 2)

$$Q = (\boldsymbol{\zeta} - \boldsymbol{\mu}_\zeta)^{\mathsf{H}}\boldsymbol{\Gamma}^{\dagger}(\boldsymbol{\zeta} - \boldsymbol{\mu}_\zeta) \sim \chi_q^2 \tag{E.51}$$

If $\boldsymbol{\Gamma}$ is invertible, then

$$Q \sim \chi_{2k}^2 . \tag{E.52}$$

If $\boldsymbol{\zeta}$ and $\boldsymbol{V}$ are deterministic vectors, the two quadratic forms in (E.50) assume the same real value. Thus, when $\boldsymbol{\zeta}$ and $\boldsymbol{V}$ are random vectors, the two quadratic forms have the same distribution. A similar result holds for (E.51) since

$$(\boldsymbol{\zeta} - \boldsymbol{\mu}_\zeta)^{\mathsf{H}}\boldsymbol{\Gamma}^{\dagger}(\boldsymbol{\zeta} - \boldsymbol{\mu}_\zeta) = (\boldsymbol{V} - \boldsymbol{\mu}_V)^{\mathsf{T}}\boldsymbol{C}_{VV}^{\dagger}(\boldsymbol{V} - \boldsymbol{\mu}_V) . \tag{E.53}$$

E.2.3.1 Sum of the squares of independent complex normal random variables

Let $Z = X + jY$ be a complex-valued scalar random variable with X and Y real-valued independent zero-mean normal random variables with different variances. That is $X \sim \mathcal{N}(0, \sigma_X^2)$, $Y \sim \mathcal{N}(0, \sigma_Y^2)$. According to the results of Section E.2.1 we have

$$\frac{X^2}{\sigma_X^2} + \frac{Y^2}{\sigma_Y^2} \sim \chi_2^2 . \tag{E.54}$$

Accounting for the results of Section E.1.1 we have

$$\frac{|Z|^2}{\sigma_Z^2} = \frac{X^2}{\sigma_X^2 + \sigma_Y^2} + \frac{Y^2}{\sigma_X^2 + \sigma_Y^2} = \frac{1}{2}\left(\frac{X^2}{\sigma_X^2} + \frac{Y^2}{\sigma_Y^2}\right). \tag{E.55}$$

Therefore, if $\sigma_X^2 = \sigma_Y^2$ it results

$$2\,\frac{|Z|^2}{\sigma_Z^2} \sim \chi_2^2 \,. \tag{E.56}$$

E.2.4 Asymptotic results

Let X_n be a sequence of random variables and X a random variable (or possibly a constant). In the following, we denote with $X_n \xrightarrow{\mathrm{d}} X$, $X_n \xrightarrow{\mathrm{p}} X$, and $X_n \xrightarrow{\mathrm{w.p.1}} X$ the convergence of X_n to X in distribution, probability, and with probability 1, respectively. We write $X_n \sim \mathcal{AN}(\mu_n, \sigma_n^2)$ if $(X_n - \mu_n)/\sigma_n \xrightarrow{\mathrm{d}} \mathcal{N}(0,1)$.

Theorem E.4. (Serfling, 1980, Sec. 3.2.1, p. 120). *Let be $X_n \sim \mathcal{AN}(\mu, \sigma_n^2)$ with $\sigma_n^2 \to 0$ as $n \to \infty$. If $\mu \neq 0$, then $X_n^2 \sim \mathcal{AN}(\mu^2, 4\mu^2\sigma_n^2)$. If $\mu = 0$, then $X_n^2/\sigma_n^2 \xrightarrow{\mathrm{d}} \chi_1^2$.* □

Theorem E.5. (Anderson, 2003, Sec. 6.5.3, p. 222). *The probability limits (limit with probability 1 or limit in probability) of sums, differences, products, and quotients of sequences of random variables are the sums, differences, products, and quotients of their probability limits, as long as the probability limit of each denominator is different from zero.* □

Theorem E.6. (Slutsky) (Serfling, 1980, p. 19). *Assume that*

$$X_n \xrightarrow{\mathrm{d}} X \qquad Y_n \xrightarrow{\mathrm{p}} c \tag{E.57}$$

where c is a real-valued finite constant. We have

$$X_n + Y_n \xrightarrow{\mathrm{d}} X + c \tag{E.58}$$

$$X_n Y_n \xrightarrow{\mathrm{d}} cX \tag{E.59}$$

$$\frac{X_n}{Y_n} \xrightarrow{\mathrm{d}} \frac{X}{c} \qquad c \neq 0 \tag{E.60}$$

□

Theorem E.7. (Serfling, 1980, p. 24). *Let $g(\cdot) : \mathbb{R}^k \to \mathbb{R}^p$ be continuous with $p_{\boldsymbol{X}}$ probability 1 and $\boldsymbol{X}_n$ a sequence of random vectors with probability limit (limit with probability 1 or limit in probability as $n \to \infty$) $\boldsymbol{X}$. We have*

$$\boldsymbol{X}_n \xrightarrow{\mathrm{w.p.1}} \boldsymbol{X} \Rightarrow g(\boldsymbol{X}_n) \xrightarrow{\mathrm{w.p.1}} g(\boldsymbol{X}) \tag{E.61}$$

$$\boldsymbol{X}_n \xrightarrow{\mathrm{p}} \boldsymbol{X} \Rightarrow g(\boldsymbol{X}_n) \xrightarrow{\mathrm{p}} g(\boldsymbol{X}) \,. \tag{E.62}$$

□

Let the deterministic vector parameter $\boldsymbol{\theta} = [\theta_1, \ldots, \theta_k]^\mathsf{T}$ be estimated by $\widehat{\boldsymbol{\theta}}^{(n)} = [\widehat{\theta}_1^{(n)}, \ldots, \widehat{\theta}_k^{(n)}]^\mathsf{T}$.

As a consequence of Theorem E.5, we have the following result.

Theorem E.8. (Serfling, 1980, Sec. 4.1.2, p. 140). *Let $\boldsymbol{\Sigma}_{\widehat{\boldsymbol{\theta}}^{(n)}}$ be an estimate of the covariance $\boldsymbol{\Sigma}_{\boldsymbol{\theta}}$ obtained starting from the estimate $\widehat{\boldsymbol{\theta}}^{(n)}$. We have that*

$$\boldsymbol{\Sigma}_{\widehat{\boldsymbol{\theta}}^{(n)}} \xrightarrow{\mathrm{p}} \boldsymbol{\Sigma}_{\boldsymbol{\theta}} \Rightarrow \boldsymbol{\Sigma}^{-1}_{\widehat{\boldsymbol{\theta}}^{(n)}} \xrightarrow{\mathrm{p}} \boldsymbol{\Sigma}^{-1}_{\boldsymbol{\theta}} \tag{E.63}$$

provided that the inverse $\boldsymbol{\Sigma}_{\boldsymbol{\theta}}^{-1}$ exists. □

In (E.63), the convergence in probability of matrices is defined through a suitable norm. For example, the Frobenius norm $\|\cdot\|_\mathrm{F}$ can be adopted. That is, the right-hand side of (E.63) is

$$\forall \epsilon > 0 \qquad \lim_{n\to\infty} \mathrm{P}\Big(\|\boldsymbol{\Sigma}^{-1}_{\widehat{\boldsymbol{\theta}}^{(n)}} - \boldsymbol{\Sigma}^{-1}_{\boldsymbol{\theta}}\|_\mathrm{F} < \epsilon\Big) = 1 \tag{E.64}$$

Note that mean-square convergence implies convergence in probability (Serfling, 1980, Sec. 1.3.2).

Assumption E.9. *The deterministic vector parameter $\boldsymbol{\theta} = [\theta_1, \ldots, \theta_k]^\mathsf{T}$ is estimated by $\widehat{\boldsymbol{\theta}}^{(n)} = [\widehat{\theta}_1^{(n)}, \ldots, \widehat{\theta}_k^{(n)}]^\mathsf{T}$ with $\widehat{\boldsymbol{\theta}}^{(n)} \sim \mathcal{AN}\Big(\boldsymbol{\theta}, \frac{1}{n}\boldsymbol{\Sigma}_{\boldsymbol{\theta}}\Big)$, that is*

$$\sqrt{n}\,(\widehat{\boldsymbol{\theta}}^{(n)} - \boldsymbol{\theta}) \sim \mathcal{N}(\mathbf{0}, \boldsymbol{\Sigma}_{\boldsymbol{\theta}}) \quad \textit{as } n \to \infty\,. \tag{E.65}$$

□

As a consequence of Theorem E.8, we have the following result.

Theorem E.10. (Serfling, 1980, Sec. 4.1.2, p. 140), (Soderstrom and Stoica, 1989, Lemma B.4). *Under Assumption E.9, let*

$$\widehat{Q} \triangleq n\,(\widehat{\boldsymbol{\theta}}^{(n)} - \boldsymbol{\theta})^\mathsf{T}\boldsymbol{\Sigma}^{-1}_{\widehat{\boldsymbol{\theta}}^{(n)}}(\widehat{\boldsymbol{\theta}}^{(n)} - \boldsymbol{\theta}) \tag{E.66}$$

$$Q \triangleq n\,(\widehat{\boldsymbol{\theta}}^{(n)} - \boldsymbol{\theta})^\mathsf{T}\boldsymbol{\Sigma}^{-1}_{\boldsymbol{\theta}}(\widehat{\boldsymbol{\theta}}^{(n)} - \boldsymbol{\theta})\,. \tag{E.67}$$

It results that

$$\widehat{Q} - Q \xrightarrow{\mathrm{p}} 0 \tag{E.68}$$

and, consequently,

$$\widehat{Q} \xrightarrow{\mathrm{d}} \chi^2_k\,. \tag{E.69}$$

Let c_α such that $\mathrm{P}(\chi^2_k > c_\alpha) = \alpha$. Then,

$$\mathrm{P}(\widehat{Q} \leqslant c_\alpha) \to \mathrm{P}(\chi^2_k \leqslant c_\alpha) = 1 - \alpha \qquad \textit{as } n \to \infty \tag{E.70}$$

□

E.3 Nonlinear transformation of two complex random variables

Let us consider a nonlinear transformation of the two complex variables $x = x_{\rm r} + j x_{\rm i}$ and $y = y_{\rm r} + j y_{\rm i}$. In the most general case we have

$$z = h(x_{\rm r}, x_{\rm i}, y_{\rm r}, y_{\rm i}) = g(x, x^*, y, y^*)\,, \tag{E.71}$$

where h is a complex-valued function of the four real-valued variables $(x_{\rm r}, x_{\rm i}, y_{\rm r}, y_{\rm i})$ and g is a complex-valued function of the four complex-valued variables (x, x^*, y, y^*), where x and x^* must be considered as independent variables and similarly y and y^* (Brandwood, 1983).

For notation convenience, let us define $\boldsymbol{\zeta} \triangleq (\zeta_1, \zeta_2, \zeta_3, \zeta_4) = (x, x^*, y, y^*)$ and $\boldsymbol{\zeta}_0 \triangleq (\zeta_{01}, \zeta_{02}, \zeta_{03}, \zeta_{04})$. If g is sufficiently smooth around the point $\boldsymbol{\zeta}_0$, it can be expanded in Taylor series around this point. We have

$$\begin{aligned} g(\boldsymbol{\zeta}) = g(\boldsymbol{\zeta}_0) &+ \sum_{i=1}^{4} \left.\frac{\partial g}{\partial \zeta_i}\right|_{\boldsymbol{\zeta}=\boldsymbol{\zeta}_0} (\zeta_i - \zeta_{0i}) \\ &+ \frac{1}{2}\sum_{i_1=1}^{4}\sum_{i_2=1}^{4} \left.\frac{\partial^2 g}{\partial \zeta_{i_1}\partial \zeta_{i_2}}\right|_{\boldsymbol{\zeta}=\boldsymbol{\zeta}_0} (\zeta_{i_1} - \zeta_{0i_1})\,(\zeta_{i_2} - \zeta_{0i_2}) \\ &+ o(|\zeta_{i_1} - \zeta_{0i_1}|\,|\zeta_{i_2} - \zeta_{0i_2}|) \end{aligned} \tag{E.72}$$

Let us consider the nonlinear function of the two complex random variables X and Y

$$\mathsf{Z} = g(\mathsf{X}, \mathsf{X}^*, \mathsf{Y}, \mathsf{Y}^*)\,, \tag{E.73}$$

and let, for notation convenience, $(\mathsf{U}_1, \mathsf{U}_2, \mathsf{U}_3, \mathsf{U}_4) = (\mathsf{X}, \mathsf{X}^*, \mathsf{Y}, \mathsf{Y}^*)$. If g is sufficiently smooth in a neighborhood of the point $(x, x^*, y, y^*) = (\eta_{\mathsf{X}}, \eta_{\mathsf{X}}^*, \eta_{\mathsf{Y}}, \eta_{\mathsf{Y}}^*)$, where $\eta_{\mathsf{X}} \triangleq \mathrm{E}\{\mathsf{X}\}$ and $\eta_{\mathsf{Y}} \triangleq \mathrm{E}\{\mathsf{Y}\}$, it can be expanded in Taylor series around $(\eta_{\mathsf{X}}, \eta_{\mathsf{X}}^*, \eta_{\mathsf{Y}}, \eta_{\mathsf{Y}}^*) = (\mathrm{E}\{\mathsf{U}_1\}, \mathrm{E}\{\mathsf{U}_2\}, \mathrm{E}\{\mathsf{U}_3\}, \mathrm{E}\{\mathsf{U}_4\}) = \mathrm{E}\{\mathsf{U}\}$. From (E.72) it follows that

$$\begin{aligned} \mathrm{E}\{\mathsf{Z}\} &= \mathrm{E}\{g(\mathsf{U}_1, \mathsf{U}_2, \mathsf{U}_3, \mathsf{U}_4)\} \\ &= g\Big(\mathrm{E}\{\mathsf{U}_1\}, \mathrm{E}\{\mathsf{U}_2\}, \mathrm{E}\{\mathsf{U}_3\}, \mathrm{E}\{\mathsf{U}_4\}\Big) \\ &\quad + \sum_{i=1}^{4} \left.\frac{\partial g}{\partial \zeta_i}\right|_{\boldsymbol{\zeta}=\mathrm{E}\{\mathsf{U}\}} \underbrace{\mathrm{E}(\mathsf{U}_i - \mathrm{E}\{\mathsf{U}_i\})}_{=0} \\ &\quad + \frac{1}{2}\sum_{i_1=1}^{4}\sum_{i_2=1}^{4} \left.\frac{\partial^2 g}{\partial \zeta_{i_1}\partial \zeta_{i_2}}\right|_{\boldsymbol{\zeta}=\mathrm{E}\{\mathsf{U}\}} \mathrm{E}\{(\mathsf{U}_{i_1} - \mathrm{E}\{\mathsf{U}_{i_1}\})\,(\mathsf{U}_{i_2} - \mathrm{E}\{\mathsf{U}_{i_2}\})\} \\ &\quad + \mathrm{E}\{o(|\mathsf{U}_{i_1} - \mathrm{E}\{\mathsf{U}_{i_1}\}|\,|\mathsf{U}_{i_2} - \mathrm{E}\{\mathsf{U}_{i_2}\}|)\} \end{aligned} \tag{E.74}$$

If g is *analytic* with respect to both x and y, then $g(x, x^*, y, y^*) = g(x, y)$ and, hence,

$$\frac{\partial g}{\partial x^*} = 0 \quad \frac{\partial g}{\partial y^*} = 0 \tag{E.75}$$

Mean and variance of Z can be approximately expressed as

$$\begin{aligned} \mathrm{E}\{\mathsf{Z}\} \simeq g(\eta_{\mathsf{X}}, \eta_{\mathsf{Y}}) + \frac{1}{2}\Bigg(& \frac{\partial^2 g}{\partial x^2}\, \mathrm{E}\{(\mathsf{X}-\eta_{\mathsf{X}})^2\} + \frac{\partial^2 g}{\partial y^2}\, \mathrm{E}\{(\mathsf{Y}-\eta_{\mathsf{Y}})^2\} + \\ & + 2\frac{\partial^2 g}{\partial x \partial y}\, \mathrm{E}\{(\mathsf{X}-\eta_{\mathsf{X}})(\mathsf{Y}-\eta_{\mathsf{Y}})\} \Bigg)\Bigg|_{\substack{x=\eta_{\mathsf{X}} \\ y=\eta_{\mathsf{Y}}}} \end{aligned} \tag{E.76}$$

and

$$\begin{aligned} \mathrm{var}\{\mathsf{Z}\} \simeq \Bigg(& \left|\frac{\partial g}{\partial x}\right|^2 \mathrm{E}\{|\mathsf{X}-\eta_{\mathsf{X}}|^2\} + \left|\frac{\partial g}{\partial y}\right|^2 \mathrm{E}\{|\mathsf{Y}-\eta_{\mathsf{Y}}|^2\} \\ & + 2\mathrm{Re}\left\{\left(\frac{\partial g}{\partial x}\right)\left(\frac{\partial g}{\partial y}\right)^* \mathrm{E}\{(\mathsf{X}-\eta_{\mathsf{X}})(\mathsf{Y}-\eta_{\mathsf{Y}})^*\}\right\}\Bigg)\Bigg|_{\substack{x=\eta_{\mathsf{X}} \\ y=\eta_{\mathsf{Y}}}} \end{aligned} \tag{E.77}$$

respectively.

In the special case $g(x, y) = y/x$ (considered in Section 9.5.2.3), it results that

$$\frac{\partial g}{\partial x} = -\frac{y}{x^2} \quad \frac{\partial^2 g}{\partial x^2} = 2\frac{y}{x^3} \quad \frac{\partial g}{\partial y} = \frac{1}{x} \quad \frac{\partial^2 g}{\partial y^2} = 0 \quad \frac{\partial^2 g}{\partial x \partial y} = \frac{\partial^2 g}{\partial y \partial x} = -\frac{1}{x^2} \tag{E.78}$$

F

Bibliographic notes

Contents

In this chapter, selected references are classified into categories and listed in chronological order.

F.1 Almost-periodic functions

Fundamental treatments

(Besicovitch, 1932), (Bohr, 1933), (Amerio and Prouse, 1971), (Corduneanu, 1989).

Generalizations

(von Neumann, 1934), (Jessen and Tornehave, 1945), (Eberlein, 1949), (Eberlein, 1956), (Kahane, 1961), (Urbanik, 1962), (Kahane, 1962), (Zhang, 1994), (Zhang, 1995), (Ait Dads and Arino, 1996), (Han and Hong, 2007), (Phong, 2007), (Zhang and Liu, 2010).

Reviews

(Andreas et al., 2006), (Chérif, 2011a), (Chérif, 2011b), (Napolitano, 2012, Sec. 1.2).

F.2 Cyclostationary signals

Fundamental treatments

(Gladyshev, 1961), (Gladyshev, 1963), (Gardner, 1985), (Gardner, 1986c), (Gardner, 1987f), (Hurd, 1989b), (Dehay and Hurd, 1994), (Gardner, 1994a), (Giannakis, 1998), (Gardner et al., 2006), (Hurd and Miamee, 2007), (Napolitano, 2016b).

Cyclostationary Processes and Time Series. https://doi.org/10.1016/B978-0-08-102708-0.00033-9

Developments and applications

(Ogura, 1971), (Hurd, 1974a), (Gardner and Franks, 1975), (Gardner, 1978), (Boyles and Gardner, 1983), (Gardner, 1986a), (Gardner, 1986b), (Brown, 1987), (Gardner, 1987c), (Gardner, 1987a), (Gardner, 1987e), (Gardner et al., 1987), (Alekseev, 1988), (Gardner, 1988a), (Gardner, 1988c), (Gardner and Chen, 1988a), (Gardner, 1988b), (Hurd, 1989a), (Agee et al., 1990), (Gardner, 1991a), (Hurd, 1991), (Hurd and Gerr, 1991), (Gardner and Chen, 1992), (Chen and Gardner, 1992), (Gardner and Spooner, 1992b), (Hurd and Leśkow, 1992b), (Hurd and Leśkow, 1992a), (Gardner, 1993), (Gardner and Spooner, 1993), (Sathe and Vaidyanathan, 1993), (Dehay and Hurd, 1994), (Franks, 1994), (Genossar et al., 1994), (Gerr and Allen, 1994a), (Schell, 1995a), (Dehay and Leśkow, 1996a), (Swift, 1996), (Akkarakaran and Vaidyanathan, 2000), (De Angelis et al., 2005), (Lii and Rosenblatt, 2006), (Dehay, 2007), (Javorskyj et al., 2007), (Wahlberg and Schreier, 2008), (Antoni, 2009), (Spooner and Nicholls, 2009), (Napolitano and Doğançay, 2011), (Napolitano, 2017a).

Higher-order cyclostationarity

(Gardner and Spooner, 1994a), (Spooner and Gardner, 1994a), (Dandawaté and Giannakis, 1994a), (Dandawaté and Giannakis, 1995), (Napolitano, 1995), (Izzo and Napolitano, 1996), (Izzo and Napolitano, 1997), (Izzo and Napolitano, 1998b), (Sadler and Dandawaté, 1998), (Flagiello et al., 2000), (Napolitano and Spooner, 2000), (Napolitano and Spooner, 2001), (Spooner and Nicholls, 2009).

F.3 Generalizations of cyclostationarity

Generalized almost-cyclostationary signals

(Izzo and Napolitano, 1998a), (Izzo and Napolitano, 2002a), (Izzo and Napolitano, 2002b), (Izzo and Napolitano, 2003), (Izzo and Napolitano, 2005), (Napolitano, 2007a), (Napolitano, 2009), (Napolitano and Tesauro, 2011). (Napolitano, 2012), (Napolitano, 2013), (Napolitano, 2015b), (Napolitano, 2016a).

Spectrally correlated signals

(Allen and Hobbs, 1992), (Genossar, 1992), (Lii and Rosenblatt, 1998), (Lii and Rosenblatt, 2002), (Napolitano, 2001), (Napolitano, 2003), (Øigård et al., 2006), (Soltani and Parvardeh, 2006), (Napolitano, 2007b), (Napolitano, 2007c), (Napolitano, 2010b), (Napolitano, 2010a), (Napolitano, 2011), (Napolitano, 2012), (Napolitano, 2013), (Napolitano, 2015b), (Napolitano, 2016a).

Oscillatory almost-cyclostationary signals

(Napolitano, 2016a), (Napolitano and Gardner, 2016), (Napolitano, 2017b), (Napolitano, 2017c).

F.4 Other nonstationary signals

(Silverman, 1957), (Kampé de Fériet and Frenkiel, 1962), (Priestley, 1965), (Mandelbrot and Van Ness, 1968), (Hurd, 1973), (Melititskiy and Radzievskii, 1973), (Martin and Flandrin, 1985), (Gardner, 1988a), (Hurd, 1988), (Flandrin, 1989), (Cohen, 1989), (Gardner, 1991c), (Gardner, 1992), (Amin, 1992), (Hlawatsch and Bourdeaux-Bartels, 1992), (Wornell, 1993), (Cohen, 1995), (Allen and Hobbs, 1997), (Matz et al., 1997), (Wu and Lev-Ari, 1997), (Hanin and Schreiber, 1998), (Wang and Zhou, 1998), (Almeida, 1994), (Flandrin, 1999), (Ozaktas et al., 2001), (Soedjack, 2002), (Hanssen and Scharf, 2003), (Hopgood and Rayner, 2003), (Scharf et al., 2005), (Dmochowski et al., 2009), (Flandrin et al., 2011), (Boashash, 2015).

F.5 Functional approach and generalized harmonic analysis

(Wiener, 1930), (Wiener, 1933), (Kac and Steinhaus, 1938), (Wiener, 1949), (Kac, 1959), (Lee, 1967), (Bass, 1971), (Bass, 1974a), (Bass, 1974b), (Pfaffelhuber, 1975), (Gardner, 1985), (Gardner, 1987f), (Gardner and Brown, 1991), (Gardner, 1994a), (Benedetto, 1996), (Leśkow and Napolitano, 2002), (Hurd and Koski, 2004b), (Leśkow and Napolitano, 2006), (Leśkow and Napolitano, 2007), (Casinovi, 2007), (Casinovi, 2009), (Dehay et al., 2013), (Dehay et al., 2018b).

F.6 Linear time-variant processing

Channels

(Bello, 1963), (Middleton, 1967), (Aiken, 1967), (Crochiere and Rabiner, 1981), (Vaidyanathan, 1990), (Varghese and Donohue, 1994a), (Munk et al., 1995), (Sadowsky and Kafedziski, 1998), (Yoo, 2008).

Doppler effect

(Kelly, 1961), (Kelly and Wishner, 1965), (Rihaczek, 1967), (Swick, 1969), (Censor, 1973), (Cooper, 1980), (Kapoulitsas, 1981), (Censor, 1984), (Chiu, 1986), (Weiss, 1994), (Jin et al., 1995), (Ho and Chan, 1998), (Ferguson, 1999), (Schlotz, 2002), (Gray and Addison, 2003), (Neipp et al., 2003), (Oberg, 2004), (Chan and Ho, 2005), (Napolitano, 2012, Chap. 7), (Napolitano, 2016a).

Time warping

(Cochran and Clark, 1990), (Xia and Zhang, 1992), (Clark and Cochran, 1993), (Azizi and Cochran, 1999), (Azizi et al., 2002).

Frequency warping

(Franaszek, 1967), (Franaszek and Liu, 1967), (Liu and Franaszek, 1969), (Rabiner et al., 1969), (Oppenheim et al., 1971), (Oppenheim and Johnson, 1972), (Braccini and Oppenheim, 1974), (Claasen and Mecklenbräuker, 1982), (Oppenheim and Schafer, 1989), (von Schroeter, 1999), (Makur and Mitra, 2001), (Franz et al., 2003).

F.7 Sampling

(Lloyd, 1959), (Zakai, 1965), (Gardner, 1972b), (Lee, 1978), (Garcia et al., 2001), (Song et al., 2007).

F.8 Complex random variables, signals, and systems

(Aiken, 1967), (Brandwood, 1983), (van den Bos, 1994), (van den Bos, 1995), (Picinbono, 1996), (Picinbono and Bondon, 1997), (Schreier and Scharf, 2003a), (Schreier and Scharf, 2003b), (Politis, 2005), (Yoo, 2008), (Schreier and Scharf, 2010), (Adali et al., 2011).

F.9 Stochastic processes

(Kolmogorov, 1933), (Cramér, 1940), (Wold, 1948), (Doob, 1953), (Grenander and Rosenblatt, 1957), (Parzen, 1957), (Leonov and Shiryaev, 1959), (Loève, 1963), (Brillinger, 1965), (Brillinger and Rosenblatt, 1967), (Billingsley, 1968), (Brillinger, 1969), (Gikhman and Skorokhod, 1969), (Prohorov and Rozanov, 1989), (Van Trees, 1971b), (Brillinger, 1974), (Rosenblatt, 1974), (Cambanis and Masry, 1976), (Brillinger, 1981), (Thomson, 1982), (Rosenblatt, 1985), (Mendel, 1991), (Papoulis, 1991), (Scharf, 1991), (Boashash et al., 1995), (Politis, 1998), (Lahiri, 2003), (Politis, 2005), (Rao, 2008).

F.10 Mathematics

(Morse and Transue, 1956), (Kolmogorov, 1960), (Gel'fand and Vilenkin, 1964), (Smirnov, 1964), (Henniger, 1970), (Kolmogorov and Fomin, 1970), (Rudin, 1987), (Zemanian, 1987), (Champeney, 1990), (Johnsonbaugh and Pfaffenberger, 2002), (NIST, 2010).

Bibliography

Aach, T., 2007. Comparative analysis of shift variance and cyclostationarity in multirate filter banks. IEEE Transactions on Circuits and Systems. I, Regular Papers (ISSN 1549-8328) 54 (5), 1077–1087. https://doi.org/10.1109/TCSI.2007.890630.

Abdelaziz, M., Chonavel, T., Aissa-El-Bey, A., Belouchrani, A., Hamadouche, M., 2013. Sea clutter texture estimation: exploiting decorrelation and cyclostationarity. IEEE Transactions on Aerospace and Electronic Systems (ISSN 0018-9251) 49 (2), 726–743. https://doi.org/10.1109/TAES.2013.6494377.

Abdulrahman, M., Falconer, D.D., 1992. Cyclostationary cross-talk suppression by decision feedback equalization on digital subscriber loops. IEEE Journal on Selected Areas in Communications 12 (4), 640–649.

Abed-Meraim, K., Xiang, Y., Manton, J.H., Hua, Y., 2001. Blind source separation using second-order cyclostationary statistics. IEEE Transactions on Signal Processing 49 (4), 694–701.

Ackroyd, M.H., 1984. Stationary and cyclostationary finite buffer behaviour computation via Levinson's method. AT & T Bell Laboratories Technical Journal 63, 2159–2170.

Adali, T., Schreier, P.J., Scharf, L.L., 2011. Complex-valued signal processing: the proper way to deal with impropriety. IEEE Transactions on Signal Processing 59 (11), 5101–5125. https://doi.org/10.1109/TSP.2011.2162954.

Adams, G.J., Goodwin, G.C., 1995. Parameter estimation for periodic ARMA models. Journal of Time Series Analysis 16, 127–146.

Agee, B.G., 1988. The baseband modulus restoral approach to blind adaptive signal demodulation. In: Proceedings of Digital Signal Processing Workshop.

Agee, B.G., 1989a. The Property Restoral Approach to Blind Adaptive Signal Extraction. Ph.D. Dissertation. Department of Electrical Engineering and Computer Science, University of California, Davis, California.

Agee, B.G., 1989b. The property-restoral approach to blind adaptive signal extraction. In: Proceedings of CSI-ARO Workshop on Advanced Topics in Communications.

Agee, B.G., 1993. Solving the near-far problem: exploitation of spectral and spatial coherence in wireless personal communication networks. In: Proceedings of Virginia Tech Third Symposium on Wireless Personal Communications, vol. 15, pp. 1–12.

Agee, B.G., 2000. Stacked-carrier discrete multiple tone communication technology and combinations with code nulling, interference cancellation, retrodirective communication adaptive antenna arrays. U.S. Patent No. 6,128,276.

Agee, B.G., 2003. Stacked carrier discrete multiple tone communication systems. U.S. Patent No. 6,512,737.

Agee, B.G., Venkataraman, S., 1989. Adaptive demodulation of PCM signals in the frequency domain. In: Proceedings of Twenty-Third Asilomar Conference on Signals, Systems and Computers. Pacific Grove, CA.

Agee, B.G., Young, D., 1990. Blind capture and geolocation of stationary waveforms using multiplatform temporal and spectral self-coherence restoral. In: Proceedings of Twenty-Fourth Asilomar Conference on Signals, Systems and Computers. Pacific Grove, California.

Agee, B.G., Schell, S.V., Gardner, W.A., 1987. Self-coherence restoral: a new approach to blind adaptation of antenna arrays. In: Proceedings of the Twenty-First Annual Asilomar Conference on Signals, Systems, and Computers. Pacific Grove, CA, pp. 589–593.

Agee, B.G., Schell, S.V., Gardner, W.A., 1988. The SCORE approach to blind adaptive signal extraction: an application of the theory of spectral correlation. In: Proceedings of the IEEE/ASSP Fourth Workshop on Spectral Estimation and Modeling. Minneapolis, MN, pp. 277–282.

Agee, B., Schell, S., Gardner, W., 1990. Spectral self-coherence restoral: a new approach to blind adaptive signal extraction using antenna arrays. Proceedings of the IEEE (ISSN 0018-9219) 78 (4), 753–767. https://doi.org/10.1109/5.54812.

Agee, B.G., Kleinman, R.J., Reed, J.H., 1996. Soft synchronization of direct sequence spread-spectrum signals. IEEE Transactions on Communications 44 (11), 1527–1536.

Agee, B.G., Kelly, P., Gerlach, D., 1997. The backtalk airlink for full exploitation of spectral and spatial diversity in wireless communication systems. In: Proceedings of Fourth Workshop on Smart Antennas in Wireless Mobile Communications.

Agee, B.G., Bromberg, M., Gerlach, D., Gibbons, D., Golden, J.T., Ho, M., Hole, E., Jesse, M., Maxwell, R.L., Mechaley Jr., R.G., Naish, R.R., Nix, D.J., Ryan, D.J., Stephenson, D., 2002. High bandwidth efficient communications. U.S. Patent No. 6,359,923.

Agostini, V., Knaflitz, M., 2012. An algorithm for the estimation of the signal-to-noise ratio in surface myoelectric signals generated during cyclic movements. IEEE Transactions on Biomedical Engineering (ISSN 0018-9294) 59 (1), 219–225. https://doi.org/10.1109/TBME.2011.2170687.

Ahmad, B., Tarczynski, A., 2011. A SARS method for reliable spectrum sensing in multiband communication systems. IEEE Transactions on Signal Processing (ISSN 1053-587X) 59 (12), 6008–6020. https://doi.org/10.1109/TSP.2011.2165060.

Aiken, R.T., 1967. Time-variant filters and analytic signals. IEEE Transactions on Information Theory IT-13, 331–333.

Ait Dads, E., Arino, O., 1996. Exponential dichotomy and existence of pseudo almost-periodic solutions of some differential equations. Nonlinear Analysis 27 (4), 369–386.

Akdi, Y., 1998. The discrete Fourier transform approximation for periodically correlated time series. Istatistik, Journal of the Turkish Statistical Association 1 (3), 47–53.

Akinshin, N.S., Rumyantsev, V.L., Mikhailov, A.V., Melititskaya, V.V., 1988. Statistical characteristics of the envelope of an additive mixture of a non-Gaussian periodically nonstationary radio signal and non-Gaussian interference. Radioelectronics and Communications Systems 31, 89–92.

Akkarakaran, S., Vaidyanathan, P.P., 2000. Bifrequency and bispectrum maps: a new look at multirate systems with stochastic inputs. IEEE Transactions on Signal Processing 48 (3), 723–736.

Aknouche, A., 2015. Explosive strong periodic autoregression with multiplicity one. Journal of Statistical Planning and Inference (ISSN 0378-3758) 161, 50–72. https://doi.org/10.1016/j.jspi.2014.12.010.

Al-Habashna, A., Dobre, O., Venkatesan, R., Popescu, D., 2012. Second-order cyclostationarity of mobile WiMAX and LTE OFDM signals and application to spectrum awareness in cognitive radio systems. IEEE Journal of Selected Topics in Signal Processing (ISSN 1932-4553) 6 (1), 26–42. https://doi.org/10.1109/JSTSP.2011.2174773.

Albuquerque, J.P.A., Shimbo, O., Ngugen, L.N., 1984. Modulation transfer noise effects from a continuous digital carrier to FDM/FM carriers in memoryless nonlinear devices. IEEE Transactions on Communications COM-32, 337–353.

Alekseev, V.O., 1987a. Extraction of the trend in a periodically correlated time series. Atmospheric and Oceanic Physics 23, 187–191.

Alekseev, V.O., 1987b. Symmetry properties of high-order spectral densities of stationary and periodic-nonstationary stochastic processes. Problems of Information Transmission 23, 210–215.

Alekseev, V.G., 1988. Estimating the spectral densities of a Gaussian periodically correlated stochastic process. Problems of Information Transmission 24, 109–115.

Alekseev, V.G., 1990. On the construction of spectral densities of a periodically correlated random process. Problemy Peredači Informacii 26, 106–108.

Alekseev, V.G., 1991a. On spectral density estimates of Gaussian periodically correlated random fields. Probability and Mathematical Statistics 11 (2), 157–167.

Alekseev, V.G., 1991b. Spectral density estimators of a periodically correlated stochastic process. Problems of Information Transmission 26, 286–288.

Allen, J.C., Hobbs, S.L., 1992. Detecting target motion by frequency-plane smoothing. In: Twenty-Sixth Annual Asilomar Conference on Signals, Systems, and Computers. Pacific Grove, CA, pp. 1042–1047.

Allen, J.C., Hobbs, S.L., 1997. Spectral estimation of non-stationary white noise. Journal of the Franklin Institute 334B (1), 99–116.

Alloin, L.F., 2007. Method and apparatus for channel equalization and cyclostationary interference rejection for ADSL-DMT modems. U.S. Patent No. 7,289,554.

Almeida, L., 1994. The fractional Fourier transform and time-frequency representations. IEEE Transactions on Signal Processing (ISSN 1053-587X) 42 (11), 3084–3091. https://doi.org/10.1109/78.330368.

Alpay, D., Chevreuil, A., Loubaton, P., 2001. An extension problem for discrete-time periodically correlated stochastic processes. Journal of Time Series Analysis 22 (1), 1–12.

Amerio, L., Prouse, 1971. Almost-Periodic Functions and Functional Equations. Van Nostrand, New York.

Amin, M.G., 1992. Time-frequency spectrum analysis and estimation for nonstationary random processes. In: Boashash, B. (Ed.), Time-Frequency Signal Analysis: Methods and Applications. Longman Cheshire.

Andel, J., 1985a. On periodic autoregression with unknown mean. Applications of Mathematics 30, 126–139.

Andel, J., 1985b. Periodic autoregression with exogenous variables and equal variances. In: Grossman, W., et al. (Eds.), Probability Theory and Mathematical Statistics with Applications. Reidel, Dordrecht, NL, Boston, MA, pp. 237–245.

Andel, J., 1989. Periodic autoregression with exogenous variables and periodic variances. Ceskoslovenska Akodemie Ved. Aplikace Motemotiky 34, 387–395.

Andel, J., Rubio, A., 1986. On interpolation on periodic autoregressive processes. Applications of Mathematics 31, 480–485.

Anderson, T., 2003. An Introduction to Multivariate Statistical Analysis, third ed. John Wiley & Sons, Inc., Hoboken, New Jersey.

Anderson, P.L., Meerschaert, M.M., 1997. Periodic moving averages of random variables with regularly varying tails. The Annals of Statistics 25, 771–785.

Anderson, P.L., Meerschaert, M.M., 2005. Parameter estimation for periodically stationary time series. Journal of Time Series Analysis 26 (4), 489–518.

Anderson, P.L., Vecchia, A.V., 1993. Asymptotic results for periodic autoregressive moving-average processes. Journal of Time Series Analysis 14, 1–18.

Anderson, R.R., Foschini, G.T., Gopinath, B., 1979. A queuing model for a hybrid data multiplexer. The Bell System Technical Journal 58, 279–300.

Anderson, P.L., Meerschaert, M.M., Vecchia, A., 1999. Innovations algorithm for periodically stationary time series. Stochastic Processes and Their Applications 83, 149–169.

Anderson, P., Meerschaert, M., Zhang, K., 2013. Forecasting with prediction intervals for periodic ARMA models. Journal of Time Series Analysis 34 (2), 187–193. https://doi.org/10.1111/jtsa.12000.

Andreas, J., Bersani, A.M., Grande, R.F., 2006. Hierarchy of almost-periodic function spaces. Rendiconti Di Matematica 26, 121–188.

Andrieu, C., Duvat, P., Doucet, A., 1996. Bayesian deconvolution of cyclostationary processes based on point processes. In: Proceedings of VIII European Signal Processing Conference. EUSIPCO'96, Trieste, Italy.

Angrisani, L., Napolitano, A., Vadursi, M., 2009. True-power measurement in digital communication systems affected by in-channel interference. IEEE Transactions on Instrumentation and Measurement (ISSN 0018-9456) 58 (12), 3985–3994. https://doi.org/10.1109/TIM.2009.2021621.

Ankarali, Z.E., Arslan, H., 2017. Cyclic feature suppression for physical layer security. Physical Communication (ISSN 1874-4907) 25, 588–597. https://doi.org/10.1016/j.phycom.2016.09.003.

Anthony, M.P., Chuang, J., Mody, A.N., Spooner, C.M., 2017. Process for tunnelized cyclostationary to achieve low-energy spectrum sensing. U.S. Patent No. 9,755,869.

Antoni, J., 2007a. Cyclic spectral analysis in practice. Mechanical Systems and Signal Processing (ISSN 0888-3270) 21 (2), 597–630. https://doi.org/10.1016/j.ymssp.2006.08.007.

Antoni, J., 2007b. Cyclic spectral analysis of rolling-element bearing signals: facts and fictions. Journal of Sound and Vibration (ISSN 0022-460X) 304, 497–529. https://doi.org/10.1016/j.jsv.2007.02.029.

Antoni, J., 2009. Cyclostationarity by examples. Mechanical Systems and Signal Processing 23 (4), 987–1036.

Antoni, J., 2016. The infogram: entropic evidence of the signature of repetitive transients. Mechanical Systems and Signal Processing (ISSN 0888-3270) 74, 73–94. https://doi.org/10.1016/j.ymssp.2015.04.034.

Antoni, J., Hanson, D., 2012. Detection of surface ships from interception of cyclostationary signature with the cyclic modulation coherence. IEEE Journal of Oceanic Engineering (ISSN 0364-9059) 37 (3), 478–493. https://doi.org/10.1109/JOE.2012.2195852.

Antoni, J., Randall, R.B., 2002. Differential diagnosis of gear and bearing faults. Journal of Vibration and Acoustics 124 (2), 165–171.

Antoni, J., Randall, R.B., 2003. A stochastic model for simulation and diagnostics of rolling element bearings with localized faults. Journal of Vibration and Acoustics 125 (3), 282–289.

Antoni, J., Daniere, J., Guillet, F., 2002a. Effective vibration analysis of IC engines using cyclostationarity. Part I–a methodology for condition monitoring. Journal of Sound and Vibration 257 (5), 815–837.

Antoni, J., Daniere, J., Guillet, F., Randall, R.B., 2002b. Effective vibration analysis of IC engines using cyclostationarity. Part II–new results on the reconstruction of the cylinder pressure. Journal of Sound and Vibration 257 (5), 839–856.

Antoni, J., Bonnardot, F., Raad, A., El Badaoui, M., 2004a. Cyclostationary modelling of rotating machine vibration signals. Mechanical Systems and Signal Processing (ISSN 0888-3270) 18 (6), 1285–1314. https://doi.org/10.1016/S0888-3270(03)00088-8.

Antoni, J., Wagstaff, P., Henrio, J.C., 2004b. H_α – a consistent estimator for frequency response function with input and output noise. IEEE Transactions on Instrumentation and Measurement 53 (2), 457–465.

Antoni, J., Guillet, F., El Badaoui, M., Bonnardot, F., 2005. Blind separation of convolved cyclostationary processes. Signal Processing 85, 51–66.

Antoni, J., Ducleaux, N., NGhiem, G., Wang, S., 2013. Separation of combustion noise in IC engines under cyclo-non-stationary regime. Mechanical Systems and Signal Processing (ISSN 0888-3270) 38 (1), 223–236. https://doi.org/10.1016/j.ymssp.2013.02.015.

Antoni, J., Abboud, D., Baudin, S., 2014. Time-angle periodically correlated processes. In: Chaari, F., Leskow, J., Napolitano, A., Sanchez-Ramirez, A. (Eds.), Cyclostationarity: Theory and Methods. In: Lecture Notes in Mechanical Engineering. Springer International Publishing, pp. 3–14.

Antoni, J., Xin, G., Hamzaoui, N., 2017. Fast computation of the spectral correlation. Mechanical Systems and Signal Processing (ISSN 0888-3270) 92, 248–277. https://doi.org/10.1016/j.ymssp.2017.01.011.

Antoniadis, I., Glossiotis, G., 2001. Cyclostationary analysis of rolling-element bearing vibration signals. Journal of Sound and Vibration 248 (5), 829–845.

Archer, T.L., Gardner, W.A., 1990. New methods for identifying the Volterra kernels of a nonlinear system. In: Proceedings of the Twenty-Fourth Annual Asilomar Conference on Signals, Systems, and Computers. Pacific Grove, CA, pp. 592–597.

Arciplani, B., Carloni, F., Marseguerra, M., 1979. Neutron counting statistics in a subcritical cyclostationary multiplying system, Part I. Nuclear Instruments and Methods 167, 465–473.

Arciplani, B., Carloni, F., Marseguerra, M., 1980. Neutron counting statistics in a subcritical cyclostationary multiplying system, Part II. Nuclear Instruments and Methods 172, 531–534.

Ariananda, D., Leus, G., 2014. Non-uniform sampling for compressive cyclic spectrum reconstruction. In: IEEE International Conference on Acoustics, Speech and Signal Processing. ICASSP 2014, pp. 41–45.

Arora, R., Sethares, W., Bucklew, J., 2008. Latent periodicities in genome sequences. IEEE Journal of Selected Topics in Signal Processing (ISSN 1932-4553) 2 (3), 332–342. https://doi.org/10.1109/JSTSP.2008.923861.

Arslan, H., Molnar, K.J., 2003. Methods of receiving co-channel signals by channel separation and successive cancellation and related receivers. U.S. Patent No. 6,574,235.

Aru, J., Aru, J., Priesemann, V., Wibral, M., Lana, L., Pipa, G., Singer, W., Vicente, R., 2015. Untangling cross-frequency coupling in neuroscience. Current Opinion in Neurobiology (ISSN 0959-4388) 31, 51–61. https://doi.org/10.1016/j.conb.2014.08.002.

Astaneh, S.A., Gazor, S., 2013. Cooperative spectrum sensing over mixture-Nakagami channels. IEEE Wireless Communications Letters (ISSN 2162-2337) 2 (3), 259–262. https://doi.org/10.1109/WCL.2013.020513.120825.

Astaneh, S.A., Gazor, S., 2014. Relay-assisted spectrum sensing. IET Communications (ISSN 1751-8628) 8, 11–18. https://doi.org/10.1049/iet-com.2013.0232.

Axell, E., Leus, G., Larsson, E.G., Poor, H.V., 2012. Spectrum sensing for cognitive radio: state-of-the-art and recent advances. IEEE Signal Processing Magazine (ISSN 1053-5888) 29 (3), 101–116. https://doi.org/10.1109/MSP.2012.2183771.

Azizi, S., Cochran, D., 1999. Reproducing kernel structure and sampling on time-warped Kramer spaces. In: Proceedings of IEEE International Conference on Acoustics, Speech, and Signal Processing, vol. 3. ICASSP 1999, pp. 1649–1652.

Azizi, S., Cochran, D., McDonald, J., 2002. Reproducing kernel structure and sampling on time-warped spaces with application to warped wavelets. IEEE Transactions on Information Theory 48 (3), 789–790.

Ballerini, R., McCormick, W.P., 1989. Extreme value theory for processes with periodic variances. Stochastic Models 5, 45–61.

Barnett, T.P., 1983. Interaction of the monsoon and Pacific trade wind system at interannual time scales. I. The equatorial zone. Monthly Weather Review III, 756–773.

Barnett, T.P., Heinz, H.D., Hasselmann, K., 1984. Statistical prediction of seasonal air temperature over Eurasia. Tellus. Series A (Sweden) 36, 132–146.

Bartolini, P., Salas, J.D., Obeysekera, J.T.B., 1988. Multivariate periodic ARMA(1, 1) processes. Water Resources Research 24, 1237–1246.

Barzilai, J., Borwein, J., 1988. Two-point step size gradient methods. IMA Journal of Numerical Analysis 8 (1), 141–148.

Basawa, I.V., Lund, R., 2001. Large sample properties of parameter estimates for periodic ARMA models. Journal of Time Series Analysis 22 (6), 651–664.

Bass, J., 1971. Cours de Mathématiques, III. Masson & Cie, Paris.

Bass, J., 1974a. Stationary functions and their applications to the theory of turbulence: I. Stationary functions. Journal of Mathematical Analysis and Applications 47 (2), 354–399. https://doi.org/10.1016/0022-247X(74)90026-2.

Bass, J., 1974b. Stationary functions and their applications to the theory of turbulence: II. Turbulent solutions of the Navier-Stokes equations. Journal of Mathematical Analysis and Applications 47 (3), 458–503. https://doi.org/10.1016/0022-247X(74)90002-X.

Bell, M.R., Grubbs, R.A., 1993. JEM modeling and measurements for radar target identification. IEEE Transactions on Aerospace 29, 73–87.

Bello, P.A., 1963. Characterization of randomly time-variant channels. IEEE Transactions on Communications Systems CS-11, 360–393.

Benedetto, J.J., 1996. Harmonic Analysis and Applications. CRC Press, New York.

Bennett, W.R., 1958. Statistics of regenerative digital transmission. The Bell System Technical Journal 37, 1501–1542.

Bensley, S., Aazhang, B., 1996. Subspace-based channel estimation for code division multiple access systems. IEEE Transactions on Communications 44 (8), 1009–1020.

Bensley, S., Aazhang, B., 1998. Maximum-likelihood synchronization of a single user for code-division multiple-access communication systems. IEEE Transactions on Communications 46 (3), 392–399.

Bentarzi, M., Hallin, M., 1993. On the invertibility of periodic moving-average models. Journal of Time Series Analysis 15, 263–268.

Bentarzi, M., Hallin, M., 1996. Locally optimal tests against periodical autocorrelation: parametric and nonparametric approaches. Econometric Theory 12, 88–112.

Bentarzi, M., Hallin, M., 1998. Spectral factorization of periodically correlated MA(1) processes. Journal of Applied Probability 35 (1), 46–54.

Berezin, Y.V., Krasheninnikov, I.V., 1978. Distribution functions of the envelope and phase of a periodically non stationary process frequently encountered in radio physics. Geomagnetism and Aeronomy 18, 309–312.

Bergmans, J.W.M., 1994. Data receiver comprising a control loop with reduced sampling frequency. U.S. Patent No. 5,311,558.

Bergmans, J.W.M., 1995. Transmission system with increased sampling rate detection. U.S. Patent No. 5,463,654.

Bergmans, J.W.M., Janssen, A.J.E.M., 1987. Robust data equalization, fractional tap spacing and the Zak transform. Philips Journal of Research 23, 351–398.

Bergmans, J.W.M., van Gerwen, P.J., Wouda, K.J., 1989. Data signal transmission system using decision feedback equalization. U.S. Patent No. 4,870,657.

Bershad, N., Eweda, E., Bermudez, J., 2014. Stochastic analysis of the LMS and NLMS algorithms for cyclostationary white Gaussian inputs. IEEE Transactions on Signal Processing (ISSN 1053-587X) 62 (9), 2238–2249. https://doi.org/10.1109/TSP.2014.2307278.

Bershad, N.J., Eweda, E., Bermudez, J.C.M., 2016. Stochastic analysis of an adaptive line enhancer/canceler with a cyclostationary input. IEEE Transactions on Signal Processing (ISSN 1053-587X) 64 (1), 104–119. https://doi.org/10.1109/TSP.2015.2486745.

Bershad, N.J., Eweda, E., Bermudez, J.C., 2018. Performance of soft limiters in the LMS algorithm for cyclostationary white Gaussian inputs. Signal Processing (ISSN 0165-1684) 152, 197–205. https://doi.org/10.1016/j.sigpro.2018.05.023.

Besicovitch, A.S., 1932. Almost Periodic Functions. Cambridge University Press, London (also Dover, New York, 1954).

Besson, O., Stoica, P., 1999. Simple test for distinguishing constant from time varying amplitude in harmonic retrieval problem. IEEE Transactions on Signal Processing 47 (4), 1137–1141.

Bhuiya, R.K., 1971. Stochastic analysis of periodic hydrologic process. American Society of Civil Engineers. Journal of the Hydraulics Division 97, 949–963.

Bianchi, P., Loubaton, P., Sirven, F., 2004. Non data-aided estimation of the modulation index of continuous phase modulations. IEEE Transactions on Signal Processing 52 (10), 2847–2861.

Bibi, A., Franq, C., 2003. Consistent and asymptotically Normal estimators for cyclically time-dependent linear models. Annals of the Institute of Statistical Mathematics 55 (1), 41–68.

Bibi, A., Lescheb, I., 2014. A note on integrated periodic GARCH processes. Statistics & Probability Letters (ISSN 0167-7152) 87, 121–124. https://doi.org/10.1016/j.spl.2014.01.007.

Biedka, T., 1993. Subspace-constrained SCORE algorithms. In: Proceedings of Twenty-Seventh Asilomar Conference on Signals, Systems and Computers. Pacific Grove, California, pp. 716–720.

Biedka, T., 1995. A method for reducing computations in cyclostationarity-exploiting beamforming. In: Proceedings of IEEE International Conference on Acoustics, Speech, and Signal Processing. ICASSP'95, pp. 1828–1831.

Biedka, T., 2001. Analysis and Development of Blind Adaptive Beamforming Algorithms. Ph.D. Dissertation. Bradley Department of Electrical Engineering, Virginia Polytechnic Institute and State University, Virginia.

Biedka, T., Agee, B.G., 1991. Subinterval cyclic MUSIC - robust DF with inaccurate knowledge of cycle frequency. In: Proceedings of Twenty-Fifth Asilomar Conference on Signals, Systems and Computers. Pacific Grove, California.

Biedka, T., Kahn, M., 1994. Methods for constraining a CMA beamformer to extract a cyclostationary signal. In: Proceedings of Second Workshop on Cyclostationary Signals. Yountville, CA.

Biedka, T., Mili, L., Reed, J., 1995a. Robust estimation of cyclic correlation in contaminated Gaussian noise. In: Conference Record of the Twenty-Ninth Asilomar Conference on Signals, Systems and Computers, vol. 1, pp. 511–515.

Biedka, T., Mili, L., Reed, J.H., 1995b. Robust estimation of cyclic correlation in contaminated gaussian noise. In: Proceedings of 29th Asilomar Conference on Signals, Systems and Computers. Pacific Grove, CA.

Bilardi, C., Pupolin, S., 1984. Spectral analysis of the powers of a PAM digital signal. Alta Frequenza 53 (2), 70–76.

Billingsley, P., 1968. Convergence in Probability Measures. Wiley, New York.

Bilyik, M.G., 1973. Up-crossings of periodically nonstationary processes. Otbor i Peredacha Informacii 36, 28–31 (in Russian).

Bilyik, M.G., 1974a. Models and level crossings of periodically nonstationary processes. Otbor i Peredacha Informacii 39, 3–7 (in Russian).

Bilyik, M.G., 1974b. On the theory of up-crossings of periodically nonstationary processes. Otbor i Peredacha Informacii 38, 15–21 (in Russian).

Bilyik, M.G., 1976. Some properties of periodically nonstationary processes and inhomogeneous fields. Otbor i Peredacha Informacii 48, 27–33 (in Russian).

Bilyik, M.G., 1977a. Periodic nonstationary processes. Otbor i Peredacha Informacii 50, 28–34 (in Russian).

Bilyik, M.G., 1977b. Statistical properties of level crossings of a periodic nonstationary process and a periodic inhomogeneous field. Otbor i Peredacha Informacii 52, 3–8 (in Russian).

Bilyik, M.G., 1979. Some transformations of periodic inhomogeneous random fields. Otbor i Peredacha Informacii 57, 21–25 (in Russian).

Bilyik, M.G., 1980. High crossings of periodically nonstationary Gaussian process. Otbor i Peredacha Informacii 59, 24–31 (in Russian).

Bittanti, S., 1986. The periodic prediction problem for cyclostationary processes–an introduction. In: Proc. of the NATO Advanced Research Workshop. Groningen, pp. 239–249.

Bittanti, S., Colaneri, P., 2008. Periodic Systems: Filtering and Control, Communications and Control Engineering. Springer. ISBN 9781848009103.

Bittanti, S., De Nicolao, G., 1993. Spectral factorization of linear periodic systems with application to the optimal prediction of periodic ARMA models. Automatica 29 (2), 517–522.

Bittanti, S., Guardabassi, G., 1982. Optimal cyclostationary control: a parameter-optimization frequency-domain approach. In: Akashi, H. (Ed.), Proceedings of the Eighth Triennial World Congress of the Internat. Fed. of Automatic Control, vol. 2. Kyoto, Japan, pp. 857–862.

Bittanti, S., Bolzern, P., De Nicolao, G., Piroddi, L., Purassanta, D., 1991a. A minimum prediction error algorithm for estimation of periodic ARMA models. In: Proceedings of the European Control Conference. Grenoble, France.

Bittanti, S., Hernandez, D.B., Zerbi, G., 1991b. The simple pendulum and the periodic LQG control problem. Journal of the Franklin Institute 328, 299–315.

Bittanti, S., Bolzern, P., De Nicolao, G., Piroddi, L., 1994. Representation, prediction, and identification of cyclostationary processes–a state-space approach. In: Gardner, W.A. (Ed.), Cyclostationarity in Communications and Signal Processing. IEEE Press, pp. 267–294.

Blachman, N.M., 1994. Beneficial effects of spectral correlation on synchronization. In: Gardner, W.A. (Ed.), Cyclostationarity in Communications and Signal Processing. IEEE Press, pp. 362–390.

Black, T.R., Donohue, K.D., 2000a. Frequency correlation analysis for periodic echoes. In: Proceedings of IEEE South-East Conference '00. Nashville, TN, pp. 131–138.

Black, T.R., Donohue, K.D., 2000b. Pitch determination of music signals using the generalized spectrum. In: Proceedings of IEEE South-East Conference '00. Nashville, TN, pp. 104–109.

Bloomfield, P., Hurd, H.L., Lund, R., 1994. Periodic correlation in stratospheric ozone time series. Journal of Time Series Analysis 15 (2), 127–150.

Boashash, B. (Ed.), 2003. Time Frequency Signal Analysis and Processing. A Comprehensive Reference. Elsevier.

Boashash, B. (Ed.), 2015. Time Frequency Signal Analysis and Processing. A Comprehensive Reference, second ed. Elsevier–Academic Press.

Boashash, B., Powers, E.J., Zoubir, A.M., 1995. Higher-Order Statistical Signal Processing. Longman, Australia.

Bohr, H., 1933. Almost Periodic Functions. Springer, Berlin (also Chelsea, New York, 1947).

Boiteau, D., Le Martret, C., 1998. A general maximum likelihood framework for modulation classification. In: Proceedings of International Conference on Acoustics, Speech, and Signal Processing, vol. 4. ICASSP'98, pp. 2165–2168.

Bölcskei, H., 2001. Blind estimation of symbol timing and carrier frequency offset in wireless OFDM systems. IEEE Transactions on Communications 49 (6), 988–999.

Bölcskei, H., Duhamel, P., Hleiss, R., 2001. A subspace-based approach to blind channel identification in pulse shaping OFDM/OQAM systems. IEEE Transactions on Signal Processing 49 (7), 1594–1598.

Bölcskei, H., Health, R.W., Paulraj, A.J., 2002. Blind channel identification and equalization in OFDM-based multiantenna systems. IEEE Transactions on Signal Processing 50 (1), 96–109.

Bollerslev, T., Ghysels, E., 1996. Periodic autoregressive conditional heteroscedasticity. American Statistical Association Journal of Business & Economic Statistics 14 (2), 139–151.

Bollig, A., Mathar, R., 2013. Dictionary-based reconstruction of the cyclic autocorrelation via ℓ_1-minimization for cyclostationary spectrum sensing. In: IEEE International Conference on Acoustics, Speech and Signal Processing. ICASSP 2013, pp. 4908–4912.

Bonani, F., Donati, S., Ghione, G., 2002. Noise source modeling for cyclostationary noise analysis in large-signal device operation. IEEE Transactions on Electron Devices 49 (9), 1640–1647.

Bonnardot, F., Randall, R., Guillet, F., 2005. Extraction of second-order cyclostationary sources: application to vibration analysis. Mechanical Systems and Signal Processing (ISSN 0888-3270) 19 (6), 1230–1244. https://doi.org/10.1016/j.ymssp.2005.07.008.

Borghesani, P., 2015. The envelope-based cyclic periodogram. Mechanical Systems and Signal Processing (ISSN 0888-3270) 58–59, 245–270. https://doi.org/10.1016/j.ymssp.2014.11.009.

Borghesani, P., Antoni, J., 2018. A faster algorithm for the calculation of the fast spectral correlation. Mechanical Systems and Signal Processing (ISSN 0888-3270) 111, 113–118. https://doi.org/10.1016/j.ymssp.2018.03.059.

Borghesani, P., Pennacchi, P., Ricci, R., Chatterton, S., 2013. Testing second order cyclostationarity in the squared envelope spectrum of non-white vibration signals. Mechanical Systems and Signal Processing (ISSN 0888-3270) 40 (1), 38–55. https://doi.org/10.1016/j.ymssp.2013.05.012.

Borgnat, P., Flandrin, P., Amblard, P.O., 2002. Stochastic discrete scale invariance. IEEE Signal Processing Letters 9 (6), 181–184.

Borgnat, P., Amblard, P., Flandrin, P., 2005. Scale invariances and Lamperti transformations for stochastic processes. Journal of Physics. A, Mathematical and General 38, 2081–2101.
Boshnakov, G.N., 1995. Recursive computation of the parameters of periodic autoregressive moving average processes. Journal of Time Series Analysis 17, 333–349.
Boshnakov, G.N., Lambert-Lacroix, S., 2012. A periodic Levinson–Durbin algorithm for entropy maximization. Computational Statistics & Data Analysis (ISSN 0167-9473) 56 (1), 15–24. https://doi.org/10.1016/j.csda.2011.07.001.
Boswijk, H.P., Franses, P.H., 1995. Testing for periodic integration. Economics Letters 48, 241–248.
Botero, F., Hasmatuchi, V., Roth, S., Farhat, M., 2014. Non-intrusive detection of rotating stall in pump-turbines. Mechanical Systems and Signal Processing (ISSN 0888-3270) 48, 162–173. https://doi.org/10.1016/j.ymssp.2014.03.007.
Bouguerriou, N., Haritopoulos, M., Capdessus, C., Allam, L., 2005. Novel cyclostationarity-based blind source separation algorithm using second order statistical properties: theory and application to the bearing defect diagnosis. Mechanical Systems and Signal Processing (ISSN 0888-3270) 19 (6), 1260–1281. https://doi.org/10.1016/j.ymssp.2005.07.007.
Bouillaut, L., Sidahmed, M., 2001. Cyclostationary approach and bilinear approach: comparison, applications to early diagnosis for helicopter gearbox and classification method based on HOCS. Mechanical Systems and Signal Processing 15 (5), 923–943.
Boungou, D., Guillet, F., El Badaoui, M., Lyonnet, P., Rosario, T., 2015. Fatigue damage detection using cyclostationarity. Mechanical Systems and Signal Processing (ISSN 0888-3270) 58–59, 128–142. https://doi.org/10.1016/j.ymssp.2014.11.010.
Boustany, R., Antoni, J., 2005. A subspace method for the blind extraction of a cyclostationary source: application to rolling element bearing diagnostics. Mechanical Systems and Signal Processing (ISSN 0888-3270) 19 (6), 1245–1259. https://doi.org/10.1016/j.ymssp.2005.08.016.
Boustany, R., Antoni, J., 2008. Blind extraction of a cyclostationary signal using reduced-rank cyclic regression: a unifying approach. Mechanical Systems and Signal Processing (ISSN 0888-3270) 22 (3), 520–541. https://doi.org/10.1016/j.ymssp.2007.09.014.
Bouzegzi, A., Ciblat, P., Jallon, P., 2010. New algorithms for blind recognition of OFDM based systems. Signal Processing (ISSN 0165-1684) 90 (3), 900–913. https://doi.org/10.1016/j.sigpro.2009.09.017.
Boyles, R., Gardner, W., 1983. Cycloergodic properties of discrete- parameter nonstationary stochastic processes. IEEE Transactions on Information Theory (ISSN 0018-9448) IT-29 (1), 105–114. https://doi.org/10.1109/TIT.1983.1056613.
Braccini, C., Oppenheim, A.V., 1974. Unequal bandwidth spectral analysis using digital frequency warping. IEEE Transactions on Acoustics, Speech, and Signal Processing ASSP-22 (4), 236–244.
Bradaric, I., Petropulu, A.P., Diamantaras, K.I., 2003. Blind MIMO FIR channel identification based on second-order spectra correlations. IEEE Transactions on Signal Processing 51 (6), 1668–1674.
Brandwood, D.H., 1983. A complex gradient operator and its application in adaptive array theory. IEE Proceedings. Parts F and H 130 (1), 11–16.
Braun, S., Seth, B., 1980. Analysis of repetitive mechanism signature. Journal of Sound and Vibration 70, 513–526.
Brelsford, W.M., 1967. Probability Predictions and Time Series With Periodic Structure. Ph.D. Dissertation. Johns Hopkins University, Baltimore, MD.
Brennan, D.G., 1961. Probability theory in communication system engineering. In: Baghdady, E.J. (Ed.), Communication System Theory. McGraw-Hill, New York.
Brillinger, D.R., 1965. An introduction to polyspectra. Annals of Mathematics 36 (5), 1351–1374.
Brillinger, D.R., 1969. Asymptotic properties of spectral estimates of second order. Biometrika 56, 375–390.
Brillinger, D.R., 1974. Fourier analysis of stationary processes. Proceedings of the IEEE 62 (12), 1628–1643.
Brillinger, D.R., 1981. Time series. Data analysis and theory. In: SIAM 2001. Holden Day, Inc., San Francisco, CA.
Brillinger, D.R., Rosenblatt, M., 1967. Asymptotic theory of estimates of kth-order spectra. In: Harris, B. (Ed.), Spectral Analysis of Time Series. Wiley, New York, pp. 153–188.
Broszkiewicz-Suwaj, E., Makagon, A., Weron, R., Wyłomańska, A., 2004. On detecting and modeling periodic correlation in financial data. In: Proceedings of the XVIII Max Born Symposium. Physica A, Statistical Mechanics and Its Applications 336 (1–2), 196–205.

Brown, W.A., 1987. On the Theory of Cyclostationary Signals. Ph.D. Dissertation. Dep. Elect. Eng. Comput. Sci., Univ. California, Davis, CA.

Brown, W.A., Loomis Jr., H.H., 1988. Digital implementations of spectral correlation analyzers. In: Proceedings of the Fourth Annual ASSP Workshop on Spectrum Estimation and Modeling. Minneapolis, MN, pp. 264–270.

Brown, W.A., Loomis Jr., H.H., 1993. Digital implementation of spectral correlation analyzers. IEEE Transactions on Signal Processing 41 (2), 703–720.

Bruyevich, A.N., 1968. Fluctuations in auto-oscillators for periodically nonstationary shot noise. Radioengineering 23, 91–96.

Buiquang, C., Zhang, T., 2018. Blind recognition of space time block code in MIMO systems. Digital Signal Processing (ISSN 1051-2004) 83, 1–8. https://doi.org/10.1016/j.dsp.2018.08.011.

Bukofzer, D., 1979. Coherent and Noncoherent Detection of Signals in Cyclostationary Noise and of Cyclostationary Signals in Stationary Noise. Ph.D. Dissertation. University of California, Davis, Department of Electrical and Computer Engineering.

Bukofzer, D., 1987. Optimum and suboptimum detector performance for signals in cyclostationary noise. IEEE Journal of Oceanic Engineering OE-12, 97–115.

Bukofzer, D.C., 1990. Characterization and applications of multiplexed PAM/PPM processes. In: Proceedings of the International Symposium on Information Theory and Applications. ISITA'90, Hawaii, USA, pp. 915–918.

Buzzi, S., Lops, M., Tulino, A.M., 2001a. Blind adaptive multiuser detection for asynchronous dual rate DS/CDMA systems. IEEE Journal on Selected Areas in Communications 19, 233–244.

Buzzi, S., Lops, M., Tulino, A.M., 2001b. A new family of MMSE multiuser receiver for interference suppression in DS/CDMA systems employing BPSK modulation. IEEE Transactions on Communications 49 (1), 154–167.

Buzzi, S., Lops, M., Tulino, A.M., 2001c. Partially blind adaptive MMSE interference rejection in asynchronous DS/CDMA networks over frequency-selective fading channels. IEEE Transactions on Communications 49 (1), 94–108.

Buzzoni, M., Antoni, J., D'Elia, G., 2018. Blind deconvolution based on cyclostationarity maximization and its application to fault identification. Journal of Sound and Vibration (ISSN 0022-460X) 432, 569–601. https://doi.org/10.1016/j.jsv.2018.06.055.

Cai, X., Wang, X., Huang, Z., Wang, F., 2016. Single channel steepest descent algorithm for the correction of cycle frequency error. IET Communications (ISSN 1751-8628) 10 (14), 1829–1838. https://doi.org/10.1049/iet-com.2015.1235.

Cambanis, S., Houdré, C., 1995. On the continuous wavelet transform of second-order random processes. IEEE Transactions on Information Theory 41 (3), 628–642.

Cambanis, S., Masry, E., 1976. Zakai's class of bandlimited functions and processes: its characterization and properties. SIAM Journal on Applied Mathematics 30 (1), 10–21.

Cambanis, S., Houdré, C.H., Hurd, H.L., Leśkow, J., 1994. Laws of large numbers for periodically and almost periodically correlated processes. Stochastic Processes and Their Applications 53, 37–54.

Campbell, J.C., Gibbs, A.J., Smith, B.M., 1983a. The cyclostationary nature of crosstalk interference from digital signals in multipair cable–Part I: fundamentals. IEEE Transactions on Communications COM-31 (5), 629–637.

Campbell, J.C., Gibbs, A.J., Smith, B.M., 1983b. The cyclostationary nature of crosstalk interference from digital signals in multipair cable–Part II: applications and further results. IEEE Transactions on Communications COM-31 (5), 638–649.

Capdessus, C., Nandi, A.K., 2011. Extraction of a cyclostationary source using a new cost function without pre-whitening. Signal Processing (ISSN 0165-1684) 91 (11), 2497–2505. https://doi.org/10.1016/j.sigpro.2011.04.029.

Capdessus, C., Sidahmed, M., Lacoume, J.L., 2000. Cyclostationary processes: application in gear faults early diagnosis. Mechanical Systems and Signal Processing 14 (3), 371–385.

Cariolaro, G., Erseghe, T., Laurenti, N., Pierobon, G., 2011. New results on the spectral analysis of multi-h CPM signals. IEEE Transactions on Communications (ISSN 0090-6778) 59 (7), 1893–1903. https://doi.org/10.1109/TCOMM.2011.050911.100631.

Carrick, M., Reed, J.H., 2017. Exploiting the cyclostationarity of radar chirp signals with time-varying filters. In: 2017 IEEE Global Conference on Signal and Information Processing. GlobalSIP, pp. 274–277.

Carrick, M., Reed, J.H., Spooner, C.M., 2018. Paramorphic multicarrier communications for interference mitigation. EURASIP Journal on Advances in Signal Processing (ISSN 1687-6180) 2018 (1). https://doi.org/10.1186/s13634-017-0523-7.

Carroll, T.L., 2005. Low-interference communications device using chaotic signals. U.S. Patent No. 6,854,058.

Cartaxo, A.V.T., Albuquerque, A.A.D., 1990. A general property of the n-order moment generating function of strict-sense cyclostationary processes. In: Arikan, E. (Ed.), Proceedings of the 1990 Bilkent International Conference on New Trends in Communication, Control and Signal Processing. Ankara, Turkey. Elsevier Science, Amsterdam, Netherlands, pp. 1548–1552.

Casinovi, G., 2007. l^1-norm convergence properties of correlogram spectral estimates. IEEE Transactions on Signal Processing 55 (9), 4354–4365.

Casinovi, G., 2009. Sampling and ergodic theorems for weakly almost periodic signals. IEEE Transactions on Information Theory 55 (4), 1883–1897.

Castedo, L., Figueiras-Vidal, A.R., 1995. An adaptive beamforming technique based on cyclostationary signal properties. IEEE Transactions on Signal Processing 43 (7), 1637–1650.

Castedo, L., Tseng, C., 1993. Behavior of adaptive beamformers based on cyclostationary signal properties in multipath environments. In: Proceedings of 27th Asilomar Conference on Signals, Systems and Computers, vol. 1. Pacific Grove, CA, pp. 653–657.

Castedo, L., Tseng, C., Griffiths, L., 1993. A new cost function for adaptive beamforming using cyclostationary signal properties. In: Proceedings of IEEE International Conference on Acoustics, Speech, and Signal Processing, vol. 4. ICASSP'93, pp. 284–287.

Castedo, L., Fueiras-Vidal, A., Tseng, C., Griffiths, L., 1994. Linearly constrained adaptive beamforming using cyclostationary signal properties. In: Proceedings of IEEE International Conference on Acoustics, Speech, and Signal Processing, vol. 4. ICASSP'94, pp. 249–252.

Celebi, S., 2004. Methods and devices for simplifying blind channel estimation of the channel impulse response for a DMT signal. U.S. Patent No. 6,782,042.

Censor, D., 1973. The generalized Doppler effect and applications. Journal of the Franklin Institute 295 (2), 103–116.

Censor, D., 1984. Theory of the Doppler effect: fact, fiction, and approximation. Radio Science 19 (4), 1027–1040.

Cerrato, L.R., Eisenstein, B.A., 1977. Deconvolution of cyclostationary signals. IEEE Transactions on Acoustics, Speech, and Signal Processing ASSP-25, 466–476.

Chaari, F., Haddar, M., 2014. Modeling of gear transmissions dynamics in non-stationary conditions. In: Chaari, F., Leskow, J., Napolitano, A., Sanchez-Ramirez, A. (Eds.), Cyclostationarity: Theory and Methods. In: Lecture Notes in Mechanical Engineering. Springer International Publishing, pp. 109–124.

Challapali, K., Dong, B., Birru, D., 2017. Detection of the presence of television signals embedded in noise using cyclostationary toolbox. U.S. Patent No. 9,654,673.

Champeney, D.C., 1990. A Handbook of Fourier Theorems. Cambridge University Press, New York.

Chan, Y.T., Ho, K.C., 2005. Joint time-scale and TDOA estimation: analysis and fast approximation. IEEE Transactions on Signal Processing 53, 2625–2634.

Chang, L., Li, G.Y., Li, J., Li, R., 2016. Blind parameter estimation of GFDM signals over frequency-selective fading channels. IEEE Transactions on Communications (ISSN 0090-6778) 64 (3), 1120–1131. https://doi.org/10.1109/TCOMM.2016.2523502.

Charge, P., Wang, Y., 2005. A root-MUSIC-like direction finding method for cyclostationary signals. EURASIP Journal on Advances in Signal Processing 2005 (1), 69–73. https://doi.org/10.1155/ASP.2005.69.

Chargé, P., Wang, Y., Sillard, J., 2003. An extended cyclic MUSIC algorithm. IEEE Transactions on Signal Processing 51 (7), 1695–1701.

Chaudhari, S., Cabric, D., 2016. Cyclic weighted centroid algorithm for transmitter localization in the presence of interference. IEEE Transactions on Cognitive Communications and Networking (ISSN 2332-7731) 2 (2), 162–177. https://doi.org/10.1109/TCCN.2016.2586078.

Chaudhari, S., Kosunen, M., Mäkinen, S., Oksanen, J., Laatta, M., Ojaniemi, J., Koivunen, V., Ryynänen, J., Valkama, M., 2016. Performance evaluation of cyclostationary-based cooperative sensing using field measurements. IEEE Transactions on Vehicular Technology (ISSN 0018-9545) 65 (4), 1982–1997. https://doi.org/10.1109/TVT.2015.2422715.

Chaudhari, S., Kosunen, M., Mäkinen, S., Ramanathan, C., Oksanen, J., Laatta, M., Ryynänen, J., Koivunen, V., Valkama, M., 2018. Spatial interpolation of cyclostationary test statistics in cognitive radio networks: methods and field measurements. IEEE Transactions on Vehicular Technology (ISSN 0018-9545) 67 (2), 1113–1129. https://doi.org/10.1109/TVT.2017.2717379.

Chen, C.K., 1989. Spectral Correlation Characterization of Modulated Signals With Application to Signal Detection and Source Location. Ph.D. Dissertation. University of California, Davis, California, Department of Electrical Engineering and Computer Science.

Chen, C.K., Gardner, W.A., 1992. Signal-selective time-difference-of-arrival estimation for passive location of manmade signal sources in highly corruptive environments. Part II: algorithms and performance. IEEE Transactions on Signal Processing 40, 1185–1197.

Chen, Y., Nikias, C.L., 1994. Identifiability of a band limited system from its cyclostationary output autocorrelation. IEEE Transactions on Signal Processing 42 (2), 483–485.

Chen, T., Qiu, L., 1997. Linear periodically time-varying discrete-time systems: aliasing and LTI approximations. Systems & Control Letters 30, 225–235.

Chen, W., Giannakis, G.B., Nandhakumar, N., 1996. Spatio-temporal approach for time-varying global image motion estimation. IEEE Transactions on Image Processing 10, 1448–1461.

Chen, Z.M., Zhu, H.C., Peng, M., 2015. Identification and localization of the sources of cyclostationary sound fields. Applied Acoustics (ISSN 0003-682X) 87, 64–71. https://doi.org/10.1016/j.apacoust.2014.06.013.

Cheong, C., Joseph, P., 2014. Cyclostationary spectral analysis for the measurement and prediction of wind turbine swishing noise. Journal of Sound and Vibration (ISSN 0022-460X) 333 (14), 3153–3176. https://doi.org/10.1016/j.jsv.2014.02.031.

Chérif, F., 2011a. Various types of almost periodic functions on Banach spaces: Part I. International Mathematical Forum 6 (19), 921–952.

Chérif, F., 2011b. Various types of almost periodic functions on Banach spaces: Part II. International Mathematical Forum 6 (20), 953–985.

Chevreuil, A., 1997. Blind Equalization in a Cyclostationary Context. Ph.D. Dissertation. ENST-Telecom Paris, Dept. Signal Processing, Paris, France (in French).

Chevreuil, A., Loubaton, P., 1997. Blind second-order identification of FIR channels: forced cyclostationarity and structured subspace methods. IEEE Signal Processing Letters 4, 204–206.

Chevreuil, A., Loubaton, P., 1999. MIMO blind second-order equalization method and conjugate cyclostationarity. IEEE Transactions on Signal Processing 47 (2), 572–578.

Chevreuil, A., Serpedin, E., Loubaton, P., Giannakis, G.B., 2000. Blind channel identification and equalization using nonredundant periodic modulation precoders: performance analysis. IEEE Transactions on Signal Processing 48 (6), 1570–1586.

Chi, Y., Yang, S., Jiao, W., He, J., Gu, X., Papatheou, E., 2019. Spectral DCS-based feature extraction method for rolling element bearing pseudo-fault in rotor-bearing system. Measurement (ISSN 0263-2241) 132, 22–34. https://doi.org/10.1016/j.measurement.2018.09.006.

Chiu, S.T., 1986. Statistical estimation of the parameters of a moving source form array data. The Annals of Statistics 14 (2), 559–578.

Cho, J.H., 2004. Joint transmitter and receiver optimization in additive cyclostationary noise. IEEE Transactions on Information Theory 50 (12), 3396–3405.

Chopra, R., Ghosh, D., Mehra, D., 2014. Spectrum sensing for cognitive radios based on space-time FRESH filtering. IEEE Transactions on Wireless Communications (ISSN 1536-1276) 13 (7), 3903–3913. https://doi.org/10.1109/TWC.2014.2314125.

Chopra, R., Ghosh, D., Mehra, D.K., 2016. FRESH filter-based spectrum sensing in the presence of cyclic frequency offset. IEEE Wireless Communications Letters (ISSN 2162-2337) 5 (2), 124–127. https://doi.org/10.1109/LWC.2015.2506908.

Chopra, R., Ghosh, D., Mehra, D., 2017. Spectrum sensing for OFDM signals using pilot induced cyclostationarity in the presence of cyclic frequency offset. Physical Communication (ISSN 1874-4907) 24, 182–194. https://doi.org/10.1016/j.phycom.2017.07.008.

Chorna, L.B., 2003. Statistical estimators of a periodically correlated random process for a voiced speech signal. The Journal of the Acoustical Society of America 113 (4), 2271. https://doi.org/10.1121/1.4780527.

Ciblat, P., 2000. Some Estimation Problems Relative to Non Co-Operative Communications. Ph.D. Dissertation. Univ. Marne-la-Vallée, Marne-la-Vallée, France (in French).

Ciblat, P., Serpedin, E., 2003. On a blind fractionally sampling-based carrier frequency offset estimator for noncircular transmissions. IEEE Signal Processing Letters 10 (4), 89–92.

Ciblat, P., Serpedin, E., 2004. A fine blind frequency offset estimator for OFDM/OQAM systems. IEEE Transactions on Signal Processing 52 (1), 291–296.

Ciblat, P., Vandendorpe, L., 2003. Blind carrier frequency offset estimation for noncircular constellation-based transmissions. IEEE Transactions on Signal Processing 51 (5), 1378–1389.

Ciblat, P., Loubaton, P., Serpedin, E., Giannakis, G.B., 2002a. Asymptotic analysis of blind cyclic correlation-based symbol-rate estimators. IEEE Transactions on Information Theory (ISSN 0018-9448) 48 (7), 1922–1934. https://doi.org/10.1109/TIT.2002.1013133.

Ciblat, P., Loubaton, P., Serpedin, E., Giannakis, G.B., 2002b. Performance analysis of blind carrier frequency offset estimators for noncircular transmissions through frequency-selective channels. IEEE Transactions on Signal Processing (ISSN 1053-587X) 50 (1), 130–140. https://doi.org/10.1109/78.972489.

Cioch, W., Knapik, O., Leskow, J., 2013. Finding a frequency signature for a cyclostationary signal with applications to wheel bearing diagnostics. Mechanical Systems and Signal Processing (ISSN 0888-3270) 38 (1), 55–64. https://doi.org/10.1016/j.ymssp.2012.12.013.

Cipra, T., 1985. Periodic moving average process. Applications of Mathematics 30, 218–229.

Cipra, T., Tusty, P., 1987. Estimation in multiple autoregressive-moving average models using periodicity. Journal of Time Series Analysis 8, 293–300.

Claasen, T.A.C.M., Mecklenbräuker, W.F.G., 1982. On stationary linear time-varying systems. IEEE Transactions on Circuits and Systems 29 (3), 169–184.

Clare, L., Rubin, I., 1983. Queueing analysis of TDMA with limited and unlimited buffer capacity. In: Proceedings of IEEE INFOCOM'83. San Diego, CA, pp. 229–238.

Clark, J., Cochran, D., 1993. Time-warped bandlimited signals: sampling, bandlimitedness, and uniqueness of representation. In: Proceedings of 1993 IEEE International Symposium on Information Theory, p. 331.

Cochran, D., Clark, J., 1990. On the sampling and reconstruction of time-warped bandlimited signals. In: IEEE International Conference on Acoustics, Speech, and Signal Processing, vol. 3. ICASSP 1990, pp. 1539–1541.

Cohen, L., 1989. Time-frequency distributions–a review. Proceedings of the IEEE 77 (7), 941–981.

Cohen, L., 1995. Time Frequency Analysis: Theory and Applications. Prentice Hall, Upper Saddle River, NJ.

Cohen, D., Eldar, Y.C., 2017. Sub-Nyquist cyclostationary detection for cognitive radio. IEEE Transactions on Signal Processing (ISSN 1053-587X) 65 (11), 3004–3019. https://doi.org/10.1109/TSP.2017.2684743.

Cohen, D., Rebeiz, E., Eldar, Y., Cabric, D., 2013. Cyclic spectrum reconstruction and cyclostationary detection from sub-Nyquist samples. In: IEEE 14th Workshop on Signal Processing Advances in Wireless Communications. SPAWC, pp. 425–429.

Cohen, D., Tsiper, S., Eldar, Y.C., 2018. Analog-to-digital cognitive radio: sampling, detection, and hardware. IEEE Signal Processing Magazine (ISSN 1053-5888) 35 (1), 137–166. https://doi.org/10.1109/MSP.2017.2740966.

Cooper, J., 1980. Scattering of electromagnetic fields by a moving boundary: the one-dimensional case. IEEE Transactions on Antennas and Propagation AP-28 (6), 791–795.

Corduneanu, C., 1989. Almost Periodic Functions. Chelsea, New York.

Cramér, H., 1940. On the theory of stationary random processes. Annals of Mathematics 1, 215–230.

Crochiere, R.E., Rabiner, L.R., 1981. Interpolation and decimation of digital signals–a tutorial review. Proceedings of the IEEE 69, 300–331.

Cui, J., Falconer, D.D., Sheikh, A.U.H., 1998. Blind adaptation of antenna arrays using a simple algorithm based on small frequency offset. IEEE Transactions on Communications 46 (1), 61–70.

Cui, W., Jiang, H., Li, J., Wang, Q., 2011. A correction of spectral correlation function expression for OFDM signals. IEEE Communications Letters (ISSN 1089-7798) 15 (4), 446–448. https://doi.org/10.1109/LCOMM.2011.012511.102253.

da Silva, C., Choi, B., 2008. Non-asymptotic performance analysis of single-cycle detectors. IEEE Transactions on Wireless Communications (ISSN 1536-1276) 7 (10), 3732–3737. https://doi.org/10.1109/T-WC.2008.070639.

Dabrowski, D., Adamczyk, J., Mora, H.P., Hashemiyan, Z., 2014. Model of the planetary gear based on multi-body method and its comparison with experiment on the basis of gear meshing frequency and sidebands. In: Chaari, F., Leskow, J., Napolitano, A., Sanchez-Ramirez, A. (Eds.), Cyclostationarity: Theory and Methods. In: Lecture Notes in Mechanical Engineering. Springer International Publishing, pp. 157–169.

Dalpiaz, G., Rivola, A., Rubini, R., 2000. Effectiveness and sensitivity of vibration processing techniques for local fault detection in gears. Mechanical Systems and Signal Processing 14 (3), 387–412.

Dandawaté, A.V., 1993. Exploiting Cyclostationarity and Higher-Order Statistics in Signal Processing. Ph.D. Dissertation. University of Virginia, Charlottesville, VA.

Dandawaté, A., Giannakis, G.B., 1993. Nonparametric cyclic-polyspectral analysis of AM signals and processes with missing observations. IEEE Transactions on Information Theory 39 (6), 1864–1876.

Dandawaté, A.V., Giannakis, G.B., 1994a. Nonparametric polyspectral estimators for kth-order (almost) cyclostationary processes. IEEE Transactions on Information Theory 40 (1), 67–84.

Dandawaté, A.V., Giannakis, G.B., 1994b. Statistical tests for presence of cyclostationarity. IEEE Transactions on Signal Processing (ISSN 1053-587X) 42 (9), 2355–2369. https://doi.org/10.1109/78.317857.

Dandawaté, A.V., Giannakis, G.B., 1995. Asymptotic theory of mixed time averages and kth-order cyclic-moment and cumulant statistics. IEEE Transactions on Information Theory 41 (1), 216–232.

Dandawaté, A., Giannakis, G.B., 1996. Modeling (almost) periodic moving average processes using cyclic statistics. IEEE Transactions on Signal Processing 44 (3), 673–684.

Daponte, P., Vito, L.D., Picariello, F., Rapuano, S., 2018. A cyclostationarity-based method for jitter measurement in waveform recorders. IEEE Transactions on Instrumentation and Measurement (ISSN 0018-9456) 67 (8), 1786–1794. https://doi.org/10.1109/TIM.2018.2806022.

D'Arco, M., De Vito, L., 2016. A novel method for phase noise measurement based on cyclic complementary autocorrelation. IEEE Transactions on Instrumentation and Measurement (ISSN 0018-9456) 65 (12), 2685–2692. https://doi.org/10.1109/TIM.2016.2610138.

Datta, J., Lin, H., 2018. Cyclostationarity-based joint sensing and equalisation of fast convolution-based DWPT block-filtered orthogonal frequency division multiplexing for fifth-generation wireless systems. IET Communications (ISSN 1751-8628) 12 (19), 2460–2469. https://doi.org/10.1049/iet-com.2018.5455.

Davis, B.J., 2007. Observable coherence theory for statistically periodic fields. Physical Review A 76, 043843. https://doi.org/10.1103/PhysRevA.76.043843.

De Angelis, V., Izzo, L., Napolitano, A., Tanda, M., 2005. Cyclostationarity-based parameter estimation of wide-band signals in mobile communications. In: IEEE Statistical Signal Processing Workshop. SSP'05, Bordeaux, France.

De Nicolao, G., Ferrari-Trecate, G., 1999. On the Wold decomposition of discrete-time cyclostationary processes. IEEE Transactions on Signal Processing 47 (7), 2041–2043.

Dehay, D., 1991. Contributions to the Spectral Analysis of Nonstationary Processes. Dissertation for the Habilitation to Direct Research. Universite des Sciences et Techniques de Lille Flandres Artois.

Dehay, D., 1994. Spectral analysis of the covariance of the almost periodically correlated processes. Stochastic Processes and Their Applications 50, 315–330.

Dehay, D., 1995. Asymptotic behavior of estimators of cyclic functional parameters for some nonstationary processes. Statistics & Decisions 13, 273–286.

Dehay, D., 2007. Discrete time observation of almost periodically correlated processes and jitter phenomena. In: XV European Signal Processing Conference. EUSIPCO 2007, Poznan, Poland.

Dehay, D., 2015a. Limiting distributions for explosive PAR(1) time series with strongly mixing innovation. In: Chaari, F., Leskow, J., Napolitano, A., Zimroz, R., Wylomanska, A., Dudek, A. (Eds.), Cyclostationarity: Theory and Methods – II, Applied Condition Monitoring, vol. 3. Springer International Publishing, pp. 105–129.

Dehay, D., 2015b. Parameter maximum likelihood estimation problem for time periodic modulated drift Ornstein Uhlenbeck processes. Statistical Inference for Stochastic Processes (ISSN 1387-0874) 18 (1), 69–98. https://doi.org/10.1007/s11203-014-9104-7.

Dehay, D., Dudek, A.E., 2015. Block bootstrap for Poisson-sampled almost periodic processes. Journal of Time Series Analysis (ISSN 1467-9892) 36 (3), 327–351. https://doi.org/10.1111/jtsa.12115.

Dehay, D., Hurd, H.L., 1994. Representation and estimation for periodically and almost periodically correlated random processes. In: Gardner, W.A. (Ed.), Cyclostationarity in Communications and Signal Processing. IEEE Press, pp. 295–328.

Dehay, D., Hurd, H., 2002. Spectral estimation for strongly periodically correlated random fields defined on R^2. Mathematical Methods of Statistics 11 (2), 135–151.

Dehay, D., Hurd, H.L., 2013. Empirical determination of the frequencies of an almost periodic time series. Journal of Time Series Analysis (ISSN 1467-9892) 34 (2), 262–279. https://doi.org/10.1111/jtsa.12009.

Dehay, D., Leśkow, J., 1996a. Functional limit theory for the spectral covariance estimator. Journal of Applied Probability 33, 1077–1092.

Dehay, D., Leśkow, J., 1996b. Testing stationarity for stock market data. Economics Letters 50, 205–212.

Dehay, D., Monsan, V., 1996. Random sampling estimation for almost-periodically correlated processes. Journal of Time Series Analysis 17 (5), 425–445.

Dehay, D., Monsan, V., 2007. Discrete periodic sampling with jitter and almost periodically correlated processes. Statistical Inference for Stochastic Processes (ISSN 1387-0874) 10 (3), 223–253. https://doi.org/10.1007/s11203-006-0004-3.

Dehay, D., Waled, K.E., 2013. Nonparametric estimation problem for a time-periodic signal in a periodic noise. Statistics & Probability Letters (ISSN 0167-7152) 83 (2), 608–615. https://doi.org/10.1016/j.spl.2012.11.008.

Dehay, D., Leśkow, J., Napolitano, A., 2013. Central limit theorem in the functional approach. IEEE Transactions on Signal Processing (ISSN 1053-587X) 61 (16), 4025–4037. https://doi.org/10.1109/TSP.2013.2266324.

Dehay, D., Dudek, A., Leskow, J., 2014. Subsampling for continuous-time almost periodically correlated processes. Journal of Statistical Planning and Inference (ISSN 0378-3758) 150, 142–158. https://doi.org/10.1016/j.jspi.2014.02.012.

Dehay, D., Dudek, A.E., El Badaoui, M., 2018a. Bootstrap for almost cyclostationary processes with jitter effect. Digital Signal Processing (ISSN 1051-2004) 73, 93–105. https://doi.org/10.1016/j.dsp.2017.11.002.

Dehay, D., Leśkow, J., Napolitano, A., 2018b. Time average estimation in the fraction-of-time probability framework. Signal Processing 153, 275–290. https://doi.org/10.1016/j.sigpro.2018.07.005.

Delvecchio, S., Antoni, J., 2012. Extraction of angle deterministic signals in the presence of stationary speed fluctuations with cyclostationary blind source separation. Mechanical Systems and Signal Processing (ISSN 0888-3270) 27, 350–361. https://doi.org/10.1016/j.ymssp.2011.09.009.

Delvecchio, S., D'Elia, G., Dalpiaz, G., 2015. On the use of cyclostationary indicators in IC engine quality control by cold tests. Mechanical Systems and Signal Processing (ISSN 0888-3270) 60–61, 208–228. https://doi.org/10.1016/j.ymssp.2014.09.015.

Demorest, P.B., 2011. Cyclic spectral analysis of radio pulsars. Monthly Notices of the Royal Astronomical Society (ISSN 1365-2966) 416 (4), 2821–2826. https://doi.org/10.1111/j.1365-2966.2011.19230.x.

Derakhshani, M., Le-Ngoc, T., Nasiri-Kenari, M., 2011. Efficient cooperative cyclostationary spectrum sensing in cognitive radios at low SNR regimes. IEEE Transactions on Wireless Communications (ISSN 1536-1276) 10 (11), 3754–3764. https://doi.org/10.1109/TWC.2011.080611.101580.

Dimc, F., Baldini, G., Kandeepan, S., 2015. Experimental detection of mobile satellite transmissions with cyclostationary features. International Journal of Satellite Communications and Networking (ISSN 1542-0981) 33 (2), 163–183. https://doi.org/10.1002/sat.1081.

Ding, Z., 1994. Blind channel identification and equalization using spectral correlation measurements, Part I: frequency-domain analysis. In: Gardner, W.A. (Ed.), Cyclostationarity in Communications and Signal Processing. IEEE Press, pp. 417–436.

Ding, H., Fahmy, M.M., 1990. Inverse of linear periodically time-varying filtering. In: Proceedings of International Conference on Acoustic, Speech, and Signal Processing, vol. 3. ICASSP'90, Albuquerque, NM, pp. 1217–1220.

Ding, H., Fahmy, M.M., 1992. Synchronised linear almost periodically time-varying adaptive filters. IEE Proceedings. Part I 139 (24), 429–436.

Ding, H., Fahmy, M.M., 1993. Performance of output-modulator-structured linear almost periodically time-varying adaptive filters. IEE Proceedings. Part I 140 (2), 114–120.

Ding, Z., Thomas, P.H., Wang, R., Tong, W., 2002. Co-channel interference reduction in wireless communications systems. U.S. Patent No. 6,396,885.

Dmochowski, J., Benesty, J., Affes, S., 2009. On spatial aliasing in microphone arrays. IEEE Transactions on Signal Processing 57 (4), 1383–1395.

Dobre, O.A., 2015. Signal identification for emerging intelligent radios: classical problems and new challenges. IEEE Instrumentation & Measurement Magazine (ISSN 1094-6969) 18 (2), 11–18. https://doi.org/10.1109/MIM.2015.7066677.

Dobre, O., Abdi, A., Bar-Ness, Y., Su, W., 2007. Survey of automatic modulation classification techniques: classical approaches and new trends. IET Communications (ISSN 1751-8628) 1 (2), 137–156. https://doi.org/10.1049/iet-com:20050176.

Dobre, O., Rajan, S., Inkol, R., 2009. Joint signal detection and classification based on first-order cyclostationarity for cognitive radios. EURASIP Journal on Advances in Signal Processing (ISSN 1687-6180) 2009 (1), 656719. https://doi.org/10.1155/2009/656719.

Dobre, O., Oner, M., Rajan, S., Inkol, R., 2012. Cyclostationarity-based robust algorithms for QAM signal identification. IEEE Communications Letters (ISSN 1089-7798) 16 (1), 12–15. https://doi.org/10.1109/LCOMM.2011.112311.112006.

Dominique, F., Reed, J.H., 1997. Estimating spectral correlations using the least mean square algorithm. Electronics Letters 33 (3), 182–184.

Donohue, K.D., Cheah, H.Y., 1995. Spectral correlation filters for flaw detection. In: Proceedings of 1995 Ultrasonics Symposium. Seattle, WA, pp. 725–728.

Donohue, K.D., Huang, L., 2001. Ultrasonic scatterer structure classification with the generalized spectrum. In: Proceedings of the IEEE International Conference on Acoustics, Speech and Signal Processing. Salt Lake City, UT.

Donohue, K.D., Bressler, J., Varghese, T., Bilgutay, N.M., 1993a. Spectral correlation in ultrasonic pulse-echo signal processing. IEEE Transactions on Ultrasonics, Ferroelectrics, and Frequency Control 40 (3), 330–337.

Donohue, K.D., Varghese, T., Bilgutay, N.M., 1993b. Spectral redundancy in characterizing scatterer structures from ultrasonic echoes. In: Review of Progress in Quantitative Non-Destructive Evaluation. Brunswick, Maine, pp. 951–958.

Donohue, K.D., Varghese, T., Forsberg, F., Halpern, E.J., 1998. The analysis and classification of small-scale tissue structures using the generalized spectrum. In: Proceedings of the International Congress on Acoustics and Acoustical Society of America. Seattle, WA, pp. 2685–2686.

Donohue, K.D., Forsberg, F., Piccoli, C.W., Goldberg, B.B., 1999a. Analysis and classification of tissue with scatterer structure templates. IEEE Transactions on Ultrasonics, Ferroelectrics, and Frequency Control 46 (2), 300–310.

Donohue, K.D., Haung, L., Genis, V., Foresberg, F., 1999b. Duct size estimation in breast tissue. In: Proceedings of 1999 Ultrasonics Symposium. Reno, NV, pp. 1353–1356.

Donohue, K.D., Huang, L., Burks, T., Forsberg, F., Piccoli, C.W., 2001. Tissue classification with generalized spectrum parameters. Ultrasound in Medicine & Biology 27 (11), 1505–1514.

Donohue, K.D., Huang, L., Georgiou, G., Cohen, F.S., Piccoli, C.W., Forsberg, F., 2003. Malignant and benign breast tissue classification performance using a scatterer structure preclassifier. IEEE Transactions on Ultrasonics, Ferroelectrics, and Frequency Control 50 (6), 724–729.

Doob, J.L., 1953. Stochastic Processes. John Wiley & Sons, Inc., New York.

Dorogovtsev, A.Y., 1990a. Necessary and sufficient conditions for existence of stationary and periodic solutions of a stochastic difference equation in Hilbert space. Computers & Mathematics with Applications (ISSN 0898-1221) 19 (1), 31–37.

Dorogovtsev, A.Y., 1990b. Stationary and periodic solutions of a stochastic difference equation in a Banach space. Teoriâ Veroâtnostej I Matematičeskoj Statistiki (ISSN 0131-6982) 42, 35–42.

Dorogovtsev, A.Y., 1991. Stationary and periodic solutions of stochastic difference and differential equations in Banach space. In: New Trends in Probability and Statistics, vol. 1. Bakuriani, 1990. VSP, Utrecht, pp. 375–390.

Dorogovtsev, A.Y., 2002. Periodicity in distribution. I. Discrete systems. International Journal of Mathematics and Mathematical Sciences 30 (2), 65–127.

Dorogovtsev, A.Y., Tkhuan, L., 1988. The existence of periodic and stationary regimes of discrete dynamical systems in a Banach space. Kibernetika (Kiev) (ISSN 0023-1274) 6, 121–123. 136.

Dowell, J., Weiss, S., Hill, D., Infield, D., 2013. A cyclo-stationary complex multichannel Wiener filter for the prediction of wind speed and direction. In: Proceedings of the 21st European Signal Processing Conference. EUSIPCO 2013.

Dragan, Y.P., 1969a. Expansion of random processes and their noncommutative transformations. Otbor i Peredacha Informacii 22, 22–27 (in Russian).

Dragan, Y.P., 1969b. Periodically correlated random processes and transformations with periodically varying parameters. Otbor i Peredacha Informacii 22, 27–33 (in Russian).

Dragan, Y.P., 1971. The spectral properties of periodically correlated stochastic processes. Otbor i Peredacha Informacii 30, 16–24 (in Russian).

Dragan, Y.P., 1972a. On foundations of the stochastic model of rhythmic phenomena. Otbor i Peredacha Informacii 31, 21–27 (in Russian).

Dragan, Y.P., 1972b. Properties of counts of periodically correlated random processes. Otbor i Peredacha Informacii 33, 9–12.

Dragan, Y.P., 1975. The representation of a periodically correlated random process by stationary components. Otbor i Peredacha Informacii 45, 7–20 (in Russian).

Dragan, Y.P., 1978. Harmonizability and spectral distribution of random processes with finite mean power. Doklady Akademii Nauk Ukrainskoj SSR 8, 679–684 (in Russian).

Dragan, Y.P., 1980. Structure and Representations of Models of Stochastic Signals, Part 4. Naukova Dumka, Kiev (in Russian).

Dragan, Y.P., 1985. Periodic and periodically nonstationary random processes. Otbor i Peredacha Informacii 72, 3–17 (in Russian).

Dragan, Y.P., 1988. Principles of a linear theory of stochastic test signals and their statistical analysis. Otbor i Peredacha Informacii 77, 2–7 (in Russian).

Dragan, Y.P., Yavorskii, I.N., 1974a. Hydroacoustic communication channels with surface scattering represented in terms of linear time-varying filters. Otbor i Peredacha Informacii 39, 23–31 (in Russian).

Dragan, Y.P., Yavorskii, I.N., 1974b. A paradox of the rhythmic model. Otbor i Peredacha Informacii 41, 11–15 (in Russian).

Dragan, Y.P., Yavorskii, I.N., 1982a. The periodic correlation-random field as a model for bidimensional ocean waves. Otbor i Peredacha Informacii 51, 15–21 (in Russian).

Dragan, Y.P., Yavorskii, I.N., 1982b. Rhythmics of Sea Waves and Underwater Acoustic Signals. Naukova Dumka, Kiev (in Russian).

Dragan, Y.P., Yavorskii, I.N., 1985. Statistical analysis of periodic random processes. Otbor i Peredacha Informacii 71, 20–29 (in Russian).

Dragan, Y.P., Yavorskii, B.I., 1986. A model of the test signal noise in channels with digital signal processing. Radioelectronics and Communications Systems 29, 17–21.

Dragan, Y.P., Mezentsev, V.P., Yavorskii, I.N., 1979. Algorithm and program for calculating estimates of the characteristics of periodically correlated random processes. Otbor i Peredacha Informacii 58, 10–15 (in Russian).

Dragan, O.P., Dragan, Y.P., Karavan, Y.V., 1980a. A probabilistic model of rhythmicity in the radiolysis of solids. Otbor i Peredacha Informacii 62, 26–31 (in Russian).

Dragan, Y.P., Mezentsev, V.P., Yavorskii, I.N., 1980b. The problem of verification of a stochastic rhythmicity model. Otbor i Peredacha Informacii 59, 3–12 (in Russian).

Dragan, Y.P., Mezentsev, V.P., Yavorskii, I.N., 1982. Symmetry of the covariance matrix of measurements of a periodically correlated random process. Otbor i Peredacha Informacii 66, 3–6 (in Russian).

Dragan, Y.P., Mezentsev, V.P., Yavorskii, I.N., 1984a. Computation of estimates for spectral characteristics of periodically correlated random processes. Otbor i Peredacha Informacii 69, 29–35 (in Russian).

Dragan, Y.P., Rozhkov, V.A., Yavorskii, I.N., 1984b. Applications of the theory of periodically correlated random processes to the probabilistic analysis of oceanological time series. In: Rozhkov, V.A. (Ed.), Probabilistic Analysis and Modeling of Oceanological Processes. Gidrometeoizdat, Leningrad, USSR, pp. 4–23.

Dragan, Y.P., Rozhkov, V.A., Yavorskii, I.N., 1987. Methods of Probabilistic Analysis of Oceanological Rhythmics. Gidrometeoizdat, Leningrad, USSR (in Russian).

Drake, C., Knapik, O., Leskow, J., 2014. EM-based inference for cyclostationary time series with missing observations. In: Chaari, F., Leskow, J., Napolitano, A., Sanchez-Ramirez, A. (Eds.), Cyclostationarity: Theory and Methods. In: Lecture Notes in Mechanical Engineering. Springer International Publishing, pp. 23–35.

Drake, C., Leskow, J., Garay, A.M., 2015. Imputation of missing observations for heavy tailed cyclostationary time series. In: Chaari, F., Leskow, J., Napolitano, A., Zimroz, R., Wylomanska, A., Dudek, A. (Eds.), Cyclostationarity: Theory and Methods – II, Applied Condition Monitoring, vol. 3. Springer International Publishing, pp. 179–187.

Du, K.L., Mow, W.H., 2010. Affordable cyclostationarity-based spectrum sensing for cognitive radio with smart antennas. IEEE Transactions on Vehicular Technology (ISSN 0018-9545) 59 (4), 1877–1886. https://doi.org/10.1109/TVT.2010.2043860.

Du, K.L., Mow, W.H., 2013. Exploiting multiple antennas for spectrum sensing in cognitive radio networks. U.S. Patent No. 8,422,958.

Dubois, D.F., Besserides, B., 1976. Nonlinear theory of parametric instabilities in plasmas. Physical Review A 14, 1869–1882.

Dubovetska, I., Moklyachuk, M., 2013. Minimax estimation problem for periodically correlated stochastic processes. Journal of Mathematics and System Science 3, 26–30.

Dubovetska, I., Moklyachuk, M., 2014. Extrapolation of periodically correlated stochastic processes observed with noise. Theory of Probability and Mathematical Statistics 88, 67–83.

Dubovetska, I., Masyutka, O., Moklyachuk, M., 2014. Filtering problems for periodically correlated isotropic random fields. Mathematics and Statistics 2 (4), 162–171. https://doi.org/10.13189/ms.2014.020402.

Dubovetska, I., Masyutka, O., Moklyachuk, M., 2015. Estimation problems for periodically correlated isotropic random fields. Methodology and Computing in Applied Probability (ISSN 1387-5841) 17 (1), 41–57. https://doi.org/10.1007/s11009-013-9339-6.

Dudek, A., 2015. Circular block bootstrap for coefficients of autocovariance function of almost periodically correlated time series. Metrika (ISSN 0026-1335) 78 (3), 313–335. https://doi.org/10.1007/s00184-014-0505-9.

Dudek, A.E., 2018. Block bootstrap for periodic characteristics of periodically correlated time series. Journal of Nonparametric Statistics 30 (1), 87–124. https://doi.org/10.1080/10485252.2017.1404060.

Dudek, A., Leskow, J., 2011. A bootstrap algorithm for data from a periodic multiplicative intensity function. Communications in Statistics. Theory and Methods 40 (8), 1468–1489. https://doi.org/10.1080/03610921003615849.

Dudek, A.E., Potorski, P., 2014. Simulation comparison of CBB and GSBB in overall mean estimation problem for PC time series. In: Chaari, F., Leskow, J., Napolitano, A., Sanchez-Ramirez, A. (Eds.), Cyclostationarity: Theory and Methods. In: Lecture Notes in Mechanical Engineering. Springer International Publishing, pp. 95–106.

Dudek, A.E., Uzar, J., 2015. Simulation study of performance of MBB in overall mean estimation problem for APC time series. In: Chaari, F., Leskow, J., Napolitano, A., Zimroz, R., Wylomanska, A., Dudek, A. (Eds.), Cyclostationarity: Theory and Methods – II, Applied Condition Monitoring, vol. 3. Springer International Publishing, pp. 1–18.

Dudek, A.E., Hurd, H., Wojtowicz, W., 2014a. Bootstrap for maximum likelihood estimates of PARMA coefficients. In: Chaari, F., Leskow, J., Napolitano, A., Sanchez-Ramirez, A. (Eds.), Cyclostationarity: Theory and Methods. In: Lecture Notes in Mechanical Engineering. Springer International Publishing, pp. 15–22.

Dudek, A.E., Leskow, J., Paparoditis, E., Politis, D.N., 2014b. A generalized block bootstrap for seasonal time series. Journal of Time Series Analysis (ISSN 1467-9892) 35 (2), 89–114. https://doi.org/10.1002/jtsa.12053.

Dudek, A.E., Maiz, S., El Badaoui, M., 2014c. Generalized seasonal block bootstrap in frequency analysis of cyclostationary signals. Signal Processing (ISSN 0165-1684) 104, 358–368. https://doi.org/10.1016/j.sigpro.2014.04.022.

Dudek, A.E., Hurd, H., Wójtowicz, W., 2015. PARMA models with applications in R. In: Chaari, F., Leskow, J., Napolitano, A., Zimroz, R., Wylomanska, A., Dudek, A. (Eds.), Cyclostationarity: Theory and Methods – II, Applied Condition Monitoring, vol. 3. Springer International Publishing, pp. 131–153.

Duel-Hallen, A., 1992. Equalizers for multiple input/multiple output channels and PAM systems with cyclostationary input sequence. IEEE Journal on Selected Areas in Communications 10 (3), 630–639.

Dunsmuir, W., 1984. Time Series Regression With Periodically Correlated Errors and Missing Data, Time Series Analysis of Irregularly Observed Data. Springer, New York-Berlin, pp. 78–107.

Dusserre-Telmon, G.F.P., Flores, D., Prieux, F., 2001. Damage detection of motor pieces. European Patent No. 1111364.

Dusserre-Telmon, G.F.P., Flores, D., Prieux, F., 2002. Process for the detection of damage to components of an engine. U.S. Patent No. 6,389,887.

Duverdier, A., Lacaze, B., 2000. On the use of periodic clock changes to implement linear periodic time-varying filters. IEEE Transactions on Circuits and Systems. 2, Analog and Digital Signal Processing 47 (11), 1152–1159.

Duverdier, A., Lacaze, B., Roviras, D., 1999. Introduction of linear cyclostationary filters to model time-variant channels. In: Proceedings of Global Telecommunications Conference. GLOBECOM'99, pp. 325–329.

Eberlein, W.F., 1949. Abstract ergodic theorems and weak almost periodic functions. Transactions of the American Mathematical Society 67, 217–240.

Eberlein, W.F., 1956. The point spectrum of weakly almost periodic functions. Michigan Journal of Mathematics 3, 137–139.

Edlund, J.A., Tinto, M., Królak, A., Nelemans, G., 2005. White-dwarf–white-dwarf galactic background in the LISA data. Physical Review. D 71, 122003. https://doi.org/10.1103/PhysRevD.71.122003.

Efimov, E., Shevgunov, T., Kuznetsov, Y., 2018. Time delay estimation of cyclostationary signals on PCB using spectral correlation function. In: 2018 Baltic URSI Symposium. URSI, pp. 184–187.

El Waled, K., 2015. Parametric estimation problem for a time periodic signal in a periodic noise. In: Chaari, F., Leskow, J., Napolitano, A., Zimroz, R., Wylomanska, A., Dudek, A. (Eds.), Cyclostationarity: Theory and Methods – II, Applied Condition Monitoring, vol. 3. Springer International Publishing, pp. 19–41.

Elgenedy, M., Sayed, M., Al-Dhahir, N., Chabaan, R.C., 2018. Cyclostationary noise mitigation for SIMO powerline communications. IEEE Access (ISSN 2169-3536) 6, 5460–5484. https://doi.org/10.1109/ACCESS.2017.2789185.

Elmirghani, J.M.H., Cryan, R.A., Clayton, F.M., 1995. Spectral analysis of timing jitter effects on the cyclostationary PPM format. Signal Processing 43 (3), 269–277.

Enserink, S., Cochran, D., 1995. On detection of cyclostationary signals. In: 1995 International Conference on Acoustics, Speech, and Signal Processing, vol. 3, pp. 2004–2007.

Ericson, T.H.E., 1981. Modulation by means of linear periodic filtering. IEEE Transactions on Information Theory IT-27, 322–327.

Esman, D.J., Ataie, V., Kuo, B.P., Temprana, E., Alic, N., Radic, S., 2017. Comb-assisted cyclostationary analysis of wideband RF signals. Journal of Lightwave Technology (ISSN 0733-8724) 35 (17), 3705–3712. https://doi.org/10.1109/JLT.2017.2715336.

Estupiñan, E., White, P., San Martin, C., 2007. A cyclostationary analysis applied to detection and diagnosis of faults in helicopter gearboxes. In: Rueda, L., Mery, D., Kittler, J. (Eds.), Progress in Pattern Recognition, Image Analysis and Applications. In: Lecture Notes in Computer Science, vol. 4756. Springer, Berlin, Heidelberg, pp. 61–70.

Eweda, E., Bershad, N.J., 2017. Stochastic analysis of the signed LMS algorithms for cyclostationary white Gaussian inputs. IEEE Transactions on Signal Processing (ISSN 1053-587X) 65 (7), 1673–1684. https://doi.org/10.1109/TSP.2016.2646666.

Eweda, E., Bershad, N.J., Bermudez, J.C.M., 2018. Stochastic analysis of the LMS and NLMS algorithms for cyclostationary white Gaussian and non-Gaussian inputs. IEEE Transactions on Signal Processing (ISSN 1053-587X) 66 (18), 4753–4765. https://doi.org/10.1109/TSP.2018.2860552.

Fehske, A., Gaeddert, J., Reed, J., 2005. A new approach to signal classification using spectral correlation and neural networks. In: First IEEE International Symposium on New Frontiers in Dynamic Spectrum Access Networks. DySPAN 2005, pp. 144–150.

Feng, Z., Zuo, M., Hao, R., Chu, F., El Badaoui, M., 2011. Gear damage assessment based on cyclic spectral analysis. IEEE Transactions on Reliability (ISSN 0018-9529) 60 (1), 21–32. https://doi.org/10.1109/TR.2010.2104017.

Ferguson, B.G., 1999. Time-delay estimation techniques applied to the acoustic detection of jet aircraft transit. The Journal of the Acoustical Society of America 106 (1), 225–264.

Fernandez, B., Salas, J.D., 1986. Periodic gamma autoregressive processes for operational hydrology. Water Resources Research 22, 1385–1396.

Fernández-Pousa, C.R., 2013. Nonstationary elementary-field light randomly triggered by Poisson impulses. Journal of the Optical Society of America. A 30 (5), 932–940. https://doi.org/10.1364/JOSAA.30.000932.

Ferrara Jr., E.R., 1985. Frequency-domain implementations of periodically time-varying filters. IEEE Transactions on Acoustics, Speech, and Signal Processing ASSP-33, 883–892.

Ferrara Jr., E.R., Widrow, B., 1981. The time-sequenced adaptive filter. IEEE Transactions on Acoustics, Speech, and Signal Processing ASSP-29, 679–683.

Ferreol, A., 2002. Process of cyclic detection in diversity of polarization of digital cyclostationary radioelectric signals. U.S. Patent No. 6,430,239.

Ferréol, A., Chevalier, P., 2000. On the behavior of current second and higher order blind source separation methods for cyclostationary sources. IEEE Transactions on Signal Processing 48 (6), 1712–1725.

Ferréol, A., Chevalier, P., Albera, L., 2004. Second-order blind separation of first- and second-order cyclostationary sources–application to AM, FSK, CPFSK, and deterministic sources. IEEE Transactions on Signal Processing 52 (4), 845–861.

Ferreol, A., Chevalier, P., Albera, L., 2008. Antenna processing method for potentially non-centered cyclostationary signals. U.S. Patent No. 7,453,399.

Fijalkow, I., Lopez de Victoria, F., Johnson, C., 1994. Adaptive fractionally spaced blind equalization. In: Proceedings of Digital Signal Processing Workshop.

Finelli, C.J., Jenkins, J.M., 1991. A cyclostationary least mean squares algorithm for discrimination of ventricular tachycardia from sinus rhythm. In: Proceedings of the Annual International Conference of the IEEE Engineering in Medicine and Biology Society, vol. 13.

Firat, U., Akgul, T., 2018. Compressive sensing for detecting ships with second-order cyclostationary signatures. IEEE Journal of Oceanic Engineering (ISSN 0364-9059) 43 (4), 1086–1098. https://doi.org/10.1109/JOE.2017.2740698.

Flagiello, F., Izzo, L., Napolitano, A., 2000. A computationally efficient and interference tolerant nonparametric algorithm for LTI system identification based on higher order cyclostationarity. IEEE Transactions on Signal Processing (ISSN 1053-587X) 48 (4), 1040–1052. https://doi.org/10.1109/78.827538.

Flandrin, P., 1989. On the spectrum of fractional Brownian motions. IEEE Transactions on Information Theory (ISSN 0018-9448) 35 (1), 197–199. https://doi.org/10.1109/18.42195.

Flandrin, P., 1999. Time-Frequency/Time-Scale Analysis. Academic Press, San Diego, CA.

Flandrin, P., Napolitano, A., Ozaktas, H.M., Thompson, D.J., 2011. Recent advances in theory and methods for nonstationary signal analysis (editorial of the special issue). EURASIP Journal on Advances in Signal Processing 2011, 963642. https://doi.org/10.1155/2011/963642.

Florian, E., Chevreuil, A., Loubaton, P., 2012. Blind source separation of convolutive mixtures of non-circular linearly modulated signals with unknown baud rates. Signal Processing (ISSN 0165-1684) 92 (3), 715–726. https://doi.org/10.1016/j.sigpro.2011.09.013.

Fontes, A.I., Rego, J.B., de M. Martins, A., Silveira, L.F., Principe, J., 2017a. Corrigendum to "Cyclostationary correntropy: Definition and applications" [Expert Systems with Applications 69 (2017) 110–117]. Expert Systems with Applications (ISSN 0957-4174) 81, 472–473. https://doi.org/10.1016/j.eswa.2017.02.001.

Fontes, A.I., Rego, J.B., de M. Martins, A., Silveira, L.F., Principe, J., 2017b. Cyclostationary correntropy: definition and applications. Expert Systems with Applications (ISSN 0957-4174) 69, 110–117. https://doi.org/10.1016/j.eswa.2016.10.029.

Franaszek, P.A., 1967. On linear systems which preserve wide sense stationarity. SIAM Journal on Applied Mathematics 15, 1481–1484.

Franaszek, P.A., Liu, B., 1967. On a class of linear time-varying filters. IEEE Transactions on Information Theory IT-13, 477–481.

Francis, B., Dasgupta, S., 1999. Signal compression by subband coding. Automatica 35, 1895–1908.

Franks, L.E., 1969. Signal Theory. Prentice-Hall, Englewood Cliffs, NJ (Chapter 8).

Franks, L.E., 1980. Carrier and bit synchronization in data communication – a tutorial review. IEEE Transactions on Communications COM-28, 1107–1121.

Franks, L.E., 1994. Polyperiodic linear filtering. In: Gardner, W.A. (Ed.), Cyclostationarity in Communications and Signal Processing. IEEE Press, pp. 240–266.

Franks, L.E., 1997. Random processes, autocorrelation, and spectral densities. In: Gibson, J. (Ed.), The Communications Handbook. CRC Press and IEEE Press, New York.

Franks, L.E., Bubrouski, J., 1974. Statistical properties of timing jitter in a PAM timing recovery scheme. IEEE Transactions on Communications COM-22, 913–930.

Franses, P.H., 1995. The effects of seasonally adjusting a periodic autoregressive process. Computational Statistics & Data Analysis 19, 683–704.

Franses, P.H., McAleer, M., 1998. Cointegration analysis of seasonal time series. Journal of Economics Survey 12, 651–678.

Franses, P.H., Paap, R., 1994. Model selection in periodic autoregressions. Oxford Bulletin of Economics and Statistics 56, 421–439.

Franz, S., Mitra, S.K., Doblinger, G., 2003. Frequency estimation using warped discrete Fourier transform. Signal Processing 83, 1661–1671.

French, C.A., Gardner, W.A., 1986. Despreading spread-spectrum signals without the code. IEEE Transactions on Communications COM-34, 404–407.

Friedman, J.S., King, J.P., Pride III, J.P., 1989. Time difference of arrival geolocation method, etc. U.S. Patent No. 4,888,593.

Fu, H., Abeywickrama, S., Zhang, L., Yuen, C., 2018. Low-complexity portable passive drone surveillance via SDR-based signal processing. IEEE Communications Magazine (ISSN 0163-6804) 56 (4), 112–118. https://doi.org/10.1109/MCOM.2018.1700424.

Fung, A., Lee, L.S., Falconer, D.D., 1989. A facility for near end crosstalk measurements on ISDN subscriber loops. In: Proceedings of GLOBECOM '89, IEEE Global Telecommunications Conference and Exhibition, vol. 3. Dallas, TX, pp. 1592–1596.

Fusco, T., Izzo, L., Napolitano, A., Tanda, M., 2006. On the second-order cyclostationarity properties of long-code DS-SS signals. IEEE Transactions on Communications (ISSN 0090-6778) 54 (10), 1741–1746. https://doi.org/10.1109/TCOMM.2006.881353.

Gajecka-Mirek, E., 2014. Subsampling for weakly dependent and periodically correlated sequences. In: Chaari, F., Leskow, J., Napolitano, A., Sanchez-Ramirez, A. (Eds.), Cyclostationarity: Theory and Methods. In: Lecture Notes in Mechanical Engineering. Springer International Publishing, pp. 37–51.

Gameiro, A., 2002. Simple receiver for systems with spectrally overlapping narrowband and broadband signals. Wireless Personal Communications 23, 311–320.

Garakyaragh, N., Shahtalebi, K., Forouzan, A.R., 2016. A cyclic frequency tracking loop for wideband spectrum sensing and FM demodulation. IEEE Communications Letters (ISSN 1089-7798) 20 (10), 2015–2018. https://doi.org/10.1109/LCOMM.2016.2591938.

Garcia, F.M., Lourtie, I.M.G., Buescu, J., 2001. $l^2(\mathbb{R})$ nonstationary processes and the sampling theorem. IEEE Signal Processing Letters 8 (4), 117–119.

Garcia-Vega, S., Alvarez-Meza, A., Castellanos-Dominguez, G., 2015. Time-series prediction based on kernel adaptive filtering with cyclostationary codebooks. In: Paredes, R., Cardoso, J.S., Pardo, X.M. (Eds.), Pattern Recognition and Image Analysis. In: Lecture Notes in Computer Science, vol. 9117, pp. 354–361.

Gardner, W.A., 1972a. Representation and Estimation of Cyclostationary Processes. Ph.D. Dissertation. University of Massachusetts, Department of Electrical and Computer Engineering (reprinted as Signal and Image Processing Lab Technical Report No. SIPL-82-1, Department of Electrical and Computer Engineering, University of California, Davis, California, 95616, 1982).

Gardner, W.A., 1972b. A sampling theorem for nonstationay random processes. IEEE Transactions on Information Theory IT-18, 808–809.

Gardner, W.A., 1973. The structure of least-mean-square linear estimators for synchronous M-ary signals. IEEE Transactions on Information Theory IT-19, 240–243.

Gardner, W.A., 1976. An equivalent linear model for marked and filtered doubly stochastic Poisson processes with application to MMSE estimation for synchronous M-ary optical data signals. IEEE Transactions on Communication Technology COM-24, 917–921.

Gardner, W.A., 1978. Stationarizable random processes. IEEE Transactions on Information Theory IT-24, 8–22.

Gardner, W.A., 1980. Cyclo-ergodicity, Part I: application to spectral extraction. In: Lainiotis, D.G., Tzannes, N.S. (Eds.), Advances in Communications. D. Reidel, Holland, pp. 203–209.

Gardner, W.A., 1982. Structural characterization of locally optimum detectors in terms of locally optimum estimators and correlators. IEEE Transactions on Information Theory IT-28, 924–932.

Gardner, W.A., 1983. Non-parametric signal detection and identification of modulation type for signals hidden in noise: an application of the new theory of cyclic spectral analysis. In: Proceedings of Ninth Symposium on Signal Processing and Its Applications, pp. 119–124.

Gardner, W.A., 1985. Introduction to Random Processes With Applications to Signals and Systems. Macmillan, New York (second ed., McGraw-Hill, New York, 1990).

Gardner, W.A., 1986a. Measurements of spectral correlation. IEEE Transactions on Acoustics, Speech, and Signal Processing ASSP-34, 1111–1123.

Gardner, W.A., 1986b. The role of spectral correlation in design and performance analysis of synchronizers. IEEE Transactions on Communications COM-34 (11), 1089–1095.

Gardner, W.A., 1986c. The spectral correlation theory of cyclostationary time series. Signal Processing 11, 13–36. (Erratum: Signal Processing, vol. 11, p. 405).

Gardner, W.A., 1987a. Common pitfalls in the application of stationary process theory to time-sampled and modulated signals. IEEE Transactions on Communications COM-35, 529–534.
Gardner, W.A., 1987b. Nonstationary learning characteristics of the LMS algorithm. IEEE Transactions on Circuits and Systems CAS-34, 1199–1207.
Gardner, W.A., 1987c. Rice's representation for cyclostationary processes. IEEE Transactions on Communications COM-35, 74–78.
Gardner, W.A., 1987d. Signal interception: a unifying theoretical framework for feature detection. In: Proceedings of the Eleventh GRETSI Symposium on Signal and Image Processing. Nice, France, pp. 69–72.
Gardner, W.A., 1987e. Spectral correlation of modulated signals. Part I – analog modulation. IEEE Transactions on Communications COM-35 (6), 584–594.
Gardner, W.A., 1987f. Statistical Spectral Analysis: A Nonprobabilistic Theory. Prentice-Hall, Englewood Cliffs, NJ.
Gardner, W.A., 1988a. Correlation estimation and time-series modeling for nonstationary processes. Signal Processing 15 (1), 31–41.
Gardner, W.A., 1988b. Signal interception: a unifying theoretical framework for feature detection. IEEE Transactions on Communications COM-36, 897–906.
Gardner, W.A., 1988c. Simplification of MUSIC and ESPRIT by exploitation of cyclostationarity. Proceedings of the IEEE 76 (7), 845–847.
Gardner, W.A., 1990a. Identification of systems with cyclostationary input and correlated input/output measurement noise. IEEE Transactions on Automatic Control 35 (4), 449–452.
Gardner, W.A., 1990b. Introduction to Random Processes With Applications to Signals and Systems, second ed. McGraw-Hill, New York (Chapter 12).
Gardner, W.A., 1990c. Spectral characterization of n-th order cyclostationarity. In: Proceedings of the Fifth IEEE/ASSP Workshop on Spectrum Estimation and Modeling. Rochester, NY, pp. 251–255.
Gardner, W.A., 1991a. Exploitation of spectral redundancy in cyclostationary signals. IEEE Signal Processing Magazine 8, 14–36.
Gardner, W.A., 1991b. A new method of channel identification. IEEE Transactions on Communications 39, 813–817.
Gardner, W.A., 1991c. On the spectral coherence of nonstationary processes. IEEE Transactions on Signal Processing 39 (2), 424–430.
Gardner, W.A., 1991d. Two alternative philosophies for estimation of the parameters of time-series. IEEE Transactions on Information Theory 37, 216–218.
Gardner, W.A., 1992. A unifying view of coherence in signal processing. Signal Processing 29, 113–140.
Gardner, W.A., 1993. Cyclic Wiener filtering: theory and method. IEEE Transactions on Communications 41 (1), 151–163.
Gardner, W.A., 1994a. An introduction to cyclostationary signals. In: Gardner, W.A. (Ed.), Cyclostationarity in Communications and Signal Processing. IEEE Press, pp. 1–90.
Gardner, R.W., 1995. Detection of a digitally modulated signal using binary line enhancement. U.S. Patent No. 5,455,846.
Gardner, W.A., 2001. Suppression of cochannel interference in GSM by pre-demodulation signal processing. In: Proceedings of the 11th Virginia Tech. Symp. on Wireless Personal Communications, pp. 217–228.
Gardner, W.A., 2018a. Cyclostationarity.com. https://cyclostationarity.com. (Accessed 1 April 2019).
Gardner, W.A., 2018b. Statistically inferred time warping: extending the cyclostationarity paradigm from regular to irregular statistical cyclicity in scientific data. EURASIP Journal on Advances in Signal Processing (ISSN 1687-6180) 2018 (1), 59. https://doi.org/10.1186/s13634-018-0564-6.
Gardner, W.A., Agee, B.G., 1993. Self-coherence restoring signal extraction apparatus and method. U.S. Patent No. 5,255,210.
Gardner, W.A., Archer, T.L., 1991. Simplified methods for identifying the Volterra kernels of nonlinear systems. In: Proceedings of the Thirty-Fourth Midwest Symposium on Circuits and Systems. Monterey, CA.
Gardner, W.A., Archer, T.L., 1993. Exploitation of cyclostationarity for identifying the Volterra kernels of nonlinear systems. IEEE Transactions on Information Theory 39 (3), 535–542.
Gardner, W.A., Brown, W.A., 1989. Frequency-shift filtering theory for adaptive co-channel interference removal. In: Conference Record, Twenty-Third Asilomar Conference on Signals, Systems and Computers, vol. 2. Pacific Grove, CA, pp. 562–567.

Gardner, W.A., Brown, W.A., 1991. Fraction-of-time probability for time-series that exhibit cyclostationarity. Signal Processing 23, 273–292.

Gardner, W.A., Chen, C.K., 1988a. Interference-tolerant time-difference-of-arrival estimation for modulated signals. IEEE Transactions on Acoustics, Speech, and Signal Processing ASSP-36, 1385–1395.

Gardner, W.A., Chen, C.K., 1988b. Selective source location by exploitation of spectral coherence. In: Proceedings of the Fourth Annual ASSP Workshop on Spectrum Estimation and Modeling. Minneapolis, MN, pp. 271–276.

Gardner, W.A., Chen, C.K., 1991. The Random Processes Tutor. McGraw-Hill, New York.

Gardner, W.A., Chen, C.K., 1992. Signal-selective time-difference-of-arrival estimation for passive location of manmade signal sources in highly corruptive environments. Part I: theory and method. IEEE Transactions on Signal Processing 40, 1168–1184.

Gardner, W.A., Franks, L.E., 1975. Characterization of cyclostationary random signal processes. IEEE Transactions on Information Theory IT-21, 4–14.

Gardner, W.A., Paura, L., 1995. Identification of polyperiodic nonlinear systems. Signal Processing 46, 75–83.

Gardner, W.A., Reed, C.W., 2001. Making the most out of spectral redundancy in GSM: cheap CCI suppression. In: Proceedings of the Thirty-Sixth Annual Asilomar Conference on Signals, Systems and Computers. Pacific Grove, CA.

Gardner, W.A., Roberts, R.S., 1993. One-bit spectral correlation algorithms. IEEE Transactions on Signal Processing 41 (1), 423–427.

Gardner, W.A., Schell, S.V., 1993. Method and apparatus for multiplexing communications signals through blind adaptive spatial filtering. U.S. Patent No. 5,260,968.

Gardner, W.A., Schell, S.V., 1998. GMSK signal processors for improved communications capacity and quality. U.S. Patent No. 5,848,105.

Gardner, W.A., Spooner, C.M., 1988. Cyclic spectral analysis for signal detection and modulation recognition. In: Proceedings of MILCOM 88. San Diego, CA, pp. 419–424.

Gardner, W.A., Spooner, C.M., 1990. Higher order cyclostationarity, cyclic cumulants, and cyclic polyspectra. In: Proceedings of the 1990 International Symposium on Information Theory and Its Applications. ISITA'90, Honolulu, HI, pp. 355–358.

Gardner, W.A., Spooner, C.M., 1992a. Comparison of autocorrelation and cross-correlation methods for signal-selective TDOA estimation. IEEE Transactions on Signal Processing 40 (10), 2606–2608.

Gardner, W.A., Spooner, C.M., 1992b. Signal interception: performance advantages of cyclic feature detectors. IEEE Transactions on Communications 40 (1), 149–159.

Gardner, W.A., Spooner, C.M., 1993. Detection and source location of weak cyclostationary signals: simplifications of the maximum-likelihood receiver. IEEE Transactions on Communications 41 (6), 905–916.

Gardner, W.A., Spooner, C.M., 1994a. The cumulant theory of cyclostationary time-series. Part I: foundation. IEEE Transactions on Signal Processing 42, 3387–3408.

Gardner, W.A., Spooner, C.M., 1994b. Cyclostationary signal processing. In: Leondes, C.T. (Ed.), Control and Dynamic Systems. Academic Press.

Gardner, W.A., Venkataraman, S., 1990. Performance of optimum and adaptive frequency-shift filters for co-channel interference and fading. In: Proceedings of the Twenty-Fourth Annual Asilomar Conference on Signals, Systems, and Computers. Pacific Grove, CA, pp. 242–247.

Gardner, W.A., Brown, W.A., Chen, C.K., 1987. Spectral correlation of modulated signals. Part II – digital modulation. IEEE Transactions on Communications COM-35 (6), 595–601.

Gardner, W.A., Schell, S.V., Agee, B.G., 1994. Self-coherence restoring signal extraction and estimation of signal direction of arrival. U.S. Patent No. 5,299,148.

Gardner, W.A., Yeung, G.K., Brown, W.A., 1995. Signal reconstruction after spectral excision. In: Proceedings of the Twenty-Ninth Annual Asilomar Conference on Signals, Systems, and Computers. Pacific Grove, CA.

Gardner, W.A., Napolitano, A., Paura, L., 2006. Cyclostationarity: half a century of research. Signal Processing 86 (4), 639–697. https://doi.org/10.1016/j.sigpro.2005.06.016.

Gardner, W.A. (Ed.), 1994b. Cyclostationarity in Communications and Signal Processing. IEEE Press.

Gartshtein, B.N., Smelyakov, V.V., 1982. Assessing characteristics of periodic random signals. Radiotekhnika 60, 116–123 (in Russian).

Gaucherel, C., 2010. Analysis of ENSO interannual oscillations using non-stationary quasi-periodic statistics: a study of ENSO memory. International Journal of Climatology (ISSN 1097-0088) 30 (6), 926–934. https://doi.org/10.1002/joc.1937.

Gefen, S., Tretiak, O.J., Piccoli, C., Donohue, K.D., Petropulu, A.P., Shankar, P.M., Dumane, V.A., Huang, L., Kutay, M.A., Genis, V., Forsberg, F., Reid, J.M., Goldberg, B.B., 2003. ROC analysis of classifiers based on ultrasonic tissue characterization features. IEEE Transactions on Medical Imaging 22 (2), 170–177.

Gel'fand, I.M., Vilenkin, N.Y., 1964. Generalized Functions, Applications of Harmonic Analysis, vol. 4. Academic Press, New York, NJ.

Gelli, G., 1994. Sensor Array-Based Parameter Estimation of Cyclostationary Signals. Ph.D. Dissertation. Università di Napoli Federico II, Napoli, Italy, Dipartimento di Ingegneria Elettronica (in Italian).

Gelli, G., 1995. Power and timing parameter estimation of multiple cyclostationary signals from sensor array data. Signal Processing 42 (1), 97–102.

Gelli, G., Izzo, L., 1997. Minimum-redundancy linear arrays for cyclostationarity-based source location. IEEE Transactions on Signal Processing 45 (10), 2605–2608.

Gelli, G., Izzo, L., 2003. Cyclostationarity-based coherent methods for wideband-signal source location. IEEE Transactions on Signal Processing 51 (10), 2471–2482.

Gelli, G., Verde, F., 2001. Adaptive minimum variance equalization with interference suppression capabilities. IEEE Communications Letters 5 (12), 491–493.

Gelli, G., Verde, F., 2002. Two-stage interference-resistant adaptive periodically time-varying CMA blind equalization. IEEE Transactions on Signal Processing 50 (3), 662–672.

Gelli, G., Verde, F., 2003. Blind FSR-LPTV equalization and interference rejection. IEEE Transactions on Communications 51 (2), 145–150.

Gelli, G., Izzo, L., Paura, L., Poggi, G., 1991a. A cyclic SVD-based algorithm for multiple source localization. In: Treizième Colloque sur le Traitement du Signal et des Images. GRETSI'91. Juan-Les-Pins, France.

Gelli, G., Napolitano, A., Paura, L., 1991b. Spectral-correlation based estimation of channel parameters by noncoherent processing. In: Proceedings of China 1991 International Conference on Circuits and Systems. Sheuzhen, China, pp. 354–357.

Gelli, G., Izzo, L., Napolitano, A., Paura, L., 1993. Multipath channel identification by an improved Prony algorithm based on spectral correlation measurements. Signal Processing 31 (1), 17–29.

Gelli, G., Izzo, L., Paura, L., 1996. Cyclostationarity-based signal detection and source location in non-Gaussian noise. IEEE Transactions on Communications 44 (3), 368–376.

Gelli, G., Paura, L., Tulino, A.M., 1998. Cyclostationarity-based filtering for narrowband interference suppression in direct-sequence spread-spectrum systems. IEEE Journal on Selected Areas in Communications 16, 1747–1755.

Gelli, G., Paura, L., Ragozzini, A.R.P., 2000. Blind widely linear multiuser detection. IEEE Communications Letters 4 (6), 187–189.

Gelli, G., Mattera, D., Paura, L., 2005. Blind wide-band spatio-temporal filtering based on higher-order cyclostationarity. IEEE Transactions on Signal Processing 53 (4), 1282–1290.

Genossar, M.J., 1992. Spectral Characterization of Nonstationary Processes. Ph.D. Dissertation. Stanford University, Stanford, CA.

Genossar, M.J., Lev-Ari, H., Kailath, T., 1994. Consistent estimation of the cyclic autocorrelation. IEEE Transactions on Signal Processing 42, 595–603.

Gerr, N.L., Allen, J.C., 1994a. The generalized spectrum and spectral coherence of harmonizable time series. Digital Signal Processing 4, 222–238.

Gerr, N.L., Allen, J.C., 1994b. Time-delay estimation for harmonizable signals. Digital Signal Processing 4, 49–62.

Ghaderi, F., Sanei, S., Makkiabadi, B., Abolghasemi, V., McWhirter, J.G., 2009. Heart and lung sound separation using periodic source extraction method. In: 16th International Conference on Digital Signal Processing, pp. 1–6.

Ghaderi, F., Nazarpour, K., McWhirter, J.G., Sanei, S., 2010. Removal of ballistocardiogram artifacts using the cyclostationary source extraction method. IEEE Transactions on Biomedical Engineering (ISSN 0018-9294) 57 (11), 2667–2676. https://doi.org/10.1109/TBME.2010.2060334.

Ghogo, M., Swami, A., Garel, B., 1999a. Performance analysis of cyclic statistics for the estimation of harmonics in multiplicative and additive noise. IEEE Transactions on Signal Processing 47, 3235–3249.

Ghogo, M., Swami, A., Nandi, A.K., 1999b. Nonlinear least-squares estimation for harmonics in multiplicative and additive noise. Signal Processing 79 (2), 43–60.

Ghozzi, M., Dohler, M., Marx, F., Palicot, J., 2006. Cognitive radio: methods for the detection of free bands. Comptes Rendus. Physique (ISSN 1631-0705) 7 (7), 794–804. https://doi.org/10.1016/j.crhy.2006.07.009.

Ghysels, E., Osborne, D.R., 2001. The Econometric Analysis of Seasonal Time Series, Themes in Modern Econometrics. Cambridge University Press, Cambridge, UK.

Ghysels, E., Hall, A., Lee, H.S., 1996. On periodic structures and testing for seasonal unit roots. Journal of the American Statistical Association 91, 1551–1559.

Gianaroli, F., Pancaldi, F., Sironi, E., Vigilante, M., Vitetta, G., Barbieri, A., 2012. Statistical modeling of periodic impulsive noise in indoor power-line channels. IEEE Transactions on Power Delivery (ISSN 0885-8977) 27 (3), 1276–1283. https://doi.org/10.1109/TPWRD.2012.2188910.

Giannakis, G.B., 1995. Polyspectral and cyclostationary approaches for identification of closed-loop systems. IEEE Transactions on Automatic Control 40 (5), 882–885.

Giannakis, G.B., 1998. Cyclostationary signal analysis. In: Madisetti, V.K., Williams, D.B. (Eds.), The Digital Signal Processing Handbook. CRC Press and IEEE Press.

Giannakis, G.B., Halford, S.D., 1997. Asymptotically optimal blind fractionally spaced channel estimation and performance analysis. IEEE Transactions on Signal Processing 45 (7), 1815–1830.

Giannakis, G.B., Serpedin, E., 1998. Blind identification of ARMA channels with periodically modulated inputs. IEEE Transactions on Signal Processing 46 (11), 3099–3104.

Giannakis, G.B., Zhou, G., 1994. Parameter estimation of cyclostationary AM time series with application to missing observations. IEEE Transactions on Signal Processing 42 (9), 2408–2419.

Giannakis, G.B., Zhou, G., 1995. Harmonics in multiplicative and additive noise: parameter estimation using cyclic statistics. IEEE Transactions on Signal Processing 43 (9), 2217–2221.

Gikhman, I.I., Skorokhod, A.V., 1969. Introduction to the Theory of Random Processes. Dover, New York.

Gini, F., Giannakis, G.B., 1998. Frequency offset and symbol timing recover in flat-fading channels: a cyclostationary approach. IEEE Transactions on Communications 46, 400–411.

Gini, F., Giannakis, G.B., 1999. Hybrid FM-polynomial phase signal modeling: parameter estimation and Cramér-Rao bounds. IEEE Transactions on Signal Processing 47 (2), 363–377.

Gini, F., Greco, M., 2002. Texture modelling, estimation and validation using measured sea clutter. IEE Proceedings. Radar, Sonar and Navigation 149 (3), 115–124.

Girault, J., Biard, M., Kouame, D., Bleuzen, A., Tranquart, F., 2006. Spectral correlation of the embolic blood Doppler signal. In: IEEE International Conference on Acoustics, Speech and Signal Processing. ICASSP, 2006, p. 2.

Girondin, V., Pekpe, K.M., Morel, H., Cassar, J.P., 2013. Bearings fault detection in helicopters using frequency readjustment and cyclostationary analysis. Mechanical Systems and Signal Processing (ISSN 0888-3270) 38 (2), 499–514. https://doi.org/10.1016/j.ymssp.2013.03.015.

Gitlin, R.D., Weinstein, S.B., 1981. Fractionally-spaced equalization: an improved digital transversal equalizer. The Bell System Technical Journal 60, 275–296.

Gladyshev, E.G., 1961. Periodically correlated random sequences. Soviet Mathematics. Doklady 2, 385–388.

Gladyshev, E.G., 1963. Periodically and almost periodically correlated random processes with continuous time parameter. Theory of Probability and Its Applications 8 (2), 173–177 (in Russian).

GNU Octave, 1998. www.gnu.org/software/octave. https://www.gnu.org/software/octave. (Accessed 1 April 2019).

Goette, J., Gobet, C., 1989. Exact noise analysis of SC circuits and an approximate computer implementation. IEEE Transactions on Circuits and Systems 36, 508–521.

Goldberger, A., Amaral, L., Glass, L., Hausdorff, J., Ivanov, P., Mark, R., Mietus, J., Moody, G., Peng, C.K., Stanley, H., 2000. PhysioBank, PhysioToolkit, and PhysioNet: components of a new research resource for complex physiologic signals. Circulation 101 (23), e215–e220.

Golden, G.D., 1992a. Use of a fractionally spaced equalizer to perform echo cancellation in a full-duplex modem. U.S. Patent No. 5,163,044.

Golden, G.D., 1992b. Use of cyclostationary signal to constrain the frequency response of a fractionally spaced equalizer. U.S. Patent No. 5,095,495.

Golden, G.D., 1994. Extended bandwidth transmitter for crosstalk channels. U.S. Patent No. 5,377,230.

Gonzalez, G.J., Gregorio, F.H., Cousseau, J., Werner, S., Wichman, R., 2013. Data-aided CFO estimators based on the averaged cyclic autocorrelation. Signal Processing (ISSN 0165-1684) 93 (1), 217–229. https://doi.org/10.1016/j.sigpro.2012.07.032.

Goodman, N.R., 1965. Statistical Test for Stationarity Within the Framework of Harmonizable Processes. Research Report AD619270. Rokeydyne, Canoga Park, CA.

Gourge, F., Roosen, F., 2005. Device enambling different spreading factors while preserving a common rambling code, in particular for transmission in a code division multiple access cellular mobile radio system. U.S. Patent No. 6,868,077.

Gournay, P., Nicolas, P., 1995. Cyclic spectral analysis and time-frequency analysis for automatic transmission classification. In: Proceedings of Quinzieme Colloque GRETSI. Juan-Les-Pins, France.

Gournay, P., Viravau, P., 1998. Theorical spectral correlation of CPM modulations–Part I: analytic results for two-states CPFSK modulation (1REC). Annales des Télécommunications 53 (7–8), 267–278 (in French).

Graef, F.K., 1983. Joint optimization of transmitter and receiver for cyclostationary random signal processes. In: Proceedings of the NATO Advanced Study Institute on Nonlinear Stochastic Problems. Reidel, Algarve, Portugal, pp. 581–592.

Graffeuil, J., Liman, R., Muraro, J.L., Llopis, O., 2010. Cyclostationary shot-noise measurements in RF Schottky-barrier diode detectors. IEEE Electron Device Letters (ISSN 0741-3106) 31 (1), 74–76. https://doi.org/10.1109/LED.2009.2035339.

Granger, C.W.J., 1978. Some new time series models: non-linear, bilinear and non-stationary. Statistician 27 (3–4), 237–254.

Gravely, T.B., Alles, M.C., Beck, A.E., Srinivasan, N., 2018. System and method for passive intermodulation identification using cyclostationary signals. U.S. Patent No. 10,027,368.

Gray, J.E., Addison, S.R., 2003. Effect of nonuniform target motion on radar backscattered waveforms. IEE Proceedings. Radar, Sonar and Navigation 150 (4), 262–270. https://doi.org/10.1049/ip-rsn:20030637.

Grenander, U., Rosenblatt, M., 1957. Statistical Analysis of Stationary Time Series. Wiley, New York.

Gryllias, K., Andre, H., Leclere, Q., Antoni, J., 2017. Condition monitoring of rotating machinery under varying operating conditions based on cyclo-non-stationary indicators and a multi-order probabilistic approach for instantaneous angular speed tracking. In: 20th IFAC World Congress. IFAC-PapersOnLine (ISSN 2405-8963) 50 (1), 4708–4713. https://doi.org/10.1016/j.ifacol.2017.08.857.

Gudzenko, L.I., 1959. On periodic nonstationary processes. Radio Engineering and Electronic Physics 4, 220–224 (in Russian).

Guglielmi, F., Luschi, C., Spalvieri, A., 1998. Fractionally spaced equalizing circuits and method. U.S. Patent No. 5,751,768.

Guidorzi, R., Diversi, R., 2003. Minimal representations of MIMO time-varying systems and realization of cyclostationary models. Automatica 39, 1903–1914.

Gul, S., Oner, M., Cirpan, H.A., 2017. Cyclostationarity based blind block timing estimation for Alamouti coded MIMO signals. IEEE Communications Letters (ISSN 1089-7798) 21 (6), 1341–1344. https://doi.org/10.1109/LCOMM.2017.2677420.

Gullapalli, K.K., 2003. Method and apparatus for analyzing small signal response and noise in nonlinear circuits. U.S. Patent No. 6,536,026.

Güner, A., Kaya, I., 2014. Periodic variation method for blind symbol rate estimation. International Journal of Communication Systems (ISSN 1099-1131) 27 (12), 4199–4209. https://doi.org/10.1002/dac.2606.

Gupta, K.N., 1972. Stationariness provided by filtration to a periodic non-stationary random process. Journal of Sound and Vibration 23, 319–329.

Gurevich, V.E., Korshunov, S.Y., 1987. Statistical characteristics of a narrow-band cyclo-stationary process. Telecommunications & Radio Engineering, Part 2 42, 87–91.

Habibi, D., Lewis, D.J.H., 1996. Fourier analysis for modelling some cyclic behaviour of networks. Computer Communications 19, 426–434.

Hajivandi, M., 1982. Tracking Performance of Stochastic Gradient Algorithms for Nonstationary Processes. Ph.D. Dissertation. University of California, Davis, CA, Department of Electrical and Computer Engineering.

Hajivandi, M., Gardner, W.A., 1990. Measures of tracking performance for the LMS algorithm. IEEE Transactions on Acoustics, Speech, and Signal Processing 38, 1953–1958.

Halmos, P., 1944. The foundations of probability. The American Mathematical Monthly 51 (9), 493–510.

Halmos, P., 1956. Lectures on Ergodic Theory. Mathematical Society of Japan (and American Mathematical Society (AMS) Chelsea Publishing, New York, 2006).

Hamlington, B.D., Leben, R.R., Nerem, R.S., Han, W., Kim, K.Y., 2011. Reconstruction sea level using cyclostationary empirical orthogonal functions. Journal of Geophysical Research: Oceans (ISSN 2156-2202) 116 (C12). https://doi.org/10.1029/2011JC007529.

Han, B.W., Cho, J.H., 2012. Capacity of second-order cyclostationary complex Gaussian noise channels. IEEE Transactions on Communications (ISSN 0090-6778) 60 (1), 89–100. https://doi.org/10.1109/TCOMM.2011.110711.100414.

Han, Y., Hong, J., 2007. Almost periodic random sequences in probability. Journal of Mathematical Analysis and Applications 336, 962–974.

Hanin, L.G., Schreiber, B.M., 1998. Discrete spectrum of nonstationary stochastic processes on groups. Journal of Theoretical Probability 11 (4).

Hanson, D., Randall, R., Antoni, J., Thompson, D., Waters, T., Ford, R., 2007a. Cyclostationarity and the cepstrum for operational modal analysis of MIMO systems–Part I: modal parameter identification. Mechanical Systems and Signal Processing (ISSN 0888-3270) 21 (6), 2441–2458. https://doi.org/10.1016/j.ymssp.2006.11.008.

Hanson, D., Randall, R., Antoni, J., Waters, T., Thompson, D., Ford, R., 2007b. Cyclostationarity and the cepstrum for operational modal analysis of MIMO systems–Part II: obtaining scaled mode shapes through finite element model updating. Mechanical Systems and Signal Processing (ISSN 0888-3270) 21 (6), 2459–2473. https://doi.org/10.1016/j.ymssp.2006.11.009.

Hanssen, A., Scharf, L.L., 2003. A theory of polyspectra for nonstationary stochastic processes. IEEE Transactions on Signal Processing 51 (5), 1243–1252.

Hardin, J.C., Miamee, A.G., 1990. Correlation autoregressive processes with application to helicopter noise. Journal of Sound and Vibration 142, 191–202.

Hardin, J.C., Makagon, A., Miamee, A.G., 1991. Correlation autoregressive sequences: a summary. In: Miamee, A.G. (Ed.), Proceeding of the Workshop on Nonstationary Stochastic Processes and Their Applications. World Scientific, Virginia, VA, pp. 165–175.

Haritopoulos, M., Capdessus, C., Nandi, A., 2010. Foetal PQRST extraction from ECG recordings using cyclostationarity-based source separation method. In: Annual International Conference of the IEEE Engineering in Medicine and Biology Society. EMBC 2010, pp. 1910–1913.

Haritopoulos, M., Roussel, J., Capdessus, C., Nandi, A.K., 2014. Cyclostationarity-Based Estimation of the Foetus Subspace Dimension From ECG Recordings. Springer International Publishing, Cham, pp. 726–729.

Hasselmann, K., Barnett, T.P., 1981. Techniques of linear prediction for systems with periodic statistics. Journal of the Atmospheric Sciences 38, 2275–2283.

Hatzinakos, D., 1994. Nonminimum phase channel deconvolution using the complex cepstrum of the cyclic autocorrelation. IEEE Transactions on Signal Processing 42 (11), 3026–3042.

Haykin, S., Thomson, D.J., Reed, J.H., 2009. Spectrum sensing for cognitive radio. Proceedings of the IEEE (ISSN 0018-9219) 97 (5), 849–877. https://doi.org/10.1109/JPROC.2009.2015711.

Heat, R.W., Giannakis, G.B., 1999. Exploiting input cyclostationarity for blind channel identification in OFDM systems. IEEE Transactions on Signal Processing 47 (3), 848–856.

Hellbourg, G., Weber, R., Capdessus, C., Boonstra, A.J., 2012. Cyclostationary approaches for spatial RFI mitigation in radio astronomy. Comptes Rendus. Physique (ISSN 1631-0705) 13 (1), 71–79. https://doi.org/10.1016/j.crhy.2011.10.010.

Henniger, J., 1970. Functions of bounded mean square, and generalized Fourier-Stieltjes transforms. Canadian Journal of Mathematics XXII (5), 1016–1034.

Herbst, L.J., 1963. Almost periodic variances. The Annals of Mathematical Statistics 34, 1549–1557.

Herbst, L.J., 1965. The statistical Fourier analysis of variances. Journal of the Royal Statistical Society, Series B 27, 159–165.

Heydari, K., Azmi, P., Arand, B.A., Heydari, A., 2018. Detection of the chirp signal features caused by Doppler phenomenon in the presence of destructive agents based on cyclostationary and Hough transform methods. IET Signal Processing (ISSN 1751-9675) 12 (4), 394–402. https://doi.org/10.1049/iet-spr.2017.0092.

Hlawatsch, F., Bourdeaux-Bartels, G.F., 1992. Linear and quadratic time-frequency signal representations. IEEE Signal Processing Magazine, 21–67.

Ho, K.C., Chan, Y.T., 1998. Optimum discrete wavelet scaling and its application to delay and Doppler estimation. IEEE Transactions on Signal Processing 46 (9), 2285–2290.

Ho, K.C., Hamelberg, D., 2016. Oscillatory diffusion and second-order cyclostationarity in alanine tripeptide from molecular dynamics simulation. Journal of Chemical Theory and Computation 12 (1), 372–382. https://doi.org/10.1021/acs.jctc.5b00876.

Hofstetter, E.M., 1964. Random processes. In: Margenau, H., Murphy, G.M. (Eds.), The Mathematics of Physics and Chemistry, II. D. Van Nostrand Co., Princeton, NJ.

Holt, C., 1987. Two-channel likelihood detectors for arbitrary linear channel distortion. IEEE Transactions on Acoustics, Speech, and Signal Processing (ISSN 0096-3518) 35 (3), 267–273. https://doi.org/10.1109/TASSP.1987.1165129.

Honda, I., 1980. Sample periodicity of periodically correlated processes. Keio Mathematical Reports 5, 13–18.

Honda, I., 1981. Spectral representation of periodically correlated stochastic processes and approximate Fourier series. Keio Mathematical Seminar Reports 6, 11–16.

Honda, I., 1982. On the spectral representation and related properties of a periodically correlated stochastic process. Transactions of the Institute of Electronics and Communication Engineers of Japan. Section E 65, 723–729.

Honda, I., 1983. A note on periodogram analysis of periodically correlated stochastic processes. Keio Mathematical Seminar Reports 8, 1–7.

Honda, I., 1990. On the ergodicity of Gaussian periodically correlated stochastic processes. Transactions of the Institute of Electronics, Information, and Communication Engineers. Section E 73 (10), 1729–1737.

Hong, S., Li, Y., He, Y.C., Wang, G., Jin, M., 2012. A cyclic correlation-based blind SINR estimation for OFDM systems. IEEE Communications Letters (ISSN 1089-7798) 16 (11), 1832–1835. https://doi.org/10.1109/LCOMM.2012.100812.122009.

Hongsheng, L., You, H., Rijie, Y., 2005. Neural blind beamformer for cyclostationary signals. Journal of Systems Engineering and Electronics 16 (3), 498–501.

Hongyi, Y., Zhang, B., 1998. Fully blind estimation of time delays and spatial signatures for cyclostationary signals. Elettronic Letters 34 (25), 2378–2380.

Hopgood, J.R., Rayner, P.J.W., 2003. Single channel nonstationary stochastic signal separation using linear time-varying filters. IEEE Transactions on Signal Processing 51 (7), 1739–1752. https://doi.org/10.1109/TSP.2003.812837.

Horstmann, S., Ramirez, D., Schreier, P.J., 2018. Joint detection of almost-cyclostationary signals and estimation of their cycle period. IEEE Signal Processing Letters (ISSN 1070-9908) 25 (11), 1695–1699. https://doi.org/10.1109/LSP.2018.2871961.

Houcke, S., Chevreuil, A., Loubaton, P., 2003. Blind equalization–case of unknown symbol period. IEEE Transactions on Signal Processing 51 (3), 781–793.

Howard, S.D., Sirianunpiboon, S., Cochran, D., 2018. The geometry of coherence and its application to cyclostationary time series. In: IEEE Workshop on Statistical Signal Processing. SSP 2018.

Huang, J.P., North, G.R., 1996. Cyclic spectral analysis of fluctuations in a GCM simulation. Journal of the Atmospheric Sciences 53 (3), 370–379. https://doi.org/10.1175/1520-0469(1996)053<0370:CSAOFI>2.0.CO;2.

Huang, G., Tugnait, J.K., 2013. On cyclostationarity based spectrum sensing under uncertain Gaussian noise. IEEE Transactions on Signal Processing (ISSN 1053-587X) 61 (8), 2042–2054. https://doi.org/10.1109/TSP.2013.2246158.

Huang, Z.T., Zhou, Y.Y., 2006. Multi-cycle estimator for time-difference-of-arrival (TDOA) and its performance. IEE Proceedings. Radar, Sonar and Navigation (ISSN 1350-2395) 153 (5), 381–388.

Huang, L., Donohue, K.D., Genis, V., Forsberg, F., 2000. Duct detection and wall spacing estimation in breast tissue. Ultrasonic Imaging 22 (3), 137–152.

Huang, Z., Zhou, Y., Jiang, W., 2002a. Direction-of-arrival of signal sources exploiting cyclostationarity property. Acta Electronica Sinica 30 (3), 372–375 (in Chinese).

Huang, Z., Zhou, Y., Jiang, W., 2002b. On cyclic correlation matched filtering. Acta Electronica Sinica 30 (12), 122–126 (in Chinese).

Huang, Z.T., Zhou, Y.Y., Jiang, W.L., Lu, Q.Z., 2002c. Joint estimation of Doppler and time-difference-of-arrival exploiting cyclostationary property. IEE Proceedings. Radar, Sonar and Navigation (ISSN 1350-2395) 149 (4), 161–165. https://doi.org/10.1049/ip-rsn:20020292.

Huang, Z., Wang, W., Zhou, Y., Jiang, W., 2003. Asymptotic analysis of estimated cyclic cross-correlation function between stationary and cyclostationary processes. Systems Engineering and Electronics 14 (3), 87–91.

Huang, Z., Zhou, Y., Jiang, W., 2008. TDOA and Doppler estimation for cyclostationary signals based on multi-cycle frequencies. IEEE Transactions on Aerospace and Electronic Systems (ISSN 0018-9251) 44 (4), 1251–1264. https://doi.org/10.1109/TAES.2008.4667707.

Huang, J., Wang, P., Wan, Q., 2011. Sidelobe suppression for blind adaptive beamforming with sparse constraint. IEEE Communications Letters (ISSN 1089-7798) 15 (3), 343–345. https://doi.org/10.1109/LCOMM.2011.012511.102215.

Hurd, H.L., 1969. An Investigation of Periodically Correlated Stochastic Processes. Ph.D. Dissertation. Duke University, Durham, North Carolina.

Hurd, H.L., 1973. Testing for harmonizability. IEEE Transactions on Information Theory IT-19, 316–320.

Hurd, H.L., 1974a. Periodically correlated processes with discontinuous correlation function. Theory of Probability and Its Applications 19, 804–808.

Hurd, H.L., 1974b. Stationarizing properties of random shifts. SIAM Journal on Applied Mathematics 26 (1), 203–212.

Hurd, H.L., 1982. Clipping degradation in a system used to detect time-periodic variance fluctuations. The Journal of the Acoustical Society of America 72, 1827–1830.

Hurd, H.L., 1988. Spectral coherence of nonstationary and transient stochastic processes. In: IEEE Fourth Annual ASSP Workshop on Spectrum Estimation and Modeling. Minneapolis, MN.

Hurd, H.L., 1989a. Nonparametric time series analysis for periodically correlated processes. IEEE Transactions on Information Theory 35, 350–359.

Hurd, H.L., 1989b. Representation of strongly harmonizable periodically correlated processes and their covariances. Journal of Multivariate Analysis 29, 53–67.

Hurd, H.L., 1991. Correlation theory of almost periodically correlated processes. Journal of Multivariate Analysis 37, 24–45.

Hurd, H.L., 1992. Almost periodically unitary stochastic processes. Stochastic Processes and Their Applications 43, 99–113.

Hurd, H.L., 1995. Spectral correlation of randomly jittered periodic functions of two variables. In: Proceedings of the Twenty-Ninth Annual Asilomar Conference on Signals, Systems, and Computers. Pacific Grove, CA.

Hurd, H., 2005. Periodically correlated sequences of less than full rank, IISA 2002 DeKalb Conference. Journal of Statistical Planning and Inference (ISSN 0378-3758) 129 (1–2), 279–303. https://doi.org/10.1016/j.jspi.2004.06.053.

Hurd, H.L., Gerr, N.L., 1991. Graphical methods for determining the presence of periodic correlation. Journal of Time Series Analysis 12, 337–350.

Hurd, H.L., Jones, C.H., 1994. Dynamical systems with cyclostationary orbits. In: Katz, R. (Ed.), The Chaos Paradigm: Developments and Applications in Engineering and Science. AIP Press.

Hurd, H.L., Kallianpur, G., 1991. Periodically correlated processes and their relationship to $L_2[0, T]$-valued stationary sequences. In: Miamee, A.G. (Ed.), Proceeding of the Workshop on Nonstationary Stochastic Processes and Their Applications. World Scientific, Virginia, VA, pp. 256–284.

Hurd, H.L., Koski, T., 2004a. Spectral theory of cyclostationary arrays. In: Krinik, A.C., Swift, R.J. (Eds.), Stochastic Processes and Functional Analysis: Recent Advances. A Volume in Honor of Professor M.M. Rao. Marcel Dekker, New York.

Hurd, H.L., Koski, T., 2004b. The Wold isomorphism for cyclostationary sequences. Signal Processing 84, 813–824.

Hurd, H.L., Leśkow, J., 1992a. Estimation of the Fourier coefficient functions and their spectral densities for ϕ-mixing almost periodically correlated processes. Statistics & Probability Letters 14, 299–306.

Hurd, H.L., Leśkow, J., 1992b. Strongly consistent and asymptotically normal estimation of the covariance for almost periodically correlated processes. Statistics & Decisions 10, 201–225.

Hurd, H.L., Miamee, A.G., 2007. Periodically Correlated Random Sequences: Spectral Theory and Practice. John Wiley & Sons Inc., New Jersey.

Hurd, H.L., Makagon, A., Miamee, A.G., 2002. On $AR(1)$ models with periodic and almost periodic coefficients. Stochastic Processes and Their Applications 100, 167–185.

Hurd, H., Kallianpur, G., Farshidi, J., 2004. Correlation and spectral theory for periodically correlated random fields indexed on Z^2. Journal of Multivariate Analysis 90 (2), 359–383.

Im, G.H., Won, H.C., 2004. Performance analysis of cyclostationary interference suppression for multiuser wired communication systems. Journal of Communication and Networks 6 (2), 93–105.

Ionescu, M., Sato, M., Thomsen, B., 2015. Cyclostationarity-based joint monitoring of symbol-rate, frequency offset, CD and OSNR for Nyquist WDM superchannels. Optics Express 23 (20), 25762–25772. https://doi.org/10.1364/OE.23.025762.

Iqelan, B., 2011. Periodically Correlated Time Series: Models and Examples. Lambert Academic Publishing. ISBN 9783844301106.

Isokawa, Y., 1982. An identification problem in almost and asymptotically almost periodically correlated processes. Journal of Applied Probability 19, 456–462.

Ivanov, A., Mihovska, A., Tonchev, K., Poulkov, V., 2018. Real-time adaptive spectrum sensing for cyclostationary and energy detectors. IEEE Aerospace and Electronic Systems Magazine (ISSN 0885-8985) 33 (5–6), 20–33. https://doi.org/10.1109/MAES.2018.170098.

Izzo, L., Napolitano, A., 1995. Cyclic higher-order statistics of decimated and interpolated digital signals. In: IEEE Signal Processing ATHOS Workshop on Higher-Order Statistics. Parador d'Aiguablava, Begur, Girona, Spain.

Izzo, L., Napolitano, A., 1996. Higher-order cyclostationarity properties of sampled time-series. Signal Processing 54, 303–307. https://doi.org/10.1016/S0165-1684(96)00157-0.

Izzo, L., Napolitano, A., 1997. Higher-order statistics for Rice's representation of cyclostationary signals. Signal Processing 56, 279–292. https://doi.org/10.1016/S0165-1684(96)00175-2.

Izzo, L., Napolitano, A., 1998a. The higher-order theory of generalized almost-cyclostationary time-series. IEEE Transactions on Signal Processing 46 (11), 2975–2989. https://doi.org/10.1109/78.726811.

Izzo, L., Napolitano, A., 1998b. Multirate processing of time series exhibiting higher order cyclostationarity. IEEE Transactions on Signal Processing 46 (2), 429–439. https://doi.org/10.1109/78.655427.

Izzo, L., Napolitano, A., 2002a. Linear time-variant transformations of generalized almost-cyclostationary signals. Part I: theory and method. IEEE Transactions on Signal Processing 50 (12), 2947–2961. https://doi.org/10.1109/TSP.2002.805499.

Izzo, L., Napolitano, A., 2002b. Linear time-variant transformations of generalized almost-cyclostationary signals. Part II: development and applications. IEEE Transactions on Signal Processing 50 (12), 2962–2975. https://doi.org/10.1109/TSP.2002.805500.

Izzo, L., Napolitano, A., 2003. Sampling of generalized almost-cyclostationary signals. IEEE Transactions on Signal Processing 51 (6), 1546–1556. https://doi.org/10.1109/TSP.2003.811236.

Izzo, L., Napolitano, A., 2005. Generalized almost-cyclostationary signals. In: Hawkes, P.W. (Ed.), Advances in Imaging and Electron Physics, vol. 135. Elsevier, pp. 103–223.

Izzo, L., Tanda, M., 1995. Delay-and-multiply detectors for signal interception in non-Gaussian noise. In: Proceedings of 15th GRTESI Symposium on Signal and Image Processing. Juan-Les-Pins, France.

Izzo, L., Paura, L., Tanda, M., 1990. Optimum and suboptimum detection of weak signals in cyclostationary non-Gaussian noise. European Transactions on Telecommunications and Related Technologies 1 (3), 233–238.

Izzo, L., Napolitano, A., Tanda, M., 1992a. Spectral-correlation based methods for multipath channel identification. European Transactions on Telecommunications and Related Technologies 3 (1), 341–348.

Izzo, L., Paura, L., Poggi, G., 1992b. An interference-tolerant algorithm for localization of cyclostationary-signal sources. IEEE Transactions on Signal Processing 40, 1682–1686.

Izzo, L., Paura, L., Tanda, M., 1992c. Signal interception in non-Gaussian noise. IEEE Transactions on Communications 40, 1030–1037.

Izzo, L., Napolitano, A., Paura, L., 1994a. Cyclostationarity-exploiting methods for multipath-channel identification. In: Gardner, W.A. (Ed.), Cyclostationarity in Communications and Signal Processing. IEEE Press, pp. 391–416.

Izzo, L., Napolitano, A., Paura, L., 1994b. MIMO Volterra system input/output relations for cyclic higher-order statistics. In: Proceedings of VII European Signal Processing Conference. EUSIPCO'94, Edinburgh, UK.

Izzo, L., Napolitano, A., Paura, L., 1994c. Modified cyclic methods for signal selective TDOA estimation. IEEE Transactions on Signal Processing (ISSN 1053-587X) 42 (11), 3294–3298. https://doi.org/10.1109/78.330403.

Jacob, C., 1990. Ergodicity in periodic autoregressive models. Comptes Rendus de l'Academie des Sciences, Serie I 310, 431–434 (in French).

Jalili, A., Sahami, S., Chi, C., Amirfattahi, R., 2018. Speech emotion recognition using cyclostationary spectral analysis. In: 2018 IEEE 28th International Workshop on Machine Learning for Signal Processing. MLSP, pp. 1–6.

Jallon, P., 2012. Method for detecting cyclostationary signals. U.S. Patent No. 8,254,497.

Jallon, P., Chevreuil, A., 2008. Estimation of the symbol rate of linearly modulated sequences of symbols. Signal Processing (ISSN 0165-1684) 88 (8), 1971–1979. https://doi.org/10.1016/j.sigpro.2008.02.001.

Jallon, P., Chevreuil, A., Loubaton, P., 2010. Separation of digital communication mixtures with the CMA: case of unknown symbol rates. Signal Processing (ISSN 0165-1684) 90 (9), 2633–2647. https://doi.org/10.1016/j.sigpro.2010.03.002.

Jang, W.M., 2014. Blind cyclostationary spectrum sensing in cognitive radios. IEEE Communications Letters (ISSN 1089-7798) 18 (3), 393–396. https://doi.org/10.1109/LCOMM.2014.012714.132507.

Janssen, A.J.E.M., 1988. The Zak transform: a signal transform for sampled time-continuous signals. Philips Journal of Research 43, 23–69.

Javorskyj, I.M., 2013. Mathematical Models and Analysis of Stochastic Oscillations. Academician of Ukrainian NAS, Lviv (in Ukrainian).

Javorskyj, I., Isayev, I., Zakrzewski, Z., Brooks, S., 2007. Coherent covariance analysis of periodically correlated random processes. Signal Processing (ISSN 0165-1684) 87 (1), 13–32. https://doi.org/10.1016/j.sigpro.2006.04.002.

Javorskyj, I., Isayev, I., Majewski, J., Yuzefovych, R., 2010. Component covariance analysis for periodically correlated random processes. Signal Processing (ISSN 0165-1684) 90 (4), 1083–1102. https://doi.org/10.1016/j.sigpro.2009.07.031.

Javorskyj, I., Leskow, J., Kravets, I., Isayev, I., Gajecka, E., 2011. Linear filtration methods for statistical analysis of periodically correlated random processes–Part II: harmonic series representation. Signal Processing (ISSN 0165-1684) 91 (11), 2506–2519. https://doi.org/10.1016/j.sigpro.2011.04.031.

Javorskyj, I., Leskow, J., Kravets, I., Isayev, I., Gajecka, E., 2012. Linear filtration methods for statistical analysis of periodically correlated random processes–Part I: coherent and component methods and their generalization. Signal Processing (ISSN 0165-1684) 92 (7), 1559–1566. https://doi.org/10.1016/j.sigpro.2011.09.030.

Javorskyj, I., Dehay, D., Kravets, I., 2014a. Component statistical analysis of second order hidden periodicities. Digital Signal Processing (ISSN 1051-2004) 26, 50–70. https://doi.org/10.1016/j.dsp.2013.12.002.

Javorskyj, I., Yuzefovych, R., Kravets, I., Matsko, I., 2014b. Methods of periodically correlated random processes and their generalizations. In: Chaari, F., Leskow, J., Napolitano, A., Sanchez-Ramirez, A. (Eds.), Cyclostationarity: Theory and Methods. In: Lecture Notes in Mechanical Engineering. Springer International Publishing, pp. 73–93.

Javors'kyj, I., Yuzefovych, R., Matsko, I., Kravets, I., 2015. The stochastic recurrence structure of geophysical phenomena. In: Chaari, F., Leskow, J., Napolitano, A., Zimroz, R., Wylomanska, A., Dudek, A. (Eds.), Cyclostationarity: Theory and Methods – II, Applied Condition Monitoring, vol. 3. Springer International Publishing, pp. 55–88.

Jerjawi, W.A., Eldemerdash, Y.A., Dobre, O.A., 2015. Second-order cyclostationarity-based detection of LTE SC-FDMA signals for cognitive radio systems. IEEE Transactions on Instrumentation and Measurement (ISSN 0018-9456) 64 (3), 823–833. https://doi.org/10.1109/TIM.2014.2357592.

Jessen, B., Tornehave, H., 1945. Mean motions and zeros of almost periodic functions. Acta Mathematica 77 (1), 137–279.

Jiang, Y., Hu, H., Yang, Y., Chen, H.H., 2012. Signal representation: wireless applications in a statistical spectrum domain. IEEE Vehicular Technology Magazine (ISSN 1556-6072) 7 (4), 88–94. https://doi.org/10.1109/MVT.2012.2203886.

Jin, Q., Wong, K.M., Luo, Z.Q., 1995. The estimation of time delay and Doppler stretch of wideband signals. IEEE Transactions on Signal Processing 43, 904–916.

Jin, F., Qiu, T., Liu, T., 2019. Robust cyclic beamforming against cycle frequency error in Gaussian and impulsive noise environments. AEÜ. International Journal of Electronics and Communications (ISSN 1434-8411) 99, 153–160. https://doi.org/10.1016/j.aeue.2018.11.035.

Johnson, W.K., 1981. The dynamic pneumocardiogram: an application of coherent signal processing to cardiovascular measurement. IEEE Transactions on Biomedical Engineering BME-28, 471–475.

Johnson, C.M., Lemke, P., Barnett, T.P., 1985. Linear prediction of sea ice anomalies. Journal of Geophysical Research 90, 5665–5675.

Johnsonbaugh, R., Pfaffenberger, W.E., 2002. Foundations of Mathematical Analysis. Dover Publications, Inc., Mineola, NY.

Jones, R.H., Brelsford, W.M., 1967. Time series with periodic structure. Biometrika 54, 403–408.

Jordan, K.L., 1961. Discrete Representations of Random Signals. Ph.D. Dissertation. Massachusetts Institute of Technology, Cambridge, MA, Department of Electrical Engineering.

Joshi, V., Falconer, D.D., 1988. Reduced state sequence estimation techniques for digital subscriber loop application. In: Proceedings of GLOBECOM '88, vol. 2. Hollywood, CA, pp. 799–803.

Joshi, V., Falconer, D.D., 1990. Sequence estimation techniques for digital subscriber loop transmission with crosstalk interference. IEEE Transactions on Communications 38, 1367–1374.

Kac, M., 1959. Statistical Independence in Probability, Analysis and Number Theory. The Mathematical Association of America, USA.

Kac, M., Steinhaus, H., 1938. Sur les foncions indépendantes IV. Studia Mathematica 7, 1–15.

Kacimov, A.R., Obnosov, Y.V., Yakimov, N.D., 1999. Groundwater flow in a medium with parquet-type conductivity distribution. Journal of Hidrology 226, 242–249.

Kaddoum, G., Gagné, S., Gagnon, F., 2011. Removing cyclostationary properties in a chaos-based communication system. Circuits, Systems, and Signal Processing (ISSN 0278-081X) 30 (6), 1391–1400. https://doi.org/10.1007/s00034-010-9232-2.

Kadiman, K., Williamson, D., 1987. Discrete minimax linear quadratic regulation of continuous-time systems. Automatica 23, 741–747.

Kahane, J.P., 1961. Sur les coefficients de Fourier-Bohr. Studia Mathematica 21, 103–106.

Kahane, J.P., 1962. Sur les fonctions presque-périodiques généralisées dont le spectre est vide. Studia Mathematica 21, 231–236.

Kahn, M., Mow, M., Gardner, W.A., Biedka, T., 1994. A recursive programmable canonical correlation analyzer. In: Proceedings of Second Workshop on Cyclostationary Signals. Yountville, CA.

Kailath, T., 1966. The complex envelope of white noise. IEEE Transactions on Information Theory 12 (3), 397–398. https://doi.org/10.1109/TIT.1966.1053905.

Kamolor, A.I., 1984. Kolmogorov diameters of a class of random processes. Doklady Akademii Nauk USSR 8, 8–10 (in Russian).

Kampé de Fériet, J., Frenkiel, F.N., 1962. Correlations and spectra for non-stationary random functions. Mathematics of Computation 16 (77), 1–21.

Kaplan, M., 1983. A single-server queue with cyclostationary arrivals and arithmetic service. Operations Research 31, 184–205.

Kapoulitsas, G.M., 1981. On the non-relativistic Doppler effect. European Journal of Physics 2, 174–177.

Kapustinskas, A.I., 1977. Estimation of the parameters of a periodically nonstationary autoregressive process. Trudy Akademii Nauk Litovskoj SSR. Series B 4 (101), 115–121 (in Russian).

Kapustinskas, A.I., 1978. Covariational estimation of the parameters of a periodically nonstationary autoregressive process. Lietuvos TSR Mokslų Akademijos Darbai. B Serija 104, 113–119 (in Russian).

Kapustinskas, A.I., 1981. Properties of periodically nonstationary processes. Trudy Akademii Nauk Litovskoj SSR. Series B 1 (122), 87–95 (in Russian).

Karabinis, P.D., 2011. Systems, methods, devices, and/or computer program products for providing communications devoid of cyclostationary features. U.S. Patent No. 8,050,337.

Karabinis, P.D., 2014. Systems and/or methods of increased privacy wireless communications. U.S. Patent No. 8,670,493.

Karami, E., Dobre, O.A., 2015. Identification of SM-OFDM and AL-OFDM signals based on their second-order cyclostationarity. IEEE Transactions on Vehicular Technology (ISSN 0018-9545) 64 (3), 942–953. https://doi.org/10.1109/TVT.2014.2326107.

Katayama, M., Yamazato, T., Okada, H., 2006. A mathematical model of noise in narrowband power line communication systems. IEEE Journal on Selected Areas in Communications (ISSN 0733-8716) 24 (7), 1267–1276. https://doi.org/10.1109/JSAC.2006.874408.

Kato, Y., Yamaguchi, S., 1990. A prediction method for probability distribution of road traffic noise at an intersection. Acoustics Australia 18, 46–50.

Kayatskas, A.A., 1968. Periodically correlated random processes. Telecommunications & Radio Engineering, Part 2 23, 136–141 (English translation of Radiotekhnika, vol. 23, pp. 91–97).

Kayatskas, A.A., 1970. Quasi-periodically correlated random processes. Telecommunications & Radio Engineering, Part 2 25, 145–146 (English translation of Radiotekhnika).

Kazemi, S., Ghorbani, A., Amindavar, H., Li, C., 2014. Cyclostationary approach to Doppler radar heart and respiration rates monitoring with body motion cancelation using radar Doppler system. Biomedical Signal Processing and Control (ISSN 1746-8094) 13, 79–88. https://doi.org/10.1016/j.bspc.2014.03.012.

Kazemi, S., Ghorbani, A., Amindavar, H., 2015. Cyclostationary modelling of amplitude and frequency modulated signals in heart and respiration monitoring Doppler radar systems. IEE Proceedings. Radar, Sonar and Navigation (ISSN 1751-8784) 9 (2), 116–124. https://doi.org/10.1049/iet-rsn.2014.0165.

Kelly, E.J., 1961. The radar measurement of range, velocity, and acceleration. IRE Transactions on Military Electronics MIL-5, 51–57.

Kelly, E.J., Wishner, R.P., 1965. Matched-filter theory for high-velocity, accelerating targets. IEEE Transactions on Military Electronics MIL-9, 56–69.

Khalaf, Z., Nafkha, A., Palicot, J., 2011a. Blind cyclostationary feature detector based on sparsity hypotheses for cognitive radio equipment. In: IEEE 54th International Midwest Symposium on Circuits and Systems. MWSCAS, pp. 1–4.

Khalaf, Z., Nafkha, A., Palicot, J., 2011b. Blind spectrum detector for cognitive radio using compressed sensing. In: IEEE Global Telecommunications Conference. GLOBECOM 2011, pp. 1–5.

Kilundu, B., Chiementin, X., Duez, J., Mba, D., 2011. Cyclostationarity of acoustic emissions (AE) for monitoring bearing defects. Mechanical Systems and Signal Processing (ISSN 0888-3270) 25 (6), 2061–2072. https://doi.org/10.1016/j.ymssp.2011.01.020.

Kim, K.Y., 2000. Statistical prediction of cyclostationary processes. Journal of Climate 13, 1098–1115.

Kim, K.Y., 2002. Investigation of ENSO variability using cyclostationary EOFs of observational data. Meteorology and Atmospheric Physics 81 (3–4), 149–168. https://doi.org/10.1007/s00703-002-0549-7.

Kim, K.Y., Kim, Y.Y., 2004. Investigation of tropical Pacific upper-ocean variability using cyclostationary EOFs of assimilated data. Ocean Dynamics 54 (5), 489–505. https://doi.org/10.1007/s10236-004-0094-7.

Kim, T., Min, M., 2018. Improved non-data-aided feedforward symbol timing estimator for low-rate sampling systems. IEEE Communications Letters (ISSN 1089-7798) 22 (5), 1010–1013. https://doi.org/10.1109/LCOMM.2018.2812728.

Kim, K.Y., North, G.R., 1992. Seasonal cycle and second-moment statistics of a simple coupled climate system. Journal of Geophysical Research 97 (20), 437–448.

Kim, K.Y., North, G.R., 1997a. EOFs of harmonizable cyclostationary processes. Journal of the Atmospheric Sciences 54 (19), 2416–2427. https://doi.org/10.1175/1520-0469(1997)054<2416:EOHCP>2.0.CO;2.

Kim, K.Y., North, G.R., 1997b. Statistical interpolation using cyclostationary EOFs. Journal of Climate 10 (11), 2931–2942. https://doi.org/10.1175/1520-0442(1997)010<2931:SIUCE>2.0.CO;2.

Kim, K.Y., Wu, Q., 1999. A comparison of study of EOF techniques: analysis of nonstationary data with periodic statistics. Journal of Climate 12 (1), 185–199. https://doi.org/10.1175/1520-0442-12.1.185.

Kim, K.Y., North, G.R., Huang, J., 1996. EOFs of one-dimensional cyclostationary time series: computations, examples and stochastic modeling. Journal of the Atmospheric Sciences 53, 1007–1017.

Kim, K., Akbar, I., Bae, K., Um, J.S., Spooner, C., Reed, J., 2007. Cyclostationary approaches to signal detection and classification in cognitive radio. In: 2nd IEEE International Symposium on New Frontiers in Dynamic Spectrum Access Networks. DySPAN 2007, pp. 212–215.

Kim, K.Y., Hamlington, B., Na, H., 2015. Theoretical foundation of cyclostationary EOF analysis for geophysical and climatic variables: concepts and examples. Earth-Science Reviews (ISSN 0012-8252) 150, 201–218. https://doi.org/10.1016/j.earscirev.2015.06.003.

Kim, M., Cho, J.H., Lehnert, J.S., 2016. Asymptotically optimal low-complexity SC-FDE with noise prediction in data-like improper-complex interference. IEEE Transactions on Wireless Communications (ISSN 1536-1276) 15 (3), 2090–2103. https://doi.org/10.1109/TWC.2015.2498604.

Kipnis, A., Goldsmith, A.J., Eldar, Y.C., 2018. The distortion rate function of cyclostationary Gaussian processes. IEEE Transactions on Information Theory (ISSN 0018-9448) 64 (5), 3810–3824. https://doi.org/10.1109/TIT.2017.2741978.

Knapp, C., Carter, G., 1976. The generalized correlation method for estimation of time delay. IEEE Transactions on Acoustics, Speech, and Signal Processing (ISSN 0096-3518) 24 (4), 320–327. https://doi.org/10.1109/TASSP.1976.1162830.

Kochel, P., 1980. Periodically stationary Markovian decision models. Elektronische Informationsverarbeitung und Kybernetik 16, 553–567 (in German).

Kochel, P., 1982. A dynamic multilocation supply model with redistribution between the stores. Mathematische Operationsforschung und Statistik. Series Optimization 13, 267–286 (in German).

Koenig, D., Boehme, J., 1994. Application of cyclostationarity and time-frequency analysis to engine car diagnostics. In: Proceedings of International Conference on Acoustics, Speech, and Signal Processing. ICASSP, Adelaide, Australia, pp. 149–152.

Kolmogorov, A.N., 1933. Foundations of the Theory of Probability (and Chelsea, New York, 1956).

Kolmogorov, A.N., 1960. On the $\phi^{(n)}$ classes of Fortet and Blanc-Lapierre. Theory of Probability and Its Applications 5, 337.

Kolmogorov, A.N., Fomin, S.V., 1970. Introductory Real Analysis. Prentice Hall, Englewood Cliffs, NJ (Dover, New York, 1975).

Kommate, J., Liang, Y.C., Zeng, Y.H., 2013. Method for spectrum sensing and communication device. U.S. Patent No. 8,615,211.

Konstantinides, D.G., Pieterbarg, V.I., 1993. Extreme values of the cyclostationary Gaussian random process. Journal of Applied Probability 30, 82–97.

Konstantinides, D.G., Pieterbarg, V.I., Stamotovic, S., 2004. Gnedenko-type limit theorems for cyclostationary χ^2-processes. Lithuanian Mathematical Journal 44 (2), 157–167.

Kosunen, M., Turunen, V., Kokkinen, K., Ryynanen, J., 2013. Survey and analysis of cyclostationary signal detector implementations on FPGA. IEEE Journal on Emerging and Selected Topics in Circuits and Systems (ISSN 2156-3357) 3 (4), 541–551. https://doi.org/10.1109/JETCAS.2013.2280810.

Kozlov, A.I., Mosionzhik, A.I., Rusinov, V.R., 1990. Statistical characteristics of polarization parameters of non-Gaussian periodically non stationary radio signals. Soviet Journal of Communications Technology & Electronics 35, 130–134.

Kozłowski, S., 2016. Implementation and verification of cyclostationary feature detector for DVB-T signals. IET Signal Processing (ISSN 1751-9675) 10 (2), 162–167. https://doi.org/10.1049/iet-spr.2014.0509.

Krein, M.G., 1978. Measurable Hermitian-positive functions. Matematičeskie Zametki 23 (1), 79–91.

Krikorian, K.V., Rosen, R.A., 2003. Technique for robust characterization of weak RF emitters and accurate time difference of arrival estimation for passive ranging of RF emitters. U.S. Patent No. 6,646,602.

Kruczek, P., Obuchowski, J., Wylomanska, A., Zimroz, R., 2019. Multiple local damage detection in gearbox by novel coherent bi-frequency map and its spatial, cycle oriented enhancement. In: Special Issue on ICTD-CMMNO'2016, ICTD: International Congress on Technical Diagnostics. Applied Acoustics (ISSN 0003-682X) 144, 23–30. https://doi.org/10.1016/j.apacoust.2017.10.006.

Kuznetsov, G.G., Chernenko, P.A., 1984. Mathematical model and device for diagnostics of periodically nonstationary random processes. Elektron. Model. 6, 45–48 (in Russian).

Lacaze, B., Roviras, D., 2002. Effect of random permutations applied to random sequences and related applications. Signal Processing 82 (6), 821–831.

Lacoume, J.L., Amblard, P.O., Common, P., 1997. Higher Order Statistics for Signal Processing. Masson, Paris (in French).

Lafon, B., Antoni, J., Sidahmed, M., Polac, L., 2011. The concept of cyclic sound intensity and its application to acoustical imaging. Journal of Sound and Vibration (ISSN 0022-460X) 330 (9), 2107–2121. https://doi.org/10.1016/j.jsv.2010.11.003.

Lahiri, S.N., 2003. A necessary and sufficient condition for independence of discrete Fourier transforms under short- and long-range dependence. The Annals of Statistics 31 (2), 613–641.

Lakkis, I., O'Shea, D., Tayebi, M.K., Hatim, B., 2002. Non-data-aided maximum likelihood based feedforward synchronization method. U.S. Patent No. 6,359,878.

Lakkis, I., O'Shea, D., Tayebi, M., 2004. Non-data aided maximum likelihood based feedforward timing synchronization method. U.S. Patent No. 6,768,780.

Lall, B., Joshi, S.D., Bhatt, R.K.P., 1999. Second-order statistical characterization of the filter bank and its elements. IEEE Transactions on Signal Processing 47 (6), 1745–1749.

Lambert, S., 1996. Extension of autocovariance coefficients sequence for periodically correlated random processes by using the partial autocorrelation function. In: Proceedings of VIII European Signal Processing Conference. EUSIPCO'96, Trieste, Italy.

Lambert-Lacroix, S., 2000. On periodic autoregressive processes estimation. IEEE Transactions on Signal Processing 48 (6), 1800–1803.

Lambert-Lacroix, S., 2005. Extension of autocovariance coefficients sequence for periodically correlated processes. Journal of Time Series Analysis 26 (3), 423–435.
Lamraoui, M., Thomas, M., El Badaoui, M., 2014. Cyclostationarity approach for monitoring chatter and tool wear in high speed milling. Mechanical Systems and Signal Processing (ISSN 0888-3270) 44, 177–198. https://doi.org/10.1016/j.ymssp.2013.05.001.
Lathi, B.P., 2002. Linear Systems and Signals. Oxford University Press, New York.
Lebedev, V.L., 1959. On random processes having nonstationarity of periodic character. Nauchnye Doklady Vysshchei Shchkoly. Seria Radiotekhnika i Elektronika 2, 32–34 (in Russian).
Leclere, Q., Hamzaoui, N., 2014. Using the moving synchronous average to analyze fuzzy cyclostationary signals. Mechanical Systems and Signal Processing (ISSN 0888-3270) 44 (1–2), 149–159. https://doi.org/10.1016/j.ymssp.2013.01.005.
Leduc, J.P., 1990. Image modelling for digital TV codecs. In: Proceedings of the SPIE. Lausanne, Switzerland, pp. 1160–1170.
Lee, Y.W., 1967. Statistical Theory of Communication. Wiley, New York.
Lee, A.J., 1978. Sampling theorems for nonstationary random processes. Transactions of the American Mathematical Society 242, 225–241.
Lee, J.H., Huang, C.C., 2010. Robust adaptive beamforming for multiple signals of interest with cycle frequency error. EURASIP Journal on Advances in Signal Processing (ISSN 1687-6180) 2010 (1), 873916. https://doi.org/10.1155/2010/873916.
Lee, J.H., Huang, C.C., 2012. Robust cyclic adaptive beamforming using a compensation method. Signal Processing (ISSN 0165-1684) 92 (4), 954–962. https://doi.org/10.1016/j.sigpro.2011.10.008.
Lee, J.H., Lee, Y.T., 1999. Robust adaptive array beamforming for cyclostationary signals under cycle frequency error. IEEE Transactions on Antennas and Propagation 47 (2), 233–241.
Lee, J.H., Lee, Y.T., 2001a. A novel direction finding method for cyclostationary signals. Signal Processing 81 (6), 1317–1323.
Lee, Y.T., Lee, J.H., 2001b. Robust adaptive array beamforming with random error in cycle frequency. IEE Proceedings. Radar, Sonar and Navigation 148 (4), 193–199.
Lee, J.H., Lee, Y.T., 2003. Robust technique for estimating the bearings of cyclostationary signals. Signal Processing 83 (5), 1035–1046.
Lee, J.H., Tung, C.H., 2002. Estimating the bearings of near-field cyclostationary signals. IEEE Transactions on Signal Processing 50 (1), 110–118.
Lee, J.H., Lee, Y.T., Shih, W.H., 2000. Efficient robust adaptive beamforming for cyclostationary signals. IEEE Transactions on Signal Processing 48 (7), 1893–1901.
Lee, D., Yeo, J., Han, B., Kim, B., Cho, J., 2016. Apparatus and method for performing properrizing frequency shift(p-fresh) vectorizing. U.S. Patent No. 9,344,310.
Lenart, L., 2011. Asymptotic distributions and subsampling in spectral analysis for almost periodically correlated time series. Bernoulli 17 (1), 290–319. https://doi.org/10.3150/10-BEJ269.
Lenart, L., 2013. Non-parametric frequency identification and estimation in mean function for almost periodically correlated time series. Journal of Multivariate Analysis (ISSN 0047-259X) 115, 252–269. https://doi.org/10.1016/j.jmva.2012.10.006.
Lenart, L., Pipien, M., 2013. Seasonality revisited – statistical testing for almost periodically correlated stochastic processes. Central European Journal of Economic Modelling and Econometrics 5 (2), 85–102.
Lenart, L., Leśkow, J., Synowiecki, R., 2008. Subsampling in testing autocovariance for periodically correlated time series. Journal of Time Series Analysis (ISSN 1467-9892) 29 (6), 995–1018. https://doi.org/10.1111/j.1467-9892.2008.00591.x.
Léonard, F., 2015. Time domain cyclostationarity signal-processing tools. Mechanical Systems and Signal Processing (ISSN 0888-3270) 62–63, 100–112. https://doi.org/10.1016/j.ymssp.2014.10.013.
Leonov, V.P., Shiryaev, A.N., 1959. On a method of calculation of semi-invariants. Theory of Probability and Its Applications 4 (3), 319–329.
Leonov, V.G., Mogilevskaya, L.Y., Khotuntsev, Y.L., Frolov, A.V., 1987. Correlation relations between the complex amplitudes of the noise components in bipolar and field-effect transistors in the presence of strong signals. Soviet Journal of Communications Technology & Electronics 32, 126–130.
Lepage, K.Q., Thomson, D.J., 2009. Spectral analysis of cyclostationary time-series: a robust method. Geophysical Journal International (ISSN 1365-246X) 179 (2), 1199–1212. https://doi.org/10.1111/j.1365-246X.2009.04339.x.

Leskow, J., 1994. Asymptotic normality of the spectral density estimators for almost periodically correlated stochastic processes. Stochastic Processes and Their Applications 52, 351–360.
Leśkow, J., Napolitano, A., 2002. Quantile prediction for time series in the fraction-of-time probability framework. Signal Processing (ISSN 0165-1684) 82 (11), 1727–1741. https://doi.org/10.1016/S0165-1684(02)00334-1.
Leśkow, J., Napolitano, A., 2006. Foundations of the functional approach for signal analysis. Signal Processing 86 (12), 3796–3825. https://doi.org/10.1016/j.sigpro.2006.03.028.
Leśkow, J., Napolitano, A., 2007. Non-relatively measurable functions for secure-communications signal design. Signal Processing 87 (11), 2765–2780. https://doi.org/10.1016/j.sigpro.2007.05.005.
Leskow, J., Synowiecki, R., 2010. On bootstrapping periodic random arrays with increasing period. Metrika (ISSN 0026-1335) 71 (3), 253–279. https://doi.org/10.1007/s00184-008-0228-x.
Leśkow, J., Weron, A., 1992. Ergodic behavior and estimation for periodically correlated processes. Statistics & Probability Letters 15, 299–304.
Leus, G., Tian, Z., 2011. Recovering second-order statistics from compressive measurements. In: 4th IEEE International Workshop on Computational Advances in Multi-Sensor Adaptive Processing. CAMSAP, pp. 337–340.
Lev-Ari, H., 1998. Adaptive RLS filtering under the cyclo-stationary regime. In: Proceedings of International Conference on Acoustics, Speech, and Signal Processing. ICASSP'98, pp. 2185–2188.
Li, H., Cheng, Q., 1997. Almost sure convergence analysis of mixed time averages and kth-order cyclic statistics. IEEE Transactions on Information Theory 43 (4), 1265–1268.
Li, Y., Ding, Z., 1994. ARMA system identification based on second-order cyclostationarity. IEEE Transactions on Signal Processing 42 (12), 3483–3494.
Li, W.K., Hui, Y.V., 1988. An algorithm for the exact likelihood of periodic autoregressive moving average models. Communications in Statistics. Simulation and Computation 17, 1483–1494.
Li, Y., Jayaweera, S., 2013. Dynamic spectrum tracking using energy and cyclostationarity-based multivariate non-parametric quickest detection for cognitive radios. IEEE Transactions on Wireless Communications (ISSN 1536-1276) 12 (7), 3522–3532. https://doi.org/10.1109/TW.2013.060413.121814.
Li, D., Jayaweera, S.K., 2017. Uncertainty modeling and price-based demand response scheme design in smart grid. IEEE Systems Journal (ISSN 1932-8184) 11 (3), 1743–1754. https://doi.org/10.1109/JSYST.2014.2369451.
Li, L., Ji, H., 2011. Radar emitter recognition based on cyclostationary signatures and sequential iterative least-square estimation. Expert Systems with Applications (ISSN 0957-4174) 38 (3), 2140–2147. https://doi.org/10.1016/j.eswa.2010.07.155.
Li, J.X., Ke, X.Z., 2010. Period estimation method for weak pulsars based on coherent statistic of cyclostationary signal. Acta Physica Sinica 59 (11), 8304–8309. https://doi.org/10.7498/aps.59.8304.
Li, L., Qu, L., 2003. Cyclic statistics in rolling bearing diagnosis. Journal of Sound and Vibration 267, 253–265.
Li, H., Cheng, Q., Yuan, B., 1998. Strong laws of large numbers for two dimensional processes. In: Proceedings of Fourth International Conference on Signal Processing. ICSP'98.
Li, T., Liang, W., Ding, Z., Tugnait, J.K., 2005. Blind multiuser detection for long-code CDMA systems with transmission-induced cyclostationarity. EURASIP Journal on Wireless Communications and Networking (ISSN 1687-1499) 2005 (2), 621515. https://doi.org/10.1155/WCN.2005.206.
Li, K., Liu, Y., Huo, K., Jiang, W., Li, X., Zhuang, Z., 2010. Estimation of micro-motion parameters based on cyclostationary analysis. IET Signal Processing (ISSN 1751-9675) 4 (3), 218–223. https://doi.org/10.1049/iet-spr.2009.0044.
Li, J.X., Ke, X.Z., Zhao, B.S., 2012. A new time-domain estimation method for period of pulsars. Acta Physica Sinica 61 (6), 69701. https://doi.org/10.7498/aps.61.069701.
Li, J., Wei, G., Ding, Y., 2013a. Adaptive beamforming based on covariance matrix reconstruction by exploiting interferences' cyclostationarity. Signal Processing (ISSN 0165-1684) 93 (9), 2543–2547. https://doi.org/10.1016/j.sigpro.2013.02.009.
Li, T., Qiu, T., Tang, H., 2013b. Optimum heart sound signal selection based on the cyclostationary property. Computers in Biology and Medicine (ISSN 0010-4825) 43 (6), 607–612. https://doi.org/10.1016/j.compbiomed.2013.03.002.
Li, T., Tang, H., Qiu, T., Park, Y., 2013c. Heart sound cancellation from lung sound record using cyclostationarity. Medical Engineering & Physics (ISSN 1350-4533) 35 (12), 1831–1836. https://doi.org/10.1016/j.medengphy.2013.05.004.

Li, S., Chu, N., Yan, P., Wu, D., Antoni, J., 2019. Cyclostationary approach to detect flow-induced effects on vibration signals from centrifugal pumps. Mechanical Systems and Signal Processing (ISSN 0888-3270) 114, 275–289. https://doi.org/10.1016/j.ymssp.2018.05.027.

Liang, Y.C., Leyman, A.R., 1998. (Almost) periodic moving average system identification using higher order cyclic-statistics. IEEE Transactions on Signal Processing 46 (3), 779–783.

Liang, Y.C., Leyman, A.R., Soong, B.H., 1997. (Almost) periodic FIR system identification using third-order cyclic-statistics. Electronics Letters 33 (5), 356–357.

Lii, K.S., Rosenblatt, M., 1998. Line spectral analysis for harmonizable processes. Proceedings of the National Academy of Sciences of the United States of America 95, 4800–4803.

Lii, K.S., Rosenblatt, M., 2002. Spectral analysis for harmonizable processes. The Annals of Statistics 30 (1), 258–297.

Lii, K.S., Rosenblatt, M., 2006. Estimation for almost periodic processes. The Annals of Statistics 34 (3), 1115–1139. https://doi.org/10.1214/009053606000000218.

Like, E., Chakravarthy, V., Ratazzi, P., Wu, Z., 2009. Signal classification in fading channels using cyclic spectral analysis. EURASIP Journal on Wireless Communications and Networking (ISSN 1687-1499) 2009 (1), 879812. https://doi.org/10.1155/2009/879812.

Lim, C.W., Wakin, M.B., 2012. CHOCS: a framework for estimating compressive higher order cyclostationary statistics. In: Proc. SPIE, vol. 8365.

Lim, C.W., Wakin, M.B., 2015. Compressive temporal higher order cyclostationary statistics. IEEE Transactions on Signal Processing (ISSN 1053-587X) 63 (11), 2942–2956. https://doi.org/10.1109/TSP.2015.2415760.

Lima, A.D., Silveira, L.F., de Souza, S.X., 2018. Spectrum sensing with a parallel algorithm for cyclostationary feature extraction. Computers & Electrical Engineering (ISSN 0045-7906) 71, 151–161. https://doi.org/10.1016/j.compeleceng.2018.07.016.

Liu, B., Franaszek, P.A., 1969. A class of time-varying digital filters. IEEE Transactions on Circuit Theory CT-16, 467–471.

Liu, P., Sohn, H., 2017. Development of nonlinear spectral correlation between ultrasonic modulation components. NDT & E International (ISSN 0963-8695) 91, 120–128. https://doi.org/10.1016/j.ndteint.2017.07.005.

Liu, H., Xu, G., Tong, L., Kailath, T., 1996. Recent developments in blind channel equalization: from cyclostationarity to subspaces. Signal Processing 50, 83–99.

Liu, Z.M., Huang, Z.T., Zhou, Y.Y., 2009. Generalized wideband cyclic MUSIC. EURASIP Journal on Advances in Signal Processing (ISSN 1687-6180) 2009 (1), 539727. https://doi.org/10.1155/2009/539727.

Liu, Y., Qiu, T., Sheng, H., 2012. Time-difference-of-arrival estimation algorithms for cyclostationary signals in impulsive noise. Signal Processing (ISSN 0165-1684) 92 (9), 2238–2247. https://doi.org/10.1016/j.sigpro.2012.02.016.

Liu, G., Sun, X., Liu, Y., Qin, Y., 2013. Low-complexity estimation of signal parameters via rotational invariance techniques algorithm for mixed far-field and near-field cyclostationary sources localisation. IET Signal Processing (ISSN 1751-9675) 7 (5), 382–388. https://doi.org/10.1049/iet-spr.2012.0394.

Liu, Y., Qiu, T., Li, J., 2015. Joint estimation of time difference of arrival and frequency difference of arrival for cyclostationary signals under impulsive noise. Digital Signal Processing (ISSN 1051-2004) 46, 68–80. https://doi.org/10.1016/j.dsp.2015.08.013.

Liu, J., Lu, Y., Zhang, Y., Wang, W., 2017. DOA estimation with enhanced DOFs by exploiting cyclostationarity. IEEE Transactions on Signal Processing (ISSN 1053-587X) 65 (6), 1486–1496. https://doi.org/10.1109/TSP.2016.2645542.

Liu, M., Wang, J., Li, B., 2018a. Non-data aided joint estimation of symbol timing offset and carrier frequency offset for OFDM/OQAM systems. AEÜ. International Journal of Electronics and Communications (ISSN 1434-8411) 87, 164–172. https://doi.org/10.1016/j.aeue.2018.02.021.

Liu, T., Qiu, T., Luan, S., 2018b. Cyclic correntropy: foundations and theories. IEEE Access (ISSN 2169-3536) 6, 34659–34669. https://doi.org/10.1109/ACCESS.2018.2847346.

Liu, Y., Zhang, Y., Qiu, T., Gao, J., Na, S., 2018c. Improved time difference of arrival estimation algorithms for cyclostationary signals in α-stable impulsive noise. Digital Signal Processing (ISSN 1051-2004) 76, 94–105. https://doi.org/10.1016/j.dsp.2018.02.010.

Lloyd, S.P., 1959. A sampling theorem for stationary (wide sense) stochastic processes. Transactions of the American Mathematical Society 92 (1), 1–12.

Loève, M., 1963. Probability Theory. Van Nostrand, Princeton, NJ.

Lohan, E., Lundén, J., Seco-Granados, G., Lopez-Salcedo, J., Koivunen, V., 2014. Cyclic frequencies of BOC-modulated GNSS signals and their potential within a cognitive positioning framework. Journal of the Institute of Navigation 61 (2), 95–114.

Lopez-Salcedo, J., Vazquez, G., 2006. Asymptotic equivalence between the unconditional maximum likelihood and the square-law nonlinearity symbol timing estimation. IEEE Transactions on Signal Processing (ISSN 1053-587X) 54 (1), 244–257. https://doi.org/10.1109/TSP.2005.859245.

Lops, M., Tulino, A.M., 2000. Simultaneous suppression of multiaccess and narrowband interference in asynchronous CDMA networks. IEEE Transactions on Vehicular Technology 49, 1705–1718.

Lops, M., Ricci, G., Tulino, A.M., 1998. Narrow-band-interference suppression in multiuser CDMA systems. IEEE Transactions on Communications 46 (9), 1163–1175.

Lugannani, R., 1973. Sample stability of periodically correlated pulse trains. Journal of the Franklin Institute 296, 179–190.

Lund, R., Basawa, I., 1997. Modeling and inference for periodically correlated time series. In: Gosh, S. (Ed.), Asymptotics, Nonparametrics and Time Series.

Lund, R., Basawa, I.V., 2000. Recursive prediction and likelihood evaluation for periodic ARMA models. Journal of Time Series Analysis 21 (1), 75–93.

Lund, R.B., Hurd, H.L., Bloomfield, P., Smith, R.L., 1995. Climatological time series with periodic correlation. Journal of Climate 11, 2787–2809.

Lunden, J., Koivunen, V., 2012. Locally optimum and suboptimum detection of complex-valued cyclostationary signals. In: 46th Annual Conference on Information Sciences and Systems. CISS 2012, pp. 1–5.

Lunden, J., Koivunen, V., 2014. Spatial sign and rank cyclic detectors. IEEE Signal Processing Letters (ISSN 1070-9908) 21 (5), 595–599. https://doi.org/10.1109/LSP.2014.2311886.

Lundén, J., Koivunen, V., Huttunen, A., Poor, H.V., 2009. Collaborative cyclostationary spectrum sensing for cognitive radio systems. IEEE Transactions on Signal Processing (ISSN 1053-587X) 57 (11), 4182–4195. https://doi.org/10.1109/TSP.2009.2025152.

Lundén, J., Kassam, S., Koivunen, V., 2010. Robust nonparametric cyclic correlation-based spectrum sensing for cognitive radio. IEEE Transactions on Signal Processing (ISSN 1053-587X) 58 (1), 38–52. https://doi.org/10.1109/TSP.2009.2029790.

Maciá-Fernández, G., Camacho, J., Magán-Carrión, R., García-Teodoro, P., Therón, R., 2018. UGR'16: a new dataset for the evaluation of cyclostationarity-based network IDSs. Computers & Security (ISSN 0167-4048) 73, 411–424. https://doi.org/10.1016/j.cose.2017.11.004.

Madhow, U., 1998. Blind adaptive interference suppression for direct-sequence CDMA. Proceedings of the IEEE 86 (10), 2049–2069.

Madhow, U., Honig, M., 1994. MMSE interference suppression for direct-sequence spread-spectrum CDMA. IEEE Transactions on Communications 42 (12), 3178–3188.

Madhow, U., Honig, M., 1999. On the average near-far resistance for MMSE detection of direct-sequence CDMA signals with random spreading. IEEE Transactions on Information Theory 45 (6), 2039–2045.

Mahdian, B., Saic, S., 2009. A cyclostationarity analysis applied to image forensics. In: Workshop on Applications of Computer Vision. WACV 2009, pp. 1–6.

Mahdian, B., Amsky, A., Saic, S., 2015. Detecting cyclostationarity in re-captured LCD screens. Journal of Forensic Research 6. https://doi.org/10.4172/21577145.1000294.

Mahmoudi, M.R., Nematollahi, A.R., Soltani, A.R., 2015. On the detection and estimation of the simple harmonizable processes. Iranian Journal of Science and Technology (ISSN 1028-6276) 39 (2), 239–242. https://doi.org/10.22099/ijsts.2015.3024.

Mahmoudi, M.R., Heydari, M.H., Avazzadeh, Z., 2018. On the asymptotic distribution for the periodograms of almost periodically correlated (cyclostationary) processes. Digital Signal Processing (ISSN 1051-2004) 81, 186–197. https://doi.org/10.1016/j.dsp.2018.07.004.

Mahmoudi, M.R., Heydari, M.H., Avazzadeh, Z., 2019. Testing the difference between spectral densities of two independent periodically correlated (cyclostationary) time series models. Communications in Statistics. Theory and Methods 48 (9), 1–9. https://doi.org/10.1080/03610926.2018.1472776.

Maiz, S., El Badaoui, M., Bonnardot, F., Serviere, C., 2014. New second order cyclostationary analysis and application to the detection and characterization of a runner's fatigue. Signal Processing (ISSN 0165-1684) 102, 188–200. https://doi.org/10.1016/j.sigpro.2014.03.024.

Majhi, S., Ho, T.S., 2015. Blind symbol-rate estimation and test bed implementation of linearly modulated signals. IEEE Transactions on Vehicular Technology (ISSN 0018-9545) 64 (3), 954–963. https://doi.org/10.1109/TVT.2014.2327985.

Makagon, A., 1999a. Induced stationary process and structure of locally square integrable periodically correlated processes. Studia Mathematica 136 (1), 71–86.

Makagon, A., 1999b. Theoretical prediction of periodically correlated sequences. Probability and Mathematical Statistics 19 (2), 287–322.

Makagon, A., 2001. Characterization of the spectra of periodically correlated processes. Journal of Multivariate Analysis 78 (1), 1–10.

Makagon, A., 2011. Stationary sequences associated with a periodically correlated sequence. Probability and Mathematical Statistics 31 (2), 263–283.

Makagon, A., Miamee, A.G., 1996. Weak law of large numbers for almost periodically correlated processes. Proceedings of the American Mathematical Society 124 (6), 1899–1902.

Makagon, A., Miamee, A.G., 1997. On the spectrum of correlation autoregressive sequences. Stochastic Processes and Their Applications 69, 179–193.

Makagon, A., Miamee, A., 2014. Structure of PC sequences and the 3rd prediction problem. In: Chaari, F., Leskow, J., Napolitano, A., Sanchez-Ramirez, A. (Eds.), Cyclostationarity: Theory and Methods. In: Lecture Notes in Mechanical Engineering. Springer International Publishing, pp. 53–72.

Makagon, A., Wylomanska, A., 2008. On the support of the spectral measure of a harmonizable sequence. Proceedings of the American Mathematical Society 136 (7), 2609–2613.

Makagon, A., Miamee, A.G., Salehi, H., 1991. Periodically correlated processes and their spectrum. In: Miamee, A.G. (Ed.), Proceeding of the Workshop on Nonstationary Stochastic Processes and Their Applications. World Scientific, Virginia, VA, pp. 147–164.

Makagon, A., Miamee, A.G., Salehi, H., 1994. Continuous time periodically correlated processes: spectrum and prediction. Stochastic Processes and Their Applications 49, 277–295.

Mäkilä, P., 2006. LTI approximation of nonlinear systems via signal distribution theory. Automatica (ISSN 0005-1098) 42 (6), 917–928. https://doi.org/10.1016/j.automatica.2006.02.018.

Mäkilä, P.M., Partington, J.R., Norlander, T., 1998. Bounded power signal spaces for robust control and modeling. SIAM Journal on Control and Optimization 37 (1), 92–117. https://doi.org/10.1137/S0363012997316664.

Makur, A., Mitra, S.K., 2001. Warped discrete-Fourier transform: theory and applications. IEEE Transactions on Circuits and Systems. I, Fundamental Theory and Applications 48 (9), 1086–1093.

Mandelbrot, B.B., Van Ness, H.W., 1968. Fractional Brownian motions, fractional noises and applications. SIAM Review 10, 422–436.

Mandyam, G.D., 2002. Digital-to-analog conversion of pulse amplitude modulated systems using adaptive quantization. Wireless Personal Communications 23, 253–281.

Mandyam, G.D., 2003. Adaptive quantization in a pulse-amplitude modulated system. U.S. Patent No. 6,535,564.

Mangalvedhe, N., 1999. Development and Analysis of Adaptive Interference Rejection Techniques for Direct Sequence Code Division Multiple Access Systems. Ph.D. Dissertation. Bradley Department of Electrical Engineering, Virginia Polytechnic Institute and State University.

Manouchehri, T., Nematollahi, A.R., 2019. On the estimation problem of periodic autoregressive time series: symmetric and asymmetric innovations. Journal of Statistical Computation and Simulation 89 (1), 71–97. https://doi.org/10.1080/00949655.2018.1535599.

Marchand, P., 1998. Detection and Classification of Digital Modulations by Higher Order Cyclic Statistics. Ph.D. Dissertation. INPG-Nat. Poly. Inst. Grenoble, Grenoble, France (in French).

Marchand, P., Amblard, P.O., Lacoume, J.L., 1995. Higher than second-order statistics for complex-valued cyclostationary signals. In: Proceedings of 15th GRETSI Symposium. Juan-Les-Pins, France, pp. 69–72.

Marchand, P., Lacoume, J.L., Le Martret, C., 1998. Multiple hypothesis modulation classification based on cyclic cumulants of different orders. In: Proceedings of International Conference on Acoustics, Speech, and Signal Processing, vol. 4. ICASSP'98, pp. 2157–2160.

Marey, M., Dobre, O.A., Inkol, R.J., 2012. Classification of space-time block codes based on second-order cyclostationarity with transmission impairments. IEEE Transactions on Wireless Communications (ISSN 1536-1276) 11 (7), 2574–2584. https://doi.org/10.1109/TWC.2012.041612.111488.

Mari, J., 1996. A counterexample in power signals space. IEEE Transactions on Automatic Control (ISSN 0018-9286) 41 (1), 115–116. https://doi.org/10.1109/9.481613.

Markelov, V.A., 1966a. Axis crossings and relative time of existence of a periodically nonstationary random process. Soviet Radiophysics 9, 440–443 (in Russian).

Markelov, V.A., 1966b. Extrusions and phases of the periodically nonstationary random process. Izvestiâ Vysših Učebnyh Zavedenij. Radiotekhnika 9 (5) (in Russian).

Markelov, V.A., 1966c. On extrusions and relative time of staying of the periodically nonstationary random process. Izvestiâ Vysših Učebnyh Zavedenij. Radiotekhnika 9 (4) (in Russian).

Martin, D.E., 1990. Estimation of the Minimal Period of Periodically Correlated Processes. Ph.D. Dissertation. University of Maryland.

Martin, D.E.K., 1999. Detection of periodic autocorrelation in time series data via zero crossing. Journal of Time Series Analysis 20 (4), 435–452.

Martin, W., Flandrin, P., 1985. Wigner-Ville spectral analysis of nonstationary processes. IEEE Transactions on Acoustics, Speech, and Signal Processing ASSP-33 (6), 1461–1470.

Martone, M., 1999. Adaptive multistage beamforming using cyclic higher order statistics. IEEE Transactions on Signal Processing 47 (10), 2867–2873.

Martone, M., 2000. Blind adaptive detection of DS/CDMA signals on time-varying multipath channels with antenna array using higher-order statistics. IEEE Transactions on Communications 48 (9), 1590–1600.

MathWorks, 1994. www.mathworks.com. https://www.mathworks.com. (Accessed 1 April 2019).

Matsko, I., Javorskyj, I., Yuzefovych, R., Zakrzewski, Z., 2018. Forced oscillations of cracked beam under the stochastic cyclic loading. Mechanical Systems and Signal Processing (ISSN 0888-3270) 104, 242–263. https://doi.org/10.1016/j.ymssp.2017.08.021.

Mattera, D., 1999. Identification of polyperiodic Volterra systems by means of input-output noisy measurements. Signal Processing 75 (1), 41–50.

Mattera, D., Paura, L., 1998. Higher-order cyclostationarity-based methods for identifying Volterra systems by input-output noisy measurements. Signal Processing 67 (1), 77–98.

Matz, G., Hlawatsch, F., Kozek, W., 1997. Generalized evolutionary spectral analysis and the Weyl spectrum of nonstationary random processes. IEEE Transactions on Signal Processing 45 (6), 1520–1534. https://doi.org/10.1109/78.599994.

Mazzenga, F., 1998. Blind multipath channel identification for wideband communication systems based on cyclostationary statistics. European Transactions on Telecommunications and Related Technologies 9 (1), 27–31.

Mazzenga, F., 2000a. Blind adaptive parameter estimation for CDMA systems using cyclostationary statistics. European Transactions on Telecommunications and Related Technologies 11 (5), 495–500.

Mazzenga, F., 2000b. On the identifiability of a channel transfer function from cyclic and conjugate cyclic statistics. European Transactions on Telecommunications and Related Technologies 11 (3), 293–296.

Mazzenga, F., Vatalaro, F., 1996. Parameter estimation in CDMA multiuser detection using cyclostationary statistics. Electronics Letters 32 (3).

McCallister, R.D., Ronald, D., Shearer III, D.D., 1990. Compensating for distortion in a communication channel. U.S. Patent No. 4,922,506.

McCormick, A.C., Nandi, A.K., 1998. Cyclostationarity in rotating machinery vibrations. Mechanical Systems and Signal Processing 12 (2), 225–242.

McLeod, A.I., 1993. Parsimony, model adequacy and periodic correlation in time series forecasting. International Statistical Review 61, 387–393.

McLeod, A.I., Noakes, D.J., Hipel, K.W., Thompstone, R.M., 1987. Combining hydrologic forecasts. Journal of Water Resources Planning and Management 113, 29–41.

McLernon, D.C., 1991. Analysis of LMS algorithm with inputs from cyclostationary random processes. Electronics Letters 27, 136–138.

McLernon, D., 1998. Inter-relationships between different structures for periodic systems. In: Proceedings of IX European Signal Processing Conference. EUSIPCO'98, Rhodes, Greece, pp. 1885–1888.

Mehr, A.S., Chen, T., 2001. On alias-component matrices of discrete-time periodically time-varying systems. IEEE Signal Processing Letters 8, 114–116.

Mehr, A.S., Chen, T., 2002. Representation of linear periodically time-varying multirate systems. IEEE Transactions on Signal Processing 50 (9), 2221–2229.

Melititskiy, V.A., Mosionzhik, A.I., 1987. A probabilistic model of non Gaussian periodically nonstationary radio signals. Soviet Journal of Communications Technology & Electronics 32, 100–106.
Melititskiy, V.A., Radzievskii, V.O., 1973. Statistical characteristics of a nonstationary Normal process and ways of its modeling. Radioelectronics and Communications Systems 16, 82–89 (English translation of Izv. VUZ Radioelektron., vol. 16, pp. 82–89).
Mendel, J.M., 1991. Tutorial on higher-order statistics (spectra) in signal processing and system theory: theoretical results and some applications. Proceedings of the IEEE 79, 278–305.
Mendoza, R., Reed, J.H., Hsia, T.C., Agee, B.G., 1991. Interference rejection using time-dependent constant modulus algorithms (CMA) and the hybrid CMA/spectral correlation discriminator. IEEE Transactions on Signal Processing 39 (9), 2108–2111.
Meng, Z., Shen, F., Zhou, W., 2017. Adaptive beamforming using the cyclostationarity of the source signals. Electronics Letters (ISSN 0013-5194) 53 (13), 858–860. https://doi.org/10.1049/el.2017.1190.
Mengali, U., Pezzani, E., 1978. Tracking properties of phase-locked loops in optical communication systems. IEEE Transactions on Communications COM-26, 1811–1818.
Mesiya, M.F., McLane, P.J., Campbell, L.L., 1978. Optimal receiver filters for BPSK transmission over a bandlimited nonlinear channel. IEEE Transactions on Communications COM-26, 12–22.
Meyer, R.A., Burrus, C.S., 1975. A unified analysis of multirate and periodically time-varying digital filters. IEEE Transactions on Circuits and Systems CAS-22, 162–168.
Mezentsev, V.P., Yavorskii, I.N., 1988. Estimation of probability characteristics of rhythmic signals as a problem of linear filtration. Radioelectronics and Communications Systems 31, 70–72.
Miamee, A.G., 1990. Periodically correlated processes and their stationary dilations. SIAM Journal on Applied Mathematics 50 (4), 1194–1199.
Miamee, A.G., 1993. Explicit formulas for the best linear predictor and predictor error matrix of a periodically correlated process. SIAM Journal on Mathematical Analysis 24, 703–711.
Miamee, A.G., 1994. On recent developments in prediction theory for cyclostationary processes. In: Gardner, W.A. (Ed.), Cyclostationarity in Communications and Signal Processing. IEEE Press, pp. 480–493.
Miamee, A.G., Hardin, J.C., 1990. On a class of nonstationary stochastic processes. Sankhya. The Indian Journal of Statistics 52, Series A, Pt. 2, 145–156.
Miamee, A.G., Salehi, H., 1980. On the prediction of periodically correlated stochastic processes. In: Krishnaiah, P.R. (Ed.), Multivariate Analysis–V. North Holland Publishing Company, Amsterdam, pp. 167–179.
Miamee, A.G., Shahkar, G.H., 2002. Shift operator for periodically correlated processes. Indian Journal of Pure and Applied Mathematics 33 (5), 705–712.
Middleton, D., 1967. A statistical theory of reverberation and similar first-order scattered fields, Part II: moments spectra, and special distributions. IEEE Transactions on Information Theory IT-13 (3), 393–414.
Mihandoost, S., Amirani, M.C., 2017. Cyclic spectral analysis of electrocardiogram signals based on GARCH model. Biomedical Signal Processing and Control (ISSN 1746-8094) 31, 79–88. https://doi.org/10.1016/j.bspc.2016.07.012.
Miller, S., 1995. An adaptive direct sequence code division multiple access receiver for multiuser interference rejection. IEEE Transactions on Communications 43 (2), 1746–1755.
Ming, Y., Chen, J., Dong, G., 2011. Weak fault feature extraction of rolling bearing based on cyclic Wiener filter and envelope spectrum. Mechanical Systems and Signal Processing (ISSN 0888-3270) 25 (5), 1773–1785. https://doi.org/10.1016/j.ymssp.2010.12.002.
Ming, Y., Chen, J., Dong, G., 2014. Application of convolved blind separation based on second-order cyclic statistics in rolling element bearing feature extraction. Journal of Vibration and Control 20 (4), 617–633. https://doi.org/10.1177/1077546312463744.
Mirbagheri, A., Plataniotis, K., Pasupathy, S., 2006. An enhanced widely linear CDMA receiver with OQPSK modulation. IEEE Transactions on Communications (ISSN 0090-6778) 54 (2), 261–272. https://doi.org/10.1109/TCOMM.2005.863766.
Mirdamadi, A., Sabbaghian, M., 2014. Spectrum sensing of interleaved SC-FDMA signals in cognitive radio networks. IEEE Transactions on Vehicular Technology (ISSN 0018-9545) 64 (4), 1633–1637. https://doi.org/10.1109/TVT.2014.2331695.
Moeneclaey, M., 1982a. Comment on 'Tracking performance of the filter and square bit synchronizer'. IEEE Transactions on Communications COM-30, 407–410.

Moeneclaey, M., 1982b. Linear phase-locked loop theory for cyclostationary input disturbances. IEEE Transactions on Communications COM-30 (10), 2253–2259.
Moeneclaey, M., 1983. The optimum closed-loop transfer function of a phase-locked loop used for synchronization purposes. IEEE Transactions on Communications COM-31, 549–553.
Moeneclaey, M., 1984. A fundamental lower bound on the performance of practical joint carrier and bit synchronizers. IEEE Transactions on Communications COM-32, 1007–1012.
Mohammadpour, M., Soltani, A.R., 2014. Forward moving average representations for MA processes of finite order: multivariate stationary and periodically correlated. Communications in Statistics. Theory and Methods 43 (1), 141–150. https://doi.org/10.1080/03610926.2012.656874.
Monin, A.S., 1963. Stationary and periodic time series in the general circulation of the atmosphere. In: Rosenblatt, M. (Ed.), Proceedings of the Symposium on Time Series Analysis. Wiley, New York, pp. 144–151.
Monti, C.M., Pierobon, G.L., 1985. Block codes for linear timing recovery in data transmission systems. IEEE Transactions on Communications COM-33, 527–534.
Morelande, M.R., Zoubir, A.M., 2002. On the performance of cyclic moment-based parameter estimators of amplitude modulated polynomial phase signals. IEEE Transactions on Signal Processing 50 (3), 590–605.
Morse, M., Transue, W., 1956. ℂ-bimeasures and their integral extensions. Annals of Mathematics 64, 480–504.
Moshrefzadeh, A., Fasana, A., 2018. The autogram: an effective approach for selecting the optimal demodulation band in rolling element bearings diagnosis. Mechanical Systems and Signal Processing (ISSN 0888-3270) 105, 294–318. https://doi.org/10.1016/j.ymssp.2017.12.009.
Mosquera, C., Scalise, S., Lopez-Valcarce, R., 2008. Non-data-aided symbol rate estimation of linearly modulated signals. IEEE Transactions on Signal Processing (ISSN 1053-587X) 56 (2), 664–674. https://doi.org/10.1109/TSP.2007.907888.
Munk, W., Worchester, P., Wunsch, C., 1995. Ocean Acoustic Tomography. Cambridge University Press, New York.
Murphy, C.D., 2005. Symbol constellations having second-order statistics with cyclostationary phase. U.S. Patent No. 6,952,445.
Murty, M.S., Shrestha, R., 2018a. Hardware implementation and VLSI design of spectrum sensor for next-generation LTE-A cognitive-radio wireless network. IET Circuits, Devices & Systems (ISSN 1751-858X) 12 (5), 542–550. https://doi.org/10.1049/iet-cds.2017.0292.
Murty, M.S., Shrestha, R., 2018b. Reconfigurable and memory-efficient cyclostationary spectrum sensor for cognitive-radio wireless networks. IEEE Transactions on Circuits and Systems. II, Express Briefs (ISSN 1549-7747) 65 (8), 1039–1043. https://doi.org/10.1109/TCSII.2018.2790952.
Nakatsuka, T., 1986. Periodic property of streetcar congestion at the first station. Journal of the Operations Research Society of Japan 29, 1–20.
Napolitano, A., 1994. Spectral-Correlation Based Methods for Multipath Channel Identification. Ph.D. Dissertation. Università di Napoli Federico II, Napoli, Italy, Dipartimento di Ingegneria Elettronica (in Italian).
Napolitano, A., 1995. Cyclic higher-order statistics: input/output relations for discrete- and continuous-time MIMO linear almost-periodically time-variant systems. Signal Processing 42 (2), 147–166. https://doi.org/10.1016/0165-1684(94)00124-I.
Napolitano, A., 2001. On the spectral correlation measurement of nonstationary stochastic processes. In: Thirty-Fifth Annual Asilomar Conference on Signals, Systems, and Computers. Pacific Grove, CA.
Napolitano, A., 2003. Uncertainty in measurements on spectrally correlated stochastic processes. IEEE Transactions on Information Theory 49 (9), 2172–2191. https://doi.org/10.1109/TIT.2003.815768.
Napolitano, A., 2004. Mean-square consistency of statistical-function estimators for generalized almost-cyclostationary processes. In: Proceedings of XII European Signal Processing Conference. EUSIPCO 2004, Vienna, Austria.
Napolitano, A., 2005. Asymptotic normality of statistical-function estimators for generalized almost-cyclostationary processes. In: Proceedings of XIII European Signal Processing Conference. EUSIPCO 2005, Antalya, Turkey.
Napolitano, A., 2007a. Estimation of second-order cross-moments of generalized almost-cyclostationary processes. IEEE Transactions on Information Theory 53 (6), 2204–2228. https://doi.org/10.1109/TIT.2007.896868.

Napolitano, A., 2007b. Generalized almost-cyclostationary processes and spectrally correlated processes: two extensions of the class of the almost-cyclostationary processes. In: IEEE-EURASIP International Symposium on Signal Processing and Its Applications. ISSPA 2007, Sharjah, United Arab Emirates.

Napolitano, A., 2007c. Mean-square consistent estimation of the spectral correlation density for spectrally correlated stochastic processes. In: IEEE International Conference on Acoustics, Speech, and Signal Processing. ICASSP 2007, Honolulu, Hawai, USA.

Napolitano, A., 2009. Discrete-time estimation of second-order statistics of generalized almost-cyclostationary processes. IEEE Transactions on Signal Processing 57 (5), 1670–1688. https://doi.org/10.1109/TSP.2009.2013889.

Napolitano, A., 2010a. Interpolation and decimation of spectrally correlated stochastic processes. In: XVIII European Signal Processing Conference. EUSIPCO 2010, Aalborg, Denmark.

Napolitano, A., 2010b. Sampling theorems for Doppler-stretched wide-band signals. Signal Processing 90 (7), 2276–2287. https://doi.org/10.1016/j.sigpro.2010.02.016.

Napolitano, A., 2011. Sampling of spectrally correlated processes. IEEE Transactions on Signal Processing 59 (2), 525–539. https://doi.org/10.1109/TSP.2010.2090873.

Napolitano, A., 2012. Generalizations of Cyclostationary Signal Processing: Spectral Analysis and Applications. John Wiley & Sons Ltd - IEEE Press.

Napolitano, A., 2013. Generalizations of cyclostationarity: a new paradigm for signal processing for mobile communications, radar, and sonar. IEEE Signal Processing Magazine (ISSN 1053-5888) 30 (6), 53–63. https://doi.org/10.1109/MSP.2013.2265101.

Napolitano, A., 2014. Wide-band moving source passive localization in highly corruptive environments. In: IEEE International Conference on Acoustics, Speech, and Signal Processing. ICASSP 2014, Firenze, Italy.

Napolitano, A., 2015a. Asymptotic normality of cyclic autocorrelation estimate with estimated cycle frequency. In: XXIII European Signal Processing Conference. EUSIPCO 2015, Nice, France.

Napolitano, A., 2015b. Characterization of cyclostationary signals and their generalizations. In: Boashash, B. (Ed.), Time-Frequency Signal Analysis and Processing: a Comprehensive Reference. Elsevier–Academic Press, pp. 562–573.

Napolitano, A., 2016a. Cyclostationarity: limits and generalizations. Signal Processing 120, 323–347. https://doi.org/10.1016/j.sigpro.2015.09.013.

Napolitano, A., 2016b. Cyclostationarity: new trends and applications. Signal Processing 120, 385–408. https://doi.org/10.1016/j.sigpro.2015.09.011.

Napolitano, A., 2016c. On cyclic spectrum estimation with estimated cycle frequency. In: XXIV European Signal Processing Conference. EUSIPCO 2016, Budapest, Hungary.

Napolitano, A., 2017a. Cyclic statistic estimators with uncertain cycle frequencies. IEEE Transactions on Information Theory 63 (1), 649–675. https://doi.org/10.1109/TIT.2016.2614321.

Napolitano, A., 2017b. On characterization and application of oscillatory almost-cyclostationary processes. In: XXV European Signal Processing Conference. EUSIPCO 2017, Kos, Greece.

Napolitano, A., 2017c. Time-warped almost-cyclostationary signals: characterization and statistical function measurements. IEEE Transactions on Signal Processing 65 (20), 5526–5541. https://doi.org/10.1109/TSP.2017.2728499.

Napolitano, A., 2018. On cyclostationarity-based signal detection. In: XXVI European Signal Processing Conference. EUSIPCO 2018, Rome, Italy.

Napolitano, A., Doğançay, K., 2011. On waveform design in interference-tolerant range-Doppler estimation for wideband multistatic radars. In: XIX European Signal Processing Conference. EUSIPCO 2011, Barcelona, Spain.

Napolitano, A., Doğançay, K., 2018. Bandpass sampling of almost-cyclostationary signals. Signal Processing 153, 266–274. https://doi.org/10.1016/j.sigpro.2018.07.003.

Napolitano, A., Gardner, W.A., 2016. Algorithms for analysis of signals with time-warped cyclostationarity. In: 50th Asilomar Conference on Signals, Systems, and Computers. Pacific Grove, California.

Napolitano, A., Perna, I., 2014a. Cyclic spectral analysis of the GPS signal. Digital Signal Processing 33, 13–33. https://doi.org/10.1016/j.dsp.2014.06.003.

Napolitano, A., Perna, I., 2014b. Effects of satellite motion on the received signal in GPS. In: Chaari, F., Leśkow, J., Napolitano, A., Sanchez-Ramirez, A. (Eds.), Cyclostationarity: Theory and Methods. In: Lecture Notes in Mechanical Engineering. Springer, pp. 125–140.

Napolitano, A., Spooner, C.M., 2000. Median-based cyclic polyspectrum estimation. IEEE Transactions on Signal Processing 48 (5), 1462–1466. https://doi.org/10.1109/78.839992.

Napolitano, A., Spooner, C.M., 2001. Cyclic spectral analysis of continuous-phase modulated signals. IEEE Transactions on Signal Processing 49 (1), 30–44. https://doi.org/10.1109/78.890336.

Napolitano, A., Tanda, M., 2001. Blind parameter estimation in multiple-access systems. IEEE Transactions on Communications 49 (4), 688–698.

Napolitano, A., Tanda, M., 2004. Doppler-channel blind identification for non-circular transmissions in multiple-access systems. IEEE Transactions on Communications 52 (12), 2172–2191.

Napolitano, A., Tesauro, M., 2011. Almost-periodic higher-order statistic estimation. IEEE Transactions on Information Theory 57 (1), 514–533. https://doi.org/10.1109/TIT.2010.2090244.

Napolitano, A., Paura, L., Tanda, M., 1992. Signal detection in cyclostationary generalized Gaussian noise with unknown parameters. European Transactions on Telecommunications and Related Technologies 3 (1), 39–43.

Napolitano, A., Ricciardelli, S., Tanda, M., 2005. Blind estimation of amplitudes, phases, time delays and frequency offsets in multiple access systems with circular transmissions. Signal Processing 85 (8), 1588–1601.

Narasimhan, S.V., Hazarataiah, M., Giridhar, P.V.S., 2005. Channel blind identification based on cyclostationarity and group delay. Signal Processing 85 (7), 1275–1286.

Narieda, S., Kageyama, T., 2013. Simple spectrum sensing techniques based on cyclostationarity detection in cognitive radio networks. Electronics Letters (ISSN 0013-5194) 49 (17), 1108–1109. https://doi.org/10.1049/el.2013.2156.

Neipp, C., Hernandez, A., Rodes, J.J., Marquez, A., Belendez, T., Belendez, A., 2003. An analysis of the classical Doppler effect. European Journal of Physics 24, 497–505.

Nematollahi, A.R., Rao, T.S., 2005. On the spectral density estimation of periodically correlated (cyclostationary) time series. Sankhya. The Indian Journal of Statistics (2003–2007) (ISSN 0972-7671) 67 (3), 568–589.

Newton, H.J., 1982. Using periodic autoregressions for multiple spectral estimation. Technometrics 24, 109–116.

Ngoya, E., 2011. On the cyclostationary noise analysis in large RF integrated circuits. IEEE Transactions on Circuits and Systems. I, Regular Papers (ISSN 1549-8328) 58 (11), 2729–2740. https://doi.org/10.1109/TCSI.2011.2143070.

Nicolas, J.J., Lim, J.S., 1995. Method and apparatus for decoding broadcast digital HDTV in the presence of quasi-cyclostationary interference. U.S. Patent No. 5,453,797.

Nie, H., Mathiopoulos, P., 2006. Adaptive prediction and cancellation digitization method for wideband multistandard software radio base-station receivers. IEEE Transactions on Vehicular Technology (ISSN 0018-9545) 55 (3), 887–902. https://doi.org/10.1109/TVT.2005.863410.

NIST, 2010. In: Olver, F.W.J., Lozier, D.W., Boisvert, R.F., Clark, C.W. (Eds.), NIST Handbook of Mathematical Functions. Cambridge University Press, New York.

Noguet, D., Biard, L., Laugeois, M., 2010. Cyclostationarity detectors for cognitive radio: architectural tradeoffs. EURASIP Journal on Wireless Communications and Networking (ISSN 1687-1499) 2010 (1), 526429. https://doi.org/10.1155/2010/526429.

Oberg, J., 2004. Titan calling. IEEE Spectrum 41, 28–33.

Obeysekera, J.T.B., Salas, J.D., 1986. Modeling of aggregated hydrologic time series. Journal of Hydrology 86, 197–219.

Obrezkov, G.V., Sizyakov, Y.A., 1985. Approximate analysis of pulsed tracking systems when there is fluctuating interference. Radioelectronics and Communications Systems 28, 63–65 (English translation of Izv. VUZ Radioelectron.).

Ogura, H., 1971. Spectral representation of a periodic nonstationary random process. IEEE Transactions on Information Theory IT-17, 143–149.

Ogura, H., Yoshida, Y., 1982. Time series analysis of a periodic stationary random process. Transactions of the Institute of Electronics and Communication Engineers of Japan 365-A, 22–29 (in Japanese).

Ohno, S., Sakai, H., 1996. Optimization of filter banks using cyclostationary spectral analysis. IEEE Transactions on Signal Processing 44 (11), 2718–2725.

Øigård, T.A., Hanssen, A., Scharf, L.L., 2006. Spectral correlations of fractional Brownian motion. Physical Review E 74, 031114. https://doi.org/10.1103/PhysRevE.74.031114.

Ojeda, O., Grajal, J., 2012. Sensitivity analysis of cyclostationarity-based and radiometric detectors for single-sensor receivers. IEEE Transactions on Aerospace and Electronic Systems (ISSN 0018-9251) 48 (1), 27–43. https://doi.org/10.1109/TAES.2012.6129619.

Okumura, M., Tanimoto, H., Itakura, T., Sugawara, T., 1993. Numerical noise analysis for nonlinear circuits with a periodic large signal excitation including cyclostationary noise sources. IEEE Transactions on Circuits and Systems. I, Fundamental Theory and Applications 40, 581–590.

Olhede, S., Ombao, H., 2013. Modeling and estimation of covariance of replicated modulated cyclical time series. IEEE Transactions on Signal Processing (ISSN 1053-587X) 61 (8), 1944–1957. https://doi.org/10.1109/TSP.2012.2237168.

Omel'chenko, V.A., Omel'chenko, A.V., Dragan, Y.P., Kolesnikov, O.A., 1988. Recognising Gaussian periodically-correlated random signals. I. Radiotekhnika 85, 75–79 (in Russian).

Oner, M., 2008a. On the cyclostationary statistics of ultra-wideband signals in the presence of timing and frequency jitter. AEÜ. International Journal of Electronics and Communications (ISSN 1434-8411) 62 (3), 174–184. https://doi.org/10.1016/j.aeue.2007.03.010.

Oner, M., 2008b. On the spectral correlation of UWB impulse radio signals. IEEE Communications Letters (ISSN 1089-7798) 12 (10), 714–716. https://doi.org/10.1109/LCOMM.2008.080941.

Oner, M., 2009. Spectral correlation of a digital pulse stream modulated by a cyclostationary sequence in the presence of timing jitter. IEEE Transactions on Communications (ISSN 0090-6778) 57 (2), 339–342. https://doi.org/10.1109/TCOMM.2009.02.070163.

Oner, M., Dobre, O., 2011. On the second-order cyclic statistics of signals in the presence of receiver impairments. IEEE Transactions on Communications (ISSN 0090-6778) 59 (12), 3278–3284. https://doi.org/10.1109/TCOMM.2011.081111.100028.

Oner, M., Jondral, F., 2007a. Air interface identification for software radio systems. AEÜ. International Journal of Electronics and Communications (ISSN 1434-8411) 61 (2), 104–117. https://doi.org/10.1016/j.aeue.2006.03.005.

Oner, M., Jondral, F., 2007b. On the extraction of the channel allocation information in spectrum pooling systems. IEEE Journal on Selected Areas in Communications (ISSN 0733-8716) 25 (3), 558–565. https://doi.org/10.1109/JSAC.2007.070406.

Oppenheim, A.V., Johnson, D.H., 1972. Discrete representations of signals. Proceedings of the IEEE 60 (6), 681–691.

Oppenheim, A.V., Schafer, R.W., 1989. Discrete-Time Signal Processing. Prentice-Hall, Englewood Cliffs, NJ.

Oppenheim, A., Johnson, D.H., Steiglitz, K., 1971. Computation of spectra with unequal resolution using the fast Fourier transform. Proceedings of the IEEE 59 (2), 299–301.

O'Reilly, J.J., 1984. Timing extraction for baseband digital transmission. In: Cattermole, K.W., O'Reilly, J.J. (Eds.), Problems of Randomness in Communication Engineering. Plymouth, London.

Ortiz, M.J., Ruiz de Elvira, A., 1985. A cyclo-stationary model of sea surface temperature in the Pacific Ocean. Tellus. Series A 37, 14–23.

Osborn, D.R., 1991. The implications of periodically varying coefficients for seasonal time-series processes. Journal of Econometrics 48, 373–384.

Osborn, D.R., Smith, J.P., 1989. The performance of periodic autoregressive models in forecasting seasonal U.K. consumption. Journal of Business and Economic Statistics 7, 117–127 (see also Journal of Applied Econometrics, vol. 3, pp. 255-266, 1989).

Osteyee, D., 1981. Testing Markov properties of time-series. In: Proceedings of Time-Series Analysis. Houston, TX, pp. 385–400.

Ozaktas, H., Zalevsky, Z., Kutay, M., 2001. The Fractional Fourier Transform With Applications in Optics and Signal Processing. John Wiley and Sons Series in Pure and Applied Optics.

Pagano, M., 1978. On periodic and multiple autoregressions. The Annals of Statistics 6, 1310–1317.

Pandharipande, A., Dasgupta, S., 1999. Subband coding of cyclostationary signals with static bit allocation. IEEE Signal Processing Letters 6 (11), 284–286.

Papoulis, A., 1972. Narrow-band systems and gaussianity. IEEE Transactions on Information Theory (ISSN 0018-9448) 18 (1), 20–27. https://doi.org/10.1109/TIT.1972.1054731.

Papoulis, A., 1977. Signal Analysis. McGraw-Hill, New York.

Papoulis, A., 1983. Random modulation: a review. IEEE Transactions on Acoustics, Speech, and Signal Processing ASSP-31, 96–105.

Papoulis, A., 1991. Probability, Random Variables, and Stochastic Processes, third ed. McGrawHill, New York.

Parzen, E., 1957. On consistent estimates of the spectrum of stationary time series. The Annals of Mathematical Statistics 28 (2), 329–348.

Parzen, E., Pagano, M., 1979. An approach to modeling seasonally stationary time-series. Journal of Econometrics 9, 137–153.

Pateros, C., Saulnier, G., 1992. Interference suppression and multipath mitigation using an adaptive correlator direct sequence spread spectrum receiver. In: Proceedings of IEEE ICC, pp. 662–666.

Pateros, C., Saulnier, G., 1996. An adaptive correlator receiver for direct-sequence spread-spectrum communication. IEEE Transactions on Communications 44, 1543–1552.

Pelkowitz, L., 1981. Frequency domain analysis of wraparound error in fast convolution algorithms. IEEE Transactions on Acoustics, Speech, and Signal Processing ASSP-29, 413–422.

Peng, C., Bo, L., 2014. Vibration signal analysis of journal bearing supported rotor system by cyclostationarity. Shock and Vibration 2014, 952958. https://doi.org/10.1155/2014/952958.

Pennacchi, P., Borghesani, P., Chatterton, S., 2015. A cyclostationary multi-domain analysis of fluid instability in Kaplan turbines. Mechanical Systems and Signal Processing (ISSN 0888-3270) 60–61, 375–390. https://doi.org/10.1016/j.ymssp.2014.08.026.

Petersen, B.R., Falconer, D.D., 1991. Minimum mean square equalization in cyclostationary and stationary interference–analysis and subscriber line calculations. IEEE Journal on Selected Areas in Communications 9 (6), 931–940.

Pfaffelhuber, E., 1975. Generalized harmonic analysis for distributions. IEEE Transactions on Information Theory IT-21, 605–611.

Philippe, A., Surgailis, D., Viano, M.C., 2006. Almost periodically correlated processes with long memory. In: Bertail, P., Soulier, P., Doukhan, P. (Eds.), Dependence in Probability and Statistics. In: Lecture Notes in Statistics, vol. 187. Springer, New York, pp. 159–194.

Phong, V.Q., 2007. A new proof and generalizations of Gearhart's theorem. Proceedings of the American Mathematical Society 135 (7), 2065–2072.

Picinbono, B., 1996. Second-order complex random vectors and Normal distributions. IEEE Transactions on Signal Processing 44, 2637–2640.

Picinbono, B., Bondon, P., 1997. Second-order statistics of complex signals. IEEE Transactions on Signal Processing 45 (2), 411–420.

Picinbono, B., Chevalier, P., 1995. Widely linear estimation with complex data. IEEE Transactions on Signal Processing (ISSN 1053-587X) 43 (8), 2030–2033. https://doi.org/10.1109/78.403373.

Pierini, S., 2014. Ensemble simulations and pullback attractors of a periodically forced double-gyre system. Journal of Physical Oceanography 44, 3245–3254. https://doi.org/10.1175/JPO-D-14-0117.1.

Piranashvili, Z.A., Kavtaradze, L.O., 1975. Certain properties of periodic nonstationary random processes. V.I. Lenin. Sakharth. Politekh. Inst. Samecn. Shrom, Math. Mehk. 3 (176), 93–105 (in Russian).

Politis, D.N., 1998. Computer intensive methods in statistical analysis. IEEE Signal Processing Magazine 15 (1), 39–55.

Politis, D.N., 2005. Complex-valued tapers. IEEE Signal Processing Letters 12 (7), 512–515.

Porotskii, S.M., Tsymbalyuk, M.Y., 1979. Extension of stochastic approximation procedures to periodically nonstationary random processes. Automation and Remote Control 40, Part 2, 606–609 (English translation of Avtom. and Telemekh., vol. 40, pp. 152–155).

Pourahmadi, M., Salehi, H., 1983. On subordination and linear transformation of harmonizable and periodically correlated processes. In: Probability Theory on Vector Spaces, III. Springer (Lublin), Berlin-New York, pp. 195–213.

Prakriya, S., Hatzinakos, D., 1997. Blind identification of linear subsystems of LTI-ZMNL-LTI models with cyclostationary inputs. IEEE Transactions on Signal Processing 45 (8), 2023–2036.

Prendergast, R., Nguyen, T., 2006. Minimum mean-squared error reconstruction for generalized undersampling of cyclostationary processes. IEEE Transactions on Signal Processing (ISSN 1053-587X) 54 (8), 3237–3242. https://doi.org/10.1109/TSP.2006.877649.

Pries, A., Ramirez, D., Schreier, P.J., 2018. LMPIT-inspired tests for detecting a cyclostationary signal in noise with spatio-temporal structure. IEEE Transactions on Wireless Communications (ISSN 1536-1276) 17 (9), 6321–6334.

Priestley, M.B., 1965. Evolutionary spectra and non-stationary processes. Journal of the Royal Statistical Society, Series B, Methodological 27 (2), 204–237.

Proakis, J., 1993. Reduced-complexity simultaneous beamforming and equalization for underwater acoustic communications. In: Proceedings of Oceans'93.

Prohorov, Y.V., Rozanov, Y.A., 1989. Probability Theory. Springer-Verlag, New York.

Punchihewa, A., Zhang, Q., Dobre, O., Spooner, C., Rajan, S., Inkol, R., 2010. On the cyclostationarity of OFDM and single carrier linearly digitally modulated signals in time dispersive channels: theoretical developments and application. IEEE Transactions on Wireless Communications (ISSN 1536-1276) 9 (8), 2588–2599. https://doi.org/10.1109/TWC.2010.061510.091080.

Punchihewa, A., Bhargava, V., Despins, C., 2011. Blind estimation of OFDM parameters in cognitive radio networks. IEEE Transactions on Wireless Communications (ISSN 1536-1276) 10 (3), 733–738. https://doi.org/10.1109/TWC.2010.010411.100276.

Pupolin, S., Tomasi, C., 1984. Spectral analysis of line regenerator time jitter. IEEE Transactions on Communications COM-32, 561–566.

Qiu, Z.H., Cassara, F.A., 1989. Steady-state analysis of the clock synchronizer for a narrowband communication system. International Journal of Electronics 66, 551–570.

Quezada Acuña, D., Molina Vicuña, C., 2015. Damage assessment of rolling element bearing using cyclostationary processing of AE signals with electromagnetic interference. In: Chaari, F., Leskow, J., Napolitano, A., Zimroz, R., Wylomanska, A., Dudek, A. (Eds.), Cyclostationarity: Theory and Methods – II, Applied Condition Monitoring, vol. 3. Springer International Publishing, pp. 43–54.

Raad, A., 2003. Contributions to the Higher-Order Cyclic Statistics: Applications to Gear Faults. Ph.D. Dissertation. Université de Technologie Compiègne, Compiègne, France (in French).

Raad, A., Antoni, J., Sidahmed, M., 2002. Third-order cyclic characterization of vibration signals in rotating machinery. In: Proceedings of XI European Signal Processing Conference. EUSIPCO'02. Toulouse, France.

Raad, A., Antoni, J., Sidahmed, M., 2008. Indicators of cyclostationarity: theory and application to gear fault monitoring. Mechanical Systems and Signal Processing (ISSN 0888-3270) 22 (3), 574–587. https://doi.org/10.1016/j.ymssp.2007.09.011.

Rabiner, L.R., Shafer, R.W., Rader, C.M., 1969. The chirp z-transform algorithm. IEEE Transactions on Audio and Electroacoustics AU-17 (2), 86–92.

Rahimzadeh, F., Shahtalebi, K., Parvaresh, F., 2018. Cyclic frequency distinction and its application in modulation parameter estimation: an adaptive approach. AEÜ. International Journal of Electronics and Communications (ISSN 1434-8411) 84, 13–20. https://doi.org/10.1016/j.aeue.2017.11.003.

Ramirez, D., Scharf, L., Via, J., Santamaria, I., Schreier, P., 2014. An asymptotic GLRT for the detection of cyclostationary signals. In: IEEE International Conference on Acoustics, Speech and Signal Processing. ICASSP 2014, pp. 3415–3419.

Ramirez, D., Schreier, P., Via, J., Santamaria, I., Scharf, L., 2015. Detection of multivariate cyclostationarity. IEEE Transactions on Signal Processing (ISSN 1053-587X) 63 (20), 5395–5408. https://doi.org/10.1109/TSP.2015.2450201.

Ramkumar, B., 2009. Automatic modulation classification for cognitive radios using cyclic feature detection. IEEE Circuits and Systems Magazine (ISSN 1531-636X) 9 (2), 27–45. https://doi.org/10.1109/MCAS.2008.931739.

Randall, R., 2011. Vibration-Based Condition Monitoring: Industrial, Aerospace and Automotive Applications. Wiley Online Library, Wiley. ISBN 9780470977583.

Randall, R.B., Antoni, J., Chobsaard, S., 2001. The relationship between spectral correlation and envelope analysis in the diagnostics of bearing faults and other cyclostationary machine signals. Mechanical Systems and Signal Processing 15 (5), 945–962.

Rao, M.M., 2008. Integral representations of second-order processes. Nonlinear Analysis 69, 979–986.

Rapajic, P., Vucetic, B., 1994. Adaptive receiver structures for asynchronous CDMA systems. IEEE Journal on Selected Areas in Communications 12 (4), 685–697.

Rebeiz, E., Jain, V., Cabric, D., 2012. Cyclostationary-based low complexity wideband spectrum sensing using compressive sampling. In: IEEE International Conference on Communications. ICC, pp. 1619–1623.

Rebeiz, E., Urriza, P., Cabric, D., 2013. Optimizing wideband cyclostationary spectrum sensing under receiver impairments. IEEE Transactions on Signal Processing (ISSN 1053-587X) 61 (15), 3931–3943. https://doi.org/10.1109/TSP.2013.2262680.

Rebeiz, E., Yuan, F.L., Urriza, P., Markovic, D., Cabric, D., 2014. Energy-efficient processor for blind signal classification in cognitive radio networks. IEEE Transactions on Circuits and Systems. I, Regular Papers (ISSN 1549-8328) 61 (2), 587–599. https://doi.org/10.1109/TCSI.2013.2278392.

Rebeiz, E., Ghadam, A., Valkama, M., Cabric, D., 2015. Spectrum sensing under RF non-linearities: performance analysis and DSP-enhanced receivers. IEEE Transactions on Signal Processing (ISSN 1053-587X) 63 (8), 1950–1964. https://doi.org/10.1109/TSP.2015.2401532.

Reed, J.H., 1987. Time-Dependent Adaptive Filters for Interference Rejection. Ph.D. Dissertation. Department of Electrical Engineering and Computer Science, University of California, Davis, California.

Reed, J.H., Hsia, T.C., 1990. The performance of time-dependent adaptive filters for interference rejection. IEEE Transactions on Acoustics, Speech, and Signal Processing 38 (8), 1373–1385.

Reed, D.E., Wickert, M.A., 1989. Nonstationary moments of a random binary pulse train. IEEE Transactions on Information Theory 35, 700–703.

Reed, J.H., Quilici, A.A., Hsia, T.C., 1988. A frequency domain time dependent adaptive filter for interference rejection. In: Proceedings of MILCOM 88, vol. 2. San Diego, CA, pp. 391–397.

Reed, J.H., Greene, C., Hsia, T.C., 1989. Demodulation of a direct sequence spread-spectrum signal using an optimal time-dependent filter. In: Proceedings of IEEE Militar Communications Conference. MILCOM'89, Boston, MA, pp. 657–662.

Reed, J.H., Yuen, N.M., Hsia, T.C., 1995. An optimal receiver using a time-dependent adaptive filter. IEEE Transactions on Communications 43 (2/3/4), 187–190.

Renard, J., Verlant-Chenet, J., Dricot, J.M., De Doncker, P., Horlin, F., 2010. Higher-order cyclostationarity detection for spectrum sensing. EURASIP Journal on Wireless Communications and Networking (ISSN 1687-1499) 2010 (1), 721695. https://doi.org/10.1155/2010/721695.

Renard, J., Lampe, L., Horlin, F., 2013a. Nonparametric cyclic polyspectrum-based spectrum sensing. IEEE Wireless Communications Letters (ISSN 2162-2337) 2 (1), 98–101. https://doi.org/10.1109/WCL.2012.120312.120749.

Renard, J., Lampe, L., Horlin, F., 2013b. Spatial sign cyclic-feature detection. IEEE Transactions on Signal Processing (ISSN 1053-587X) 61 (18), 4521–4531. https://doi.org/10.1109/TSP.2013.2271483.

Renger, A., 1971. Behaviour of linear oscillatory systems which are energized by stochastic impulse trains. Maschinenbau Tech. 20, 596–600 (in German).

Renger, A., 1977. Spectral analysis of periodically nonstationary stochastic impulse processes. Zeitschrift für Angewandte Mathematik und Mechanik 57, 681–692 (in German).

Restrepo, A., Chacon, L.P., 1998. On the period of sums of discrete periodic signals. IEEE Signal Processing Letters (ISSN 1070-9908) 5 (7), 164–166. https://doi.org/10.1109/97.700917.

Reuven, A.M., Weiss, A.J., 2009. Direct position determination of cyclostationary signals. In: Special Section: Visual Information Analysis for Security. Signal Processing (ISSN 0165-1684) 89 (12), 2448–2464. https://doi.org/10.1016/j.sigpro.2009.04.009.

Riba, J., Vazquez, G., 1994. Bayesian recursive estimation of frequency and timing exploiting the cyclostationarity property. Signal Processing 40, 21–37.

Riba, J., Villares, J., Vazquez, G., 2010. A nondata-aided SNR estimation technique for multilevel modulations exploiting signal cyclostationarity. IEEE Transactions on Signal Processing (ISSN 1053-587X) 58 (11), 5767–5778. https://doi.org/10.1109/TSP.2010.2059017.

Riba, J., Font-Segura, J., Villares, J., Vazquez, G., 2014. Frequency-domain GLR detection of a second-order cyclostationary signal over fading channels. IEEE Transactions on Signal Processing (ISSN 1053-587X) 62 (8), 1899–1912. https://doi.org/10.1109/TSP.2014.2303433.

Rihaczek, A.W., 1967. Radar resolution of moving targets. IEEE Transactions on Information Theory IT-13, 51–56.

Roberts, R.S., 1989. Architectures for Digital Cyclic Spectral Analysis. Ph.D. Dissertation. University of California, Davis, California, Department of Electrical Engineering and Computer Science.

Roberts, R.S., Loomis Jr., H.H., 1995. Parallel computation structures for a class of cyclic spectral analysis algorithms. Journal of VLSI Signal Processing 10, 25–40.

Roberts, R.S., Brown, W.A., Loomis Jr., H.H., 1991. Computationally efficient algorithms for cyclic spectral analysis. IEEE Signal Processing Magazine 8, 38–49.

Roberts, R.S., Brown, W.A., Loomis Jr., H.H., 1994. A review of digital spectral correlation analysis: theory and implementation. In: Gardner, W.A. (Ed.), Cyclostationarity in Communications and Signal Processing. IEEE Press, pp. 455–479.

Rocha, F.L., 1980. Adaptive deconvolution of cyclostationary signals. I. Revista Telegrafica Electronica 68, 1026–1030 (in Spanish).

Rodriguez, P.M., Fernandez, Z., Torrego, R., Lizeaga, A., Mendicute, M., Val, I., 2017. Low-complexity cyclostationary-based modulation classifying algorithm. AEÜ. International Journal of Electronics and Communications (ISSN 1434-8411) 74, 176–182. https://doi.org/10.1016/j.aeue.2017.02.008.

Rootenberg, J., Ghozati, S.A., 1977. Stability properties of periodic filters. International Journal of Systems Science 8, 953–959.

Rootenberg, J., Ghozati, S.A., 1978. Generation of a class of nonstationary random processes. International Journal of Systems Science 9, 935–947.

Rosenblatt, M., 1974. Random Processes. Springer-Verlag, New York.

Rosenblatt, M., 1985. Stationary Sequences and Random Fields. Birkhäuser, Boston.

Rostaing, P., 1996. Detection of Modulated Signals by Exploiting Their Cyclostationarity Properties: Application to Sonar Signals. Ph.D. Dissertation. University of Nice-Sophia Antipolis, Nice, France (in French).

Rostaing, P., Thierry, E., Pitarque, T., 1999. Asymptotic performance analysis of cyclic detectors. IEEE Transactions on Communications 47 (1), 10–13.

Roussel, J., Haritopoulos, M., Ravier, P., Buttelli, O., 2013. Cyclostationary analysis of electromyographic signals. In: Proceedings of the 21st European Signal Processing Conference. EUSIPCO 2013.

Roussel, J., Ravier, P., Haritopoulos, M., Farina, D., Buttelli, O., 2017. Decomposition of multi-channel intramuscular EMG signals by cyclostationary-based blind source separation. IEEE Transactions on Neural Systems and Rehabilitation Engineering (ISSN 1534-4320) 25 (11), 2035–2045. https://doi.org/10.1109/TNSRE.2017.2700890.

Roy, S., Yang, J., Kumar, P.S., 1994. Joint transmitter/receiver optimization for multiuser communications. In: Gardner, W.A. (Ed.), Cyclostationarity in Communications and Signal Processing. IEEE Press, pp. 329–361.

Roychowdhury, J.S., 2004. Apparatus and method for reduced-order modeling of time-varying systems and computer storage medium containing the same. U.S. Patent No. 6,687,658.

Roychowdhury, J., Long, D., Feldmann, P., 1998. Cyclostationary noise analysis of large RF circuits with multitone excitations. IEEE Journal of Solid-State Circuits 33 (3), 324–336.

Roychowdhury, J., Feldmann, P., Long, D.E., 2000. Method of making an integrated circuit including noise modeling and prediction. U.S. Patent No. 6,072,947.

Rozenvasser, E.N., 1973. Periodically Nonstationary Control Systems. Izdat. Nauka, Moskow (in Russian).

Rudin, W., 1987. Real and Complex Analysis, third ed. McGraw-Hill.

Ruiz-Medina, M.D., 2011. Spatial autoregressive and moving average Hilbertian processes. Journal of Multivariate Analysis (ISSN 0047-259X) 102 (2), 292–305. https://doi.org/10.1016/j.jmva.2010.09.005.

Rybakov, B.S., 1980. Optimum detection of a class of random Gaussian pulsed signals in the presence of noise with a priori parametric uncertainty. Radio Engineering and Electronic Physics 25, 62–66 (English translation of Radiotekhnika i Electronika, vol. 25, pp. 1198–1202).

Rybakov, B.S., 1981. Correlation properties of the envelope and phase of a periodically nonstationary Normal process. Telecommunications & Radio Engineering 35–36, 96–97 (English translation of Elektrosvyaz and Radiotekhnika).

Rybakov, B.S., 1986. Correlation coefficients of the envelope and phase of a signal for periodically nonstationary Gaussian noise. Radiotekhnika 41, 3–7 (in Russian).

Rytov, S.M., Kratsov, Y.A., Tatarskii, V.I., 1988. Principles of Statistical Radiophysics. Correlation Theory of Random Processes, vol. 2. Springer-Verlag (Chapter 4).

Sabat, S., Srinu, S., Kumar, N., Udgata, S., 2010. FPGA realization of spectrum sensing techniques for cognitive radio network. In: International Workshop on Cognitive Radio. IWCR 2010, pp. 1–5.

Sabharwal, A., Mitra, U., Moses, R., 2001. MMSE receivers for multirate DS-CDMA systems. IEEE Transactions on Communications 49 (12), 2184–2197.

Sabri, K., El Badaoui, M., Guillet, F., Adib, A., Aboutajdine, D., 2009. A frequency domain-based approach for blind MIMO system identification using second-order cyclic statistics. Signal Processing (ISSN 0165-1684) 89 (1), 77–86. https://doi.org/10.1016/j.sigpro.2008.07.011.

Sabri, K., El Badaoui, M., Guillet, F., Belli, A., Millet, G., Morin, J.B., 2010. Cyclostationary modeling of ground reaction force signals. Signal Processing (ISSN 0165-1684) 90 (4), 1146–1152. https://doi.org/10.1016/j.sigpro.2009.09.027.

Sadeghi, H., Azmi, P., 2014. Performance analysis of linear cooperative cyclostationary spectrum sensing over Nakagami-m fading channels. IEEE Transactions on Vehicular Technology (ISSN 0018-9545) 63 (9), 4748–4756. https://doi.org/10.1109/TVT.2014.2312110.

Sadeghi, B., Yu, R., 2016. Shift-variance and nonstationarity of linear periodically shift-variant systems and applications to generalized sampling-reconstruction processes. IEEE Transactions on Signal Processing (ISSN 1053-587X) 64 (6), 1493–1506. https://doi.org/10.1109/TSP.2015.2502545.

Sadeghi, H., Azmi, P., Arezumand, H., 2012a. Cyclostationarity-based cooperative spectrum sensing over imperfect reporting channels. AEÜ. International Journal of Electronics and Communications (ISSN 1434-8411) 66 (10), 833–840. https://doi.org/10.1016/j.aeue.2012.02.004.

Sadeghi, H., Azmi, P., Arezumand, H., 2012b. Cyclostationarity-based soft cooperative spectrum sensing for cognitive radio networks. IET Communications (ISSN 1751-8628) 6 (1), 29–38. https://doi.org/10.1049/iet-com.2011.0269.

Sadler, B.M., 1995. Acousto-optic cyclostationary signal processing. Applied Optics 34 (23), 5091–5099.

Sadler, B.M., Dandawaté, A.V., 1998. Nonparametric estimation of the cyclic cross spectrum. IEEE Transactions on Information Theory 44, 351–358. https://doi.org/10.1109/18.651065.

Sadowsky, J.S., Kafedziski, V., 1998. On the correlation and scattering functions of the WSS channel for mobile communications. IEEE Transactions on Vehicular Technology 47 (1), 270–282.

Safatly, L., Aziz, B., Nafkha, A., Louet, Y., Nasser, Y., El-Hajj, A., Kabalan, K., 2014. Blind spectrum sensing using symmetry property of cyclic autocorrelation function: from theory to practice. EURASIP Journal on Wireless Communications and Networking (ISSN 1687-1499) 2014 (1), 26. https://doi.org/10.1186/1687-1499-2014-26.

Saghafian, S., Hopp, W.J., Van Oyen, M.P., Desmond, J.S., Kronick, S.L., 2014. Complexity-augmented triage: a tool for improving patient safety and operational efficiency. Manufacturing & Service Operations Management 16 (3), 329–345. https://doi.org/10.1287/msom.2014.0487.

Sahmoudi, M., Yang, C., Calmettes, V., 2010. The merits of the cyclostationarity of BOC signals for a cognitive GNSS receiver design. In: IEEE/ION Position Location and Navigation Symposium. PLANS 2010, pp. 1181–1188.

Sakai, H., 1982. Circular lattice filtering using Pagano's method. IEEE Transactions on Acoustics, Speech, and Signal Processing ASSP-30, 279–287.

Sakai, H., 1983. Covariance matrices characterization by a set of scalar partial autocorrelation coefficients. The Annals of Statistics 11, 337–340.

Sakai, H., 1990. Spectral analysis and lattice filter for periodic autoregressive processes. Electronics & Communications in Japan. Part 3 73, 9–15.

Sakai, H., 1991. On the spectral density matrix of a periodic ARMA process. Journal of Time Series Analysis 12, 73–82.

Sakai, H., Sakaguchi, F., 1990. Simultaneous confidence bands for the spectral estimate of two-channel autoregressive processes. Journal of Time Series Analysis 11, 49–56.

Salas, J.D., Smith, R.A., 1980. Correlation properties of periodic $AR(p)$ models. In: Proceedings of the 3rd International Symposium on Stochastic Hydraulics. Tokyo, pp. 107–115.

Salas, J.D., Boes, D.C., Smith, R.A., 1982. Estimation of ARMA models with seasonal parameters. Water Resources Research 18, 1006–1010.

Sankar, B., Tugbay, N., 1988. The use of the Wigner-Ville distribution (WVD) for the classification and the analysis of modulation techniques. In: Proceedings of ISCIS III. Cesme, Turkey, pp. 75–82.

Sarnaglia, A.J.Q., Reisen, V.A., Lévy-Leduc, C., 2010. Robust estimation of periodic autoregressive processes in the presence of additive outliers. Journal of Multivariate Analysis (ISSN 0047-259X) 101 (9), 2168–2183. https://doi.org/10.1016/j.jmva.2010.05.006.

Sathe, V.P., Vaidyanathan, P.P., 1993. Effects of multirate systems on statistical properties of random signals. IEEE Transactions on Signal Processing 41 (1), 131–146.

Savic, M., 1994. Detection of cholesterol deposits in arteries. U.S. Patent No. 5,327,893.

Scharf, L.L., 1991. Statistical Signal Processing. Detection, Estimation, and Time Series Analysis. Addison-Wesley Publishing Company, New York.

Scharf, L.L., Schreier, P.J., Hanssen, A., 2005. The Hilbert space geometry of the Rihaczek distribution for stochastic analytic signals. IEEE Signal Processing Letters 12 (4), 297–300.

Schell, S.V., 1990. Exploitation of Spectral Correlation for Signal-Selective Direction Finding. Ph.D. Dissertation. University of California, Davis, CA, Department of Electrical Engineering and Computer Science.

Schell, S.V., 1994a. An overview of sensor array processing for cyclostationary signals. In: Gardner, W.A. (Ed.), Cyclostationarity in Communications and Signal Processing. IEEE Press, pp. 168–239.
Schell, S.V., 1994b. Performance analysis of the cyclic MUSIC method of direction estimation for cyclostationary signals. IEEE Transactions on Signal Processing 42 (11), 3043–3050.
Schell, S.V., 1995a. Asymptotic moments of estimated cyclic correlation matrices. IEEE Transactions on Signal Processing 43 (1), 173–180.
Schell, S.V., 1995b. Higher-order cyclostationarity properties of coded communication signals. In: Proceedings of the Twenty-Ninth Annual Asilomar Conference on Signals, Systems, and Computers. Pacific Grove, CA.
Schell, S.V., Agee, B.G., 1988. Application of the SCORE algorithm and SCORE extensions to sorting in the rank-L spectral self-coherence environment. In: Conference Record, Twenty-Second Asilomar Conference on Signals, Systems and Computers. Pacific Grove, CA, pp. 274–278.
Schell, S.V., Gardner, W.A., 1989. Signal-selective direction finding for fully correlated signals. In: Abstracts of Sixth Multidimensional Signal Processing Workshop of the IEEE/ASSP Society. Pacific Grove, CA, pp. 139–140.
Schell, S.V., Gardner, W.A., 1990a. The Cramer-Rao lower bound for parameters of Gaussian cyclostationary signals. In: Proceedings of International Symposium on Information Theory and Its Applications. ISITA'90, Honolulu, HI, pp. 255–258.
Schell, S.V., Gardner, W.A., 1990b. Detection of the number of cyclostationary signals in unknown interference and noise. In: Proceedings of the Twenty-Fourth Annual Asilomar Conference on Signals, Systems, and Computers. Pacific Grove, CA, pp. 473–477.
Schell, S.V., Gardner, W.A., 1990c. Progress on signal-selective direction finding. In: Proceedings of the Fifth IEEE/ASSP Workshop on Spectrum Estimation and Modeling. Rochester, NY, pp. 144–148.
Schell, S.V., Gardner, W.A., 1990d. Signal-selective high-resolution direction finding in multipath. In: Proceedings of International Conference on Acoustic, Speech, and Signal Processing. ICASSP'90, Albuquerque, NM, pp. 2667–2670.
Schell, S.V., Gardner, W.A., 1992. The Cramér-Rao lower bound for parameters of Gaussian cyclostationary signals. IEEE Transactions on Information Theory 38 (4), 1418–1422.
Schell, S.V., Gardner, W.A., 1993a. Blind adaptive spatio-temporal filtering for wideband cyclostationary signals. IEEE Transactions on Signal Processing 41, 1961–1964.
Schell, S.V., Gardner, W.A., 1993b. High resolution direction finding. In: Bose, N.K., Rao, C.R. (Eds.), Handbook of Statistics. Elsevier, Amsterdam, NL.
Schell, S.V., Gardner, W.A., 1994. Spatio-temporal filtering and equalization for cyclostationary signals. In: Leondes, C.T. (Ed.), Control and Dynamic Systems. Academic Press.
Schell, S.V., Gardner, W.A., 1995. Programmable canonical correlation analysis: a flexible framework for blind adaptive spatial filtering. IEEE Transactions on Signal Processing 43 (12), 2898–2908.
Schell, S.V., Calabretta, R.A., Gardner, W.A., Agee, B.G., 1989. Cyclic MUSIC algorithms for signal-selective direction estimation. In: Proceedings of the International Conference on Acoustics, Speech, and Signal Processing. ICASSP'89, Glasgow, UK, pp. 2278–2281.
Schkoda, R., Lund, R., Wagner, J., 2014. Clustering of cyclostationary signals with applications to climate station sitings, eliminations, and merges. IEEE Journal of Selected Topics in Applied Earth Observations and Remote Sensing (ISSN 1939-1404) 7 (5), 1754–1762. https://doi.org/10.1109/JSTARS.2013.2293713.
Schlotz, R.A., 2002. Problems in modeling UWB channels. In: Thirty-Sixth Asilomar Conference on Signals, Systems, and Computers. Pacific Grove, CA.
Schoonover, R.W., Davis, B.J., Bartels, R.A., Carney, P.S., 2008. Optical interferometry with pulsed fields. Journal of Modern Optics 55 (10), 1541–1556. https://doi.org/10.1080/09500340701847895.
Schoonover, R.W., Davis, B.J., Bartels, R.A., Carney, P.S., 2009a. Propagation of spatial coherence in fast pulses. Journal of the Optical Society of America. A 26 (9), 1945–1953. https://doi.org/10.1364/JOSAA.26.001945.
Schoonover, R.W., Davis, B.J., Carney, P.S., 2009b. The generalized Wolf shift for cyclostationary fields. In: Frontiers in Optics 2009/Laser Science XXV/Fall 2009 OSA Optics & Photonics Technical Digest. Optical Society of America, p. FWH3.
Schoonover, R., Lavarello, R., Oelze, M., Carney, P., 2011. Observation of generalized Wolf shifts in short pulse spectroscopy. Applied Physics Letters (ISSN 0003-6951) 98 (25), 251107. https://doi.org/10.1063/1.3599842.

Schreier, P.J., Scharf, L.L., 2003a. Second-order analysis of improper complex random vectors and processes. IEEE Transactions on Signal Processing 51 (3), 714–725.
Schreier, P.J., Scharf, L.L., 2003b. Stochastic time-frequency analysis using the analytic signal: why the complementary distribution matters. IEEE Transactions on Signal Processing 51 (12), 3071–3079.
Schreier, P.J., Scharf, L.L., 2010. Statistical Signal Processing of Complex-Valued Data: The Theory of Improper and Noncircular Signals. Cambridge University Press.
Scott, K.E., Olasz, E.B., 1995. Simultaneous clock phase and frequency offset estimation. IEEE Transactions on Communications 43 (7), 2263–2270.
Scott, K.E., Kaube, M., Anvari, K., 1994. Timing and automatic frequency control of digital receiver using the cyclic properties of a non-linear operation. U.S. Patent No. 5,282,228.
Seco-Granados, G., Lopez-Salcedo, J., Jimenez-Banos, D., Lopez-Risueno, G., 2012. Challenges in indoor global navigation satellite systems: unveiling its core features in signal processing. IEEE Signal Processing Magazine (ISSN 1053-5888) 29 (2), 108–131.
Semenov, V.A., Smirnov, A.I., 1983. On the stability of linear stochastic systems with periodically nonstationary parametric excitation. Mechanics of Solids 18, 14–20 (English translation of Izvestiya Akademii Nauk SSR, Mekhanika Tverdogo Tela).
Sepidband, P., Entesari, K., 2015. A CMOS spectrum sensor based on quasi-cyclostationary feature detection for cognitive radios. IEEE Transactions on Microwave Theory and Techniques (ISSN 0018-9480) 63 (12), 4098–4109. https://doi.org/10.1109/TMTT.2015.2496239.
Serena, P., Bononi, A., 2015. A time-domain extended Gaussian noise model. Journal of Lightwave Technology (ISSN 0733-8724) 33 (7), 1459–1472. https://doi.org/10.1109/JLT.2015.2398873.
Serfling, R.J., 1980. Approximation Theorems of Mathematical Statistics. John Wiley & Sons, Inc., USA.
Serpedin, E., Giannakis, G.B., 1998. Blind channel identification and equalization with modulation induced cyclostationarity. IEEE Transactions on Signal Processing 46 (7), 1930–1944.
Serpedin, E., Chevreuil, A., Giannakis, G.B., Loubaton, P., 2000. Blind channel and carrier frequency offset estimation using periodic modulation precoders. IEEE Transactions on Signal Processing 48 (8), 2389–2405.
Serpedin, E., Panduru, F., Sari, I., Giannakis, G.B., 2005. Bibliography on cyclostationarity. Signal Processing (ISSN 0165-1684) 85 (12), 2233–2303. https://doi.org/10.1016/j.sigpro.2005.05.002.
Shaked, R., Shlezinger, N., Dabora, R., 2018. Joint estimation of carrier frequency offset and channel impulse response for linear periodic channels. IEEE Transactions on Communications (ISSN 0090-6778) 66 (1), 302–319. https://doi.org/10.1109/TCOMM.2017.2741943.
Shamsunder, S., Giannakis, G.B., 1994. Signal selective localization of nonGaussian cyclostationary sources. IEEE Transactions on Signal Processing 42 (10), 2860–2864.
Shamsunder, S., Giannakis, G.B., Friedlander, B., 1995. Estimating random amplitude polynomial phase signals: a cyclostationary approach. IEEE Transactions on Signal Processing 43 (2), 492–505.
Shao, Q., Lund, R., 2004. Computation and characterization of autocorrelations and partial autocorrelations in periodic ARMA models. Journal of Time Series Analysis 25 (3), 359–372.
Shell, S., Biowas, K.K., Sihka, A.K., 1981. Modelling and predictions of stochastic processes involving periodicities. Applied Mathematical Modelling 5, 241–245.
Shen, J.C., Alsusa, E., 2013a. An efficient multiple lags selection method for cyclostationary feature based spectrum-sensing. IEEE Signal Processing Letters (ISSN 1070-9908) 20 (2), 133–136. https://doi.org/10.1109/LSP.2012.2233471.
Shen, J.C., Alsusa, E., 2013b. Joint cycle frequencies and lags utilization in cyclostationary feature spectrum sensing. IEEE Transactions on Signal Processing (ISSN 1053-587X) 61 (21), 5337–5346. https://doi.org/10.1109/TSP.2013.2278810.
Shevgunov, T., 2019. A comparative example of cyclostationary description of a non-stationary random process. Journal of Physics. Conference Series 1163, 012037. https://doi.org/10.1088/1742-6596/1163/1/012037.
Shevgunov, T., Efimov, E., Zhukov, D., 2018. Averaged absolute spectral correlation density estimator. In: 2018 Moscow Workshop on Electronic and Networking Technologies. MWENT, pp. 1–4.
Shi, M., Bar-Ness, Y., Su, W., 2009. Revisiting the timing and frequency offset estimation based on cyclostationarity with new improved method. IEEE Communications Letters (ISSN 1089-7798) 13 (7), 537–539. https://doi.org/10.1109/LCOMM.2009.090149.

Shiktorov, P., Starikov, E., Gružinskis, V., Reggiani, L., Varani, L., Vaissière, J.C., Gonzáles, S.P.T., 2003. Monte Carlo simulation in electronic noise in semiconductor materials and devices operating under cyclostationary conditions. Journal of Computational Electronics 2, 455–458.
Shishebor, Z., Nematollahi, A.R., Soltani, A.R., 2006. On covariance generating functions and spectral densities of periodically correlated autoregressive processes. Journal of Applied Mathematics and Stochastic Analysis 2006, 94746. https://doi.org/10.1155/JAMSA/2006/94746.
Shlezinger, N., Dabora, R., 2014. Frequency-shift filtering for OFDM signal recovery in narrowband power line communications. IEEE Transactions on Communications (ISSN 0090-6778) 62 (4), 1283–1295. https://doi.org/10.1109/TCOMM.2014.020514.130421.
Shlezinger, N., Dabora, R., 2015. On the capacity of narrowband PLC channels. IEEE Transactions on Communications (ISSN 0090-6778) 63 (4), 1191–1201. https://doi.org/10.1109/TCOMM.2015.2408318.
Shlezinger, N., Todros, K., 2019. Performance analysis of LMS filters with non-gaussian cyclostationary signals. Signal Processing (ISSN 0165-1684) 154, 260–271. https://doi.org/10.1016/j.sigpro.2018.08.008.
Shlezinger, N., Todros, K., Dabora, R., 2017. Adaptive filtering based on time-averaged MSE for cyclostationary signals. IEEE Transactions on Communications (ISSN 0090-6778) 65 (4), 1746–1761. https://doi.org/10.1109/TCOMM.2017.2655526.
Silaiev, A.M., Yakimov, A.V., 1979. Periodic nonstationarity of noise in harmonic systems. Radiotekhnika i Elektronika 24, 1806–1811 (in Russian).
Sills, J.A., Black, Q.R., 2004. Signal recognizer for communications signals. U.S. Patent No. 6,690,746.
Silva, I., Moody, G., 2014. An open-source toolbox for analysing and processing PhysioNet Databases in MATLAB and Octave. Journal of Open Research Software 2 (1), e27.
Silverman, R.A., 1957. Locally stationary random processes. IRE Transactions on Information Theory 3 (3), 182–187.
Sirianunpiboon, S., Howard, S.D., Cochran, D., 2018. Detection of cyclostationarity using generalized coherence. In: IEEE International Conference on Acoustics, Speech and Signal Processing. ICASSP 2018, Calgary, Alberta, Canada.
Smirnov, V.I., 1964. A Course of Higher Mathematics, I. Pergamon, Oxford, UK.
Smith, B.M., Potter, P.G., 1986. Design criteria for crosstalk interference between digital signals in multipair cable. IEEE Transactions on Communications COM-34, 593–599.
Socheleau, F.X., Houcke, S., Ciblat, P., Aissa-El-Bey, A., 2011. Cognitive OFDM system detection using pilot tones second and third-order cyclostationarity. Signal Processing (ISSN 0165-1684) 91 (2), 252–268. https://doi.org/10.1016/j.sigpro.2010.07.003.
Soderstrom, T., Stoica, P., 1989. System Identification. Prentice-Hall Int., New York.
Soedjack, H., 2002. Consistent estimation of the bispectral density function of a harmonizable process. Journal of Statistical Planning and Inference 100, 159–170.
Soltani, A., Azimmohseni, M., 2007. Periodograms asymptotic distributions in periodically correlated processes and multivariate stationary processes: an alternative approach. Journal of Statistical Planning and Inference (ISSN 0378-3758) 137 (4), 1236–1242. https://doi.org/10.1016/j.jspi.2006.01.015.
Soltani, A., Hashemi, M., 2011a. Periodically correlated autoregressive Hilbertian processes. Statistical Inference for Stochastic Processes (ISSN 1387-0874) 14 (2), 177–188. https://doi.org/10.1007/s11203-011-9056-0.
Soltani, A., Hashemi, M., 2011b. Periodically correlated autoregressive Hilbertian processes of order p. In: Ferraty, F. (Ed.), Recent Advances in Functional Data Analysis and Related Topics, Contributions to Statistics. Physica-Verlag HD, pp. 301–306.
Soltani, A., Parvardeh, A., 2005. Decomposition of discrete time periodically correlated and multivariate stationary symmetric stable processes. Stochastic Processes and Their Applications (ISSN 0304-4149) 115 (11), 1838–1859. https://doi.org/10.1016/j.spa.2005.06.005.
Soltani, A., Parvardeh, A., 2006. Simple random measures and simple processes. Theory of Probability and Its Applications 50 (3), 448–462. https://doi.org/10.1137/S0040585X9798186X.
Soltani, A., Shishebor, Z., 2007. On infinite dimensional discrete time periodically correlated processes. The Rocky Mountain Journal of Mathematics 37 (3), 1043–1058. https://doi.org/10.1216/rmjm/1182536176.
Soltani, A., Nematollahi, A., Mahmoudi, M., 2019. On the asymptotic distribution of the periodograms for the discrete time harmonizable simple processes. Statistical Inference for Stochastic Processes 22 (2), 307–322. https://doi.org/10.1007/s11203-018-9189-5.

Song, Z., Sun, W., Zhou, X., Hou, Z., 2007. An average sampling theorem for bandlimited stochastic processes. IEEE Transactions on Information Theory 53 (12), 4798–4800.

Soriano-Irigaray, F.J., Fernandez-Prat, J.S., Lopez-Martinez, F.J., Martos-Naya, E., Cobos-Morales, O., Entrambasaguas, J.T., 2018. Adaptive self-interference cancellation for full duplex radio: analytical model and experimental validation. IEEE Access (ISSN 2169-3536) 6, 65018–65026. https://doi.org/10.1109/ACCESS.2018.2878448.

Spooner, C.M., 1992. Theory and Application of Higher-Order Cyclostationarity. Ph.D. Dissertation. University of California, Davis, CA, Department of Electrical and Computer Engineering.

Spooner, C.M., 1994. Higher-order statistics for nonlinear processing of cyclostationary signals. In: Gardner, W.A. (Ed.), Cyclostationarity in Communications and Signal Processing. IEEE Press, pp. 91–167.

Spooner, C.M., 1995. Classification of co-channel communication signals using cyclic cumulants. In: Twenty-Ninth Asilomar Conference on Signals, Systems and Computers, vol. 1. Pacific Grove, CA, USA, pp. 531–536.

Spooner, C.M., 2001. On the utility of sixth-order cyclic cumulants for RF signal classification. In: Proceedings of the Thirty-Fifth Annual Asilomar Conference on Signals, Systems, and Computers. Pacific Grove, CA.

Spooner, C.M., 2015. Cyclostationary signal processing: understanding and using the statistics of communication signals. https://cyclostationary.blog. (Accessed 1 April 2019).

Spooner, C.M., Gardner, W.A., 1991a. Estimation of cyclic polyspectra. In: Proceedings of Twenty-Fifth Annual Asilomar Conference on Signals, Systems, and Computers. Pacific Grove, CA.

Spooner, C.M., Gardner, W.A., 1991b. An overview of higher-order cyclostationarity. In: Miamee, A.G. (Ed.), Proceeding of the Workshop on Nonstationary Stochastic Processes and Their Applications. World Scientific, Virginia, VA, pp. 110–125.

Spooner, C.M., Gardner, W.A., 1994a. The cumulant theory of cyclostationary time-series. Part II: development and applications. IEEE Transactions on Signal Processing 42, 3409–3429.

Spooner, C.M., Gardner, W.A., 1994b. Robust feature detection for cyclostationary signals. IEEE Transactions on Communications 42 (5), 2165–2173.

Spooner, C.M., Mody, A.N., 2018. Wideband cyclostationary signal processing using sparse subsets of narrowband subchannels. IEEE Transactions on Cognitive Communications and Networking (ISSN 2332-7731) 4 (2), 162–176. https://doi.org/10.1109/TCCN.2018.2790971.

Spooner, C.M., Nicholls, R.B., 2009. Spectrum sensing based on spectral correlation. In: Fette, B. (Ed.), Cognitive Radio Technology. Elsevier.

Spooner, C.M., Chen, C.K., Gardner, W.A., 1989. Maximum likelihood two-sensor detection and TDOA estimation for cyclostationary signals. In: Abstracts of Sixth Multidimensional Signal Processing Workshop of the IEEE/ASSP Society. Pacific Grove, CA, pp. 119–120.

Spooner, C.M., Brown, W.A., Yeung, G.K., 2000. Automatic radio-frequency environment analysis. In: Proceedings of the Thirty-Fourth Annual Asilomar Conference on Signals, Systems, and Computers. Pacific Grove, CA.

Spooner, C., Mody, A., Chuang, J., Anthony, M., 2013. Tunnelized cyclostationary signal processing: a novel approach to low-energy spectrum sensing. In: IEEE Military Communications Conference. MILCOM 2013, pp. 811–816.

Spooner, C.M., Mody, A.N., Chuang, J., Petersen, J., 2017. Modulation recognition using second- and higher-order cyclostationarity. In: 2017 IEEE International Symposium on Dynamic Spectrum Access Networks. DySPAN, pp. 1–3.

Spurbeck, M.S., Schreier, P.J., 2007. Causal Wiener filter banks for periodically correlated time series. Signal Processing (ISSN 0165-1684) 87 (6), 1179–1187. https://doi.org/10.1016/j.sigpro.2006.10.008.

Stefanakos, C.N., 2008. Investigation of the long-term wind and wave spectral climate of the Mediterranean sea. Proceedings of the Institution of Mechanical Engineers, Part M: Journal of Engineering for the Maritime Environment 222 (1), 27–39. https://doi.org/10.1243/14750902JEME64.

Stojanovic, M., Catipovic, J., Proakis, J., 1993a. Adaptive multichannel combining and equalization for underwater acoustic communications. The Journal of the Acoustical Society of America 94, 1621–1631.

Stojanovic, M., Catipovic, J., Proakis, J., 1993b. Adaptive receivers for underwater acoustic communications: their relation to beamforming and diversity combining. In: Proceedings of COMCON 4.

Stratonovich, R.L., 1967. Topics in the Theory of Random Noise, vol. I and II. Gordon and Breach, New York (revised English edition translated by R.A. Silverman).

Streight, D.A., Lott, G.K., Brown, W.A., 2000. Maximum likelihood estimates of the time and frequency differences of arrival of weak cyclostationary digital communications signals. In: 21st Century Military Communications Conference Proceedings, vol. 2. MILCOM 2000, pp. 957–961.

Strom, E., Miller, S., 1999. Properties of the single-bit single-user MMSE receiver for DS-CDMA systems. IEEE Transactions on Communications 47 (3), 416–425.

Strom, T., Signell, S., 1977. Analysis of periodically switched linear circuits. IEEE Transactions on Circuits and Systems CAS-24, 531–541.

Strom, E., Parkvall, S., Miller, S., Ottersten, B., 1996. Propagation delay estimation in asynchronous direct-sequence code-division multiple access systems. IEEE Transactions on Communications 44 (1), 84–93.

Subotic, N.S., Thelen, B.J., Carrara, D.A., 1998. Cyclostationary signal models for the detection and characterization of vibrating objects in SAR data. In: Proceedings of Thirty-Second Asilomar Conference on Signals, Systems, and Computers. Pacific Grove, California, pp. 1304–1308.

Sullivan, M.C., 1999. Wireless geolocation system. U.S. Patent No. 5,999,131.

Sullivan, M.C., Wegman, E.J., 1995. Estimating spectral correlations with simple nonlinear transformations. IEEE Transactions on Signal Processing 43 (6), 1525–1526.

Sun, Z., Liu, R., Wang, W., 2013. Joint time-frequency domain cyclostationarity-based approach to blind estimation of OFDM transmission parameters. EURASIP Journal on Wireless Communications and Networking (ISSN 1687-1499) 2013 (1), 117. https://doi.org/10.1186/1687-1499-2013-117.

Sun, Z., Chen, Y., Liu, S., Wang, W., 2014. Cyclostationarity-based joint domain approach to blind recognition of SCLD and OFDM signals. EURASIP Journal on Advances in Signal Processing (ISSN 1687-6180) 2014 (1), 5. https://doi.org/10.1186/1687-6180-2014-5.

Sun, H., Yuan, S., Luo, Y., 2018. Cyclic spectral analysis of vibration signals for centrifugal pump fault characterization. IEEE Sensors Journal (ISSN 1530-437X) 18 (7), 2925–2933. https://doi.org/10.1109/JSEN.2018.2804908.

Sutton, P., Nolan, K., Doyle, L., 2008. Cyclostationary signatures in practical cognitive radio applications. IEEE Journal on Selected Areas in Communications (ISSN 0733-8716) 26 (1), 13–24. https://doi.org/10.1109/JSAC.2008.080103.

Sutton, P., Ozgul, B., Doyle, L., 2014. Cyclostationary signatures for LTE Advanced and beyond. Physical Communication (ISSN 1874-4907) 10, 179–189. https://doi.org/10.1016/j.phycom.2013.11.006.

Swick, D.A., 1969. A Review of Wideband Ambiguity Functions. Technical Report 6994. Naval Research Laboratory.

Swift, R.J., 1996. Almost periodic harmonizable processes. Georgian Mathematical Journal 3 (3), 275–292.

Synowiecki, R., 2007. Consistency and application of moving block bootstrap for non-stationary time series with periodic and almost periodic structure. Bernoulli 13 (4), 1151–1178. https://doi.org/10.3150/07-BEJ102.

Tan, Z., Hashorva, E., 2013a. Exact asymptotics and limit theorems for supremum of stationary-processes over a random interval. Stochastic Processes and Their Applications (ISSN 0304-4149) 123 (8), 2983–2998. https://doi.org/10.1016/j.spa.2013.03.009.

Tan, Z., Hashorva, E., 2013b. Limit theorems for extremes of strongly dependent cyclo-stationary χ-processes. Extremes (ISSN 1386-1999) 16 (2), 241–254. https://doi.org/10.1007/s10687-013-0170-9.

Tanda, M., 1992. Cyclostationarity-Based Algorithms for Signal Interception in Non-Gaussian Noise. Ph.D. Dissertation. Università di Napoli Federico II, Napoli, Italy, Dipartimento di Ingegneria Elettronica (in Italian).

Tandra, R., Sahai, A., 2007. SNR walls for feature detectors. In: 2nd IEEE International Symposium on New Frontiers in Dynamic Spectrum Access Networks. DySPAN 2007, pp. 559–570.

Tani, A., Fantacci, R., 2010. A low-complexity cyclostationary-based spectrum sensing for UWB and WiMAX coexistence with noise uncertainty. IEEE Transactions on Vehicular Technology (ISSN 0018-9545) 59 (6), 2940–2950. https://doi.org/10.1109/TVT.2010.2049511.

Tani, A., Fantacci, R., Marabissi, D., 2016. A low-complexity cyclostationary spectrum sensing for interference avoidance in femtocell LTE-A-Based networks. IEEE Transactions on Vehicular Technology (ISSN 0018-9545) 65 (4), 2747–2753. https://doi.org/10.1109/TVT.2015.2419877.

Tardu, S., 2003. Characterization of unsteady time periodical turbulent flows. Comptes Rendus. Mécanique 331, 767–774 (in French).

Teng, W., Ding, X., Zhang, Y., Liu, Y., Ma, Z., Kusiak, A., 2017. Application of cyclic coherence function to bearing fault detection in a wind turbine generator under electromagnetic vibration. Mechanical Systems and Signal Processing (ISSN 0888-3270) 87, 279–293. https://doi.org/10.1016/j.ymssp.2016.10.026.

Teplitsky, M., Yeredor, A., 2009. TDOA estimation for cyclostationary sources: new correlations-based bounds and estimators. In: IEEE International Conference on Acoustics, Speech and Signal Processing. ICASSP 2009, pp. 3309–3312.

Thibault, J., Chevalier, P., Pipon, F., Monot, J.J., Multedo, G., 2001. Method and device for space division multiplexing of radio signals transmitted in cellular radio communications. U.S. Patent No. 6,240,098.

Thompstone, R.M., Hipel, K.W., McLeod, A.I., 1985. Grouping of periodic autoregressive models. In: Anderson, O.D., Ord, J.K., Robinson, E.A. (Eds.), Time Series Analysis: Theory and Practice 6. Elsevier, Amsterdam, Holland, pp. 35–49.

Thomson, D.J., 1982. Spectrum estimation and harmonic analysis. Proceedings of the IEEE 70, 1055–1096.

Thomson, D., 2011. Some problems in the analysis of possibly cyclostationary data. In: Conference Record of the Forty Fifth Asilomar Conference on Signals, Systems and Computers. ASILOMAR 2011, pp. 2040–2044.

Tian, C.J., 1988. A limiting property of sample autocovariances of periodically correlated processes with application to period determination. Journal of Time Series Analysis 9, 411–417.

Tian, Z., 2011. Cyclic feature based wideband spectrum sensing using compressive sampling. In: IEEE International Conference on Communications. ICC, pp. 1–5.

Tian, Z., Tafesse, Y., Sadler, B., 2012. Cyclic feature detection with sub-Nyquist sampling for wideband spectrum sensing. IEEE Journal of Selected Topics in Signal Processing (ISSN 1932-4553) 6 (1), 58–69. https://doi.org/10.1109/JSTSP.2011.2181940.

Tiao, G.C., Grupe, M.R., 1980. Hidden periodic autoregressive-moving average models in time series data. Biometrika 67, 365–373.

Todd, R.M., 1990. Periodic linear-quadratic models of seasonality. Journal of Economic Dynamics and Control 14, 763–796.

Tong, L., Xu, G., Kailath, T., 1994a. Blind channel identification and equalization using spectral correlation measurements, Part II: a time-domain approach. In: Gardner, W.A. (Ed.), Cyclostationarity in Communications and Signal Processing. IEEE Press, pp. 437–454.

Tong, L., Xu, G., Kailath, T., 1994b. Blind identification and equalization based on second order statistics: a time domain approach. IEEE Transactions on Information Theory 40 (2), 340–349.

Tong, L., Xu, G., Hassibi, B., Kailath, T., 1995. Blind channel identification based on second order statistics: a frequency domain approach. IEEE Transactions on Information Theory 41 (1), 329–334.

Tong, H., Pourrostam, J., Zekavat, S.A., 2007. LCMV beamforming for a novel wireless local positioning system: nonstationarity and cyclostationarity analysis. EURASIP Journal on Advances in Signal Processing (ISSN 1687-6180) 2007 (1), 098243. https://doi.org/10.1155/2007/98243.

Touati, S., Pesquet, J.C., 2001. Statistical properties of the wavelet decomposition of cyclostationary processes. In: Proceedings of the IEEE-EURASIP Workshop on Nonlinear Signal and Image Processing. Baltimore, MD.

Trans, F., 2003. Means and method for increasing performance of interference-suppression based receivers. U.S. Patent No. 6,553,085.

Troutman, B.M., 1979. Some results in periodic autoregression. Biometrika 66, 219–228.

Tsatsanis, M., Giannakis, G.B., 1997. Transmitter induced cyclostationarity for blind channel equalization. IEEE Transactions on Signal Processing 45 (7), 1785–1794.

Tugnait, J.K., 1995. On blind identifiability of multipath channels using fractional sampling and second-order cyclostationary statistics. IEEE Transactions on Information Theory 41 (1), 308–311.

Tugnait, J.K., 1996. Blind equalization and estimation of FIR communication channels using fractional sampling. IEEE Transactions on Communications 44 (3), 324–336.

Tugnait, J.K., Luo, W., 2002. Linear prediction error method for blind identification of periodically time-varying channels. IEEE Transactions on Signal Processing 50 (12), 3070–3082.

Tulino, A.M., 1999. Interference Suppression in CDMA Systems. Ph.D. Dissertation. Seconda Università di Napoli, Aversa, Italy, Dipartimento di Ingegneria dell'Informazione (in Italian).

Tulino, A.M., Verdù, S., 2001. Asymptotic analysis of improved linear receivers for BPSK-CDMA subject to fading. IEEE Journal on Selected Areas in Communications 19, 1544–1555.

Turunen, V., Kosunen, M., 2013. Cyclostationary signal detection in the angular domain for communication systems. U.S. Patent No. 8,526,520.

Ula, T.A., 1990. Periodic covariance stationarity of multivariate periodic autoregressive moving average processes. Water Resources Research 26, 855–861.

Ula, T.A., 1993. Forecasting of multivariate periodic autoregressive moving-average processes. Journal of Time Series Analysis 14 (6).

Urbanek, J., Barszcz, T., Antoni, J., 2013. Time-frequency approach to extraction of selected second-order cyclostationary vibration components for varying operational conditions. Measurement 46 (4), 1454–1463. https://doi.org/10.1016/j.measurement.2012.11.042.

Urbanik, K., 1962. Fourier analysis in Marcinkiewicz spaces. Studia Mathematica 21, 93–102.

Urriza, P., Rebeiz, E., Cabric, D., 2013. Multiple antenna cyclostationary spectrum sensing based on the cyclic correlation significance test. IEEE Journal on Selected Areas in Communications (ISSN 0733-8716) 31 (11), 2185–2195. https://doi.org/10.1109/JSAC.2013.131118.

Uzun, N., Haddad, R.A., 1995. Cyclostationary modeling, analysis, and optimization compensation of quantization errors in subband codecs. IEEE Transactions on Signal Processing 43 (9), 2109–2119.

Vaidyanathan, P.P., 1990. Multirate digital filters, filter banks, polyphase networks, and applications: a tutorial review. Proceedings of the IEEE 78, 56–93.

van den Bos, A., 1994. Complex gradient and Hessian. IEE Proceedings. Vision, Image and Signal Processing 141 (6), 380–382.

van den Bos, A., 1995. The multivariate complex Normal distribution–a generalization. IEEE Transactions on Information Theory 41, 537–539.

van Der Wurf, P., 1974. On the spectral density of a cyclostationary process. IEEE Transactions on Communications 22 (10), 1727–1730.

Van Trees, H.L., 1971a. Detection, Estimation, and Modulation Theory. Part I. John Wiley & Sons, Inc., New York.

Van Trees, H.L., 1971b. Detection, Estimation, and Modulation Theory. Part III. John Wiley & Sons, Inc., New York.

Varghese, T., Donohue, K.D., 1994a. Characterization of tissue microstructure with spectral crosscorrelation. Ultrasound Imaging 15, 238–254.

Varghese, T., Donohue, K.D., 1994b. Mean-scatter spacing estimates with spectral correlation. The Journal of the Acoustical Society of America 96 (6), 3504–3515.

Varghese, T., Donohue, K.D., 1995a. Estimating mean scatterer spacing with frequency-smoothed spectral autocorrelation function. IEEE Transactions on Ultrasonics, Ferroelectrics, and Frequency Control 42 (3), 451–463.

Varghese, T., Donohue, K.D., 1995b. Spectral redundancy in tissue characterization. In: Twentieth International Symposium on Ultrasonic Imaging and Tissue Characterization.

Varghese, T., Donohue, K.D., Chatterjee, J.P., 1995. Specular echo imaging with spectral correlation. In: Proceedings of 1995 Ultrasonics Symposium. Seattle, WA, pp. 1315–1318.

Vazquez-Padin, D., Mosquera, C., Perez-Gonzalez, F., 2010. Two-dimensional statistical test for the presence of almost cyclostationarity on images. In: 17th IEEE International Conference on Image Processing. ICIP 2010, pp. 1745–1748.

Vecchia, A.V., 1983. Aggregation and Estimation for Periodic Autoregressive Moving Average Models. Ph.D. Dissertation. Colorado State University, Fort Collins, Colorado.

Vecchia, A.V., 1985a. Maximum likelihood estimation for periodic autoregressive moving average models. Technometrics 27, 375–384.

Vecchia, A.V., 1985b. Periodic autoregressive-moving average (PARMA) modeling with applications to water resources. Water Resources Bulletin 21, 721–730.

Vecchia, A.V., Ballerini, R., 1991. Testing for periodic autocorrelations in seasonal time series data. Biometrika 78 (1), 53–63.

Verde, F., 2001. Equalization and Interference Suppression Techniques for High-Capacity Digital Systems. Ph.D. Dissertation. Università di Napoli Federico II, Napoli, Italy, Dipartimento di Ingegneria Elettronica e delle Telecomunicazioni (in Italian).

Vicuña, C.M., Acuña, D.Q., 2014. Cyclostationary processing of vibration and acoustic emissions for machine failure diagnosis. In: Chaari, F., Leskow, J., Napolitano, A., Sanchez-Ramirez, A. (Eds.), Cyclostationarity: Theory and Methods. In: Lecture Notes in Mechanical Engineering. Springer International Publishing, pp. 141–156.

Viravau, P., Gournay, P., 1998. Theorical spectral correlation of CPM modulations–Part II: general calculation method and analysis. Annales des Télécommunications 53 (7–8), 279–288 (in French).

Viswanathan, R., 2006. On the autocorrelation of complex envelope of white noise. IEEE Transactions on Information Theory 52 (9), 4298–4299. https://doi.org/10.1109/TIT.2006.880018.

Vito, L.D., Dobre, O.A., 2017. A compressive sampling-based method for classification and parameter estimation of FSK signals. Measurement (ISSN 0263-2241) 98, 439–444. https://doi.org/10.1016/j.measurement.2015.12.038.

Vokurka, K., 1982. Application of the group pulse processes in the theory of Barkhausen noise. Czechoslovak Journal of Physics 32, 1384–1398.

Voloshin, V.I., 1971. Frequency fluctuations in isochronous self-oscillators at periodically nonstationary random actions. Radiophysics and Quantum Electronics (Izv. Vuz Radiofiz.) 14, 1383–1391.

von Neumann, J., 1934. Almost periodic functions in a group I. Transactions of the American Mathematical Society 36, 445–492.

von Schroeter, T., 1999. Frequency warping with arbitrary allpass maps. IEEE Signal Processing Letters 6 (5), 116–118.

Vourinen, O., Seppanen, T., Roning, J., Lilleberg, J., Kolehmainen, T., 2005. Method for distinguishing signals from one another, and filter. U.S. Patent No. 6,847,689.

Voychishin, K.S., Dragan, Y.P., 1971. A simple stochastic model of the natural rhythmic processes. Otbor i Peredacha Informacii 2, 7–15 (in Russian).

Voychishin, K.S., Dragan, Y.P., 1972a. The elimination of rhythm from periodically correlated random processes. Otbor i Peredacha Informacii 33, 12–16 (in Russian).

Voychishin, K.S., Dragan, Y.P., 1972b. Some properties of the stochastic model of systems with periodically varying parameters. In: Algorithms in Production Processes. Kiev, pp. 11–20 (in Russian).

Voychishin, K.S., Dragan, Y.P., 1973. Example of formation of periodically correlated random processes. Radio Engineering and Electronic Physics 18, 1426–1429 (English translation of Radiotekhnika i Electronika, vol. 18, pp. 1957–1960).

Voychishin, K.S., Dragan, Y.P., et al., 1972. Coll. Algorithmization of Industrial Processes. Cybernetics Institute, AN USSR (in Russian).

Vucic, D., Eric, M., 2009. Cyclic spectral analysis of UWB-IR signals. IEEE Signal Processing Letters (ISSN 1070-9908) 16 (8), 723–726. https://doi.org/10.1109/LSP.2009.2023949.

Vučić, D., Obradović, M., 1998. Matrix-based stochastic method for the spectral correlation characterization of digital modulation. Facta Universitatis (NIŠ), Series and Energetics 11 (3), 271–284.

Vučić, D., Obradović, M., 1999. Spectral correlation evaluation of MSK and offset QPSK modulation. Signal Processing 78, 363–367.

Vučić, D., Vukotić, S., Erić, M., 2017. Cyclic spectral analysis of OFDM/OQAM signals. AEÜ. International Journal of Electronics and Communications (ISSN 1434-8411) 73, 139–143. https://doi.org/10.1016/j.aeue.2017.01.005.

Wagstaff, A., 2008. Logarithmic cyclic frequency domain profile for automatic modulation recognition. IET Communications (ISSN 1751-8628) 2 (8), 1009–1015. https://doi.org/10.1049/iet-com:20070634.

Wahlberg, P., Schreier, P., 2008. Spectral relations for multidimensional complex improper stationary and (almost) cyclostationary processes. IEEE Transactions on Information Theory (ISSN 0018-9448) 54 (4), 1670–1682. https://doi.org/10.1109/TIT.2008.917626.

Wang, C.L., Chen, H.W., 2014. A new signal structure for active sensing in cognitive radio systems. IEEE Transactions on Communications (ISSN 0090-6778) 62 (3), 822–835. https://doi.org/10.1109/TCOMM.2014.011614.120732.

Wang, X., Poor, H.V., 1998. Blind multiuser detection: a subspace approach. IEEE Transactions on Information Theory 44 (2), 30–44.

Wang, W., Zhao, H., 2017. Diffusion signed LMS algorithms and their performance analyses for cyclostationary white Gaussian inputs. IEEE Access (ISSN 2169-3536) 5, 18876–18894. https://doi.org/10.1109/ACCESS.2017.2733766.

Wang, W., Zhao, H., 2018. Performance analysis of diffusion LMS algorithm for cyclostationary inputs. Signal Processing (ISSN 0165-1684) 150, 33–50. https://doi.org/10.1016/j.sigpro.2018.03.019.

Wang, Y., Zhou, G., 1998. On the use of higher-order ambiguity function for multicomponent polynomial phase signals. Signal Processing 65, 283–296.

Wang, Y., Ciblat, P., Serpedin, E., Loubaton, P., 2002. Performance analysis of a class of nondata-aided frequency offset and symbol timing estimators for flat-fading channels. IEEE Transactions on Signal Processing 50 (9), 2295–2305.

Wang, Y., Serpedin, E., Ciblat, P., 2003a. An alternative blind feedforward symbol timing estimator using two samples per symbol. IEEE Transactions on Communications 51 (9), 1451–1455.

Wang, Y., Serpedin, E., Ciblat, P., 2003b. Optimal blind carrier recovery for MPSK burst transmissions. IEEE Transactions on Communications 51 (9), 1571–1581.

Wang, W., Jafari, M.G., Sanei, S., Chambers, J.A., 2004. Blind separation of convolutive mixtures of cyclostationary signals. International Journal of Adaptive Control and Signal Processing 18 (3), 279–298.

Wang, J., Chen, T., Huang, B., 2005. On spectral theory of cyclostationary signals in multirate systems. IEEE Transactions on Signal Processing (ISSN 1053-587X) 53 (7), 2421–2431. https://doi.org/10.1109/TSP.2005.849192.

Wang, H., Dobre, O.A., Li, C., Popescu, D.C., 2015. Blind cyclostationarity-based symbol period estimation for FSK signals. IEEE Communications Letters (ISSN 1089-7798) 19 (7), 1149–1152. https://doi.org/10.1109/LCOMM.2015.2436915.

Weiss, L.G., 1994. Wavelets and wideband correlation processing. IEEE Signal Processing Magazine 11 (1), 13–22.

Wen, F., Wan, Q., 2013. Maximum likelihood and signal-selective TDOA estimation for noncircular signals. Journal of Communications and Networks (ISSN 1229-2370) 15 (3), 245–251. https://doi.org/10.1109/JCN.2013.000046.

Wiener, N., 1930. Generalized harmonic analysis. Acta Mathematica 55, 117–258.

Wiener, N., 1933. The Fourier Integral and Certain of Its Applications. Cambridge University Press, London (also Dover, New York, 1958).

Wiener, N., 1949. Extrapolation, Interpolation, and Smoothing of Stationary Time Series. MIT Press, Cambridge, MA.

Wilbur, J., Bono, J., 1991. Wigner/cycle spectrum analysis of spread spectrum and diversity transmissions. IEEE Journal of Oceanic Engineering 16 (1), 98–106.

Wilbur, J., McDonald, R.J., 1992. Nonlinear analysis of cyclically correlated spectral spreading in modulated signals. The Journal of the Acoustical Society of America 92, 219–230.

Williamson, D., 1991. Digital Control and Implementation. Prentice-Hall, Inc., Englewood Cliffs, NJ.

Williamson, D., Kadiman, K., 1985. Cyclostationarity in the digital regulation of continuous time systems. In: Byrnes, C.I., Lindquist, A. (Eds.), Frequency Domain and State Space Methods for Linear Systems. North-Holland, Amsterdam, pp. 297–309.

Willis, D.M., 1964. The statistics of a particular non-homogeneous Poisson process. Biometrika 51, 399–404.

Win, M.Z., 1998. On the power spectral density of digital pulse streams generated by M-ary cyclostationary sequences in the presence of stationary timing jitter. IEEE Transactions on Communications 46 (9), 1135–1145.

Wisdom, S., Atlas, L., Pitton, J., 2014. Extending coherence time for analysis of modulated random processes. In: IEEE International Conference on Acoustics, Speech, and Signal Processing. ICASSP 2014, Firenze, Italy.

Woerdeman, H.J., 2002. The Carathéodory-Toeplitz problem with partial data. Linear Algebra and Its Applications 342, 149–161.

Wohlbrandt, A., Kissner, C., Guérin, S., 2018. Impact of cyclostationarity on fan broadband noise prediction. Journal of Sound and Vibration (ISSN 0022-460X) 420, 142–164. https://doi.org/10.1016/j.jsv.2018.01.039.

Wold, H.O.A., 1948. On prediction in stationary time series. The Annals of Mathematical Statistics 19, 558–567.

Wong, H.E., Chambers, J.A., 1996. Two-stage interference immune blind equaliser which exploits cyclostationary statistics. Electronics Letters 32 (19), 1763–1764.

Wornell, G.W., 1993. Wavelet-based representations for the $1/f$ family of fractal processes. Proceedings of the IEEE 81 (10), 1428–1450. https://doi.org/10.1109/5.241506.

Wouters, N., Serafim, M., Valayer, P., Kirkup, B., Wang, Y., Cabral, H., 2016. Early warning signals as indicators of cyclostationarity in three-species hierarchies. Ecological Indicators (ISSN 1470-160X) 60, 586–593. https://doi.org/10.1016/j.ecolind.2015.06.038.

Wu, P.W., Lev-Ari, H., 1997. Optimized estimation of moments for nonstationary signals. IEEE Transactions on Signal Processing 45 (5), 1210–1221. https://doi.org/10.1109/78.575695.

Wu, Q., Wong, M., 1996. Blind adaptive beamforming for cyclostationary signals. IEEE Transactions on Signal Processing 44 (11), 2757–2767.

Wu, Q., Wong, K.M., Ho, R., 1994. Fast algorithm for adaptive beamforming of cyclic signals. IEE Proceedings. Radar, Sonar and Navigation 141 (6), 312–318.

Wu, Y.C., Halvorsen, E., Lallart, M., Richard, C., Guyomar, D., 2015. Stochastic modeling in the frequency domain for energy harvester with switching electronic interface. IEEE/ASME Transactions on Mechatronics (ISSN 1083-4435) 20 (1), 50–60. https://doi.org/10.1109/TMECH.2014.2308930.

Wylomanska, A., Obuchowski, J., Zimroz, R., Hurd, H., 2014. Periodic autoregressive modeling of vibration time series from planetary gearbox used in bucket wheel excavator. In: Chaari, F., Leskow, J., Napolitano, A., Sanchez-Ramirez, A. (Eds.), Cyclostationarity: Theory and Methods. In: Lecture Notes in Mechanical Engineering. Springer International Publishing, pp. 171–186.

Wylomanska, A., Obuchowski, J., Zimroz, R., Hurd, H., 2015. Influence of different signal characteristics on PAR model stability. In: Chaari, F., Leskow, J., Napolitano, A., Zimroz, R., Wylomanska, A., Dudek, A. (Eds.), Cyclostationarity: Theory and Methods – II, Applied Condition Monitoring, vol. 3. Springer International Publishing, pp. 89–104.

Xia, X.G., Zhang, Z., 1992. On a conjecture on time-warped band-limited signals. IEEE Transactions on Signal Processing 40 (1), 252–254. https://doi.org/10.1109/78.157208.

Xiang, Y., Peng, D., Ubhayaratne, I., Rolfe, B., Pereira, M., 2017. Second-order cyclostationary statistics-based blind source extraction from convolutional mixtures. IEEE Access (ISSN 2169-3536) 5, 2011–2019. https://doi.org/10.1109/ACCESS.2017.2664978.

Xin, J., Sano, A., 2001. Linear prediction approach to direction estimation of cyclostationary signals in multipath environments. IEEE Transactions on Signal Processing 49 (4), 710–720.

Xin, J., Tsuji, H., Sano, A., 2000. Higher-order cyclostationarity based direction estimation of coherent narrow-band signals. IEICE Transactions on Fundamentals E83-A (8), 1624–1633.

Xu, G., Kailath, T., 1989. Array signal processing via exploitation of spectral correlation–a combination of temporal and spatial processing. In: Conference Record, Twenty-Third Asilomar Conference on Signals, Systems and Computers. Pacific Grove, CA, pp. 945–949.

Xu, G., Kailath, T., 1990. A simple and effective algorithm for estimating time delay of communication signals. In: Proceedings of International Symposium on Information Theory and Its Applications. ISITA'90, pp. 267–270.

Xu, G., Kailath, T., 1992. Direction-of-arrival estimation via exploitation of cyclostationarity–a combination of temporal and spatial filtering. IEEE Transactions on Signal Processing 40 (7), 1775–1785.

Xu, T., Chen, H., Hu, H., 2014. A low-complexity detection method for statistical signals in OFDM systems. IEEE Communications Letters (ISSN 1089-7798) 18 (4), 632–635. https://doi.org/10.1109/LCOMM.2014.030614.140193.

Yaglom, A.M., 1986. Correlation Theory of Stationary and Related Random Functions, vol. I and II. Springer-Verlag, New York.

Yamaguchi, S., Kato, Y., 1989. A practical method of predicting noise produced by road traffic controlled by traffic signals. The Journal of the Acoustical Society of America 86, 2206–2214.

Yamaguchi, S., Kato, Y., 1991. A statistical study for determining the minimum sample size for L_{eq} estimation of periodic nonstationary random noise. Applied Acoustics 32, 35–48.

Yan, H., Fan, H., 2005a. On source association of DOA estimation under multipath propagation. IEEE Signal Processing Letters (ISSN 1070-9908) 12 (10), 717–720. https://doi.org/10.1109/LSP.2005.855541.

Yan, H., Fan, H.H., 2005b. On improvements of cyclic MUSIC. EURASIP Journal on Advances in Signal Processing (ISSN 1687-6180) 2005 (1), 760297. https://doi.org/10.1155/ASP.2005.61.

Yan, H., Fan, H.H., 2005c. Wideband cyclic MUSIC algorithm. Signal Processing 85, 643–649.

Yan, H., Fan, H., 2007. Signal-selective DOA tracking for wideband cyclostationary sources. IEEE Transactions on Signal Processing (ISSN 1053-587X) 55 (5), 2007–2015. https://doi.org/10.1109/TSP.2007.893204.

Yang, D., Si, J., Li, Z., Beaulieu, N.C., Zhu, J., Zhou, F., Hao, B., 2016. Blind carrier frequency offset estimation for single carrier and orthogonal frequency division multiplexing signals using least-order cyclic moments. IET Communications (ISSN 1751-8628) 10 (6), 669–676. https://doi.org/10.1049/iet-com.2015.0862.

Yang, J.X., Ser, W., Lei, L., 2018. Multiple cycle frequencies estimation under cochannel interference. IEEE Signal Processing Letters (ISSN 1070-9908) 25 (11), 1645–1649. https://doi.org/10.1109/LSP.2018.2867730.

Yavorskii, I.N., 1983. Properties of estimators of mathematical expectation and correlation function of periodically-correlated random processes. Otbor i Peredacha Informacii 67, 22–28 (in Russian).

Yavorskii, I.N., 1985. Statistical analysis of periodically correlated random processes. Radiotekhnika i Elektronika 30, 1096–1104 (in Russian).

Yavorskii, I.N., 1986. Component estimates of the probability characteristics of periodically correlated random processes. Avtomatika 19, 44–48 (in Russian).

Yavorskii, I.N., 1987a. Interpolation of estimates of probability characteristics of periodically correlated random processes. Avtometrika 1, 36–41 (in Russian).

Yavorskii, I.N., 1987b. Statistical analysis of periodically correlated vector random processes. Otbor i Peredacha Informacii 76, 3–12.

Yavorskii, I.N., 1989. Statistical analysis of poly- and nearly periodically-correlated random processes. Otbor i Peredacha Informacii 79, 1–10 (in Russian).

Yavorskyj, I., Yuzefovych, R., Matsko, I., Zakrzewski, Z., 2014. Discrete estimations of cross-correlation components of periodically correlated random signals. Radioelectronics and Communications Systems (ISSN 0735-2727) 57 (2), 78–91. https://doi.org/10.3103/S0735272714020034.

Yellepeddi, A., Preisig, J.C., 2018. Efficient system tracking with decomposable graph-structured inputs and application to adaptive equalization with cyclostationary inputs. IEEE Transactions on Signal Processing (ISSN 1053-587X) 66 (10), 2645–2658. https://doi.org/10.1109/TSP.2018.2811745.

Yen, C.P., Tsai, Y., Wang, X., 2013. Wideband spectrum sensing based on sub-Nyquist sampling. IEEE Transactions on Signal Processing (ISSN 1053-587X) 61 (12), 3028–3040. https://doi.org/10.1109/TSP.2013.2251342.

Yeo, J.H., Cho, J.H., 2012. Properization of second-order cyclostationary random processes and its application to signal presence detection. In: IEEE Military Communications Conference. MILCOM 2012, pp. 1–6.

Yeo, J., Cho, J.H., 2014. Asymptotic frequency-shift properizer for block processing of improper-complex second-order cyclostationary random processes. IEEE Transactions on Information Theory (ISSN 0018-9448) 60 (7), 4083–4100. https://doi.org/10.1109/TIT.2014.2320938.

Yeo, J., Cho, J.H., Lehnert, J., 2015a. Joint transmitter and receiver optimization for improper-complex second-order stationary data sequence. Journal of Communications and Networks (ISSN 1229-2370) 17 (1), 1–11. https://doi.org/10.1109/JCN.2015.000003.

Yeo, J., Han, B.W., Cho, J.H., Lehnert, J.S., 2015b. Capacity of an orthogonal overlay channel. IEEE Transactions on Wireless Communications (ISSN 1536-1276) 14 (11), 6111–6124. https://doi.org/10.1109/TWC.2015.2448538.

Yeung, G.K., Gardner, W.A., 1996. Search-efficient methods of detection of cyclostationary signals. IEEE Transactions on Signal Processing 44 (5), 1214–1223.

Yeung, G.K., Brown, W.A., Gardner, W.A., 1994. A new algorithm for blind-adaptive frequency-shift filtering. In: 28th Annual Conference on Information Science and Systems. Princeton, NJ.

Yoo, D.S., 2008. Equivalent complex baseband representations of linear time-variant systems and signals. IEEE Transactions on Signal Processing 56 (8), 3775–3778.

Yoo, Y.G., Cho, J.H., 2010. Asymptotically optimal low-complexity SC-FDE in data-like co-channel interference. IEEE Transactions on Communications (ISSN 0090-6778) 58 (6), 1718–1728. https://doi.org/10.1109/TCOMM.2010.06.090242.

Yoo, D.S., Lim, J., Kang, M.H., 2014. ATSC digital television signal detection with spectral correlation density. Journal of Communications and Networks (ISSN 1229-2370) 16 (6), 600–612. https://doi.org/10.1109/JCN.2014.000106.

Yoon, C.Y., Lindsay, W.C., 1982. Phase-locked loop performance in the presence of CW interference and additive noise. IEEE Transactions on Communications COM-30, 2305–2311.

Yu, S.J., Ueng, F.B., 2000. Implementation of cyclostationary signal-based adaptive arrays. Signal Processing 80, 2249–2254.

Yu, L., Antoni, J., Wu, H., Jiang, W., 2019a. Reconstruction of cyclostationary sound source based on a back-propagating cyclic Wiener filter. Journal of Sound and Vibration (ISSN 0022-460X) 442, 787–799. https://doi.org/10.1016/j.jsv.2018.11.034.

Yu, L., Wu, H., Antoni, J., Jiang, W., 2019b. Extraction and imaging of aerodynamically generated sound field of rotor blades in the wind tunnel test. Mechanical Systems and Signal Processing (ISSN 0888-3270) 116, 1017–1028. https://doi.org/10.1016/j.ymssp.2018.07.042.

Yun, Y.H., Cho, J.H., 2010. An optimal orthogonal overlay for a cyclostationary legacy signal. IEEE Transactions on Communications (ISSN 0090-6778) 58 (5), 1557–1567. https://doi.org/10.1109/TCOMM.2010.05.090001.

Yuran, S.I., 1980. Methods for the instrumental analysis of periodically nonstationary random processes. Measurement & Technology 23, 1069–1071 (English translation of Izmer. Tekh., vol. 23, pp. 16–17).

Zak, G., Wylomanska, A., Zimroz, R., 2019. Periodically impulsive behavior detection in noisy observation based on generalized fractional order dependency map. In: Special Issue on ICTD-CMMNO'2016, ICTD: International Congress on Technical Diagnostics. Applied Acoustics (ISSN 0003-682X) 144, 31–39. https://doi.org/10.1016/j.apacoust.2017.05.003.

Zakai, M., 1965. Band-limited functions and the sampling theorem. Information and Control 8, 143–158.

Zalud, V., Kulesov, V.N., 1980. Low-Noise Semiconductor Circuits. SNTL, Prague (Section 1.18) (in Czech).

Zaytsev, V.A., Pechinkin, A.V., 1979. Optimal control of servicing in a multichannel system. Engineering Cybernetics 17, 36–42 (English translation of Tekh. Kibern, vol. 17, pp. 57–64).

Zemanian, A.H., 1987. Distribution Theory and Transform Analysis. Dover, New York.

Zeng, H.H., Tong, L., 1997. Blind channel estimation using the second-order statistics: algorithms. IEEE Transactions on Signal Processing 45 (8), 1919–1930.

Zeng, Y., Koh, C.L., Liang, Y.C., 2013. Method for determining as to whether a received signal includes a data signal. U.S. Patent No. 8,433,009.

Zerhouni, K., Elassali, R., Elbahhar, F., Elbaamrani, K., Idboufker, N., 2018. On the cyclostationarity of universal filtered multi-carrier UFMC. AEÜ. International Journal of Electronics and Communications (ISSN 1434-8411) 89, 174–180. https://doi.org/10.1016/j.aeue.2018.03.016.

Zhang, C., 1994. Pseudo almost periodic solutions of some differential equations. Journal of Mathematical Analysis and Applications 181, 62–76.

Zhang, C., 1995. Pseudo almost periodic solutions of some differential equations, II. Journal of Mathematical Analysis and Applications 192, 543–561.

Zhang, H., 1997. Maximum entropy modeling of periodically correlated processes. IEEE Transactions on Information Theory 43 (6), 2033–2035.

Zhang, H., Abdi, A., 2009. Cyclostationarity-based Doppler spread estimation in mobile fading channels. IEEE Transactions on Communications (ISSN 0090-6778) 57 (4), 1061–1067. https://doi.org/10.1109/TCOMM.2009.04.070255.

Zhang, H., Abdi, A., 2012. Mobile speed and Doppler frequency estimation using cyclostationarity. U.S. Patent No. 8,209,150.

Zhang, C., Liu, J., 2010. Two unsolved problems on almost periodic type functions. Applied Mathematics Letters 23, 1133–1136.

Zhang, J., Wong, K.M., Luo, Z.Q., Ching, P.C., 1999. Blind adaptive FRESH filtering for signal extraction. IEEE Transactions on Signal Processing 47 (5), 1397–1402.

Zhang, J., Liao, G., Wang, J., 2005. Robust direction finding for cyclostationary signals with cycle frequency error. Signal Processing (ISSN 0165-1684) 85 (12), 2386–2393. https://doi.org/10.1016/j.sigpro.2005.01.015.

Zhang, Hai-Bin, Jiang, Wei-Kang, Wan, Quan, 2008. Nearfield acoustic holography based on wave superposition algorithm for cyclostationary sound filed. Acta Physica Sinica 57 (1), 313–320. https://doi.org/10.7498/aps.57.313.

Zhang, H., Le Ruyet, D., Terre, M., 2010. Spectral correlation of multicarrier modulated signals and its application for signal detection. EURASIP Journal on Advances in Signal Processing (ISSN 1687-6180) 2010 (1), 794246. https://doi.org/10.1155/2010/794246.

Zhang, Q., Dobre, O.A., Eldemerdash, Y.A., Rajan, S., Inkol, R.J., 2013. Second-order cyclostationarity of BT-SCLD signals: theoretical developments and applications to signal classification and blind parameter estimation. IEEE Transactions on Wireless Communications 12 (4), 1501–1511. https://doi.org/10.1109/TWC.2013.021213.111888.

Zhang, Y., Fu, X., Zhang, Q., Feng, Z., Liu, X., 2015. Novel schemes to optimize sampling rate for compressed sensing. Journal of Communications and Networks (ISSN 1229-2370) 17 (5), 517–524. https://doi.org/10.1109/JCN.2015.000090.

Zhang, M., Haritopoulos, M., Nandi, A.K., 2016. Fetal ECG subspace estimation based on cyclostationarity. In: 2016 24th European Signal Processing Conference. EUSIPCO, pp. 2060–2064.

Zhang, Z., Qu, Y., Wu, Z., Nowak, M.J., Ellinger, J., Wicks, M.C., 2018. RF steganography via LFM chirp radar signals. IEEE Transactions on Aerospace and Electronic Systems (ISSN 0018-9251) 54 (3), 1221–1236. https://doi.org/10.1109/TAES.2017.2776698.

Zhao, Y., Su, Y., 2018. Cyclostationary phase analysis on micro-Doppler parameters for radar-based small UAVs detection. IEEE Transactions on Instrumentation and Measurement (ISSN 0018-9456) 67 (9), 2048–2057. https://doi.org/10.1109/TIM.2018.2811256.

Zhao, L., Xu, H., Bai, C., 2018. Blind identification of the number of sub-carriers for orthogonal frequency division multiplexing-based elastic optical networking. Optics Communications (ISSN 0030-4018) 411, 101–107. https://doi.org/10.1016/j.optcom.2017.10.076.

Zhong, G., Guo, J., Zhao, Z., Qu, D., 2010. Cyclostationarity based multi-antenna spectrum sensing in cognitive radio networks. In: IEEE 71st Vehicular Technology Conference. VTC 2010-Spring, pp. 1–5.

Živanović, G.D., 1991. On the instantaneous frequency of cyclostationary random signals. IEEE Transactions on Signal Processing 39 (7), 1604–1610.

Živanović, G.D., Gardner, W.A., 1991. Degrees of cyclostationarity and their application to signal detection and estimation. Signal Processing 22, 287–297.

Zhou, G., 1995. Random Amplitude and Polynomial Phase Modeling of Nonstationary Processes Using Higher-Order and Cyclic Statistics. Ph.D. Dissertation. University of Virginia, Charlottesville, VA.

Zhou, G., Giannakis, G.B., 1995a. Harmonics in multiplicative and additive noise: performance analysis of cyclic estimators. IEEE Transactions on Signal Processing 43 (6), 1445–1460.

Zhou, G., Giannakis, G.B., 1995b. Retrieval of self-coupled harmonics. IEEE Transactions on Signal Processing 43 (5), 1173–1185.

Zhou, G., Giannakis, G.B., 1996. Polyspectral analysis of mixed processes and coupled harmonics. IEEE Transactions on Information Theory 42 (3), 943–958.

Zhou, L., Man, H., 2013. Distributed automatic modulation classification based on cyclic feature via compressive sensing. In: IEEE Military Communications Conference. MILCOM 2013, pp. 40–45.

Zhu, Z., Feng, Z., Kong, F., 2005. Cyclostationarity analysis for gearbox condition monitoring: approaches and effectiveness. Mechanical Systems and Signal Processing 19 (3), 467–482. https://doi.org/10.1016/j.ymssp.2004.02.007.

Zhu, L., Luo, C.E., McClellan, J.H., 2013. Cyclostationarity-based wideband spectrum sensing using random sampling. In: IEEE Global Conference on Signal and Information Processing. GlobalSIP 2013, pp. 1202–1205.

Zhu, Y., Liu, J., Feng, Z., Zhang, P., 2014. Sensing performance of efficient cyclostationary detector with multiple antennas in multipath fading and lognormal shadowing environments. Journal of Communications and Networks (ISSN 1229-2370) 16 (2), 162–171. https://doi.org/10.1109/JCN.2014.000027.

Zoltowski, M.D., 2012. Open loop cyclostationarity based timing recovery for accelerated timing acquisition in frequency selective channels. U.S. Patent No. 8,233,575.

Index

9780081027080